Data Conversion Handbook

Data Conversion Handbook

Walt Kester, Editor

with the technical staff of Analog Devices

A Volume in the Analog Devices Series

ELSEVIER

AMSTERDAM • BOSTON • HEIDELBERG • LONDON
NEW YORK • OXFORD • PARIS • SAN DIEGO
SAN FRANCISCO • SINGAPORE • SYDNEY • TOKYO

Newnes is an imprint of Elsevier

Newnes

Newnes is an imprint of Elsevier
30 Corporate Drive, Suite 400, Burlington, MA 01803, USA
Linacre House, Jordan Hill, Oxford OX2 8DP, UK

 Recognizing the importance of preserving what has been written, Elsevier
prints its books on acid-free paper whenever possible.

Library of Congress Cataloging-in-Publication Data

(Application submitted.)

British Library Cataloguing-in-Publication Data
A catalogue record for this book is available from the British Library.

ISBN-13: 978-0-7506-7841-4   ISBN-10: 0-7506-7841-0

For information on all Newnes publications
visit our Web site at www.books.elsevier.com

Transferred to Digital Printing, 2010
Printed and bound in the United Kingdom

Contents

Foreword

The signal-processing products of Analog Devices (and its worthy competitors) have always had broad applications, but in a special way: they tend to be used in critical roles making possible—and at the same time limiting—the excellence in performance of the device, instrument, apparatus, or system using them.

Think about the *op amp*—how it can play a salient role in amplifying an ultrasound wave from deep within a human body, or measure and help reduce the error of a feedback system; the *data converter*—and its critical position in translating rapidly and accurately between the world of tangible physics and the world of abstract digits; the *digital signal processor*—manipulating the transformed digital data to extract information, provide answers, and make crucial instant-by-instant decisions in control systems; *transducers*, such as the life-saving MEMS accelerometers and gyroscopes; and even *control chips*, such as the one that empowers the humble thermometric junction placed deep in the heart of a high-performance—but very vulnerable—microcomputer chip.

From its founding two human generations ago, in 1965, Analog Devices has been committed to a leadership role in designing and manufacturing products that meet the needs of the existing market, anticipate the near-term needs of present and future users, and envision the needs of users yet unknown—and perhaps *unborn*—who will create the markets of the future. These existing, anticipated and envisioned "needs" must perforce include far more than just the design, manufacture and timely delivery of a physical device that performs a function reliably to a set of specifications at a competitive price.

We've always called a product that satisfies these needs "the augmented product," but what does this mean?

The *physical* product is a highly technological product that, above all, requires *knowledge* of its possibilities, limitations and subtleties. But when the earliest generations—and to some extent later generations—of such a product appear in the marketplace, there exist few (if any) school courses that have produced graduates proficient in its use. There are few knowledgeable designers who can foresee its possibilities. So we have the huge task of creating awareness; teaching about principles, performance measures, and existing applications; and providing ideas to stimulate the imagination of those creative users who will provide our next round of challenges.

This problem is met by deploying people and publications. The *people* are Applications Engineers, who can deal with user questions arriving via phone, fax, and e-mail—as well as working with users in the field to solve particular problems. These experts also spread the word by giving seminars to small and large groups for purposes from inspiring the creative user to imbuing the system, design, and components engineer with the nuts-and-bolts of practice. The *publications*—both in hard copy and on-line—range from authoritative handbooks, such as the present volume, comprehensive data sheets, application notes, hardware and software manuals, to periodic publications, such as "Solutions Bulletins" and our unique *Analog Dialogue*—the sole survivor among its early peers—currently approaching its 39th year of continuous publication in print and its 7th year of regular publication on the Internet.

This book is the ultimate expression of product "augmentation" as it relates to data converters. It can be considered a direct descendant of the Analog Devices 1972 *Analog-Digital Conversion Handbook*, edited by the undersigned. This timely publication was seminal in the early days of the mini- and microcomputer era—advocating the understanding and use of data converters and their links to an IC computer market that was then on the verge of explosive growth. Its third—and most recent—edition was published nearly 20 years ago, in 1986.

Data converters have been marketed as board-mountable components since the mid- to late 1960s, and practical IC D/A and A/D converters have been available since the mid-'70s. Yet, a third of a century later, there is still a need for a book that embraces the many aspects of conversion technology—one that is thorough in its technical content, that looks forward to tomorrow's uses and back to the principles and applications that still make data converters a vital necessity today. This is indeed such a book, and I am delighted that Walt Kester continues the practice of "augmenting" our data converters in such an interesting and accessible form.

Dan Sheingold, August 24, 2004
Norwood, Massachusetts

Preface

This book is written for the practicing design engineer who must routinely use data converters and related support circuitry. We have therefore included many practical design suggestions. Much of the material has been taken—and updated where necessary—from previous popular Analog Devices' seminar books. Most of the tutorial sections have undergone several revisions over the years to ensure their accuracy and clarity. Various highly experienced members of the Analog Devices' technical staff have contributed to the material, and they are recognized at the beginning of each major section in the book.

Chapter 1, *Data Converter History*, covers the chronological history of data converters. The chapter covers the period starting from the invention of the telegraph to the present time, and focuses on the hardware evolution of data converters. The history of data converter architectures is covered in Chapter 3, Data Converter Architectures, along with the descriptions of the various architectures themselves. A history of data converter processes is included in Chapter 4, Data Converter Process Technology, and the history of data converter testing is included as part of Chapter 5, Testing Data Converters.

Chapter 2, *Fundamentals of Sampled Data Systems*, contains the basics of coding, sampling, and quantizing. The various static and dynamic data converter error sources as well as specification definitions are also included in this chapter.

Chapter 3, *Data Converter Architectures*, includes not only architecture descriptions but a historical perspective on the popular DAC and ADC topologies. This chapter includes descriptions of both high and low speed ADC architectures as well as sigma-delta data converters.

Chapter 4, *Data Converter Process Technology*, takes a look at the various processes used to produce data converters, including a brief historical perspective. The chapter concludes with a discussion on "smart partitioning" of systems to achieve the maximum performance at the lowest cost.

Chapter 5, *Testing Data Converters*, includes both classic and modern techniques for testing data static and dynamic data converter performance, including a brief history of data converter testing. Of particular interest to modern applications is the relatively non-mathematical section on FFT testing.

Chapter 6, *Interfacing to Data Converters*, discusses solutions to various interface problems associated with data converters. The chapter begins with a study of the ADC analog inputs, and the various methods available for achieving optimum performance. Details of the digital interface to data converters are treated in a general manner, and various DAC output buffer configurations are presented. The chapter ends with a discussion of the critical issue of generating low jitter sampling clocks.

Chapter 7, *Data Converter Support Circuits*, covers various external components required to support data converters and data acquisition systems, including voltage references, regulators, analog switches and multiplexers, and sample-and-holds. Even though many of these function are incorporated into modern data converters, a fundamental understanding of them is useful to the system design engineer.

Chapter 8, *Data Converter Applications*, concentrates on specific data converter applications, including precision measurement and sensor conditioning, multichannel data acquisition systems, digital potentiometers, digital audio, digital video and display electronics, software radio and IF sampling, direct digital synthesis, and precision analog microcontrollers.

The material contained in Chapter 9, *Hardware Design Techniques*—the longest chapter in the book—has always been the most popular topic in the Analog Devices seminar series and probably contains the most important material in the entire book regarding practical issues in using data converters and their related components. The chapter begins with a thorough discussion of various pitfalls and solutions relating to the non-ideal qualities of passive components—capacitors, resistors, and inductors.

The section of Chapter 9, PC Board Design Issues, covers the important topics of grounding, layout, and decoupling. The recommendations presented reflect the collective inputs of a large number of experienced design and applications engineers at Analog Devices whose experience in data converters spans several decades.

Another key section in Chapter 9 is Analog Power Supply Systems, where the issues of generating clean analog supply voltages are addressed. Considerations for both linear and switching regulators are covered in the section.

Circuits used in data acquisition systems are often connected to external sensors and are therefore subject to overvoltage conditions. The section on Overvoltage Protection in Chapter 9 includes a discussion of the impact of overvoltage on ICs as well as methods for protecting critical analog circuits. The section concludes with a brief description of electrostatic discharge (ESD), including models and testing.

The Thermal Management section in Chapter 9 is especially important when dealing with devices which dissipate more than a few hundred milliwatts of power. The section covers the basics of thermal calculations using the traditional thermal resistance analysis. Methods for heatsinking high power devices, such as series pass transistors used in high current linear voltage regulators, are also discussed.

In the section titled EMI/RFI Considerations, the basics principles of EMI/RFI are first discussed, including EMI/RFI mechanisms, noise sources, coupling paths, near and far-field interference. The next part of the section contains a discussion of how passive components can be used to minimize EMI/RFI problems. A review of shielding concepts follows, and the general issue of RFI rectification is covered. PC board layout techniques useful in combating EMI/RFI problems conclude the section.

The section on Low Voltage Logic Interfacing deals with the common problem of interfacing devices in systems which operate on multiple supply voltages.

Chapter 9 concludes with a discussion of breadboarding and prototyping techniques as well as the general use of manufacturer's evaluation boards.

Acknowledgments

Thanks are due the many technical staff members of Analog Devices in Engineering, Marketing, and Applications who provided invaluable inputs during this project. Particular credit is give to the individual authors whose names appear at the beginning of their material.

Dan Sheingold graciously provided material from his classic 1986, Analog-Digital Conversion Handbook. Over the years, Dan has truly set the standards for technical publication quality, not only with Analog Dialogue, but with many other Analog Devices' technical books.

Special thanks also go to Brad Brannon, Wes Freeman, Walt Jung, Bob Marwin, Hank Zumbahlen, and Scott Wayne who reviewed the material for accuracy.

Judith Douville compiled the index and also offered many helpful manuscript comments.

A thank you also goes to ADI management, especially Dave Kress, for encouragement and support of the project.

Walt Kester, May 2004
Central Applications Department, Analog Devices

Direct questions, corrections, and comments to Linear.Apps@analog.com, with a subject line of "Data Conversion Handbook."

CHAPTER 1

Data Converter History

Data Converter History

Walt Kester

Chapter Preface

This chapter was inspired by Walt Jung's treatment of op amp history in his book, *Op Amp Applications Handbook* (Reference 1). His writing on the subject contains references to hundreds of interesting articles, patents, etc., which taken as a whole, paints a fascinating picture of the development of the operational amplifier—from Harold Black's early feedback amplifier sketch to modern high performance IC op amps.

We have attempted to do the same for the history of data converters. In considering the scope of this effort—and the somewhat chaotic and fragmented development of data converters—we were faced with a difficult challenge in organizing the material. Rather than putting all the historical material in this single chapter, we have chosen to disperse some of it throughout the book. For instance, most of the historical material related to data converter architectures is included in Chapter 3 *(Data Converter Architectures)*. along with the individual converter architectural descriptions. Likewise, Chapter 4 *(Data Converter Process Technology)* includes most of the key events related to data converter process technology. Chapter 5 *(Testing Data Converters)* touches on some of the key historical developments relating to data converter testing.

In an effort to make each chapter of this book stand on its own as much as possible, some of the historical material is repeated in several places—therefore, the reader should realize that this repetition is intentional and not the result of careless editing.

Early History

It is difficult to determine exactly when the first data converter was made or what form it took. The earliest recorded binary DAC known to the authors of this book is not electronic at all, but hydraulic. Turkey, under the Ottoman Empire, had problems with its public water supply, and sophisticated systems were built to meter water. One of these is shown in Figure 1.1 and dates to the 18th century. An example of an actual dam using this metering system was the Mahmud II dam built in the early 19th century near Istanbul and described in Reference 2.

The metering system used reservoirs (labeled *header tank* in the diagrams) maintained at a constant depth (corresponding to the reference potential) by means of a spillway over which water *just* trickled (the criterion was sufficient flow to float a straw). This is illustrated in Figure 1.1A. The water output from the header tank is controlled by gated binary-weighted *nozzles* submerged 96 mm below the surface of the water. The output of the nozzles feeds an *output trough* as shown in Figure 1.1B. The nozzle sizes corresponded to flows of binary multiples and submultiples of the basic unit of 1 lüle (= 36 l/min or 52 m³/day). An eight-lüle nozzle was known as a "sekizli lüle," a four lüle nozzle a "dörtlü lüle," a ¼ lüle nozzle a "kamuş," an eighth lüle a "masura," and a thirty-second lüle a "çuvaldiz." Details of the metering system using the binary weighted nozzles are shown in Figure 1.1C. Functionally this is an 8-bit DAC with manual (rather than digital, no doubt) input and a wet output, and it may be the oldest DAC in the world. There are probably other examples of early data converters, but we will now turn our attention to those based on more familiar electronic techniques.

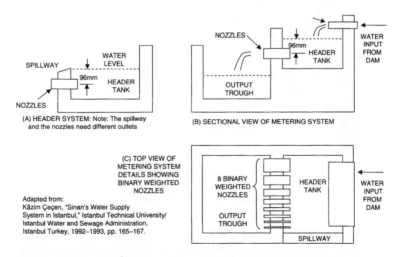

Adapted from:
Kâzim Çeçen, "Sinan's Water Supply System in Istanbul," Istanbul Technical University/ Istanbul Water and Sewage Administration, Istanbul Turkey, 1992–1993, pp. 165–167.

Figure 1.1: Early 18th Century Binary Weighted Water Metering System

Probably the single largest driving force behind the development of electronic data converters over the years has been the field of communications. The telegraph led to the invention of the telephone, and the subsequent formation of the Bell System. The proliferation of the telegraph and telephone, and the rapid demand for more capacity, led to the need for multiplexing more than one channel onto a single pair of copper conductors. While time division multiplexing (TDM) achieved some measure of popularity, frequency division multiplexing (FDM) using various carrier-based systems was by far the most successful and widely used. It was pulse code modulation (PCM), however, that put data converters on the map, and understanding its evolution is where we begin.

The material in the following sections has been extracted from a number of sources, but K. W. Cattermole's classic 1969 book, *Principles of Pulse Code Modulation* (Reference 3), is by far the most outstanding source of historical material for both PCM and data converters. In addition to the historical material, the book has excellent tutorials on sampling theory, data converter architectures, and many other topics relating to the subject. An extensive bibliography cites the important publications and patents behind the major developments. In addition to Cattermole's book, the reader is also referred to an excellent series of books published by the Bell System under the title of *A History of Engineering and Science in the Bell System* (References 4 through 8). These Bell System books are also excellent sources for background material on the entire field of communications.

The Early Years: Telegraph to Telephone

According to Cattermole (Reference 3), the earliest proposals for the electric telegraph date from about 1753, but most actual development occurred from about 1825–1875. Various ideas for binary and ternary numbers, codes of length varying inversely with probability of occurrence (Schilling, 1825), reflected-binary (Elisha Gray, 1878—now referred to as the *Gray* code), and chain codes (Baudot, 1882) were explored. With the expansion of telegraphy came the need for more capacity, and multiplexing more than one signal on a single pair of conductors. Figure 1.2 shows a typical telegraph key and some highlights of telegraph history.

- Telegraph proposals: Started 1753
- Major telegraph development: 1825–1875
- Various binary codes developed
- Experiments in multiplexing for increased channel capacity

- Telephone invented: 1875 by A. G. Bell while working on a telegraph multiplexing project

- Evolution:
  - Telegraph: Digital
  - Telephone: Analog
  - Frequency division multiplexing (FDM): Analog
  - Pulse code modulation (PCM): Back to Digital

Figure 1.2: The Telegraph

The invention of the telephone in 1875 by Alexander Graham Bell (References 9 and 10) was probably the most significant event in the entire history of communications. It is interesting to note, however, that Bell was actually experimenting with a telegraph multiplexing system (Bell called it the *harmonic* telegraph) when he recognized the possibility of transmitting the voice itself as an analog signal.

Figure 1.3 shows a diagram from Bell's original patent which puts forth his basic proposal for the telephone. Sound vibrations applied to the transmitter *A* cause the membrane *a* to vibrate. The vibration of *a* causes a vibration in the armature *c* which induces a current in the wire *e* via the electromagnet *b*. The current in *e* produces a corresponding fluctuation in the magnetic field of electromagnet *f*, thereby vibrating the receiver membrane *i*.

The proliferation of the telephone generated a huge need to increase channel capacity by multiplexing. It is interesting to note that studies of multiplexing with respect to telegraphy led to the beginnings of information theory. Time division multiplexing (TDM) for telegraph was conceived as early as 1853 by a little known American inventor, M. B. Farmer; and J. M. E. Baudot put it into practice in 1875 using rotating mechanical commutators as multiplexers.

UNITED STATES PATENT OFFICE.

ALEXANDER GRAHAM BELL, OF SALEM, MASSACHUSETTS.

IMPROVEMENT IN TELEGRAPHY.

Specification forming part of Letters Patent No. **174,465**, dated March 7, 1876; application filed February 14, 1876.

Fig. 7

Extracted from U.S. Patent 174,465,
Filed February 14, 1876, Issued March 7, 1876

Figure 1.3: The Telephone

In a 1903 patent (Reference 11), Willard M. Miner describes experiments using this type of electromechanical rotating commutator to multiplex several analog telephone conversations onto a single pair of wires as shown in Figure 1.4. Quoting from his patent, he determined that each channel must be sampled at

> *"… a frequency or rapidity approximating the frequency or average frequency of the finer or more complex vibrations which are characteristic of the voice or of articulate speech, …, as high as 4320 closures per second, at which rate I find that the voice with all its original timbre and individuality may be successfully reproduced in the receiving instrument. … I have also succeeded in getting what might be considered as commercial results by using rates of closure that, comparatively speaking, are as low as 3500 closures per second, this being practically the rate of the highest note which characterizes vowel sounds."*

At higher sampling rates, Miner found no perceptible improvement in speech quality, probably because of other artifacts and errors in his rather crude system.

There was no follow-up to Miner's work on sampling and TDM, probably because there were no adequate electrical components available to make it practical. FDM was well established by the time adequate components did arrive.

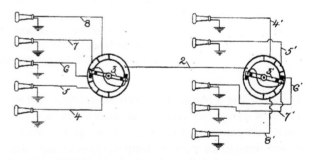

Extracted from: Williard M. Miner, "Multiplex Telephony," U.S. Patent 745,734,
Filed February 26, 1903, Issued December 1, 1903

Figure 1.4: One of the Earliest References to a Criteria for Determining the Sampling Rate

The Invention of PCM

Pulse code modulation was first disclosed in a relatively obscure patent issued to Paul M. Rainey of Western Electric in 1921 (Reference 12). The patent describes a method to transmit facsimile information in coded form over a telegraph line using 5-bit PCM. The figure from the patent is shown in Figure 1.5 (additional labels have been added for clarity).

Rainey proposed that a light beam be focused on the transparency of the material to be transmitted. A photocell is placed on the other side of the transparency to gather the light and produce a current proportional to the intensity of the light. This current drives a galvanometer which in turn moves another beam of light which activates one of 32 individual photocells, depending upon the amount of galvanometer deflection. Each individual photocell output activates a corresponding relay. The five relay outputs are connected in such a way as to generate the appropriate code corresponding to the photocell location. The digital code is thus generated from an "m-hot out of 32 code," similar to modern flash converters. The output of this simple electo-optical-mechanical "flash" converter is then transmitted serially using a rotating electromechanical commutator, called a distributor.

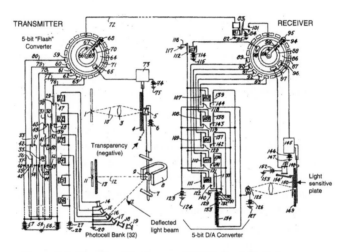

Figure 1.5: The First Disclosure of PCM: Paul M. Rainey, "Facsimile Telegraph System," U.S. Patent 1,608,527, Filed July 20, 1921, Issued November 30, 1926

The serial data is transmitted, received, and converted into a parallel format using a second distributor and a bank of relays. The received code determines the combination of relays to be activated, and the relay outputs are connected across appropriate taps of a resistor which is in series with the receiving lamp. The current through the receiving lamp therefore changes depending upon the received code, thereby varying its intensity proportionally to the received code and performing the digital-to-analog conversion. The receiving lamp output is focused on a photographically sensitive receiving plate, thus reproducing the original image in quantized form.

Rainey's patent illustrates several important concepts: quantization using a flash A/D converter, serial data transmission, and reconstruction of the quantized data using a D/A converter. These are the fundamentals of PCM. However, his invention aroused little interest at the time and was, in fact, forgotten by Bell System engineers. His patent was discovered years later after many other PCM patents had already been issued.

The Mathematical Foundations of PCM

In the mid-1920's, Harry Nyquist studied telegraph signaling with the objective of finding the maximum signaling rate that could be used over a channel with a given bandwidth. His results are summarized in two classic papers published in 1924 (Reference 13) and 1928 (Reference 14), respectively.

In his model of the telegraph system, he defined his signal as:

$$s(t) = \sum_k a_k f(t - kT).$$

Eq. 1.1

In the equation, $f(t)$ is the basic pulse shape, a_k is the amplitude of the kth pulse, and T is the time between pulses. DC telegraphy fits this model if f(t) is assumed to be a rectangular pulse of duration T, and a_k equal to 0 or 1. A simple model is shown in Figure 1.6. The signal is bandlimited to a frequency W by the transmission channel.

His conclusion was that the pulse rate, 1/T, could not be increased beyond 2 W pulses per second. Another way of stating this conclusion is *if a signal is sampled instantaneously at regular intervals at a rate at least twice the highest significant signal frequency, then the samples contain all the information in the original signal*. This is clear from Figure 1.6 if the filtered rectangular pulses are each represented by a sinx/x response. The sinx/x time domain impulse response of an ideal lowpass filter of bandwidth W has zeros at intervals of 1 /2W. Therefore, if the output waveform is sampled at the points indicated in the diagram, there will be no interference from adjacent pulses, provided T ≥ 1 /2W (or more commonly expressed as: $f_s \geq 2$ W), and the amplitude of the individual pulses can be uniquely recovered.

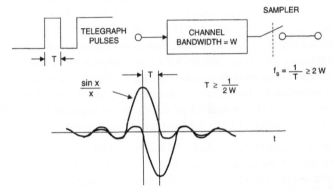

- Up to 2W pulses per second can be transmitted over a channel that has a bandwidth W.

- If a signal is sampled instantaneously at regular intervals at a rate at least twice the highest significant signal frequency, the samples contain all the information in the original signal.

Figure 1.6: Harry Nyquist's Classic Theorem: 1924

Except for a somewhat general article by Hartley in 1928 (Reference 15), there were no significant additional publications on the specifics of sampling until 1948 in the classic papers by Shannon, Bennett, and Oliver (References 16–19) which solidified PCM theory for all time. A summary of the classic papers on PCM is shown in Figure 1.7.

- Multiplexing experiments such as Williard Miner, "Multiplex Telephony," U.S. Patent 745,734, filed February 26, 1903, issued December 1, 1903.
- H. Nyquist, "Certain Factors Affecting Telegraph Speed," Bell System Technical Journal, Vol. 3, April 1924, pp. 324–346.
- H. Nyquist, Certain Topics in Telegraph Transmission Theory, A.I.E.E. Transactions, Vol. 47, April 1928, pp. 617–644.
- R.V.L. Hartley, "Transmission of Information," Bell System Technical Journal, Vol. 7, July 1928, pp. 535–563.
- Note: Shannon's classic paper was written in 1948, well after the invention of PCM:
- C. E. Shannon, "A Mathematical Theory of Communication," Bell System Technical Journal, Vol. 27, July 1948, pp. 379–423, and October 1948, pp. 623–656.
- W. R. Bennett, "Spectra of Quantized Signals," Bell System Technical Journal, Vol. 27, July 1948, pp. 446–471.
- B. M. Oliver, J. R. Pierce, C. E. Shannon, "The Philosophy of PCM," IRE Proceedings, Vol. 36, November 1948, pp. 1324–1331.
- W. R. Bennett, "Noise in PCM Systems," Bell Labs Record, Vol. 26, December 1948, pp. 495–499.

Figure 1.7: Mathematical Basis of PCM

The PCM Patents of Alec Harley Reeves

By 1937, frequency division multiplexing (FDM) based on vacuum tube technology was widely used in the telephone industry for long-haul routes. However, noise and distortion were the limiting factors in expanding the capacity of these systems. Although wider bandwidths were becoming available on microwave links, the additional noise and distortion made them difficult to adapt to FDM signals.

Alec Harley Reeves had studied analog-to-time conversion techniques using pulse time modulation (PTM) during the beginning of his career in the 1920s. In fact, he was one of the first to make use of counter chains to accurately define time bases using bistable multivibrators invented by Eccles and Jordan a few years earlier. In PTM, the amplitude of the pulses is constant, and the analog information is contained in the relative timing of the pulses. This technique gave better noise immunity than strictly analog transmission, but Reeves was shortly to invent a system that would completely revolutionize communications from that point forward.

It was therefore the need for a system with noise immunity similar to the telegraph system that led to the (re-) invention of pulse code modulation (PCM) by Reeves at the Paris labs of the International Telephone and Telegraph Corporation in 1937. The very first PCM patent by Reeves was filed in France, but was immediately followed by similar patents in Britain and the United States, all listing Reeves as the inventor (Reference 20). These patents were very comprehensive and covered the far-reaching topics of (1) general principles of quantization and encoding, (2) the choice of resolution to suit the noise and bandwidth of the transmission medium, (3) transmission of signals in digital format serially, in parallel, and as modulated carriers, and (4) a counter-based design for the required 5-bit ADCs and DACs. Unlike the previous PCM patent by Rainey in 1926, Reeves took full advantage of existing vacuum tube technology in his design.

The ADC and DAC developed by Reeves deserves some further discussion, since they represent one of the first all-electronic data converters on record. The ADC technique (Figure 1.8) basically uses a sampling pulse to take a sample of the analog signal, set an R/S flip-flop, and simultaneously start a controlled ramp voltage. The ramp voltage is compared with the input, and when they are equal, a pulse is generated that resets the R/S flip-flop. The output of the flip-flop is a pulse whose width is proportional to the analog signal at the sampling instant. This pulse width modulated (PWM) pulse controls a gated oscillator, and the number of pulses out of the gated oscillator represents the quantized value of the analog signal. This pulse train can be easily converted to a binary word by driving a counter. In Reeves' system, a master clock of 600 kHz is used, and a 100:1 divider generates the 6 kHz sampling pulses. The system uses a 5-bit counter, and 31 counts (out of the 100 counts between sampling pulses) therefore represents a full-scale signal.

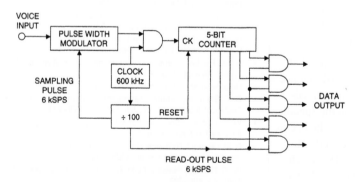

Adapted from: Alec Harley Reeves, "Electric Signaling System,"
U.S. Patent 2,272,070, Filed November 22, 1939, Issued February 3, 1942

Figure 1.8: A. H. Reeves' 5-Bit Counting ADC

The DAC uses a similar counter and clock source as shown in Figure 1.9. The received binary code is first loaded into the counter, and the R/S flip-flop is reset. The counter is then allowed to count upward by applying the clock pulses. When the counter overflows and reaches 00000, the clock source is disconnected, and the R/S flip-flop is set. The number of pulses counted by the encoding counter is thus the complement of the incoming data word. The output of the R/S flip-flop is a PWM signal whose analog value is the complement of the input binary word. Reeves uses a simple low-pass filter to recover the analog signal from the PWM output. The phase inversion in the DAC is easily corrected in either the logic or in an amplifier further down the signal chain.

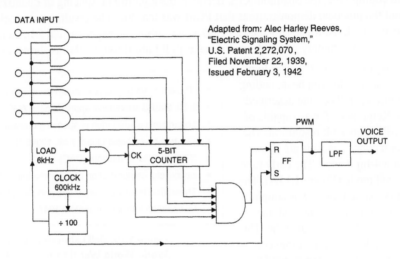

Figure 1.9: A. H. Reeves' 5-Bit Counting DAC

Reeves' patents covered all the essentials of PCM: sampling, quantizing, and coding the digitized samples for serial, parallel, phase-modulated, and other transmission methods. On the receiving end, Reeves proposed a suitable decoder to reconstruct the original analog signal. In spite of the significance of his work, it is interesting to note that after the patent disclosures, Reeves shifted his attention to the shortwave transmission of speech using pulse-amplitude modulation, pulse-duration modulation, and pulse-position modulation, rather than pursuing PCM techniques.

PCM and the Bell System: World War II through 1948

Under a cross-licensing arrangement with International Telephone and Telegraph Corporation, Bell Telephone Laboratories' engineers reviewed Reeves' circuit descriptions and embarked upon their own pursuit of PCM technology. Starting in about 1940 and during World War II, studies were conducted on a speech secrecy system that made PCM techniques mandatory.

The highly secret "Project-X" to develop a speech secrecy system was started in 1940 by Bell Labs and is described in detail in Reference 6 (pp. 296–317). It used a complex technique based on vacuum tube technology that made use of the previously developed "vocoder," PCM techniques, and a unique data scrambling technique utilizing a phonograph recording containing the electronic "key" to the code. This system was designed at Bell Labs and put into production by Western Electric in late 1942. By April, 1943, several terminals were completed and installed in Washington, London, and North Africa. Shortly thereafter, additional terminals were installed in Paris, Hawaii, Australia, and the Philippines.

By the end of the war, several groups at Bell Labs were studying PCM; however, most of the wartime results were not published until several years later because of secrecy issues. The work of H. S. Black, J. O. Edson, and W. M. Goodall were published in 1947–1948. (References 21, 22, and 23). Their emphasis was on speech encryption systems based on PCM techniques, and many significant developments came out of their work. A PCM system which digitized the entire voice band to 5-bits, sampling at 8 kSPS using a successive approximation ADC was described by Edson and Black (Reference 21 and 22). W. M. Goodall described an experimental PCM system in his classic paper based on similar techniques (Reference 23).

Some of the significant developments that came out of this work were the successive approximation ADC, the electron beam coding tube, the Shannon-Rack decoder, the logarithmic spacing of quantization levels (companding), and the practical demonstrations that PCM was feasible. The results were nicely summarized in a 1948 article by L. A. Meacham and E. Peterson describing an experimental 24-channel PCM system (Reference 24). A summary of PCM work done at Bell Labs through 1948 is shown in Figure 1.10.

A significant development in ADC technology during the period was the electron beam coding tube shown in Figure 1.11. The tube described by R. W. Sears in Reference 25 was capable of sampling at 96 kSPS with 7-bit resolution. The basic electron beam coder concepts are shown in Figure 1.11 for a 4-bit device. The early tubes operated in the serial mode (Figure 1.11A). The analog signal is first passed through a sample-and-hold, and during the "hold" interval, the beam is swept horizontally across the tube. The Y-deflection for a single sweep therefore corresponds to the value of the analog signal from the sample-and-hold. The shadow mask is coded

- "Project-X" voice secrecy system using PCM, 1940–1943.
- 5-bit, 8kSPS successive approximation ADC
- Logarithmic quantization of speech (companding)
- Electron beam coding tube, 7-bit, 100kSPS
- "Shannon-Rack" decoder (DAC)
- Successful demonstration of experimental PCM terminals
- Theoretical PCM work expanded and published by Shannon
- Germanium transistor invented: 1947

Figure 1.10: Bell Laboratories' PCM Work: World War II through 1948.

to produce the proper binary code, depending on the vertical deflection. The code is registered by the collector, and the bits are generated in serial format. Later tubes used a fan-shaped beam (shown in Figure 1.11B), creating the first electronic "flash" converter delivering a parallel output word.

Early electron tube coders used a binary-coded shadow mask, and large errors could occur if the beam straddled two adjacent codes and illuminated both of them. The way these errors occur is illustrated in Figure 1.12A, where the horizontal line represents the beam sweep at the midscale transition point (transition between code 0111 and code 1000). For example, an error in the most significant bit (MSB) produces an error of ½ scale. These errors were minimized by placing fine horizontal sensing wires across the boundaries of each of the quantization levels. If the beam initially fell on one of the wires, a small voltage was added to the vertical deflection voltage which moved the beam away from the transition region.

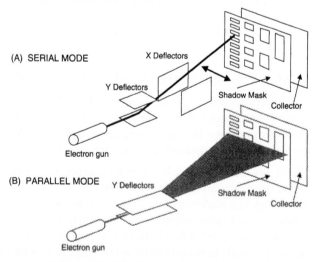

Figure 1.11: The Electron Beam Coder

The errors associated with binary shadow masks were eliminated by using a Gray code shadow mask as shown in Figure 1.12B. This code was originally called the "reflected binary" code, and was invented by Elisha Gray in 1878, and later re-invented by Frank Gray in 1949 (see Reference 26). The Gray code has the property that adjacent levels differ by only one digit in the corresponding Gray-coded word. Therefore, if there is an error in a bit decision for a particular level, the corresponding error after conversion to binary code is only one least significant bit (LSB). In the case of midscale, note that only the

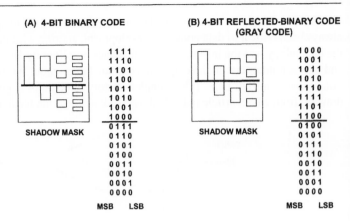

(A) 4-BIT BINARY CODE		(B) 4-BIT REFLECTED-BINARY CODE (GRAY CODE)	
	1111		1000
	1110		1001
	1101		1011
	1100		1010
	1011		1110
	1010		1111
	1001		1101
	1000		1100
SHADOW MASK	0111	SHADOW MASK	0100
	0110		0101
	0101		0111
	0100		0110
	0011		0010
	0010		0011
	0001		0001
	0000		0000
MSB LSB		MSB LSB	

Figure 1.12: Electron Beam Coder Shadow Masks for Binary and Gray Code

MSB changes. It is interesting to note that this same phenomenon can occur in modern comparator-based flash converters due to comparator metastability. With small overdrive, there is a finite probability that the output of a comparator will generate the wrong decision in its latched output, producing the same effect if straight binary decoding techniques are used. In many cases, Gray code, or "pseudo-Gray" codes are used to decode the comparator bank output before finally converting to a binary code output (refer to Chapter 3 for further architectural descriptions).

In spite of the many mechanical and electrical problems relating to beam alignment, electron tube coding technology reached its peak in the mid-1960s with an experimental 9-bit coder capable of 12 MSPS sampling rates (Reference 27). Shortly thereafter, however, advances in solid-state ADC techniques quickly made the electron tube converter technology obsolete.

Op Amps and Regenerative Repeaters: Vacuum Tubes to Solid-State

Except for early relatively inefficient electro-mechanical amplifiers (see Reference 5), electronic amplifier development started with the invention of the vacuum tube by Lee de Forest in 1906 (References 28 and 29). A figure from the original de Forest patent is shown in Figure 1.13.

Figure 1.13: The Invention of the Vacuum Tube: 1906

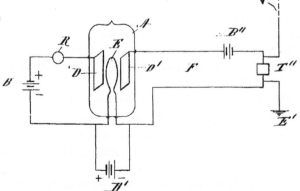

Extracted from: Lee De Forest, "Device for Amplifying Feeble Electrical Currents," U.S. Patent 841,387, Filed October 25, 1906, Issued January 15, 1907

By 1914, vacuum tube amplifiers had been introduced into the telephone plant. Amplifier development has always been critical to data converter development, starting with these early vacuum tube circuits. Key to the technology was the invention of the feedback amplifier by Harold S. Black in 1927 (References 30, 31, and 32). Amplifier circuit development continued throughout World War II, and many significant contributions came from Bell Labs. (The complete history of op amps is given in Reference 1). Figure 1.14 shows a drawing from a later article published by Black defining the feedback amplifier.

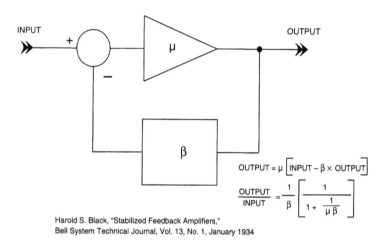

$$\text{OUTPUT} = \mu \left[\text{INPUT} - \beta \times \text{OUTPUT} \right]$$

$$\frac{\text{OUTPUT}}{\text{INPUT}} = \frac{1}{\beta} \left[\frac{1}{1 + \frac{1}{\mu\beta}} \right]$$

Harold S. Black, "Stabilized Feedback Amplifiers,"
Bell System Technical Journal, Vol. 13, No. 1, January 1934

Figure 1.14: Harold Black's Feedback Amplifier of 1927

The invention of the germanium transistor in 1947 (References 33, 34, and 35) was key to the development of PCM and all other electronic systems. In order for PCM to be practical, regenerative repeaters had to be placed periodically along the transmission lines. Vacuum tube repeaters had been somewhat successfully designed and used in the telegraph and voice network for a number of years prior to the development of the transistor, but suffered from obvious reliability problems. However, the solid state regenerative repeater designed by L. R. Wrathall in 1956 brought the PCM research phase to a dramatic conclusion (Reference 36). This repeater was demonstrated on an experimental cable system using repeater spacings of 2.3 miles on 19-gauge cable, and 0.56 miles on 32-gauge cable. A schematic diagram of the repeater is reproduced in Figure 1.15.

The Wrathall repeater used germanium transistors designed by Bell Labs and built by Western Electric. The silicon transistor was invented in 1954 by Gordon Teal at Texas Instruments and gained wide commercial acceptance because of the increased temperature performance and reliability. Finally, the invention of the integrated circuit (References 37 and 38) in 1958 followed by the planar process in 1959 (Reference 39) set the stage for future PCM developments. These key solid state developments are summarized in Figure 1.16 and discussed in greater depth in Chapter 4 of this book.

With the development of the Wrathall repeater, it was therefore clear in 1956 that PCM could be effectively used to increase the number of voice channels available on existing copper cable pairs. This was especially attractive in metropolitan areas where many cable conduits were filled to capacity. Many of these pairs were equipped with loading coils at a spacing of 1.8 kM to improve their response in the voice band. It was natural to consider replacing the loading coils with solid-state repeaters and to extend the capacity from 1 to 24 channels by using PCM.

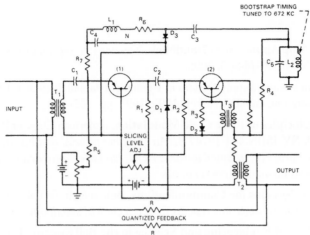

From: L.R. Wrathall, "Transistorized Binary Pulse Regenerator,"
Bell System Technical Journal, Vol. 35, September 1956, pp. 1059–1084

**Figure 1.15: L. R. Wrathall's Solid State
PCM Repeater: 1956**

- Invention of the (Germanium) transistor at Bell Labs: John
  Bardeen, Walter Brattain, and William Shockley in 1947.

- Silicon Transistor: Gordon Teal, Texas Instruments, 1954.

- Birth of the Integrated Circuit:
  - Jack Kilby, Texas Instruments, 1958 (used bond wires for
    interconnections).
  - Robert Noyce, Fairchild Semiconductor, 1959 (used
    metallization for interconnections).

- The Planar Process: Jean Hoerni, Fairchild Semiconductor, 1959.

Figure 1.16: Key Solid-state Developments: 1947–1959

For these reasons, a decision was made at Bell Labs to develop a PCM carrier system, and a prototype 24-channel system was designed and tested during 1958 and 1959 on a link between Summit, New Jersey and South Orange, New Jersey. This system, called the T-1 carrier system, transmitted 24 voice channels using a 1.544 MHz pulse train in a bipolar code. The system used 7-bit logarithmic encoding with 26 dB of companding, and was later expanded to 8-bit encoding. The solid-state repeaters were spaced at 1.8 kM intervals, corresponding to the placement of the existing loading coils. The first T-1 operating link went into service in 1962, and by 1984 there were more than 200 million circuit-kilometers of T-1 carrier in the United States.

References:

1.1 Early History

1. Walter G. Jung, **Op Amp Applications Handbook**, Newnes (an imprint of Elsevier Science and Technology Books), ISBN 0-7506-7844-5, 2005.

2. Kâzim Çeçen, "Sinan's Water Supply System in Istanbul," Istanbul Technical University/Istanbul Water and Sewage Administration, Istanbul Turkey, 1992–1993, pp. 165–167.

3. K. W. Cattermole, **Principles of Pulse Code Modulation**, American Elsevier Publishing Company, Inc., 1969, New York NY, ISBN 444-19747-8. *(An excellent tutorial and historical discussion of data conversion theory and practice, oriented towards PCM, but covers practically all aspects. This one is a must for anyone serious about data conversion.)*

4. Editors, **Transmission Systems for Communications**, Bell Telephone Laboratories, 1964. *(Excellent discussion of Bell System transmission systems from a technical standpoint.)*

5. M. D. Fagen, **A History of Engineering and Science in the Bell System: The Early Years (1875–1925)**, Bell Telephone Laboratories, 1975.

6. M. D. Fagen, **A History of Engineering and Science in the Bell System: National Service in War and Peace (1925–1975)**, Bell Telephone Laboratories, 1978, ISBN 0-932764-00-2.

7. S. Millman, **A History of Engineering and Science in the Bell System: Communications Sciences (1925–1980)**, AT&T Bell Laboratories, 1984, ISBN 0-932764-06-1.

8. E. F. O'Neill, **A History of Engineering and Science in the Bell System: Transmission Technology (1925–1975)**, AT&T Bell Telephone Laboratories, 1985.

9. Alexander Graham Bell, "Improvement in Telegraphy," **U.S. Patent 174,465**, filed February 14, 1876, issued March 7, 1876. *(This is the original classic patent on the telephone.)*

10. Alexander Graham Bell, "Improvement in Electric Telegraphy," **U.S. Patent 186,787**, filed January 15, 1877, issued January 30, 1877. *(This and the preceding patent formed the basis of the Bell System patents.)*

11. Willard M. Miner, "Multiplex Telephony," **U.S. Patent 745,734**, filed February 26, 1903, issued December 1, 1903. *(A relatively obscure patent on electro-mechanical multiplexing of telephone channels in which experiments describing voice quality versus sampling frequency are mentioned.)*

12. Paul M. Rainey, "Facimile Telegraph System," **U.S. Patent 1,608,527**, filed July 20, 1921, issued November 30, 1926. *(Although A. H. Reeves is generally credited with the invention of PCM, this patent discloses an electro-mechanical PCM system complete with A/D and D/A converters. The patent was largely ignored and forgotten until many years after the various Reeves' patents were issued in 1939–1942.)*

13. H. Nyquist, "Certain Factors Affecting Telegraph Speed," **Bell System Technical Journal**, Vol. 3, April 1924, pp. 324–346.

14. H. Nyquist, "Certain Topics in Telegraph Transmission Theory," **A.I.E.E. Transactions**, Vol. 47, April 1928, pp. 617–644.

15. R.V.L. Hartley, "Transmission of Information," **Bell System Technical Journal**, Vol. 7, July 1928, pp. 535–563.

16. C. E. Shannon, "A Mathematical Theory of Communication," **Bell System Technical Journal**, Vol. 27, July 1948, pp. 379–423 and October 1948, pp. 623–656.

17. W. R. Bennett, "Spectra of Quantized Signals," **Bell System Technical Journal**, Vol. 27, July 1948, pp. 446–471.

18. B. M. Oliver, J. R. Pierce, and C. E. Shannon, "The Philosophy of PCM," **Proceedings IRE**, Vol. 36, November 1948, pp. 1324–1331.

19. W. R. Bennett, "Noise in PCM Systems," **Bell Labs Record**, Vol. 26, December 1948, pp. 495–499.

20. Alec Harley Reeves, "Electric Signaling System," **U.S. Patent 2,272,070**, filed November 22, 1939, issued February 3, 1942. Also **French Patent 852,183** issued 1938, and **British Patent 538,860** issued 1939.

21. H. S. Black and J. O. Edson, "Pulse Code Modulation," **AIEE Transactions**, Vol. 66, 1947, pp. 895–899.

22. H. S. Black, "Pulse Code Modulation," **Bell Labs Record**, Vol. 25, July 1947, pp. 265–269.

23. W. M. Goodall, "Telephony by Pulse Code Modulation," **Bell System Technical Journal**, Vol. 26, pp. 395–409, July 1947. *(Describes an experimental PCM system using a 5-bit, 8 kSPS successive approximation ADC based on the subtraction of binary weighted charges from a capacitor to implement the internal DAC function.)*

24. L. A. Meacham and E. Peterson, "An Experimental Multichannel Pulse Code Modulation System of Toll Quality," **Bell System Technical Journal**, Vol 27, No. 1, January 1948, pp. 1–43. *(Describes the culmination of much work leading to this 24-channel experimental PCM system.)*

25. R. W. Sears, "Electron Beam Deflection Tube for Pulse Code Modulation," **Bell System Technical Journal**, Vol. 27, pp. 44–57, Jan. 1948. *(Describes an electon-beam deflection tube 7-bit,100 kSPS flash converter for early experimental PCM work.)*

26. Frank Gray, "Pulse Code Communication," **U.S. Patent 2,632,058**, filed November 13, 1947, issued March 17, 1953. *(Detailed patent on the Gray code and its application to electron beam coders.)*

27. J. O. Edson and H. H. Henning, "Broadband Codecs for an Experimental 224Mb/s PCM Terminal," **Bell System Technical Journal**, Vol. 44, pp. 1887–1940, Nov. 1965. *(Summarizes experiments on ADCs based on the electron tube coder as well as a bit-per-stage Gray code 9-bit solid state ADC. The electron beam coder was 9-bits at 12 MSPS, and represented the fastest of its type.)*

28. Lee de Forest, "Device for Amplifying Feeble Electrical Currents," **U.S. Patent 841,387**, filed October 25, 1906, issued January 15, 1907.

29. Lee de Forest, "Space Telegraphy," **U.S. Patent 879,532**, filed January 29, 1907, issued February 18, 1908.

30. H. S. Black, "Wave Translation System," **U.S. Patent 2,102,671**, filed August 8, 1928, issued December 21, 1937. *(The basis of feedback amplifier systems.)*

31. H. S. Black, "Stabilized Feedback Amplifiers," **Bell System Technical Journal**, Vol. 13, No. 1, January 1934, pp. 1–18. *(A practical summary of feedback amplifier systems.)*

32. Harold S. Black, "Inventing the Negative Feedback Amplifier," **IEEE Spectrum**, December, 1977. *(Inventor's 50th anniversary story on the invention of the feedback amplifier.)*

33. C. Mark Melliar-Smith et al, "Key Steps to the Integrated Circuit," **Bell Labs Technical Journal**, Vol. 2, #4, Autumn 1997.

34. J. Bardeen, W. H. Brattain, "The Transistor, a Semi-Conductor Triode," **Physical Review**, Vol. 74, No. 2, July 15, 1947 pp. 230–231. *(The invention of the germanium transistor.)*

35. W. Shockley, "The Theory of p-n Junctions in Semiconductors and p-n Junction Transistors," **Bell System Technical Journal**, Vol. 28, No. 4, July 1949, pp. 435–489. *(Theory behind the germanium transistor.)*

36. L. R. Wrathall, "Transistorized Binary Pulse Regenerator," **Bell System Technical Journal**, Vol. 35, September 1956, pp. 1059–1084.

37. J. S. Kilby, "Invention of the Integrated Circuit," **IRE Transactions on Electron Devices**, Vol. ED-23, No. 7, July 1976, pp. 648–654. *(Kilby's IC invention at TI.)*

38. Robert N. Noyce, "Semiconductor Device-and-Lead Structure," **U.S. Patent 2,981,877**, filed July 30, 1959, issued April 25, 1961. *(Noyce's IC invention at Fairchild.)*

39. Jean Hoerni, "Planar Silicon Diodes and Transistors," **IRE Transactions on Electron Devices**, Vol. 8, March 1961, p. 168.

40. Jean A. Hoerni, "Method of Manufacturing Semiconductor Devices," **U.S. Patent 3,025,589**, filed May 1, 1959, issued March 20, 1962. *(The planar process— a manufacturing means of protecting and stabilizing semiconductors.)*

Data Converters of the 1950s and 1960s
Walt Kester

Commercial Data Converters: 1950s

Up until the mid-1950s, data converters were primarily developed and used within specialized applications, such as the Bell System work on PCM, and message encryption systems of World War II. Because of vacuum tube technology, the converters were very expensive, bulky, and dissipated lots of power. There was practically no commercial usage of these devices.

The digital computer was a significant early driving force behind commercial ADC development. The ENIAC computer development project was started in 1942 and was revealed to the general public in February 1946. The ENIAC led to the development of the first commercially available digital computer, the UNIVAC, by Eckert and Mauchly. The first UNIVAC was delivered to the United States Census Bureau in June 1951.

Military applications, such as ballistic trajectory computation, were early driving forces behind the digital computer, but as time went on, the possibilities of other applications in the area of data analysis and industrial process control created more general interest in digital processing, and hence the need for data converters. In 1953 Bernard M. Gordon, a pioneer in the field of data conversion, founded a company called Epsco Engineering in his basement in Concord MA. Gordon had previously worked on the UNIVAC computer, and saw the need for commercial data converters. In 1954 Epsco introduced an 11-bit, 50 kSPS vacuum-tube based ADC. This converter is believed to be the first commercial offering of such a device.

The Epsco "Datrac" converter dissipated 500 watts, was designed for rack mounting (19" × 15" × 26") and sold for $8,000 to $9,000 (see Reference 1). A photograph of the instrument is shown in Figure 1.17. The Datrac was the first commercially offered ADC to utilize the shift-programmable successive approximation architecture, and Gordon was granted a patent on the logic required to perform the conversion algorithm (Reference 2). Because it had a sample-and-hold function, the Epsco Datrac was the first commercial ADC suitable for digitizing ac waveforms, such as speech.

During the same period, a few other companies manufactured lower speed ADCs suitable for digital voltmeter measurement applications, and there were

- 19" × 15" × 26"
- 500W
- 150 lbs
- $8,500.00

Courtesy,
Analogic Corporation
8 Centennial Drive
Peabody, MA 01960

www.analogic.com

Figure 1.17: 1954 "DATRAC" 11-Bit, 50 kSPS Vacuum Tube ADC Designed by Bernard M. Gordon at EPSCO

offerings of optical converters based on coded discs for measuring the angular position of shafts in avionics applications (see Reference 1). Converters of the mid to late 1950s used a combination of vacuum tubes, solid-state diodes, and transistors to implement the conversion process. A few of the companies in the data converter business at the time were Epsco, Non-Linear Systems, Inc., J.B. Rea, and Adage. In order to gain further insight to the converters of the 1950s, References 1, 3, 4, 5, and 6 are excellent sources.

Commercial Data Converter History: 1960s

During the mid 1950s through the early 1960s, electronic circuit designs began to migrate from vacuum tubes to transistors, thereby opening up many new possibilities in data conversion products. As was indicated earlier, the silicon transistor was responsible for the increased interest in solid state designs. There was more and more interest in data converter products, as indicated in two survey articles published in 1964 (Reference 5) and 1967 (Reference 6). Because these devices were basically unfamiliar to new customers, efforts were begun to define specifications and testing requirements for converter products (References 7–16).

The IBM-360 mainframe computer and solid state minicomputers (such as the DEC PDP-series starting in 1963) added to the general interest in data analysis applications. Other driving forces requiring data converters in the 1960s were industrial process control, measurement, PCM, and military systems.

Efforts continued during the 1960s at Bell Telephone Labs to develop high speed converters (e.g., 9 bits, 5 MSPS) for PCM applications (Reference 17), and the military division of Bell Labs began work on the development of hardware and software for an anti-ballistic missile (ABM) system.

In 1958, the U.S. Army began the development of the Nike-Zeus anti-ballistic missile system, with Bell Laboratories responsible for much of the hardware design. This program was replaced by Nike-X in 1963, which was the first program to propose a digitally controlled phased array radar for guiding the short and long-range interceptor missiles. The objective of the system was to intercept and destroy incoming Soviet nuclear warheads above the atmosphere and thereby protect U.S. population centers.

In 1967, President Lyndon Johnson and Secretary of Defense Robert McNamara redefined the ABM program and changed the name to Sentinel. This system used basically the same hardware as Nike-X, but the threat definition was changed from the Soviet Union to China, where work was underway on less sophisticated ICBMs, and nuclear capability had been demonstrated. This program provoked large scale public protests when it became clear that nuclear tipped interceptor missiles would be deployed very close to the cities they were meant to defend.

Richard Nixon became President in 1969, and the ABM program objective and name was once again changed for political reasons, but still using basically the same hardware. This time the program was called Safeguard, and the new objective was to protect Minuteman ICBM fields, Strategic Air Command bases, and Washington DC. The system would be deployed at up to 12 sites and utilize both short- and long-range missiles.

The Safeguard program became entangled in the politics of the SALT talks with the Soviet Union, and was eventually scaled back significantly. In the end, the Grand Forks, North Dakota site was the only site ever built; it became operational on October 1, 1975. On October 2, 1975, the House of Representatives voted to deactivate the Safeguard program.

Key to the Nike-X/Sentinel/Safeguard systems was the use of digital techniques to control the phased array radar and perform other command and control tasks. The logic was resistor-transistor-logic (RTL), and was mounted in hybrid packages. Also important to the system were the high speed ADCs used in the phased array radar receiver. Early prototypes for the required 8-bit 10-MSPS ADC were developed by John M. Eubanks and Robert C. Bedingfield at Bell Labs between 1963 and 1965. In 1966, these two pioneers in high speed data conversion left Bell Labs and founded Computer Labs—a Greensboro, NC based company—and introduced a commercial version of this ADC.

The 8-bit, 10 MSPS converter was rack-mounted, contained its own linear power supply, dissipated nearly 150 watts, and sold for approximately $10,000 (Figure 1.18). The same technology was used to produce 9-bit, 5 MSPS and 10-bit 3 MSPS versions. Although the next generation of Computer Labs' designs would take advantage of modular op amps (Computer Labs OA-125 and FS-125), ICs such as the Fairchild μA710/711 comparators, as well as 7400 TTL logic, the first ADCs offered used all discrete devices.

The early high speed ADCs produced by Computer Labs were primarily used in research and development projects associated with radar receiver development by companies such as Raytheon, General Electric, and MIT Lincoln Labs.

In the mid-1960s, development of lower speed instrument, PC-board, and modular ADCs was pioneered by such companies as Analogic (founded by Bernard M. Gordon) and Pastoriza Electronics (founded by James Pastoriza). Other companies in data converter business were Adage, Burr Brown, General Instrument Corp, Radiation, Inc., Redcor Corporation, Beckman Instruments, Reeves Instruments, Texas Instruments, Raytheon Computer, Preston Scientific, and Zeltex, Inc. Many of the data converters of the 1960s were in the form of digital voltmeters which used integrating architectures, although Adage introduced an 8-bit, 1-MSPS sampling ADC, the Voldicon VF7, in the early 1960s (Reference 5).

In addition to the widespread proliferation of discrete transistor circuits in the 1960s, various integrated circuit building blocks became available which led to size and power reductions in data converters. In 1964 and 1965, Fairchild introduced two famous Bob Widlar IC designs: the μA709 op amp and the μA710/μA711 comparator. These were quickly followed by a succession of linear ICs from Fairchild and other manufacturers. The 7400-series of transistor-transistor-logic (TTL) and high speed emitter-coupled-logic (ECL) also emerged during this period as well as the 4000-series CMOS logic from RCA in 1968. In addition to these devices, Schottky diodes, Zener reference diodes, FETs suitable for switches, and matched dual JFETs also made up some of the building blocks required in data converter designs of the period.

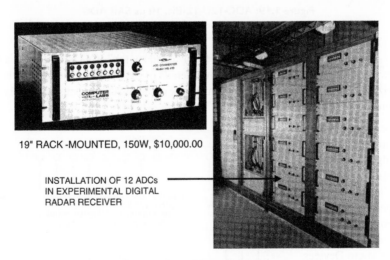

19" RACK -MOUNTED, 150W, $10,000.00

INSTALLATION OF 12 ADCs
IN EXPERIMENTAL DIGITAL
RADAR RECEIVER

Figure 1.18: HS-810, 8-bit, 10-MSPS ADC
Released by Computer Labs, Inc. in 1966

In 1965, Ray Stata and Matt Lorber founded Analog Devices, Inc. (ADI) in Cambridge, MA. The initial product offerings were high performance modular op amps, but in 1969 ADI acquired Pastoriza Electronics, a leader in data converter products, thereby making a solid commitment to both data acquisition and linear products.

Pastoriza had a line of data acquisition products, and Figure 1.19 shows a photograph of a 1969 12-bit, 10 µs general-purpose successive approximation ADC, the ADC-12U, that sold for approximately $800.00. The architecture was successive approximation, and the ADC-12U utilized a µA710 comparator, a modular 12-bit "MiniDAC," and 14 7400 series logic packages to perform the successive approximation conversion algorithm.

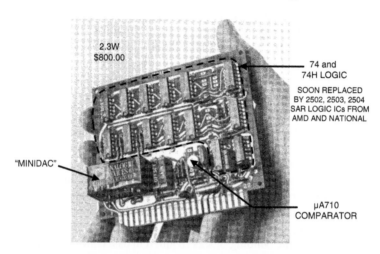

**Figure 1.19: ADC-12U 12-Bit, 10 µs SAR ADC
from Pastoriza Division of Analog Devices, 1969**

The "MiniDAC" module was actually constructed from "quad switch" ICs (AD550) and a thin film network (AD850) as shown in Figure 1.20. Figure 1.21 shows details of the famous quad switch that was patented by James Pastoriza (Reference 18). Chapter 3 of this book contains more discussion on the quad switch and other DAC architectures.

Notice that in the ADC-12U, the implementation of the successive approximation algorithm required 14 logic packages. In 1958, Bernard M. Gordon had filed a patent on the logic to perform the successive approximation algorithm (Reference 19), and in the early 1970s, Advanced Micro Devices and National Semiconductor introduced commercial *successive approximation register* logic ICs: the 2502 (8-bit, serial, not expandable), 2503 (8-bit, expandable) and 2504 (12-bit, serial, expandable).

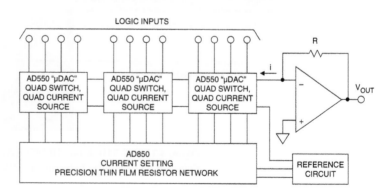

**Figure 1.20: A 1969 Vintage 12-bit "MiniDAC" Using
Quad Current Switches, a Thin Film Resistor Network,
a Voltage Reference, and an Op Amp**

These were designed specifically to perform the register and control functions in successive approximation ADCs. These became standard building blocks in many modular and hybrid data converters. In fact, the acronym *SAR* actually stands for *successive approximation register*, and hence the term *SAR* ADC. Earlier converters utilizing the successive approximation architecture were simply referred to as *sequential coders*, *feedback coders*, or *feedback subtractor coders*.

Data Converter Architectures

Details of the history of the various data converter architectures are contained in Chapter 3 of this book so, for now, we will just summarize the key developments.

The basic algorithm used in the successive approximation (initially called *feedback subtraction*) ADC conversion process can be traced back to the 1500s relating to the solution of a certain mathematical puzzle regarding the determination of an unknown weight by a minimal sequence of weighing operations (Reference 20). In this problem, as stated, the object is to determine the least number of weights that would serve to weigh an integral number of pounds from 1 lb to 40 lb using a balance scale. One solution put forth by the mathematician Tartaglia in 1556, was to use the series of weights 1 lb, 2 lb, 4 lb, 8 lb, 16 lb, and 32 lb. The proposed weighing algorithm is the same as used in modern successive approximation ADCs. (It should be noted that this solution will actually measure unknown weights up to 63 lb rather than 40 lb as stated in the problem). The algorithm is shown in Figure 1.22 where the unknown weight is 45 lbs. The balance scale analogy is used to demonstrate the algorithm. The electronic implementation of the successive approximation ADC is shown in Figure 1.23.

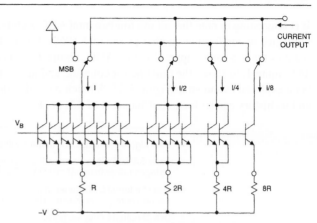

Figure 1.21: "Quad Switch" DAC Building Block with External Thin Film Resistor Network

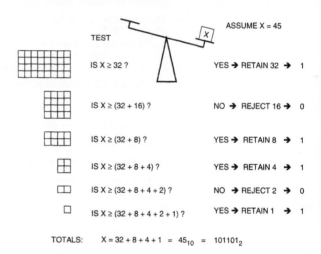

Figure 1.22: Successive Approximation ADC Algorithm Analogy Using Binary Weights

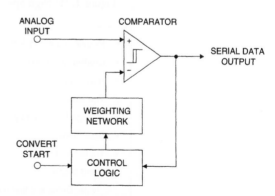

Figure 1.23: Basic Successive Approximation ADC (Feedback Subtraction ADC)

It is interesting to note that all the fundamental ADC architectures used today had been discovered and published in one form or another by the mid-1960s. Figure 1.24 shows a detailed timeline for the development of the successive approximation ADC. Figure 1.25 shows the timeline high speed ADC architectures, and Figure 1.26 shows the timeline for counting and integrating ADCs. Even the Σ-Δ ADC architecture had been explored as shown in Figure 1.27. A much more detailed discussion of each individual architecture and its history can be found in Chapter 3 of this book.

- SAR algorithm dates back to the 1500's
- Early SAR ADCs used individually switched binary reference voltages rather than internal DAC (Schelling: 1946, Goodall: 1947)
- Use of internal DAC rather than switched reference voltages to perform conversion (Kaiser: 1953, B. D. Smith: 1953)
- Use of nonuniformly weighted DAC to produce companding transfer function (B. D. Smith: 1953)
- First commercial vacuum tube SAR ADC, 11-bits, 50kSPS (Bernard M. Gordon, Epsco: 1954)
- Design of specific logic function to perform SAR algorithm (Gordon: 1958)
  Led to popular SAR logic ICs: 2503, 2504 from National Semiconductor and Advanced Micro Devices in the early 1970's

Figure 1.24: SAR ADC Development Summary

• Reeve's counting ADC	1939
• Successive approximation	1946
• Flash (electron tube coders)	1948
• Bit-per-stage (binary and Gray)	1956
• Subranging	1956
• Subranging with error correction	1964
• Pipeline with error correction	1966
Note: Dates are first publications or patent filings	

Figure 1.25: High Speed ADC Architecture Timeline

• Reeve's counting ADC	1939
• Charge run-down:	1946
• Ramp run-up	1951
• Tracking	1950
• Voltage-to-frequency converter (VFC)	1952
• Dual Slope	1957
• Triple Slope	1967
• Quad Slope	1973
Note: Dates are first publications or patent filings	

Figure 1.26: Counting and Integrating ADC Architecture Timeline

By the end of the 1960s, the key architectures and building blocks were available to allow for modular and ultimately hybrid converters, and significant work was already underway to produce the first monolithic converters which were to appear in the early 1970s.

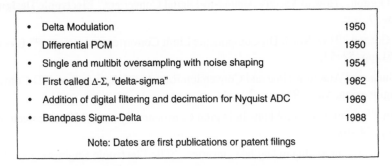

• Delta Modulation	1950
• Differential PCM	1950
• Single and multibit oversampling with noise shaping	1954
• First called Δ-Σ, "delta-sigma"	1962
• Addition of digital filtering and decimation for Nyquist ADC	1969
• Bandpass Sigma-Delta	1988

Note: Dates are first publications or patent filings

Figure 1.27: Sigma-Delta ADC Architecture Timeline

References:
1.2 Data Converters of the 1950s and 1960s

1. G.G. Bowers, "Analog to Digital Converters," **Control Engineering**, April 1957, pp. 107–118.

2. Bernard M. Gordon and Robert P. Talambiras, "Signal Conversion Apparatus," **U.S. Patent 3,108,266**, filed July 22, 1955, issued October 22, 1963. *(Classic patent describing Gordon's 11-bit, 50 kSPS vacuum tube successive approximation ADC done at Epsco. The internal DAC represents the first known use of equal currents switched into an R/2R ladder network.)*

3. G. J. Herring, "High Speed Analogue to Digital Conversion," **Journal British IRE**, August 1957, pp. 407–420.

4. G. J. Herring, "Electronic Digitizing Techniques," **Journal British IRE**, July 1960, pp. 513–517.

5. F.M. Young, "Factors Limiting A/D Conversion State of the Art," **Data Systems Engineering**, May 1964, pp. 35–39.

6. George Flynn, "Analog to Digital Converters," **Electronic Products**, Vol. 10, No. 5, October 1967, pp. 18–48.

7. Bernard M. Gordon, "Definition of Accuracy of Voltage to Digital Converters," **Instruments and Control Systems**, May 1959, p. 710.

8. W. M. Gaines, "Terminology for Functional Characteristics of Analog to Digital Converters," **Control Engineering**, February 1961.

9. Bernard M. Gordon, "Designing Sampled Data Systems," **Control Engineering**, April 1961, pp. 127–132.

10. Bernard M. Gordon, "How to Specify Analog-to-Digital Converters," **Electronic Design**, May 10 1961, pp. 36–39.

11. Bernard M. Gordon, "How Much Do Components Limit Converter Performance?" **Electronic Design**, June 21, 1961, pp. 52–53.

12. P. Barr, "Influence of Aperture Time and Conversion Rate on the Accuracy of A/D Converters," **Data Systems Engineering**, May 1964, pp. 30–34.

13. A. Van Doren, "Solving Error Problems in Digital Conversion Systems," **Electromechanical Design**, April 1966, pp. 44–46.

14. J. Freeman, "Specifying Analog to Digital Converters," **The Electronic Engineer**, June 1968, pp. 44–48.

15. Bernard M. Gordon, "Speaking Out on Analog to Digital Converters," **EEE Magazine**, December 1968.

16. Bernard M. Gordon, "Bernard Gordon of Analogic Speaks Out on What's Wrong with A/D Converter Specs," **EEE Magazine**, February 1969, pp. 54–61.

17. J. O. Edson and H. H. Henning, "Broadband Codecs for an Experimental 224Mb/s PCM Terminal," **Bell System Technical Journal**, Vol. 44, pp. 1887–1940, Nov. 1965.

18. James J. Pastoriza, "Solid State Digital-to-Analog Converter," **U.S. Patent 3,747,088**, filed December 30, 1970, issued July 17, 1973. *(The first patent on the quad switch approach to building high resolution DACs.)*

19. Bernard M. Gordon and Evan T. Colton, "Signal Conversion Apparatus," **U.S. Patent 2,997,704**, filed February 24, 1958, issued August 22, 1961. *(Classic patent describes the logic to perform the successive approximation algorithm in a SAR ADC.)*

20. W. W. Rouse Ball and H. S. M. Coxeter, **Mathematical Recreations and Essays**, Thirteenth Edition, Dover Publications, 1987, pp. 50, 51. *(Describes a mathematical puzzle for measuring unknown weights using the minimum number of weighing operations. The solution proposed in the 1500's is the same basic successive approximation algorithm used today.)*

Note on Anti-Ballistic Missile (ABM) system history:

Although most of the information regarding the various ABM systems was classified at the time, today the information is in the public domain. There are several excellent websites which can provide further information and references:

- www.ucsusa.org is a site maintained by the Union of Concerned Scientists, 2 Brattle Square, Cambridge, MA 02238. There is a section on the site under the security section titled "From Nike-Zeus to Safeguard: US Defenses Against ICBMs, 1958–1976."

- www.painless.id.au/missiles is a site maintained by an individual with an excellent collection of references about the ABM programs. A good whitepaper on the site is titled "Nuclear ABM Defense of the USA."

In addition to the websites, the following book published by the Bell System provides a good discussion of the development of the ABM system:

- M. D. Fagen, **A History of Engineering and Science in the Bell System: National Service in War and Peace (1925–1975)**, Bell Telephone Laboratories, 1978, ISBN 0-932764-00-2.

Data Converters of the 1970s

Introduction

The year 1970 began one of the most exciting decades in the history of data converters. The ADC/DAC market was driven by a number of applications, including high resolution digital voltmeters, industrial process control, digital video, military phased array radar, medical imaging, vector scan displays, and raster scan displays. Most of these systems had formerly utilized conventional analog signal processing techniques, and the increased availability of low cost computing technology generated a desire to take advantage of the increased performance and flexibility offered by digital signal processing and analysis—and of course, the need for compatible data converters.

As a result, a number of companies entered the data converter field, including Analog Devices, Analogic Corporation (initially Epsco and later Gordon Engineering), Burr Brown, Computer Labs, Datel, Hybrid Systems, ILC/Data Device Corporation, Micronetworks, National Semiconductor, Teledyne Philbrick, and Zeltex.

Integrated circuit building blocks, as well as complete IC data converters of the 1970s, came from Analog Devices, Advanced Micro Devices, Fairchild, Signetics, Intersil, Micro Power Systems, Motorola, National Semiconductor, TRW (LSI Division), and Precision Monolithics.

The data converters of the 1970s maximized utilization of all the available technologies: monolithic, modular, and hybrid, with the modular and hybrid products typically offering higher resolution and faster speed than the existing monolithic parts.

An often overlooked fact is that both customer education and high quality application support is required to take full advantage of an emerging technology such as data conversion. From the beginning, Analog Devices has always realized the importance of excellent application material, starting with a set of tutorial articles on op amps by its founder, Ray Stata (Reference 1). These articles were published in 1965—the same year Analog Devices was founded—and are still recognized as classic tutorials on basic op amp theory and applications.

The ADI continuing thread of customer support through applications information was enhanced considerably in 1967, when the *Analog Dialogue* magazine was launched (see Reference 2). The initial charter for the magazine was stated as "A Journal for the Exchange of Operational Amplifier Technology," later on this was broadened to "A Journal for the Exchange of Analog Technology."

Dissemination of analog circuit information is what the early *Analog Dialogue* did, and did well. The premier issue featured an op amp article by Ray Stata that is still available as an application note (see Reference 3). A similar comment can also be made for a subsequent Ray Stata article (see Reference 4).

A milestone in the life of the young magazine was the arrival of Dan Sheingold as editor, in 1969 (see Reference 5). Already highly experienced as a skilled op amp expert and editorial writer from vacuum tube and early solid-state years at George A. Philbrick Researches (GAP/R), Dan Sheingold brought a unique set of skills to the task of editorial guidance for *Analog Dialogue*. Dan's leadership as editor continues today, in 2004. For more than 35 years his high technical communication standards have been an industry benchmark.

Realizing the need for a comprehensive book on the newly emerging field of data conversion, Analog Devices published the first edition of the *Analog-Digital Conversion Handbook* in 1972, under the editorship of Dan Sheingold (Reference 6). An interim revision, *Analog-Digital Conversion Notes*, was published in 1977 (Reference 7). In 1986, in conjunction with Prentice-Hall, Sheingold published a third revision, again titled *Analog-Digital Conversion Handbook* (Reference 8). All of these books provided detailed information on data converter architectures, specifications, design, and applications, and helped in the adoption of uniform terminology and performance metrics throughout the industry.

Monolithic Data Converters of the 1970s

Bipolar Process IC DACs of the 1970s

The earliest monolithic DACs were made using bipolar process technology; they included only the basic core of a complete DAC—the array of switches and resistors to set the weight of each bit. An example is the 1408 and a later higher speed derivative, the DAC08, introduced in 1975 and shown in Figure 1.28.

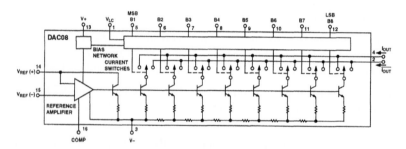

Figure 1.28: DAC08 8-Bit 85 ns IC DAC, 1975

These converters were produced by several manufacturers and were available at low cost. However, they required many additional external components in order to be usable in a system design. These external components included several resistors, a voltage reference, a latch, an output op amp, possibly a compensation capacitor, and usually one or more trimming potentiometers.

Converters like the 1408 and DAC08 were limited to 8-bit accuracy by the matching and tracking limitations of the diffused resistors. When higher accuracy is required, lower tempco resistors are needed, and some means of post-fabrication adjustment is desirable.

Thin film resistors exhibit low tempcos and can be trimmed with a laser—they are well suited for use in data converters. By the mid-1970s, Analog Devices had developed considerable expertise, not only the deposition of thin film resistors, but also trimming at the wafer level.

The Analog Devices' AD562, designed by Bob Craven and introduced in 1974, was originally a "compound monolithic" manufactured using two IC chips mounted in the same package, without the traditional hybrid substrate for mounting and interconnection. Instead, the two chips were designed so that a set of wire bonds between the two chips (in addition to the usual ones to the package pins) were all that were necessary to assemble a 12-bit accurate DAC in an IC package. In the original AD562, one chip contained the thin film resistor network (including the bit weight-setting resistors and output gain-setting resistors), and the other chip contained the reference control amplifier and the current switches for the 12 bits. As the processing matured, the manufacture of larger chips became more practical. The two chips of the original AD562 were later merged into a single-chip version. It became the first 12-bit DAC qualified by the U.S. Department of Defense under MIL-M-38510.

While the AD562 was the first 12-bit IC DAC and embodied the solution to some extremely difficult design problems, it was still really only a building block, since it lacked *buffer latches*, a *voltage reference*, and an *output amplifier*. Shortly after the two-chip AD562 was introduced (Reference 9, 10), a version with a third chip was developed. The third chip was a 2.5 V bandgap reference (designed by Paul Brokaw and described in detail in Reference 11 and 12). This made the DAC function much more complete. The resulting product, known as the AD563, also became quite popular and eventually made the transition to a completely monolithic single-chip device.

Another problem with the AD562 was that, while reasonably fast, it lacked sufficient speed for many applications—its settling time was approximately 1 μs. Later advances in switch design and Zener diode fabrication led to a higher speed DAC, the AD565, introduced in 1978 (later followed by the AD565A in 1981). A simplified diagram of the AD565 is shown in Figure 1.29.

The bit switches used in this design were much smaller than those used in the AD562, allowing a substantial reduction in chip area and increasing the yield of good chips per wafer. The new switches yielded a settling time of 200 ns to ½ LSB. The AD565 used a buried Zener reference which had less noise than the bandgap reference used in the AD563.

The AD565 retained the same pinout as the earlier AD563, allowing drop-in replacement with improved performance and lower price.

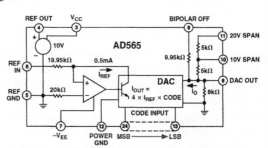

Figure 1.29: AD565, 12-Bit, 200 ns DAC, 1978

It should be noted that the first monolithic single-chip 10-bit DAC with thin film laser wafer trimmed (LWT) resistors and internal reference was the AD561, designed by Peter Holloway, and was introduced by Analog Devices in 1976 (Reference 13). This DAC utilized a method of compensating for errors produced by operating the internal current source transistors at different current densities. This idea is patented by Paul Brokaw and is one of the most widely referenced patents in data conversion (Reference 14).

CMOS IC DACs of the 1970s

As we have seen, the early commercially available monolithic DACs were principally processed by conventional bipolar linear processing techniques. Before 1974, when the AD7520 CMOS DAC was introduced, 10-bit conversion had been difficult to obtain with good yields (and low cost) because of the finite β of switching devices, the V_{BE}-matching requirement, the matching and tracking requirements on the diffused resistor ladders, and the tracking limitations caused by the thermal gradients produced by high internal power dissipation.

Most of these problems were solved or avoided with CMOS devices. The CMOS transistors have nearly infinite current gain, eliminating β problems. There is no equivalent in CMOS circuitry to a bipolar transistor's V_{BE} drop; instead, a CMOS switch in the ON condition is almost purely resistive, with the resistance value controllable by device geometry. The temperature problems of diffused resistors were eliminated by using thin film resistors instead. A simplified diagram of the AD7520 10-bit, 500 ns CMOS multiplying DAC introduced in 1974 is shown in Figure 1.30.

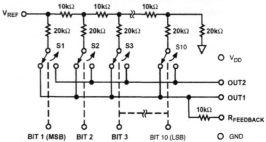

Figure 1.30: AD7520, the First Monolithic CMOS Multiplying 10-Bit DAC, 1974

The AD7520 architecture is a standard current-mode R-2R (also referred to as an "inverted R-2R") and is described further in Chapter 3 of this book. The output drives the inverting input of an external op amp connected as an I/V converter. The 10 kΩ feedback resistor for the op amp is internal to the AD7520 to provide good tracking. The key to the linearity of the AD7520 is that the geometries of the switches corresponding to the first six bits are tapered so as to obtain ON resistances that are related in binary fashion.

The AD7520 architecture was extended to 12-bit resolution in the AD7541 by merely adding additional switch cells and resistors. However, in order to achieve 12-bit linearity, laser trimming at the wafer level was required. The AD7541, introduced in 1978, was the first 12-bit CMOS multiplying DAC. Settling time to ½ LSB was 1 μs.

The AD7520 and AD7541 were the beginning of an entire product line of general purpose multiplying CMOS DACs from Analog Devices. Some of these products will be discussed in Section 1.3, *Data Converters of the 1980s*.

Another useful feature for a DAC is the addition of an on-chip latch (generally referred to as a "buffered" DAC). The latch allows the DAC to be connected to a microprocessor data bus. The AD7524, shown in Figure 1.31, was an 8-bit multiplying CMOS DAC which had an on-chip latch. Data is loaded into the DAC by first asserting the CHIP SELECT pin. Data is then written into the latch by asserting the WRITE pin. Returning the WRITE pin to 0 disconnects the latch from the data bus, and the bus can then be connected to another device if desired.

Future DAC products added a second latch, and are referred to as a "double-buffered" DAC. The input latch is used to load the data (either serial, parallel, or in bytes), and when the second parallel "DAC latch" is clocked, the DAC output is updated.

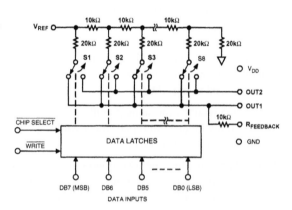

Figure 1.31: AD7524 8-Bit Buffered μP-Compatible DAC, 1978

Data converter designers soon realized the importance of making them easy to interface to microprocessors and DSPs, and this trend added additional functionality to the devices and continues to this day.

A summary of the monolithic DAC developments during the 1970s is shown in Figure 1.32.

Figure 1.32: Summary: Monolithic DACs of the 1970s

- Bipolar:
  - AD550 "μDAC" Building Block Quad Switch, 1970
  - AD562 12-Bit, 1.5μs (2-chip, compound monolithic) DAC, 1974
  - 1408 8-Bit 250ns DAC, 1975
  - DAC08 8-Bit, 80ns DAC, 1976
  - AD561 10-Bit, 250ns, LWT Current-Output DAC with Reference, 1976
  - AD565 12-Bit, 200ns, LWT Current-Output DAC with Reference, 1978

- CMOS:
  - AD7520 10-Bit, 500ns, Multiplying DAC, 1974
  - AD7541 12-Bit, 1μs, LWT Multiplying DAC, 1978
  - AD7524, 8-Bit, 150ns LWT Multiplying DAC with DAC buffer latch, μP interface, 1978

Monolithic ADCs of the 1970s

Although most of the ADCs of the early 1970s were modular or hybrid, there was considerable effort by data converter manufacturers to produce an all-monolithic ADC. An early attempt was the AD7570 10-bit, 20 μs CMOS SAR ADC introduced in 1975. However, due to the difficulty of designing good comparators, amplifiers, and references on the early CMOS process, the AD7570 required an external LM311 comparator as well as a voltage reference.

The integrating ADC architecture was suitable for the early CMOS processes, and in 1976, Analog Devices introduced the 13-bit AD7550 which utilized a unique architecture called "quad slope." The architecture was patented by Ivar Wold (Reference 15).

The first complete monolithic ADC was the 10-bit, 25 μs AD571 SAR ADC introduced in 1978 and designed by Paul Brokaw (Reference 16). The AD571 was designed on a bipolar process with LWT thin film resistors. In order to implement the logic functions required in the SAR ADC, integrated-injection logic (I²L) was added to the bipolar process. This process allowed reasonably dense low voltage logic to be included on the same chip as high breakdown precision linear circuitry.

The I²L process was particularly useful in manufacturing ADCs, because only a single additional diffusion step was required beyond those used in the standard linear process. Furthermore, this diffusion did not significantly interfere with the other steps in the process, so the analog circuitry was relatively unaffected by the addition of the logic.

The 10-bit AD571 (and the 8-bit AD570) were completely self-contained monolithic ADCs with internal clock, buried Zener voltage reference, laser-trimmed DAC (based on the design in Reference 14), and three-state output buffers. A simplified diagram of the AD571 is shown in Figure 1.33.

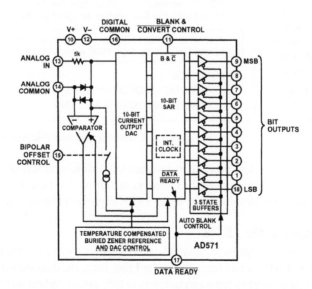

Figure 1.33: AD571 Complete 10-Bit, 25 μs IC ADC, 1978

Probably the most significant SAR ADC ever introduced was the 12-bit, 35 μs AD574 in 1978. The AD574 represents a complete solution, including buried Zener reference, timing circuits, and three-state output buffers for direct interfacing to an 8-, 12-, or 16-bit microprocessor bus. In its introductory form, the AD574 was manufactured using compound monolithic construction, based on two chips—one an AD565 12-bit current-output DAC, including reference and thin film scaling resistors; and the other containing the successive approximation register (SAR) and microprocessor interface logic functions as well as a precision latching comparator. The AD574 soon emerged as the industry-standard 12-bit ADC in the early 1980s. In 1985, the device became available in single-chip monolithic form for the first time; thereby making low-cost commercial plastic packaging possible. A simplified block diagram of the AD574 is shown in Figure 1.34.

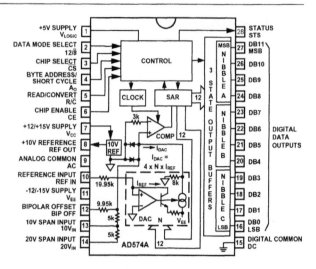

Figure 1.34: The Industry-Standard AD574 12-Bit, 35 μs IC ADC, 1978

As the 1970s came to a close, the first high speed video flash ADCs emerged, starting with the TDC-1007J 8-bit, 30 MSPS from the LSI division of TRW in 1979 (Reference 17). TRW also introduced a lower power 6-bit version, the TDC-1014J. Also in 1979, Advanced Micro Devices introduced the AM6688 4-bit, 100 MSPS flash

- AD7570, 10-bit, 20μs CMOS SAR ADC, (required external comparator, reference) 1975
- AD7550, 13-bit integrating ADC based on "quad slope" technique, 1976
- AD571, Complete 10-bit, 25μs monolithic SAR ADC, *with reference,* using I²L and LWT thin film resistors, designed by Paul Brokaw, 1978

- AD574, 12-bit, 35μs two-chip, compound monolithic complete SAR ADC: AD565 DAC + logic chip), 1978; single-chip version 1985

- Flash ADCs:
  - TRW TDC-1007J/TDC-1016J, 8-bit/6-bit 30MSPS ADCs, 1979
  - AM6688 4-bit, 100MSPS ADC, 1979

Figure 1.35: Summary: Monolithic ADCs of the 1970s

ADC, designed by Jim Giles, who had previously designed the AM685 and AM687 fast ECL comparators. There is more discussion on the history of flash converters in *Section 1.4, Data Converters of the 1980s*. A summary of key monolithic ADC developments in the 1970s is given in Figure 1.35.

Hybrid Data Converters of the 1970s

Although a few more instrument-type rack-mounted data converters were developed in the early 1970s (such as the VHS-series and 7000-series from Computer Labs, Inc.), the demand for lower cost, more compact high-performance data converters led manufacturers to turn toward hybrid and modular techniques—as the monolithic technology of the period was not yet capable of supporting the high-end converter functions in single-chip form.

Hybrid and modular data converter designers of the 1970s had a virtual smorgasbord of components from which to choose, including IC op amps, IC DACs, comparators, discrete transistors, various logic chips, etc. Figure 1.36 lists some of the more popular hybrid and modular building blocks for the 1970s.

- Quad switches (AD550µDAC)
- Precision thin film resistor networks (AD850)
- IC DACs: AD562, AD563, AD565, 1408, DAC08
- IC Comparators: µA710, µA711, NE521, LM311, LM361, MC1650, AM685, AM687
- Successive approximation registers (SARs): 2502, 2503, 2504
- IC and hybrid op amps
- IC voltage references, Zenerreferences
- Fast PNP and fast NPN discrete transistors
- Matched monolithic dual FETs
- Monolithic transistor arrays (RCA CA-series)
- Schottky diodes
- CMOS and DMOS switches
- TTL, CMOS, ECL logic
- 4-, 6-, 8-bit monolithic flash ADCs (starting in 1979)

**Figure 1.36: Building Block Components for
1970s Hybrid and Modular Data Converters**

Hybrids generally utilize ceramic substrates with either thick or thin film conductors. Individual die are bonded to the substrate (usually with epoxy), and wire bonds make the connections between the bond pads and the conductors. The hybrid is usually hermetically sealed in some sort of ceramic or metal package. Accuracy was achieved by trimming thick or thin film resistors after assembly and interconnection, but before sealing. Manufacturers used thin film networks, discrete thin film resistors, deposited thick or thin film resistors, or some combination of the above.

Although the chip-and-wire hybrid was certainly more expensive to manufacture than an IC, it allowed performance levels that could not be achieved with existing monolithic technology of the period. The popular hybrid circuits followed an evolutionary path to monolithic form, generally over a 5-to-10-year period, depending upon the particular device. One of the most popular 12-bit DACs of the 1970s was the DAC80, originally introduced in the mid-1970s as a hybrid device consisting of 11 chips: three quad switch arrays, two op amps, two resistor networks, a Zener diode, two clamp diodes, and a chip capacitor (see Figure 1.37A). In 1978, when monolithic technology had progressed to the point where it was possible to combine the switch and resistor network into a single chip, a three-chip DAC80 was introduced as shown in Figure 1.37B. The three chips in this design included a voltage reference, an output op amp, and the switch/resistor/control-amplifier chip.

The newer design offered performance identical to that of the original DAC80, but with a tremendous improvement in reliability and at a much lower cost. Then, in 1983, the first single-chip DAC80 became available (Figure 1.37C). It, of course, provided further cost reduction and reliability improvement, compared to the 3- and 11-chip versions. Finally, in

Figure 1.37: DAC80 12-Bit DAC Evolution

1984, this popular device was offered in a low-cost plastic DIP package. Thus, in approximately 10 years, the DAC80 evolved from a relatively high cost hybrid to a high volume commodity IC.

Another excellent example of hybrid technology was the AD572 12-bit, 25 μs SAR ADC introduced in 1977. The AD572 was complete with internal clock, voltage reference, comparator, and input buffer amplifier. The SAR register was the popular 2504. The internal DAC was comprised of a 12-bit switch chip and an actively trimmed thin film ladder network (separately packaged as the two-chip AD562 DAC). The AD572 was the first military-approved 12-bit ADC processed to MIL-STD-883B, and specified over the full operating temperature range of –55°C to +125°C. A photograph of the AD572 is shown in Figure 1.38.

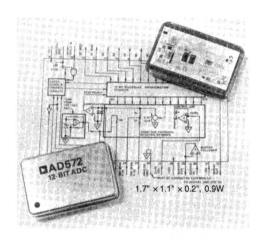

Figure 1.38: AD572 12-Bit, 25 μs
Military-Approved Hybrid ADC, 1977

Many hybrid circuits were introduced in the 1970s, and a few of the key ones are listed in Figure 1.39. Toward the end of the 1970s, Computer Labs, Inc., introduced a number of very fast hybrid data converters based on thick film laser trimmed resistor technology (Computer Labs was acquired by Analog Devices in 1978). The thick film resistor technology developed at Computer Labs during the 1970s was capable of 12-bit accuracy, which was quite remarkable, as most hybrid manufacturers used more expensive thin film resistors for 12-bit devices.

- DAC80, 12-bit DAC, 1975

- ADC80, 12-bit, 25μs SAR ADC, 1975

- AD572, 12-bit, 25μs Military-Approved ADC, 1977

- HDS-1250 12-bit, 35ns DAC (also 8-, 10-bit versions), 1979

- HAS-1202, 12-bit, 2.2μs SAR ADC (also 8-, 10-bit versions), 1979

- HTC-0300, 300ns SHA; HTS-0025, 25ns SHA, 1979

Figure 1.39: Hybrid ADC and DAC Milestones of the 1970s

Both the HDS-series DACs and the HAS-series ADCs utilized actively trimmed thick film resistors and discrete PNP transistor switches for the internal DACs.

It should be noted at this point that none of the monolithic or hybrid ADCs of the 1970s were sampling ADCs with internal sample-and-holds (SHAs). In order to process ac signals, a separate SHA had to be connected to the ADC (with the appropriate interface and timing circuits). This generated the need for hybrid SHAs, such as the HTC-0300 and the faster HTS-0025.

Modular Data Converters of the 1970s

Designers of modular data converters in the 1970s had even more flexibility than hybrid designers. In fact, modular technology began in the late 1960s before hybrids became popular, and still exists to this day for certain products. Figure 1.40 shows two of the early popular modular ADCs: the ADC-12QZ and the MAS-1202. The modular technology was quite straightforward—components were mounted on a small PC board and encapsulated in a potted module after trimming (usually done with manually selected resistors). The potting compound served to distribute the heat throughout the module, provided some degree of thermal tracking between critical components, and made it somewhat more difficult for competitors to reverse engineer the circuits.

ADC-12QZ GENERAL PURPOSE 12-BIT,
40µs SAR ADC INTRODUCED IN 1972

MAS-1202 12-BIT, 2µs
SAR ADC INTRODUCED IN 1975

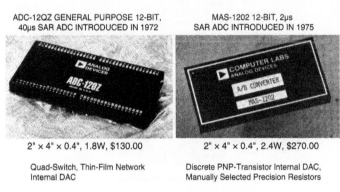

2" × 4" × 0.4", 1.8W, $130.00

2" × 4" × 0.4", 2.4W, $270.00

Quad-Switch, Thin-Film Network
Internal DAC

Discrete PNP-Transistor Internal DAC,
Manually Selected Precision Resistors

Figure 1.40: Early Modular ADCs of the 1970s

The ADC-12QZ 12-bit, 40 µs SAR ADC was introduced by Analog Devices in 1972, an outgrowth of the Pastoriza Electronics' early converter product line acquired in 1969. The ADC-12QZ utilized a DAC made up of three "quad switch" building blocks previously discussed and a matching pretrimmed thin film resistor network. A SAR logic chip, comparator, a reference, and miscellaneous other components completed the parts list.

The MAS-1202 12-bit, 2 µs SAR ADC designed at Computer Labs, Inc., was released in 1975 and utilized an internal DAC based on fast PNP transistor switches and manually-selected precision resistors.

The ADC-12QZ and the MAS-1202 each dissipated approximately 2 W, and commanded premium prices of $130.00 and $270.00, respectively.

One of the first complete modular *sampling* ADCs of the 1970s utilized an open-card construction with a combination of hybrid, IC, and discrete building blocks. The MOD-815 8-bit, 15 MSPS ADC was introduced by Computer Labs in 1976. The design utilized two 4-bit flash converters in a subranging architecture (see Chapter 3 of this book). Each 4-bit flash converter was constructed of eight dual AM687 ECL comparators. The MOD-815 was one of the first commercial ADCs used in the rapidly emerging field of digital television.

Other popular card-level modules included the MOD-1205 12-bit, 5 MSPS ADC and the MOD-1020 10-bit, 20 MSPS ADC introduced by Analog Devices/Computer Labs in 1979. These products made use of the first flash converters on the market: the Advanced Micro Devices' AM6688 4-bit, 100 MSPS flash and the TRW TDC-1007J (8-bit) and TDC-1014J (6-bit) flash converters. Figure 1.41 shows a photo of the MOD-1020 with the key devices labeled. The architecture was subranging, with two "stacked" AM6688s providing the first 5-bit conversion, and the 6-bit TDC-1014J the second 6-bit conversion. The extra bit was used for error correction. The MOD-1020 utilized quite a bit of ECL logic and dissipated a total of 21 W. Due to the high level of performance and the large number of costly hybrid and IC building blocks, the MOD-1020 commanded a premium price of $3,500.00.

Starting in the mid-1970s, most of the modular sampling converters were tested using FFT techniques to measure SNR, ENOB, and distortion. (See Chapter 5 of this book, *Testing Data Converters*).

A summary of some of the popular modular ADCs and DACs of the 1970s is given in Figure 1.42.

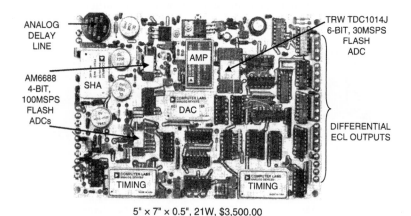

5" × 7" × 0.5", 21W, $3,500.00

Figure 1.41: MOD-1020, 10-bit, 20 MSPS Sampling ADC Introduced in 1979.

- DAC-12QZ, 12-bit DAC, 1970
- ADC-12QZ, 12-bit, 40μs SAR ADC, 1972
- MAS-1202, 12-bit, 2μs SAR ADC (also 8- 10- bit versions) 1975
- ADC1130, 14-bit, 12μs SAR ADC, 1975
- MDS-1250, 12-bit, 50ns DAC, (also 8-, 10-bit versions) 1975
- THS-0300, 300ns SHA; THS-0025, 25ns SHA, 1975
- MOD-815, 8-bit, 15MSPS Video, sampling ADC, 1976
- SDC1700, Synchro-to-Digital Converter, 1977
- DAC1138, 18-bit DAC, (most accurate DAC for 10 years), 1977
- MOD-1205, 12-bit, 5MSPS, sampling ADC, 1979
- MOD-1020, 10-bit, 20MSPS, sampling ADC, 1979

Figure 1.42: Modular ADC and DAC Milestones of the 1970s

References:
1.3 Data Converters of the 1970s

1. Ray Stata, "Operational Amplifiers – Parts I and II," **Electromechanical Design**, September, November, 1965.

2. **Analog Dialogue**, Vol. 1, No. 1, April 1967. *(The premier issue of Analog Dialogue—'A Journal for the Exchange of Operational Amplifier Technology'.)*

3. Ray Stata, "Operational Integrators," **Analog Dialogue**, Vol. 1, No. 1, April, 1967. *(Reprinted as ADI AN357.)*

4. Ray Stata, "User's Guide to Applying and Measuring Operational Amplifier Specifications," **Analog Dialogue**, Vol. 1, No. 3. *(Reprinted as ADI AN356).*

5. **Analog Dialogue**, Vol. 3, No. 1, March 1969. *(The first issue of Analog Dialogue under the editorial guidance of Dan Sheingold.)*

6. Dan Sheingold, **Analog-Digital Conversion Handbook**, First Edition, Analog Devices, 1972.

7. Dan Sheingold, **Analog-Digital Conversion Notes**, Analog Devices, 1977.

8. Dan Sheingold, **Analog-Digital Conversion Handbook**, Analog Devices/Prentice-Hall, 1986, ISBN-0-13-032848-0.

9. Robert B. Craven, "An Integrated Circuit 12-bit D/A Converter, **IEEE International Solid-State Circuits Conference Digest of Technical Papers**, February 1975, pp. 40–41.

10. Robert B. Craven, "Solid-State Digital-to-Analog Converter, **U.S. Patent 3,961,326**, filed September 12, 1974, issued June 1, 1976.

11. Paul Brokaw, "A Simple Three-Terminal IC Bandgap Voltage Reference," **IEEE Journal of Solid State Circuits**, Vol. SC-9, December, 1974.

12. Paul Brokaw, "More About the AD580 Monolithic IC Voltage Regulator," **Analog Dialogue**, 9-1, 1975.

13. Peter Holloway and Mark Norton, "A High-Yield Second-Generation 10-Bit Monolithic DAC," **ISSCC Digest of Technical Papers**, February 1976, pp. 106–107.

14. Adrian Paul Brokaw, "Digital-to-Analog Converter with Current Source Transistors Operated Accurately at Different Current Densities," **U.S. Patent 3,940,760**, filed March 21, 1975, issued February 24, 1976.

15. Ivar Wold, "Integrating Analog-to-Digital Converter Having Digitally Derived Offset Error Compensation and Bipolar Operation without Zero Discontinuity," **U.S. Patent 3,872,466**, filed July 19, 1973, issued March 18, 1975.

16. A. Paul Brokaw, "A Monolithic 10-Bit A/D Using I^2L and LWT Thin-Film Resistors," **IEEE Journal of Solid-State Circuits**, Vol. SC-13, pp. 736–745, December 1978.

17. J. Peterson, "A Monolithic Video A/D Converter," **IEEE Journal of Solid-State Circuits**, Vol. SC-14, No. 6, December 1979, pp. 932–937.

Data Converters of the 1980s
Walt Kester

Introduction

The 1980s represented high growth years for both IC, hybrid, and modular data converters. The driving market forces were instrumentation, data acquisition, medical imaging, professional and consumer audio/video, computer graphics, and a host of others. The increased availability of relatively low cost microprocessors, high speed memory, DSPs, and the emergence of the IBM-compatible PC increased the interest in all areas of signal processing. The emphasis in ADCs began to rapidly shift to include ac performance and wide dynamic range; and hence, a great demand for *sampling* ADCs at all frequencies. Specifications such as signal-to-noise ratio (SNR), signal-to-noise and distortion (SINAD), effective number of bits (ENOB), noise power ratio (NPR), spurious free dynamic range (SFDR), aperture time jitter, etc., began to appear on most ADC data sheets; and glitch impulse area, SFDR, etc., on DAC data sheets.

There was a proliferation of high speed bipolar and CMOS flash ADCs in the 1980s, with 4-, 6-, 8-, 9-, and 10-bit at sampling rates from 20 MSPS to 100 MSPS. Digital video was a chief driving force for the 8-, 9-, and 10-bit devices. In the graphics display area, high speed video RAM-DACs emerged, with CMOS being the ideal process for these memory-intensive devices.

Voiceband and audio signal processing led to the demand for 16- and 18-bit ADCs and DACs, and the emergence of the compact disk (CD) player fueled the need for low cost audio DACs.

In the 1980s, general-purpose ADCs and DACs began to offer more resolution, more functionality, and more complete solutions to the problems of data acquisition and distribution, including multichannel ADCs and DACs. The development of linear-compatible CMOS processes (such as the Analog Devices' LC^{2}MOS and BiCMOS II in the mid-1980s) allowed data converter designers to provide more functionality by the addition of features such as on-chip voltage references and buffer amplifiers, for example.

Another important process development in the mid-1980s was the introduction of Analog Devices' first-generation complementary bipolar (CB) process, which offered high-speed, high performance matching PNP and NPN transistors. The high speed op amps produced on the CB process made excellent drivers for many of the new ADCs, and the CB process eventually yielded some extremely high performance IF-sampling ADCs in the 1990s. Refer to Chapter 4 of this book, *Data Converter Process Technology*, for more discussion regarding this topic.

Monolithic DACs of the 1980s

A listing of key IC DAC product introductions during the 1980s is shown in Figure 1.43. Rather than discuss each one individually, we will simply point out examples that illustrate the general trends in the product line.

- AD558, 8-bit, 1µs, *µP-compatible*, *voltage output*, Bipolar/I²L DAC, 1980
- AD7528, *Dual* 8-bit, *buffered*,CMOS MDAC, 1981
- AD7546, *16-bit*, *segmented*, CMOS voltage mode DAC (required external amps), 1982
- AD7545, *Buffered*, 12-bit CMOS MDAC, 1982
- AD390, *Quad* 12-bit voltage output DAC (compound monolithic), 1982
- AD7240, *12-bit*, *voltage mode*, CMOS DAC, 1983
- AD7226, *Quad* 8-bit, *double-buffered*, *voltage output*, *LC²MOS* DAC, 1984
- AD9700, 125MSPS, 8-bit *Video* ECL DAC, 1984
- AD7535, *14-bit*, double-buffered LC²MOS MDAC, 1985
- AD569, *16-bit*, *segmented*, double-buffered voltage output, *BiCMOS* DAC, 1986
- AD7245, 12-bit, double-buffered, voltage output, LC²MOS DAC, *with reference*, 1987
- AD1856/AD1860, 16-/18-bit *audio* BiCMOS DACs for CD players, 1988
- ADV453/ADV471/ADV476/ADV478 CMOS *Video RAM-DACs*, 1988
- AD7840, *14-bit*, double-buffered, voltage output, LC²MOS DAC, *with reference*, 1989
- AD7846, *16-bit*, *segmented*, *voltage output*, LC²MOS DAC, 1989

Figure 1.43: Monolithic DACs of the 1980s

The AD558 8-bit, 1 µs CMOS DAC introduced in 1980 illustrates the trend toward microprocessor-compatible interfaces, which are almost universal today for general-purpose DACs and ADCs. Later products would utilize double-buffered digital inputs, where an input register accepted parallel, serial, or byte-wide data, and a second parallel latch was used to actually update the DAC switches.

The AD7546, although requiring two external op amps to perform the complete 16-bit DAC function, illustrates the trend toward higher resolution. The segmented architecture used in the AD7546 was later utilized in the 16-bit AD569, which integrated the entire function onto one chip.

The introduction of multiple DACs in the same package is illustrated by the dual AD7528, the quad AD390 (compound monolithic), and the quad AD7226 (single-chip).

The initial CMOS DACs were current-output, requiring an external op amp to perform the current-to-voltage conversion; but with the advent of LC²MOS and BiCMOS processes, a number of DACs provided voltage-output capability. These same processes also allowed the integration of the voltage reference on-chip, thereby providing a more complete solution.

Audio and video monolithic DACs began to appear in the mid 1980s. The AD1856/AD1860 16-/18-bit audio DACs were targeted toward the emerging compact disk (CD) player market. The AD9700 DAC, introduced in 1984, was the first monolithic IC DAC designed for raster scan graphics applications which allowed the *sync*, *blanking*, *10% white*, and *reference white* levels to be set with separate internal switches. This allowed the full 8-bit range to be dedicated to the *active* video region. The AD9700 was based on the HDG-series of hybrid DACs previously introduced in 1980. The AD9700 would later be replaced by the ADV series of CMOS video RAM-DACs which included on-chip color palette memory as well as the basic DAC function.

Monolithic ADCs of the 1980s

The two-chip AD574, introduced in 1978, was well on its way to becoming an industry standard converter by the time of the introduction of the single-chip AD574 12-bit, 35 μs SAR ADC in 1985. Since that time, this converter did become an industry standard, and is still in production today.

Figure 1.44 lists some of the important monolithic ADCs introduced during the 1980s. It is interesting to note the emergence of the *sampling* monolithic ADC starting in the mid-1980s. The addition of sample-and-holds, references, and buffer amplifiers was made considerably easier with the addition of bipolar capability to the CMOS process (LC²MOS and BiCMOS). Another trend which emerged was the addition of front-end multiplexers to the basic ADC, as in the case of the 4-channel AD7582 in 1984, thereby providing a more complete data acquisition solution.

Although the basic Σ-Δ ADC architecture had been well known since the 1950s and 1960s, the first actual commercial offering of a monolithic Σ-Δ ADC was in 1988 by Crystal Semiconductor (CSZ5316). The device had 16-bit resolution and an effective throughput rate of 20 kSPS, making it suitable for voiceband digitization. The digital audio market (both professional and consumer) generated a demand for Σ-Δ ADCs with higher throughput rates and greater resolution; and the precision measurement market required 20+ bit resolution, although at much lower throughput rates. Both these needs would be addressed during the 1990s by an explosion of ADCs and DACs using the Σ-Δ architecture.

- AD574, 12-bit, 35μs, industry-standard, single-chip ADC, 1985
- AD673, 8-bit, complete ADC, 1983
- AD7582, *4-channel* muxed input 12-bit CMOS ADC, 1984
- AD670, 8-bit, 10μs ADCPORT, 1984
- AD7820, 8-bit, 1.36μs half-flash, *sampling* ADC, 1985
- AD7572, 12-bit, 5μs SAR LC²MOS ADC with reference, 1986
- AD7575, 8-bit, 5μs SAR LC²MOS *sampling* ADC, 1986
- AD7579, 10-bit, 50kSPS, LC²MOS SAR *sampling* ADC with AC specs, 1987
- AD7821, 8-bit, 1MSPS half-flash *sampling* ADC with AC specs, 1988
- AD674, 12-bit,15μs ADC, 1988
- AD7870, 12-bit, 100kSPS LC²MOS SAR *sampling* ADC with AC specs, 1989
- AD7871, 14-bit, 83kSPS LC²MOS SAR *sampling* ADC with AC specs, 1989
- First commercial 16-bit *sigma-delta* ADC, Crystal Semiconductor, 1988

Figure 1.44: Monolithic ADCs of the 1980s

Monolithic Flash ADCs of the 1980s

As previously mentioned, the rapidly growing digital video market, coupled with the introduction of the TRW TDC-1007J 8-bit, 30 MSPS flash ADC in 1979, spurred many IC manufacturers to develop similar, but lower powered flash ADCs with resolutions ranging from 4 to 10 bits, and sampling rates as high as high as 500 MSPS. Most, but certainly not all, are listed in Figure 1.45 which covers the period from approximately 1979 to 1990, the peak years for flash converters.

- TDC1007J, 8-bit, 30MSPS, (TRW, LSI), 1979
- TDC1016J, 6-bit, 30MSPS, (TRW, LSI), 1979
- AM6688, 4-bit, 100MSPS, (AMD), 1979
- SDA6020, 6-bit, 50MSPS, (Siemens) 1980
- TLM1070, 7-bit, 20MSPS, CMOS (Telmos), 1982
- MP7684, 8-bit, 20MSPS, CMOS (Micro Power), 1983
- TDC1048, 8-bit, 30MSPS, (TRW, LSI), 1983
- AD9000, 6-bit, 75MSPS, 1984
- AD9002, 8-bit, 150MSPS, 1987
- AD770, 8-bit, 200MSPS, 1988
- AD9048, 8-bit, 35MSPS, 1988
- AD9006/AD9016, 6-bit, 500MSPS, 1989
- AD9012, 8-bit, 100MSPS, TTL, 1988
- AD9028/AD9038 8-bit, 300MSPS, 1989
- AD9020, 10-bit, 60MSPS, 1990
- AD9058, Dual 8-bit, 50MSPS, 1990
- AD9060, 10-bit, 75MSPS, 1990

Figure 1.45: Monolithic Flash ADCs of the 1980s

Both bipolar and CMOS technology was used to produce these devices, with the CMOS converters offering lower power, but at the expense of somewhat inferior performance—especially in the early offerings. The chief problem with the early CMOS flash converters were error codes known as "sparkle codes" produced by comparator metastability (see *Data Converter Architectures*, Chapter 3 for details). The bipolar process comparator designs were less susceptible because they typically had much higher regenerative gain. Today, these metastability problems have largely been overcome in CMOS devices designed on submicron processes; however, they still may occur if care is not taken in the design.

Although bipolar and CMOS flash converter designs were used in most of the 6- to 10-bit video ADCs of the 1980s, the lower power subranging and pipelined architectures became prevalent in the 1990s, as ADC designers gained experience with faster CMOS and BiCMOS processes. Today, the flash converter architecture is widely used as a building block within pipelined ADCs. There are, however, a few GaAs flash converters of 6- or 8-bits resolutions which serve the relatively niche markets requiring sampling rates of 1 GSPS or greater.

Hybrid and Modular DACs and ADCs of the 1980s

The demand for hybrid and modular DACs and ADCs peaked in the 1980s, primarily because of the 3-to-5-year lead time hybrids and modules held compared to single-chip monolithic converters with equivalent performance. In addition, the large number of flash converters and other components served as building blocks for higher resolution subranging ADCs. Some examples of important hybrid and modular data converters introduced in the 1980s are shown in Figure 1.46.

- Hybrids:
  - HDS-1240E, 12-bit, 40ns ECL DAC, 1980
  - HDG-Series, 4-, 6-, 8-bit, 5ns *video* ECL DACs, 1980
  - HAS-1409, 14-bit, 1.25MSPS *sampling* ADC, 1983
  - HAS-1201, 12-bit, 1MSPS *sampling* ADC, 1984
  - AD376, 16-bit, 20µs SAR ADC, 1985
  - AD1332, 12-bit, 125kSPS *sampling* ADC with 32-word FIFO, 1988
  - AD9003, 12-bit, 1MSPS *sampling* ADC, 1988
  - AD9005, 12-bit, 10MSPS *sampling* ADC, 1988
  - AD1377, 16-bit, 10µs SAR ADC, 1989
- Modules:
  - ADC1140, 16-bit, 35µs SAR ADC, 1982
  - CAV-1220, 12-bit, 20MSPS *sampling* ADC, 1986
  - CAV-1040, 10-bit, 40MSPS *sampling* ADC, 1986
  - AD1175, *22-bit integrating* ADC, 1987

**Figure 1.46: High Performance Hybrid and
Modular DACs and ADCs of the 1980s**

An interesting thick film hybrid family, introduced by Analog Devices in 1980, was the HDG-series 4-, 6-, and 8-bit video ECL DACs. Designed for raster scan RGB graphics displays, these DACs had an 8-bit settling time of approximately 5 ns. In addition to fast settling, these were among the first video DACs to allow the *sync, blanking, 10% white*, and *reference white* levels to be set with separate internal switches. This allowed the full 8-bit DAC range to be dedicated to the *active video* region. The HDG-series was a precursor to the fully monolithic CMOS video DACs and RAM-DACs which were introduced later in the 1980s (ADV-series in Figure 1.43).

High performance hybrid subranging ADCs also appeared during the 1980s, most utilizing high speed flash converters as internal building blocks. Most were *sampling* devices, complete with ac specifications, culminating in the 12-bit 10-MSPS AD9005 introduced in 1988.

Hybrids of the 1980s also achieved resolutions not yet attainable in monolithic technology, such as the AD1377 16-bit, 10 µs SAR ADC introduced in 1989.

Also worthy of notice were several modules introduced in the 1980s which also pushed the speed and resolution envelope. The CAV-1220 12-bit, 20 MSPS ADC and the CAV-1040 10-bit, 40 MSPS ADC introduced in 1986 set new standards in high-speed dynamic range performance, while the AD1175 22-bit integrating ADC set the standard for high resolution when it was introduced in 1987.

Data Converters of the 1990s
Walt Kester

Introduction

The markets influencing data converters in the 1990s were even more diverse and demanding than those of the 1980s. Some of the major applications were industrial process control, measurement, instrumentation, medical imaging, audio, video, and computer graphics. In addition, communications became an even bigger driving force for low cost, low power, high performance data converters in modems, cell phone handsets, and wireless infrastructure (basestations).

Other trends were the emphasis on lower power and single-supply voltages for portable battery-powered applications. While the reduced supply voltages were compatible with the higher speed, lower voltage processes, the reduced signal range and headroom made the converter designs more sensitive to noise. Packaging trends also changed in the 1990s from the traditional DIP to smaller surface-mount packages suitable for high volume automatic mass-assembly manufacturing techniques. These included both leaded types and nonleaded types such as ball grid array (BGA) and chip-scale packages (CSP).

Even in the general-purpose data converter market, there was a demand for more analog and digital functionality, such as putting an entire data acquisition system on a chip including the input multiplexer, programmable gain amplifier (PGA), sample-and-hold, and the ADC function. Many applications required both the ADC and DAC function; and this led to the integration of both on a single chip, called a coder-decoder, or CODEC. Specially designed analog front ends (AFEs) and mixed-signal front ends (MxFE™) were included with the basic ADC function in applications such as CCD image processors and IF sampling receivers.

In the TxDAC® family, digital functions such as interpolation filters and digital modulators were integrated with a high speed, low distortion CMOS DAC core. Data converter designers made more use of "core" designs to yield several products with various options, such as serial or parallel output parts, etc. The TxDAC series is a good illustration of the application of this concept, where a variety of resolutions, update rates, and internal digital processing are offered across a broad number of individual products, all of which use essentially the same DAC core.

Because of the increase in frequency-domain signal processing applications, there was even greater emphasis on dynamic range and ac performance in practically all data converters, and a large number of monolithic *sampling* ADCs were introduced to meet the demand. The pipelined subranging architecture virtually replaced the higher power flash ADCs of the 1980s, and some of the key ac specifications were SNR, SINAD, ENOB, and SFDR.

In the 1990s, CMOS became the process-of-choice for general-purpose data converters, with BiCMOS reserved for the high end devices. In a few cases, high speed complementary bipolar processes were utilized for ultrahigh performance data converters. CMOS is also the ideal process for the Σ-Δ architecture, which became the topology of choice for ADCs and DACs used in voiceband and audio applications, as well as in higher resolution, low frequency measurement converters.

A major process technology shift occurred in the 1990s, when parasitics ultimately became the performance-limiting factor for high speed chip-and-wire hybrid data converters. The newer ICs, with their smaller feature size and reduced parasitics, allowed them to achieve higher levels of performance than attainable in a chip-and-wire hybrid or a module—a reversal of the situation that existed throughout the 1970s and most of the 1980s.

In the following sections, we will examine the data converter trends of the 1990s and 2000s, using a few example products as illustrations. It would be impossible to cover the individual products in as much detail as we did for the 1970s and 1980s, simply because of the large number of individual data converter introductions in the 1990s. Many of these products are discussed in Chapter 8 of this book, *Data Converter Applications*.

Monolithic DACs of the 1990s

A significant trend in general-purpose DACs of the 1990s was toward more functionality in all areas, especially with respect to the input structure. DACs were specifically designed to handle parallel, serial, and byte-wide loading, and the inputs were generally double-buffered. The serial interface became popular for interfacing to microprocessors and DSPs. In many cases, the same core DAC design was utilized to provide these various options as separate products in the appropriate packages. Obviously, this required the introduction of many individual DAC products to cover all the desired options.

A variety of options existed with respect to the output structure also. Audio and video DACs tended to use current outputs, while some of the more general-purpose DACs offered either current or voltage output options.

The trend toward multiple DACs is illustrated by an early example of an octal DAC, the AD7568 12-bit LC²MOS DAC shown in Figure 1.47, introduced in 1991. This DAC utilized the popular multiplying architecture and provided current outputs designed to drive an op amp connected as an I/V converter. Notice that the DAC is double buffered—the input shift register accepts the serial input data and loads it into the appropriate input latch, and asserting $\overline{LDAC}$ simultaneously latches the data into the eight parallel individual DAC latches.

Consumer audio compact disk players spurred the market for low distortion 16-bit DACs in the late 1980s, and the first audio DACs were parallel linear DACs capable of oversampling at rates of 8 times or 16 times the basic CD update rate of 44.1 kSPS. Resolutions ranged from 16 bits with the early audio DACs to 18 and 20 bits with later versions. For example, the AD1865 dual 18-bit stereo DAC was introduced in 1991 and was capable of 16 times oversampling.

By the mid-1990s, the Σ-Δ architecture began to replace the parallel DACs in audio applications. Sigma-delta offered much higher oversampling ratios, thereby relaxing output filter requirements, as well as providing higher dynamic range with lower distortion. Some early offerings were the AD1857, AD1858, and AD1859 introduced in 1996. These DACs offered resolutions ranging from 16 bits to 20 bits, used serial interfaces, and were single-supply devices.

Although there were a few bipolar process high speed ECL DACs introduced in the early 1990s, such as the AD9712 12-bit, 100 MSPS DAC and the AD9720 10-bit,

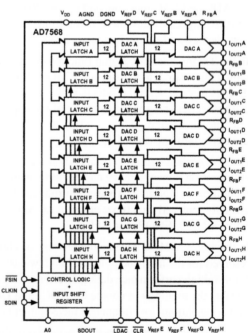

Figure 1.47: AD7568 Octal 12-Bit LC²MOS MDAC, 1991

400 MSPS DAC, the majority of video and communications DACs were low power, low glitch, low distortion CMOS devices. The ADV series of video CMOS RAM-DACs continued to expand during the 1990s, and an entire family of 8-, 10-, 12-, 14-, and 16-bit transmit-DACs (TxDAC) for communications was started in 1996, with new introductions continuing to this day (AD976x, AD977x, and AD978x series).

Direct digital synthesis (DDS) systems on a single chip emerged in the 1990s, largely because of the relative ease with which digital logic could be added to a high performance CMOS DAC core. The first of these was the AD7008, 10-bit, 50 MSPS DDS introduced in 1993 (see Figure 1.48). This was soon followed by additional offerings, such as the 10-bit, 125 MSPS AD9850 in 1996. Later DDS systems added phase and frequency modulation capability, on-chip clock multipliers, more resolution, and update rates as high as 1 GHz.

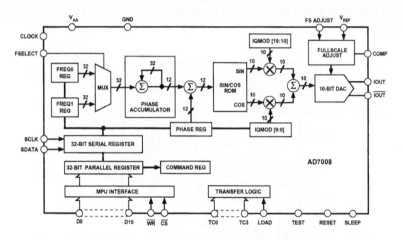

Figure 1.48: AD7008 10-Bit, 50 MSPS Complete CMOS DDS, 1993

The *digital potentiometer*, another highly popular component today, had its origins in 1989 with the release of the AD8800 TrimDAC®, the first in a series. The TrimDACs were basically 8-bit voltage-output DACs designed and optimized to replace mechanical potentiometers. The TrimDAC family became popular, and in 1995, the first DigiPOTs® were introduced. The basic concept behind the digital pot was simply to use a CMOS "string DAC" as a variable resistor. The AD8402 2-channel (8 bits), AD8403 4-channel (8 bits) were the first offerings in 1995. Since then the product line has been expanded to include many more products including some with nonvolatile memory (AD51xx and AD52xx series).

Figure 1.49 summarizes some of the key DAC developments during the 1990s.

> - Multiple DACs: AD7568, 12-bit octal single 5V supply CMOS MDAC, 1991
> - Audio DACs
>   - Parallel 8×, 16× oversampling, early 1990s
>   - Sigma-delta, starting with the AD1857, AD1858, AD1859 in 1996
> - Video RAM-DACs-continued expansion of product line
> - Transmit DACs (TxDACs) for communications, 1996
> - Direct Digital Synthesis (DDS) Systems, AD9008, 1993
> - TrimDACs, 1989
> - Digital potentiometers, 1995

Figure 1.49: Summary: Monolithic DACs of the 1990s

Monolithic ADCs of the 1990s

During the decade of the 1990s, monolithic ADC performance overtook that of modular and hybrid converters, primarily because of the greatly reduced parasitics in the new IC processes. The AD1674 12-bit, 100 kSPS sampling SAR ADC introduced in 1990 was pin-compatible with the industry-standard AD574 introduced a decade earlier. A simplified block diagram of the AD1674 is shown in Figure 1.50. This represented a trend toward sampling ADCs that was to continue throughout the 1990s, because of the increased interest in the digital processing of ac signals.

Progress was also being made in CMOS sampling ADCs, as illustrated by the introduction of the AD7880 12-bit, 66 kSPS ADC in 1990. A simplified block diagram is shown in Figure 1.51. Although the AD7880 required an external reference, it still represented a breakthrough because of its low power (25 mW) and single 5 V operation.

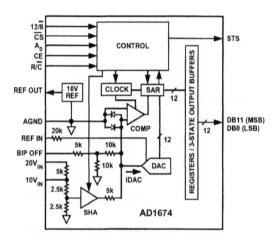

Figure 1.50: AD1674, 12-Bit, 100 kSPS
Sampling ADC (AD574 Pin-Compatible), 1990

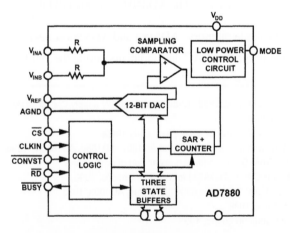

Figure 1.51: AD7880 12-Bit, 66 kSPS, Single 5 V,
LC²MOS Sampling SAR ADC, 1990

In 1992, the 12-bit, 1.25 MSPS AD1671 BiCMOS sampling ADC was introduced. The basic subranging pipelined architecture utilized in the AD1671 was based on a previous nonsampling version, the 2 MSPS AD671, which had been introduced in 1990. A simplified block diagram of the AD1671 is shown in Figure 1.52.

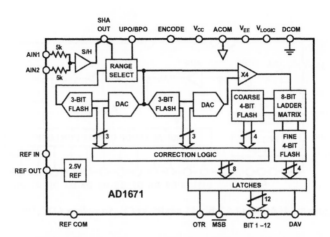

Figure 1.52: AD1671 12-Bit, 1.25 MSPS Sampling BiCMOS ADC, 1992

A significant breakthrough in speed and performance was also achieved in 1992 with the release of the AD872 12-bit, 10 MSPS BiCMOS sampling ADC, shown in Figure 1.53. This ADC also used the pipelined architecture with error correction.

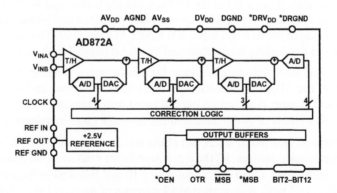

**Figure 1.53: AD872 12-Bit 10 MSPS
BiCMOS Sampling ADC, 1992**

In 1996, three single supply (5 V) CMOS ADCs were released using an architecture similar to the 12-bit AD872: the AD9220 (10 MSPS), the AD9221 (1 MSPS), and AD9223 (3 MSPS). All three parts utilized the same basic design, with the operating current scaled to yield the three sampling rate options. Power dissipation for the three devices is 250 mW (AD9220, 10 MSPS), 60 mW (AD9221, 1 MSPS), and 100 mW (AD9223, 3 MSPS).

A true breakthrough in wide dynamic range IF sampling ADCs occurred in 1995 with the release of the 12-bit, 41 MSPS AD9042. A functional diagram of the AD9042 is shown in Figure 1.54. This converter was the first to achieve greater than 80 dB SFDR for signals over a 20 MHz Nyquist bandwidth. It was fabricated on the Analog Devices' XFCB high speed complementary bipolar process.

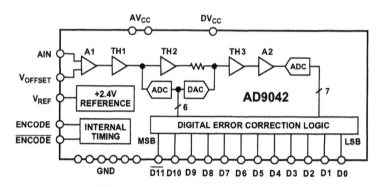

Figure 1.54: AD9042 12-Bit 41 MSPS XFCB, Sampling ADC,1995

The concept of a complete data acquisition system on a chip saw fruition in the 1990s with the introduction of the AD789x-series of LC²MOS SAR single-supply (5 V) ADCs in 1993. These parts offer up to eight channels of multiplexed inputs and sampling rates of 100 kSPS to 600 kSPS. In order to accommodate the more traditional industry-standard bipolar inputs of ±10 V and ±5 V, the series provides input thin film resistor attenuators/level shifters to match the input range of the internal SAR ADCs.

A significant technology change occurred in the 1990s, when the high-power 8-, 9-, and 10-bit flash converters of the 1980s were gradually replaced by lower power pipelined and folding architectures. This was typified by the introduction of the AD9054 8-bit, 200 MSPS ADC in 1997. The AD9054 utilized a unique architecture consisting of five folding stages followed by a 3-bit parallel flash stage.

CMOS Σ-Δ ADCs became the architecture of choice for measurement, voiceband, and audio ADCs beginning in the early 1990s. The AD7001 was the first GSM baseband converter and was introduced in 1990. Sigma-delta was utilized in many other voiceband and audio converters as well as high-resolution measurement ADCs. The AD771x family of 24-bit measurement converters was introduced in 1992.

These converters include on-chip multiplexers and PGAs and are designed for direct interfacing to many sensors, such as thermocouples, bridges, RTDs, etc. A significant product in the family is the AD7730, introduced in 1997, which allows a load cell output with a 10 mV full-scale voltage to be digitized to over 80,000 noise-free codes (16.5 bits). A simplified block diagram of the AD7730 is shown in Figure 1.55.

Another application of Σ-Δ technology appeared in the late 1990s, with the introduction of the ADE775x series of energy metering ICs. These parts measure the instantaneous current and voltage of the power mains and calculate power usage, thereby replacing mechanical devices.

A summary of key monolithic ADCs of the 1990s is given in Figure 1.56.

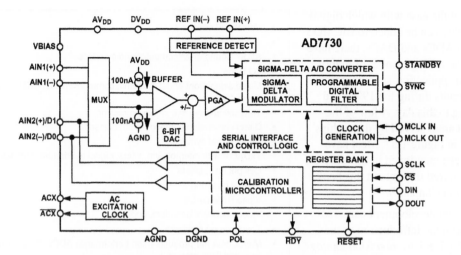

Figure 1.55: AD7730 24-Bit Signal-Conditioning Σ-Δ ADC, 1997

- AD1674, 12-bit, 100kSPS, sampling ADC, AD574A pin-compatible, 1990
- AD7880, 12-bit, 66kSPS LC²MOS sampling ADC, 1990
- AD7001, CMOS GSM baseband converter, 1990
- AD771x-series 24-bit Σ-Δ measurement ADCs, 1992
- AD1671, 12-bit, 1.25MSPS BiCMOS sampling ADC, 1992
- AD872, 12-bit, 10MSPS BiCMOS sampling ADC, 1992
- AD9220/AD9221/AD9223, 12-bit, 10/1/3 MSPS, CMOS sampling ADCs, 1996
- AD9042, 12-bit, 41MSPS sampling ADC, 80dB SFDR, 1995
- AD7730, 24-bit bridge transducer measurement ADC, 1997
- AD9054, 8-bit, 200MSPS sampling ADC, 1997
- ADuC812 MicroConverter® (precision ADCs, DACs, 8051-core, flash memory, 1999

Figure 1.56: Summary: Monolithic ADCs of the 1990s

Because of the ease with which digital functionality can be added to BiCMOS or CMOS ADCs and DACs, there have been an increasing number of highly integrated application-specific integrated circuits during the 1990s and continuing to this day. A few of the important areas served by these chips are listed in Figure 1.57. Most have already been mentioned in this chapter, and are covered throughout this book, especially in Chapter 8.

An important development occurred in 1999 with the introduction of the first Analog Devices' *precision analog microcontroller*, the ADuC812 MicroConverter. The MicroConverter includes not only precision signal conditioning circuitry (ADCs, DACs, multiplexers, etc.) but also flash memory and an 8051-based microprocessor core. Later MicroConverter products have included higher resolution Σ-Δ ADCs (see Chapter 8 of this book). This level of integration represents an optimum solution for many general-purpose sensor conditioning and signal processing applications.

- Voiceband codecs
- Audio codecs
- AC'97 SoundMAX® computer audio codecs
- I/O Ports
- Mixed signal front ends: modems, communications, CCD imaging, flat panel displays
- Transmit and receive signal processors
- Direct conversion (Othello® radio) chipsets
- Direct digital synthesis
- TxDACs® with interpolation, filters, digital quadrature modulation, etc.
- Cell phone chipsets
- Energy metering
- Video RAM-DACs
- Video encoders/decoders, codecs
- Touchscreen digitizers
- MicroConverter products (high performance ADCs, DACs, + 8051 µP core and flash memory)

Figure 1.57: Summary: Integrated Functions of the 1990s

Hybrid and Modular DACs and ADCs of the 1990s

Although monolithic data converter solutions largely replaced the chip-and-wire hybrids and modules of the 1970s and 1980s, there were a few significant introductions early in the 1990s. The AD9014 14-bit, 10 MSPS modular ADC introduced in 1990 represented a major breakthrough in dynamic performance. The part achieved 90 dB SFDR over the Nyquist bandwidth, and utilized a number of proprietary monolithic building blocks which were later used in the fully integrated design of the AD9042 12-bit, 41 MSPS ADC introduced in 1995.

The AD1382 and AD1385 hybrid 16-bit, 500 kSPS sampling ADCs, introduced in 1992, represented state-of-the-art performance at the time, and the AD1385 was one of the first ADCs utilizing autocalibration to maintain its linearity.

Later in the 1990s, multichip module (MCM) technology became an excellent alternative to the expensive modules and chip-and-wire hybrids. Lower-cost packaging techniques allow high performance monolithic ADCs, such as the AD9042, to be packaged as duals along with front-end conditioning circuits. For instance, the AD10242 introduced in 1996 (dual 12-bit, 41 MSPS AD9042s) offered an attractive cost-effective solution in applications requiring dual high performance 12-bit ADCs and more functionality in the analog front end.

Data Converters of the 2000s
Walt Kester

The data converter trends started in the 1990s, shown in Figures 1.56 and 1.57, have continued into the 2000s. Power dissipation has dropped, and along with it, power supply voltages. Supplies of 5 V, 3.3 V, 2.5 V, and 1.8 V parts have followed as CMOS line spacings shrank to 0.6 μm, 0.35 μm, 0.25μm, and 0.18 μm. Smaller surface-mount and chip-scale packages have also emerged as the modern replacement for the nearly obsolete DIP packages of the 1970s and 1980s.

Although the trend toward more highly integrated functions continues, data converter manufacturers are realizing that "smart partitioning" can offer a higher performance and more cost-effective solution than simply always adopting a "system-on-a-chip" philosophy. This topic is explored in much more detail in Chapter 4 of this book, *Data Converter Process Technology*.

The Analog Devices' portfolio of 16- and 18-bit SAR sampling ADCs has grown to over 30 models, including the latest offerings in the breakthrough high-resolution Pulsar® series. For example, the AD7664 16-bit, 570 kSPS ADC was introduced in 2000, the AD7677 16-bit, 1 MSPS ADC in 2001, and the AD7674 18-bit, 800 kSPS ADC in 2003, and the AD7621 18-bit, 3 MSPS ADC in 2003.

In the 2000s, general-purpose multiple DAC offerings expanded to include 16 channels (AD5390, AD5391), 32-channels (AD5382, AD5383) and 40 channels (AD5380, AD5381). High speed DACs reached 1 GSPS update rates with the AD9858 10-bit DDS system.

Turning to IF sampling data converters, the AD6645 14-bit 80 MSPS/105 MSPS ADC was introduced in 2000, and the AD9430 12-bit, 210 MSPS ADC in 2002. Both converters represent breakthroughs in sampling rate and dynamic range.

Significant multichip module (MCM) introductions were the AD10678 16-bit, 65/80/105 MSPS ADC introduced in 2003, and the AD12400 12-bit, 400 MSPS ADC also introduced in 2003. Both devices use high performance IC sampling ADCs as building blocks followed by proprietary digital post processing.

In 2002 and 2003, the ADuC-series of MicroConverter products was expanded to include 16- and 24-bit on-chip Σ-Δ ADCs. In addition, the SAR-based MicroConverter product line was expanded. Some of the future MicroConverter products to be introduced starting in 2004 will utilize the highly popular ARM7®-microcontroller core.

These data converter highlights of the 2000s are summarized in Figure 1.58.

Many other things could be said about the history of data converters, and many other examples of modern ADCs and DACs are given throughout the remaining chapters of this book. Looking to the future, we can expect many new breakthroughs, not only in sheer performance, but in levels of integration.

- Continued expansion of Analog Front Ends (AFEs) and Multiplexed Front Ends (MxFEs®)
- 16-, 18-bit Pulsar® series of switched capacitor SAR sampling ADCs
  - AD7674, 18-bit, 800kSPS ADC, 2003
  - AD7621, 16-bit, 3MSPS ADC, 2003
- Multiple DACs: 16-channel (AD5390, AD5391), 32-channel (AD5382/AD5383), 40-channel (AD5380/AD5381)
- IF-Sampling ADCs
  - AD6645 14-bit, 105MSPS ADC, 2000
  - AD9430 12-bit, 210MSPS ADC, 2002
- Multichip modules (MCMs):
  - AD12400, 12-bit, 400MSPS ADC, 2003
  - AD10678, 16-bit, 65/80/105 MSPS ADC, 2003
- AD9858 10-bit, 1GSPS DDS system, 2003
- 16-/24-bit Σ-Δ MicroConverter products, 2002, 2003
- ARM7-based MicroConverter products, 2004

Figure 1.58: Data Converter Highlights of the 2000s

CHAPTER 2

Fundamentals of Sampled Data Systems

Fundamentals of Sampled Data Systems

Coding and Quantizing
Walt Kester, Dan Sheingold, James Bryant

Analog-to-digital converters (ADCs) translate analog quantities, which are characteristic of most phenomena in the "real world," to digital language, used in information processing, computing, data transmission, and control systems. Digital-to-analog converters (DACs) are used in transforming transmitted or stored data, or the results of digital processing, back to "real-world" variables for control, information display, or further analog processing. The relationships between inputs and outputs of DACs and ADCs are shown in Figure 2.1.

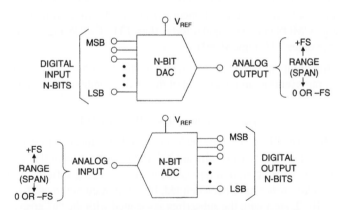

Figure 2.1: Digital-to-Analog Converter (DAC) and Analog-to-Digital Converter (ADC) Input and Output Definitions

Analog input variables, whatever their origin, are most frequently converted by transducers into voltages or currents. These electrical quantities may appear (1) as fast or slow "dc" continuous direct measurements of a phenomenon in the time domain, (2) as modulated ac waveforms (using a wide variety of modulation techniques), (3) or in some combination, with a spatial configuration of related variables to represent shaft angles. Examples of the first are outputs of thermocouples, potentiometers on dc references, and analog computing circuitry; of the second, "chopped" optical measurements, ac strain gage or bridge outputs, and digital signals buried in noise; and of the third, synchros and resolvers.

The analog variables to be dealt with in this chapter are those involving voltages or currents representing the actual analog phenomena. They may be either wideband or narrowband. They may be either scaled from the direct measurement, or subjected to some form of analog preprocessing, such as linearization, combination, demodulation, filtering, sample-hold, etc.

As part of the process, the voltages and currents are "normalized" to ranges compatible with assigned ADC input ranges. Analog output voltages or currents from DACs are direct and in normalized form, but they may be subsequently post-processed (e.g., scaled, filtered, amplified, etc.).

Information in digital form is normally represented by arbitrarily fixed voltage levels referred to "ground," either occurring at the outputs of logic gates, or applied to their inputs. The digital numbers used are all basically binary; that is, each "bit," or unit of information has one of two possible states. These states are "off," "false," or "0," and "on," "true," or "1." It is also possible to represent the two logic states by two different levels of current; however, this is much less popular than using voltages. There is also no particular reason why the voltages need be referenced to ground—as in the case of emitter-coupled-logic (ECL), positive-emitter-coupled-logic (PECL) or low-voltage-differential-signaling logic (LVDS) for example.

Words are groups of levels representing digital numbers; the levels may appear simultaneously in *parallel,* on a bus or groups of gate inputs or outputs, *serially* (or in a time sequence) on a single line, or as a sequence of parallel bytes (i.e., "byte-serial") or nibbles (small bytes). For example, a 16-bit word may occupy the 16 bits of a 16-bit bus, or it may be divided into two sequential bytes for an 8-bit bus, or four 4-bit nibbles for a 4-bit bus.

Although there are several systems of logic, the most widely used choice of levels are those used in TTL (transistor-transistor logic) and, in which positive *true,* or 1, corresponds to a minimum output level of 2.4 V (inputs respond unequivocally to "1" for levels greater than 2.0 V); and *false,* or 0, corresponds to a maximum output level of 0.4 V (inputs respond unequivocally to "0" for anything less than 0.8 V). It should be noted that even though CMOS is more popular today than TTL, CMOS logic levels are generally made to be compatible with the older TTL logic standard.

A unique parallel or serial grouping of digital levels, or a *number,* or *code,* is assigned to each analog level which is quantized (i.e., represents a unique portion of the analog range). A typical digital code would be this array:

$$a_7\, a_6\, a_5\, a_4\, a_3\, a_2\, a_1\, a_0 = 1\ 0\ 1\ 1\ 1\ 0\ 0\ 1$$

It is composed of eight bits. The "1" at the extreme left is called the "most significant bit" (MSB, or Bit 1), and the one at the right is called the "least significant bit" (LSB, or bit *N:* 8 in this case). The meaning of the code, as either a number, a character, or a representation of an analog variable, is unknown until the *code* and the *conversion relationship* have been defined. It is important not to confuse the designation of a particular bit (i.e., Bit 1, Bit 2, etc.) with the subscripts associated with the "a" array. The subscripts correspond to the power of 2 associated with the weight of a particular bit in the sequence.

The best-known code (other than base 10) is *natural or straight binary* (base 2). Binary codes are most familiar in representing integers; i.e., in a natural binary integer code having N bits, the LSB has a weight of 2^0 (i.e., 1), the next bit has a weight of 2^1 (i.e., 2), and so on up to the MSB, which has a weight of 2^{N-1} (i.e., $2^N/2$). The value of a binary number is obtained by adding up the weights of all non-zero bits. When the weighted bits are added up, they form a unique number having any value from 0 to $2^N - 1$. Each additional trailing zero bit, if present, essentially doubles the size of the number.

In converter technology, full-scale (abbreviated *FS*) is independent of the number of bits of resolution, N. A more useful coding is *fractional* binary, which is always normalized to full-scale. Integer binary can be interpreted as fractional binary if all integer values are divided by 2^N. For example, the MSB has a weight of ½ (i.e., $2^{(N-1)}/2^N = 2^{-1}$), the next bit has a weight of ¼ (i.e., 2^{-2}), and so forth down to the LSB, which has a weight of $1/2^N$ (i.e., 2^{-N}). When the weighted bits are added up, they form a number with any of 2^N values, from 0 to $(1 - 2^{-N})$ of full-scale. Additional bits simply provide more fine structure without affecting full-scale range. The relationship between base-10 numbers and binary numbers (base 2) are shown in Figure 2.2 along with examples of each.

WHOLE NUMBERS:

$$\text{Number}_{10} = a_{N-1}2^{N-1} + a_{N-2}2^{N-2} + ... + a_1 2^1 + a_0 2^0$$

$\uparrow$ MSB $\qquad\qquad\qquad\qquad$ LSB

Example: $1011_2 = (1 \times 2^3) + (0 \times 2^2) + (1 \times 2^1) + (1 \times 2^0)$
$\qquad\qquad = \quad 8 \quad + \quad 0 \quad + \quad 2 \quad + \quad 1 \quad = 11_{10}$

FRACTIONAL NUMBERS:

$$\text{Number}_{10} = a_{N-1}2^{-1} + a_{N-2}2^{-2} + ... + a_1 2^{-(N-1)} + a_0 2^{-N}$$

$\uparrow$ MSB $\qquad\qquad\qquad\qquad$ LSB

Example: $0.1011_2 = (1 \times 0.5) + (0 \times 0.25) + (1 \times 0.125) + (1 \times 0.0625)$
$\qquad\qquad = \quad 0.5 \quad + \quad 0 \quad + \quad 0.125 \quad + \quad 0.0625 \quad = 0.6875_{10}$

Figure 2.2: Representing a Base-10 Number with a Binary Number (Base 2)

Unipolar Codes

In data conversion systems, the coding method must be related to the analog input range (or span) of an ADC or the analog output range (or span) of a DAC. The simplest case is when the input to the ADC or the output of the DAC is always a unipolar positive voltage (current outputs are very popular for DAC outputs, much less for ADC inputs). The most popular code for this type of signal is *straight binary* and is shown in Figure 2.3 for a 4-bit converter. Notice that there are 16 distinct possible levels, ranging from the all-zeros code 0000, to the all-ones code 1111. It is important to note that the analog value represented by the all-ones code is not full-scale (abbreviated FS), but FS – 1 LSB. This is a common convention in data conversion notation and applies to both ADCs and DACs. Figure 2.3 gives the base-10 equivalent number, the value of the base-2 binary code relative to full-scale (FS), and also the corresponding voltage level for each code (assuming a 10 V full-scale converter. The Gray code equivalent is also shown, and will be discussed shortly.

BASE 10 NUMBER	SCALE	+10V FS	BINARY	GRAY
+15	+FS – 1LSB = +15/16 FS	9.375	1 1 1 1	1 0 0 0
+14	+7/8 FS	8.750	1 1 1 0	1 0 0 1
+13	+13/16 FS	8.125	1 1 0 1	1 0 1 1
+12	+3/4 FS	7.500	1 1 0 0	1 0 1 0
+11	+11/16 FS	6.875	1 0 1 1	1 1 1 0
+10	+5/8 FS	6.250	1 0 1 0	1 1 1 1
+9	+9/16 FS	5.625	1 0 0 1	1 1 0 1
+8	+1/2 FS	5.000	1 0 0 0	1 1 0 0
+7	+7/16 FS	4.375	0 1 1 1	0 1 0 0
+6	+3/8 FS	3.750	0 1 1 0	0 1 0 1
+5	+5/16 FS	3.125	0 1 0 1	0 1 1 1
+4	+1/4 FS	2.500	0 1 0 0	0 1 1 0
+3	+3/16 FS	1.875	0 0 1 1	0 0 1 0
+2	+1/8 FS	1.250	0 0 1 0	0 0 1 1
+1	1LSB = +1/16 FS	0.625	0 0 0 1	0 0 0 1
0	0	0.000	0 0 0 0	0 0 0 0

Figure 2.3: Unipolar Binary Codes, 4-Bit Converter

Figure 2.4 shows the transfer function for an ideal 3-bit DAC with straight binary input coding. Notice that the analog output is zero for the all-zeros input code. As the digital input code increases, the analog output increases 1 LSB (1/8 scale in this example) per code. The most positive output voltage is 7/8 FS, corresponding to a value equal to FS – 1 LSB. The midscale output of 1/2 FS is generated when the digital input code is 100.

The transfer function of an ideal 3-bit ADC is shown in Figure 2.5. There is a range of analog input voltage over which the ADC will produce a given output code; this range is the *quantization uncertainty* and is equal to 1 LSB. Note that the width of the transition regions between adjacent codes is zero for an ideal ADC. In practice, however, there is always transition noise associated with these levels, and therefore the width is non-zero. It is customary to define the analog input corresponding to a given code by the *code center* which lies halfway between two adjacent transition regions (illustrated by the black dots in the diagram). This requires that the first transition region occur at ½ LSB. The full-scale analog input voltage is defined by 7/8 FS, (FS – 1 LSB).

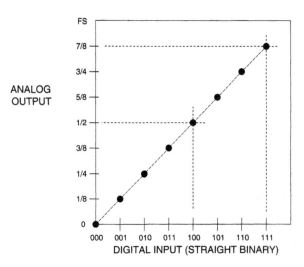

Figure 2.4: Transfer Function for Ideal Unipolar 3-Bit DAC

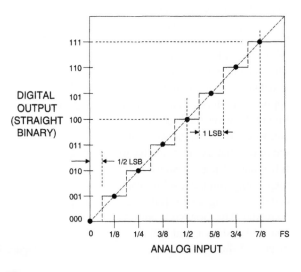

Figure 2.5: Transfer Function for Ideal Unipolar 3-Bit ADC

Gray Code

Another code worth mentioning at this point is the *Gray* code (or *reflective-binary*) which was invented by Elisha Gray in 1878 (Reference 1) and later re-invented by Frank Gray in 1949 (see Reference 2). The Gray code equivalent of the 4-bit straight binary code is also shown in Figure 2.3. Although it is rarely used in computer arithmetic, it has some useful properties which make it attractive to A/D conversion. Notice that in Gray code, as the number value changes, the transitions from one code to the next involve only one bit at a time. Contrast this to the binary code where all the bits change when making the transition between 0111 and 1000. Some ADCs make use of it internally and then convert the Gray code to a binary code for external use.

One of the earliest practical ADCs to use the Gray code was a 7-bit, 100 kSPS electron beam encoder developed by Bell Labs and described in a 1948 reference (Reference 3).

The basic electron beam coder concepts for a 4-bit device are shown in Figure 2.6. The early tubes operated in the serial mode (A). The analog signal is first passed through a sample-and-hold, and during the "hold" interval, the beam is swept horizontally across the tube. The Y-deflection for a single sweep therefore corresponds to the value of the analog signal from the sample-and-hold. The shadow mask is coded to produce the proper binary code, depending on the vertical deflection. The code is registered by the collector, and the bits are generated in serial format. Later tubes used a fan-shaped beam (shown in Figure 2.6B), creating a "flash" converter delivering a parallel output word.

Early electron tube coders used a binary-coded shadow mask, and large errors can occur if the beam straddles two adjacent codes and illuminates both of them. The way these errors occur is illustrated in Figure 2.7A, where the horizontal line represents the beam sweep at the midscale transition point (transition between code 0111 and code 1000). For example, an error in the most significant bit (MSB) produces an error of ½ scale. These errors were minimized by placing fine horizontal sensing wires across the boundaries of each of the quantization levels. If the beam initially fell on one of the wires, a small voltage was added to the vertical deflection voltage which moved the beam away from the transition region.

The errors associated with binary shadow masks were eliminated by using a Gray

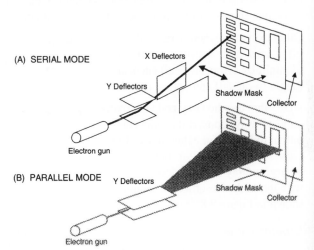

Figure 2.6: The Electron Beam Coder: (A) Serial Mode and (B) Parallel or "Flash" Mode

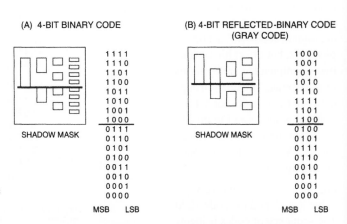

Figure 2.7: Electron Beam Coder Shadow Masks for Binary Code (A) and Gray Code (B)

code shadow mask as shown in Figure 2.7B. As mentioned above, the Gray code has the property that adjacent levels differ by only one digit in the corresponding Gray-coded word. Therefore, if there is an error in a bit decision for a particular level, the corresponding error after conversion to binary code is only one least significant bit (LSB). In the case of midscale, note that only the MSB changes. It is interesting to note that this same phenomenon can occur in modern comparator-based flash converters due to comparator metastability. With small overdrive, there is a finite probability that the output of a comparator will generate the wrong decision in its latched output, producing the same effect if straight binary decoding techniques are used. In many cases, Gray code, or "pseudo-Gray" codes are used to decode the comparator bank. The Gray code output is then latched, converted to binary, and latched again at the final output.

As a historical note, in spite of the many mechanical and electrical problems relating to beam alignment, electron tube coding technology reached its peak in the mid-1960s with an experimental 9-bit coder capable of 12 MSPS sampling rates (Reference 4). Shortly thereafter, however, advances in all solid-state ADC techniques made the electron tube technology obsolete.

Other examples where Gray code is often used in the conversion process to minimize errors are shaft encoders (angle-to-digital) and optical encoders.

ADCs that use the Gray code internally almost always convert the Gray code output to binary for external use. The conversion from Gray-to-binary and binary-to-Gray is easily accomplished with the exclusive-or logic function as shown in Figure 2.8.

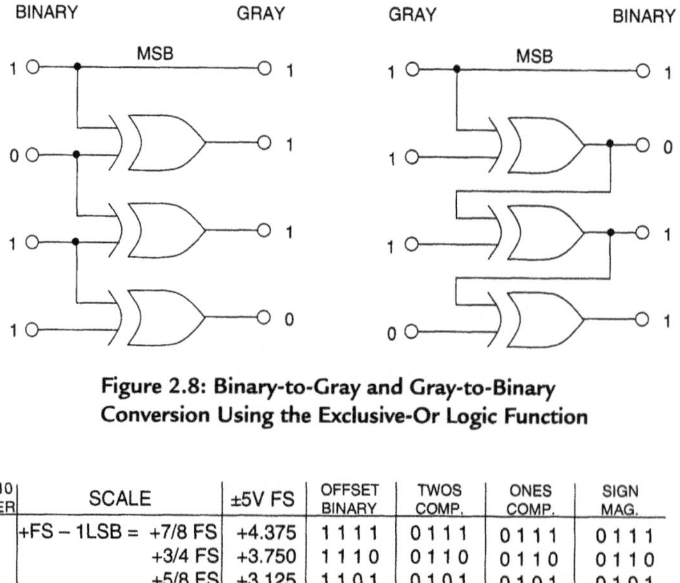

Figure 2.8: Binary-to-Gray and Gray-to-Binary Conversion Using the Exclusive-Or Logic Function

Bipolar Codes

In many systems, it is desirable to represent both positive and negative analog quantities with binary codes. Either *offset binary*, *two's complement*, *one's complement*, or *sign magnitude* codes will accomplish this, but offset binary and two's complement are by far the most popular. The relationships between these codes for a 4-bit systems is shown in Figure 2.9. Note that the values are scaled for a ±5 V full-scale input/output voltage range.

For *offset binary*, the zero signal value is assigned the code 1000. The sequence of codes is identical to that of straight binary.

BASE 10 NUMBER	SCALE	±5V FS	OFFSET BINARY	TWOS COMP.	ONES COMP.	SIGN MAG.
+7	+FS – 1LSB = +7/8 FS	+4.375	1 1 1 1	0 1 1 1	0 1 1 1	0 1 1 1
+6	+3/4 FS	+3.750	1 1 1 0	0 1 1 0	0 1 1 0	0 1 1 0
+5	+5/8 FS	+3.125	1 1 0 1	0 1 0 1	0 1 0 1	0 1 0 1
+4	+1/2 FS	+2.500	1 1 0 0	0 1 0 0	0 1 0 0	0 1 0 0
+3	+3/8 FS	+1.875	1 0 1 1	0 0 1 1	0 0 1 1	0 0 1 1
+2	+1/4 FS	+1.250	1 0 1 0	0 0 1 0	0 0 1 0	0 0 1 0
+1	+1/8 FS	+0.625	1 0 0 1	0 0 0 1	0 0 0 1	0 0 0 1
0	0	0.000	1 0 0 0	0 0 0 0	*0 0 0 0	*1 0 0 0
–1	– 1/8 FS	–0.625	0 1 1 1	1 1 1 1	1 1 1 0	1 0 0 1
–2	– 1/4 FS	–1.250	0 1 1 0	1 1 1 0	1 1 0 1	1 0 1 0
–3	– 3/8 FS	–1.875	0 1 0 1	1 1 0 1	1 1 0 0	1 0 1 1
–4	–1/2 FS	–2.500	0 1 0 0	1 1 0 0	1 0 1 1	1 1 0 0
–5	–5/8 FS	–3.125	0 0 1 1	1 0 1 1	1 0 1 0	1 1 0 1
–6	–3/4 FS	–3.750	0 0 1 0	1 0 1 0	1 0 0 1	1 1 1 0
–7	– FS + 1LSB = –7/8 FS	–4.375	0 0 0 1	1 0 0 1	1 0 0 0	1 1 1 1
–8	– FS	–5.000	0 0 0 0	1 0 0 0		

NOT NORMALLY USED IN COMPUTATIONS (SEE TEXT)

	ONES COMP.	SIGN MAG.
* 0+	0 0 0 0	0 0 0 0
0–	1 1 1 1	1 0 0 0

Figure 2.9: Bipolar Codes, 4-Bit Converter

The only difference between a straight and offset binary system is the half-scale offset associated with analog signal. The most negative value (–FS + 1 LSB) is assigned the code 0001, and the most positive value (+FS – 1 LSB) is assigned the code 1111. Note that in order to maintain perfect symmetry about midscale, the all-zeros code (0000) representing negative full-scale (–FS) is not normally used in computation. It can be used to represent a negative off-range condition or simply assigned the value of the 0001 (–FS + 1 LSB).

The relationship between the offset binary code and the analog output range of a bipolar 3-bit DAC is shown in Figure 2.10. The analog output of the DAC is zero for the zero-value input code 100. The most negative output voltage is generally defined by the 001 code (–FS + 1 LSB), and the most positive by 111 (+FS – 1 LSB). The output voltage for the 000 input code is available for use if desired, but makes the output nonsymmetrical about zero and complicates the mathematics.

The offset binary output code for a bipolar 3-bit ADC as a function of its analog input is shown in Figure 2.11. Note that zero analog input defines the center of the midscale code 100. As in the case of bipolar DACs, the most negative input voltage is generally defined by the 001 code (–FS + 1 LSB), and the most positive by 111 (+FS – 1 LSB). As discussed above, the 000 output code is available for use if desired, but makes the output nonsymmetrical about zero and complicates the mathematics.

Two's complement is identical to offset binary with the most-significant-bit (MSB) complemented (inverted). This is obviously very easy to accomplish in a data converter, using a simple inverter or taking the complementary output of a "D" flip-flop. The popularity of two's complement coding lies in the ease with which mathematical operations can be performed in computers and DSPs. Two's complement, for conversion purposes, consists of a binary code for positive magnitudes (0 sign bit), and the two's complement of each positive number to represent its negative. The two's complement is formed arithmetically by complementing the number and adding 1 LSB. For example, –3/8 FS is obtained by taking the two's complement of +3/8 FS. This is done by first complementing +3/8 FS, 0011 obtaining 1100. Adding 1 LSB, we obtain 1101.

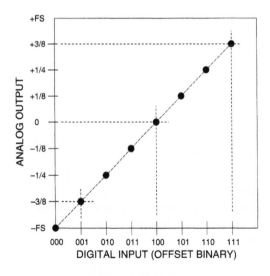

Figure 2.10: Transfer Function for Ideal Bipolar 3-Bit DAC

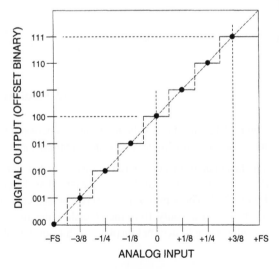

Figure 2.11: Transfer Function for Ideal Bipolar 3-Bit ADC

Two's complement makes subtraction easy. For example, to subtract 3/8 FS from 4/8 FS, add 4/8 to –3/8, or 0100 to 1101. The result is 0001, or 1/8, disregarding the extra carry.

One's complement can also be used to represent negative numbers, although it is much less popular than two's complement and rarely used today. The one's complement is obtained by simply complementing all of a positive number's digits. For instance, the one's complement of 3/8 FS (0011) is 1100. A one's complemented code can be formed by complementing each positive value to obtain its corresponding negative value. This includes zero, which is then represented by either of two codes, 0000 (referred to as 0+) or 1111 (referred to as 0–). This ambiguity must be dealt with mathematically, and presents obvious problems relating to ADCs and DACs for which there is a single code that represents zero.

Sign-magnitude would appear to be the most straightforward way of expressing signed analog quantities digitally. Simply determine the code appropriate for the magnitude and add a polarity bit. Sign-magnitude BCD is popular in bipolar digital voltmeters, but has the problem of two allowable codes for zero. It is therefore unpopular for most applications involving ADCs or DACs.

To Convert From → To ↓	Sign Magnitude	Two's Complement	Offset Binary	One's Complement
Sign Magnitude	No Change	If MSB = 1, complement other bits, add 00...01	Complement MSB If new MSB = 1, complement other bits, add 00...01	If MSB = 1, complement other bits
Two's Complement	If MSB = 1, complement other bits, add 00...01	No Change	Complement MSB	If MSB = 1, add 00...01
Offset Binary	Complement MSB If new MSB = 0 complement other bits, add 00...01	Complement MSB	No Change	Complement MSB If new MSB = 0, add 00...01
One's Complement	If MSB = 1, complement other bits	If MSB = 1, add 11...11	Complement MSB If new MSB = 1, add 11...11	No Change

Figure 2.12: Relationships Among Bipolar Codes

Figure 2.12 summarizes the relationships between the various bipolar codes: offset binary, two's complement, one's complement, and sign-magnitude, and shows how to convert between them.

The last code to be considered in this section is *binary-coded-decimal (BCD)*, where each base-10 digit (0 to 9) in a decimal number is represented as the corresponding 4-bit straight binary word as shown in Figure 2.13. The minimum digit 0 is represented as 0000, and the digit 9 by 1001. This code is relatively inefficient, since only 10 of the 16 code states for each decade are used. It is, however, a very useful code for interfacing to decimal displays such as in digital voltmeters.

BASE 10 NUMBER	SCALE	+10V FS	DECADE 1	DECADE 2	DECADE 3	DECADE 4
+15	+FS – 1LSB = +15/16 FS	9.375	1001	0011	0111	0101
+14	+7/8 FS	8.750	1000	0111	0101	0000
+13	+13/16 FS	8.125	1000	0001	0010	0101
+12	+3/4 FS	7.500	0111	0101	0000	0000
+11	+11/16 FS	6.875	0110	1000	0111	0101
+10	+5/8 FS	6.250	0110	0010	0101	0000
+9	+9/16 FS	5.625	0101	0110	0010	0101
+8	+1/2 FS	5.000	0101	0000	0000	0000
+7	+7/16 FS	4.375	0100	0011	0111	0101
+6	+3/8 FS	3.750	0011	0111	0101	0000
+5	+5/16 FS	3.125	0011	0001	0010	0101
+4	+1/4 FS	2.500	0010	0101	0000	0000
+3	+3/16 FS	1.875	0001	1000	0111	0101
+2	+1/8 FS	1.250	0001	0010	0101	0000
+1	1LSB = +1/16 FS	0.625	0000	0110	0010	0101
0	0	0.000	0000	0000	0000	0000

Figure 2.13: Binary Coded Decimal (BCD) Code

Complementary Codes

Some forms of data converters (for example, early DACs using monolithic NPN quad current switches), require standard codes such as natural binary or BCD, but with all bits represented by their complements. Such codes are called *complementary codes*. All the codes discussed thus far have complementary codes which can be obtained by this method. A *complementary* code should not be confused with a *one's complement* or a *two's complement* code.

In a 4-bit complementary-binary converter, 0 is represented by 1111, half-scale by 0111, and FS – 1 LSB by 0000. In practice, the complementary code can usually be obtained by using the complementary output of a register rather than the true output, since both are available.

Sometimes the complementary code is useful in inverting the analog output of a DAC. Today many DACs provide differential outputs which allow the polarity inversion to be accomplished without modifying the input code. Similarly, many ADCs provide differential logic inputs which can be used to accomplish the polarity inversion.

DAC and ADC Static Transfer Functions and DC Errors

The most important thing to remember about both DACs and ADCs is that either the input or output is digital, and therefore the signal is quantized. That is, an N-bit word represents one of 2^N possible states, and therefore an N-bit DAC (with a fixed reference) can have only 2^N possible analog outputs, and an N-bit ADC can have only 2^N possible digital outputs. As previously discussed, the analog signals will generally be voltages or currents.

The resolution of data converters may be expressed in several different ways: the weight of the Least Significant Bit (LSB), parts per million of full-scale (ppm FS), millivolts (mV), etc. Different devices (even from the same manufacturer) will be specified differently, so converter users must learn to translate between the different types of specifications if they are to compare devices successfully. The size of the least significant bit for various resolutions is shown in Figure 2.14.

RESOLUTION N	2^N	VOLTAGE (10V FS)	ppm FS	% FS	dB FS
2-bit	4	2.5 V	250,000	25	– 12
4-bit	16	625 mV	62,500	6.25	– 24
6-bit	64	156 mV	15,625	1.56	– 36
8-bit	256	39.1 mV	3,906	0.39	– 48
10-bit	1,024	9.77 mV (10 mV)	977	0.098	– 60
12-bit	4,096	2.44 mV	244	0.024	– 72
14-bit	16,384	610 µV	61	0.0061	– 84
16-bit	65,536	153 µV	15	0.0015	– 96
18-bit	262,144	38 µV	4	0.0004	– 108
20-bit	1,048,576	9.54 µV (10 µV)	1	0.0001	– 120
22-bit	4,194,304	2.38 µV	0.24	0.000024	– 132
24-bit	16,777,216	596 nV*	0.06	0.000006	– 144

*600nV is the Johnson Noise in a 10kHz BW of a 2.2kΩ Resistor @ 25°C

Remember: 10 Bits and 10V FS yields an LSB of 10mV, 1000ppm, or 0.1%.
All other values may be calculated by powers of 2.

Figure 2.14: Quantization: The Size of a Least Significant Bit (LSB)

Before we can consider the various architectures used in data converters, it is necessary to consider the performance to be expected, and the specifications which are important. The following sections will consider the definition of errors and specifications used for data converters. This is important in understanding the strengths and weaknesses of different ADC/DAC architectures.

The first applications of data converters were in measurement and control where the exact timing of the conversion was usually unimportant, and the data rate was slow. In such applications, the dc specifications of converters are important, but timing and ac specifications are not. Today many, if not most, converters are used in *sampling* and *reconstruction* systems where ac specifications are critical (and dc ones may not be)—these will be considered in Section 2.3 of this chapter.

Figure 2.15 shows the ideal transfer characteristics for a 3-bit unipolar DAC and a 3-bit unipolar ADC. In a DAC, both the input and the output are quantized, and the graph consists of eight points. While it is reasonable to discuss the line through these points, it is very important to remember that the actual transfer characteristic is *not* a line, but a number of discrete points.

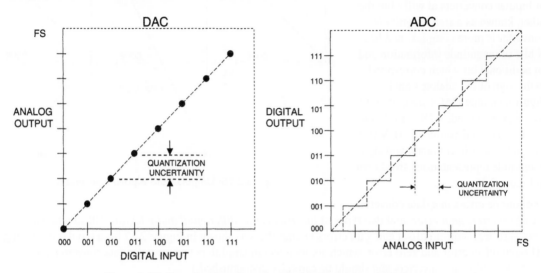

Figure 2.15: Transfer Functions for Ideal 3-Bit DAC and ADC

The input to an ADC is analog and is not quantized, but its output is quantized. The transfer characteristic therefore consists of eight horizontal steps. When considering the offset, gain and linearity of an ADC we consider the line joining the midpoints of these steps—often referred to as the *code centers*.

For both DACs and ADCs, digital full-scale (all "1"s) corresponds to 1 LSB below the analog full-scale (FS). The (ideal) ADC transitions take place at ½ LSB above zero, and thereafter every LSB, until 1½ LSB below analog full-scale. Since the analog input to an ADC can take any value, but the digital output is quantized, there may be a difference of up to ½ LSB between the actual analog input and the exact value of the digital output. This is known as the *quantization error* or *quantization uncertainty* as shown in Figure 2.15. In ac (sampling) applications this quantization error gives rise to *quantization noise* which will be discussed in Section 2.3 of this chapter.

As previously discussed, there are many possible digital coding schemes for data converters: *straight binary*, *offset binary*, *one's complement*, *two's complement*, sign *magnitude*, *gray code*, *BCD* and others. This section, being devoted mainly to the *analog* issues surrounding data converters, will use simple *binary* and *offset binary* in its examples and will not consider the merits and disadvantages of these, or any other forms of digital code.

The examples in Figure 2.15 use *unipolar* converters, whose analog port has only a single polarity. These are the simplest type, but *bipolar* converters are generally more useful in real-world applications. There are two types of bipolar converters: the simpler is merely a unipolar converter with an accurate 1 MSB of negative offset (and many converters are arranged so that this offset may be switched in and out so that they can be used as either unipolar or bipolar converters at will), but the other, known as a *sign-magnitude* converter is more complex, and has N bits of magnitude information and an additional bit which corresponds to the sign of the analog signal. Sign-magnitude DACs are quite rare, and sign-magnitude ADCs are found mostly in digital voltmeters (DVMs). The unipolar, offset binary, and sign-magnitude representations are shown in Figure 2.16.

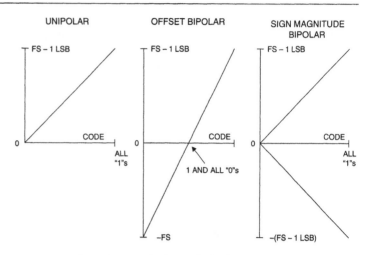

Figure 2.16: Unipolar and Bipolar Converters

The four dc errors in a data converter are *offset error*, *gain error*, and two types of *linearity error (differential and integral)*. Offset and gain errors are analogous to offset and gain errors in amplifiers as shown in Figure 2.17 for a bipolar input range. (Though offset error and zero error, which are identical in amplifiers and unipolar data converters, are not identical in bipolar converters and should be carefully distinguished.)

The transfer characteristics of both DACs and ADCs may be expressed as a straight line given by $D = K + GA$, where D is the digital code, A is the analog signal, and K and G are constants. In a unipolar converter, the ideal value of K is zero; in an offset bipolar converter it is –1 MSB. The offset error is the amount by which the actual value of K differs from its ideal value.

The gain error is the amount by which G differs from its ideal value, and is generally expressed as the percentage difference between the two, although it may be defined as the gain error contribution (in mV or LSB) to the total error at full scale. These errors can usually be trimmed by the data converter user. Note, however, that amplifier offset is trimmed at zero input, and then the gain is trimmed near to full scale. The trim algorithm for a bipolar data converter is not so straightforward.

The integral linearity error of a converter is also analogous to the linearity error of an amplifier, and is defined as the maximum deviation of the actual transfer characteristic of the converter from a straight line, and is generally expressed as a percentage of full scale (but may be given in LSBs). For an ADC, the most popular convention is to draw the straight line through the mid-points of the codes, or the code centers. There are two common ways of choosing the straight line: *end point* and *best straight line* as shown in Figure 2.18.

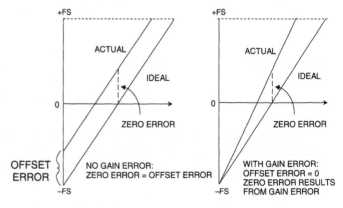

Figure 2.17: Bipolar Data Converter Offset and Gain Error

In the *end point* system, the deviation is measured from the straight line through the origin and the full-scale point (after gain adjustment). This is the most useful integral linearity measurement for measurement and control applications of data converters (since error budgets depend on deviation from the ideal transfer characteristic, not from some arbitrary "best fit"), and is the one normally adopted by Analog Devices, Inc.

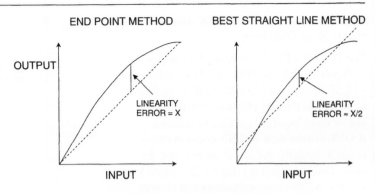

Figure 2.18: Method of Measuring Integral Linearity Errors (Same Converter on Both Graphs)

The *best straight line*, however, does give a better prediction of distortion in ac applications, and also gives a lower value of "linearity error" on a data sheet. The best fit straight line is drawn through the transfer characteristic of the device using standard curve-fitting techniques, and the maximum deviation is measured from this line. In general, the integral linearity error measured in this way is only 50% of the value measured by end point methods. This makes the method good for producing impressive data sheets, but it is less useful for error budget analysis. For ac applications it is better to specify distortion than dc linearity, so it is rarely necessary to use the best straight line method to define converter linearity.

The other type of converter nonlinearity is *differential nonlinearity* (DNL). This relates to the linearity of the code transitions of the converter. In the ideal case, a change of 1 LSB in digital code corresponds to a change of exactly 1 LSB of analog signal. In a DAC, a change of 1 LSB in digital code produces exactly 1 LSB change of analog output, while in an ADC there should be exactly 1 LSB change of analog input to move from one digital transition to the next. Differential linearity error is defined as the maximum amount of deviation of any quantum (or LSB change) in the entire transfer function from its ideal size of 1 LSB.

Where the change in analog signal corresponding to 1 LSB digital change is more or less than 1 LSB, there is said to be a DNL error. The DNL error of a converter is normally defined as the maximum value of DNL to be found at any transition across the range of the converter. Figure 2.19 shows the nonideal transfer

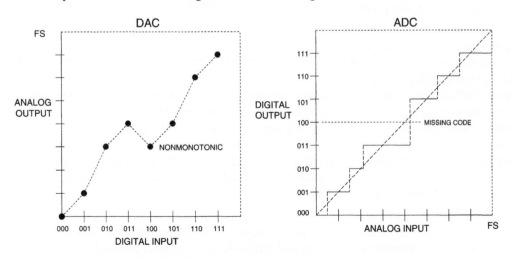

Figure 2.19: Transfer Functions for Non-Ideal 3-Bit DAC and ADC

functions for a DAC and an ADC and shows the effects of the DNL error.

The DNL of a DAC is examined more closely in Figure 2.20. If the DNL of a DAC is less than −1 LSB at any transition, the DAC is *nonmonotonic* i.e., its transfer characteristic contains one or more localized maxima or minima. A DNL greater than +1 LSB does not cause nonmonotonicity, but is still undesirable. In many DAC applications (especially closed-loop systems where nonmonotonicity can change negative feedback to positive feedback), it is critically important that DACs are monotonic. DAC monotonicity is often explicitly specified on data sheets, although if the DNL is guaranteed to be less than 1 LSB (i.e., |DNL| ≤ 1 LSB), the device must be monotonic, even without an explicit guarantee.

In Figure 2.21, the DNL of an ADC is examined more closely on an expanded scale. ADCs can be nonmonotonic, but a more common result of excess DNL in ADCs is *missing codes*. Missing codes in an ADC are as objectionable as nonmonotonicity in a DAC. Again, they result from DNL < −1 LSB.

Not only can ADCs have missing codes, they can also be nonmonotonic as shown in Figure 2.22. As in the case of DACs, this can present major problems—especially in servo applications.

In a DAC, there can be no missing codes—each digital input word will produce a corresponding analog output.

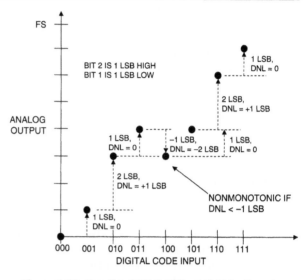

Figure 2.20: Details of DAC Differential Nonlinearity

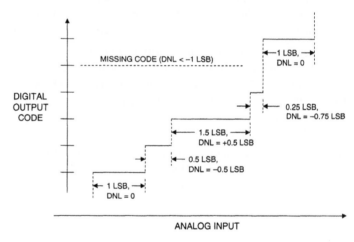

Figure 2.21: Details of ADC Differential Nonlinearity

However, DACs can be nonmonotonic as previously discussed. In a straight binary DAC, the most likely place a nonmonotonic condition can develop is at midscale between the two codes: 011…11 and 100…00. If a nonmonotonic conditions occurs here, it is generally because the DAC is not properly calibrated or trimmed. A successive approximation ADC with an internal nonmonotonic DAC will generally produce missing codes but remain monotonic. However it is possible for an ADC to be nonmonotonic—again depending on the particular conversion architecture. Figure 2.22 shows the transfer function of an ADC which is nonmonotonic and has a missing code.

ADCs that use the *subranging* architecture divide the input range into a number of coarse segments, and each coarse segment is further divided into smaller segments—and ultimately the final code is derived. This process is described in more detail in Chapter 4 of this book. An improperly trimmed subranging ADC may exhibit nonmonotonicity, wide codes, or missing codes at the subranging points as shown in Figure 2.23 A, B, and C, respectively. This type of ADC should be trimmed so that drift due to aging or temperature produces wide codes at the sensitive points rather than nonmonotonic or missing codes.

Defining missing codes is more difficult than defining nonmonotonicity. All ADCs suffer from some inherent transition noise as shown in Figure 2.24 (think of it as the flicker between adjacent values of the last digit of a DVM). As resolutions and bandwidths become higher, the range of input over which transition noise occurs may approach, or even exceed, 1 LSB. High resolution wideband ADCs generally have internal noise sources that can be reflected to the input as effective input noise summed with the signal. The effect of this noise, especially if combined with a negative DNL error, may be that there are some (or even all) codes where transition noise is present for the whole range of inputs. There are therefore some codes for which there is *no* input that will *guarantee* that code as an output, although there may be a range of inputs that will *sometimes* produce that code.

For low resolution ADCs, it may be reasonable to define *no missing codes* as a combination of transition noise and DNL which guarantees some level (perhaps 0.2 LSB) of noise-free code for all codes. However, this is impossible to achieve at the very high resolutions achieved by modern sigma-delta ADCs, or even at lower resolutions in wide bandwidth

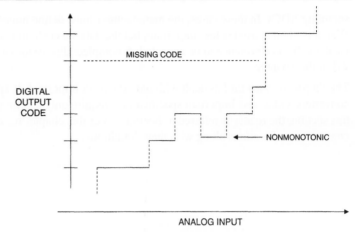

Figure 2.22: Nonmonotonic ADC with Missing Code

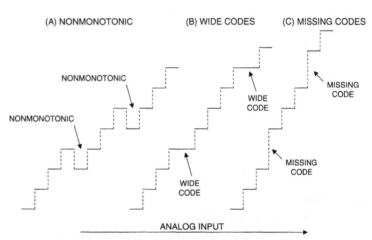

Figure 2.23: Errors Associated with Improperly Trimmed Subranging ADC

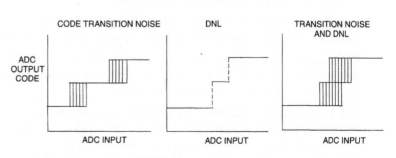

Figure 2.24: Combined Effects of Code Transition Noise and DNL

sampling ADCs. In these cases, the manufacturer must define noise levels and resolution in some other way. Which method is used is less important, but the data sheet should contain a clear definition of the method used and the performance to be expected. A complete discussion of effective input noise follows in Section 2.3 of this chapter.

The discussion thus far has dealt with only the most important dc specifications associated with data converters. Other less important specifications require only a definition. For specifications not covered in this section, the reader is referred to Section 2.5 of this chapter for a complete alphabetical listing of data converter specifications along with their definitions.

References:
2.1 Coding And Quantization

1. K. W. Cattermole, **Principles of Pulse Code Modulation**, American Elsevier Publishing Company, Inc., 1969, New York NY, ISBN 444-19747-8. *(An excellent tutorial and historical discussion of data conversion theory and practice, oriented towards PCM, but covers practically all aspects. This one is a must for anyone serious about data conversion.)*

2. Frank Gray, "Pulse Code Communication," **U.S. Patent 2,632,058**, filed November 13, 1947, issued March 17, 1953. *(Detailed patent on the Gray code and its application to electron beam coders.)*

3. R. W. Sears, "Electron Beam Deflection Tube for Pulse Code Modulation," **Bell System Technical Journal**, Vol. 27, pp. 44–57, Jan. 1948. *(Describes an electon-beam deflection tube 7-bit,100 kSPS flash converter for early experimental PCM work.)*

4. J. O. Edson and H. H. Henning, "Broadband Codecs for an Experimental 224Mb/s PCM Terminal," **Bell System Technical Journal**, Vol. 44, pp. 1887–1940, Nov. 1965. *(Summarizes experiments on ADCs based on the electron tube coder as well as a bit-per-stage Gray code 9-bit solid state ADC. The electron beam coder was 9 bits at 12 MSPS, and represented the fastest of its type.)*

5. Dan Sheingold, **Analog-Digital Conversion Handbook, 3rd Edition**, Analog Devices and Prentice-Hall, 1986, ISBN-0-13-032848-0. *(The defining and classic book on data conversion.)*

Sampling Theory
Walt Kester

This section discusses the basics of sampling theory. A block diagram of a typical real-time sampled data system is shown in Figure 2.25. Prior to the actual analog-to-digital conversion, the analog signal usually passes through some sort of signal conditioning circuitry which performs such functions as amplification, attenuation, and filtering. The low-pass/bandpass filter is required to remove unwanted signals outside the bandwidth of interest and prevent aliasing.

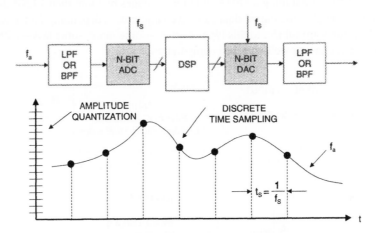

Figure 2.25: Sampled Data System

The system shown in Figure 2.25 is a real-time system; i.e., the signal to the ADC is continuously sampled at a rate equal to f_s, and the ADC presents a new sample to the DSP at this rate. In order to maintain real-time operation, the DSP must perform all its required computation within the sampling interval, $1/f_s$, and present an output sample to the DAC before arrival of the next sample from the ADC. An example of a typical DSP function would be a digital filter.

In the case of FFT analysis, a block of data is first transferred to the DSP memory. The FFT is calculated at the same time a new block of data is transferred into the memory, in order to maintain real-time operation. The DSP must calculate the FFT during the data transfer interval so it will be ready to process the next block of data.

Note that the DAC is required only if the DSP data must be converted back into an analog signal (as would be the case in a voiceband or audio application, for example). There are many applications where the signal remains entirely in digital format after the initial A/D conversion. Similarly, there are applications where the DSP is solely responsible for generating the signal to the DAC. If a DAC is used, it must be followed by an analog anti-imaging filter to remove the image frequencies. Finally, there are slower speed industrial process control systems where sampling rates are much lower—regardless of the system, the fundamentals of sampling theory still apply.

There are two key concepts involved in the actual analog-to-digital and digital-to-analog conversion process: *discrete time sampling* and *finite amplitude resolution due to quantization*. An understanding of these concepts is vital to data converter applications.

The Need for a Sample-and-Hold Amplifier (SHA) Function

The generalized block diagram of a sampled data system, shown in Figure 2.25, assumes some type of ac signal at the input. It should be noted that this does not necessarily have to be so, as in the case of modern digital voltmeters (DVMs) or ADCs optimized for dc measurements, but for this discussion assume that the input signal has some upper frequency limit f_a.

Most ADCs today have a built-in sample-and-hold function, thereby allowing them to process ac signals. This type of ADC is referred to as a *sampling ADC*. However many early ADCs, such as Analog Devices' industry-standard AD574, were not of the sampling type, but simply *encoders* as shown in Figure 2.26. If the input signal to a SAR ADC (assuming no SHA function) changes by more than 1 LSB during the conversion time (8 μs in the example), the output data can have large errors, depending on the location of the code. Most ADC architectures are subject to this type of error—some more, some less—with the possible exception of flash converters having well-matched comparators.

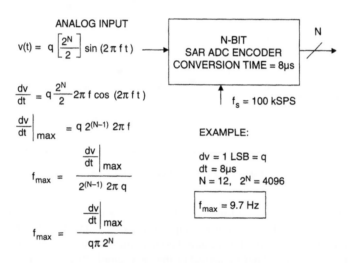

**Figure 2.26: Input Frequency Limitations
of Nonsampling ADC (Encoder)**

Assume that the input signal to the encoder is a sinewave with a full-scale amplitude ($q2^N/2$), where q is the weight of 1 LSB.

$$v(t) = q \ (2^N/2) \sin (2\pi \ f \ t). \qquad \text{Eq. 2.1}$$

Taking the derivative:

$$dv/dt = q \ 2\pi f \ (2^N/2) \cos (2\pi \ f \ t). \qquad \text{Eq. 2.2}$$

The maximum rate of change is therefore:

$$dv/dt \ |_{max} = q \ 2\pi f \ (2^N/2). \qquad \text{Eq. 2.3}$$

Solving for f:

$$f = (dv/dt \big|_{max})/(q \pi 2^N).$$

Eq. 2.4

If N = 12, and 1 LSB change (dv = q) is allowed during the conversion time (dt = 8 μs), the equation can be solved for f_{max}, the maximum full-scale signal frequency that can be processed without error:

$$f_{max} = 9.7 \text{ Hz}.$$

This implies any input frequency greater than 9.7 Hz is subject to conversion errors, even though a sampling frequency of 100 kSPS is possible with the 8 μs ADC (this allows an extra 2 μs interval for an external SHA to reacquire the signal after coming out of the hold mode).

To process ac signals, a sample-and-hold function is added as shown in Figure 2.27. The ideal SHA is simply a switch driving a hold capacitor followed by a high input impedance buffer. The input impedance of the buffer must be high enough so that the capacitor is discharged by less than 1 LSB during the hold time. The SHA samples the signal in the *sample* mode, and holds the signal constant during the *hold* mode. The timing is adjusted so that the encoder performs the conversion during the hold time. A sampling ADC can therefore process fast signals—the upper frequency limitation is determined by the SHA aperture jitter, bandwidth, distortion, etc., not the encoder. In the example shown, a good sample-and-hold could acquire the signal in 2 μs, allowing a sampling frequency of 100 kSPS, and the capability of processing input frequencies up to 50 kHz. A complete discussion of the SHA function including these specifications follows later in this chapter.

It is important to understand a subtle difference between a true *sample-and-hold* amplifier (SHA) and a *track-and-hold* amplifier (T/H, or THA). Strictly speaking, the output of a sample-and-hold is not defined during the sample mode, however the output of a track-and-hold tracks the signal during the sample or *track* mode. In practice, the function is generally implemented as a track-and-hold, and the terms *track-and-hold* and *sample-and-hold* are often used interchangeably. The waveforms shown in Figure 2.27 are those associated with a track-and-hold.

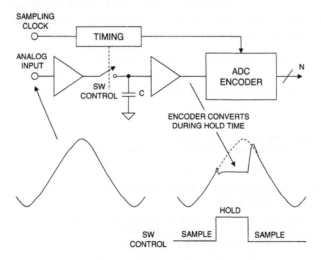

**Figure 2.27: Sample-and-Hold Function
Required for Digitizing AC Signals**

In order to better understand the types of ac errors an ADC can make without a sample-and-hold function, consider Figure 2.28. The photos show the reconstructed output of an 8-bit ADC (flash converter) with and without the sample-and-hold function. In an ideal flash converter the comparators are perfectly matched, and no sample-and-hold is required. In practice, however, there are timing mismatches between the comparators that cause high frequency inputs to exhibit nonlinearities and missing codes as shown in the right-hand photos. The data was taken by driving a DAC with the ADC output. The DAC output is a low frequency aliased sinewave corresponding to the difference between the sampling frequency (20 MSPS) and the ADC input frequency (19.98 MHz). In this case, the alias frequency is 20 kHz. (Aliasing is explained in detail in the next section.)

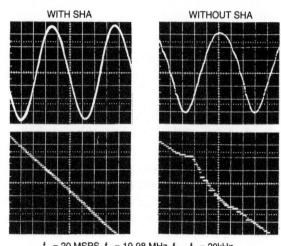

WITH SHA   WITHOUT SHA

$f_s = 20$ MSPS, $f_a = 19.98$ MHz, $f_s - f_a = 20$kHz

Figure 2.28: 8-Bit, 20 MSPS Flash ADC with and without Sample-and-Hold

The Nyquist Criteria

A continuous analog signal is sampled at discrete intervals, $t_s = 1/f_s$, which must be carefully chosen to ensure an accurate representation of the original analog signal. It is clear that the more samples taken (faster sampling rates), the more accurate the digital representation; however, if fewer samples are taken (lower sampling rates), a point is reached where critical information about the signal is actually lost. The mathematical basis of sampling was set forth by Harry Nyquist of Bell Telephone Laboratories in two classic papers published in 1924 and 1928, respectively. (See References 1 and 2 as well as Chapter 1 of this book.) Nyquist's original work was shortly supplemented by R. V. L. Hartley (Reference 3). These papers formed the basis for the PCM work to follow in the 1940s, and in 1948 Claude Shannon wrote his classic paper on communication theory (Reference 4).

Simply stated, the Nyquist criteria requires that the sampling frequency be at least twice the highest frequency contained in the signal, or information about the signal will be lost. If the sampling frequency is less than twice the maximum analog signal frequency, a phenomena known as aliasing will occur.

- A signal with a maximum frequency f_a must be sampled at a rate $f_s > 2f_a$ or information about the signal will be lost because of aliasing.

- Aliasing occurs whenever $f_s < 2 f_a$

- The concept of aliasing is widely used in communications applications such as direct IF-to-digital conversion.

- A signal which has frequency components between f_a and f_b must be sampled at a rate $f_s > 2 (f_b - f_a)$ in order to prevent alias components from overlapping the signal frequencies.

Figure 2.29: Nyquist's Criteria

In order to understand the implications of *aliasing* in both the time and frequency domain, first consider the case of a time domain representation of a single tone sinewave sampled as shown in Figure 2.30. In this example, the sampling frequency f_s is not at least 2 f_a, but only slightly more than the analog input frequency f_a—the Nyquist criteria is violated. Notice that the pattern of the actual samples produces an *aliased* sinewave at a lower frequency equal to $f_s - f_a$.

The corresponding frequency domain representation of this scenario is shown in Figure 2.31B. Now consider the case of a single frequency sinewave of frequency f_a sampled at a frequency f_s by an ideal impulse sampler (see Figure 2.31A). Also assume that $f_s > 2f_a$ as shown. The frequency-domain output of the sampler shows *aliases* or *images* of the original signal around every multiple of f_s; i.e., at frequencies equal to $|\pm Kf_s \pm f_a|$, K = 1, 2, 3, 4, ...

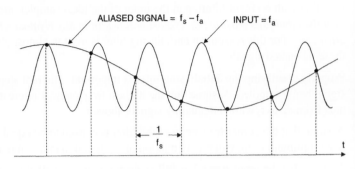

ALIASED SIGNAL = $f_s - f_a$ INPUT = f_a

$\frac{1}{f_s}$

NOTE: f_a IS SLIGHTLY LESS THAN f_s

Figure 2.30: Aliasing in the Time Domain

The *Nyquist* bandwidth is defined to be the frequency spectrum from dc to $f_s/2$. The frequency spectrum is divided into an infinite number of *Nyquist zones*, each having a width equal to 0.5 f_s as shown. In practice, the ideal sampler is replaced by an ADC followed by an FFT processor. The FFT processor only provides an output from dc to $f_s/2$, i.e., the signals or aliases that appear in the first Nyquist zone.

Now consider the case of a signal that is outside the first Nyquist zone (Figure 2.31B). The signal frequency is only slightly less than the sampling frequency, corresponding to the condition shown in the time domain representation in Figure 2.30. Notice that even though the signal is outside the first Nyquist zone, its image (or *alias*), $f_s - f_a$, falls inside. Returning to Figure 2.31A, it is clear that if an unwanted signal appears at any of the image frequencies of f_a, it will also occur at f_a, thereby producing a spurious frequency component in the first Nyquist zone.

This is similar to the analog mixing process and implies that some filtering ahead of the sampler (or ADC) is required to remove frequency components that are outside the Nyquist bandwidth, but whose aliased components fall inside it. The filter performance will depend on how close the out-of-band signal is to $f_s/2$ and the amount of attenuation required.

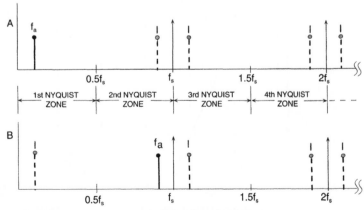

Figure 2.31: Analog Signal f_a Sampled @ f_s Using Ideal Sampler Has Images (Aliases) at $|\pm Kf_s \pm f_a|$, K = 1, 2, 3, . . .

Baseband Antialiasing Filters

Baseband sampling implies that the signal to be sampled lies in the first Nyquist zone. It is important to note that with no input filtering at the input of the ideal sampler, *any frequency component (either signal or noise) that falls outside the Nyquist bandwidth in any Nyquist zone will be aliased back into the first Nyquist zone*. For this reason, an antialiasing filter is used in almost all sampling ADC applications to remove these unwanted signals.

Properly specifying the antialiasing filter is important. The first step is to know the characteristics of the signal being sampled. Assume that the highest frequency of interest is f_a. The antialiasing filter passes signals from dc to f_a while attenuating signals above f_a.

Assume that the corner frequency of the filter is chosen to be equal to f_a. The effect of the finite transition from minimum to maximum attenuation on system dynamic range is illustrated in Figure 2.32A.

Assume that the input signal has full-scale components well above the maximum frequency of interest, f_a. The diagram shows how full-scale frequency components above $f_s - f_a$ are aliased back into the bandwidth dc to f_a. These aliased components are indistinguishable from actual signals and therefore limit the dynamic range to the value on the diagram which is shown as *DR*.

Some texts recommend specifying the antialiasing filter with respect to the Nyquist frequency, $f_s/2$, but this assumes that the signal bandwidth of interest extends from dc to $f_s/2$ which is rarely the case. In the example shown in Figure 2.32A, the aliased components between f_a and $f_s/2$ are not of interest and do not limit the dynamic range.

The antialiasing filter transition band is therefore determined by the corner frequency f_a, the stopband frequency $f_s - f_a$, and the desired stopband attenuation, DR. The required system dynamic range is chosen based on the requirement for signal fidelity.

Filters become more complex as the transition band becomes sharper, all other things being equal. For instance, a Butterworth filter gives 6 dB attenuation per octave for each filter pole (as do all filters). Achieving 60 dB attenuation in a transition region between 1 MHz and 2 MHz (1 octave) requires a minimum of 10 poles—not a trivial filter, and definitely a design challenge.

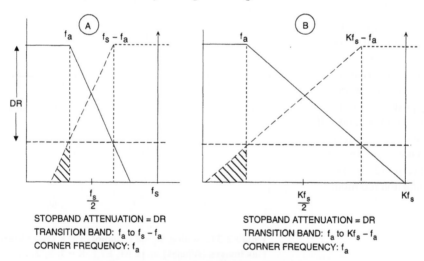

STOPBAND ATTENUATION = DR
TRANSITION BAND: f_a to $f_s - f_a$
CORNER FREQUENCY: f_a

STOPBAND ATTENUATION = DR
TRANSITION BAND: f_a to $Kf_s - f_a$
CORNER FREQUENCY: f_a

Figure 2.32: Oversampling Relaxes Requirements on Baseband Antialiasing Filter

Therefore, other filter types are generally more suited to applications where the requirement is for a sharp transition band and in-band flatness coupled with linear phase response. Elliptic filters meet these criteria and are a popular choice. A number of companies specialize in supplying custom analog filters; TTE is an example of such a company (Reference 5).

From this discussion, we can see how the sharpness of the antialiasing transition band can be traded off against the ADC sampling frequency. Choosing a higher sampling rate (oversampling) reduces the requirement on transition band sharpness (hence, the filter complexity) at the expense of using a faster ADC and processing data at a faster rate. This is illustrated in Figure 2.32B which shows the effects of increasing the sampling frequency by a factor of K, while maintaining the same analog corner frequency, f_a, and the same dynamic range, DR, requirement. The wider transition band (f_a to $Kf_s - f_a$) makes this filter easier to design than for the case of Figure 2.32A.

The antialiasing filter design process is started by choosing an initial sampling rate of 2.5 to 4 times f_a. Determine the filter specifications based on the required dynamic range and see if such a filter is realizable within the constraints of the system cost and performance. If not, consider a higher sampling rate which may require using a faster ADC. It should be mentioned that sigma-delta ADCs are inherently highly oversampled converters, and the resulting relaxation in the analog antialiasing filter requirements is therefore an added benefit of this architecture.

The antialiasing filter requirements can also be relaxed somewhat if it is certain that there will never be a full-scale signal at the stopband frequency $f_s - f_a$. In many applications, it is improbable that full-scale signals will occur at this frequency. If the maximum signal at the frequency $f_s - f_a$ will never exceed X dB below full-scale, then the filter stopband attenuation requirement can be reduced by that same amount. The new requirement for stopband attenuation at $f_s - f_a$ based on this knowledge of the signal is now only DR – X dB. When making this type of assumption, be careful to treat any noise signals that may occur above the maximum signal frequency f_a as unwanted signals that will also alias back into the signal bandwidth.

As an example, the normalized response of the TTE, Inc., LE1182 11-pole elliptic antialiasing filter is shown in Figure 2.33. Notice that this filter is specified to achieve at least 80 dB attenuation between f_c and 1.2 f_c. The corresponding pass band ripple, return loss, delay, and phase response are also shown in Figure 2.33. This custom filter is available in corner frequencies up to 100 MHz and in a choice of PC board, BNC, or SMA with compatible packages.

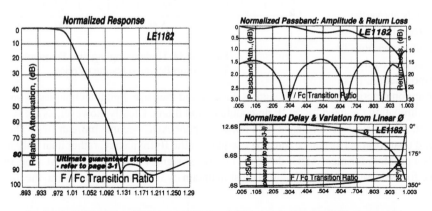

Reprinted with Permission of TTE, Inc., 11652 Olympic Blvd., Los Angeles CA 90064, www.tte.com

Figure 2.33: Characteristics of 11-Pole Elliptical Filter (TTE, Inc., LE1182 Series)

Undersampling (Harmonic Sampling, Bandpass Sampling, IF Sampling, Direct IF-to-Digital Conversion)

Thus far we have considered the case of baseband sampling, where all the signals of interest lie within the first Nyquist zone. Figure 2.34A shows such a case, where the band of sampled signals is limited to the first Nyquist zone, and images of the original band of frequencies appear in each of the other Nyquist zones.

Consider the case shown in Figure 2.34B, where the sampled signal band lies entirely within the second Nyquist zone. The process of sampling a signal outside the first Nyquist zone is often referred to as *under-sampling*, or *harmonic sampling*. Note that the image which falls in the first Nyquist zone contains all the information in the original signal, with the exception of its original location (the order of the frequency components within the spectrum is reversed, but this is easily corrected by re-ordering the output of the FFT).

Figure 2.34C shows the sampled signal restricted to the third Nyquist zone. Note that the image that falls into the first Nyquist zone has no frequency reversal. In fact, the sampled signal frequencies may lie in *any* unique Nyquist zone, and the image falling into the first Nyquist zone is still an accurate representation (with the exception of the frequency reversal tjat occurs when the signals are located in even Nyquist zones). At this point we can clearly restate the Nyquist criteria:

*A signal must be sampled at a rate equal to or greater than twice its **bandwidth** in order to preserve all the signal information.*

Notice that there is no mention of the absolute *location* of the band of sampled signals within the frequency spectrum relative to the sampling frequency. The only constraint is that the band of sampled signals be restricted to a *single* Nyquist zone, i.e., the signals must not overlap any multiple of $f_s/2$ (this, in fact, is the primary function of the antialiasing filter).

Sampling signals above the first Nyquist zone has become popular in communications because the process is equivalent to analog demodulation. It is

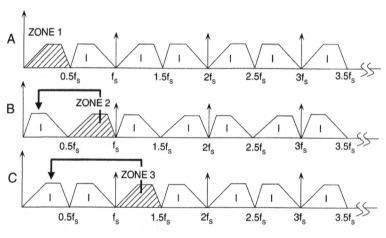

Figure 2.34: Undersampling and Frequency Translation Between Nyquist Zones

becoming common practice to sample IF signals directly and then use digital techniques to process the signal, thereby eliminating the need for an IF demodulator and filters. Clearly, however, as the IF frequencies become higher, the dynamic performance requirements on the ADC become more critical. The ADC input bandwidth and distortion performance must be adequate at the IF frequency, rather than only baseband. This presents a problem for most ADCs designed to process signals in the first Nyquist zone, therefore an ADC suitable for undersampling applications must maintain dynamic performance into the higher order Nyquist zones.

Antialiasing Filters in Undersampling Applications

Figure 2.35 shows a signal in the second Nyquist zone centered around a carrier frequency, f_c, whose lower and upper frequencies are f_1 and f_2. The antialiasing filter is a bandpass filter. The desired dynamic range is DR, which defines the filter stopband attenuation. The upper transition band is f_2 to $2f_s - f_2$, and the lower is f_1 to $f_s - f_1$. As in the case of baseband sampling, the antialiasing filter requirements can be relaxed by proportionally increasing the sampling frequency, but f_c must also be increased so that it is always centered in the second Nyquist zone.

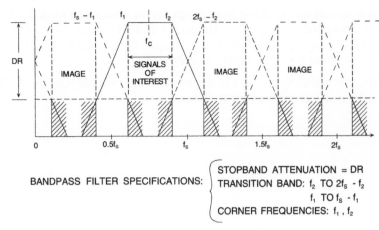

BANDPASS FILTER SPECIFICATIONS:
$\left\{ \begin{array}{l} \text{STOPBAND ATTENUATION = DR} \\ \text{TRANSITION BAND: } f_2 \text{ TO } 2f_s - f_2 \\ \qquad\qquad\qquad f_1 \text{ TO } f_s - f_1 \\ \text{CORNER FREQUENCIES: } f_1 , f_2 \end{array} \right.$

Figure 2.35: Antialiasing Filter for Undersampling

Two key equations can be used to select the sampling frequency, f_s, given the carrier frequency, f_c, and the bandwidth of its signal, Δf. The first is the Nyquist criteria:

$$f_s > 2\Delta f \qquad\qquad \text{Eq. 2.5}$$

The second equation ensures that f_c is placed in the center of a Nyquist zone:

$$f_s = \frac{4f_c}{2NZ - 1} \qquad\qquad \text{Eq. 2.6}$$

where NZ = 1, 2, 3, 4, ... and NZ corresponds to the Nyquist zone in which the carrier and its signal fall (see Figure 2.36).

NZ is normally chosen to be as large as possible while still maintaining $f_s > 2\Delta f$. This results in the minimum required sampling rate. If NZ is chosen to be odd, then f_c and its signal will fall in an odd Nyquist zone, and the image frequencies in the first Nyquist zone will not be reversed. Trade-offs can be made between the sampling frequency and the complexity of the antialiasing filter by choosing smaller values of NZ (hence a higher sampling frequency).

As an example, consider a 4 MHz wide signal centered around a carrier frequency of 71 MHz. The minimum required sampling frequency is therefore 8 MSPS. Solving Eq. 2.6 for NZ using f_c = 71 MHz and f_s = 8 MSPS yields NZ = 18.25. However, NZ must be an integer, so we round 18.25 to the next lowest integer, 18. Solving Eq. 2.6 again for f_s yields f_s = 8.1143 MSPS. The final values are therefore f_s = 8.1143 MSPS, f_c = 71 MHz, and NZ = 18.

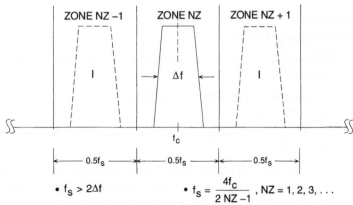

Figure 2.36: Centering an Undersampled Signal within a Nyquist Zone

Now assume that we desire more margin for the antialiasing filter, and we select f_s to be 10 MSPS. Solving Eq. 2.6 for NZ, using f_c = 71 MHz and f_s = 10 MSPS yields NZ = 14.7. We round 14.7 to the next lowest integer, giving NZ = 14. Solving Eq. 2.6 again for f_s yields f_s = 10.519 MSPS. The final values are therefore f_s = 10.519 MSPS, f_c = 71 MHz, and NZ = 14.

The above iterative process can also be carried out starting with f_s and adjusting the carrier frequency to yield an integer number for NZ.

References:
2.2 Sampling Theory

1. H. Nyquist, "Certain Factors Affecting Telegraph Speed," **Bell System Technical Journal**, Vol. 3, April 1924, pp. 324–346.

2. H.. Nyquist, Certain Topics in Telegraph Transmission Theory, **A.I.E.E. Transactions**, Vol. 47, April 1928, pp. 617–644.

3. R.V.L. Hartley, "Transmission of Information," **Bell System Technical Journal**, Vol. 7, July 1928, pp. 535–563.

4. C. E. Shannon, "A Mathematical Theory of Communication," **Bell System Technical Journal**, Vol. 27, July 1948, pp. 379–423 and October 1948, pp. 623–656.

5. TTE, Inc., 11652 Olympic Blvd., Los Angeles, CA 90064, www.tte.com.

Data Converter AC Errors
Walt Kester, James Bryant

This section examines the ac errors associated with data converters. Many of the errors and specifications apply equally to ADCs and DACs, while some are more specific to one or the other. All possible specifications are not discussed here, only the most common ones. Section 2.4 of this chapter contains a comprehensive listing of converter specifications as well as their definitions, including some not discussed in this section.

Theoretical Quantization Noise of an Ideal N-Bit Converter

The only errors (dc or ac) associated with an ideal N-bit data converter are those related to the sampling and quantization processes. The maximum error an ideal converter makes when digitizing a signal is ±½ LSB. The transfer function of an ideal N-bit ADC is shown in Figure 2.37. The quantization error for any ac signal that spans more than a few LSBs can be approximated by an uncorrelated sawtooth waveform having a peak-to-peak amplitude of q, the weight of an LSB. Although this analysis is not precise, it is accurate enough for most applications. W. R. Bennett of Bell Laboratories analyzed the actual spectrum of quantization noise in his classic 1948 paper (Reference 1). With certain simplifying assumptions, his detailed mathematical analysis simplifies to that of Figure 2.37. Other significant papers on converter noise (References 2–5) followed Bennett's classic publication.

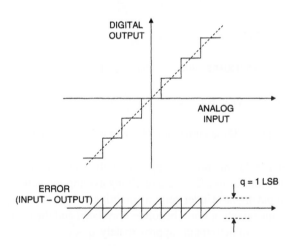

Figure 2.37: Ideal N-bit ADC Quantization Noise

The quantization error as a function of time is shown in Figure 2.38. Again, a simple sawtooth waveform provides a sufficiently accurate model for analysis. The equation of the sawtooth error is given by

$$e(t) = st, \quad -q/2s < t < +q/2s \qquad \text{Eq. 2.7}$$

The mean-square value of e(t) can be written:

$$\overline{e^2(t)} = \frac{s}{q} \int_{-q/2s}^{+q/2s} (st)^2 \, dt \qquad \text{Eq. 2.8}$$

Performing the simple integration and simplifying,

$$\overline{e^2(t)} = \frac{q^2}{12} \qquad \text{Eq. 2.9}$$

The root-mean-square quantization error is therefore

$$\text{rms quantization noise} = \sqrt{r^2(t)} = \frac{q}{\sqrt{12}} \qquad \text{Eq. 2.10}$$

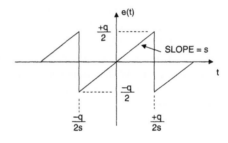

- ERROR = $e(t) = st, \quad \frac{-q}{2s} < t < \frac{+q}{2s}$

- MEAN-SQUARE ERROR = $\overline{e^2(t)} = \frac{s}{q} \int_{-q/2s}^{+q/2s} (st)^2 \, dt = \frac{q^2}{12}$

- ROOT-MEAN-SQUARE ERROR = $\sqrt{\overline{e^2(t)}} = \frac{q}{\sqrt{12}}$

Figure 2.38: Quantization Noise as a Function of Time

As Bennett points out (Reference 1), this noise is approximately Gaussian and spread more or less uniformly over the Nyquist bandwidth dc to $f_s/2$. The underlying assumption here is that the quantization noise is uncorrelated to the input signal. Under certain conditions where the sampling clock and the signal are harmonically related, the quantization noise becomes correlated and the energy is concentrated at the harmonics of the signal—the rms value remains approximately $q/\sqrt{12}$.

The theoretical signal-to-noise ratio can now be calculated assuming a full-scale input sinewave:

$$\text{Input FS Sinewave} = v(t) = \frac{q2^N}{2} \sin(2\pi ft) \qquad \text{Eq. 2.11}$$

The rms value of the input signal is therefore

$$\text{rms value of FS input} = \frac{q2^N}{2\sqrt{2}} \qquad \text{Eq. 2.12}$$

The rms signal-to-noise ratio for an ideal N-bit converter is therefore

$$SNR = 20 \log_{10} \frac{\text{rms value of FS input}}{\text{rms value of quantization noise}} \qquad \text{Eq. 2.13}$$

$$SNR = 20 \log_{10} \left[\frac{q 2^N / 2\sqrt{2}}{q / \sqrt{12}} \right] = 6.02N + 1.76 \text{ dB, over dc to } f_s/2 \text{ bandwidth} \qquad \text{Eq. 2.14}$$

These relationships are summarized in Figure 2.39.

- FS INPUT = $v(t) = \left[\frac{q \, 2^N}{2} \right] \sin(2\pi f t)$

- RMS Value of FS Sinewave = $\frac{q \, 2^N}{2?2}$

- RMS Value of Quantization Noise = $\frac{q}{\sqrt{12}}$

- SNR = $20 \log_{10} \left[\frac{\text{RMS Value of FS Sinewave}}{\text{RMS Value of Quantization Noise}} \right] = 20 \log_{10} 2^N + 20 \log_{10} \sqrt{\frac{3}{2}}$

> SNR = 6.02N + 1.76dB
>
> (Measured over the Nyquist Bandwidth : DC to $f_s/2$)

Figure 2.39: Theoretical Signal-to-Quantization Noise Ratio of an Ideal N-Bit Converter

Bennett's paper shows that although the actual spectrum of the quantization noise is quite complex to analyze—the simplified analysis which leads to Eq. 2.14 is accurate enough for most purposes. However, it is important to emphasize again that the rms quantization noise is measured over the full Nyquist bandwidth, dc to $f_s/2$. In many applications, the actual signal of interest occupies a smaller bandwidth, BW. If digital filtering is used to filter out noise components outside the bandwidth BW, then a correction factor (called *process gain*) must be included in the equation to account for the resulting increase in SNR. The process of sampling a signal at a rate greater than twice its bandwidth is often referred to as *oversampling*. In fact, oversampling in conjunction with quantization noise shaping and digital filtering is a key concept in sigma-delta converters.

$$SNR = 6.02N + 1.76 \text{ dB} + 10 \log_{10} \frac{f_s}{2 \times BW}, \text{ over bandwidth BW} \qquad \text{Eq. 2.15}$$

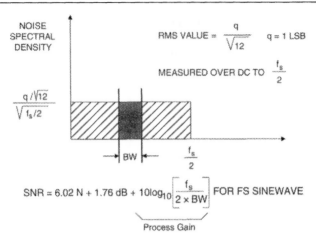

Figure 2.40: Quantization Noise Spectrum

The significance of process gain can be seen from the following example. In many digital basestations or other wideband receivers the signal bandwidth is composed of many individual channels, and a single ADC is used to digitize the entire bandwidth. For instance, the analog cellular radio system (AMPS) in the U.S. consists of 416 30-kHz-wide channels, occupying a bandwidth of approximately 12.5 MHz. Assume a 65 MSPS sampling frequency, and that digital filtering is used to separate the individual 30 kHz channels. The process gain due to oversampling is therefore given by:

$$\text{Process Gain} = 10 \log_{10} \frac{f_s}{2 \times BW} = 10 \log_{10} \frac{65 \times 10^6}{2 \times 30 \times 10^3} = 30.3 \text{ dB} \qquad \text{Eq. 2.16}$$

The process gain is added to the ADC SNR specification to yield the actual SNR in the 30 kHz bandwidth. In the above example, if the ADC SNR specification is 65 dB (dc to $f_s/2$), then it is increased to 95.3 dB in the 30 kHz channel bandwidth (after appropriate digital filtering).

Figure 2.41 shows an application that combines oversampling and undersampling. The signal of interest has a bandwidth BW and is centered around a carrier frequency f_c. The sampling frequency can be much less than f_c and is chosen such that the signal of interest is centered in its Nyquist zone. Analog and digital filtering removes the noise outside the signal bandwidth of interest, and therefore results in process gain per Eq. 2.16.

Figure 2.41: Undersampling and Oversampling Combined Results in Process Gain

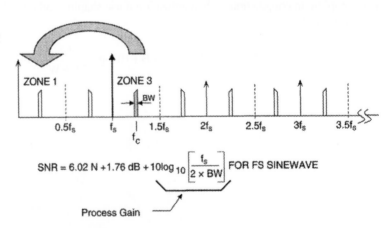

Although the rms value of the noise is accurately approximated by $q/\sqrt{12}$, its frequency domain content may be highly correlated to the ac input signal. For instance, there is greater correlation for low amplitude periodic signals than for large amplitude random signals. Quite often, the assumption is made that the theoretical quantization noise appears as white noise, spread uniformly over the Nyquist bandwidth dc to $f_s/2$. Unfortunately, this is not true in all cases. In the case of strong correlation, the quantization noise appears concentrated at the various harmonics of the input signal, just where you don't want them. Bennett (Reference 1) has an extensive analysis of the frequency content contained in the quantization noise spectrum in his classic 1948 paper.

In most practical applications, the input to the ADC is a band of frequencies (always summed with some unavoidable system noise), so the quantization noise tends to be random. In spectral analysis applications (or in performing FFTs on ADCs using spectrally pure sinewaves—see Figure 2.42), however, the correlation between the quantization noise and the signal depends upon the ratio of the sampling frequency to the input signal. This is demonstrated in Figure 2.43, where the output of an ideal 12-bit ADC is analyzed using a 4096-point FFT. In the left-hand FFT plot, the ratio of the sampling frequency to the input frequency was chosen to be exactly 32, and the worst harmonic is about 76 dB below the fundamental. The right hand diagram shows the effects of slightly offsetting the ratio to 4096/127 = 32.25196850394, showing a relatively random noise spectrum, where the SFDR is now about 92 dBc. In both cases, the rms value of all the noise components is approximately $q/\sqrt{12}$, but in the first case, the noise is concentrated at harmonics of the fundamental.

Note that this variation in the apparent harmonic distortion of the ADC is an artifact of the sampling process and the correlation of the quantization error with the input frequency. In a practical ADC application, the quantization error generally appears as random noise because of the random nature of the wideband input signal and the additional fact that there is a usually a small amount of system noise which acts as a *dither* signal to further randomize the quantization error spectrum.

It is important to understand the above point, because single-tone sinewave FFT testing of ADCs is one of the universally accepted methods of performance evaluation. In order to accurately measure the harmonic distortion of an ADC, steps must be taken to ensure that the test setup truly measures the ADC distortion, not the artifacts due to quantization noise correlation. This is done by properly choosing the frequency ratio and sometimes by injecting a small amount of noise (dither)

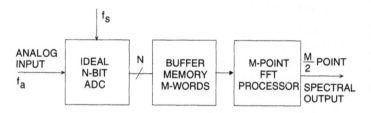

Figure 2.42: Dynamic Performance Analysis of an Ideal N-bit ADC

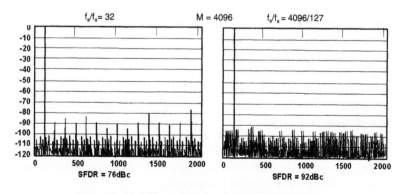

Figure 2.43: Effect of Ratio of Sampling Clock to Input Frequency on SFDR for Ideal 12-Bit ADC

with the input signal. The exact same precautions apply to measuring DAC distortion with an analog spectrum analyzer.

Figure 2.44 shows the FFT output for an ideal 12-bit ADC. Note that the average value of the noise floor of the FFT is approximately 100 dB below full-scale, but the theoretical SNR of a 12-bit ADC is 74 dB. The FFT noise floor is *not* the SNR of the ADC, because the FFT acts like an analog spectrum analyzer with a bandwidth of f_s/M, where M is the number of points in the FFT. The theoretical FFT noise floor is therefore $10\log_{10}(M/2)$ dB below the quantization noise floor due to the *processing gain* of the FFT. In the case of an ideal 12-bit ADC with

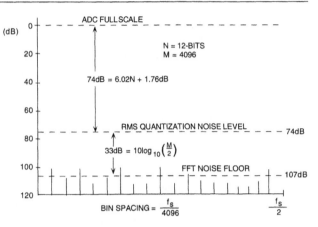

Figure 2.44: Noise Floor for an Ideal 12-Bit ADC Using 4096-Point FFT

an SNR of 74 dB, a 4096-point FFT would result in a processing gain of $10\log_{10}(4096/2) = 33$ dB, thereby resulting in an overall FFT noise floor of $74 + 33 = 107$ dBc. In fact, the FFT noise floor can be reduced even further by going to larger and larger FFTs; just as an analog spectrum analyzer's noise floor can be reduced by narrowing the bandwidth. When testing ADCs using FFTs, it is important to ensure that the FFT size is large enough that the distortion products can be distinguished from the FFT noise floor itself.

Noise in Practical ADCs

A practical sampling ADC (one that has an integral sample-and-hold), regardless of architecture, has a number of noise and distortion sources as shown in Figure 2.45. The wideband analog front-end buffer has wideband noise, nonlinearity, and also finite bandwidth. The SHA introduces further nonlinearity, bandlimiting, and aperture jitter. The actual quantizer portion of the ADC introduces quantization noise, and both integral and differential nonlinearity. In this discussion, assume

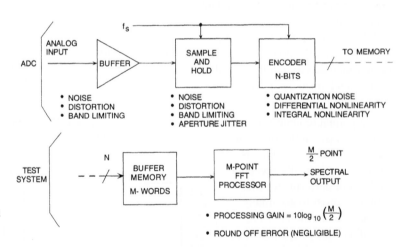

Figure 2.45: ADC Model Showing Noise and Distortion Sources

that sequential outputs of the ADC are loaded into a buffer memory of length M and that the FFT processor provides the spectral output. Also assume that the FFT arithmetic operations themselves introduce no significant errors relative to the ADC. However, when examining the output noise floor, the FFT processing gain (dependent on M) must be considered.

Equivalent Input Referred Noise

Wideband ADC internal circuits produce a certain amount of rms noise due to resistor noise and "kT/C" noise. This noise is present even for dc input signals, and accounts for the fact that the output of most wideband (or high resolution) ADCs is a distribution of codes, centered around the nominal value of a dc input (Figure 2.46). To measure its value, the input of the ADC is either grounded or connected to a heavily decoupled voltage source, and a large number of output samples are collected and plotted as a histogram (sometimes referred to as a *grounded-input* histogram). Since the noise is approximately Gaussian, the standard deviation of the histogram is easily calculated (Reference 6), corresponding to the effective input rms noise. It is common practice to express this rms noise in terms of LSBs rms, although it can be expressed as an rms voltage referenced to the ADC full-scale input range.

Noise-Free (Flicker-Free) Code Resolution

The *noise-free code resolution* of an ADC is the number of bits beyond which it is impossible to distinctly resolve individual codes. The cause is the effective input noise (or input-referred noise) associated with all ADCs and described above. This noise can be expressed as an rms quantity, usually having the units of *LSBs rms*. Multiplying by a factor of 6.6 converts the rms noise into

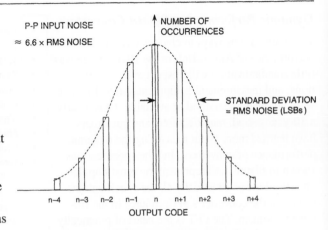

Figure 2.46: Effect of Input-Referred Noise on ADC "Grounded Input" Histogram

Effective Input Noise $= e_{n\,rms}$

Peak-to-Peak Input Noise $= 6.6\ e_{n\,rms}$

$$\text{Noise-Free Code Resolution} = \log_2 \left[\frac{\text{Peak-to-Peak Input Range}}{\text{Peak-to-Peak Input Noise}} \right]$$

$$= \log_2 \left[\frac{2^N}{\text{Peak-to-Peak Input Noise (LSBs)}} \right]$$

$$\text{"Effective Resolution"} = \log_2 \left[\frac{\text{Peak-to-Peak Input Range}}{\text{RMS Input Noise}} \right]$$

$$= \log_2 \left[\frac{2^N}{\text{RMS Input Noise (LSBs)}} \right]$$

$$= \text{Noise-Free Code Resolution} + 2.7\ \text{bits}$$

Figure 2.47: Calculating Noise-Free (Flicker-Free) Code Resolution from Input-Referred Noise

peak-to-peak noise (expressed in *LSBs peak-to-peak*). The total range of an N-bit ADC is 2^N LSBs. The noise-free (or flicker-free) resolution can be calculated using the equation:

$$\text{Noise-Free Code Resolution} = \log_2 (2^N/\text{Peak-to-Peak Noise}) \qquad \text{Eq. 2.17}$$

The specification is generally associated with high-resolution sigma-delta measurement ADCs, but is applicable to all ADCs.

The ratio of the FS range to the *rms* input noise is sometimes used to calculate resolution. In this case, the term *effective resolution* is used. Note that under identical conditions, effective resolution is larger than noise-free code resolution by $\log_2(6.6)$, or approximately 2.7 bits.

$$\text{Effective Resolution} = \log_2 (2^N/\text{RMS Input Noise}) \qquad \text{Eq. 2.18}$$

$$\text{Effective Resolution} = \text{Noise-Free Code Resolution} + 2.7\ \text{bits} \qquad \text{Eq. 2.19}$$

The calculations are summarized in Figure 2.47.

Dynamic Performance of Data Converters

There are various ways to characterize the ac performance of ADCs. Before the 1970s, there was little standardization with respect to ac specifications, and measurement equipment and techniques were not well understood or available. Over nearly a 30-year period, manufacturers and customers have learned more about measuring the dynamic performance of converters, and the specifications shown in Figure 2.48 represent the most popular ones used today. Practically all the specifications represent the converter's performance in the frequency domain. The FFT is the heart of practically all these measurements and is discussed in more detail in Chapter 6 of this book.

- Harmonic Distortion
- Worst Harmonic
- Total Harmonic Distortion (THD)
- Total Harmonic Distortion Plus Noise (THD + N)
- Signal-to-Noise-and-Distortion Ratio (SINAD, or S/N + D)
- Effective Number of Bits (ENOB)
- Signal-to-Noise Ratio (SNR)
- Analog Bandwidth (Full-Power, Small-Signal)
- Spurious Free Dynamic Range (SFDR)
- Two-Tone IntermodulationDistortion
- Multitone Intermodulation Distortion
- Noise Power Ratio (NPR)
- Adjacent Channel Leakage Ratio (ACLR)
- Noise Figure
- Settling Time, Overvoltage Recovery Time

Figure 2.48: Quantifying Data Converter Dynamic Performance

Integral and Differential Nonlinearity Distortion Effects

One of the first things to realize when examining the nonlinearities of data converters is that the transfer function of a data converter has artifacts that do not occur in conventional linear devices such as op amps or gain blocks. The overall integral nonlinearity of an ADC is due to the integral nonlinearity of the front-end and SHA as well as the overall integral nonlinearity in the ADC transfer function. However, *differential nonlinearity is due exclusively to the encoding process* and may vary considerably, dependent on the ADC encoding architecture. Overall integral nonlinearity produces distortion products whose amplitude varies as a function of the input signal amplitude. For instance, second-order intermodulation products increase 2 dB for every 1 dB increase in signal level, and third-order products increase 3 dB for every 1 dB increase in signal level.

The differential nonlinearity in the ADC transfer function produces distortion products which not only depend on the amplitude of the signal but the positioning of the differential nonlinearity errors along the ADC transfer function. Figure 2.49 shows two ADC transfer functions having differential nonlinearity. The left-hand diagram shows an error that occurs at midscale. Therefore, for both large and small signals, the signal crosses through this point producing a distortion product which is relatively independent of the signal amplitude. The right-hand diagram shows another ADC transfer function which has differential nonlinearity errors at 1/4 and 3/4 full-scale. Signals above 1/2 scale peak-to-peak will exercise these codes and produce distortion, while those less than 1/2 scale peak-to-peak will not.

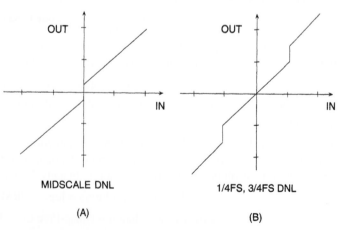

Figure 2.49: Typical ADC/ DAC DNL Errors (Exaggerated)

Most high-speed ADCs are designed so that differential nonlinearity is spread across the entire ADC range. Therefore, for signals that are within a few dB of full-scale, the overall integral nonlinearity of the transfer function determines the distortion products. For lower level signals, however, the harmonic content becomes dominated by the differential nonlinearities and does not generally decrease proportionally with decreases in signal amplitude.

Harmonic Distortion, Worst Harmonic, Total Harmonic Distortion (THD), Total Harmonic Distortion Plus Noise (THD + N)

There are a number of ways to quantify the distortion of an ADC. An FFT analysis can be used to measure the amplitude of the various harmonics of a signal. The harmonics of the input signal can be distinguished from other distortion products by their location in the frequency spectrum. Figure 2.50 shows a 7 MHz input signal sampled at 20 MSPS and the location of the first nine harmonics. Aliased harmonics of f_a fall at frequencies equal to $|\pm Kf_s \pm nf_a|$, where n is the order of the harmonic, and K = 0, 1, 2, 3,.... The second and third harmonics are generally the only ones specified on a data sheet because they tend to be the largest, although some data sheets may specify the value of the *worst* harmonic.

Harmonic distortion is normally specified in dBc (decibels below *carrier*), although at audio frequencies it may be specified as a percentage. Harmonic distortion is generally specified with an input signal near full-scale (generally 0.5 to 1 dB below full-scale to prevent clipping), but it can be specified at any level. For signals much lower than full-scale, other distortion products due to the DNL of the converter (not direct harmonics) may limit performance.

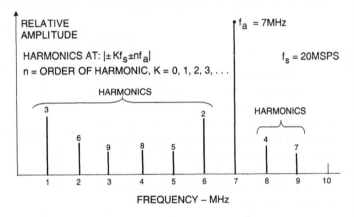

Total harmonic distortion (THD) is the ratio of the rms value of the fundamental signal to the mean value of the root-sum-square of its harmonics (generally, only the first five are significant). THD of an ADC is also generally specified with the input signal close to full-scale, although it can be specified at any level.

Figure 2.50: Location of Distortion Products: Input Signal = 7 MHz, Sampling Rate = 20 MSPS

Total harmonic distortion plus noise (THD + N) is the ratio of the rms value of the fundamental signal to the mean value of the root-sum-square of its harmonics plus all noise components (excluding dc). The bandwidth over which the noise is measured must be specified. In the case of an FFT, the bandwidth is dc to $f_s/2$. (If the bandwidth of the measurement is dc to $f_s/2$, THD + N is equal to SINAD—see below).

Signal-to-Noise-and-Distortion Ratio (SINAD), Signal-to-Noise Ratio (SNR), and Effective Number of Bits (ENOB)

SINAD and SNR deserve careful attention, because there is still some variation between ADC manufacturers as to their precise meaning. Signal-to-Noise-and Distortion (SINAD, or S/(N + D) is the ratio of the rms signal amplitude to the mean value of the root-sum-square (rss) of all other spectral components, *including harmonics*, but excluding dc (Figure 2.50). SINAD is a good indication of the overall dynamic performance of an ADC as a function of input frequency because it includes all components which make up noise (including thermal noise) and distortion. It is often plotted for various input amplitudes. SINAD is equal to THD + N if the bandwidth for the noise measurement is the same. A typical plot for the AD9226 12-bit, 65 MSPS ADC is shown in Figure 2.52.

The SINAD plot shows where the ac performance of the ADC degrades due to high-frequency distortion and is usually plotted for frequencies well above the Nyquist frequency so that performance in undersampling applications can be evaluated. SINAD is often converted to *effective-number-of-bits* (ENOB) using the relationship for the theoretical SNR of an ideal N-bit ADC: SNR = 6.02N + 1.76 dB. The equation is solved for N, and the value of SINAD is substituted for SNR:

- SINAD (Signal-to-Noise-and-Distortion Ratio):
  - The ratio of the rms signal amplitude to the mean value of the root-sum-squares (RSS) of all other spectral components, including harmonics, but excluding dc
- ENOB (Effective Number of Bits):

$$ENOB = \frac{SINAD - 1.76dB}{6.02}$$

- SNR (Signal-to-Noise Ratio, or Signal-to-Noise Ratio Without Harmonics:
  - The ratio of the rms signal amplitude to the mean value of the root-sum-squares (RSS) of all other spectral components, excluding the first five harmonics and dc

Figure 2.51: SINAD, ENOB, and SNR

$$ENOB = \frac{SINAD - 1.76 \text{ dB}}{6.02} \qquad \text{Eq. 2.20}$$

Signal-to-noise ratio (SNR, or *SNR-without-harmonics*) is calculated the same as SINAD except that the signal harmonics are excluded from the calculation, leaving only the noise terms. In practice, it is only necessary to exclude the first five harmonics since they dominate. The SNR plot will degrade at high frequencies, but not as rapidly as SINAD because of the exclusion of the harmonic terms.

Many current ADC data sheets somewhat loosely refer to SINAD as SNR, so the engineer must be careful when interpreting these specifications.

Analog Bandwidth

The analog bandwidth of an ADC is that frequency at which the spectral output of the *fundamental* swept frequency (as determined by the FFT analysis) is reduced by 3 dB. It may be specified

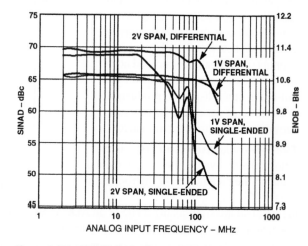

Figure 2.52: AD9226 12-Bit, 65 MSPS ADC SINAD and ENOB for Various Input Full-Scale Spans (Range)

for either a small signal (SSBW—*small signal bandwidth*), or a full-scale signal (FPBW—*full power bandwidth*), so there can be a wide variation in specifications between manufacturers.

Like an amplifier, the analog bandwidth specification of a converter does not imply that the ADC maintains good distortion performance up to its bandwidth frequency. In fact, the SINAD (or ENOB) of most ADCs will begin to degrade considerably before the input frequency approaches the actual 3 dB bandwidth frequency. Figure 2.53 shows ENOB and full-scale frequency response of an ADC with a FPBW of 1 MHz, however, the ENOB begins to drop rapidly above 100 kHz.

Spurious Free Dynamic Range (SFDR)

Probably the most significant specification for an ADC used in a communications application is its *spurious free dynamic range* (SFDR). SFDR of an ADC is defined as the ratio of the rms signal amplitude to the rms value of the *peak spurious spectral content* measured over the bandwidth of interest. Unless otherwise stated, the bandwidth is assumed to be the Nyquist bandwidth dc to $f_s/2$.

Occasionally the frequency spectrum is divided into an *in-band* region (containing the signals of interest) and an *out-of-band* region (signals here are filtered out digitally). In this case there may be an *in-band SFDR* specification and an *out-of-band SFDR* specification, respectively.

SFDR is generally plotted as a function of signal amplitude and may be expressed relative to the signal amplitude (dBc) or the ADC full-scale (dBFS) as shown in Figure 2.54.

For a signal near full-scale, the peak spectral spur is generally determined by one of the first few harmonics of the fundamental. However, as the signal falls several dB below full-scale, other spurs generally occur which are not direct harmonics of the input signal. This is because of the differential nonlinearity of the ADC transfer function as discussed earlier. Therefore, SFDR considers *all* sources of distortion, regardless of their origin.

The AD6645 is a 14-bit, 80 MSPS wideband ADC designed for communications applications where high SFDR is important. The single-tone SFDR for a 69.1 MHz input and a sampling frequency of 80 MSPS is shown in Figure 2.55. Note that a minimum of 89 dBc SFDR is obtained over the entire first Nyquist zone (dc to 40 MHz).

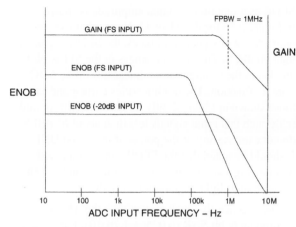

Figure 2.53: ADC Gain (Bandwidth) and ENOB versus Frequency Shows Importance of ENOB Specification

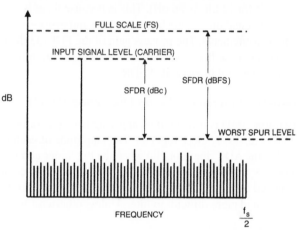

Figure 2.54: Spurious Free Dynamic Range (SFDR)

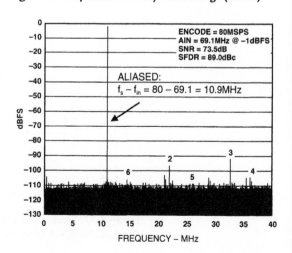

Figure 2.55: AD6645 14-Bit, 80/105 MSPS ADC SFDR for 69.1 MHz Input

SFDR as a function of signal amplitude is shown in Figure 2.56 for the AD6645. Notice that over the entire range of signal amplitudes, the SFDR is greater than 90 dBFS. The abrupt changes in the SFDR plot are due to the differential nonlinearities in the ADC transfer function. The nonlinearities correspond to those shown in Figure 2.49B, and are offset from mid-scale such that input signals less than about 65 dBFS do not exercise any of the points of increased DNL. It should be noted that the SFDR can be improved by injecting a small out-of-band dither signal—at the expense of a slight degradation in SNR.

SFDR is generally much greater than the ADCs theoretical N-bit SNR (6.02N + 1.76 dB). For example, the AD6645 is a 14-bit ADC with an SFDR of 90 dBc and a typical SNR of 73.5 dB (the theoretical SNR for 14 bits is 86 dB). This is because there is a fundamental distinction between noise and distor-

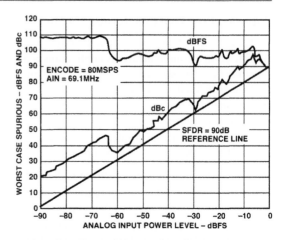

Figure 2.56: AD6645 14-Bit, 80/105 MSPS ADC SFDR versus Input Power Level for 69.1 MHz Input

tion measurements. The process gain of the FFT (33 dB for a 4096-point FFT) allows frequency spurs well below the noise floor to be observed. Adding extra resolution to an ADC may serve to increase its SNR but may or may not increase its SFDR.

Two-Tone Intermodulation Distortion (IMD)

Two-tone IMD is measured by applying two spectrally pure sinewaves to the ADC at frequencies f_1 and f_2, usually relatively close together. The amplitude of each tone is set slightly more than 6 dB below full scale so that the ADC does not clip when the two tones add in-phase. The location of the second- and third-order products are shown in Figure 2.57. Notice that the second-order products fall at frequencies that can be removed by digital filters. However, the third-order products, $2f_2 - f_1$ and $2f_1 - f_2$, are close to the original signals and more difficult to filter. Unless otherwise specified, two-tone IMD refers to these third-order

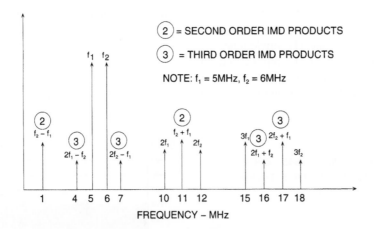

Figure 2.57: Second and Third-Order Intermodulation Products for f_1 = 5 MHz, f_2 = 6 MHz

94

products. The value of the IMD product is expressed in dBc relative to the value of *either* of the two original tones, and not to their sum.

Note, however, that if the two tones are close to $f_s/4$, the aliased third harmonics of the fundamentals can make the identification of the actual $2f_2 - f_1$ and $2f_1 - f_2$ products difficult. This is because the third harmonic of $f_s/4$ is $3f_s/4$, and the alias occurs at $f_s - 3f_s/4 = f_s/4$. Similarly, if the two tones are close to $f_s/3$, the aliased second harmonics may interfere with the measurement. The same reasoning applies here; the second harmonic of $f_s/3$ is $2f_s/3$, and its alias occurs at $f_s - 2f_s/3 = f_s/3$.

Second- and Third-Order Intercept Points, 1 dB Compression Point

Third-order IMD products are especially troublesome in multichannel communications systems where the channel separation is constant across the frequency band. Third-order IMD products can mask out small signals in the presence of larger ones.

In amplifiers, it is common practice to specify the third-order IMD products in terms of the *third-order intercept* point, as shown by Figure 2.58. Two spectrally pure tones are applied to the system. The output signal power in a single tone (in dBm) as well as the relative amplitude of the third-order products (referenced to a single tone) are plotted as a function of input signal power. The fundamental is shown by the *slope = 1* curve in the diagram. If the system nonlinearity is approximated by a power series expansion, it can be shown that second-order IMD amplitudes increase 2 dB for every 1 dB of signal increase, as represented by *slope = 2* curve in the diagram.

Similarly, the third-order IMD amplitudes increase 3 dB for every 1 dB of signal increase, as indicated by the *slope = 3* plotted line. With a low level two-tone input signal, and two data points, one can draw the second- and third-order IMD lines as they are shown in Figure 2.58 (using the principle that a point and a slope define a straight line).

Once the input reaches a certain level however, the output signal begins to soft-limit, or compress. A parameter of interest here is the *1 dB compression point*. This is the point where the output signal is compressed 1 dB from an ideal input/output transfer function. This is shown in Figure 2.58 within the region where the ideal slope = 1 line becomes dotted, and the actual response exhibits compression (solid).

Nevertheless, both the second- and third-order intercept lines may be extended, to intersect the (dotted) extension of the ideal output signal line. These intersections are called the *second-* and *third-order intercept points*, respectively, or IP2 and IP3. These power level values are usually referenced to the output power of the device delivered to a matched load (usually, but not necessarily 50 Ω) expressed in dBm.

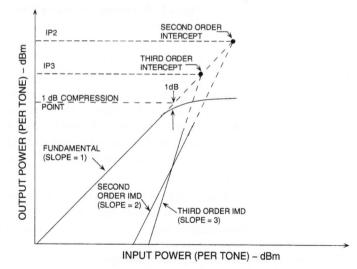

Figure 2.58: Definition of Intercept Points and 1 dB Compression Points for Amplifiers

It should be noted that IP2, IP3, and the 1 dB compression point are all a function of frequency and, as one would expect, the distortion is worse at higher frequencies.

For a given frequency, knowing the third-order intercept point allows calculation of the approximate level of the third-order IMD products as a function of output signal level.

The concept of *second- and third-order intercept points* is not valid for an ADC, because tne distortion products do not vary in a predictable manner (as a function of signal amplitude). The ADC does not gradually begin to compress signals approaching full scale (there is no 1 dB compression point); it acts as a *hard limiter* as soon as the signal exceeds the ADC input range, thereby suddenly producing extreme amounts of distortion because of clipping. On the other hand, for signals much below full scale, the distortion floor remains relatively constant and is independent of signal level. This is shown graphically in Figure 2.59.

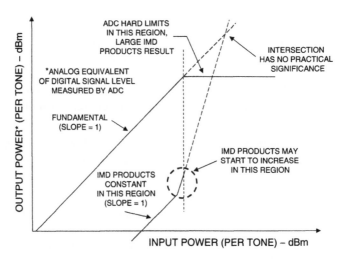

**Figure 2.59: Intercept Points for Data Converters
Have No Practical Significance**

The IMD curve in Figure 2.59 is divided into three regions. For low level input signals, the IMD products remain relatively constant regardless of signal level. This implies that as the input signal increases 1 dB, the ratio of the signal to the IMD level will also increase 1 dB. When the input signal is within a few dB of the ADC full-scale range, the IMD may start to increase (but it might not in a very well-designed ADC). The exact level at which this occurs is dependent on the particular ADC under consideration—some ADCs may not exhibit significant increases in the IMD products over their full input range, however, most will. As the input signal continues to increase beyond full scale, the ADC should function to act as an ideal limiter, and the IMD products become very large.

For these reasons, the second and third order IMD intercept points are not specified for ADCs. It should be noted that essentially the same arguments apply to DACs. In either case, the single- or multitone SFDR specification is the most accepted way to measure data converter distortion.

Multitone Spurious Free Dynamic Range

Two-tone and multitone SFDR is often measured in communications applications. The larger number of tones more closely simulates the wideband frequency spectrum of cellular telephone systems such as

AMPS or GSM. Figure 2.60 shows the two-tone intermodulation performance of the AD6645 14-bit, 80/105 MSPS ADC. The input tones are at 55.25 MHz and 56.25 MHz and are located in the second Nyquist Zone.

The aliased tones therefore occur at 23.75 MHz and 24.75 MHz in the first Nyquist Zone. High SFDR increases the receiver's ability to capture small signals in the presence of large ones, and prevents the small signals from being masked by the intermodulation products of the larger ones. Figure 2.61 shows the AD6645 two-tone SFDR as a function of input signal amplitude for the same input frequencies.

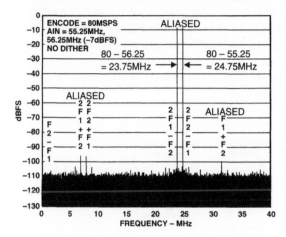

Figure 2.60: Two-Tone SFDR for AD6645 14-Bit, 80/105 MSPS ADC, Input Tones: 55.25 MHz and 56.25 MHz

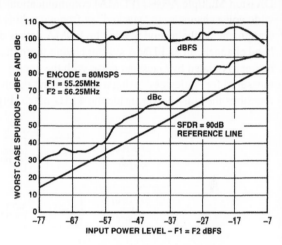

Figure 2.61: Two-Tone SFDR versus Input Amplitude for AD6645 14-Bit, 80/105 MSPS ADC

Wideband CDMA (WCDMA) Adjacent Channel Power Ratio (ACPR) and Adjacent Channel Leakage Ratio (ADLR)

A wideband CDMA channel has a bandwidth of approximately 3.84 MHz, and channel spacing is 5 MHz. The ratio in dBc between the measured power within a channel relative to its adjacent channel is defined as the *adjacent channel power ratio* (ACPR).

The ratio in dBc between the measured power within the channel bandwidth relative to the noise level in an adjacent empty carrier channel is defined as *adjacent channel leakage ratio* (ACLR).

Figure 2.62 shows a single wideband CDMA channel centered at 140 MHz sampled at a frequency of 76.8 MSPS using the AD6645. This is a good example of undersampling (direct IF-to-digital conversion). The signal lies within

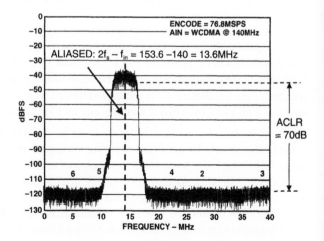

Figure 2.62: Wideband CDMA (WCDMA) Adjacent Channel Leakage Ratio (ACLR)

the fourth Nyquist zone: $3f_s/2$ to $2f_s$ (115.2 MHz to 153.6 MHz). The aliased signal within the first Nyquist zone is therefore centered at $2f_s - f_a = 153.6 - 140 = 13.6$ MHz. The diagram also shows the location of the aliased harmonics. For example, the second harmonic of the input signal occurs at $2 \times 140 = 280$ MHz, and the aliased component occurs at $4f_s - 2f_a = 4 \times 76.8 - 280 = 307.2 - 280 = 27.2$ MHz.

Noise Power Ratio (NPR)

Noise power ratio has been used extensively to measure the transmission characteristics of Frequency Division Multiple Access (FDMA) communications links (Reference 7). In a typical FDMA system, 4 kHz wide voice channels are "stacked" in frequency bins for transmission over coaxial, microwave, or satellite equipment. At the receiving end, the FDMA data is demultiplexed and returned to 4 kHz individual base-band channels. In an FDMA system having more than approximately 100 channels, the FDMA signal can be approximated by Gaussian noise with the appropriate bandwidth. An individual 4 kHz channel can be measured for "quietness" using a narrow-band notch (band-stop) filter and a specially tuned receiver which measures the noise power inside the 4 kHz notch (Figure 2.63).

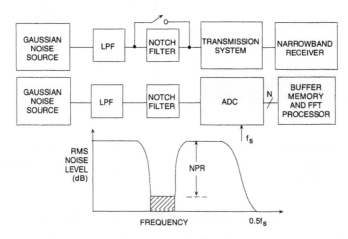

Figure 2.63: Noise Power Ratio (NPR) Measurements

Noise Power Ratio (NPR) measurements are straightforward. With the notch filter out, the rms noise power of the signal inside the notch is measured by the narrowband receiver. The notch filter is then switched in, and the residual noise inside the slot is measured. The ratio of these two readings expressed in dB is the NPR. Several slot frequencies across the noise bandwidth (low, midband, and high) are tested to character-ize the system adequately. NPR measurements on ADCs are made in a similar manner except the analog receiver is replaced by a buffer memory and an FFT processor.

The NPR is plotted as a function of rms noise level referred to the peak range of the system. For very low noise loading level, the undesired noise (in nondigital systems) is primarily thermal noise and is indepen-dent of the input noise level. Over this region of the curve, a 1 dB increase in noise loading level causes a 1 dB increase in NPR. As the noise loading level is increased, the amplifiers in the system begin to over-load, creating intermodulation products that cause the noise floor of the system to increase. As the input noise increases further, the effects of "overload" noise predominate, and the NPR is dramatically reduced. FDMA systems are usually operated at a noise loading level a few dB below the point of maximum NPR.

In a digital system containing an ADC, the noise within the slot is primarily quantization noise when low levels of noise input are applied. The NPR curve is linear in this region. As the noise level increases, there is a one-for-one correspondence between the noise level and the NPR. At some level, however, "clipping" noise caused by the hard-limiting action of the ADC begins to dominate. A theoretical curve for 10-, 11-, and 12-bit ADCs is shown in Figure 2.64 (References 8 and 21).

Figure 2.65 shows the maximum theoretical NPR and the noise loading level at which the maximum value occurs for 8- to 16-bit ADCs. The ADC input range is 2 V_O peak-to-peak. The rms noise level is σ, and the noise-loading factor k (crest factor) is defined as V_O/σ, the peak-to-rms ratio (k is expressed either as numerical ratio or in dB).

In multichannel high frequency communication systems, where there is little or no phase correlation between channels, NPR can also be used to simulate the distortion caused by a large number of individual channels, similar to an FDMA system. A notch filter is placed between the noise source and the ADC, and an FFT output is used in place of the analog receiver. The width of the notch filter is set for several MHz as shown in Figure 2.66 for the AD9430 12-bit 170/210 MSPS ADC. The notch is centered at 19 MHz, and the NPR is the "depth" of the notch. An ideal ADC will only generate quantization noise inside the notch; however, a practical one has additional noise components due to additional noise and intermodulation distortion caused by ADC imperfections. Notice that the NPR is about 57 dB compared to 62.7 dB theoretical.

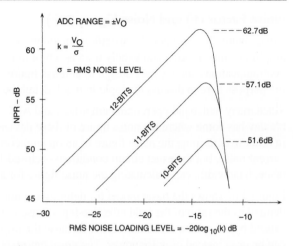

Figure 2.64: Theoretical NPR for 10-, 11-, 12-bit ADCs

BITS	k OPTIMUM	k(dB)	MAX NPR (dB)
8	3.92	11.87	40.60
9	4.22	12.50	46.05
10	4.50	13.06	51.56
11	4.76	13.55	57.12
12	5.01	14.00	62.71
13	5.26	14.41	68.35
14	5.49	14.79	74.01
15	5.72	15.15	79.70
16	5.94	15.47	85.40

ADC Range = $\pm V_O$
k = V_O/σ
σ = RMS Noise Level

Figure 2.65: Theoretical Maximum NPR for 8- to 16-bit ADCs

Figure 2.66: AD9430 12-Bit, 170/210 MSPS ADC NPR Measures 57 dB (62.7 dB Theoretical)

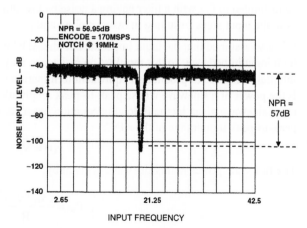

Noise Factor (F) and Noise Figure (NF)

Noise figure (NF) is a popular specification among RF system designers. It is used to characterize RF amplifiers, mixers, etc., and widely used as a tool in radio receiver design. Many excellent textbooks on communications and receiver design treat noise figure extensively (see Reference 9, for example)—it is not the purpose here to discuss the topic in much detail, but only how it applies to data converters.

Since many wideband operational amplifiers and ADCs are now being used in RF applications, the inevitable day has come where the noise figure of these devices becomes important. As discussed in Reference 10, in order to determine the noise figure of an op amp correctly, one must not only know op amp voltage and current noise, but the exact circuit conditions—closed-loop gain, gain-setting resistor values, source resistance, bandwidth, etc. Calculating the noise figure for an ADC is even more of a challenge as will be seen.

Figure 2.67 shows the basic model for defining the noise figure of an ADC. The *noise factor*, F, is simply defined as the ratio of the total effective input noise power of the ADC to the amount of that noise power caused by the source resistance alone. Because the impedance is matched, the square of the voltage noise can be used instead of noise power. The *noise figure*, NF, is simply the noise factor expressed in dB, $NF = 10\log_{10}F$.

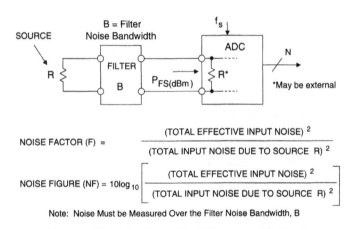

$$NOISE\ FACTOR\ (F)\ =\ \frac{(TOTAL\ EFFECTIVE\ INPUT\ NOISE)^2}{(TOTAL\ INPUT\ NOISE\ DUE\ TO\ SOURCE\ R)^2}$$

$$NOISE\ FIGURE\ (NF) = 10\log_{10}\left[\frac{(TOTAL\ EFFECTIVE\ INPUT\ NOISE)^2}{(TOTAL\ INPUT\ NOISE\ DUE\ TO\ SOURCE\ R)^2}\right]$$

Note: Noise Must be Measured Over the Filter Noise Bandwidth, B

Figure 2.67: Noise Figure for ADCs: Use with Caution

This model assumes the input to the ADC comes from a source having a resistance, R, and that the input is band-limited to $f_s/2$ with a filter having a noise bandwidth equal to $f_s/2$. It is also possible to further band-limit the input signal resulting in oversampling and process gain, and this condition will be discussed shortly.

It is also assumed that the input impedance to the ADC is equal to the source resistance. Many ADCs have a high input impedance, so this termination resistance may be external to the ADC or used in parallel with the internal resistance to produce an equivalent termination resistance equal to R. The full-scale input power is the power of a sinewave whose peak-to-peak amplitude fills the entire ADC input range. The full-scale input sinewave given by the following equation has a peak-to-peak amplitude of $2V_O$ corresponding to the peak-to-peak input range of the ADC:

$$v(t) = V_O \sin 2\pi ft \qquad\qquad \text{Eq. 2.21}$$

The full-scale power in this sinewave is given by:

$$P_{FS} = \frac{(V_O/\sqrt{2})^2}{R} = \frac{V_O^2}{2R} \qquad\qquad \text{Eq. 2.22}$$

It is customary to express this power in dBm (referenced to 1 mW) as follows:

$$P_{FS(dBm)} = 10 \log_{10} \left[\frac{P_{FS}}{1\,mW} \right]$$

Eq. 2.23

The *noise bandwidth* of a nonideal brick wall filter is defined as the bandwidth of an ideal brick wall filter which will pass the same noise power as the nonideal filter. Therefore, the noise bandwidth of a filter is always greater than the 3 dB bandwidth of the filter by a factor which depends upon the sharpness of the cutoff region of the filter. Figure 2.68 shows the relationship between the noise bandwidth and the 3 dB bandwidth for Butterworth filters up to five poles. Note that for two poles, the noise bandwidth and 3 dB bandwidth are within 11% of each other, and beyond that the two quantities are essentially equal.

NUMBER OF POLES	NOISE BW/3dB BW
1	1.57
2	1.11
3	1.05
4	1.03
5	1.02

Figure 2.68: Relationship between Noise Bandwidth and 3 dB Bandwidth for Butterworth Filter

The first step in the NF calculation is to calculate the effective input noise of the ADC from its SNR. The SNR of the ADC is given for a variety of input frequencies, so be sure and use the value corresponding to the input frequency of interest. Also, make sure that the harmonics are not included in the SNR number—some ADC data sheets may confuse SINAD with SNR. Once the SNR is known, the equivalent input rms voltage noise can be calculated starting from the equation:

$$SNR = 20 \log_{10} \left[\frac{V_{FS\,RMS}}{V_{NOISE\,RMS}} \right]$$

Eq. 2.24

Solving for $V_{NOISE\,RMS}$:

$$V_{NOISE\,RMS} = V_{FS\,RMS} \times 10^{-SNR/20}$$

Eq. 2.25

This is the total effective input rms noise voltage at the carrier frequency measured over the Nyquist bandwidth, dc to $f_s/2$. Note that this noise includes the source resistance noise. These results are summarized in Figure 2.69.

- Start with the SNR of the ADC measured at the carrier frequency (Note: this SNR value does not include the harmonics of the fundamental and is measured over the Nyquist bandwidth, dc to $f_s/2$)

$$SNR = 20 \log_{10} \frac{V_{FS\text{-}RMS}}{V_{NOISE\text{-}RMS}}$$

$$V_{NOISE\text{-}RMS} = V_{FS\text{-}RMS}\, 10^{-SNR/20}$$

- This is the total ADC effective input noise at the carrier frequency measured over the Nyquist bandwidth, dc to $f_s/2$

Figure 2.69: Calculating ADC Total Effective Input Noise from SNR

The next step is to actually calculate the noise figure. In Figure 2.70 notice that the amount of the input voltage noise due to the source resistance is the voltage noise of the source resistance $\sqrt{(4kTBR)}$ divided by two, or $\sqrt{(kTBR)}$ because of the 2:1 attenuator formed by the ADC input termination resistor.

The expression for the noise factor F can be written:

$$F = \frac{V_{NOISE\ RMS}^2}{kTRB} = \left[\frac{V_{FS\ RMS}^2}{R}\right]\left[\frac{1}{kT}\right]\left[10^{-SNR/10}\right]\left[\frac{1}{B}\right] \qquad \text{Eq. 2.26}$$

The noise figure is obtained by converting F into dB and simplifying:

$$NF = 10_{10}\log F = P_{FS(dBm)} + 174\ dBm - SNR - 10_{10}\log B, \qquad \text{Eq. 2.27}$$

Where SNR is in dB, B in Hz, T = 300 K, k = 1.38×10^{-23} J/K.

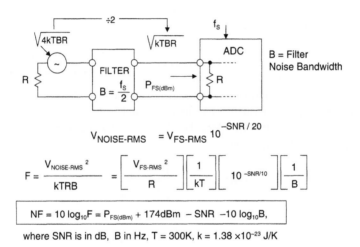

Figure 2.70: ADC Noise Figure in Terms of SNR, Sampling Rate, and Input Power

Oversampling and filtering can be used to decrease the noise figure as a result of the process gain as has been previously discussed. In this case, the signal bandwidth B is less than $f_s/2$. Figure 2.71 shows the correction factor which results in the following equation:

$$NF = 10_{10}\log F = P_{FS(dBm)} + 174\ dBm - SNR - 10\log_{10}[f_s/2B] - 10\log_{10}B. \qquad \text{Eq. 2.28}$$

Figure 2.72 shows an example NF calculation for the AD6645 14-bit, 80 MSPS ADC. A 52.3 Ω resistor is added in parallel with the AD6645 input impedance of 1 kΩ to make the net input impedance 50 Ω. The ADC is operating under Nyquist conditions, and the SNR of 74 dB is the starting point for the calculations using Eq. 2.28 above. A noise figure of 34.8 dB is obtained.

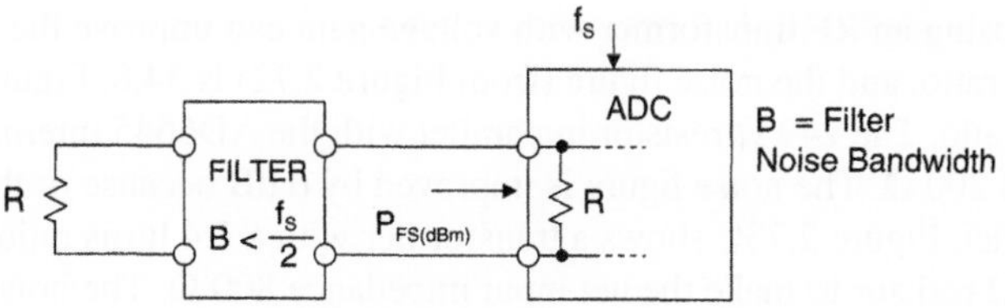

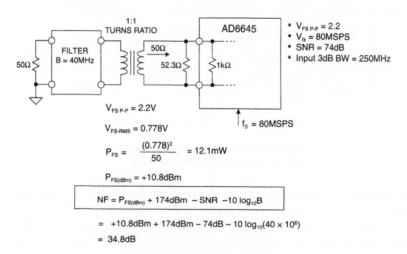

$$NF = P_{FS(dBm)} + 174dBm - SNR - 10 \log_{10} \left[\frac{f_s / 2}{B} \right] - 10 \log_{10} B,$$

Measured DC to $f_s / 2$

Process Gain

where SNR is in dB, B in Hz, T = 300K, k = 1.38×10^{-23} J/K

Figure 2.71: Effect of Oversampling and Process Gain on ADC Noise Figure

$V_{FS\,P-P} = 2.2V$

$V_{FS-RMS} = 0.778V$

$P_{FS} = \dfrac{(0.778)^2}{50} = 12.1mW$

$P_{FS(dBm)} = +10.8dBm$

$$NF = P_{FS(dBm)} + 174dBm - SNR - 10 \log_{10} B$$

$= +10.8dBm + 174dBm - 74dB - 10 \log_{10}(40 \times 10^6)$

$= 34.8dB$

Figure 2.72: Example Calculation of Noise Figure Under Nyquist Conditions for AD6645

Figure 2.73 shows how using an RF transformer with voltage gain can improve the noise figure. Figure 2.73A shows a 1:1 turns ratio, and the noise figure (from Figure 2.72) is 34.8. Figure 2.73B shows a transformer with a 1:2 turns ratio. The 249 Ω resistor in parallel with the AD6645 internal resistance results in a net input impedance of 200 Ω. The noise figure is improved by 6 dB because of the "noise-free" voltage gain of the transformer. Figure 2.73C shows a transformer with a 1:4 turns ratio. The AD6645 input is paralleled with a 4.02 kΩ resistor to make the net input impedance 800 Ω. The noise figure is improved by another 6 dB. Transformers with higher turns ratios are not generally practical because of bandwidth and distortion limitations.

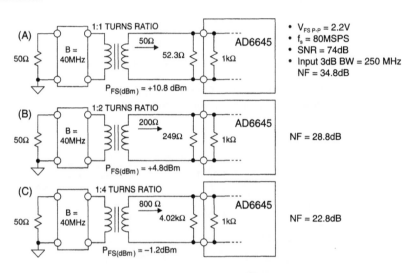

Figure 2.73: Using RF Transformers to Improve Overall ADC Noise Figure

Even with the 1:4 turns ratio transformer, the overall noise figure for the AD6645 was still 22.8 dB, still relatively high by RF standards. The solution is to provide low noise high gain stages ahead of the ADC. Figure 2.74 shows how the Friis equation is used to calculate the noise factor for cascaded gain stages. Notice that high gain in the first stage reduces the contribution of the noise factor of the second stage—the noise factor of the first stage dominates the overall noise factor.

Figure 2.74: Cascaded Noise Figure Using the Friis Equation

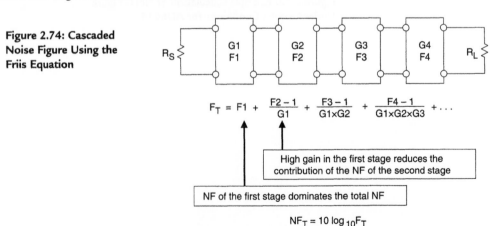

$$F_T = F1 + \frac{F2-1}{G1} + \frac{F3-1}{G1\times G2} + \frac{F4-1}{G1\times G2\times G3} + \cdots$$

High gain in the first stage reduces the contribution of the NF of the second stage

NF of the first stage dominates the total NF

$$NF_T = 10 \log_{10} F_T$$

Figure 2.75 shows the effects of a high-gain (25 dB) low-noise (NF = 4 dB) stage placed in front of a relatively high NF stage (30 dB)—the noise figure of the second stage is typical of high performance ADCs. The overall noise figure is 7.53 dB, only 3.53 dB higher than the first stage noise figure of 4 dB.

In summary, applying the noise figure concept to characterize wideband ADCs must be done with extreme caution to prevent misleading results. Simply trying to minimize the noise figure using the equations can actually increase circuit noise.

For instance, NF decreases with increasing source resistance according to the calculations, but increased source resistance increases circuit noise. Also, NF decreases with increasing ADC input bandwidth if there is no input filtering. This is also contradictory, because widening the bandwidth increases noise. In both these cases, the circuit noise increases, and the NF decreases. The reason NF decreases is that the source noise makes up a larger component of the total noise (which remains relatively constant because the ADC noise is much greater than the source noise); therefore, according to the calculation, NF decreases, but actual circuit noise increases.

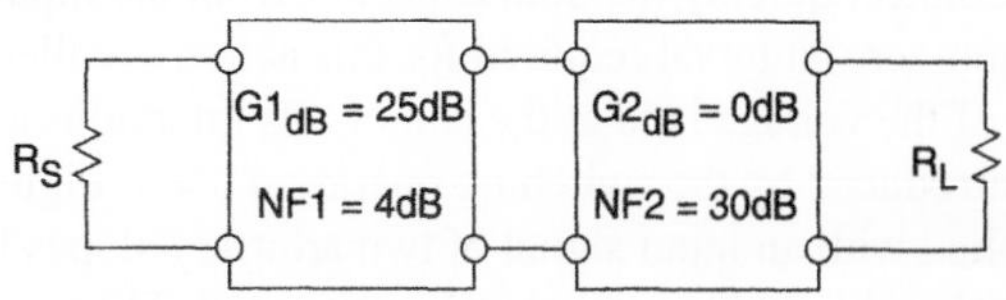

$$G1 = 10^{25/10} = 10^{2.5} = 316, \quad F1 = 10^{4/10} = 10^{0.4} = 2.51$$
$$G2 = 1, \qquad\qquad\qquad\qquad F2 = 10^{30/10} = 10^{3} = 1000$$

$$F_T = F1 + \frac{F2 - 1}{G1} = 2.51 + \frac{1000 - 1}{316} = 2.51 + 3.16 = 5.67$$

$$NF_T = 10 \log_{10} 5.67 = 7.53 dB$$

- The first stage dominates the overall NF
- It should have the highest gain possible with the lowest NF possible

Figure 2.75: Example of Two-Stage Cascaded Network

It is true that on a standalone basis ADCs have relatively high noise figures compared to other RF parts such as LNAs or mixers. In the system the ADC should be preceded with low noise gain blocks as shown in the example of Figure 2.75. Noise figure considerations for ADCs are summarized in Figure 2.76.

- NF decreases with increasing source resistance.
- NF decreases with increasing ADC input bandwidth if there is no input filtering.
- In both cases, the circuit noise increases, and the NF decreases.
- The reason NF decreases is that the source noise makes up a larger component of the total noise (which remains relatively constant because the ADC noise is much greater than the source noise).
- In practice, input filtering is used to limit the input noise bandwidth and reduce overall system noise.
- ADCs have relatively high NF compared to other RF parts. In the system the ADC should be preceded with low-noise gain blocks.
- Exercise caution when using NF.

Figure 2.76: Noise Figure Considerations for ADCs: Summary and Caution

Aperture Time, Aperture Delay Time, and Aperture Jitter

Perhaps the most misunderstood and misused ADC and sample-and-hold (or track-and-hold) specifications are those that include the word *aperture*. The most essential dynamic property of a SHA is its ability to disconnect quickly the hold capacitor from the input buffer amplifier as shown in Figure 2.77. The short (but non-zero) interval required for this action is called *aperture time (or sampling aperture)*, t_a. The actual value of the voltage held at the end of this interval is a function of both the input signal slew rate and the errors introduced by the switching operation itself. Figure 2.77 shows what happens when the hold command is applied with an input signal of two arbitrary slopes labeled as 1 and 2. For clarity, the sample-to-hold pedestal and switching transients are ignored. The value that is finally held is a delayed version of the input signal, averaged over the aperture time of the switch as shown in Figure 2.77. The first-order model assumes that the final value of the voltage on the hold capacitor is approximately equal to the average value of the signal applied to the switch over the interval during which the switch changes from a low to high impedance (t_a).

The model shows that the finite time required for the switch to open (t_a) is equivalent to introducing a small delay (t_e) in the sampling clock driving the SHA. This delay is constant and may be either positive or negative. The diagram shows that the same value of t_e works for the two signals, even though the slopes are different. This delay is called *effective aperture delay time, aperture delay time*, or simply *aperture delay*, t_e. In an ADC, the aperture delay time is referenced to the input of the converter, and the effects of the analog propagation delay through the input buffer, t_{da} and the digital delay through the switch driver, t_{dd}, must be considered. Referenced to the ADC inputs, aperture time, t_e', is defined as the time difference between the analog propagation delay of the front-end buffer, t_{da}, and the switch driver digital delay, t_{dd}, plus one-half the aperture time, $t_a/2$.

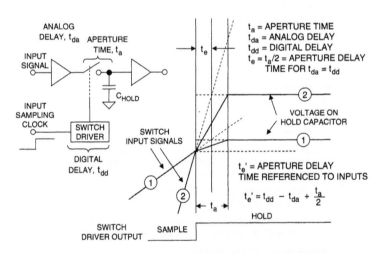

Figure 2.77: Sample-and-Hold Waveforms and Definitions

The effective aperture delay time is usually positive, but may be negative if the sum of one-half the aperture time, $t_a/2$, and the switch driver digital delay, t_{dd}, is less than the propagation delay through the input buffer, t_{da}. The aperture delay specification thus establishes when the input signal is actually sampled with respect to the sampling clock edge.

Aperture delay time can be measured by applying a bipolar sinewave signal to the ADC and adjusting the synchronous sampling clock delay such that the output of the ADC is midscale (corresponding to the zero-crossing of the sinewave). The relative delay between the input sampling clock edge and the actual zero-crossing of the input sinewave is the aperture delay time (see Figure 2.78).

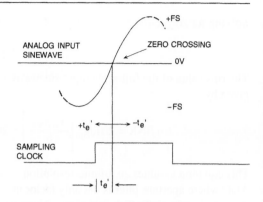

Figure 2.78: Effective Aperture Delay Time Measured with Respect to ADC Input

Aperture delay produces no errors (assuming it is relatively short with respect to the hold time), but acts as a fixed delay in either the sampling clock input or the analog input (depending on its sign). However, in simultaneous sampling applications or in direct I/Q demodulation where two or more ADCs must be well matched, variations in the aperture delay between converters can produce errors on fast slewing signals. In these applications, the aperture delay mismatches must be removed by properly adjusting the phases of the individual sampling clocks to the various ADCs.

If, however, there is *sample-to-sample* variation in aperture delay (*aperture jitter*), a corresponding voltage error is produced as shown in Figure 2.79. This sample-to-sample variation in the instant the switch opens is called *aperture uncertainty*, or *aperture jitter* and is usually measured in rms picoseconds. The amplitude of the associated output error is related to the rate-of-change of the analog input. For any given value of aperture jitter, the aperture jitter error increases as the input dv/dt increases. The effects of phase jitter on the external sampling clock (or the analog input for that matter) produce exactly the same type of error.

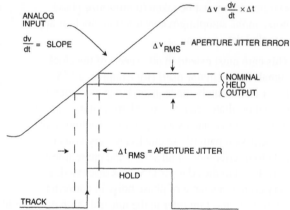

Figure 2.79: Effects of Aperture Jitter and Sampling Clock Jitter

The effects of aperture and sampling clock jitter on an ideal ADC's SNR can be predicted by the following simple analysis. Assume an input signal given by

$$v(t) = V_O \sin 2\pi ft \qquad \text{Eq. 2.29}$$

The rate of change of this signal is given by:

$$dv/dt = 2\pi fV_O \cos 2\pi ft \qquad \text{Eq. 2.30}$$

The rms value of dv/dt can be obtained by dividing the amplitude, $2\pi fV_O$, by $\sqrt{2}$:

$$dv/dt\big|_{rms} = 2\pi fV_O/\sqrt{2} \qquad \text{Eq. 2.31}$$

Now let Δv_{rms} = the rms voltage error and Δt = the rms aperture jitter t_j, and substitute:

$$\Delta v_{rms}/t_j = 2\pi fV_O/\sqrt{2} \qquad \text{Eq. 2.32}$$

Solving for Δv_{rms}:

$$\Delta v_{rms} = 2\pi f V_o t_j / \sqrt{2}$$

Eq. 2.33

The rms value of the full-scale input sinewave is $V_o / \sqrt{2}$, therefore the rms signal to rms noise ratio is given by

$$SNR = 20 \log_{10} \left[\frac{V_o / \sqrt{2}}{\Delta v_{rms}} \right] = 20 \log_{10} \left[\frac{V_o / \sqrt{2}}{2\pi f V_o t_j / \sqrt{2}} \right] = 20 \log_{10} \left[\frac{1}{2\pi f\, t_j} \right]$$

Eq. 2.34

This equation assumes an infinite-resolution ADC where aperture jitter is the only factor in determining the SNR. This equation is plotted in Figure 2.80 and shows the serious effects of aperture and sampling clock jitter on SNR, especially at higher input/output frequencies. Therefore, extreme care must be taken to minimize phase noise in the sampling/reconstruction clock of any sampled data system.

This care must extend to all aspects of the clock signal: the oscillator itself (for example, a 555 timer is absolutely inadequate, but even a quartz crystal oscillator can give problems if it uses an active device that shares a chip with noisy logic); the transmission path (these clocks are very vulnerable to interference of all sorts), and phase noise introduced in the ADC or DAC. As discussed, a very common source of phase noise in converter circuitry is aperture jitter in the integral sample-and-hold (SHA) circuitry; however, the total rms jitter will be composed of a number of components—the actual SHA aperture jitter often being the least of them.

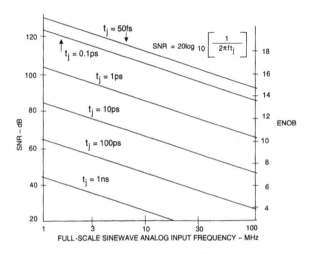

Figure 2.80: Theoretical SNR and ENOB Due to Jitter versus Full-scale Sinewave Input Frequency

A Simple Equation for the Total SNR of an ADC

A relatively simple equation for the ADC SNR in terms of sampling clock and aperture jitter, DNL, effective input noise, and the number of bits of resolution is shown in Figure 2.81. The equation combines the various error terms on an rss basis. The average DNL error, ε, is computed from histogram data. This equation is used in Figure 2.82 to predict the SNR performance of the AD6645 14-bit, 80 MSPS ADC as a function of sampling clock and aperture jitter.

Before the 1980s, most sampling ADCs were generally built up from a separate SHA and ADC. Interface design was difficult, and a key parameter was aperture jitter in the SHA. Today, almost all sampled data systems use *sampling* ADCs that contain an integral SHA. The aperture jitter of the SHA may not be specified as such, but this is not a cause of concern if the SNR or ENOB is clearly specified, since a guarantee of a specific SNR is an implicit guarantee of an adequate aperture jitter specification. However, the use of an additional high-performance SHA will sometimes improve the high frequency ENOB of even the best sampling ADC by presenting "dc" to the ADC, and may be more cost effective than replacing the ADC with a more expensive one.

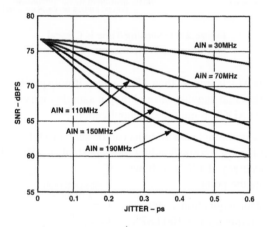

$$\text{SNR} = -20\log_{10}\left[\overbrace{(2\pi \times f_a \times t_{j\,rms})^2}^{\substack{\text{SAMPLING}\\\text{CLOCK JITTER}}} + \overbrace{\frac{2}{3}\left[\frac{1+\varepsilon}{2^N}\right]^2}^{\substack{\text{QUANTIZATION}\\\text{NOISE, DNL}}} + \overbrace{\left[\frac{2 \times \sqrt{2} \times V_{NOISErms}}{2^N}\right]^2}^{\substack{\text{EFFECTIVE}\\\text{INPUT NOISE}}}\right]^{-\frac{1}{2}}$$

f_a =   Analog input frequency of full-scale input sinewave

$t_{j\,rms}$ =   Combined rms jitter of internal ADC and external clock

ε =   Average DNL of the ADC (typically 0.41 LSB for AD6645)

N   =   Number of bits in the ADC

$V_{NOISErms}$ =   Effective input noise of ADC (typically 0.9LSB rms for AD6645)

If $t_j = 0$, $\varepsilon = 0$, and $V_{NOISErms} = 0$, the above equation reduces to the familiar:

$$\text{SNR} = 6.02\,N + 1.76\text{dB}$$

Figure 2.81: Relationship Between SNR, Sampling Clock Jitter, Quantization Noise, DNL, and Input Noise

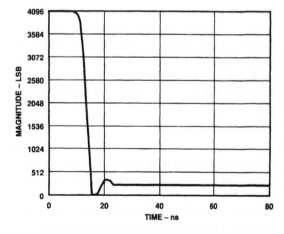

Figure 2.82: AD6645 SNR Versus Jitter

ADC Transient Response and Overvoltage Recovery

Most high-speed ADCs designed for communications applications are specified primarily in the frequency domain. However, in general-purpose data acquisition applications the transient response (or settling time) of the ADC is important. The *transient response* of an ADC is the time required for the ADC to settle to rated accuracy (usually 1 LSB) after the application of a full-scale step input. The typical response of a general-purpose 12-bit, 10 MSPS ADC is shown in Figure 2.83, showing a 1 LSB settling time of less than 40 ns. The settling time specification is critical in the typical data acquisition system application where

Figure 2.83: ADC Transient Response (Settling Time)

the ADC is being driven by an analog multiplexer as shown in Figure 2.84. The multiplexer output can deliver a full-scale sample-to-sample change to the ADC input. If both the multiplexer and the ADC have not settled to the required accuracy, channel-to-channel crosstalk will result, even though only dc or low frequency signals are present on the multiplexer inputs.

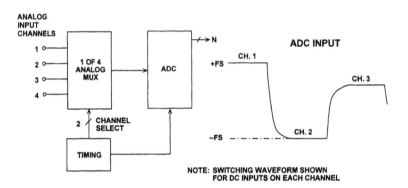

**Figure 2.84: Settling Time is Critical
in Multiplexed Applications**

Most ADCs have settling times which are less than $1/f_{s\,max}$, even if not specified. However sigma-delta ADCs have a built-in digital filter that can take several output sample intervals to settle. This should be kept in mind when using sigma-delta ADCs in multiplexed applications.

The importance of settling time in multiplexed systems can be seen in Figure 2.85, where the ADC input is modeled as a single-pole filter having a corresponding time constant, $\tau = RC$. The required number of time constants to settle to a given accuracy (1 LSB) is shown. A simple example will illustrate the point.

Assume a multiplexed 16-bit data acquisition system uses an ADC with a sampling frequency $f_s = 100$ kSPS. The ADC must settle to 16-bit accuracy for a full-scale step function input in less than $1/f_s = 10$ μs. The chart shows that 11.09 time constants are required to settle to 16-bit accuracy. The input filter time constant must therefore be less than $\tau = 10$ μs/11.09 = 900 ns. The corresponding rise time $t_r = 2.2\tau = 1.98$ μs. The required ADC full power input bandwidth can now be calculated from BW = $0.35/t_r = 177$ kHz. This neglects the settling time of the multiplexer and second-order settling time effects in the ADC.

RESOLUTION, # OF BITS	LSB (%FS)	# OF TIME CONSTANTS
6	1.563	4.16
8	0.391	5.55
10	0.0977	6.93
12	0.0244	8.32
14	0.0061	9.70
16	0.00153	11.09
18	0.00038	12.48
20	0.000095	13.86
22	0.000024	15.25

**Figure 2.85: Settling Time as a Function of
Time Constant for Various Resolutions**

Overvoltage recovery time is defined as that amount of time required for an ADC to achieve a specified accuracy, measured from the time the overvoltage signal re-enters the converter's range, as shown in Figure 2.86. This specification is usually given for a signal that is some stated percentage outside the ADC's input range. Needless to say, the ADC should act as an ideal limiter for out-of-range signals and should produce either the positive full-scale code or the negative full-scale code during the overvoltage condition. Some converters provide over- and underrange flags to allow gain-adjustment circuits to be activated. Care should always be taken to avoid overvoltage signals that will damage an ADC input.

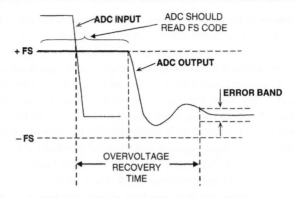

Figure 2.86: Overvoltage Recovery Time

ADC Sparkle Codes, Metastable States, and Bit Error Rate (BER)

A primary concern in the design of many digital communications systems using ADCs is the bit error rate (BER). Unfortunately, ADCs contribute to the BER in ways that are not predictable by simple analysis. This section describes the mechanisms within the ADCs that can contribute to the error rate, ways to minimize the problem, and methods for measuring the BER.

Random noise, regardless of the source, creates a finite probability of errors (deviations from the expected output). Before describing the error code sources, however, it is important to define what constitutes an ADC error code. Noise generated prior to or inside the ADC can be analyzed in the traditional manner. Therefore, an ADC error code is any deviation from the expected output that is not attributable to the equivalent input noise of the ADC. Figure 2.87 illustrates an exaggerated output of a low amplitude sinewave applied to an ADC that has error codes. Note that the noise of the ADC creates some uncertainty in the output. These anomalies are not considered error codes, but are simply the result of ordinary noise and quantization. The large errors are more significant and are not expected. These errors are random and so infrequent that an SNR test of the ADC will rarely detect them. These types of errors plagued a few of the early ADCs for video applications, and were given the name *sparkle codes* because of their appearance on a TV screen as small white dots or "sparkles" under certain test conditions. These errors have also been called *rabbits* or *flyers*. In digital communications applications, this type of error increases the overall system bit error rate (BER).

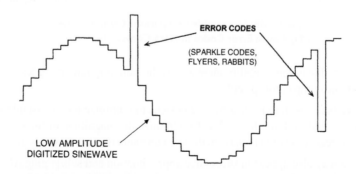

Figure 2.87: Exaggerated Output of ADC Showing Error Codes

In order to understand the causes of the error codes, we will first consider the case of a simple flash converter. The comparators in a flash converter are latched comparators usually arranged in a master-slave configuration. If the input signal is in the center of the threshold of a particular comparator, that comparator will balance, and its output will take a longer period of time to reach a valid logic level after the application of the latch strobe than the outputs of its neighboring comparators which are being overdriven. This phenomenon is known as *metastability* and occurs when a balanced comparator cannot reach a valid logic level in the time allowed for decoding. If simple binary decoding logic is used to decode the thermometer code, a metastable comparator output may result in a large output code error. Consider the case of a simple 3-bit flash converter shown in Figure 2.88. Assume that the input signal is exactly at the threshold of Comparator 4 and random noise is causing the comparator to toggle between a "1" and a "0" output each time a latch strobe is applied. The corresponding binary output should be interpreted as either 011 or 100. If, however, the comparator output is in a metastable state, the simple binary decoding logic shown may produce binary codes 000, 011, 100, or 111. The codes 000 and 111 represent a one-half scale departure from the expected codes.

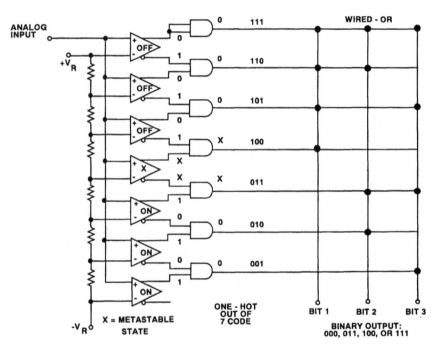

Figure 2.88: Metastable Comparator Output States
May Cause Error Codes in Data Converters

The probability of errors due to metastability increases as the sampling rate increases because less time is available for a metastable comparator to settle.

Various measures have been taken in flash converter designs to minimize the metastable state problem. Decoding schemes described in References 12 to 15 minimize the magnitude of these errors. Optimizing comparator designs for regenerative gain and small time constants is another way to reduce these problems.

Metastable state errors may also appear in successive approximation and subranging ADCs that make use of comparators as building blocks. The same concepts apply, although the magnitudes and locations of the errors may be different.

The test system shown in Figure 2.89 may be used to test for BER in an ADC. The analog input to the ADC is provided by a high stability low noise sinewave generator. The analog input level is set slightly greater than full-scale, and the frequency such that there is always slightly less than 1 LSB change between samples as shown in Figure 2.90.

The test set uses series latches to acquire successive codes A and B. A logic circuit determines the absolute difference between A and B. This difference is then compared to the error limit, chosen to allow for expected random noise spikes and ADC quantization errors. Errors that cause the difference to be larger than the limit will increment the counters. The number of errors, E, are counted over a period of time, T. The error rate is then calculated as BER = E/2Tf$_s$. The factor of 2 in the denominator is required because the hardware records a second error when the output returns to the correct code after making the initial error. The error counter is therefore incremented twice for each error. It should be noted that the same function can be accomplished in software if the ADC outputs are stored in a memory and analyzed by a computer program.

The input frequency must be carefully chosen such that at least one sample is taken per code.

Assume a full-scale input sinewave having an amplitude of $2^N/2$:

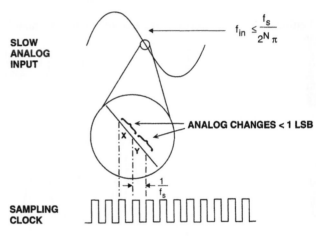

$$E = \text{Number of Errors in Interval } T$$

$$BER = \frac{E}{2\,T\,f_s}$$

Figure 2.89: ADC Bit Error Rate Test Setup

Figure 2.90: ADC Analog Signal for Low Frequency BER Test

$$v(t) = \frac{2^N}{2} \sin 2\pi ft \qquad \text{Eq. 2.35}$$

The maximum rate of change of this signal is

$$\left. \frac{dv}{dt} \right]_{max} \le 2^N \pi f \qquad \text{Eq. 2.36}$$

Letting $dv = 1$ LSB, $dt = 1/f_s$, and solving for the input frequency:

$$f_{in} \le \frac{f_s}{2^N \pi} \qquad \text{Eq. 2.37}$$

Choosing an input frequency less than this value will ensure that there is at least one sample per code.

The same test can be conducted at high frequencies by applying an input frequency slightly offset from $f_s/2$ as shown in Figure 2.91. This causes the ADC to slew full-scale between conversions. Every other conversion is compared, and the "beat" frequency is chosen such that there is slightly less than 1 LSB change between alternate samples. The equation for calculating the proper frequency for the high frequency BER test is derived as follows.

Assume an input full-scale sinewave of amplitude $2^N/2$ whose frequency is slightly less than $f_s/2$ by a frequency equal to Δf.

$$v(t) = \frac{2^N}{2} \sin\left[2\pi\left(\frac{f_s}{2} - \Delta f\right)t\right]$$

Eq. 2.38

The maximum rate of change of this signal is

$$\left.\frac{dv}{dt}\right]_{max} \leq 2^N \pi\left(\frac{f_s}{2} - \Delta f\right)$$

Eq. 2.39

Letting $dv = 1$ LSB and $dt = 2/f_s$, and solving for the input frequency Δf:

$$\Delta f \leq \frac{f_s}{2}\left(1 - \frac{1}{2\times2^N\pi}\right)$$

Eq. 2.40

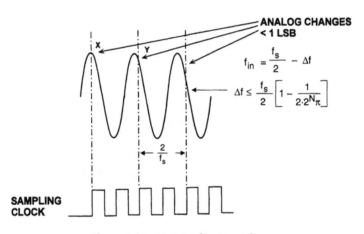

Figure 2.91: ADC Analog Input for
High Frequency BER Test

Establishing the BER of a well-behaved ADC is a difficult, time-consuming task; a single unit can sometimes be tested for days without an error. For example, tests on a typical 8-bit flash converter operating at a sampling rate of 75 MSPS yield a BER of approximately 3.7×10^{-12} (1 error per hour) with an error limit of 4 LSBs. Meaningful tests for longer periods of time require special attention to EMI/RFI effects (possibly requiring a shielded screen room), isolated power supplies, isolation from soldering irons with mechanical thermostats, isolation from other bench equipment, etc. Figure 2.92 shows the average time between errors as a function of BER for a sampling frequency of 75 MSPS. This illustrates the difficulty in measuring low BER because the long measurement times increase the probability of power supply transients, noise, etc., causing an error.

Bit Error Rate (BER)	Average Time Between Errors
1×10^{-8}	1.3 seconds
1×10^{-9}	13.3 seconds
1×10^{-10}	2.2 minutes
1×10^{-11}	22 minutes
1×10^{-12}	3.7 hours
1×10^{-13}	1.5 days
1×10^{-14}	15 days

Figure 2.92: Average Time between Errors versus BER when Sampling at 75 MSPS

DAC Dynamic Performance

The ac specifications most likely to be important with DACs are *settling time, glitch impulse area, distortion,* and *Spurious Free Dynamic Range (SFDR).*

DAC Settling Time

The input to output settling time of a DAC is the time from a change of digital code (t = 0) to when the output comes within *and remains within* some error band as shown in Figure 2.93. With amplifiers, it is hard to make comparisons of settling time, since their specified error bands may differ from amplifier to amplifier, but with DACs the error band will almost invariably be specified as ±1 or ±½ LSB. Note that in some cases, the *output* settling time may be of more interest, in which case it is referenced to the time the output first leaves the error band.

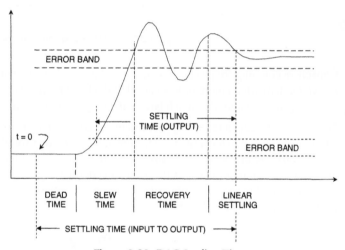

Figure 2.93: DAC Settling Time

The input to output settling time of a DAC is made up of four different periods: the *switching time* or *dead time* (during which the digital switching, but not the output, is changing), the *slewing time* (during which the rate of change of output is limited by the slew rate of the DAC output), the *recovery time* (when the DAC is recovering from its fast slew and may overshoot), and the *linear settling time* (when the DAC output approaches its final value in an exponential or near-exponential manner). If the slew time is short compared to the other three (as is usually the case with current output DACs), the settling time will largely be independent of the output step size. On the other hand, if the slew time is a significant part of the total, the larger the step, the longer the settling time.

Settling time is especially important in video display applications. For example a standard 1024×768 display updated at a 60 Hz refresh rate must have a pixel rate of $1024 \times 768 \times 60$ Hz = 47.2 MHz with no overhead. Allowing 35% overhead time increases the pixel frequency to 64 MHz corresponding to a pixel duration of $1/(64 \times 10^6) = 15.6$ ns. In order to accurately reproduce a single fully-white pixel located between two black pixels, the DAC settling time should be less than the pixel duration time of 15.6 ns.

Higher resolution displays require even faster pixel rates. For example, a 2048×2048 display requires a pixel rate of approximately 330 MHz at a 60 Hz refresh rate.

Glitch Impulse Area

Ideally, when a DAC output changes it should move from one value to its new one monotonically. In practice, the output is likely to overshoot, undershoot, or both (Figure 2.94). This uncontrolled movement of the DAC output during a transition is known as a *glitch*. It can arise from two mechanisms: capacitive coupling of digital transitions to the analog output, and the effects of some switches in the DAC operating more quickly than others and producing temporary spurious outputs.

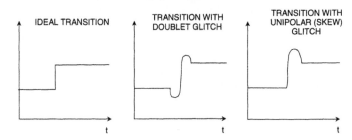

Figure 2.94: DAC Transitions (Showing Glitch)

Capacitive coupling frequently produces roughly equal positive and negative spikes (sometimes called a *doublet* glitch) which more or less cancel in the longer term. The glitch produced by switch timing differences is generally unipolar, much larger, and of greater concern.

Glitches can be characterized by measuring the *glitch impulse area*, sometimes inaccurately called glitch energy. The term *glitch energy* is a misnomer, since the unit for glitch impulse area is volt-seconds (or more probably μV-sec or pV-sec. The *peak glitch area* is the area of the largest of the positive or negative glitch areas. The glitch impulse area is the net area under the voltage-versus-time curve and can be estimated by approximating the waveforms by triangles, computing the areas, and subtracting the negative area from the positive area as shown in Figure 2.95.

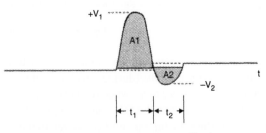

- PEAK GLITCH IMPULSE AREA $= A1 \approx \dfrac{V_1 \times t_1}{2}$

- NET GLITCH IMPULSE AREA $= A1 - A2 \approx \dfrac{V_1 \times t_1}{2} - \dfrac{V_2 \times t_2}{2}$

Figure 2.95: Calculating Net Glitch Impulse Area

The midscale glitch produced by the transition between the codes 0111...111 and 1000...000 is usually the worst glitch because all switches are changing states. Glitches at other code transition points (such as 1/4 and 3/4 full scale) are generally less. Figure 2.96 shows the midscale glitch for a fast low glitch DAC. The peak and net glitch areas are estimated using triangles as described above. Settling time is measured from the time the waveform leaves the initial 1 LSB error band until it enters and remains within the final 1 LSB error band. The step size between the transition regions is also 1 LSB.

DAC SFDR and SNR

DAC settling time is important in applications such as RGB raster scan video display drivers, but frequency-domain specifications such as SFDR are generally more important in communications.

If we consider the spectrum of a waveform reconstructed by a DAC from digital data, we find that in addition to the expected spectrum (which will contain one or more frequencies, depending on the nature of the reconstructed waveform), there will also be noise and distortion products. Distortion may be specified in terms of harmonic distortion, Spurious Free Dynamic Range (SFDR),

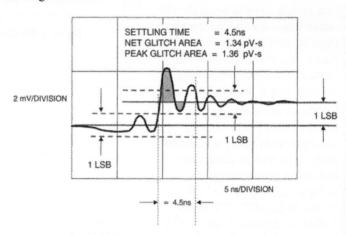

Figure 2.96: DAC Midscale Glitch Shows 1.34 pV-s Net Impulse Area and Settling Time of 4.5 ns

intermodulation distortion, or all of the above. Harmonic distortion is defined as the ratio of harmonics to fundamental when a (theoretically) pure sine wave is reconstructed, and is the most common specification. Spurious free dynamic range is the ratio of the worst spur (usually, but not necessarily always a harmonic of the fundamental) to the fundamental.

Code-dependent glitches will produce both out-of-band and in-band harmonics when the DAC is reconstructing a digitally generated sinewave as in a Direct Digital Synthesis (DDS) system. The midscale glitch occurs twice during a single cycle of a reconstructed sinewave (at each midscale crossing), and will therefore produce a second harmonic of the sinewave, as shown in Figure 2.97. Note that the higher order harmonics of the sinewave, which alias back into the Nyquist bandwidth (dc to $f_s/2$), cannot be filtered.

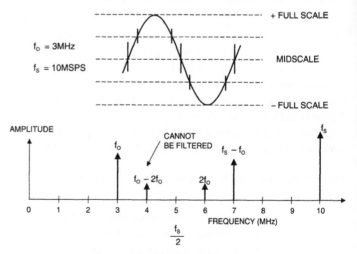

Figure 2.97: Effect of Code-Dependent Glitches on Spectral Output

It is difficult to predict the harmonic distortion or SFDR from the glitch area specification alone. Other factors, such as the overall linearity of the DAC, also contribute to distortion as shown in Figure 2.98. In addition, certain ratios between the DAC output frequency and the sampling clock cause the quantization noise to concentrate at harmonics of the fundamental thereby increasing the distortion at these points.

- Resolution
- Integral Nonlinearity
- Differential Nonlinearity
- Code-Dependent Glitches
- Ratio of Clock Frequency to Output Frequency (Even in an Ideal DAC)
- Mathematical Analysis is Difficult

Figure 2.98: Contributors to DDS DAC Distortion

It is therefore customary to test reconstruction DACs in the frequency domain (using a spectrum analyzer) at various clock rates and output frequencies as shown in Figure 2.99. Typical SFDR for the 16-bit AD9777 Transmit TxDAC is shown in Figure 2.100. The clock rate is 160 MSPS, and the output frequency is swept to 50 MHz. As in the case of ADCs, quantization noise will appear as increased harmonic distortion if the ratio between the clock frequency and the DAC output frequency is an integer number. These ratios should be avoided when making the SFDR measurements.

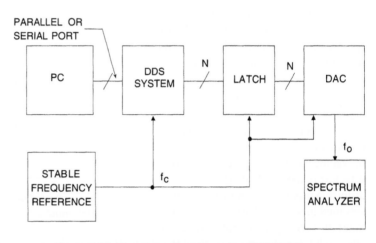

Figure 2.99: Test Setup for Measuring DAC SFDR

There is nearly an infinite combination of possible clock and output frequencies for a low distortion DAC, and SFDR is generally specified for a limited number of selected combinations. For this reason, Analog Devices offers fast turnaround on customer-specified test vectors for the Transmit TxDAC family. A test vector is a combination of amplitudes, output frequencies, and update rates specified directly by the customer for SFDR data on a particular DAC.

Measuring DAC SNR with an Analog Spectrum Analyzer

Analog spectrum analyzers are used to measure the distortion and SFDR of high performance DACs. Care must be taken that the front end of the analyzer is not overdriven by the fundamental signal. If overdrive is a problem, a band-stop filter can be used to filter out the fundamental signal so the spurious components can be observed.

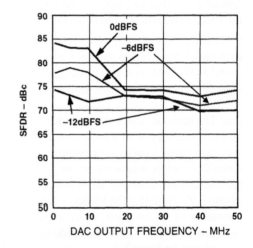

Figure 2.100: AD9777 16-bit TxDAC SFDR, Data Update Rate = 160 MSPS

Spectrum analyzers can also be used to measure the SNR of a DAC provided attention is given to bandwidth considerations. SNR of an ADC is normally defined as the signal-to-noise ratio measured over the Nyquist bandwidth dc to $f_s/2$. However, spectrum analyzers have a resolution bandwidth less than $f_s/2$—this therefore lowers the analyzer noise floor by the process gain equal to $10 \log_{10}[f_s/(2 \times BW)]$, where BW is the resolution noise bandwidth of the analyzer (Figure 2.101).

It is important that the noise bandwidth (not the 3-dB bandwidth) be used in the calculation; however, from Figure 2.68 the error is small assuming that the analyzer narrowband filter is at least two poles. The ratio of the noise bandwidth to the 3-dB bandwidth of a one-pole Butterworth filter is 1.57 (causing an error of 1.96 dB in the process gain calculation). For a two-pole Butterworth filter, the ratio is 1.11 (causing an error of 0.45 dB in the process gain calculation).

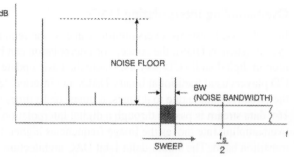

- BW = ANALYZER RESOLUTION NOISE BANDWIDTH

- $SNR = NOISE\ FLOOR - 10 \log_{10}\left[\dfrac{f_s/2}{BW}\right]$

Figure 2.101: Measuring DAC SNR with an Analog Spectrum Analyzer

DAC Output Spectrum and sin (x)/x Frequency Roll-off

The output of a reconstruction DAC can be represented as a series of rectangular pulses whose width is equal to the reciprocal of the clock rate as shown in Figure 2.102. Note that the reconstructed signal amplitude is down 3.92 dB at the Nyquist frequency, $f_c/2$. An inverse sin(x)/x filter can be used to compensate for this effect in most cases. The images of the fundamental signal occur as a result of the sampling function and are also attenuated by the sin(x)/x function.

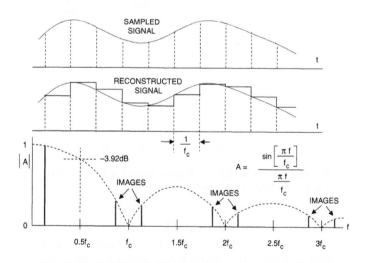

Figure 2.102: DAC sin (x)/x Roll Off (Amplitude Normalized)

Oversampling Interpolating DACs

In ADC-based systems, oversampling can ease the requirements on the antialiasing filter. In a DAC-based system (such as DDS), the concept of interpolation can be used in a similar manner. This concept is common in digital audio CD players, where the basic update rate of the data from the CD is 44.1 kSPS. Early CD players used traditional binary DACs and inserted "Zeros" into the parallel data, thereby increasing the effective update rate to 4 times, 8 times, or 16 times the fundamental throughput rate. The 4×, 8×, or 16× data stream is passed through a digital interpolation filter that generates the extra data points. The high oversampling rate moves the image frequencies higher, thereby allowing a less complex filter with a wider transition band. The sigma-delta 1-bit DAC architecture uses a much higher oversampling rate and represents the ultimate extension of this concept and has become popular in modern CD players.

The same concept of oversampling can be applied to high speed DACs used in communications applications, relaxing the requirements on the output filter as well as increasing the SNR due to process gain.

Assume a traditional DAC is driven at an input word rate of 30 MSPS (Figure 2.103A). Assume the DAC output frequency is 10 MHz. The image frequency component at $30 - 10 = 20$ MHz must be attenuated by the analog antialiasing filter, and the transition band of the filter is 10 to 20 MHz. Assume that the image frequency must be attenuated by 60 dB. The filter must therefore go from a pass band of 10 MHz to 60 dB stopband attenuation over the transition band lying between 10 and 20 MHz (one octave). A filter gives 6 dB attenuation per octave for each pole. Therefore, a minimum of 10 poles is required to provide the desired attenuation. Filters become even more complex as the transition band becomes narrower.

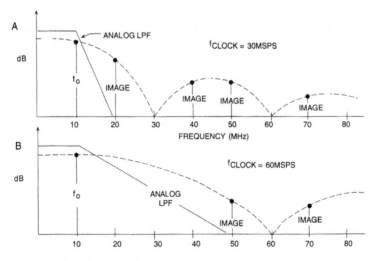

Figure 2.103: Analog Filter Requirements for f_o = 10 MHz:
(A) f_c = 30 MSPS, and (B) f_c = 60 MSPS

Assume that the DAC update rate is increased to 60 MSPS and insert a "zero" between each original data sample. The parallel data stream is now 60 MSPS, but must now be determined the value of the zero-value data points. This is done by passing the 60 MSPS data stream with the added zeros through a digital interpolation filter that computes the additional data points. The response of the digital filter relative to the 2 times oversampling frequency is shown in Figure 2.103B. The analog antialiasing filter transition zone is now 10 to 50 MHz (the first image occurs at $2f_c - f_o = 60 - 10 = 50$ MHz). This transition zone is a little greater than two octaves, implying that a 5- or 6-pole filter is sufficient.

The AD9773/AD9775/AD9777 (12-/14-/16-bit) series of Transmit DACs (TxDAC) are selectable 2×, 4×, or 8× oversampling interpolating dual DACs, and a simplified block diagram is shown in Figure 2.104. These devices are designed to handle 12-/14-/16-bit input word rates up to 160 MSPS. The output word rate is 400 MSPS maximum. For an output frequency of 50 MHz, an input update rate of 160 MHz, and an oversampling ratio of 2×, the image frequency occurs at 320 MHz – 50 MHz = 270 MHz. The transition band for the analog filter is therefore 50 MHz to 270 MHz. Without 2× oversampling, the image frequency occurs at 160 MHz – 50 MHz = 110 MHz, and the filter transition band is 50 MHz to 110 MHz.

Notice also that an oversampling interpolating DAC allows both a lower frequency input clock and input data rate, which are much less likely to generate noise within the system.

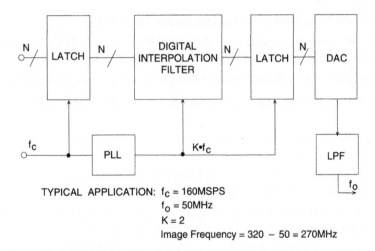

Figure 2.104: Oversampling Interpolating TxDAC
Simplified Block Diagram

References:
2.3 Data Converter AC Errors

1. W. R. Bennett, "Spectra of Quantized Signals," **Bell System Technical Journal**, Vol. 27, July 1948, pp. 446–471.

2. B. M. Oliver, J. R. Pierce, and C. E. Shannon, "The Philosophy of PCM," **Proceedings IRE**, Vol. 36, November 1948, pp. 1324–1331.

3. W. R. Bennett, "Noise in PCM Systems," **Bell Labs Record**, Vol. 26, December 1948, pp. 495–499.

4. H. S. Black and J. O. Edson, "Pulse Code Modulation," **AIEE Transactions**, Vol. 66, 1947, pp. 895–899.

5. H. S. Black, "Pulse Code Modulation," **Bell Labs Record**, Vol. 25, July 1947, pp. 265–269.

6. Steve Ruscak and Larry Singer, *Using Histogram Techniques to Measure A/D Converter Noise*, **Analog Dialogue**, Vol. 29-2, 1995.

7. M.J. Tant, **The White Noise Book**, Marconi Instruments, July 1974.

8. G.A. Gray and G.W. Zeoli, *Quantization and Saturation Noise due to A/D Conversion*, **IEEE Trans. Aerospace and Electronic Systems**, Jan. 1971, pp. 222–223.

9. Kevin McClaning and Tom Vito, **Radio Receiver Design**, Noble Publishing, 2000, ISBN 1-88-4932-07-X.

10. Walter G. Jung, editor, **Op Amp Applications**, Analog Devices, Inc., 2002, ISBN 0-916550-26-5, pp. 6.144–6.152.

11. Brad Brannon, *Aperture Uncertainty and ADC System Performance*, **Application Note AN-501**, Analog Devices, Inc., January 1998. (available for download at www.analog.com)

12. Christopher W. Mangelsdorf, *A 400 MHz Input Flash Converter with Error Correction*, **IEEE Journal of Solid-State Circuits**, Vol. 25, No. 1, February 1990, pp. 184–191.

13. Charles E. Woodward, *A Monolithic Voltage-Comparator Array for A/D Converters*, **IEEE Journal of Solid State Circuits**, Vol. SC-10, No. 6, December 1975, pp. 392–399.

14. Yukio Akazawa et. al., *A 400MSPS 8 Bit Flash A/D Converter*, **1987 ISSCC Digest of Technical Papers**, pp. 98–99.

15. A.. Matsuzawa *et al.*, *An 8b 600 MHz Flash A/D Converter with Multistage Duplex-gray Coding*, **Symposium VLSI Circuits, Digest of Technical Papers**, May 1991, pp. 113–114.

16. Ron Waltman and David Duff, *Reducing Error Rates in Systems Using ADCs,* **Electronics Engineer**, April 1993, pp. 98–104.

17. K. W. Cattermole, **Principles of Pulse Code Modulation**, American Elsevier Publishing Company, Inc., 1969, New York NY, ISBN 444-19747-8. *(An excellent tutorial and historical discussion of data conversion theory and practice, oriented towards PCM, but covers practically all aspects. This one is a must for anyone serious about data conversion.*

18. Robert A. Witte, *Distortion Measurements Using a Spectrum Analyzer*, **RF Design**, September, 1992, pp. 75–84.

19. Walt Kester, *Confused About Amplifier Distortion Specs?*, **Analog Dialogue**, 27-1, 1993, pp. 27–29.

20. Dan Sheingold, Editor, **Analog-to-Digital Conversion Handbook, Third Edition**, Prentice-Hall, 1986.

21. Fred H. Irons, "The Noise Power Ratio—Theory and ADC Testing," **IEEE Transactions on Instrumentation and Measurement**, Vol. 49, No. 3, June 2000, pp. 659–665.

General Data Converter Specifications
James Bryant

Overall Considerations

Data converters, as we have observed, have a digital port and an analog port and, like all integrated circuits, they require power supplies and will draw current from those supplies. Data converter specifications will therefore include the usual specifications common to any integrated circuit, including supply voltage and supply current, logic interfaces, power on and standby timing, package and thermal issues and ESD. We shall not consider these at any length, but there are some issues that may require a little consideration.

An over-riding piece of advice here is *read the data sheet*. There is no excuse for being unaware of the specifications of a device for which one owns a data sheet—and it is often possible to deduce extra information that is not printed on it by understanding the issues and conventions involved in preparing it.

Traditional precision analog integrated circuits (which include amplifiers, converters, and other devices) were designed for operation from supplies of ±15 V, and many (but not all—it is important to check with the data sheet) would operate within specification over quite a wide range of supply voltages. Today the processes used for many, but by no means all, modern converters have low breakdown voltages and absolute maximum ratings of only a few volts. Converters built with these processes may only work to specification over a narrow range of supply voltages.

It is therefore important when selecting a data converter to check both the absolute maximum supply voltage(s) and the range of voltages where correct operation can be expected. Some low-voltage devices work equally well with both 5-V and 3.3-V supplies, others are sold in 5-V and 3.3-V versions with different suffixes on their part numbers—with these it is important to use the correct one.

Absolute maximum ratings are ratings that can never be exceeded without grave risk of damage to the device concerned—they are not safe operating limits, but they are conservative. Integrated circuit manufacturers try to set absolute maximum ratings so that every device they manufacture will survive brief exposures to absolute maximum conditions. As a result many devices will, in fact, appear to operate safely and continuously outside the permitted limits. Good engineers do not take advantage of this for three reasons: (1) components are not tested outside their absolute maximum limits so, although they may be operating, they may not be operating at their specified accuracy. Also the damage done by incorrect operation may not be immediately fatal, but may cause low levels of disruption which, in turn, may (2) shorten the device's life, or (3) may affect its subsequent accuracy even when it is operated within specification again. None of these effects is at all desirable and absolute maximum ratings should always be respected.

The supply current in a data converter specification is usually the no-load current—i.e., the current consumption when the data converter output is driving a high impedance or open-circuit load. CMOS logic, and to a lesser extent some other types, have current consumption that is proportional to clock speed so a CMOS data converter current may be defined at a specific clock frequency and will be higher if the clock runs faster. Current consumption will also be higher when the output (or the reference output if there is one, or both) is loaded. There may be another figure for "standby" current—the current that flows when the data converter is connected to a power supply but is internally shut down into a non-operational low power state to conserve current.

When power is first applied to some data converters they may take several tens, hundreds, or even thousands of microseconds for their reference and amplifier circuitry to stabilize and, although this is less common, some may even take a long time to "wake up" from a power saving standby mode. It is therefore important to ensure that data converters that have such delays are not used in applications where full functionality is required within a short time of power-up or wake-up.

All integrated circuits are vulnerable to electrostatic discharge (ESD), but precision analog circuits are, on the whole, more vulnerable than some other types. This is because the technologies available for minimizing such damage also tend to degrade the performance of precision circuitry, and there is a necessary compromise between robustness and performance. It is always a good idea to ensure that when handling amplifiers, converters and other vulnerable circuits the necessary steps are taken to avoid ESD.

Specifications of packages, operating temperature ranges, and similar issues, although important, do not need further discussion here.

Logic Interface Issues

As it is important to read and understand power supply specifications, so it is equally important to read and understand logic specifications. In the past most integrated circuit logic circuitry (with the exception of emitter-coupled logic or ECL) operated from 5 V supplies and had compatible logic levels—with a few exceptions 5 V logic would interface with other 5 V logic. Today, with the advent of low voltage logic operating with supplies of 3.3 V, 2.7 V, or even less, it is important to ensure that logic interfaces are compatible. There are several issues which must be considered—absolute maximum ratings, worst-case logic levels, and timing. The logic inputs of integrated circuits generally have absolute maximum ratings, as do most other inputs, of 300 mV outside the power supply. Note that these are instantaneous ratings. If an IC has such a rating and is currently operating from a +5 V supply, the logic inputs may be between –0.3 V and +5.3 V—but if the supply is not present, that input must be between +0.3 V and –0.3 V, not the –0.3 V to +5.3 V which are the limits once the power is applied—ICs cannot predict the future.

The reason for the rating of 0.3 V is to ensure that no parasitic diode on the IC is ever turned on by a voltage outside the IC's absolute maximum rating. It is quite common to protect an input from such overvoltage with a Schottky diode clamp. At low temperatures the clamp voltage of a Schottky diode may be a little more than 0.3 V, and so the IC may see voltages just outside its absolute maximum rating. Although, strictly speaking, this subjects the IC to stresses outside its absolute maximum ratings and so is forbidden, this is an acceptable exception to the general rule provided the Schottky diode is at a temperature similar to the IC it is protecting (say within ±10°C).

Some low voltage devices, however, have inputs with absolute maximum ratings that are substantially greater than their supply voltage. This allows such circuits to be driven by higher voltage logic without additional interface or clamp circuitry. But it is important to read the data sheets and ensure that both logic levels and absolute maximum voltages are compatible for all combinations of high and low supplies.

This is the general rule when interfacing different low voltage logic circuitry—it is always necessary to check that at the lowest value of its power supply (a) the Logic 1 output from the driving circuit applied to its worst-case load is greater than the specified minimum Logic 1 input for the receiving circuit, and (b) with its output sinking maximum allowed current, the logic 0 output is less than the specified Logic 0 input of the receiver. If the logic specifications of the chosen devices do not meet these criteria it will be necessary to select different devices, use different power supplies, or use additional interface circuitry to ensure that the required levels are available. Note that additional interface circuitry introduces extra delays in timing.

It is not sufficient to build an experimental setup and test it. In general, logic thresholds are generously specified and usually logic circuits will work correctly well outside their specified limits—but it is not possible to rely on this in a production design. At some point a batch of devices near the limit on low output swing will be required to drive some devices needing slightly more drive than usual—and will be unable to do so.

Data Converter Logic: Timing and other Issues

It is not the purpose of this brief section to discuss logic architectures, so we shall not define the many different data converter logic interface operations and their timing specifications except to note that data converter logic interfaces may be more complex than expected—*read the data sheet*. Do not expect that because there is a pin with the same name on memory and interface chips it will behave in exactly the same way in a data converter. Also, some data converters reset to a known state on power-up but many more do not.

It is very necessary to consider general timing issues. The new low voltage processes used for many modern data converters have a number of desirable features. One that is often overlooked by users (but not by converter designers) is their higher logic speed. DACs built on older processes frequently had logic that was orders of magnitude slower than the microprocessors with which they interfaced, and it was sometimes necessary to use separate buffers, or multiple WAIT instructions, to make the two compatible. Today it is much more common for the write times of DACs to be compatible with those of the fast logic with which they interface.

Nevertheless, not all DACs are speed compatible with all logic interfaces, and it is still important to ensure that minimum data setup times and write pulsewidths are observed. Again, experiments will often show that devices work with faster signals than their specification requires—but at the limits of temperature or supply voltage some may not, and interfaces should be designed on the basis of specified rather than measured timing.

Defining the Specifications
Dan Sheingold, Walt Kester

The following list, in alphabetical order, should prove helpful regarding specifications and their definitions. Some of the most popular ones are discussed in other places in the text as well as here. Many of the application-specific specifications are defined where they are mentioned in the text and are not repeated here. The original source for these definitions was provided by Dan Sheingold from Chapter 11 in his classic book *Analog-to-Digital Conversion Handbook, Third Edition*, Prentice-Hall, 1986.

Accuracy, Absolute. Absolute accuracy error of a *DAC* is the difference between actual analog output and the output that is expected when a given digital code is applied to the converter. Error is usually commensurate with resolution, i.e., less 1/2 LSB of full-scale, for example. However, accuracy may be much better than resolution in some applications; for example, a 4-bit DAC having only 16 discrete digitally chosen levels would have a resolution of 1/16, but might have an accuracy to within 0.01 % of each ideal value.

Absolute accuracy error of an *ADC* at a given output code is the difference between the actual and the theoretical analog input voltages required to produce that code. Since the code can be produced by any analog voltage in a finite band (see *Quantizing Uncertainty*), the "input required to produce that code" is usually defined as the midpoint of the band of inputs that will produce that code. For example, if 5 V, ±1.2 mV, will theoretically produce a 12-bit half-scale code of 1000 0000 0000, then a converter for which any voltage from 4.997 V to 4.999 V will produce that code will have absolute error of $(1/2)(4.997 + 4.999) - 5$ V = +2 mV.

Sources of error include gain (calibration) error, zero error, linearity errors, and noise. Absolute accuracy measurements should be made under a set of standard conditions with sources and meters traceable to an internationally accepted standard.

Accuracy, Logarithmic DACs. The difference (measured in dB) between the actual transfer function and the ideal transfer function, as measured after calibration of gain error at 0 dB.

Accuracy, Relative. Relative accuracy error, expressed in %, ppm, or fractions of 1 LSB, is the deviation of the analog value at any code (relative to the full analog range of the device transfer characteristic) from its theoretical value (relative to the same range), after the full-scale range (FSR) has been calibrated (see *Full-Scale Range*).

Since the discrete analog values that correspond to the digital values ideally lie on a straight line, the specified worst-case relative accuracy error of a linear ADC or DAC can be interpreted as a measure of end-point nonlinearity (see *Linearity*).

The "discrete points" of a DAC transfer characteristic are measured by the actual analog outputs. The "discrete points" of an ADC transfer characteristic are the midpoints of the quantization bands at each code (see *Accuracy, Absolute*).

Acquisition Time. The acquisition time of a track-and-hold circuit for a step change is the time required by the output to reach its final value, within a specified error band, after the track command has been given. Included are switch delay time, the slewing interval, and settling time for a specified output voltage change.

Adjacent Channel Power Ratio (ACPR). The ratio in dBc between the measured power within a channel relative to its adjacent channel. See *Adjacent Channel Leakage Ratio* (ACLR).

Adjacent Channel Leakage Ratio (ACLR). The ratio in dBc between the measured power within the carrier bandwidth relative to the noise level in an adjacent empty carrier channel. Both ACPR and ACLR are Wideband CDMA (WCDMA) specifications. The channel bandwidth for WCDMA is approximately 3.84 MHz with 5 MHz spacing between channels.

Aliasing. A signal within a *bandwidth* f_a must be sampled at a rate $f_s > 2f_a$ in order to avoid the loss of information. If $f_s < 2f_a$, a phenomenon called aliasing, inherent in the spectrum of the sampled signal, will cause a frequency equal to $f_s - f_a$, called an alias, to appear in the Nyquist bandwidth, dc to $f_s/2$. For example, if $f_s = 4$ kSPS and $f_a = 3$ kHz, a 1 kHz alias will appear. Note also that for $f_a = 1$ kHz (within the dc to $f_s/2$ bandwidth), an alias will occur at 3 kHz (outside the dc to $f_s/2$ bandwidth). Since noise is also aliased, it is essential to provide low-pass (or band-pass) filtering prior to the sampling stage to prevent out-of-band noise on the input signal from being aliased into the signal range and thereby degrading the SNR.

Analog Bandwidth. For an ADC, the analog input frequency at which the spectral power of the fundamental frequency (as determined by the FFT analysis) is reduced by 3 dB. This can be specified as full power bandwidth, or small signal bandwidth. (See also (*Bandwidth, Full Linear* and *Bandwidth, Full Power.*)

Analog Bandwidth, 0.1 dB. For an ADC, the analog input frequency at which the spectral power of the fundamental frequency (as determined by the FFT analysis) is reduced by 0.1 dB. This is a popular video specification. (See also *Bandwidth, Full Linear* and *Bandwidth, Full Power.*)

Aperture Time (classic definition). Aperture time in a sample-and-hold is defined as the time required for the internal switch to switch from the closed position (zero resistance) to the fully open position (infinite resistance). A first-order analysis that neglects nonlinear effects assumes that the input signal is averaged over this time interval to produce the final output signal. The analysis shows that this does not introduce an error as long as the switch opens in a repeatable fashion, and as long as the aperture time is reasonably short with respect to the hold time. There exists an effective sampling point in time that will cause an ideal sample-and-hold to produce the same held voltage. The difference between this effective sampling point and the actual sampling point is defined as effective aperture delay time.

Aperture Delay Time, or *Effective Aperture Delay Time.* In a sample-and-hold or track-and-hold, there exists an effective sampling point in time that will cause an ideal sample-and-hold to produce the same held voltage. The difference between this effective sampling point and the actual sampling point is defined as the aperture delay time or effective aperture delay time. In a sampling ADC, aperture delay time can be measured by sampling the zero crossing of a sinewave with a sampling clock locked to the sinewave. The phase of the sampling clock is adjusted until the output of the ADC is 100…00. The time difference between the leading edge of the sampling clock and the zero crossing of the sinewave—referenced to the analog input—is the effective aperture delay time. A dual trace oscilloscope can be used to make the measurement.

Aperture Uncertainty (or Aperture Jitter). The sample-to-sample variation in the sampling point because of jitter. Aperture jitter is expressed as an rms quantity and produces a corresponding rms voltage error in the sample-and-hold output. In an ADC it is caused by internal noise and jitter in the sampling clock path from the sampling clock input pin to the internal switch. Jitter in the external sampling clock produces the same type of error.

Automatic Zero. To achieve zero stability in many integrating-type converters, a time interval is provided during each conversion cycle to allow the circuitry to compensate for drift errors. The drift error in such converters is substantially zero. A similar function exists in many high resolution sigma-delta ADCs.

Bandwidth, Full-Linear. The full-linear bandwidth of an ADC is the input frequency at which the slew-rate limit of the sample-and-hold amplifier is reached. Up to this point, the amplitude of the reconstructed fundamental signal will have been attenuated by less than 0.1 dB. Beyond this frequency, distortion of the sampled input signal increases significantly.

Bandwidth, Full-Power (FPBW). The full-power bandwidth is that input frequency at which the amplitude of the reconstructed fundamental signal (measured using FFTs) is reduced by 3 dB for a full-scale input. In order to be meaningful, the FPBW must be examined in conjunction with the signal-to-noise ratio (SNR), signal-to-noise-plus-distortion ratio (SINAD), effective number of bits (ENOB), and harmonic distortion in order to ascertain the true dynamic performance of the ADC at the FPBW frequency.

Bandwidth, Analog Input Small-Signal. Analog input bandwidth is measured similarly to FPBW at a reduced analog input amplitude. This specification is similar to the small signal bandwidth of an op amp. The amplitude of the input signal at which the small signal bandwidth is measured should be specified on the data sheet.

Bandwidth, Effective Resolution (ERB). Some ADC manufacturers define the frequency at which SINAD drops 3 dB as the *effective resolution bandwidth (ERB)*. This is the same frequency at which the ENOB drops ½ bit. This specification is a misnomer, however, since bandwidth normally is associated with signal amplitude.

Bias Current. The zero-signal dc current required from the signal source by the inputs of many semiconductor circuits. The voltage developed across the source resistance by bias current constitutes an (often negligible) offset error. When an instrumentation amplifier performs measurements of a source that is remote from the amplifier's power-supply, there *must* be a return path for bias currents. If it does not already exist and is not provided, those currents will charge stray capacitances, causing the output to drift uncontrollably or to saturate. Therefore, when amplifying outputs of "floating" sources, such as transformers, insulated thermocouples, and ac-coupled circuits, there must be a high impedance dc leakage path from each input to common, or to the driven-guard terminal (if present). If a dc return path is impracticable, an *isolator* must be used.

Bipolar Mode. (See Offset.)

Bipolar Offset. (See Offset.)

Bus. A bus is a parallel path of binary information signals—usually 4, 8, 16, 32, or 64-bits wide. Three common types of information usually found on buses are data, addresses, and control signals. Three-state output switches (inactive, high, and low) permit many sources—such as ADCs—to be connected to a bus, while only one is active at any time.

Byte. A byte is a binary digital word, usually 8 bits wide. A byte is often part of a longer word that must be placed on an 8-bit bus in two stages. The byte containing the MSB is called the *high byte*; that containing the LSB is called the *low byte*. A 4-bit byte is called a *nibble* on an 8-bit or greater bus.

Channel-to-Channel Isolation. In multiple DACs, the proportion of analog input signal from one DAC's reference input that appears at the output of the other DAC, expressed logarithmically in dB. See also *crosstalk*.

Charge Transfer, Charge Injection (or *Offset Step*). The principal component of *sample-to-hold offset* (or *pedestal),* is the small charge transferred to the storage capacitor via interelectrode capacitance of the switch and stray capacitance when switching to the *hold* mode. The offset step is directly proportional to this charge:

$$\text{Offset error} = \text{Incremental Charge/Capacitance} = \Delta Q/C$$

It can be reduced somewhat by lightly coupling an appropriate polarity version of the *hold* signal to the capacitor for first-order cancellation. The error can also be reduced by increasing the capacitance, but this increases *acquisition time.*

Code Width. This is a fundamental quantity for ADC specifications. In an ADC where the code transition noise is a fraction of an LSB, it is defined as the range of analog input values for which a given digital output code will occur. The nominal value of a code width (for all but the first and last codes) is the voltage equivalent of 1 least significant bit (LSB) of the full-scale range, or 2.44 mV out of 10 V for a 12-bit ADC. Because the full-scale range is fixed, the presence of excessively wide codes implies the existence of narrow and perhaps even missing codes. Code transition noise can make the measurement of code width difficult or impossible. In wide bandwidth and high resolution ADCs additional noise modulates the effective code width and appears as input-referred noise. Many ADCs have input-referred noise that spans several code widths, and histogram techniques must be used to accurately measure differential linearity.

Common-Mode Range. Common-mode rejection usually varies with the magnitude of the range through which the input signal can swing, determined by the sum of the common-mode and the differential voltage. *Common-mode range* is that range of *total* input voltage over which specified common-mode rejection is maintained. For example, if the common-mode signal is ±5 V and the differential signal is ±5 V, the common-mode range is ±10 V.

Common-Mode Rejection (CMR). A measure of the change in output voltage when both inputs are changed by equal amounts of ac and/or dc voltage. Common-mode rejection is usually expressed either as a ratio (e.g., CMRR = 1,000,000:1) or in decibels: CMR = $20\log_{10}$CMRR; if CMRR = 10^6, CMR = 120 dB. A CMRR of 10^6 means that 1 V of common mode is processed by the device as though it were a differential signal of 1 µV at the input.

CMR is usually specified for a full range common-mode voltage change (CMV), at a given frequency, and a specified imbalance of source impedance (e.g., 1 kΩ source unbalance, at 60 Hz). In amplifiers, the common-mode rejection ratio is defined as the ratio of the signal gain, G, to the common-mode gain (the ratio of common-mode signal appearing at the output to the CMV at the input.

Common-Mode Voltage (CMV). A voltage that appears in common at both input terminals of a differential-input device, with respect to its output reference (usually "ground"). For inputs, V_1 and V_2, with respect to ground, CMV = ½($V_1 + V_2$). An ideal differential-input device would ignore CMV. *Common-mode error (CME)* is any error at the output due to the common-mode input voltage. The errors due to supply voltage variation, an internal common-mode effect, are specified separately.

Compliance-Voltage Range. For a current-output DAC, the maximum range of (output) terminal voltage for which the device will maintain the specified current-output characteristics.

Conversion Complete. An ADC digital output signal which indicates the end of conversion. When this signal is in the opposite state, the ADC is considered to be "busy." Also called *end-of-conversion (EOC)*, *data ready*, or *status* in some converters.

Conversion Time and *Conversion Rate.* For an ADC without a sample-and-hold, the time required for a complete measurement is called *conversion time.* For most converters (assuming no significant additional systemic delays), conversion time is essentially identical with the inverse of *conversion rate.* For simple sampling ADCs, however, the conversion rate is the inverse of the conversion time plus the sample-and-hold's acquisition time. However, in many high speed converters, because of pipelining, new conversions are initiated before the results of prior conversions have been determined; thus, there can one, two, three, or more clock cycles of conversion delay (plus a fixed delay in some cases). Once a train of conversions has been initiated, as in signal-processing applications, the conversion rate can therefore be much faster than the conversion time would imply.

Crosstalk. Leakage of signals, usually via capacitance between circuits or channels of a multichannel system or device, such as a multiplexer, multiple input ADC, or multiple DAC. Crosstalk is usually determined by the impedance parameters of the physical circuit, and actual values are frequency-dependent. See also *channel-to-channel isolation.*

Multiple DACs have a *digital crosstalk* specification: the spike (sometimes called a glitch) impulse appearing at the output of one converter due to a change in the digital input code of another of the converters. It is specified in nanovolt- or picovolt-seconds and measured at $V_{REF} = 0$ V.

Data Ready. (See Conversion Complete.)

Deglitcher (See Glitch.) A device that removes or reduces the effects of time-skew in D/A conversion. A deglitcher normally employs a track-and-hold circuit, often specifically designed as part of the DAC. When the DAC is updated, the deglitcher holds the output of the DAC's output amplifier constant at the previous value until the switches reach equilibrium, then acquires and tracks the new value.

DAC Glitch. A glitch is a switching transient appearing in the output during a code transition. The worst-case DAC glitch generally occurs when the DAC is switched between the 011...111 and 100...000 codes. The net area under the glitch is referred to as *glitch impulse area* and is measured in millivolt-nanoseconds, nanovolt-seconds, or picovolt-seconds. Sometimes the term *glitch energy* is used to describe the net area under the glitch—this terminology is incorrect because the unit of measurement is not energy.

Differential Analog Input Resistance, Differential Analog Input Capacitance, and *Differential Analog Input Impedance.* The real and complex impedances measured at each analog input port of an ADC. The resistance is measured statically and the capacitance and differential input impedances are measured with a network analyzer.

Differential Analog Input Voltage Range. The peak-to-peak differential voltage that must be applied to the converter to generate a full-scale response. Peak differential voltage is computed by observing the voltage on a single pin and subtracting the voltage from the other pin, which is 180 degrees out of phase. Peak-to-peak differential is computed by rotating the inputs phase 180 degrees and taking the peak measurement again. The difference is then computed between both peak measurements.

Differential Gain (ΔG). A video specification that measures the variation in the amplitude (in percent) of a small amplitude color subcarrier signal as it is swept across the video range from black to white.

Differential Phase ($\Delta \varphi$). A video specification that measures the phase variation (in degrees) of a small amplitude color subcarrier signal as it is swept across the video range from black to white.

Droop Rate. When a sample-and-hold circuit using a capacitor for storage is in *hold,* it will not hold the information forever. Droop rate is the rate at which the output voltage changes (by increasing or decreasing), and hence gives up information. The change of output occurs as a result of leakage or bias currents flowing through the storage capacitor. The polarity of change depends on the sources of leakage within a given device. In integrated circuits with external capacitors, it is usually specified as a *(droop* or *drift)* current, in ICs having internal capacitors, a rate of change. Note: dv/dt (volts/second) = I/C (picoamperes/picofarads).

Dual-Slope Converter. An integrating ADC in which the unknown signal is converted to a proportional time interval, which is then measured digitally. This is done by integrating the unknown for a predetermined length of time. A reference input is then switched to the integrator, which integrates "down" from the level determined by the unknown until the starting level is reached. The time for the second integration process, as determined by the counter, is proportional to the average of the unknown signal level over the predetermined integrating period. The counter provides the digital readout.

Effective Input Noise. (See Input-Referred Noise.)

Effective Number of Bits (ENOB). With a sinewave input, Signal-to-Noise-and-Distortion (SINAD) can be expressed in terms of the number of bits. Rewriting the theoretical SNR formula for an ideal N-bit ADC and solving for N:

$$N = (SNR - 1.76\text{ dB})/6.02$$

The actual ADC SINAD is measured using FFT techniques, and ENOB is calculated from:

$$ENOB = (SINAD - 1.76\text{ dB})/6.02$$

Effective Resolution. (See Noise-Free Code Resolution.)

Encode Command. (See Encode, Sampling Clock.)

Encode (Sampling Clock) Pulsewidth/Duty Cycle. Pulsewidth high is the minimum amount of time that the ENCODE pulse should be left in Logic 1 state to achieve rated performance; pulsewidth low is the minimum time ENCODE or pulse should be left in low state. See timing implications of changing the width in the text of high speed ADC data sheets. At a given clock rate, these specs define an acceptable ENCODE duty cycle.

Feedthrough. Undesirable signal coupling around switches or other devices that are supposed to be turned off or provide isolation, *e.g., feedthrough error* in a sample-and-hold, multiplexer, or multiplying DAC. Feedthrough is variously specified in percent, dB, parts per million, fractions of 1 LSB, or fractions of 1 V, with a given set of inputs, at a specified frequency.

In a multiplying *DAC, feedthrough* error is caused by capacitive coupling from an ac V_{REF} to the output, with all switches off. In a *sample-and-hold, feedthrough* is the fraction of the input signal variation or ac input waveform that appears at the output in *hold.* It is caused by stray capacitive coupling from the input to the storage capacitor, principally across the open switch.

Flash Converter. A converter in which all the bit choices are made at the same time. It requires $2^N - 1$ voltage-divider taps and comparators and a comparable amount of priority encoding logic. A scheme that gives extremely fast conversion, it requires large numbers of nearly identical components, hence it is well suited to integrated-circuit form for resolutions up to eight bits. Several flash converters are often used in multistage *subranging converters,* to provide high resolution at somewhat slower speed than pure flash conversion.

Four-Quadrant. In a multiplying DAC, "four quadrant" refers to the fact that both the reference signal and the number represented by the digital input may be of either positive or negative polarity. Such a DAC can be thought of as a gain control for ac signals ("reference" input) with a range of positive and negative digitally controlled gains. A four-quadrant multiplier is expected to obey the rules of multiplication for algebraic sign.

Frequency-to-Voltage Conversion (FVC). The input of an FVC device is an ac waveform—usually a train of pulses (in the context of conversion); the output is an analog voltage, proportional to the number of pulses occurring in a given time. FVC is usually performed by a voltage-to-frequency converter in a feedback loop. Important specifications, in addition to the accuracy specs typical of VFCs (see *Voltage-to-Frequency conversion),* include *output ripple* (for specified input frequencies), *threshold* (for recognition that another cycle has been initiated, and for versatility in interfacing several types of sensors directly), *hysteresis,* to provide a degree of insensitivity to noise superimposed on a slowly varying input waveform, and *dynamic response* (important in motor control).

Full-Scale Input Power (ADC). Expressed in dBm (power level referenced to 1 mW). Computed using the following equation, where $V_{Full\ Scale\ rms}$ is in volts, and Z_{input} is in Ω.

$$Power_{Full\ Scale} = 10\log_{10}\left[\frac{\dfrac{V^2_{Full\ Scale\ rms}}{|Z|_{Input}}}{0.001}\right].$$

Full-Scale Range (FSR). For binary ADCs and DACs, that magnitude of voltage, current, or—in a multiplying DAC—gain, of which the MSB is specified to be exactly one-half or for which any bit or combination of bits is tested against its (their) prescribed ideal ratio(s). FSR is independent of resolution; the value of the LSB (voltage, current, or gain) is 2^{-N} FSR. There are several other terms, with differing meanings, that are often used in the context of discussions or operations involving full-scale range. They are:

Full-scale—similar to full-scale range, but pertaining to a single polarity. Thus, full-scale for a unipolar device is twice the prescribed value of the MSB and has the same polarity. For a bipolar device, *positive or negative full-scale* is that positive or negative value, of which the next bit after the polarity bit is tested to be one-half.

Span—the scalar voltage or current range corresponding to FSR.

All-1s—All bits on, the condition used, in conjunction with *all-zeros,* for gain adjustment of an ADC or DAC, in accordance with the manufacturer's instructions. Its magnitude, for a binary device, is $(1 - 2^{-N})$ FSR. *All-1s* is a *positive-true* definition of a specific magnitude relationship; for complementary coding the "all-1s" code will actually be all zeros. To avoid confusion, all-1s should never be called "full-scale"; FSR and FS are independent of the number of bits, all-1s isn't.

All-0s—All bits off, the condition used in offset (and gain) adjustment of a DAC or ADC, according to the manufacturer's instructions. All-0s corresponds to zero output in a unipolar DAC and negative full-scale in an offset bipolar DAC with positive output reference. In a sign-magnitude device, all-0s refers to all bits after the sign bit. Analogous to "all-1s," "all-0s" is a *positive-true* definition of the *all-bits-off* condition; in a complementary-coded device, it is expressed by all ones. To avoid confusion, all-0s should not be called "zero" unless it accurately corresponds to true analog zero output from a DAC.

Gain. The "gain" of a converter is that analog scale factor setting that establishes the nominal conversion relationship, e.g., 10 V full-scale. In a multiplying DAC or ratiometric ADC, it is indeed a gain. In a device with fixed internal reference, it is expressed as the full-scale magnitude of the output parameter (e.g., 10 V or 2 mA). In a fixed-reference converter, where the use of the internal reference is optional, the converter gain and the reference may be specified separately. Gain and zero adjustment are discussed under *zero.*

Glitch. Transients associated with code changes generally stem from several sources. Some are spikes, known as digital-to-analog feedthrough, or charge transfer, coupled from the digital signal (clock or data) to the analog output, defined with zero reference. These spikes are generally fast, fairly uniform, code-independent, and hence filterable. However, there is a more insidious form of transient, code-dependent, and difficult to filter, known as the "glitch."

If the output of a counter is applied to the input of a DAC to develop a "staircase" voltage, the number of bits involved in a code change between two adjacent codes establish "major" and "minor" transitions. The most major transition is at ½-scale, when the DAC switches all bits, i.e., from 011...111 to 100...000. If, for digital inputs having no skew, the switches are faster to switch *off* than *on,* this means that, for a short time, the DAC will seek zero output, and then return to the required 1 LSB above the previous reading. This large transient spike is commonly known as a "glitch." The better matched the input transitions and the switching times, the faster the switches, the smaller will be the area of the glitch. Because the size of the glitch is not proportional to the signal change, linear filtering may be unsuccessful and may, in fact, make matters worse. *(See also Deglitcher.)*

The severity of a glitch is specified by *glitch impulse area,* the product of its duration and its average magnitude, i.e., the net area under the curve. This product will be recognized as the physical quantity, *impulse* (electromotive *force* × Δ*time); however,* it has also been incorrectly termed "glitch energy" and "glitch charge." Glitch impulse area is usually expressed, for fast converters, in units of pV-s or mV-ns.

The glitch can be minimized through the use of fast, nonsaturating logic, such as ECL, LVDS, matched latches, and nonsaturating CMOS switches.

Glitch Charge, Glitch Energy, Glitch Impulse, Glitch Impulse Area. (See Glitch.)

Harmonic Distortion, 2nd. The ratio of the rms signal amplitude to the rms value of the second harmonic component, reported in dBc.

Harmonic Distortion, 3rd. The ratio of the rms signal amplitude to the rms value of the third harmonic component, reported in dBc.

Harmonic Distortion, Total (THD). The ratio of the rms signal amplitude to the rms sum of all harmonics (neglecting noise components). In most cases, only the first five harmonics are included in the measurement because the rest have negligible contribution to the result. The THD can be derived from the FFT of the ADC's output spectrum. For harmonics that are above the Nyquist frequency, the aliased component is used.

Harmonic Distortion, Total, Plus Noise (THD + N). Total harmonic distortion plus noise (THD + N) is the ratio of the rms signal amplitude to the rms sum of all harmonics and noise components. THD + N can be derived from the FFT of the ADC's output spectrum and is a popular specification for audio applications.

Impedance, Input. The dynamic load of an ADC presented to its input source. In unbuffered CMOS switched-capacitor ADCs, the presence of current transients at the converter's clock frequency mandates that the converter be driven from a low impedance (at the frequencies contained in the transients) in order to accurately convert. For buffered-input ADCs, the input impedance is generally represented by a resistive and capacitive component.

Input-Referred Noise (Effective Input Noise). Input-referred noise can be viewed as the net effect of all internal ADC noise sources referred to the input. It is generally expressed in *LSBs rms,* but can also be expressed as a voltage. It can be converted to a peak-to-peak value by multiplying by the factor 6.6. The peak-to-peak input-referred noise can then be used to calculate the *noise-free code resolution. (See Noise-Free Code Resolution).*

Intermodulation Distortion (IMD). With inputs consisting of sinewaves at two frequencies, f_1 and f_2, any device with nonlinearities will create distortion products of order (m + n), at sum and difference frequencies of $mf_1 \pm nf_2$, where m, n = 0, 1, 2, 3, ... Intermodulation terms are those for which m or n is not equal to zero. For example, the second-order terms are $(f_1 + f_2)$ and $(f_2 - f_1)$, and the third-order terms are $(2f_1 + f_2)$, $(2f_1 - f_2)$, $(f_1 + 2f_2)$, and $(f_1 - 2f_2)$. The IMD products are expressed as the dB ratio of the rms sum of the distortion terms to the rms sum of the measured input signals.

Latency. (See Pipelining.)

Leakage Current, Output. Current that appears at the output terminal of a DAC with all bits "off." For a converter with two complementary outputs (for example, many fast CMOS DACs), output leakage current is the current measured at OUT 1, with all digital inputs *low—and* the current measured at OUT 2, with all digital inputs *high.*

Least-Significant Bit (LSB). In a system in which a numerical magnitude is represented by a series of binary (i.e., two-valued) digits, the *least-significant bit* is that digit (or "bit") that carries the smallest value, or weight. For example, in the natural binary number 1101 (decimal 13, or $(1 \times 2^3) + (1 \times 2^2) + (0 \times 2^1) + (1 \times 2^0)$), the rightmost digit is the LSB. Its analog weight, in relation to full-scale (see *Full-Scale Range*), is 2^{-N}, where N is the number of binary digits. It represents the smallest analog change that can be resolved by an n-bit converter.

In data converter nomenclature, the LSB is bit N; in bus nomenclature (integer binary), it is Data Bit 0.

Left-Justified Data. When a 12-bit word is placed on an 8-bit bus in two bytes, the high byte contains the 4 or 8 most-significant bits. If 8, the word is said to be left justified; if 4 (plus filled-in leading sign bits), the word is said to be right justified.

Linearity. (See also Nonlinearity.) Linearity error of a converter *(also, integral nonlinearity—see Linearity, Differential),* expressed in % or parts per million of full-scale range, or (sub)multiples of 1 LSB, is a deviation of the analog values, in a plot of the measured conversion relationship, from a straight line. The straight line can be either a "best straight line," determined empirically by manipulation of the gain and/or offset to equalize maximum positive and negative deviations of the actual transfer characteristic from this straight line; or, it can be a straight line passing through the end points of the transfer characteristic after they have been calibrated, sometimes referred to as "end- point" linearity. "End-point" nonlinearity is similar to relative accuracy error *(see Accuracy, Relative).* It provides an easier method for users to calibrate a device, and it is a more conservative way to specify linearity.

For multiplying DACs, the *analog* linearity error, at a specified analog gain (digital code), is defined in the same way as for analog multipliers, i.e., by deviation from a "best straight line" through the plot of the analog output-input response.

Linearity, Differential. In a DAC, any two adjacent digital codes should result in measured output values that are exactly 1 LSB apart (2^{-N} of full-scale for an N-bit converter). Any positive or negative deviation of the measured "step" from the ideal difference is called *differential nonlinearity,* expressed in (sub)multiples of 1 LSB. It is an important specification, because a differential linearity error more negative than –1 LSB can lead to nonmonotonic response in a DAC and missed codes in an ADC using that DAC.

Similarly, in an ADC, midpoints between code transitions should be 1 LSB apart. Differential nonlinearity is the deviation between the actual difference between midpoints and 1 LSB, for adjacent codes. If this deviation is equal to or more negative than –1 LSB, a code will be missed (See *Missing Codes.)*

Often, instead of a maximum differential nonlinearity specification, there will be a simple specification of "monotonicity" or "no missing codes," which implies that the differential nonlinearity cannot be more negative than –1 for any adjacent pair of codes. However, the differential linearity error may still be more positive than +1 LSB.

Linearity, Integral. (See Linearity.) While *differential linearity* deals with errors in step size, *integral linearity* has to do with deviations of the overall shape of the conversion response. Even converters that are not subject to differential linearity errors.(e.g., integrating types) have integral linearity (sometimes just "linearity") errors.

Maximum Conversion Rate. The maximum sampling (encode) rate at which parametric testing is performed.

Minimum Conversion (Sampling) Rate. The encode rate at which the SNR of the lowest analog signal frequency drops by no more than 3 dB below the guaranteed limit.

Missing Codes. An ADC is said to have missing codes when a transition from one quantum of the analog range to the adjacent one does not result in the adjacent digital code, but in a code removed by one or more counts. Missing codes can be caused by large negative differential linearity errors, noise, or changing inputs during conversion. A converter's proclivity towards missing codes is also a function of the architecture and temperature.

Monotonicity. An DAC is said to be *monotonic* if its output either increases or remains constant as the digital input increases, with the result that the output will always be a single-valued function of the input. The condition "monotonic" requires that the derivative of the transfer function never change sign. Monotonic behavior requires that the differential nonlinearity be more positive than −1 LSB. The same basic definition applies to an ADC—the digital output code either increases or remains constant as the digital input increases. In practice, however, noise will cause the ADC output code to oscillate between two code transitions over a small range of analog input. Input-referred noise can make this effect worse, so histogram techniques are often used to measure ADC monotonicity in these situations.

Most Significant Bit (MSB). In a system in which a numerical magnitude is represented by a series of binary (i.e., two-valued) digits, the *most-significant bit* is that digit (or "bit") that carries the greatest value or weight. For example, in the natural binary number 1101 (decimal 13, or $(1 \times 2^3) + (1 \times 2^2) + (0 \times 2^1) + (1 \times 2^0)$), the leftmost "1" is the MSB, with a weight of ½ nominal peak-to-peak full-scale (full-scale range). In bipolar devices, the sign bit is the MSB.

In converter nomenclature, the MSB is bit 1; in bus nomenclature, it is Data Bit $(N - 1)$.

Multiplying DAC. A multiplying DAC differs from the conventional fixed-reference DAC in being designed to operate with varying (or ac) reference signals. The output signal of such a DAC is proportional to the product of the "reference" (i.e., analog input) voltage and the fractional equivalent of the digital input number. (See also *Four-Quadrant.*)

Multitone Spurious Free Dynamic Range (SFDR). The ratio of the rms value of an input tone to the rms value of the peak spurious component. The peak spurious component may or may not be an intermodulation distortion (IMD) product. May be reported in dBc (dB relative to the carrieror in dBFS (dB relative to full-scale). The amplitudes of the individual tones are equal and chosen such that the ADC is not overdriven when they add in-phase.

Noise-Free (Flicker-Free) Code Resolution. The noise-free code resolution of an ADC is the number of bits beyond which it is impossible to distinctly resolve individual codes. The cause is the effective input noise (or input-referred noise) associated with all ADCs. This noise can be expressed as an rms quantity, usually having the units of *LSBs rms*. Multiplying by a factor of 6.6 converts the rms noise into peak-to-peak noise (expressed in *LSBs peak-to-peak*). The total range of an N-bit ADC is 2^N. The noise-free (or flicker-free) resolution can be calculated using the equation:

Noise-Free Code Resolution $= \log_2(2^N/\text{Peak-to-Peak Noise})$

The specification is generally associated with high-resolution sigma-delta measurement ADCs, but is applicable to all ADCs.

The ratio of the FS range to the *rms* input noise is sometimes used to calculate resolution. In this case, the term *effective resolution* is used. Note that effective resolution is larger than noise-free code resolution by $\log_2(6.6)$, or approximately 2.7 bits.

Effective Resolution = $\log_2(2^N/\text{RMS Input Noise})$

Noise, Peak and RMS. Internally generated random noise is not a major factor in DACs, except at extreme resolutions and dynamic ranges. Random noise is characterized by rms specifications for a given bandwidth, or as a spectral density (current or voltage per root hertz); if the distribution is Gaussian, the probability of peak-to-peak values exceeding $6.6 \times$ the rms value is less than 0.1%.

Of much greater importance in DACs is interference, in the form of high amplitude, low energy (hence low rms) spikes appearing at a DAC's output, caused by coupling of digital signals in a surprising variety of ways; they include coupling via stray capacitance, via power supplies, via inadequate ground systems, via feedthrough, and by glitch-generation (see *Glitch*). Their presence underscores the necessity for maximum application of the designer's art, including layout, shielding, guarding, grounding, bypassing, and deglitching.

Noise in ADCs in effect narrows the region between transitions. Sources of noise include the input sample-and-hold, resistor noise, "KT/C" noise, the reference, the analog signal itself, and pickup in infinite variety.

Noise Power Ratio (NPR). In this measurement, wideband Gaussian noise (bandwidth $< f_s/2$) is applied to an ADC through a narrowband notch filter. The notch filter removes all noise within its bandwidth. The output of the ADC is examined with a large FFT. The ratio of the rms noise level to the rms noise level inside the notch (due to quantization noise, thermal noise, and intermodulation distortion) is defined as the *noise power ratio (NPR)*. The rms noise level at the input to the ADC is generally adjusted to give the best NPR value.

No Missing Codes Resolution. (See *Resolution, No Missing Codes.*)

Nonlinearity (or "gain nonlinearity") The deviation from a straight line on the plot of output versus input. The magnitude of linearity error is the maximum deviation from a "best straight line," with the output swinging through its full-scale range. Nonlinearity is usually specified in percent of full-scale output range.

Normal Mode. For an amplifier used in instrumentation, the *normal-mode* signal is the actual difference signal being measured. This signal often has noise associated with it. Signal conditioning systems and digital panel instruments usually contain input filtering to remove high frequency and line frequency noise components. *Normal-mode rejection* (NMR), is a logarithmic measure of the attenuation of normal-mode noise components at specified frequencies in dB.

Offset, Bipolar. For the great majority of bipolar converters (e.g., ± 10 V output), negative currents are not actually generated to correspond to negative numbers; instead, a unipolar DAC is used, and the output is offset by half full-scale (1 MSB). For best results, this offset voltage or current is derived from the same reference supply that determines the gain of the converter.

Because of nonlinearity, a device with perfectly calibrated end points may have offset error at analog zero.

Offset Step. (See *Pedestal.*)

Output Propagation Delay. For an ADC having a single-ended sampling (or ENCODE) clock input, the delay between the 50% point of the sampling clock and the time when all output data bits are within valid logic levels. For an ADC having differential sampling clock inputs, the delay is measured with respect to the zero crossing of the differential sampling clock signal.

Output Voltage Tolerance. For a reference, the maximum deviation from the normal output voltage at 25°C and specified input voltage, as measured by a device traceable to a recognized fundamental voltage standard.

Overload. An input voltage exceeding the ADC's full-scale input range producing an overload condition.

Overvoltage Recovery Time. Overvoltage recovery time is defined as the amount of time required for an ADC to achieve a specified accuracy after an overvoltage (usually 50% greater than full-scale range), measured from the time the overvoltage signal reenters the converter's range. The ADC should act as an ideal limiter for out-of-range signals, producing a positive or negative full-scale code during the overvoltage condition. Some ADCs provide over- and underrange flags to allow gain-adjustment circuits to be activated.

Overrange, Overvoltage. An input signal that exceeds the full-scale input range of an ADC, but is less than an overload.

Pedestal, or *Sample-to-Hold Offset Step.* In sample/track-and-hold amplifiers, a shift in level between the last value in *sample* and the value settled-to in *hold;* in devices having fixed internal capacitors, it includes *charge transfer,* or *offset step.* However, for devices that may use external capacitors, it is often defined as the residual step error after the *charge transfer* is accounted for and/or cancelled. Since it is unpredictable in magnitude and may be a function of the signal, it is also known as *offset nonlinearity.*

Pipelining. A pipelined converter is a multistage converter capable of accepting a new signal before it has completed the conversion of one or more previous ones. A new signal arrives while others are still "in the pipeline." This is a technique used where a fast conversion rate is desired and the latency of individual conversions is relatively unimportant.

Power-Supply Rejection Ratio (PSRR). The ratio of a change in dc power supply voltage to the resulting change in the specified device error, expressed in percentage, parts per million, or fractions of 1 LSB. It may also be expressed logarithmically, in dB, PSR $= 20 \log_{10}$ (PSRR).

Quad-Slope Converter. This is an integrating analog-to-digital converter that goes through two cycles of *dual-slope* conversion, once with zero input and once with the analog input being measured. The errors determined during the first cycle are subtracted digitally from the result in the second cycle. The scheme can result in high-accuracy conversion.

Quantizing Uncertainty (or " *Quantization Error"*). The analog continuum is partitioned into 2^N discrete ranges for N-bit conversion and processing. All analog values within a given quantum are represented by the same digital code, usually assigned to the nominal midrange value. There is, therefore, an inherent quantization uncertainty of $\pm\frac{1}{2}$ LSB, in addition to the actual conversion errors. In integrating ADCs, this "error" is often expressed as "± 1 count." Depending on the system context, it may be interpreted as a truncation (round-off) error or as noise.

Ratiometric. The output of an ADC is a digital number proportional to the *ratio* of (some measure of) the input to a reference voltage. Most requirements for conversions call for an absolute measurement, i.e., against a fixed reference; but this presumes that the signal applied to the converter is either reference-independent or in some way derived from another fixed reference. However, real references are not truly fixed; the references for both the converter and the signal source vary with time, temperature, loading, etc. Therefore, if the converter is used with signal sources that also rely on references (for example, strain-gage bridges, RTDs, thermistors), it makes sense to replace this multiplicity of references by a single system reference. In this case, reference-caused errors will tend to cancel out. This can be done by using the converter's internal reference (if it has one) as the system reference. Another way is to use a separate external system reference, which also becomes the reference for a *ratiometric* converter. For instance, if a bridge is excited with the same voltage used for the ADC reference, ratiometric operation is achieved, and the ADC output code is not a function of the reference. This is because the bridge output signal is proportional to the same voltage which defines the ADC input range.

Resolution. An N-bit binary converter has N digital data inputs (DAC) or N digital data outputs (ADC). A converter that satisfies this criterion is said to have a *resolution* of N bits.

Resolution, No Missing Codes. The *no missing code resolution* of an ADC is the maximum number of bits of resolution beyond which the ADC will have missing codes. For instance, if an 18-bit ADC has a no missing code resolution of 16 bits, there will be no missing codes if only the 16 MSBs are utilized. Codes may be missed at the 17- and 18-bit level.

The smallest output change that can be resolved by a linear DAC is 2^{-N} of the full-scale span. Thus, for example, the resolution of an 8-bit DAC would be 2^{-8}, or 1/256. On the other hand, a nonlinear device, such as the AD7111 LOGDAC™, can ideally achieve a dynamic range of 89.625 dB, or 30,000:1, in 0.375 dB steps, using only 8 bits of digital resolution.

Right-Justified Data. When a 12-bit word is placed on an 8-bit bus in two stages, the high byte contains the 4 or 8 most-significant bits. If 8, the word is said to be left justified; if 4 (plus filled-in leading sign bits), the word is said to be right justified.

Sample-to-Hold Offset.(See Pedestal.)

Sampling ADC. A sampling ADC includes a sample-and-hold function that acquires the input value at a given instant and holds it throughout the conversion time (or until the converter is ready for the next sample point). Flash ADCs and sigma-delta ADCs are inherently sampling devices.

Sampling Clock. (See Encode Command).

Sampling Frequency. The rate at which an ADC converts an analog input signal into digital outputs, not to be confused with *conversion time*.

Serial Output. A bit-serial output consists of a series of bits clocked out on a single line. There must be some means of identifying the beginning and ends of words; this can be accomplished via an additional clock line, by using synchronized clocks, and/or by providing a consistent identifying signature for the beginning of a word. Byte-serial consists of a series of bytes transmitted in sequence on a bus. (See *Byte*.)

Settling Time—ADC. The time required, following an analog input step change (usually full-scale), for the digital output of the ADC to reach and remain within a given fraction (usually ± ½LSB).

Settling Time—DAC. The time required, following a prescribed data change, for the output of a DAC to reach and remain within an error band (usually ±½ LSB) of the final value. Typical prescribed changes are full-scale, 1 MSB, and 1 LSB at a major carry. Settling time of current-output DACs is quite fast. The major share of settling time of a voltage-output DAC is usually contributed by the settling time of the output op-amp. DAC settling time can also be defined with respect to the output. Output settling time is the time measured from the point the output signal leaves an error band referenced to the initial output value until the time the signal enters and remains within the error band referenced to the final output value.

Signal-to-Noise-and-Distortion Ratio (SINAD). The ratio of the rms signal amplitude (set 1 dB below full-scale to prevent overdrive) to the rms value of the sum of all other spectral components, including harmonics but excluding dc.

Signal-to-Noise Ratio (without Harmonics). The ratio of the rms signal amplitude (set at 1 dB below full-scale to prevent overdrive) to the rms value of the sum of all other spectral components, excluding the first five harmonics and dc. Technically, all harmonics should be excluded, but in practice, only the first five are generally significant.

Single-Slope Conversion. In the single-slope converter, a reference voltage is integrated until the output of the integrator is equal to the input voltage. The time period required for the integrator to go from zero to the level of the input is proportional to the magnitude of the input voltage and is measured by an internal clock. Measurement accuracy is sensitive to clock speed and integrating capacitance, as well as the reference accuracy.

Slew(ing) Rate. A limitation in the rate of change of output voltage, usually imposed by some basic circuit consideration, such as limited current to charge a capacitor. The output slewing speed of a voltage-output DAC is usually limited by the slew rate of the amplifier used at its output.

Spurious-Free Dynamic Range (SFDR). The ratio of the rms signal amplitude to the rms value of the peak spurious spectral component. The peak spurious component may or may not be a harmonic. May be reported in dBc (i.e., degrades as signal level is lowered) or dBFS (related back to converter full-scale).

Stability. In a well-designed, intelligently applied converter, *dynamic stability* is not an important question. The term stability usually applies to the insensitivity of the converter's characteristics to time, temperature, etc. All measurements of stability are difficult and time consuming, but stability versus temperature is sufficiently critical in most applications to warrant universal inclusion in tables of specifications (see *Temperature Coefficient*).

Staircase. A voltage or current, increasing in equal increments as a function of time and having the appearance of a staircase (in a time plot); it is generated by applying a pulse train to a counter, and the output of the counter to the input of a DAC.

Subranging ADCs. In this type of converter, a fast converter produces the most-significant portion of the output word. This portion is stored in a holding register and also converted back to analog with a fast, high-accuracy DAC. The analog result is subtracted from the input, and the resulting residue is amplified, converted to digital at high speed, and combined with the results of the earlier conversion to form the output word. In *digitally corrected subranging* (DCS) ADCs, the two conversions are combined in a manner that corrects for the error of the LSB of the most significant bits. For example, using 8-bit and 5-bit conversion, plus this technique and a great deal of video-speed converter expertise, a full-accuracy high speed 12-bit ADC can be built. Many pipelined subranging ADCs use more than two stages with error correction between each stage.

Successive Approximation. Successive approximation is a method of conversion by comparing an unknown against a group of weighted references. The operation of a successive-approximation ADC is generally similar to the orderly weighing of an unknown quantity on a precision balance, using a set of weights, such as 1 gram ½ gram, ¼ gram, etc. The weights are tried in order, starting with the largest. Any weight that tips the scale is removed. At the end of the process, the sum of the weights remaining on the scale will be within 1 LSB of the actual weight (±½ LSB, if the scale is properly biased—see *Zero*). The successive approximation ADC is often called a SAR ADC, because the logic block that controls the conversion process is known as a successive approximation register (SAR).

Switching Time. In a DAC, the switching time is the time taken for an analog switch to change to a new state from the previous one. It includes propagation delay time, and rise time from 10% to 90%, but does not include settling time.

Temperature Coefficient. In general, temperature instabilities are expressed as %/°C, ppm/°C, fractions of 1 LSB per degree C, or as a change in a parameter over a specified temperature range. Measurements are usually made at room temperature (25°C) and at the extremes of the specified range, and the temperature coefficient (tempco, TC) is defined as the change in the parameter, divided by the corresponding temperature change. Parameters of interest include gain, linearity, offset (bipolar), and zero.

    a.   *Gain Tempco:* Two factors principally affect converter gain stability with temperature. In fixed-reference converters, the reference voltage will vary with temperature. The reference circuitry and switches (and comparator in ADCs) will also contribute to the overall gain TC.

    b.   *Linearity Tempco:* Sensitivity of linearity (integral and/or differential linearity) to temperature, in % FSR/°C or ppm FSR/°C, over the specified range. Monotonic behavior in DACs is achieved if the differential nonlinearity is less than 1 LSB at any temperature in the range of interest. The *differential nonlinearity temperature coefficient* may be expressed as a ratio, as a maximum change over a temperature range, and/or implied by a statement that the device is monotonic over the specified temperature range. To avoid missing codes in noiseless ADCs, it is sufficient that the differential nonlinearity error be less than −1 LSB at any temperature in the range of interest. The differential nonlinearity temperature coefficient is often implied by the statement that there are no missed codes when operating within a specified temperature range. In DACs, the differential nonlinearity TC is often implied by the statement that the DAC is monotonic over a specified temperature range.

c. *Zero TC (unipolar converters):* The temperature stability of a unipolar fixed-reference DAC, measured in % FSR/°C or ppm FSR/°C, is principally affected by current leakage (current-output DAC), and offset voltage and bias current of the output op amp (voltage-output DAC). The zero stability of an ADC is dependent on the zero stability of the DAC or integrator and/or the input buffer and the comparator. It is typically expressed in µV/°C or in percent or ppm of full-scale range (FSR) per degree C.

d. *Offset Tempco:* The temperature coefficient of the all-DAC-switches-off (minus full-scale) point of a bipolar converter (in % FSR/°C or ppm FSR/°C) depends on three major factors—the tempco of the reference source, the voltage zero-stability of the output amplifier, and the tracking capability of the bipolar-offset resistors and the gain resistors. In an ADC, the corresponding tempco of the negative full-scale point depends on similar quantities—the tempco of the reference source, the voltage stability of the input buffer and the sample-and-hold, and the tracking capabilities of the bipolar offset resistors and the gain-setting resistors.

Thermal Tail. The slow drift of an amplifier having a thermally induced offset due to self-heating as it settles to a final electrical equilibrium value corresponding to internal thermal equilibrium.

Total Unadjusted Error. A comprehensive specification on some devices which includes full-scale error, relative-accuracy and zero-code errors, under a specified set of conditions.

Transient Response. (See *Settling Time.*)

Two-Tone SFDR. The ratio of the rms value of either input tone to the rms value of the peak spurious component. The peak spurious component may or may not be an intermodulation distortion (IMD) product. May be reported in dBc (i.e., degrades as signal level is lowered) or in dBFS (always related back to converter full-scale).

Worst Other Spur. The ratio of the rms signal amplitude to the rms value of the worst spurious component (excluding the second and third harmonic) reported in dBc.

General References on Data Conversion and Related Topics:

1. Alfred K. Susskind, **Notes on Analog-Digital Conversion Techniques**, John Wiley, 1957.

2. David F. Hoeschele, Jr., **Analog-to-Digital/Digital-to-Analog Conversion Techniques**, John Wiley and Sons, 1968.

3. K. W. Cattermole, **Principles of Pulse Code Modulation**, American Elsevier Publishing Company, Inc., 1969, New York NY, ISBN 444-19747-8.

4. Hermann Schmid, **Electronic Analog/Digital Conversions**, Van Nostrand Reinhold Co., 1970.

5. Dan Sheingold, **Analog-Digital Conversion Handbook**, First Edition, Analog Devices, 1972.

6. Donald B. Bruck, **Data Conversion Handbook**, Hybrid Systems Corporation, 1974.

7. Eugene R. Hnatek, **A User's Handbook of D/A and A/D Converters**, John Wiley, New York, 1976, ISBN 0-471-40109-9.

8. Nuggehally S. Jayant, **Waveform Quantizing and Coding**, John Wiley-IEEE Press, 1976, ISBN 0-87942-074-X.

9. Dan Sheingold, **Analog-Digital Conversion Notes**, Analog Devices, 1977.

10. C. F. Kurth, editor, **IEEE Transactions on Circuits and Systems Special Issue on Analog/Digital Conversion**, CAS-25, No. 7, July 1978.

11. Daniel J. Dooley, **Data Conversion Integrated Circuits**, John Wiley-IEEE Press, 1980, ISBN 0-471-08155-8.

12. Bernard M. Gordon, **The Analogic Data-Conversion Systems Digest**, Fourth Edition, Analogic Corporation, 1981.

13. Eugene L. Zuch, **Data Acquisition and Conversion Handbook**, Datel-Intersil, 1982.

14. Frank F. E. Owen, **PCM and Digital Transmission Systems**, McGraw-Hill, 1982, ISBN 0-07-047954-2.

15. Dan Sheingold, **Analog-Digital Conversion Handbook**, Analog Devices/Prentice-Hall, 1986, ISBN 0-13-032848-0.

16. Matthew Mahoney, **DSP-Based Testing of Analog and Mixed-Signal Circuits**, IEEE Computer Society Press, 1987, ISBN 0-8186-0785-8.

17. Michael J. Demler, **High-Speed Analog-to-Digital Conversion**, Academic Press, Inc., 1991, ISBN 0-12-209048-9.

18. J. C. Candy and Gabor C. Temes, **Oversampling Delta-Sigma Data Converters**, IEEE Press, ISBN 0-87942-258-8, 1992.

19. David F. Hoeschele, Jr., **Analog-to-Digital and Digital-to-Analog Conversion Techniques**, Second Edition, John Wiley and Sons, 1994, ISBN-0-471-57147-4.

20. Rudy van de Plassche, **Integrated Analog-to-Digital and Digital-to-Analog Converters**, Kluwer Academic Publishers, 1994, ISBN 0-7923-9436-4.

21. David A. Johns and Ken Martin, **Analog Integrated Circuit Design**, John Wiley, 1997, ISBN 0-471-14448-7.

22. Mikael Gustavsson, J. Jacob Wikner, and Nianxiong Nick Tan, **CMOS Data Converters for Communications**, Kluwer Academic Publishers, 2000, ISBN 0-7923-7780-X.

23. R. Jacob Baker, **CMOS Circuit Design Volumes I and II**, John Wiley-IEEE Computer Society, 2002, ISBN 0-4712-7256-6.

24. Rudy van de Plassche, **CMOS Integrated Analog-to-Digital and Digital-to-Analog Converters, Second Edition**, Kluwer Academic Publishers, 2003, ISBN 1-4020-7500-6.

Analog Devices' Seminar Series:

25. Walt Kester, **Practical Analog Design Techniques**, Analog Devices, 1995, ISBN 0-916550-16-8, available for download at www.analog.com.

26. Walt Kester, **High Speed Design Techniques**, Analog Devices, 1996, ISBN 0-916550-17-6, available for download at www.analog.com.

27. Walt Kester, **Practical Design Techniques for Power and Thermal Management**, Analog Devices, 1998, ISBN 0-916550-19-2, available for download at www.analog.com.

28. Walt Kester, **Practical Design Techniques for Sensor Signal Conditioning**, Analog Devices, 1999, ISBN 0-916550-20-6, available for download at www.analog.com.

29. Walt Kester, **Mixed-Signal and DSP Design Techniques**, Analog Devices, 2000, ISBN 0-916550-22-2, available for download at www.analog.com.

30. Walt Kester, **Mixed-Signal and DSP Design Techniques**, Analog Devices and Newnes (An Imprint of Elsevier Science), ISBN 0-75067-611-6, 2003.

31. Walter G. Jung, **Op Amp Applications**, Analog Devices, 2002, ISBN 0-916550-26-5.

CHAPTER 3
Data Converter Architectures

Data Converter Architectures

DAC Architectures
James Bryant, Walt Kester

Introduction

Those unfamiliar with DACs regard them simply as devices with digital input and analog output. But the analog output depends on the presence of that analog input known as the reference, and the accuracy of the reference is almost always the limiting factor on the absolute accuracy of a DAC. We shall consider the various architectures of DACs, and the forms which the reference may take, later in this section.

Some DACs use external references (see Figure 3.1) and have a reference input terminal, while others have an output from an internal reference. The simplest DACs, of course, have neither—the reference is on the DAC chip and has no external connections.

If a DAC has an internal reference, the overall accuracy of the DAC is specified when using that reference. If such a DAC is used with a perfectly accurate external reference, its absolute accuracy may actually be worse than when it is operated with its own internal reference. This is because it is trimmed for absolute accuracy when working with its own actual reference voltage, not with the nominal value. Twenty years ago it was common for converter references to have accuracies as poor as ±5% since these references were trimmed for low temperature coefficient rather than absolute accuracy, and the inaccuracy of the reference was compensated in the gain trim of the DAC itself. Today the problem is much less severe, but it is still important to check for possible loss of absolute accuracy when using an external reference with a DAC with one that is built-in.

DACs that have reference terminals must, of course, specify their behavior and parameters. If there is a reference input, the first specification will be the reference input voltage—and of course this has two values, the absolute maximum rating, and the range of voltages over which the DAC performs correctly.

Most DACs require that their reference voltage be within quite a narrow range whose maximum value is less than or equal to the DAC's V_{DD}, but some DACs, called multiplying DACs (or "MDACs"), will work over a wide range of reference voltages that may go well outside their power

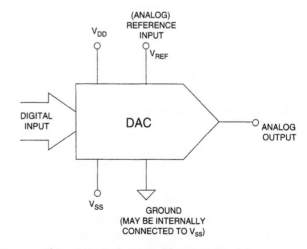

Figure 3.1: Basic DAC with External Reference

supplies. The AD7943 multiplying DAC, for example, has an absolute maximum rating on its V_{DD} terminal of +6 V but a rating of ±15 V on its reference input, and it works perfectly well with positive, negative, or ac references. (The generally-accepted definition of an MDAC is that its reference voltage range includes zero. But some authorities prefer a looser definition, "a DAC with a reference voltage range greater than 5:1." In this chapter we shall use the term "semi-multiplying DAC" for devices of this type.) MDACs that work with ac references have a "reference bandwidth" specification which defines the maximum practical frequency at the reference input.

The reference input terminal of a DAC may be buffered as shown in Figure 3.2, in which case it has input impedance (usually high) and bias current (usually low) specifications, or it may

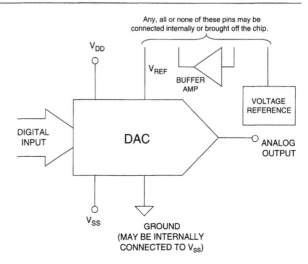

Figure 3.2: DAC with Reference and Buffer

connect directly to the DAC. In this case the input impedance specification may become more complicated since some DAC structures have an input impedance that varies substantially with the digital code applied to the DAC. In such cases the (usually simplified) structure of the DAC is shown on the data sheet, and the nominal values of resistance are given. Where the reference input impedance does not vary with code, the input impedance should be specified.

Surprisingly for such an accurate circuit, the reference input impedance of a resistive DAC network is rarely very well defined. For example, the AD7943 has a nominal input impedance of 9 kΩ, but the data sheet limits are 6 kΩ and 12 kΩ, a variation of ±33%. The reasons for this are discussed later in this book (see Chapter 4). In addition, the reference input impedance is code-dependent for the voltage-mode R-2R architecture.

Where a DAC has a reference output terminal, it will carry a defined reference voltage, with a specified accuracy. There may also be specifications of temperature coefficient and long-term stability.

The reference output (if available) may be buffered or unbuffered. If it is buffered the maximum output current will be specified. In general such a buffer will have a unidirectional output stage which sources current but does not allow current to flow into the output terminal. If the buffer does have a push-pull output stage, the output current will probably be defined as ±(SOME VALUE) mA. If the reference output is unbuffered, the output impedance may be specified, or the data sheet may simply advise the use of a high input impedance external buffer.

DAC Output Considerations

The output of a DAC may be a voltage or a current. In either case it may be important to know the output impedance. If the voltage output is buffered, the output impedance will be low. Both current outputs and unbuffered voltage outputs will be high(er) impedance and may well have a reactive component specified as well as a purely resistive one. Some DAC architectures have output structures where the output impedance is a function of the digital code on the DAC—this should be clearly noted on the data sheet.

In theory, current outputs should be connected to zero ohms at ground potential. In real life they will work with nonzero impedances and voltages. Just how much deviation they will tolerate is defined under the heading "compliance" and this specification should be heeded when terminating current-output DACs.

Basic DAC Structures

It is reasonable to consider a changeover switch (a single-pole, double-throw, SPDT switch), switching an output between a reference and ground or between equal positive and negative reference voltages, as a 1-bit DAC as shown in Figure 3.3. Such a simple device is a component of many more complex DAC structures, and is used, with oversampling, as the basic analog component in many of the sigma-delta DACs we shall discuss later. Nevertheless, it is a little too simple to require detailed discussion, and it is more rewarding to consider more complex structures.

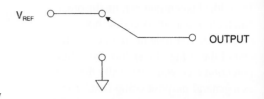

Figure 3.3: 1-Bit DAC: Changeover Switch (Single-Pole, Double Throw, SPDT)

The Kelvin Divider (String DAC)

The simplest DAC structure of all, after the changeover switch mentioned above, is the Kelvin divider or *string DAC* as shown in Figure 3.4. An N-bit version of this DAC simply consists of 2^N equal resistors in series and 2^N switches (usually CMOS), one between each node of the chain and the output. The output is taken from the appropriate tap by closing just one of the switches (there is some slight digital complexity involved in decoding to 1 of 2^N switches from N-bit data, but digital circuitry is cheap). The origins of this DAC date back to Lord Kelvin in the mid-1800s, and it was first implemented using resistors and relays, and later with vacuum tubes in the 1920s (See References 1, 2, 3).

This architecture is simple, has a voltage output (but a code-dependent output impedance) and is inherently monotonic—even if a resistor is accidentally short-circuited, output n cannot exceed output n + 1. It is linear if all the resistors are equal, but may be made deliberately nonlinear if a nonlinear DAC is required. Since only two switches operate during a transition, it is a low-glitch architecture. Also, the switching glitch is not code-dependent, making it ideal for low distortion applications. Because the glitch is constant regardless of the code transition, the frequency content of the glitch is at the DAC update rate and its harmonics—not at the harmonics of the fundamental DAC output frequency. The major drawback of the thermometer DAC is the large number of resistors and switches required for high resolution, and as a result it was not commonly used as a simple DAC architecture until the recent advent of very small IC feature sizes made it very practical for low and medium resolution DACs. Today the architecture is quite widely used in simple DACs, such as digital potentiometers and, as we shall see later, it is also used as a component in more complex high resolution DAC structures.

As we have already mentioned, the output of a DAC for an all 1s code is 1 LSB below the reference, so a string DAC intended for use as a general-purpose DAC has a resistor between the reference terminal and the first switch as shown in Figure 3.4.

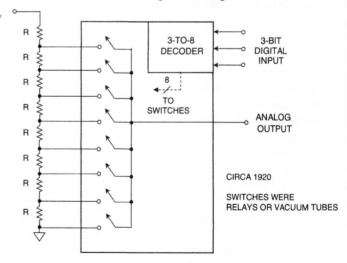

Figure 3.4: Simplest Voltage-Output Thermometer DAC: The Kelvin Divider ("String DAC")

In an ideal potentiometer, on the other hand, all 0s and all 1s codes should connect the variable tap to one or other end of the string of resistors. So a digital potentiometer, while basically the same as a general-purpose string DAC, has one fewer resistor, and neither end of the string has any other internal connection. A simple digital potentiometer is shown in Figure 3.5.

The simplest digital potentiometers are no more complex than this, and none of the potentiometer terminals may be at a potential outside the 5 V or 3 V logic supply. But others have more complex decoders with level shifters and additional high voltage supply terminals, so that while the logic control levels are low (3 V or 5 V), the potentiometer termi-

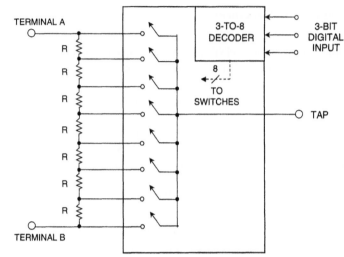

Figure 3.5: A Slight Modification to a String DAC Yields a "Digital Potentiometer"

nals have a much greater range—up to ±15 V in some cases. Digital potentiometers frequently incorporate nonvolatile logic so that their settings are retained when they are turned off.

It is evident that string DACs have a large number of resistors (2^N for an N-bit DAC as we have already seen). It is not practical to trim every resistor in a string DAC to obtain perfect DNL and INL, partly because they are too many, and partly because they are too small to trim, and mainly because it's too costly.

If required, it is still possible to trim the INL of a string DAC. The method is shown in Figure 3.6—a second string of four equal resistors is connected in parallel with the main string. These resistors are made physically large enough to laser trim. The three internal nodes of this string are connected by buffer amplifiers to the ¼, ½, and ¾ points of the main string. The trimmable string is adjusted so that these points on the main string are at the correct potentials. Typically, INL can be reduced by a factor of four by this technique, at quite a small cost in complexity. However, in many modern string DACs trimming is not required, because the resistors are well-matched, and the current drawn by the CMOS switches is negligible.

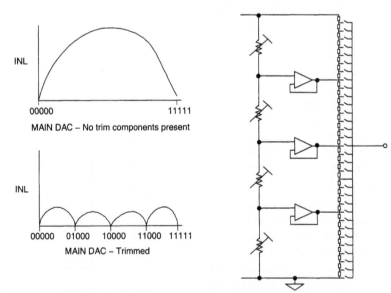

Figure 3.6: Trimming the INL of a String DAC (If Required)

Thermometer (Fully-Decoded) DACs

There is a current-output DAC analogous to a string DAC that consists of 2^N-1 switchable current sources (which may be resistors and a voltage reference or may be active current sources) connected to an output terminal, which must be at, or close to, ground. Figure 3.7 shows a thermometer DAC which uses resistors connected to a reference voltage to generate the currents.

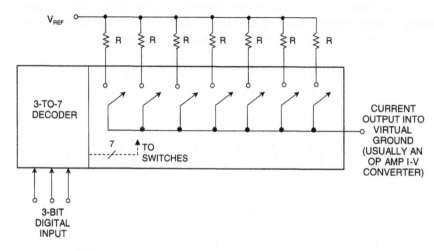

Figure 3.7: The Simplest Current-Output Thermometer (Fully-Decoded) DAC

If active current sources are used as shown in Figure 3.8, the output may have more compliance, and a resistive load used to develop an output voltage. The load resistor must be chosen so that at maximum output current the output terminal remains within its rated compliance voltage.

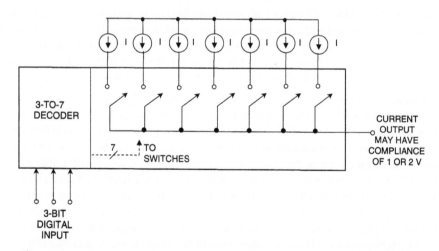

Figure 3.8: Current Sources Improve the Basic Current-Output Thermometer DAC

Once a current in a thermometer DAC is switched into the circuit by increasing the digital code, any further increases do not switch it out again. The structure is thus inherently monotonic, irrespective of inaccuracies in the currents. Again, like the Kelvin divider only the advent of high density IC processes has made this architecture practical for general-purpose medium resolution DACs, although a slightly more complex version—shown in the next diagram—is quite widely used in high speed applications. Unlike the Kelvin divider, this type of current-mode DAC does not have a unique name, although both types may be referred to as *fully-decoded* DACs or *thermometer* DACs.

A DAC where the currents are switched between two output lines—one of which is often grounded, but may, in the more general case, be used as the inverted output—is more suitable for high speed applications because switching a current between two outputs is far less disruptive, and so causes a far lower glitch than simply switching a current on and off. This architecture is shown in Figure 3.9.

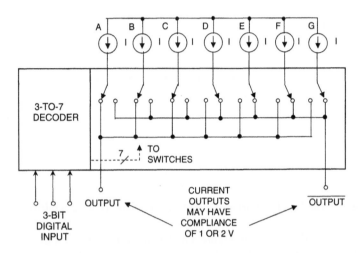

Figure 3.9: High Speed Thermometer DAC with Complementary Current Outputs

But the settling time of this DAC still varies with initial and final code, giving rise to *intersymbol interference* (ISI). This can be addressed with even more complex switching where the output current is returned to zero before going to its next value. Note that although the current in the output is returned to zero it is not "turned off"—the current is dumped when it is not being used, rather than being switched on and off. The techniques involved are too complex to discuss in detail here but can be found in Reference 4.

In the normal (linear) version of this DAC, all the currents are nominally equal. Where it is used for high speed reconstruction, its linearity can be improved by dynamically changing the order in which the currents are switched by ascending code. Instead of code 001 always turning on current A; code 010 always turning on currents A & B, code 011 always turning on currents A, B & C; etc. the order of turn-on relative to ascending code changes for each new data point. This can be done quite easily with a little extra logic in the decoder. The simplest way of achieving it is with a counter which increments with each clock cycle so that the order advances: ABCDEFG, BCDEFGA, CDEFGAB, etc., but this algorithm may give rise to spurious tones in the DAC output. A better approach is to set a new pseudo-random order on each clock cycle—this requires a little more logic, but, as we have pointed out, even complex logic is now very cheap and easily implemented on CMOS processes. There are other, even more complex, techniques that involve using the data itself to select bits and thus turn current mismatch into shaped noise. Again, they are too complex for a book of this sort. (See References 4 and 5 for a more detailed discussion.)

Binary-Weighted DACs

One of the earliest reference to an electro-mechanical binary-weighted DAC can be found in Paul M. Rainey's 1921 (filing date) patent for a PCM-based facsimile transmission system (Reference 6). This system is discussed in more detail in Chapter 1 of this book, and the 5-bit reconstruction DAC is shown in Figure 3.10.

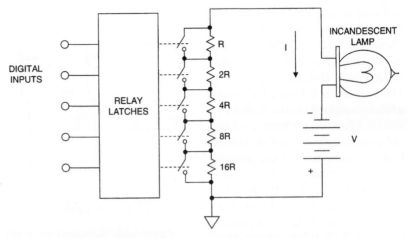

Adapted from: Paul M. Rainey, "Facsimile Telegraph System,"
U.S. Patent 1,608,527, Filed July 20, 1921, Issued November 30, 1926

Figure 3.10: Paul M. Rainey's 5-Bit Binary-Weighted DAC

The objective of Rainey's DAC was to control the intensity of the light from an incandescent lamp located in the receiver. By connecting various combinations of parallel shorting switches, 32 possible values of series resistance can be obtained ranging from 0 to 31·R, and hence 32 possible levels of light intensity. In the PCM facsimile application, the lamp output was focused on a photosensitive receiving film designed to reproduce the image digitized at the transmitter.

Another example of an early vacuum tube binary DAC can be found in John Schelleng's 1946 (filing date) patent for a 6-bit PCM system (Reference 7). Schelleng uses binary-weighted switched voltage sources whose outputs are summed together with a resistor network as shown in Figure 3.11.

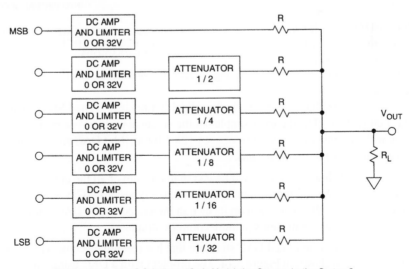

Adapted from: John C.Schelleng, "Code Modulation Communication System,"
U.S. Patent 2,453,461, Filed June 19, 1946, Issued November 9, 1948

Figure 3.11: John Schelleng's 6-Bit Binary-Weighted DAC

In a truly elegant 1953 paper on successive approximation ADCs (Reference 8), B. D. Smith of Melpar proposed two classic binary-weighted voltage-mode DAC architectures. The first, shown in Figure 3.12, uses a binary-weighted resistor network switched between a reference voltage and ground as the basis of the DAC. A transistorized version of this approach was later described in 1958 by B. K. Smith (Reference 9). B. D. Smith's second approach (shown later in Figure 3.16), is one of the first reported uses of an R-2R ladder network in a voltage-mode switching DAC (see again, Reference 8).

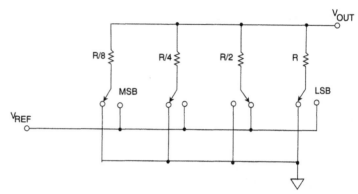

Adapted from: B. D. Smith, "Coding by Feedback Methods," Proceedings of the I. R. E., Vol. 41, August 1953, pp. 1053–1058

Figure 3.12: Voltage-Mode Binary-Weighted Resistor DAC

The voltage-mode binary-weighted resistor DAC shown in Figure 3.12 is usually the simplest textbook example of a DAC. However, this DAC is not inherently monotonic and is actually quite hard to manufacture successfully at high resolutions. In addition, the output impedance of the voltage-mode binary DAC changes with the input code.

Current-mode binary DACs are shown in Figure 3.13A (resistor-based), and Figure 3.13B (current-source based). An N-bit DAC of this type consists of N weighted current sources (which

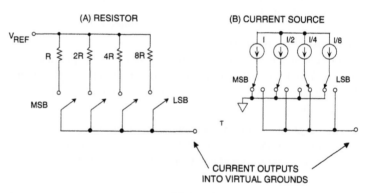

- DIFFICULT TO FABRICATE IN IC FORM DUE TO LARGE RESISTOR OR CURRENT RATIOS FOR HIGH RESOLUTIONS

Figure 3.13: Current-Mode Binary-Weighted DACs

may simply be resistors and a voltage reference) in the ratio 1:2:4:8:.....:2^{N-1}. The LSB switches the 2^{N-1} current, the MSB the 1 current, etc. The theory is simple but the practical problems of manufacturing an IC of an economical size with current or resistor ratios of even 128:1 for an 8-bit DAC are enormous, especially as they must have matched temperature coefficients.

If the MSB current is slightly low in value, it will be less than the sum of all the other bit currents, and the DAC will not be monotonic (the differential nonlinearity of most types of DAC is worst at major bit transitions). This architecture is virtually never used on its own in integrated circuit DACs, although, again, 3- or 4-bit versions have been used as components in more complex structures.

However, there is another binary-weighted DAC structure that has recently become widely used. This uses binary-weighted capacitors as shown in Figure 3.14. The problem with a DAC using capacitors is that leakage causes it to lose its accuracy within a few milliseconds of being set. This may make capacitive DACs unsuitable for general-purpose DAC applications, but it is not a problem in successive approximation ADCs, since the conversion is complete in a few μs or less—long before leakage has any appreciable effect.

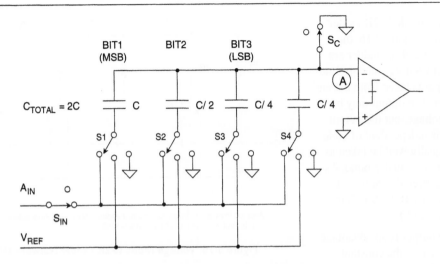

SWITCHES SHOWN IN TRACK (SAMPLE) MODE

**Figure 3.14: Capacitive Binary-Weighted
DAC in Successive Approximation ADC**

The successive approximation ADC has a very simple structure, low power, and reasonably fast conversion times. It is probably the most widely used general-purpose ADC architecture, but in the mid-1990s the subranging ADC was starting to overtake the successive approximation type in popularity because the R-2R thin-film resistor DAC in the successive approximation ADC made the chip larger and more expensive than that of a subranging ADC, even though the subranging types tend to use more power. The development of submicron CMOS processes made possible very small (and therefore cheap), and very accurate switched capacitor DACs. These enabled a new generation of successive approximation ADCs to be made small, cheap, low power and precise, and thus to regain their popularity. (See further discussion in Section 3.2 of this chapter.)

The use of capacitive charge redistribution DACs offers another advantage as well—the DAC itself behaves as a sample-and-hold circuit (SHA), so neither an external SHA nor allocation of chip area for a separate integral SHA are required.

R-2R DACs

One of the most common DAC building-block structures is the R-2R resistor ladder network shown in Figure 3.15. It uses resistors of only two different values, and their ratio is 2:1. An N-bit DAC requires 2N resistors, and they are quite easily trimmed. There are also relatively few resistors to trim.

There are two ways in which the R-2R ladder network may be used as a DAC—known respectively as the *voltage mode* and the *current mode* (they are sometimes called "normal" mode and "inverted" mode, but as there is no consensus on whether the voltage mode or the current mode is the "normal" mode for a ladder network this nomenclature can be misleading). Each mode has its advantages and disadvantages.

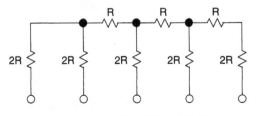

Figure 3.15: 4-Bit R-2R Ladder Network

In the voltage mode R-2R ladder DAC shown in Figure 3.16, the "rungs" or arms of the ladder are switched between V_{REF} and ground, and the output is taken from the end of the ladder. The output may be taken as a voltage, but the output impedance is independent of code, so it may equally well be taken as a current into a virtual ground. As mentioned earlier, this architecture was proposed by B. D. Smith in 1953 (Reference 8).

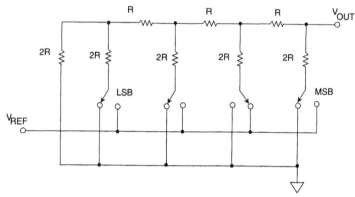

Adapted from: B. D. Smith, "Coding by Feedback Methods," Proceedings of the I. R. E., Vol. 41, August 1953, pp. 1053–1058

Figure 3.16: Voltage-Mode R-2R Ladder Network DAC

The voltage output is an advantage of this mode, as is the constant output impedance, which eases the stabilization of any amplifier connected to the output node. Additionally, the switches switch the arms of the ladder between a low impedance V_{REF} connection and ground, which is also, of course, low impedance, so capacitive glitch currents tend not to flow in the load. On the other hand, the switches must operate over a wide voltage range (V_{REF} to ground). This is difficult from a design and manufacturing viewpoint, and the reference input impedance varies widely with code, so that the reference input must be driven from a very low impedance. In addition, the gain of the DAC cannot be adjusted by means of a resistor in series with the V_{REF} terminal.

In the current-mode R-2R ladder DAC shown in Figure 3.17, the gain of the DAC may be adjusted with a series resistor at the V_{REF} terminal, since in the current mode, the end of the ladder, with its code-independent impedance, is used as the V_{REF} terminal; and the ends of the arms are switched between ground (or, sometimes, an "inverted output" at ground potential) and an output line that must be held at ground potential. The normal connection of a current-mode ladder network output is to an op amp configured as current-to-voltage (I/V) converter, but stabilization of this op amp is complicated by the DAC output impedance variation with digital code.

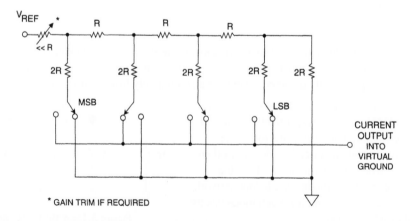

* GAIN TRIM IF REQUIRED

Figure 3.17: Current-Mode R-2R Ladder Network DAC

Current-mode operation has a larger switching glitch than voltage mode since the switches connect directly to the output line(s). However, since the switches of a current-mode ladder network are always at ground potential, their design is less demanding and, in particular, their voltage rating does not affect the reference voltage rating. If switches capable of carrying current in either direction (such as CMOS devices) are used, the reference voltage may have either polarity, or may even be ac. Such a structure is one of the most common types used as a multiplying DAC (MDAC). These will be discussed later in this section.

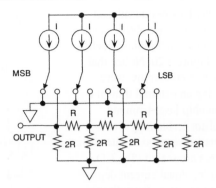

Adapted from: Bernard M. Gordon and Robert P. Talambiras, "Signal Conversion Apparatus," U.S. Patent 3,108,266, filed July 22, 1955, issued October 22, 1963

Figure 3.18: Equal Current Sources Switched into an R-2R Ladder Network

Since the switches are always at, or very close to, ground potential, the maximum reference voltage may greatly exceed the logic voltage, provided the switches are make-before-break—which they are in this type of DAC. It is not uncommon for a CMOS MDAC to accept a ±30 V reference (or even a 60 V peak-to-peak ac reference) while working from a single 5 V supply.

Another popular form of R-2R DAC switches equal currents into the R-2R network as shown in Figure 3.18. This architecture was first implemented by Bernard M. Gordon at EPSCO (now Analogic, Inc.) in a vacuum tube 11-bit, 50 kSPS successive approximation ADC. Gordon's 1955 patent application (Reference 10) describes the ADC, which was the first commercial offering of a complete converter (see Section 3.2 of this chapter and Chapter 1 for more details). In this architecture the output impedance of the DAC is equal to R, and this structure is often used in high speed video DACs. A distinct advantage is that only a 2:1 resistor ratio is required regardless of the resolution. In some applications, however, the relatively low output impedance can be a disadvantage.

Figure 3.19 shows a DAC using binary-weighted currents switched into a load. The output impedance is high, and this architecture generally has a volt or so of output compliance. The main problem with all of the binary-weighted DACs discussed thus far is that high resolutions require large resistor ratios, making manufacture very difficult.

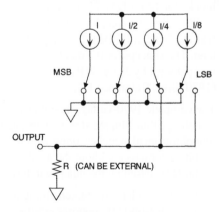

Figure 3.19: Binary-Weighted Current Sources Switched into a Load

In 1970 Analog Devices introduced the AD550 "μDAC" monolithic quad (4-bit) current switch building block IC shown in Figure 3.20. Notice that the binary-weighted currents were generated using an external thin-film network—on-chip laser-trimmed thin-film resistor technology was not developed until several years later. The transistor areas are scaled (8:4:2:1), thereby ensuring equal current densities in all the transistors for optimum V_{BE} matching.

An alternative method of developing the binary-weighted currents in the quad switch is shown in Figure 3.21, where an R-2R ladder network connected to the transistor emitters accomplishes the binary current division.

Figure 3.22 shows how three AD550 quad switches with 16:1 inter-stage attenuators are connected to form a 12-bit current-output DAC. Note that the maximum required resistor ratio of 16:1 is manageable. This monolithic "quad switch" (AD550 μDAC) along with a thin-film resistor network (AD850), voltage reference, and an op amp formed the popular building blocks for 12-bit DACs in the early 1970s before the complete function was available in IC form several years later. The concept for the quad switch was patented by James J. Pastoriza (1970 filing, Reference 11).

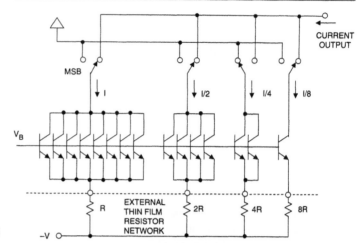

Figure 3.20: Binary-Weighted 4-Bit DAC, the AD550 "μDAC" Quad Switch

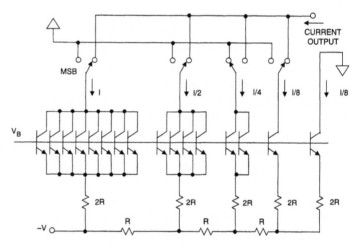

Figure 3.21: Binary-Weighted 4-Bit DAC: R/2R Ladder Network Current Setting Resistors

The complete 1970-vintage 12-bit DAC solution, shown in Figure 3.23, consists of three monolithic quad switches, a thin-film resistor network, an op amp, and a voltage reference. The matching provided by the monolithic quad switches along with the accuracy and tracking of the external thin-film network provided 12-bit performance without the need for additional trimming. An interesting and complete analysis of this 12-bit DAC based on the quad switches can be found in Reference 12.

One of the problems in implementing a completely monolithic 12-bit DAC using the quad switch approach is that each 4-bit DAC requires emitter areas scaled 8:4:2:1. This requires a total of 15 unit emitter areas, and consumes a fairly large chip area. A few years after the introduction of the quad switch building block, Paul Brokaw of Analog Devices invented a technique in which only the first two current sources have an emitter scaling of 2:1. Subsequent current sources have the same unit emitter area, but operate at different

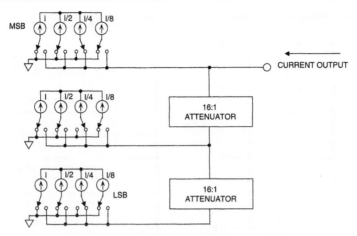

**Figure 3.22: 12-Bit Current-Output DAC Using
Cascaded Binary "Quad Switches"**

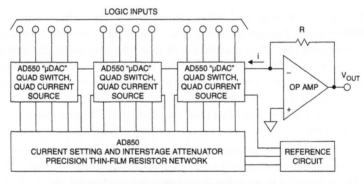

James J. Pastoriza, "Solid State Digital-to-Analog Converter,"
U.S. Patent 3,747,088, filed December 30, 1970, issued July 17, 1973

**Figure 3.23: A 1970 Vintage 12-Bit DAC Using Quad Current Switches,
Thin-Film Resistor Network, Op Amp, and Zener Diode Voltage Reference**

current densities—while still maintaining stable currents over temperature. Paul Brokaw's classic patent (filed in 1975) describes this technique in detail, and this particular patent is probably the most referenced and cited patent relating to data converters (Reference 13).

Segmented DACs

So far we have considered basic DAC architectures. When we are required to design a DAC with a specific performance, it may well be that no single architecture is ideal. In such cases, two or more DACs may be combined in a single higher resolution DAC to give the required performance. These DACs may be of the same type or of different types and need not each have the same resolution.

In principle, one DAC handles the MSBs, another handles the LSBs, and their outputs are added in some way. The process is known as "segmentation," and these more complex structures are called "segmented DACs." There are many different types of segmented DACs and some, but by no means all, of them will be illustrated in the next few diagrams.

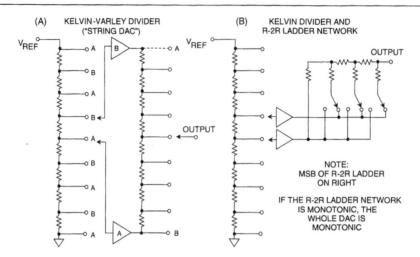

Figure 3.24: Segmented Voltage-Output DACs

Figure 3.24 shows two varieties of segmented voltage-output DACs. The architecture in Figure 3.24A is sometimes called a Kelvin-Varley Divider, or "string DAC." Since there are buffers between the first and second stages, the second string DAC does not load the first, and the resistors in this string do not need to have the same value as the resistors in the other one. All the resistors in each string, however, do need to be equal to each other or the DAC will not be linear. The examples shown have 3-bit first and second stages but for the sake of generality, let us refer to the first (MSB) stage resolution as M-bits and the second (LSB) as K-bits for a total of $N = M + K$ bits. The MSB DAC has a string of 2^M equal resistors, and a string of 2^K equal resistors in the LSB DAC.

Buffer amplifiers have offset, of course, and this can cause nonmonotonicity in a buffered segmented string DAC. In the simpler configuration of a buffered Kelvin-Varley divider buffer (Figure 3.24A), buffer A is always "below" (at a lower potential than) buffer B, and the extra tap labeled "A" on the LSB string DAC is not necessary. The data decoding is just two priority encoders. In this configuration, however, buffer offset can cause nonmonotonicity.

If the decoding of the MSB string DAC is made more complex, so that buffer A can only be connected to the taps labeled "A" in the MSB string DAC, and buffer B to the taps labeled "B," it is not possible for buffer offsets to cause nonmonotonicity. Of course, the LSB string DAC decoding must change direction each time one buffer "leapfrogs" the other, and taps A and B on the LSB string DAC are alternately not used—but this involves a fairly trivial increase in logic complexity and is justified by the increased performance.

Rather than using a second string of resistors, a binary DAC can be used to generate the three LSBs as shown in Figure 3.24B. It is quite hard to manufacture very high resolution R-2R ladder networks—to be more accurate, it is hard to trim them to monotonicity. So it is quite common to make high resolution DACs with a ladder network for the LSBs, and some other structure for two to five of the MSBs. This voltage-output DAC (Figure 3.24B) consists of a 3-bit string DAC followed by a 3-bit buffered voltage-mode ladder network.

An unbuffered version of the segmented string DAC is shown in Figure 3.25. This version is more clever in concept (and, of course, can be manufactured on CMOS processes which make resistors and switches but not amplifiers, so it may be cheaper as well). It is intrinsically monotonic. Here, the resistors in the two strings must be equal, except that the top resistor in the MSB string must be smaller—$1/2^K$ of the value of the others—and the LSB string has $2^K - 1$ resistors rather than 2^K. Because there are no buffers, the LSB

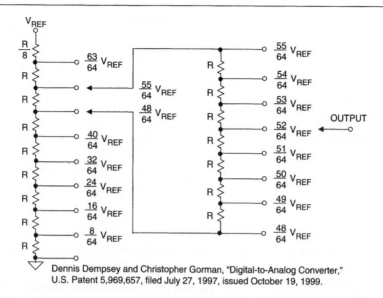

Dennis Dempsey and Christopher Gorman, "Digital-to-Analog Converter,"
U.S. Patent 5,969,657, filed July 27, 1997, issued October 19, 1999.

Figure 3.25: Segmented Unbuffered String DACs Use Patented Architecture

string appears in parallel with the resistor in the MSB string that it is switched across and loads it. This drops the voltage across that MSB resistor by 1 LSB of the LSB DAC—which is exactly what is required. The output impedance of this DAC, being unbuffered, varies with changing digital code.

In order to understand this clever concept better, the actual voltages at each of the taps has been worked out and labeled for the 6-bit segmented DAC composed of two 3-bit string DACs shown in Figure 3.25. The reader is urged to go through this simple analysis with the second string DAC connected across any other resistor in the first string DAC and verify the numbers. A detailed mathematical analysis of the unbuffered segmented string DAC can be found in the relevant patent filed by Dennis Dempsey and Christopher Gorman of Analog Devices in 1997 (Reference 14).

Very high speed DACs for video, communications, and other HF reconstruction applications are often built with arrays of fully decoded current sources. The two or three LSBs may use binary-weighted current sources. It is extremely important that such DACs have low distortion at high frequency, and there are several important issues to be considered in their design.

First of all, currents are never turned on and off—they are steered to one place or another. Turning a current off at high speed frequently involves inductive spikes and, in general, because of capacitance charging, it takes longer than current steering.

Secondly, it is important that the voltage change on the chip required to switch the current should be kept as small as possible. A voltage change causes more charge to flow in stray capacitances and a larger charge-coupled glitch.

Finally, the decoding must be done before the new data is applied to the DAC so that all the data is ready and can be applied simultaneously to all the switches in the DAC. This is generally implemented by using separate parallel latches for the individual switches in a fully decoded array. If all switches were to change state instantaneously and simultaneously there would be no skew glitch—by very careful design of propagation delays around the chip and time constants of switch resistance and stray capacitance the update synchronization can be made very good, and hence the glitch-related distortion is very small.

Two examples of segmented current-output DAC structures are shown in Figure 3.26. Figure 3.26A shows a resistor-based approach for the 7-bit DAC where the 3 MSBs are fully decoded, and the 4 LSBs are derived from an R-2R network. Figure 3.26B shows a similar implementation using current sources. The current source implementation is by far the most popular for today's high speed reconstruction DACs.

It is also often desirable to utilize more than one fully-decoded thermometer section to make up the total DAC. Figure 3.27 shows a 6-bit DAC constructed from two fully-decoded 3-bit DACs. As previously discussed, these current switches must be driven simultaneously from parallel latches in order to minimize the output glitch.

The AD9775 14-bit, 160 MSPS (input)/400-MSPS (output) TxDAC uses three sections of segmentation as shown in Figure 3.28. Other members of the AD977x family and the AD985x-family also use this same basic core.

The first 5 bits (MSBs) are fully decoded and drive 31 equally weighted current switches, each supplying 512 LSBs of current. The next 4 bits are decoded into 15 lines which drive 15 current switches, each supplying 32 LSBs of current. The 5 LSBs are latched and drive a traditional binary-weighted DAC which supplies 1 LSB per output level. A total of 51 current switches and latches are required to implement this ultra low glitch architecture.

The basic current switching cell in the TxDAC family is made up of

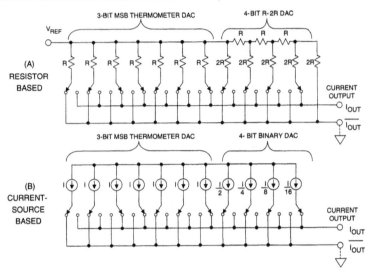

Figure 3.26: Segmented Current-Output DACs:
(A) Resistor-Based, (B) Current-source based

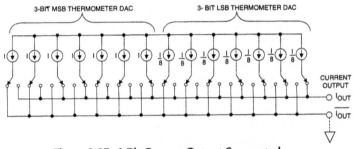

Figure 3.27: 6-Bit Current-Output Segmented DAC Based on Two 3-Bit Thermometer DACs

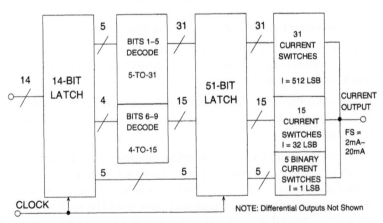

Figure 3.28: AD9775 TxDAC 14-Bit CMOS DAC Core

a differential PMOS transistor pair as shown in Figure 3.29. The differential pairs are driven with low level logic to minimize switching transients and time skew. The DAC outputs are symmetrical differential currents, which help to minimize even-order distortion products (especially when driving a differential output such as a transformer or an op amp differential current-to-voltage converter).

The overall architecture of the AD977x TxDAC family and the AD985x-DDS family is an excellent trade-off between power/performance, and allows the entire DAC function to be implemented on a standard CMOS process with no thin-film resistors. Single-supply operation on 3.3 V or 5 V make the devices extremely attractive for portable and low power applications.

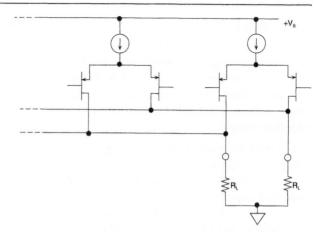

Figure 3.29: PMOS Transistor Current Switches

Oversampling Interpolating DACs

The basic concept of an oversampling/interpolating DAC is shown in Figure 3.30. The N-bits of input data are received at a rate of f_s. The digital interpolation filter is clocked at an oversampling frequency of Kf_s, and inserts the extra data points. The effects on the output frequency spectrum are shown in Figure 3.30. In the Nyquist case (A), the requirements on the analog anti-imaging filter can be quite severe. By oversampling and interpolating, the requirements on the filter are greatly relaxed as shown in (B). Also, since the quantization noise is spread over a wider region with respect to the original signal bandwidth, an improvement in the signal-to-noise ratio is also achieved. By doubling the original sampling rate (K = 2), an improvement of 3 dB is obtained, and by making K = 4, an improvement of 6 dB is obtained. Early CD players took advantage of this, and generally carried the arithmetic in the digital filter to more than N-bits. Today, most DACs in CD players are sigma-delta types.

One of the earliest publications on the oversampling/interpolating DAC concept was by Ritchie, Candy, and Ninke in 1974 (Reference 16) and followed by a 1981 patent (filing date) by Mussman and Korte (Reference 17).

Interpolation filters are not only used in very high speed DACs—they are also found in the last type of DAC we shall discuss later in Section 3.3 of this chapter—the Σ-Δ DAC. Where high resolutions are required and the output bandwidth is less than a few hundred kHz, then Σ-Δ technology is well suited for the application.

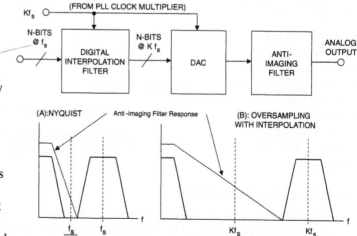

Figure 3.30: Oversampling Interpolating DAC

Multiplying DACs

In many DACs, the voltage reference is built-in—sometimes it may be varied a little as a gain adjustment, sometimes it is inaccessible. Other DACs may require an external reference voltage source, but can accept only a narrow range of reference voltages.

In all DACs, the output is the product of the reference voltage and the digital code, so in that sense, all DACs are multiplying DACs. But some DACs use an external reference voltage which may be varied over a wide range. These are "Multiplying DACs" or MDACs where the analog output is the product of the analog input and the digital code as shown in Figure 3.31. They are extremely useful in many different applications. A strict definition of an MDAC is that it will continue to work correctly as its reference is reduced to zero, but sometimes the term is used less stringently for DACs which work with a reference range of 10:1 or even 6:1—a better name for devices of this type might be "semi-multiplying" DACs.

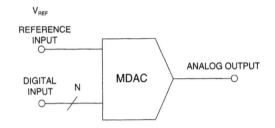

ANALOG OUTPUT = V$_{REF}$ × DIGITAL INPUT × CONSTANT

DIFFICULT TO MAINTAIN DIFFERENTIAL LINEARITY AS V$_{REF}$ APPROACHES ZERO

Figure 3.31: Multiplying DAC ("MDAC")

While some types of multiplying DACs will work only with references of one polarity (*two quadrant*) others handle bipolar (positive or negative) references, and can work with an ac signal as a reference as well.

A bipolar DAC that will work with bipolar reference voltages is known as a *four-quadrant* multiplying DAC. Some types of MDACs are so configured that they can work with reference voltages substantially greater than their supply voltage.

Current-mode ladder networks and CMOS switches permit positive, negative, and ac V$_{REF}$ as shown in Figure 3.32. While this is a simple implementation of an MDAC, several others are possible.

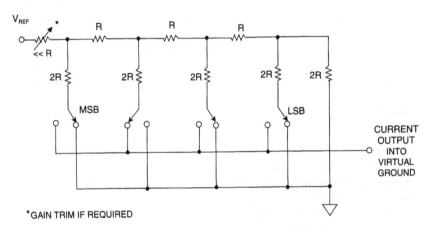

*GAIN TRIM IF REQUIRED

V$_{REF}$ CAN BE AC, ±, ALLOWING FOUR-QUADRANT OPERATION

Figure 3.32: Multiplying DAC Using Current-Mode R-2R Ladder Network and CMOS switches

Intentionally Nonlinear DACs

Thus far, we have emphasized the importance of maintaining good differential and integral linearity. However, there are situations where ADCs and DACs that have intentionally been made nonlinear (but maintain good differential linearity) are useful, especially when processing signals having a wide dynamic range. One of the earliest uses of nonlinear data converters was in the digitization of voiceband signals for pulse code modulation (PCM) systems. Major contributions were made at Bell Labs during the development of the T1 carrier system. The motive for the nonlinear ADCs and DACs was to reduce the total number of bits (and therefore the serial transmission rate) required to digitize voice channels. Straight linear encoding of a voice channel required 11 or 12 bits and a sampling rate of 8 kSPS per. In the 1960s Bell Labs determined that 7-bit nonlinear encoding was sufficient; and later in the 1970s they went to 8-bit nonlinear encoding for better performance (References 18–23).

The nonlinear transfer function allocates more quantization levels out of the total range for small signals and fewer for large amplitude signals. In effect, this reduces the quantization noise associated with small signals (where it is most noticeable) and increases the quantization noise for larger signals (where it is less noticeable). The term *companding* is generally used to describe this form of encoding.

The logarithmic transfer function chosen is referred to as the "Bell μ-255" standard, or simply "μ-law." A similar standard developed in Europe is referred to as "A-law." The Bell μ-law allows a dynamic range of about 4000:1 using 8 bits, whereas an 8-bit linear data converter provides a range of only 256:1.

The first generation channel bank (D1) generated the logarithmic transfer function using temperature-controlled resistor-diode networks for "compressors" ahead of a 7-bit linear ADC in the transmitter. Corresponding resistor-diode "expandors" having an inverse transfer function followed the 7-bit linear DAC in the receiver. The next generation D2 channel banks used nonlinear ADCs and DACs to accomplish the compression/expansion functions in a much more reliable and cost-effective manner, and eliminated the need for the temperature-controlled diode networks.

In his 1953 classic paper, B. D. Smith proposed that the transfer function of a successive approximation ADC utilizing a nonlinear internal DAC in the feedback path would be the inverse transfer function of the DAC (Reference 8). The same basic DAC could therefore be used in the ADC and also for the reconstruction DAC. Later in the 1960s and early 1970s, nonlinear ADC and DAC technology using piecewise linear approximations of the desired transfer function allowed low cost, high volume implementations (References 18–23). These nonlinear 8-bit, 8 kSPS data converters became popular telecommunications building blocks.

The nonlinear transfer function of the 8-bit DAC is first divided into 16 segments (chords) of different slopes—the slopes are determined by the desired nonlinear transfer function. The 4 MSBs determine the segment containing the desired data point, and the individual segment is further subdivided into 16 equal quantization levels by the 4 LSBs of the 8-bit word. This is shown in Figure 3.33 for a 6-bit DAC, where the first three bits identify one of the eight possible chords, and each chord is further subdivided into eight

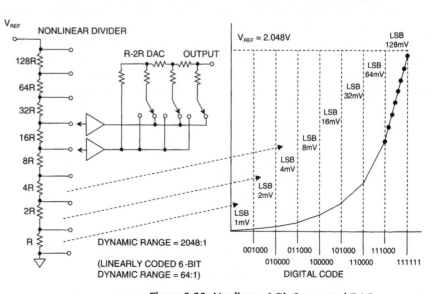

Figure 3.33: Nonlinear 6-Bit Segmented DAC

equal levels defined by the 3 LSBs. The 3 MSBs are generated using a nonlinear string DAC, and the 3 LSBs are generated using a 3-bit binary R-2R DAC.

In 1982, Analog Devices introduced the LOGDAC AD7111 monolithic multiplying DAC featuring wide dynamic range using a logarithmic transfer function. The basic DAC in the LOGDAC is a linear 17-bit voltage-mode R-2R DAC preceded by an 8-bit input decoder (Figure 3.34). The LOGDAC can attenuate an analog input signal, V_{IN}, over the range 0 dB to 88.5 dB in 0.375 dB steps. The degree of attenuation across the DAC is determined by an nonlinear-coded 8-bit word applied to the onboard decode logic. This 8-bit word is mapped into the appropriate 17-bit word, which is then applied to a 17-bit R-2R ladder. A functional diagram of the LOGDAC is shown in Figure 3.34. In addition to providing the logarithmic transfer function, the LOGDAC also acts as a full four-quadrant multiplying DAC.

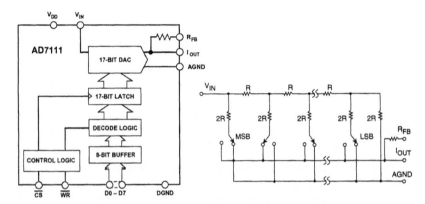

Figure 3.34: AD7111 LOGDAC (Released 1982)

With the introduction of high resolution linear ADCs and DACs, the method used in the LOGDAC is widely used today to implement various nonlinear transfer functions such as the μ-law and A-law companding functions required for telecommunications and other applications. Figure 3.35 shows a general block diagram of the modern approach. The μ-law or A-law companded input data is mapped into data points on the transfer function of a high resolution DAC. This mapping can be easily accomplished by a simple lookup table in either hardware, software, or firmware. A similar nonlinear ADC can be constructed by digitizing the analog input signal using a high resolution ADC and mapping the data points into a shorter word using the appropriate transfer function. A big advantage of this method is that the transfer curve does not have to be approximated with straight line segments as in the earlier method, thereby providing more accuracy.

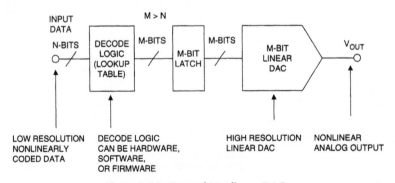

Figure 3.35: General Nonlinear DAC

Counting, Pulsewidth-Modulated (PWM) DACs

Although much less popular than the parallel input DACs previously described, various types of DACs can be constructed using counters to generate an output voltage proportional to a digital input word. A. H. Reeves' classic 1939 PCM patent (Reference 24) describes a 5-bit counting ADC and DAC, and the DAC circuit is shown in Figure 3.36. The operation is simple. A sampling clock starts the counter, which is loaded with the digital word and simultaneously sets an R/S flip-flop. The counter counts upward at a fast rate, and when it reaches all "ones," the R/S flip-flop is reset. The output of the R/S flip-flop is therefore a pulsewidth-modulated (PWM) pulse whose width is proportional to the complement of the binary word. In a variation on the method, the sampling clock starts a ramp generator, and the reset pulse activates a sample-and-hold which stores the output of the ramp generator.

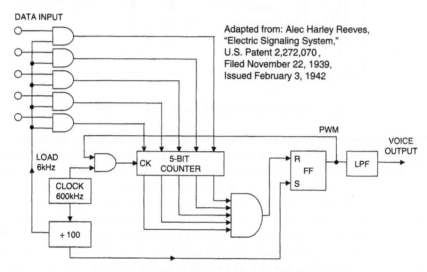

Figure 3.36: A.H. Reeves' 5-Bit Counting DAC

Resolution must be traded for update rate in a counting DAC, because the counter must cycle through all possible outputs in the sampling interval. Counting DACs do have the advantage, however, that they are inherently monotonic.

Cyclic Serial DACs

Cyclic serial DACs are rarely used today, but in the early days of PCM they were somewhat attractive because they took advantage of the serial nature of the PCM pulse stream. An example of a 4-bit implementation is shown in Figure 3.37 and is based on 1948 patent filing (Reference 25). Proper operation of this DAC depends on receiving the PCM data in the proper order: the LSB is first, and the MSB is last.

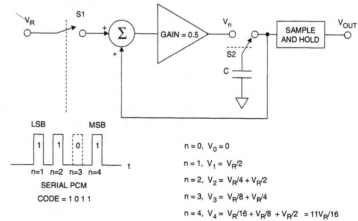

$n = 0, V_0 = 0$

$n = 1, V_1 = V_R/2$

$n = 2, V_2 = V_R/4 + V_R/2$

$n = 3, V_3 = V_R/8 + V_R/4$

$n = 4, V_4 = V_R/16 + V_R/8 + V_R/2 = 11V_R/16$

Figure 3.37: 4-Bit Cyclic Serial DAC

167

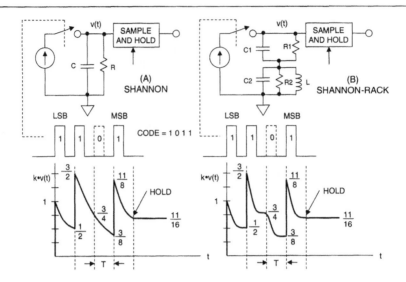

Figure 3.38: Shannon and Shannon-Rack Decoder

Assume that the initial charge on the capacitor is zero and that the serial PCM data represents the digital code 1011. The receipt of a pulse in Position 1 ($n = 1$) closes S1 and connects S2 to the output of the $G = 0.5$ amplifier. The voltage $V_R/2$ is stored on the capacitor, and S2 is then connected to the input of the summer. The receipt of a pulse in Position 2 ($n = 2$) closes S2 and connects V_R to the summer, whose other input is $V_R/2$. S2 is then connected to the amplifier output, and the charge on the capacitor is now $V_R/4 + V_R/2$. The receipt of no pulse in Position 3 ($n = 3$) simply causes the capacitor output to be divided by two, leaving $V_R/8 + V_R/4$ on the amplifier output. This voltage is transferred to the capacitor by S2. In the final cycle, the receipt of a pulse in Position 4 ($n = 4$) adds V_R to $V_R/8 + V_R/4$ which is then divided by two, leaving a final voltage on the capacitor of $V_R/16 + V_R/8 + V_R/2 = 11\,V_R/16$. The final output voltage is then sampled by a sample-and-hold which holds the output voltage until the completion of the next cycle.

It should be noted that this architecture can be made to handle PCM data which has the MSB first, by using a $G = 2$ amplifier and making a few other minor modifications relating to signal scaling.

A truly elegant serial PCM DAC architecture—for its time—was developed by C. E. Shannon and A. J. Rack in 1948 (References 26, 27). The original concept was Shannon's, but Rack added an improvement that made the DAC less sensitive to timing jitter in the PCM pulse stream. The concept circuits are shown in Figure 3.38.

In Figure 3.38A, the serial PCM pulses (LSB first, MSB last) control a switch that is closed for a small amount of time if a pulse is present (representing a Logic "1"), thereby injecting a fixed charge into the capacitor. If no pulse is present in a given pulse position (representing a Logic "0"), the switch remains open, and no additional charge is injected. The RC time constant is chosen such that the capacitor voltage discharges to exactly one-half its initial value in the time interval between PCM pulses, T. The equation that must be satisfied is simply $RC = T/\ln2$.

The diagram shows the capacitor voltage for the binary code 1011. The vertical axis is normalized so that unity represents the voltage change produced by a single switch closure. At the end of the fourth pulse position, the voltage on the capacitor is 11/16, which corresponds to the binary code of 1011 with an LSB weight of 1/16. The sample-and-hold is activated at the end of the fourth pulse position to acquire and hold the capacitor voltage until the next PCM word is completed.

Notice that any jitter in the PCM pulses or the sample-and-hold clock will produce an error in the final held output voltage. A. J. Rack devised an elegant solution to this problem, as shown in Figure 3.38B. Rack added a second capacitor, shunted by both a resistor and an inductor, in series with the original R1-C1 network. The values of the second capacitor, C2, and the inductor, L, used with it are such as to make the circuit resonant at the PCM pulse frequency, 1/T. The second resistor, R2, is adjusted so that the oscillation developed across the resonant circuit is reduced to exactly one-half amplitude between each pulse period. The resulting composite waveform has regions of zero-slope spaced one code period, T, apart, thereby making the circuit much less sensitive to timing jitter in either the PCM pulse train or the sample-and-hold clock. The Shannon-Rack encoder was implemented in a experimental late-1940s Bell Labs PCM system described in Reference 27. The resolution was 7 bits, the sampling rate was 8 kSPS, and the frequency of the PCM pulses was 672 kHz.

Other Low Distortion Architectures

Modern low glitch segmented DACs are capable of very low levels of distortion. However, in some cases, further distortion improvements can be obtained using a technique called *deglitching*. The concept requires a track-and-hold and is illustrated in Figure 3.39.

Just prior to latching new data into the DAC, the track-and-hold is put into the *hold* mode so that the DAC switching glitches are isolated from the output. The switching transients produced by the SHA are code-independent and occur at the clock frequency, and hence are easily filtered. However, great care must be taken so that the relative timing between the track-and-hold clock and the DAC update clock is optimum. In addition, the distortion performance of the track-and-hold must be at least 6 dB to 10 dB better than the DAC, or no improvement in SFDR

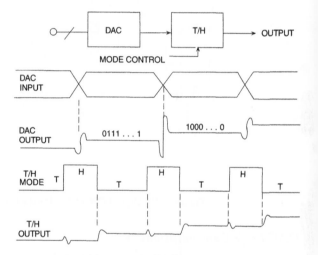

Figure 3.39: Deglitching DAC Outputs Using a Track-and-Hold

will be realized. Achieving good results using an external track-and-hold deglitcher becomes increasingly more difficult as clock frequencies approach 100 MSPS. In most cases, designers should try and use self-contained devices that are fully specified for low distortion without the requirement of additional external circuitry, but the technique may still be of use in some applications.

Digital audio applications require DACs with resolutions of over 16 bits and extremely low total harmonic distortion (THD). One way to make them is to use the segmented architecture with several decoded MSBs as previously described, but this is likely to have comparatively large DNL at the MSB transition, which is just where low DNL is needed for low level audio distortion. This problem can be avoided by using a digital adder to put a digital offset in the DAC code, so that the MSB transition of the input code is well offset from the mid-point of the DAC transfer characteristic, and then using an analog offset on the DAC output to restore the dc level at the crossover. This technique, of course, renders part of the DAC's range unusable, but it does minimize midscale distortion.

Figure 3.40 shows the AD1862 20-bit DAC (introduced in 1990) where a digital offset of 1/16th full-scale is added to the incoming 20-bit binary word. The DAC fully decodes the 3 MSBs and generates the 17 LSBs using a binary R-2R DAC section. In order to prevent clipping at the positive end of the range, the carry output of the adder drives an additional current switch having a weight equal to Bit 4. Finally, an offset current equal to 1/16th full-scale is subtracted from the DAC output to compensate for the constant digital offset. It should be noted that since the introduction of the AD1862 in 1990, most modern low distortion audio DACs today utilize the sigma-delta architecture almost exclusively.

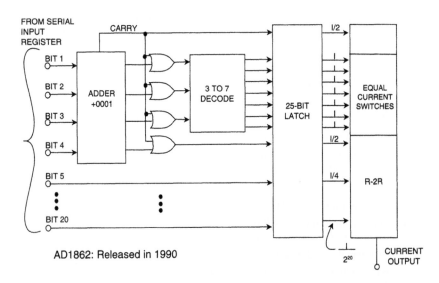

Figure 3.40: Digital Offset Minimizes Midscale Distortion of Small Signals

DAC Logic Considerations

The earliest monolithic DACs contained little, if any, logic circuitry, and parallel data had to be maintained on the digital input to maintain the digital output. Today almost all DACs are latched and data need only be written to them, not maintained. Some even have nonvolatile latches and remember settings while turned off.

There are innumerable variations of DAC input structure, which will not be discussed here, but nearly all are described as "double-buffered." A double-buffered DAC has two sets of latches. Data is initially latched in the first rank and subsequently transferred to the second as shown in Figure 3.41. There are two reasons why this arrangement is useful.

The first is that it allows data to enter the DAC in many different ways. A DAC without a latch, or with a single latch, must be loaded in parallel with all bits at once, since otherwise its output during loading may be totally different from what it was, or what it is to become. A double-buffered DAC, on the other

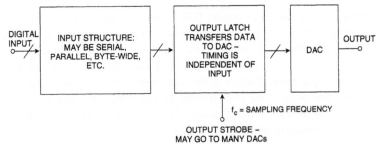

Figure 3.41: Double-Buffered DAC Permits Complex Input Structures and Simultaneous Update

hand, may be loaded with parallel data, or with serial data, with 4-bit or 8-bit words, or whatever, and the output will be unaffected until the new data is completely loaded and the DAC receives its update instruction.

The other convenience of the double-buffered structure is that many DACs may be updated simultaneously: data is loaded into the first rank of each DAC in turn, and when all is ready, the output buffers of all the DACs are updated at once. There are many DAC applications where the output of a number of DACs must change simultaneously, and the double-buffered structure allows this to be done very easily.

Most early monolithic high resolution DACs had parallel or byte-wide data ports and tended to be connected to parallel data buses and address decoders and addressed by microprocessors as if they were very small write-only memories. (Some parallel DACs are not write-only, but can have their contents read as well—this is convenient for some applications, but is not very common.) A DAC connected to a data bus is vulnerable to capacitive coupling of logic noise from the bus to the analog output, and therefore many DACs today have serial data structures. These are less vulnerable to such noise (since fewer noisy pins are involved), use fewer pins and therefore take less board space, and are frequently more convenient for use with modern microprocessors, most of which have serial data ports. Some, but not all, of such serial DACs have both data outputs and data inputs so that several DACs may be connected in series, with data clocked to all of them from a single serial port. This arrangement is often referred to as "daisy-chaining."

Of course, serial DACs cannot be used where high update rates are involved, since the clock rate of the serial data would be too high. Some very high speed DACs actually have two parallel data ports, and use them alternately in a multiplexed fashion (sometimes this is called a "ping-pong" input) to reduce the data rate on each port as shown in Figure 3.42. The alternate loading (ping-pong) DAC in the diagram loads from port A and port B alternately on the rising and falling edges of the clock, which must have a mark-space ratio close to 50:50. The internal clock multiplier ensures that the DAC itself is updated with data A and data B alternately at exactly 50:50 time ratio, even if the external clock is not so precise.

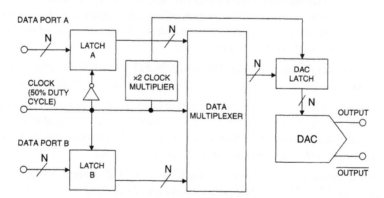

**Figure 3.42: Alternate Loading
(Ping-Pong) High Speed DAC**

References:
3.1 DAC Architectures

1. Peter I. Wold, "Signal-Receiving System," **U.S. Patent 1,514,753**, filed November 19, 1920, issued November 11, 1924. *(Thermometer DAC using relays and vacuum tubes.)*

2. Clarence A. Sprague, "Selective System," **U.S. Patent 1,593,993**, filed November 10, 1921, issued July 27, 1926. *(Thermometer DAC using relays and vacuum tubes.)*

3. Leland K. Swart, "Gas-Filled Tube and Circuit Therefor," **U.S. Patent 2,032,514**, filed June 1, 1935, issued March 3, 1936. *(A thermometer DAC based on vacuum tube switches.)*

4. Robert Adams, Khiem Nguyen, and Karl Sweetland, "A 113 dB SNR Oversampling DAC with Segmented Noise-Shaped Scrambling, " **ISSCC Digest of Technical Papers**, vol. 41, 1998, pp. 62, 63, 413. *(Describes a segmented audio DAC with data scrambling.)*

5. Robert W. Adams and Tom W. Kwan, "Data-Directed Scrambler for Multibit Noise-Shaping D/A Converters," **U.S. Patent 5,404,142**, filed August 5, 1993, issued April 4, 1995. *(Describes a segmented audio DAC with data scrambling.)*

6. Paul M. Rainey, "Facimile Telegraph System," **U.S. Patent 1,608,527**, filed July 20, 1921, issued November 30, 1926. *(The first PCM patent. Also shows a relay-based 5-bit electro-mechanical flash converter and a binary DAC using relays and multiple resistors.)*

7. John C. Schelleng, "Code Modulation Communication System," **U.S. Patent 2,453,461**, Filed June 19, 1946, Issued November 9, 1948. *(Vacuum tube binary DAC using binary weighted voltages summed into load resistor with equal resistor weights.)*

8. B. D. Smith, "Coding by Feedback Methods," **Proceedings of the I. R. E.**, Vol. 41, August 1953, pp. 1053–1058. *(Smith uses an internal binary weighted DAC and also points out that a nonlinear transfer function can be achieved by using a DAC with nonuniform bit weights, a technique which is widely used in today's voiceband ADCs with built-in companding. He was also one of the first to propose using an R/2R ladder network within the DAC core.)*

9. Bruce K. Smith, "Digital Attenuator," **U.S. Patent 1,976,527**, filed July 17, 1958, issued March 21, 1961. *(Describes a transistorized voltage output DAC similar to B. D. Smith above.)*

10. Bernard M. Gordon and Robert P. Talambiras, "Signal Conversion Apparatus," **U.S. Patent 3,108,266**, filed July 22, 1955, issued October 22, 1963. *(Classic patent describing Gordon's 11-bit, 50 kSPS vacuum tube successive approximation ADC done at Epsco. The internal DAC represents the first known use of equal currents switched into an R/2R ladder network.)*

11. James J. Pastoriza, "Solid State Digital-to-Analog Converter," **U.S. Patent 3,747,088**, filed December 30, 1970, issued July 17, 1973. *(The first patent on the quad switch approach to building high resolution DACs.)*

12. Eugene R. Hnatek, **A User's Handbook of D/A and A/D Converters**, John Wiley, New York, 1976, ISBN 0-471-40109-9, pp. 282–295. *(Contains an excellent description of the Analog Devices' AD550 monolithic μDAC quad current switch, and AD850 thin film network—building blocks for 12-bit DACs introduced in 1970.)*

13. Adrian Paul Brokaw, "Digital-to-Analog Converter with Current Source Transistors Operated Accurately at Different Current Densities," **U.S. Patent 3,940,760**, filed March 21, 1975, issued February 24, 1976. *(The most referenced data converter patent ever issued.)*

14. Dennis Dempsey and Christopher Gorman, "Digital-to-Analog Converter," **U.S. Patent 5,969,657**, filed July 27, 1997, issued October 19, 1999. *(Describes an elegant solution for segmented unbuffered string DACs.)*

15. John A. Schoeff, "An Inherently Monotonic 12 Bit DAC," **IEEE Journal of Solid State Circuits**, Vol. SC-14, No. 6, December 1979, pp. 904–911. *(Describes one of the first monolithic DACs to use segmentation.)*

16. G. R. Ritchie, J. C. Candy, and W. H. Ninke, "Interpolative Digital-to-Analog Converters," **IEEE Transactions on Communications**, Vol. COM-22, November 1974, pp. 1797–1806. *(One of the earliest papers written on oversampling interpolative DACs.)*

17. H. G. Musmann and W. W. Korte, "Generalized Interpolative Method for Digital/Analog Conversion of PCM Signals," **U.S. Patent 4,467,316**, filed June 3, 1981, issued August 21, 1984. *(A description of interpolating DACs.)*

18. B. Smith, "Instantaneous Companding of Quantized Signals, **Bell System Technical Journal**, Vol. 36, May 1957, pp. 653–709. *(One of the first papers written about using nonlinear coding techniques for speech signals in PCM.)*

19. H. Kaneko and T. Sekimoto, "Logarithmic PCM Encoding Without Diode Compandor," **IEEE Transactions on Communications Systems**, Vol. 11, No. 3, September 1963, pp. 296–307. *(Describes several methods for nonlinear encoding speech directly without the need for diode compandors.)*

20. C. L. Dammann, "An Approach to Logarithmic Coders and Decoders," **NEREM Record**, Boston MA, November 2-4, 1966, pp. 196–197. *(More discussions on nonlinear coders and decoders for PCM.)*

21. H. Kaneko, "A Unified Formulation of Segment Companding Laws and Synthesis of Codecs and Digital Compandors," **Bell System Technical Journal**, Vol. 49, September 1970, pp. 1555–1558. *(Discusses the piecewise linear approximation to the logarithmic transfer companding function.)*

22. M. R. Aaron and H. Kaneko, "Synthesis of Digital Attenuators for Segment Companded PCM Codes," **Transactions on Communications Technology**, COM-19, December 1971, pp. 1076–1087. *(More on nonlinear coding.)*

23. C. L. Dammann, L. D. McDaniel, and C. L. Maddox, "D2 Channel Bank: Multiplexing and Coding," **Bell System Technical Journal**, Vol. 51, October 1972, pp. 1675–1699. *(Still more on nonlinear coding.)*

24. A.H. Reeves, "Electric Signaling System," **U.S. Patent 2,272,070**, filed November 22, 1939, issued February 3, 1942. *(Reeves' classic PCM patent which describes a 5-bit PWM counting ADC and DAC.)*

25. R. L. Carbrey, "Decoder for Pulse Code Modulation," **U.S. Patent 2,579,302**, filed January 17, 1948, issued December 18, 1951. *(Describes cyclic or sequential attenuation DAC.)*

26. R. L. Carbrey, "Decoding in PCM,", **Bell Labs Record**, November 1948, pp. 451–456. *(Describes the Shannon-Rack PCM DAC.)*

27. L.A. Meacham and E. Peterson, "An Experimental Multichannel Pulse Code Modulation System of Toll Quality," **Bell System Technical Journal**, Vol. 27, No. 1, January 1948, pp. 1–43. *(Further details of the Shannon-Rack decoder as part of the experimental PCM system.)*

ADC Architectures
Walt Kester, James Bryant

Introduction

As in the case of DACs, the relationship between the digital output and the analog input of an ADC depends upon the value of the reference, and the accuracy of the reference is almost always the limiting factor on the absolute accuracy of a ADC. We shall consider the various architectures of ADCs, and the forms the reference may take, later in this section.

Similar to DACs, many ADCs use external references (see Figure 3.43) and have a reference input terminal, while others have an output from an internal reference. The simplest ADCs, of course, have neither—the reference is on the ADC chip and has no external connections.

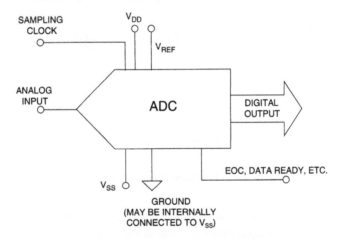

Figure 3.43: Basic ADC with External Reference

If an ADC has an internal reference, its overall accuracy is specified when using that reference. If such an ADC is used with a perfectly accurate external reference, its absolute accuracy may actually be worse than when it is operated with its own internal reference. This is because it is trimmed for absolute accuracy when working with its own actual reference voltage, not with the nominal value. Twenty years ago it was common for converter references to have accuracies as poor as ±5% since these references were trimmed for low temperature coefficient rather than absolute accuracy, and the inaccuracy of the reference was compensated in the gain trim of the ADC itself. Today the problem is much less severe, but it is still important to check for possible loss of absolute accuracy when using an external reference with an ADC that has a built-in one.

ADCs that have reference terminals must, of course, specify their behavior and parameters. If there is a reference input the first specification will be the reference input voltage—and of course this has two values, the absolute maximum rating, and the range of voltages over which the ADC performs correctly.

Most ADCs require that their reference voltage is within quite a narrow range whose maximum value is less than or equal to the ADC's V_{DD}. Notice that this is unlike DACs, where many allow the reference to be varied over a wide range (as in the MDACs or semi-multiplying DACs previously discussed in Section 3.1 of this chapter).

The reference input terminal of an ADC may be buffered as shown in Figure 3.44, in which case it has input impedance (usually high) and bias current (usually low) specifications, or it may connect directly to the ADC. In either case, the transient currents developed on the reference input due to the internal conversion process need good decoupling with external low inductance capacitors. Most ADC data sheets recommend appropriate decoupling networks.

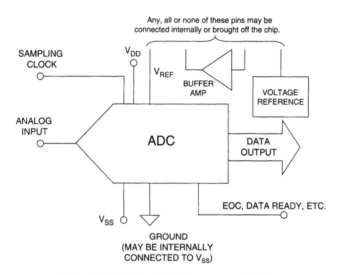

Figure 3.44: ADC with Reference and Buffer

Where an ADC has an internal reference, it will carry a defined reference voltage, with a specified accuracy. There may also be specifications of temperature coefficient and long-term stability.

The reference input may be buffered or unbuffered. If it is buffered, the maximum output current will probably be specified. In general such a buffer will have a unidirectional output stage which sources current but does not allow current to flow into the output terminal. If the buffer does have a push-pull output stage, the output current will probably be defined as ±(SOME VALUE) mA. If the reference output is unbuffered, the output impedance may be specified, or the data sheet may simply advise the use of a high input impedance external buffer.

The *sampling clock* input (sometimes called *convert-start* or *encode command*) is a critical function in an ADC and a source of some confusion. Many of the early integrated circuit ADCs (such as the industry-standard AD574) did not have a built-in sample-and-hold function, and were known simply as *encoders*. These converters required an external clock to begin the conversion process. In the case of the AD574, the application of the external clock initiated an internal high-speed clock oscillator which in turn controlled the actual conversion process.

Most modern ADCs have the sample-and-hold function on-chip and require an external sampling clock to initiate the conversion. In some ADCs, only a single sampling clock is required—in others, both a high frequency clock as well as a lower speed sampling clock are required. Regardless of the ADC, it is extremely important to read the data sheet and determine exactly what the external clock requirements are because they can vary widely from one ADC to another, since there is no standard.

At some point after the assertion of the sampling clock, the output data is valid. This data may be in parallel or serial format depending upon the ADC. Early successive approximation ADCs such as the AD574 simply provided a STATUS output (STS) which went high during the conversion, and returned to the low state when the output data was valid. In other ADCs, this line is variously called *busy, end-of-conversion (EOC), data ready,* etc. Regardless of the ADC, there must be some method of knowing when the output data is valid—and again, the data sheet is where this information can always be found.

There are one or two other practical points worth remembering about the logic of ADCs. On power-up, many ADCs do not have logic reset circuitry and may enter an anomalous logical state. One or two conversions may be necessary to restore their logic to proper operation so: (a) the first and second conversions after power-up should never be trusted, and (b) control outputs (EOC, data ready, etc.) may behave in unexpected ways at this time (and not necessarily in the same way at each power-up), and (c) care should be taken to ensure that such anomalous behavior cannot cause system latch-up. For example, EOC (end-of-conversion) should not be used to initiate conversion if there is any possibility that EOC will not occur until the first conversion has taken place, as otherwise initiation will never occur.

Some low power ADCs now have power-saving modes of operation variously called *standby, power-down, sleep,* etc. When an ADC comes out of one of these low power modes, there is a certain recovery time required before the ADC can operate at its full specified performance. The data sheet should therefore be carefully studied when using these modes of operation, and there may be several different levels of power-down.

Another detail that can cause trouble is the difference between EOC and DRDY (data ready). EOC indicates that conversion has finished, DRDY that data is available at the output. In some ADCs, EOC functions as DRDY—in others, data is not valid until several tens of nanoseconds *after* the EOC has become valid, and if EOC is used as a data strobe, the results will be unreliable.

As a final example, some ADCs use CS (Chip Select) edges to reset internal logic, and it may not be possible to perform another conversion without asserting or reasserting CS (or it may not be possible to read the same data twice, or both).

For more detail, it is important to read the whole data sheet before using an ADC since there are innumerable small logic variations from type to type. Unfortunately, many data sheets are not as clear as one might wish, so it is also important to understand the general principles of ADCs in order to interpret data sheets correctly. That is one of the purposes of this section.

We are now ready to discuss the various architectures for ADCs. Because it is a fundamental building block used in all ADCs, the comparator (a 1-bit ADC) is treated first. It is logical to follow this with the flash converter architecture because it is somewhat analogous to the fully-decoded (thermometer) DAC architecture previously discussed. The successive approximation ADC architecture is treated next followed by subranging and pipelined architectures. The folding (Gray-code) architecture completes the primary architectures used in so-called high speed ADCs.

The last part of this sections discusses the various counting and integrating architectures which are generally more suited to high resolution lower speed ADCs. Sigma-Delta (Σ-Δ) ADCs and DACs are treated in a separate section which concludes this chapter.

The Comparator: A 1-Bit ADC

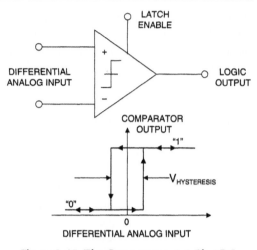

As a changeover switch is a 1-bit DAC, so a comparator is a 1-bit ADC (see Figure 3.45). If the input is above a threshold, the output has one logic value, below it has another. Moreover, there is no ADC architecture that does not use at least one comparator of some sort.

The most common comparator has some resemblance to an operational amplifier in that it uses a differential pair of transistors or FETs as its input stage, but unlike an op amp, it does not use external negative feedback, and its output is a logic level indicating which of the two inputs is at the higher potential. Op amps are not designed for use as comparators—they may saturate if overdriven and recover slowly. Many op amps have input stages that behave in unexpected ways when used with large differential voltages, and their outputs are rarely compatible with standard logic levels. There are cases, however, when it may be desirable to use an op amp as a comparator, and an excellent treatment of this subject can be found in Reference 1.

Figure 3.45: The Comparator: A 1-Bit ADC

Comparators used as building blocks in ADCs need good resolution which implies high gain. This can lead to uncontrolled oscillation when the differential input approaches zero. In order to prevent this, *hysteresis* is often added to comparators using a small amount of positive feedback. Figure 3.45 shows the effects of hysteresis on the overall transfer function. Many comparators have a millivolt or two of hysteresis to encourage "snap" action and to prevent local feedback from causing instability in the transition region. Note that the resolution of the comparator can be no less than the hysteresis, so large values of hysteresis are generally not useful.

Early comparators were designed with vacuum tubes and were often used in radio receivers—where they were called *discriminators*, not comparators. Most modern comparators used in ADCs include a built-in latch which makes them sampling devices suitable for data converters. A typical structure is shown in Figure 3.46 for the AM685 ECL (emitter-coupled-logic) latched comparator introduced in 1972 by Advanced Micro Devices, Inc. (see Reference 2). The input stage preamplifier drives a cross-coupled latch. The latch locks the output in the logic state it was in at the instant the latch was enabled. The latch thus performs a track-and-hold function, allowing short input signals to be detected and held for further processing. Because the latch operates directly on the input stage, the signal suffers no additional delays—signals

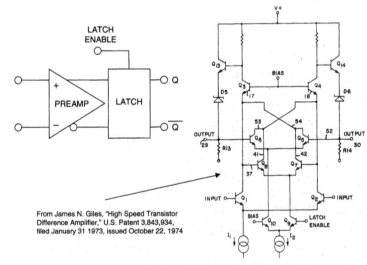

From James N. Giles, "High Speed Transistor Difference Amplifier," U.S. Patent 3,843,934, filed January 31 1973, issued October 22, 1974

Figure 3.46: The AM685 ECL Comparator (1972)

only a few nanoseconds wide can be acquired and held. The latched comparator is also less sensitive to instability caused by local feedback than an unlatched one.

Where comparators are incorporated into IC ADCs, their design must consider resolution, speed, overload recovery, power dissipation, offset voltage, bias current, and the chip area occupied by the architecture chosen. There is another subtle but troublesome characteristic of comparators which can cause large errors in ADCs if not understood and dealt with effectively. This error mechanism is the occasional inability of a comparator to resolve a small differential input into a valid output logic level. This phenomenon is known as *metastability*—the ability of a comparator to balance right at its threshold for a short period of time.

The metastable state problem is illustrated in Figure 3.47. Three conditions of differential input voltage are illustrated: (1) large differential input voltage, (2) small differential input voltage, and (3) zero differential input voltage. The approximate equation which describes the output voltage, $V_O(t)$ is given by:

$$V_O(t) = \Delta V_{IN} A e^{t/\tau}$$

Eq. 3.1

Where ΔV_{IN} = the differential input voltage at the time of latching, A = the gain of the preamp at the time of latching, τ = regeneration time constant of the latch, and t = the time that has elapsed after the comparator output is latched (see References 3 and 4).

For small differential input voltages, the output takes longer to reach a valid logic level. If the output data is read when it lies between the "valid Logic 1" and the "valid Logic 0" region, the data can be in error. If the differential input voltage is exactly zero, and the comparator is perfectly balanced at the time of latching, the time required to reach a valid logic level can be quite long (theoretically infinite). However, hysteresis and noise on the input makes this condition highly unlikely. The effects of invalid logic levels out of the

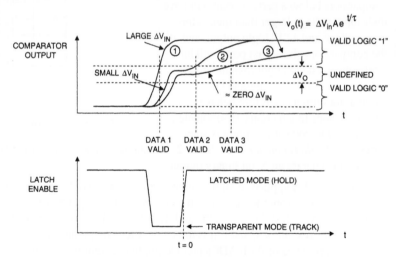

Figure 3.47: Comparator Metastable State Errors

comparator are different, depending upon how the comparator is used in the actual ADC.

From a design standpoint, comparator metastability can be minimized by making the gain, A, high, minimizing the regeneration time constant, τ, by increasing the gain-bandwidth of the latch, and allowing sufficient time, t, for the output of the comparator to settle to a valid logic level. It is not the purpose of this discussion to analyze the complex trade-offs between speed, power, and circuit complexity when optimizing comparator designs, but an excellent treatment of the subject can be found in References 3 and 4.

From a user standpoint, the effect of comparator metastability (if it affects the ADC performance at all) is in the *bit error rate* (BER)—which is not usually specified on most ADC data sheets. A discussion of this specification can be found in Chapter 2 of this book. The resulting errors are often referred to as *sparkle codes*, *rabbits*, or *flyers*.

Bit error rate should not be a problem in a properly designed ADC in most applications, however the system designer should be aware that the phenomenon exists. An application example where it can be a problem is when the ADC is used in a digital oscilloscope to detect small-amplitude single-shot randomly occurring events. The ADC can give false indications if its BER is not sufficiently small.

High Speed ADC Architectures

Flash Converters

Flash ADCs (sometimes called *parallel* ADCs) are the fastest type of ADC and use large numbers of comparators. An N-bit flash ADC consists of 2^N resistors and 2^N-1 comparators arranged as in Figure 3.48. Each comparator has a reference voltage from the resistor string which is 1 LSB higher than that of the one below it in the chain. For a given input voltage, all the comparators below a certain point will have their input voltage larger than their reference voltage and a "1" logic output, and all the comparators above that point will have a reference voltage larger than the input voltage and a "0" logic output. The 2^N-1 comparator outputs therefore behave in a way analogous to a mercury thermometer, and the output code at this point is some-

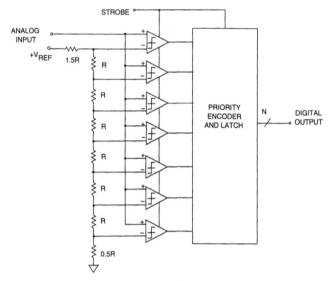

Figure 3.48: 3-Bit All-Parallel (Flash) Converter

times called a *thermometer* code. Since 2^N-1 data outputs are not really practical, they are processed by a decoder to generate an N-bit binary output.

The input signal is applied to all the comparators at once, so the thermometer output is delayed by only one comparator delay from the input, and the encoder N-bit output by only a few gate delays on top of that, so the process is very fast. However, the architecture uses large numbers of resistors and comparators and is limited to low resolutions, and if it is to be fast, each comparator must run at relatively high power levels. Hence, the problems of flash ADCs include limited resolution, high power dissipation because of the large number of high speed comparators (especially at sampling rates greater than 50 MSPS), and relatively large (and therefore expensive) chip sizes. In addition, the resistance of the reference resistor chain must be kept low to supply adequate bias current to the fast comparators, so the voltage reference has to source quite large currents (typically > 10 mA).

The first documented flash converter was part of Paul M. Rainey's electro-mechanical PCM facsimile system described in a relatively ignored patent filed in 1921 (Reference 5—see further discussions in Chapter 1 of this book). In the ADC, a current proportional to the intensity of light drives a galvanometer which in turn moves another beam of light which activates one of 32 individual photocells, depending upon the amount of galvanometer deflection (see Figure 3.49). Each individual photocell output activates part of a relay network which generates the 5-bit binary code.

A significant development in ADC technology during the period was the electron beam coding tube developed at Bell Labs and shown in Figure 3.50. The tube described by R. W. Sears in Reference 6 was capable

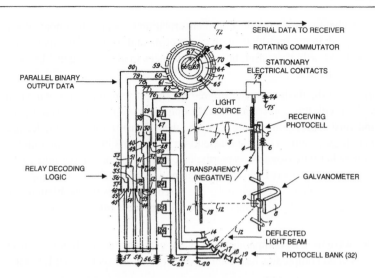

Figure 3.49: A 5-Bit Flash ADC Proposed by Paul Rainey
Adapted from Paul M. Rainey, "Facsimile Telegraph System,"
U.S. Patent 1,608,527, Filed July 20, 1921, Issued November 30, 1926

of sampling at 96 kSPS with 7-bit resolution. The basic electron beam coder concepts are shown in Figure 3.50 for a 4-bit device. The tube used a fan-shaped beam creating a "flash" converter delivering a parallel output word.

Early electron tube coders used a binary-coded shadow mask (Figure 3.50A), and large errors can occur if the beam straddles two adjacent codes and illuminates both of them. The errors associated with binary shadow masks were later eliminated by using a Gray code shadow mask as shown in Figure 3.50B. This code was originally called the "reflected binary" code, and was invented by Elisha Gray in 1878, and later re-invented by Frank Gray in 1949 (see Reference 7). The Gray code has the property that adjacent levels differ by only one digit in the corresponding Gray-coded word. Therefore, if there is an error in a bit decision for a particular level, the corresponding error after conversion to binary code is only one least significant bit (LSB). In the case of midscale, note that only the MSB changes. It is interesting to note that this same phenomenon can occur in modern comparator-based flash converters due to comparator metastability. With small overdrive, there is a finite probability that the output of a comparator will generate the wrong decision in its latched output, producing the same effect if straight binary decoding techniques are used. In many cases, Gray code, or "pseudo-Gray" codes are used to decode the comparator bank output before finally converting to a binary code output.

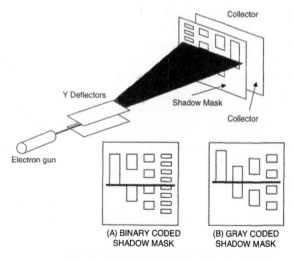

Figure 3.50: The Electron Beam Coder from Bell Labs (1948)

In spite of the many mechanical and electrical problems relating to beam alignment, electron tube coding technology reached its peak in the mid-1960s with an experimental 9-bit coder capable of 12-MSPS sampling rates (Reference 8). Shortly thereafter, however, advances in all solid-state ADC techniques made the electron tube technology obsolete.

It was soon recognized that the flash converter offered the fastest sampling rates compared to other architectures, but the problem with this approach is that the comparator circuit itself is quite bulky using discrete transistor circuits and very cumbersome using vacuum tubes. Constructing a single latched comparator cell using either technology is quite a task, and extending it to even 4 bits of resolution (15 comparators required) makes it somewhat unreasonable. Nevertheless, work was done in the mid 1950s and early 1960s as shown in Robert Staffin and Robert D. Lohman's patent which describes a subranging architecture using both tube and transistor technology (Reference 9). The patent discusses the problem of the all-parallel approach and points out the savings by dividing the conversion process into a coarse conversion followed by a fine conversion.

Tunnel (Esaki) diodes were used as comparators in several experimental early flash converters in the 1960s as an alternative to a latched comparator based solely on tubes or transistors (see References 10–13).

In 1964 Fairchild introduced the first IC comparators, the μA711/712, designed by Bob Widlar. The same year, Fairchild also introduced the first IC op amp, the μA709—another Widlar design. Other IC comparators soon followed, including the Signetics 521, National LM361, Motorola MC1650 (1968), AM685/687 (1972/1975). With the introduction of these building block comparators and the availability of TTL and ECL logic ICs, 6-bit rack-mounted discrete flash converters were introduced by Computer Labs, Inc., including the VHS-630 (6-bit, 30 MSPS in 1970) and the VHS-675 (6-bit, 75 MSPS in 1975). The VHS-675 shown in Figure 3.51 used 63 AM685 ECL comparators preceded by a high-speed track-and-hold, ECL decoding

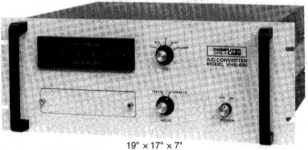

19" × 17" × 7"

VHS-630
- 6-Bits, 30 MSPS
- 32 dual MC1650 MECL III Comparators
- 100 watts (linear power supplies included)

VHS-675
- 6-Bits, 75 MSPS
- 64 AM685 Comparators
- 130 watts (linear power supplies included)

Figure 3.51: VHS-Series ADCs from Computer Labs, Inc. VHS-630 (1970), VHS-675 (1975)

logic, contained a built-in linear power supply (ac line powered), and dissipated a total of 130 W (sale price was about $10,000 in 1975). Instruments such as these found application in early high speed data acquisition applications including military radar receivers.

The AM685 comparator was also used as a building block in the 4-bit 100 MSPS board-level flash ADC, the MOD-4100, introduced in 1975 and shown in Figure 3.52.

The first integrated circuit 8-bit video-speed 30 MSPS flash converter, the TDC1007J, was introduced by TRW LSI division in 1979 (References 14 and 15). A 6-bit version of the same design, the TDC1014J followed shortly. Also in 1979, Advanced Micro Devices, Inc. introduced the AM6688, a 4-bit 100 MSPS IC flash converter.

Flash converters became very popular in the 1980s for high speed 8-bit video applications as well as building blocks for higher resolution subranging card-level, modular, and hybrid ADCs. Many were fabricated on CMOS processes for lower power dissipation. Recently, however, the subranging pipeline architecture has become popular for 8-bit ADCs up to about 250 MSPS. For instance, the AD9480 8-bit 250 MSPS ADC is fabricated on a high speed BiCMOS process and dissipates less than 400 mW compared to the several watts required for a full flash implementation on a similar process.

In practice, IC flash converters are currently available up to 10 bits, but they more commonly have 8 bits of resolution. Their maximum sampling rate can be as high as 1 GHz (these are generally made on Gallium Arse-

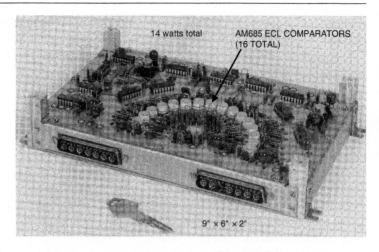

Figure 3.52: MOD-4100 4-Bit, 100 MSPS Flash Converter, Computer Labs, 1975

nide processes with several watts of power dissipation), with input full-power bandwidths in excess of 300 MHz.

As mentioned earlier, full-power bandwidths are not necessarily full-resolution bandwidths. Ideally, the comparators in a flash converter are well-matched both for dc and ac characteristics. Because the sampling clock is applied to all the comparators simultaneously, the flash converter is inherently a sampling converter. In practice, there are delay variations between the comparators and other ac mismatches which cause a degradation in the effective number of bits (ENOBs) at high input frequencies. This is because the inputs are slewing at a rate comparable to the comparator conversion time. For this reason, track-and-holds are often required ahead of flash converters to achieve high SFDR on high frequency input signals.

The input to a flash ADC is applied in parallel to a large number of comparators. Each has a voltage-variable junction capacitance, and this signal-dependent capacitance results in most flash ADCs having reduced ENOB and higher distortion at high input frequencies. For this reason, most flash converters must be driven with a wideband op amp that is tolerant to the capacitive load presented by the converter as well as high speed transients developed on the input.

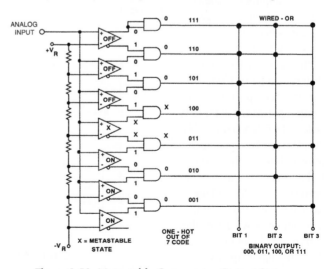

Comparator metastability in a flash converter can severely impact the bit error rate (BER). Figure 3.53 shows a simple flash converter with one stage of binary decoding logic. The two-input AND gates convert the thermometer code output of the parallel comparators into a "one-hot out of 7" code. The decoding logic is simply a "wired-or" array, a technique popular with ECL logic.

Figure 3.53: Metastable Comparator Output States May Cause Error Codes in Data Converters

Assume that the comparator labeled "X" has metastable outputs labeled "X." The desired output code should be either 011 or 100, but note that the 000 code (both gate outputs high) and the 111 code (both gate outputs low) are also possible due to the metastable states, representing a ½ FS error.

Metastable state errors in flash converters can be reduced by several techniques, one of which involves decoding the comparator outputs in Gray code followed by a Gray-to-binary conversion as in the Bell Labs electron beam encoder previously described. The advantage of Gray code decoding is that a metastable state in any of the comparators can produce only a 1 LSB error in the Gray code output. The Gray code is latched and then converted into a binary code which, in turn, will only have a maximum of 1 LSB error as shown in Figure 3.54. (This is described in more detail in Chapter 1 of this book in the section on coding.)

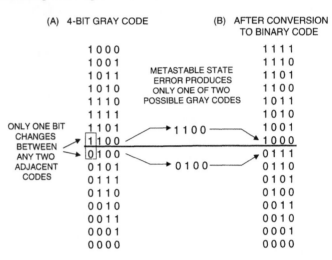

Figure 3.54: Gray Code Decoding Reduces Amplitude of Metastable State Errors

The same principles have been applied to several modern IC flash converters to minimize the effects of metastable state errors as described in References 3, 16, 17, for example.

Power dissipation is always a big consideration in flash converters, especially at resolutions above 8 bits. A clever technique was used in the AD9410 10-bit, 210-MSPS ADC called *interpolation* to minimize the number of preamplifiers in the flash converter comparators and also reduce the power (2.1 W). The method is shown in Figure 3.55 (see Reference 18).

The preamplifiers (labeled "A1," "A2," etc.) are low gain g_m stages whose bandwidth is proportional to the tail currents of the differential pairs. Consider the case for a positive-going ramp input which is initially below the reference to AMP A1, V1. As the input signal approaches V1, the differential output of A1 approaches zero (i.e., A = Ā), and the decision point is reached. The output of A1 drives the differential input of LATCH 1. As the input signals continue to go positive, A continues to go positive, and

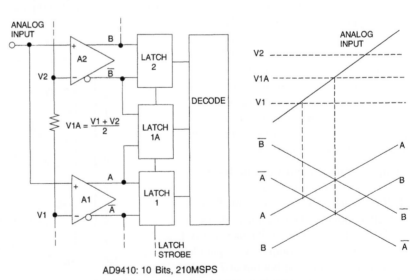

Figure 3.55: "Interpolating" Flash Reduces the Number of Preamplifiers by Factor of Two

$\overline{B}$ begins to go negative. The interpolated decision point is determined when $A = \overline{B}$. As the input continues positive, the third decision point is reached when $B = \overline{B}$. This novel architecture reduces the ADC input capacitance and thereby minimizes its change with signal level and the associated distortion. The AD9410 also uses an input sample-and-hold circuit for improved ac linearity.

Successive Approximation ADCs

The successive approximation ADC has been the mainstay of data acquisition for many years. Recent design improvements have extended the sampling frequency of these ADCs into the megahertz region. The Analog Devices PulSAR® family of SAR ADCs uses internal switched capacitor techniques along with auto calibration techniques to extend the resolution of these ADCs to 18 bits on CMOS processes without the need for expensive thin-film laser trimming.

The basic successive approximation ADC is shown in Figure 3.56. It performs conversions on command. On the assertion of the CONVERT START command, the sample-and-hold (SHA) is placed in the *hold* mode, and all the bits of the successive approximation register (SAR) are reset to "0" except the MSB, which is set to "1." The SAR output drives the internal DAC. If the DAC output is greater than the analog input, this bit in the SAR is reset, otherwise it is left set. The next most significant bit is then set to "1." If the DAC output is greater than the analog input, this bit in the SAR is reset, otherwise it is left set. The process is repeated with each bit in turn. When all the bits have been set, tested, and reset or not as appropriate, the contents of the SAR correspond to the value of the analog input, and the conversion is complete. These bit "tests" can form the basis of a serial output version SAR-based ADC.

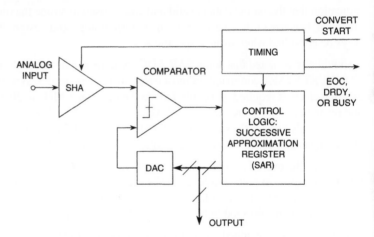

Figure 3.56: Basic Successive Approximation ADC (Feedback Subtraction ADC)

The fundamental timing diagram for a typical SAR ADC is shown in Figure 3.57. The end of conversion is generally indicated by an end-of-convert (EOC), data-ready (DRDY), or a busy signal (actually, *not*-BUSY indicates end of conversion). The polarities and name of this signal may be different for different SAR ADCs, but the fundamental concept is the same. At the beginning of the conversion interval, the signal goes high (or low) and remains in that state until the conversion is completed, at which time it goes low (or high). The trailing

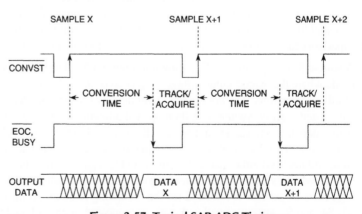

Figure 3.57: Typical SAR ADC Timing

edge is generally an indication of valid output data, but the data sheet should be carefully studied—in some ADCs additional delay is required before the output data is valid.

An N-bit conversion takes N steps. It would seem on superficial examination that a 16-bit converter would have twice the conversion time of an 8-bit one, but this is not the case. In an 8-bit converter, the DAC must settle to 8-bit accuracy before the bit decision is made, whereas in a 16-bit converter, it must settle to 16-bit accuracy, which takes a lot longer. In practice, 8-bit successive approximation ADCs can convert in a few hundred nanoseconds, while 16-bit ones will generally take several microseconds.

While there are some variations, the fundamental timing of most SAR ADCs is similar and relatively straightforward. The conversion process is generally initiated by asserting a CONVERT START signal. The CONVST signal is a negative-going pulse whose positive-going edge actually initiates the conversion. The internal sample-and-hold (SHA) amplifier is placed in the hold mode on this edge, and the various bits are determined using the SAR algorithm. The negative-going edge of the $\overline{\text{CONVST}}$ pulse causes the EOC or BUSY line to go high. When the conversion is complete, the BUSY line goes low, indicating the completion of the conversion process. In most cases the trailing edge of the BUSY line can be used as an indication that the output data is valid and can be used to strobe the output data into an external register. However, because of the many variations in terminology and design, the individual data sheet should always be consulted when using a specific ADC.

It should also be noted that some SAR ADCs require an external high frequency clock in addition to the CONVERT START command. In most cases, there is no need to synchronize the two. The frequency of the external clock, if required, generally falls in the range of 1 MHz to 30 MHz depending on the conversion time and resolution of the ADC. Other SAR ADCs have an internal oscillator THAT is used to perform the conversions and only require the CONVERT START command. Because of their architecture, SAR ADCs generally allow single-shot conversion at any repetition rate from dc to the converter's maximum conversion rate.

Notice that the overall accuracy and linearity of the SAR ADC is determined primarily by the internal DAC. Until recently, most precision SAR ADCs used laser-trimmed thin-film DACs to achieve the desired accuracy and linearity. The thin-film resistor trimming process adds cost, and the thin-film resistor values may be affected when subjected to the mechanical stresses of packaging.

For these reasons, switched-capacitor (or charge-redistribution) DACs have become popular in newer SAR ADCs. The advantage of the switched capacitor DAC is that the accuracy and linearity is primarily determined by high accuracy photolithography, which in turn, controls the capacitor plate area and the capacitance as well as matching. In addition, small capacitors can be placed in parallel with the main capacitors which can be switched in and out under control of autocalibration routines to achieve high accuracy and linearity without the need for thin-film laser trimming. Temperature tracking between the switched capacitors can be better than 1 ppm/°C, thereby offering a high degree of temperature stability.

A simple 3-bit capacitor DAC is shown in Figure 3.58. The switches are shown in the *track*, or *sample* mode where the analog input voltage, A_{IN}, is constantly charging and discharging the parallel combination of all the capacitors. The *hold* mode is initiated by opening S_{IN}, leaving the sampled analog input voltage on the capacitor array. Switch S_C is then opened allowing the voltage at node A to move as the bit switches are manipulated. If S1, S2, S3, and S4 are all connected to ground, a voltage equal to $-A_{\text{IN}}$ appears at node A. Connecting S1 to V_{REF} adds a voltage equal to $V_{\text{REF}}/2$ to $-A_{\text{IN}}$. The comparator then makes the MSB bit decision, and the SAR either leaves S1 connected to V_{REF} or connects it to ground depending on the comparator output (which is high or low, depending on whether the voltage at node A is negative or positive, respectively). A similar process is followed for the remaining two bits. At the end of the conversion interval, S1, S2, S3, S4, and S_{IN} are connected to A_{IN}, S_C is connected to ground, and the converter is ready for another cycle.

Note that the extra LSB capacitor (C/4 in the case of the 3-bit DAC) is required to make the total value of the capacitor array equal to 2C so that binary division is accomplished when the individual bit capacitors are manipulated.

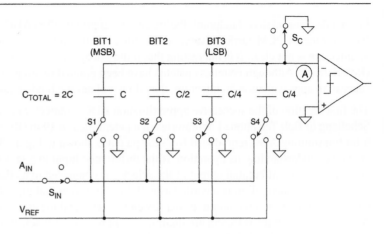

Figure 3.58: 3-Bit Switched Capacitor DAC

The operation of the capacitor DAC (cap DAC) is similar to an R-2R resistive DAC. When a particular bit capacitor is switched to V_{REF}, the voltage divider created by the bit capacitor and the total array capacitance (2C) adds a voltage to node A equal to the weight of that bit. When the bit capacitor is switched to ground, the same voltage is subtracted from node A.

The basic algorithm used in the successive approximation (initially called *feedback subtraction*) ADC conversion process can be traced back to the 1500s relating to the solution of a certain mathematical puzzle regarding the determination of an unknown weight by a minimal sequence of weighing operations (Reference 1). In this problem, as stated, the object is to determine the least number of weights which would serve to weigh an integral number of pounds from 1 lb to 40 lb using a balance scale. One solution put forth by the mathematician Tartaglia in 1556, was to use the series of weights 1 lb, 2 lb, 4 lb, 8 lb, 16 lb, and 32 lb. The proposed weighing algorithm is the same as used

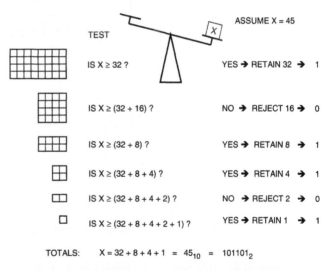

Figure 3.59: Successive Approximation ADC Algorithm

in modern successive approximation ADCs. (It should be noted that this solution will actually measure unknown weights up to 63 lb rather than 40 lb as stated in the problem). The algorithm is shown in Figure 3.59 where the unknown weight is 45 lbs. The balance scale analogy is used to demonstrate the algorithm.

Early implementations of the successive approximation ADC did not use either DACs or SARs and implemented similar functions in a variety of ways. In fact, early SAR ADCs were referred to as *sequential coders*, *feedback coders*, or *feedback subtractor coders*. The term *SAR ADC* came about in the 1970s when commercial successive approximation register logic ICs such as the 2503 and 2504 became available from National Semiconductor and Advanced Micro Devices. These devices were specifically designed to perform the register and control functions in successive approximation ADCs and were standard building blocks in many modular and hybrid data converters.

From a data conversion standpoint, the successive approximation ADC architecture formed the building block for the T1 PCM carrier system and is still a popular architecture today, but the exact origin of this architecture is not clear. It is interesting that it did not appear in Reeves' otherwise comprehensive patent (Reference 2). Although countless patents have been granted relating to refinements and variations on the successive approximation architecture, they do not claim the fundamental principle.

The first mention of the successive approximation ADC architecture in the context of PCM was by J. C. Schelleng of Bell Telephone Laboratories in a patent filed in 1946 (Reference 21). A block diagram of the 6-bit transmitting ADC reproduced from the patent is shown in Figure 3.60. The blocks labeled *selectors* are key to understanding its operation. If the differential input to a selector is greater than its designated voltage, then the differential output of the selector is connected to its designated reference voltage, and the corresponding binary digit is recorded as a "1." If the differential input to a selector is less than its designated voltage, the differential output of the selector is zero, and the corresponding binary digit is recorded as a "0." Notice that all the selectors are floating except for the 32 V selector. For that reason, the *digit grounders* are required to level shift the outputs of the floating selectors and reference them to system ground.

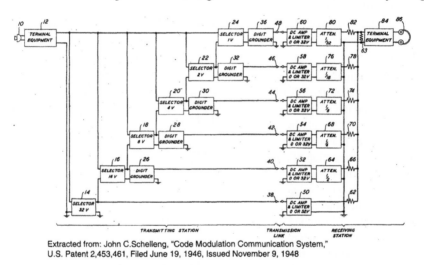

Extracted from: John C.Schelleng, "Code Modulation Communication System,"
U.S. Patent 2,453,461, Filed June 19, 1946, Issued November 9, 1948

Figure 3.60: J. C. Schelleng's 1946 Successive Approximation ADC

The operation of the ADC follows the fundamental successive approximation algorithm. The input signal is first tested by the 32 V selector. If it is greater than 32 V, the selector output is set to 32 V, and a "1" is recorded for the MSB. If it is less than 32 V, the selector output is set to 0 V, and a "0" is recorded for the MSB. The process is continued sequentially for the remaining bits. The selectors are "stacked," i.e., the output of a given selector is connected to one input of the following selector. Therefore the output of the LSB selector represents a 5-bit analog approximation to the input signal.

Details of the rather cumbersome and somewhat impractical vacuum tube design of the selectors and digit grounders are shown in Figure 3.61, also reproduced from the patent. The battery with the label *98* is 32 V, and the battery with the label *130* is 16 V, etc., thereby constituting the binary weighted voltage set required to perform the conversion algorithm.

A much more elegant implementation of the successive approximation ADC is described by Goodall of Bell Telephone Labs in a 1947 article (Reference 22). This ADC has 5-bit resolution and samples the voice channel at a rate of 8 kSPS. The voice signal is first sampled, and the corresponding voltage stored on a capacitor. It is then compared to a reference voltage which is equal to ½ the full-scale voltage. If it is greater than the reference voltage, the MSB is registered as a "1," and an amount of charge equal to ½ scale is subtracted from the storage capacitor. If the voltage on the capacitor is less than ½ scale, no charge is removed, and the bit is registered as a "0." After the MSB decision is completed, the cycle continues for the second bit, but with the reference voltage now equal to ¼ scale.

Both the Schelleng and the Goodall ADCs use a process of addition/subtraction of binary weighted reference voltages to perform the SAR algorithm. Although the DAC function is there, it is not performed using a traditional binary weighted DAC. The ADCs described by H. R. Kaiser et. al. (Reference 23) and B. D. Smith (Reference 24) in 1953 use an actual binary weighted DAC to generate the analog approximation to the input signal, similar to modern SAR ADCs. Smith also points out that nonlinear ADC transfer functions can be achieved by using a nonuniformly weighted DAC. This technique has become the basis of modern companding voiceband codecs. Before this nonlinear ADC technique was developed, linear ADCs were used, and the compression and expansion functions were performed by diode/resistor networks which had to be individually calibrated and held at a constant temperature to prevent drift errors (Reference 25).

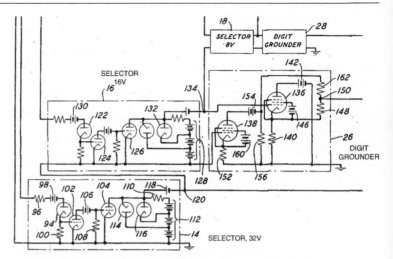

Extracted from: John C.Schelleng, "Code Modulation Communication System," U.S. Patent 2,453,461, Filed June 19, 1946, Issued November 9, 1948

Figure 3.61: Details of the Selectors and Digit Grounders

Of course, no discussion on ADC history would be complete without crediting the truly groundbreaking work of Bernard M. Gordon at EPSCO (now Analogic, Incorporated). Gordon's 1955 patent application (Reference 26) describes an all-vacuum tube 11-bit, 50 kSPS successive approximation ADC—representing the first commercial offering of a complete converter (see Figure 3.62). The DATRAC was offered in a 19" × 26" × 15" housing, dissipated several hundred watts, and sold for approximately $8000.00.

- 19" × 15" × 26"
- 150 lbs
- $8,500.00

Courtesy,
Analogic Corporation
8 Centennial Drive
Peabody, MA 01960

www.analogic.com

Figure 3.62: 1954 "DATRAC" 11-Bit, 50-kSPS SAR ADC Designed by Bernard M. Gordon at EPSCO

In a later patent (Reference 27), Gordon describes the details of the logic block required to perform the successive approximation algorithm. The SAR logic function was later implemented in the 1970s by National Semiconductor and Advanced Micro Devices—the popular 2502/2503/2504 family of IC logic chips. These chips were to become an integral building block of practically all modular and hybrid successive approximation ADCs of the 1970s and 1980s.

	RESOLUTION	SAMPLING RATE	POWER	CHANNELS
AD7482	12 BITS	3.0MSPS	80mW	1
AD7484	14 BITS	3.0MSPS	80mW	1
AD7490	12 BITS	1.0MSPS	6mW	16
AD7928	12 BITS	1.0MSPS	5.4mW	8
AD974	16 BITS	0.2MSPS	120mW	4
AD7677*	16 BITS	1.0MSPS	130mW	1
AD7621*	16 BITS	3.0MSPS	100mW	1
AD7674*	18 BITS	0.8MSPS	120mW	1

* PulSAR SERIES

Figure 3.63: Resolution/Conversion Time Comparison for Representative Single-Supply SAR ADCs

Because of their popularity, successive approximation ADCs are available in a wide variety of resolutions, sampling rates, input and output options, package styles, and costs. It would be impossible to attempt to list all types, but Figure 3.63 shows a number of recent Analog Devices' SAR ADCs which are representative. Note that many devices are complete data acquisition systems with input multiplexers which allow a single ADC core to process multiple analog channels.

An example of modern charge redistribution successive approximation ADCs is Analog Devices' PulSAR series. The AD7677 is a 16-bit, 1 MSPS, PulSAR, fully differential, ADC that operates from a single 5 V power supply (see Figure 3.64). The part contains a high speed 16-bit sampling ADC, an internal conversion clock, error correction circuits, and both serial and parallel system interface ports. The AD7677 is hardware factory calibrated and comprehensively tested to ensure such ac parameters as signal-to-noise ratio (SNR) and total harmonic distortion (THD), in addition to the more traditional dc parameters of gain, offset, and linearity. It features a very high sampling rate mode (Warp) and, for asynchronous conversion rate applications, a fast mode

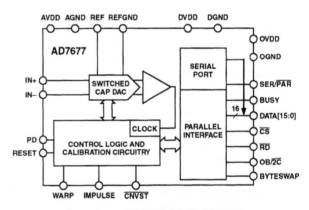

Figure 3.64: AD7677 16-Bit 1 MSPS Switched Capacitor PulSAR ADC

(Normal) and, for low power applications, a reduced power mode (Impulse) where the power is scaled with the throughput. There are three speed versions of the device, the AD7675 (100 kSPS), AD7676 (500 kSPS), and the AD7677 (1 MSPS). The latest addition to the 16-bit family is 3 MSPS AD7621. An 18-bit family of PulSARs is also available: the AD7678 (100 kSPS), AD7679 (570 kSPS) and the AD7674 (800 kSPS).

Subranging, Error Corrected, and Pipelined ADCs

Because of the sheer complexity of constructing an all-parallel flash converter using either vacuum tubes, transistors, or tunnel diodes, the early work such as in the Staffin and Lohman 1956 (filed) patent (Reference 9) used subranging to simplify the conversion process. However, in order to make the subranging ADC practical, a suitable fast sample-and-hold was required. Early subranging ADCs using vacuum tube technology were limited by the sample-and-hold performance, but by 1964 transistors were widely available and Gray and Kitsopoulos of Bell Labs describe pioneering work on the classic diode-bridge sample-and-hold in their 1964 paper (Reference 28).

A basic two-stage N-bit subranging ADC is shown in Figure 3.65. The ADC is based on two separate conversions—a coarse conversion (N1 bits) in the MSB sub-ADC

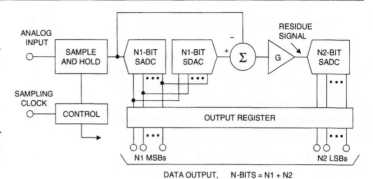

DATA OUTPUT, N-BITS = N1 + N2

See: R. Staffin and R. Lohman , "Signal Amplitude Quantizer,"
U.S. Patent 2,869,079, Filed December 19, 1956, Issued January 13, 1959

Figure 3.65: N-bit Two-Stage Subranging ADC

(SADC) followed by a fine conversion (N2 bits) in the LSB sub-ADC. Early subranging ADCs nearly always used flash converters as building blocks, but a number of recent ADCs utilize other architectures for the individual ADCs.

The conversion process begins placing the sample-and-hold in the hold mode followed by a coarse N1-bit sub-ADC (SADC) conversion of the MSBs. The digital outputs of the MSB converter drive an N1-bit sub-DAC (SDAC) which generates a coarsely quantized version of the analog input signal. The N1-bit SDAC output is subtracted from the held analog signal, amplified, and applied to the N2-bit LSB SADC. The amplifier provides gain, G, sufficient to make the "residue" signal exactly fill the input range of the N2 SADC. The output data from the N1 SADC and the N2 SADC are latched into the output registers yielding the N-bit digital output code, where N = N1 + N2.

In order for this simple subranging architecture to work satisfactorily, both the N1 SADC and SDAC (although they only have N1 bits of resolution) must be better than N-bits accurate. The residue signal offset and gain must be adjusted such that it precisely fills the range of the N2 SADC as shown in Figure 3.66A. If the residue signal drifts by more than 1 LSB (referenced to the N2 SADC), there will be missing codes as shown in Figure 3.66B where the residue signal enters the out-of-range regions labeled "X" and "Y." Any nonlinearity or drift in the N1 SADC will also cause missing codes if it exceeds 1 LSB referenced to N-bits. In practice, an 8-bit subranging ADC with

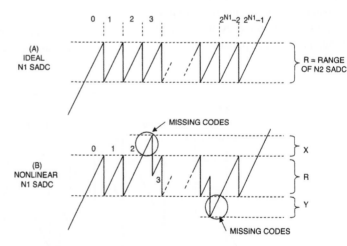

Figure 3.66: Residue Waveforms at Input of N2 Sub-ADC

N1 = 4 bits and N2 = 4 bits represents a realistic limit to this architecture in order to maintain no missing codes over a reasonable operating temperature range.

When the interstage alignment is not correct, missing codes will appear in the overall ADC transfer function as shown in Figure 3.67. If the residue signal goes into positive overrange (the "X" region), the output first "sticks" on a code and then "jumps" over a region leaving missing codes. The reverse occurs if the residue signal is negative overrange.

Figure 3.68 shows a popular 8-bit 15 MSPS subranging ADC manufactured by Computer Labs, Inc. in the mid-1970s. This converter was a basic two-stage subranging ADC with two 4-bit flash converters—each composed of eight dual AM687 high speed comparators. The interstage offset adjustment potentiometer allowed the transfer function to be optimized in the field. This ADC was popular in early digital video products such as frame stores and time base correctors.

In order to reliably achieve higher than 8-bit resolution using the subranging approach, a technique generally referred to as *digital corrected subranging, digital error correction, overlap bits, redundant bits*, etc. is utilized. This method was referred to in literature as early as 1964 by T. C. Verster (Reference 29) and quickly became widely known and utilized (References 30–33). The fundamental concept is illustrated in Figure 3.69.

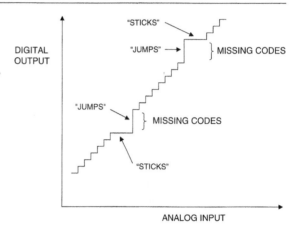

Figure 3.67: Missing Codes Due to MSB SADC Nonlinearity or Interstage Misalignment

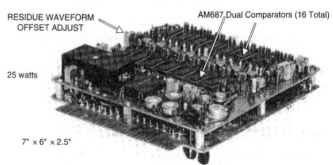

Figure 3.68: MOD-815, 8-Bit, 15 MSPS 4 × 4 Subranging ADC, 1976, Computer Labs, Inc.

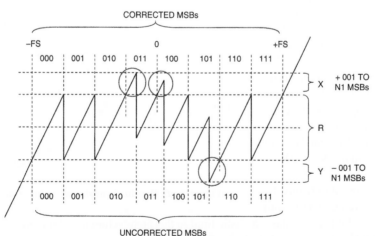

Figure 3.69: Error Correction Using Added Quantization Levels for N1 = 3

A residue waveform is shown for the specific case where N1 = 3 bits. In a standard subranging ADC, the residue waveform must exactly fill the input range of the N2 SADC—it must stay within the region designated R. The missing code problem is solved by adding extra quantization levels in the positive overrange region X and the negative overrange region Y. These additional levels require additional comparators in the basic N2 flash SADC. The scheme works as follows. As soon as the residue enters the X region, the N2 SADC should return to all zeros and start counting up again. In addition, the code 001 is added to the output of the N1 SADC to make the MSBs read the correct code. The figure labels the uncorrected MSB regions on the lower part of the waveform and the corrected MSB regions on the upper part of the waveform. A similar situation occurs when the residue waveform enters the negative overrange region Y. Here, the first quantization level in the Y region should generate the all-ones code, and the additional overrange comparators should cause the count to decrease. In the Y region, the code 001 must be subtracted from the MSBs to produce the corrected MSB code. It is key to understand that in order for this correction method to work properly, the N1 SDAC must be more accurate than the total resolution of the ADC. Nonlinearity or gain errors in the N1 SDAC affect the amplitude of the vertical portions of the residue waveform and therefore can produce missing codes in the output.

Horna, in a 1972 paper (Reference 32), describes an experimental 8-bit 15 MSPS error corrected subranging ADC using Motorola MC1650 dual ECL comparators as the flash converter building blocks. Horna adds additional comparators in the second flash converter and describes this procedure in more detail. He points out that the correction logic can be greatly simplified by adding an appropriate offset to the residue waveform so that there is never a negative overrange condition. This eliminates the need for the subtraction function—only an adder is required. The MSBs are either passed through unmodified, or 1 LSB (relative to the N1 SADC) is added to them, depending on whether the residue signal is in range or overrange.

Modern digitally corrected subranging ADCs generally obtain the additional quantization levels by using an internal ADC with higher resolution for the N2 SADC. For instance, if one additional bit is added to the N2 SADC, its range is doubled and the residue waveform can go outside either end of the range by ½ LSB referenced to the N1 SADC. Adding two extra bits to N2 allows the residue waveform to go outside either end of the range by 1½ LSBs referenced to the N1 SADC. The residue waveform is offset using Horna's technique such that only a simple adder is required to perform the correction logic. The details of how all this works are not immediately obvious, and can best be explained by going through an actual example of a 6-bit ADC with a 3-bit MSB SADC and a 4-bit LSB SADC providing one bit of error correction. The block diagram of the example ADC is shown in Figure 3.70.

After passing through an input sample-and-hold, the signal is digitized by the 3-bit SADC, reconstructed by a 3-bit SDAC, subtracted from the held analog signal and then amplified and applied to the second 4-bit SADC. The gain of the amplifier, G, is chosen so that the residue waveform occupies ½ the input

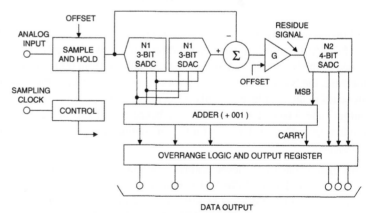

SEE: T. C. Verster, "A Method to Increase the Accuracy of Fast Serial-Parallel Analog-to-Digital Converters," IEEE Transactions on Electronic Computers, EC-13, 1964, pp. 471–473

Figure 3.70: 6-Bit Subranging Error Corrected ADC, N1 = 3, N2 = 4

range of the 4-bit SADC. The 3 LSBs of the 6-bit output data word go directly from the second SADC to the output register. The MSB of the 4-bit SADC controls whether or not the adder adds 001 to the 3 MSBs. The carry output of the adder is used in conjunction with some simple overrange logic to prevent the output bits from returning to the all-zeros state when the input signal goes outside the positive range of the ADC.

The residue waveform for a full-scale ramp input will now be examined in more detail to explain how the correction logic works. Figure 3.71 shows the ideal residue waveform assuming perfect linearity in the first ADC and perfect alignment between the two stages. Notice that the residue waveform occupies exactly one-half the range of the N2 SADC. The 4-bit digital output of the N2 SADC is shown on the left-hand side of the figure. The regions defined by the 3-bit uncorrected N1 SADC are shown on the bottom of the figure. The regions defined by the 3-bit corrected N1 ADC are shown are shown at the top of the figure.

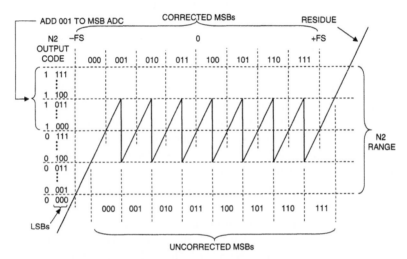

Figure 3.71: 6-Bit Error Corrected Subranging ADC N1 = 3, N2 = 4, Ideal MSB SADC

Following the residue waveform from left to right—as the input first enters the overall ADC range at –FS, the N2 SADC begins to count up, starting at 0000. When the N2 SADC reaches the 1000 code, 001 is added to the N1 SADC output causing it to change from 000 to 001. As the residue waveform continues to increase, the N2 SADC continues to count up until it reaches the code 1100, at which point the N1 SADC switches to the next level, the SDAC switches and causes the residue waveform to jump down to the 0100 output code. The adder is now disabled because the MSB of the N2 SADC is zero, so the N1 SADC output remains 001. The residue waveform then continues to pass through each of the remaining regions until +FS is reached.

This method has some clever features worth mentioning. First, the overall transfer function is offset by ½ LSB referred to the MSB SADC (which is 1/16th FS referred to the overall ADC analog input). This is easily corrected by injecting an offset into the input sample-and-hold. It is well known that the points at which the internal N1 SADC and SDAC switch are the most likely to have additional noise and are the most likely to create differential nonlinearity in the overall ADC transfer function. Offsetting them by 1/16th FS ensures that low level signals (less than ±1/16th FS) near zero volts analog input do not exercise the critical switching points and gives low noise and excellent DNL where they are most needed in communications applications. Finally, since the ideal residue signal is centered within the range of the N2 SADC, the extra range provided by the N2 SADC allows up to a ±1/16th FS error in the N1 SADC conversion while still maintaining no missing codes.

Figure 3.72 shows a residue signal where there are errors in the N1 SADC. Notice that there is no effect on the overall ADC linearity provided the residue signal remains within the range of the N2 SADC. As long as

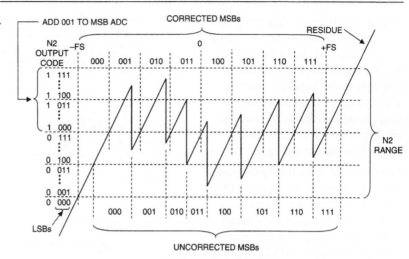

Figure 3.72: 6-Bit Error Corrected Subranging ADC 1 = 3, N2 = 4, Nonlinear MSB SADC

this condition is met, the error correction method described corrects for the following errors: *sample-and-hold droop error, sample-and-hold settling time error, N1 SADC gain error, N1 SADC offset error, N1 SDAC offset error, N1 SADC linearity error, residue amplifier offset error*. In spite of its ability to correct all these errors, it should be emphasized that this method does not correct for gain and linearity errors associated with the N1 SDAC or gain errors in the residue amplifier. The errors in these parameters must be kept less than 1 LSB referred to the N-bits of the overall subranging ADC. Another way to look at this requirement is to realize that the amplitude of the vertical transitions of the residue waveform, corresponding to the N1 SADC and SDAC changing levels, must remain within ±½ LSB referenced to the N2 SADC input in order for the correction to prevent missing codes.

Figure 3.73 shows two methods that can be used to design a pipeline stage in a subranging ADC. Figure 3.73A shows two pipelined stages that use an interstage T/H in order to provide interstage gain and give each stage the maximum possible amount of time to process the signal at its input. In Figure 3.73B a multiplying DAC is used to provide the appropriate amount of interstage gain as well as the subtraction function.

The term "pipelined" architecture refers to the ability of one stage to process data from the previous stage during any given clock cycle. At the end of each phase of a particular clock cycle, the output of a given stage is

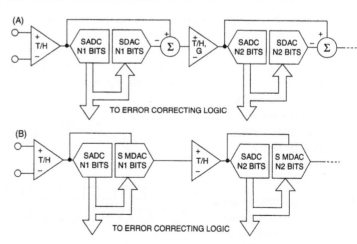

Figure 3.73: Generalized Pipeline Stages in a Subranging ADC with Error Correction

passed on to the next stage using the T/H functions and new data is shifted into the stage. Of course this means that the digital outputs of all but the last stage in the "pipeline" must be stored in the appropriate number of shift registers so that the digital data arriving at the correction logic corresponds to the same sample.

Figure 3.74 shows a timing diagram of a typical pipelined subranging ADC. Notice that the phases of the clocks to the T/H amplifiers are alternated from stage to stage such that when a particular T/H in the ADC enters the hold mode it holds the sample from the preceding T/H, and the preceding

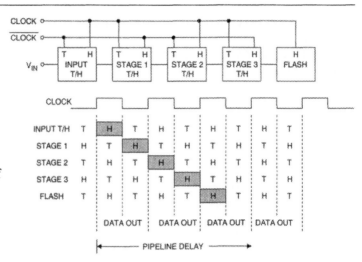

Figure 3.74: Clock Issues in Pipelined ADCs

T/H returns to the track mode. The held analog signal is passed along from stage to stage until it reaches the final stage in the pipelined ADC—in this case, a flash converter. When operating at high sampling rates, it is critical that the differential sampling clock be kept at a 50% duty cycle for optimum performance. Duty cycles other than 50% affect all the T/H amplifiers in the chain—some will have longer than optimum track times and shorter than optimum hold times; while others suffer exactly the reverse condition. Several newer pipelined ADCs including the 12-bit, 65-MSPS AD9235 and the 12-bit, 170 MSPS/210 MSPS AD9430 have on-chip clock conditioning circuits to control the internal duty cycle while allowing some variation in the external clock duty cycle.

The effects of the "pipeline" delay (sometimes called latency) in the output data are shown in Figure 3.75 for the AD9235 12-bit 65 MSPS ADC where there is a 7-clock cycle pipeline delay.

Note that the pipeline delay is a function of the number of stages and the particular architecture of the ADC under consideration—the data sheet should always be consulted for the exact details of the relationship between the sampling clock and the output data timing. In many applications the pipeline delay will not be a problem, but if the ADC is inside a feedback loop the pipeline delay may cause instability. The pipeline delay can also be troublesome in multiplexed applications or when operating the ADC in a "single-shot" mode. Other ADC architectures—such as successive approximation—may be better suited to these types of applications.

It is often erroneously assumed that all subranging ADCs are pipelined, and that all pipelined ADCs are subranging. While it is true that most modern subranging ADCs are

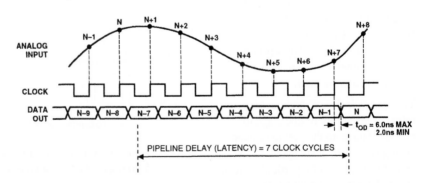

Figure 3.75: Typical Pipelined ADC Timing for AD9235 12-Bit, 65 MSPS ADC

pipelined in order to achieve the maximum possible sampling rate, they don't necessarily have to be pipelined if designed for use at much lower speeds. For instance, the leading edge of the sampling clock could initiate the conversion process, and any additional clock pulses required to continue the conversion could be generated internal to the ADC using an on-chip timing circuit. At the end of the conversion process, an end-of-conversion or data-ready signal could be generated as an external indication that the data corresponding to that particular sampling edge is valid.

Conversely, some ADCs use architectures other than subranging and are pipelined. For instance, many flash converters use an extra set of output latches (in addition to the latch associated with the parallel comparators) which introduces pipeline delay in the output data. Another example of a nonsubranging architecture that generally has quite a bit of pipeline delay is sigma-delta which will be covered in more detail in the next section of this chapter. Note, however, that it is possible to modify the timing of a normal sigma-delta ADC, reduce the output data rate, and make a "no latency" sigma-delta ADC.

The pipelined error-correcting ADC has become very popular in modern ADCs requiring wide dynamic range and low levels of distortion. There are many possible ways to design a pipelined ADC, and we will now look at just a few of the trade-offs. Figure 3.76A shows a pipelined ADC designed with identical stages of k-bits each. This architecture uses the same core hardware in each stage, offers a few other advantages, but does necessarily optimize the ADC for best possible performance. Figure 3.76B shows the simplest form of this architecture where k = 1.

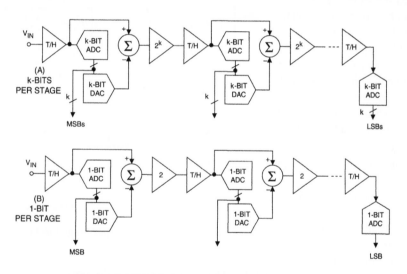

Figure 3.76: Basic Pipelined ADC with Identical Stages

In order to optimize performance at the 12-bit level, for example, 1-bit per stage pipeline is more commonly used with a multibit front-end and back-end ADC as shown in Figure 3.77.

Error correction is used in practically all pipelined ADCs, including the simple 1-bit stage. Figure 3.78 shows how an ADC constructed of uncorrected cascaded 1-bit stages will ultimately result in missing codes unless each stage is nearly ideal.

Error correction can be added to the simple 1-bit stage by adding a single extra comparator—resulting in what is commonly referred to as a "1.5-bit" stage as shown in Figure 3.79. Details of this architecture can be found in Reference 34. The two comparators have three possible output codes: 00, 01, and 10. Note that three parallel comparators form a complete 2-bit stage—which would be required for the final stage in a pipelined 1.5-bit ADC, as one additional output level is required to generate the 11 code.

It would appear at first glance that simply adding a single comparator would not allow the error correction scheme to work as previously illustrated for the multibit cases. The explanation is as follows. The thresholds of the first 1.5-bit stage are designated M1, M2, and the thresholds of the second 1.5-bit stage are designated L1, L2. The residue waveform driving the second 1.5-bit stage is shown for an ideal first stage. The "corrected" codes for the first stage are obtained by simply adding the output code of the second stage to that of the first stage. Note that the correct output code is obtained as long as the residue signal falls within the dotted box—a third comparator is not required (except for the last stage which must be 2 bits or greater). The final bit decision for the first stage is made based on the corrected bits as follows: a 00, 01, or 10 code indicates a "0," and a 11 code indicates a "1."

Figure 3.80 shows the effects of errors made in the first stage converter. Errors of up to ±½ LSB in the first stage can be corrected by this method. Figure 3.81 shows the effects of error correction for a different set of first stage errors which still fall within the ±½ LSB correction range.

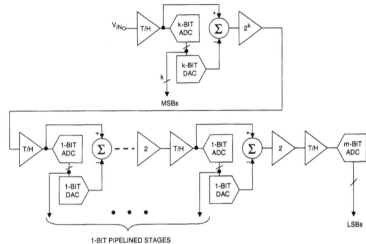

Figure 3.77: Multibit and 1-Bit Pipelined Core Combined

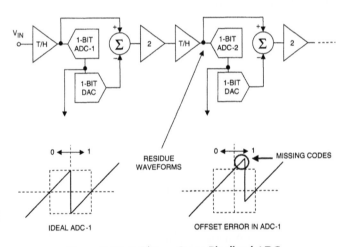

Figure 3.78: 1-Bit-per-Stage Pipelined ADC

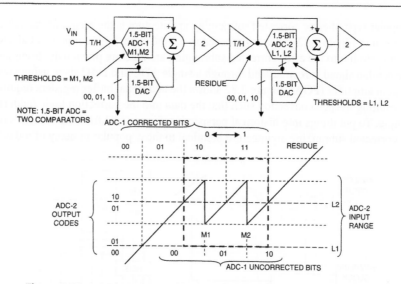

Figure 3.79: 1.5-Bit-per-Stage Pipeline Showing Error Corrected Range

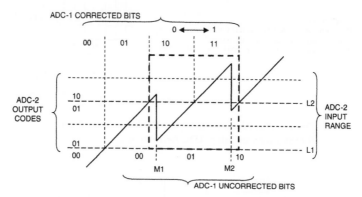

Figure 3.80: Residue Waveform for 1.5-Bit ADC-2 Stage Input with Nonlinear ADC-1 Stage, Case 1

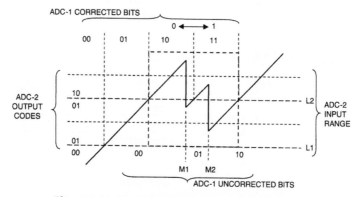

Figure 3.81: Residue Waveform for 1.5-Bit ADC-2 Stage Input with Nonlinear ADC-1 Stage, Case 2

Another less popular type of error corrected subranging architecture is the *recirculating* subranging ADC. This is shown in Figure 3.82 and was proposed in a 1966 paper by Kinniment, et.al. (Reference 31). The concept is similar to the error corrected subranging architecture previously discussed, but in this architecture the residue signal is recirculated through a single ADC and DAC stage using switches and a programmable gain amplifier (PGA). Figure 3.82 shows the additional buffer registers required to store the pipelined data resulting in each conversion such that the data into the correction logic (adder) corresponds to the same sample. To put things into historical perspective, Figure 3.83 from Kinniment's paper shows a pipelined error corrected subranging architecture identical to those popular in many of today's ADCs.

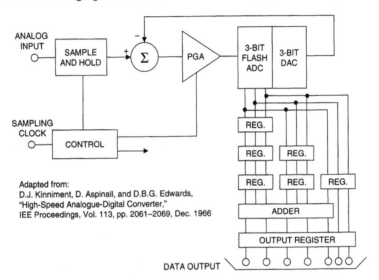

Figure 3.82: Kinniment, et. al., 1966 Pipelined 7-Bit,
9 MSPS Recirculating ADC Architecture

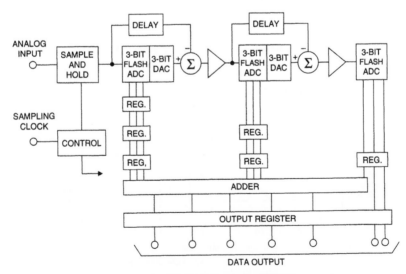

Figure 3.83: Kinniment, et. al., 1966 Proposed
Pipelined 7-Bit ADC Architecture

200

The discussion of error corrected pipelined ADCs concludes with a few examples of modern integrated circuit implementations of the popular architecture. These examples show the flexibility of the technique in optimizing ADC performance at different resolutions, sampling rates, power dissipation, etc.

The demand for wide dynamic range high speed ADCs suitable for communications applications led to the development of a breakthrough product in 1995, the AD9042 12-bit, 41 MSPS ADC (see Reference 35). A block diagram of the converter is shown in Figure 3.84.

The AD9042 uses an error-corrected subranging architecture composed of a 6-bit MSB ADC/DAC followed by a 7-bit LSB ADC and uses one bit of error correction. The AD9042 yields 80 dB SFDR performance over the Nyquist bandwidth at a sampling rate of 41 MSPS. Fabricated on a high speed complementary bipolar process, the device dissipates 600 mW and operates on a single 5 V supply.

In order to meet the need for lower cost, lower power devices, Analog Devices initiated a family of CMOS high performance ADCs such as the AD9225 12-bit, 25 MSPS ADC released in 1998. The AD9225 dissipates 280 mW, has 85 dB SFDR, and operates on a single 5 V supply. A simplified diagram of the AD9225 is shown in Figure 3.85.

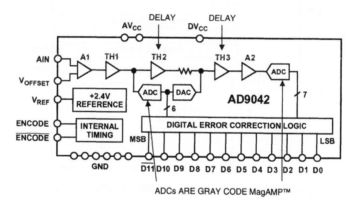

Figure 3.84: AD9042 12-Bit 41 MSPS ADC, 1995

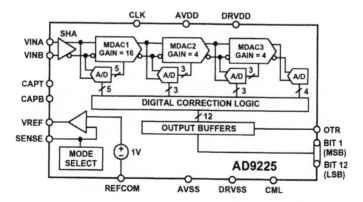

Figure 3.85: AD9225 12-Bit 25 MSPS CMOS ADC, 1998

The AD9235 12-bit 65 MSPS CMOS ADC released in 2001 shows the progression of CMOS high performance converters. The AD9235 operates on a single 3 V supply, dissipates 300 mW (at 65 MSPS), and has a 90 dB SFDR over the Nyquist bandwidth. The ADC uses eight stages of 1.5-bit converters (two comparators) previously described in this section of the book. A simplified diagram of the AD9235 is shown in Figure 3.86.

The 12-bit 210 MSPS AD9430 released in 2002 is shown in Figure 3.87 and is fabricated on a BiCMOS process, has 80 dB SFDR up to 70 MHz inputs, operates on a single 3 V supply and dissipates 1.3 W at 210 MSPS. Output data is provided on two ports at 105 MSPS each in the CMOS mode or on a single port at 210 MSPS in the LVDS (low voltage digital signal) mode.

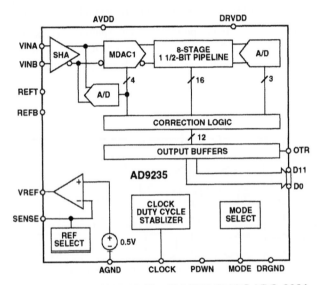

Figure 3.86: AD9235 12-Bit, 65 MSPS CMOC ADC, 2001

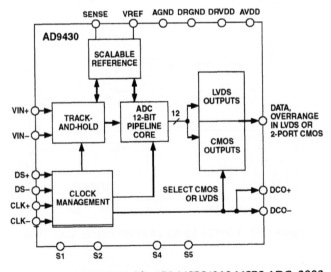

Figure 3.87: AD9430 12-Bit, 170 MSPS/210 MSPS ADC, 2002

The final example (certainly not to imply the conclusion of a complete listing) is the 14-bit 105 MSPS AD6645 ADC released in 2003 and fabricated on a high speed complementary bipolar process (XFCB), has 90 dB SFDR, operates on a single 5 V supply and dissipates 1.5 W. The 80 MSPS version of the AD6645 was released in 2002. A simplified diagram of the AD6645 is shown in Figure 3.88.

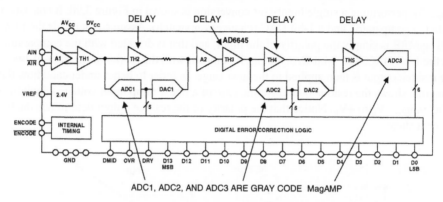

ADC1, ADC2, AND ADC3 ARE GRAY CODE MagAMP

Figure 3.88: AD6645 14-Bit, 105 MSPS ADC, 2003 (80 MSPS Version Released in 2002)

Achieving the level of performance shown in the above examples is by no means easy. A variety of circuit and process design trade-offs must be made. Circuit design requires fast-settling op amps, track-and-hold amplifiers, linear DACs with low glitch, careful attention to chip layout, etc. Bit shuffling using a thermometer DAC architecture is often used in the internal DAC structures as well as the addition of dither signals to increase the SFDR performance. In addition to the design of the actual IC, the system designer must use extremely good layout, grounding, and decoupling techniques in order to achieve specified performance. Many of these important hardware design techniques are covered in detail in Chapters 6 and 9 of this book.

Serial Bit-Per-Stage Binary and Gray Coded (Folding) ADCs

Various architectures exist for performing A/D conversion using one stage per bit. Figure 3.89 shows the overall concept. In fact, a multistage subranging ADC with one bit per stage and no error correction is one form previously discussed. In this approach, the input signal must be held constant during the entire conversion cycle. There are N stages, each of which has a bit output and a *residue* output. The residue output of one stage is the input to the next. The last bit is detected with a single comparator as shown.

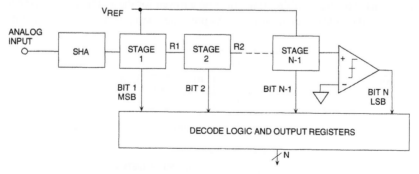

B. D. Smith, "An Unusual Electronic Analog-Digital Conversion Method,"
IRE Transactions on Instrumentation, June 1956, pp. 155–160.

Figure 3.89: Generalized Bit-Per-Stage ADC Architecture

One of the first references to these architectures appeared in an article by B. D. Smith in 1956 (Reference 36). Smith indicates, however, that previous work had been done at M.I.T. by R. P. Sallen in a 1949 thesis. In the article, Smith describes both the binary and the Gray (or folding) transfer functions required to implement the A/D conversion.

The basic stage for performing a single binary bit conversion is shown in Figure 3.90. It consists of a gain-of-two amplifier, a comparator, and a 1-bit DAC (changeover switch). Assume that this is the first stage of the ADC. The MSB is simply the polarity of the input, and that is detected with the comparator which also controls the 1-bit DAC. The 1-bit DAC output is summed with the output of the gain-of-two amplifier. The resulting residue output is then applied to the next stage. In order to better understand how the circuit works, the diagram shows the residue output for the case of a linear ramp input voltage which traverses the entire ADC range, $-V_R$ to $+V_R$. Notice that the polarity of the residue output determines the binary bit output of the next stage.

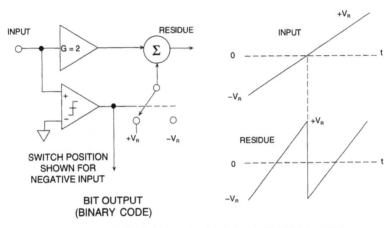

Figure 3.90: Single-Stage Transfer Function for Binary ADC

A simplified 3-bit serial-binary ADC is shown in Figure 3.91, and the residue outputs are shown in Figure 3.92. Again, the case is shown for a linear ramp input voltage whose range is between $-V_R$ and $+V_R$. Each residue output signal has discontinuities that correspond to the point where the comparator changes state and causes the DAC to switch. The fundamental problem with this architecture is the discontinuity in the residue output waveforms. Adequate settling time must be allowed for these transients to propagate through all the stages and settle at the final comparator input. As presented here, the prospects of making this

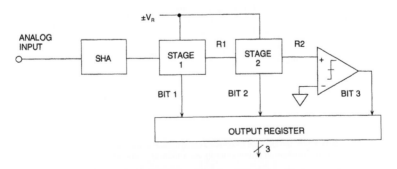

Figure 3.91: 3-Bit Serial ADC with Binary Output

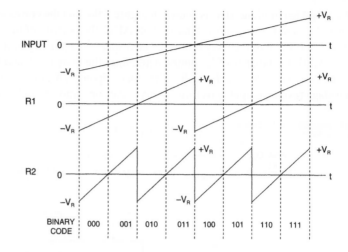

Figure 3.92: Input and Residue Waveforms of 3-Bit Binary Ripple ADC

architecture operate at high speed are dismal. However, using the 1.5-bit-per-stage pipelined architecture previously discussed in this section makes it much more attractive at high speeds.

Although the binary method is discussed in his paper, B. D. Smith also describes a much preferred bit-per-stage architecture based on absolute value amplifiers (magnitude amplifiers, or simply *MagAMPs*). This scheme has often been referred to as *serial-Gray* (since the output coding is in Gray code), or *folding* converter because of the shape of the transfer function. Performing the conversion using a transfer function that produces an initial Gray code output has the advantage of minimizing discontinuities in the residue output waveforms and offers the potential of operating at much higher speeds than the binary approach.

The basic folding stage is shown functionally in Figure 3.93 along with its transfer function. The input to the stage is assumed to be a linear ramp voltage whose range is between $-V_R$ and $+V_R$. The comparator detects the polarity of the input signal and provides the Gray bit output for the stage. It also determines whether the overall stage gain is +2 or −2. The reference voltage V_R is summed with the switch output to generate the residue signal which is applied to the next stage. The polarity of the residue signal determines the Gray bit for the next stage. The transfer function for the folding stage is also shown in Figure 3.93.

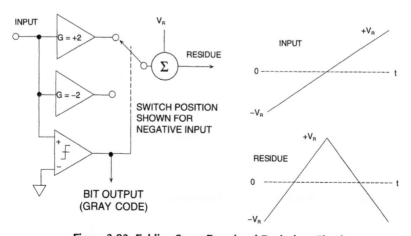

Figure 3.93: Folding Stage Functional Equivalent Circuit

205

A 3-bit MagAMP folding ADC is shown in Figure 3.94, and the corresponding residue waveforms in Figure 3.95. As in the case of the binary ripple ADC, the polarity of the residue output signal of a stage determines the value of the Gray bit for the next stage. The polarity of the input to the first stage determines the Gray MSB; the polarity of R1 output determines the Gray bit-2; and the polarity of R2 output determines the Gray bit-3. Notice that unlike the binary ripple ADC, there is no abrupt transition in any of the folding stage residue output waveforms. This makes operation at high speeds quite feasible.

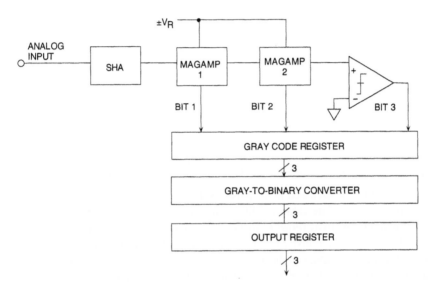

Figure 3.94: 3-Bit Folding ADC Block Diagram

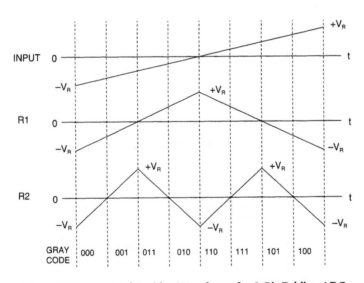

Figure 3.95: Input and Residue Waveforms for 3-Bit Folding ADC

The key to operating this architecture at high speeds is the folding stage. N. E. Chasek of Bell Telephone Labs describes a circuit for generating the folding transfer function using nested diode bridges in a patent filed in 1960 (Reference 37). This circuit made use of solid-state devices, but required different reference voltages for each stage (see Figure 3.96). Chasek's circuit also suffered from loss of headroom and gain when several stages were cascaded to form higher resolution converters as shown in Figure 3.97. What is really needed to make the folding ADC work at high resolutions is nearly ideal voltage or current rectification.

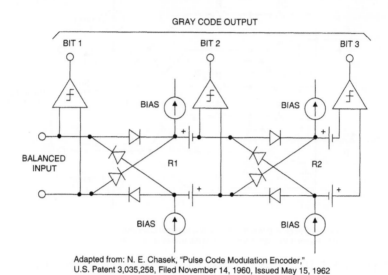

Adapted from: N. E. Chasek, "Pulse Code Modulation Encoder,"
U.S. Patent 3,035,258, Filed November 14, 1960, Issued May 15, 1962

Figure 3.96: 3-Bit Folding ADC Based on N. E. Chasek's Design

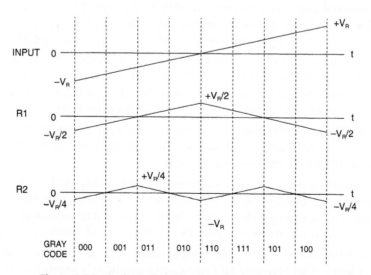

Figure 3.97: Single-Ended Waveforms in Chasek's Folding ADC

F. D. Waldhaur of Bell Telephone Labs remedied the problems of Chasek's nested diode bridge circuits in a classic patent filed in 1962 (Reference 38). Figure 3.98 shows Waldhaur's elegant implementation of the folding transfer function using solid state op amps with diodes in the feedback loop. The gain-of-two op amps allow the same reference voltages to be used for each stage and maintain the same signal level at each residue output with nearly ideal rectification.

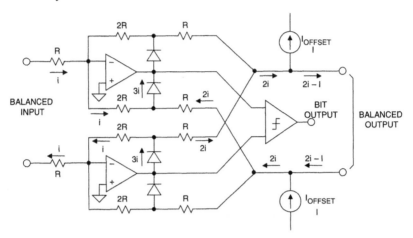

Extracted from: F. D.Waldhauer, "Analog-to-digital Converter,"
U.S. Patent 3,187,325, Filed July 2, 1962, Issued June 1, 1965

Figure 3.98: F. D. Waldhaur's Classic Folding Stage using Rectifier Amplifiers

J. O. Edson and H. H. Henning describe the operation and performance of this type of ADC in greater detail in a 1965 *Bell System Technical Journal* article (Reference 39). An operational 9-bit, 6 MSPS ADC of this type was used in experimental studies on 224-Mbit/second PCM terminals. These terminals were supposed to handle data as well as voice signals. The voiceband objective was to digitize an entire 600-channel, 2.4 MHz FDM band, therefore requiring a minimum sampling rate of approximately 6 MSPS.

It is interesting to note that the experimental terminal was also supposed to handle video as well, which required a higher sampling rate of approximately 12 MSPS. For this requirement, the latest (and final) generation Bell Labs' electron beam coder was needed to meet the ADC requirement, as the solid-state coder based on Waldhaur's patent did not have the necessary accuracy at the higher sampling rates.

The first commercial ADC using Waldhaur's Gray code architecture was the 8-bit, 10 MSPS HS-810 from Computer Labs, Inc., in 1966 (see Chapter 1 of this book). The instrument used all discrete transistor circuits (no ICs) and was designed to be mounted in a 19" rack.

The folding Gray code architecture was used in a few instrument and modular ADCs in the early 1970s, but was largely replaced by the error-corrected subranging architecture. With improvements in IC processes, there was renewed interest in the folding architecture in the late 1970s and throughout the 1980s—with quite a number of designs reported in the various journals over the period (References 40–44).

Analog Devices developed the first high speed fully complementary bipolar (CB) process in the mid-1980s, and in 1994 Frank Murden and Carl Moreland filed patents on a significantly improved current-steering architecture for a Gray code MagAMP-based ADC (References 45–49). The technique was first implemented for building block cores in the AD9042 12-bit, 41 MSPS ADC released in 1995, and refinements of

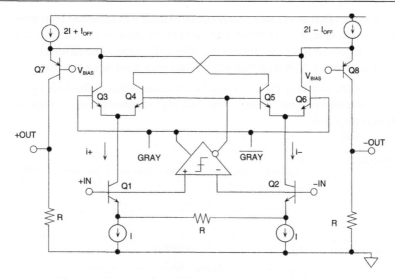

Figure 3.99: A Modern Current-Steering MagAMP Stage

the technique and a higher speed CB process, XFCB, (References 50 and 51) pushed the core technology to 14 bits with the release of the AD6644 14-bit 65 MSPS ADC in 1999, the AD6645 14-bit 80 MSPS ADC in 2001, and a 105 MSPS version of the AD6645 in 2003. Although these ADCs use the error-corrected subranging architecture, the internal building block core ADCs utilize the MagAMP architecture.

Modern IC circuit designs implement the transfer function using current-steering open-loop gain techniques that can be made to operate much faster. Fully differential stages (including the SHA) also provide speed, lower distortion, and yield 8-bit accurate folding stages with no requirement for thin film resistor laser trimming.

An example of a fully differential gain-of-two MagAMP folding stage is shown in Figure 3.99 (see References 45, 46, 48). The differential input signal is applied to the degenerated-emitter differential pair Q1,Q2 and the comparator. The differential input voltage is converted into a differential current which flows in the collectors of Q1, Q2. If +IN is greater than –IN, cascode-connected transistors Q3, Q6 are on, and Q4, Q6 are off. The differential signal currents therefore flow through the collectors of Q3, Q6 into level-shifting transistors Q7, Q8 and into the output load resistors, developing the differential output voltage between +OUT and –OUT. The overall differential voltage gain of the circuit is two.

If +IN is less than –IN (negative differential input voltage), the comparator changes state and turns Q4, Q5 on and Q3, Q6 off. The differential signal currents flow from Q5 to Q7 and from Q4 to Q8, thereby maintaining the same relative polarity at the differential output as for a positive differential input voltage. The required offset voltage is developed by adding a current I_{OFF} to the emitter current of Q7 and subtracting it from the emitter current of Q8.

The differential residue output voltage of the stage drives the next stage input, and the comparator output represents the Gray code output for the stage.

The MagAMP architecture offers lower power and can be extended to sampling rates previously dominated by flash converters. For example, the AD9054A 8-bit, 200 MSPS ADC is shown in Figure 3.100. The first five bits (Gray code) are derived from five differential MagAMP stages. The differential residue output of the fifth MagAMP stage drives a 3-bit flash converter, rather than a single comparator.

The Gray-code output of the five MagAMPs and the binary-code output of the 3-bit flash are latched, all converted into binary, and latched again in the output data register. Because of the high data rate, a demultiplexed output option is provided.

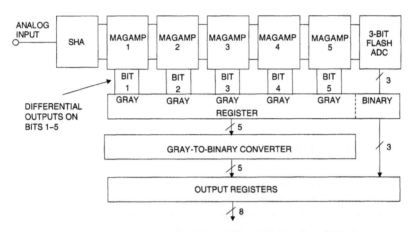

Figure 3.100: AD9054A 8-Bit, 200 MSPS ADC Functional Diagram

A summary of the timeline for the popular high speed ADC architectures discussed in this section is shown in Figure 3.101 to put things into a historical perspective.

• Reeve's counting ADC	1939
• Successive approximation	1946
• Flash (electron tube coders)	1948
• Bit-per-stage (binary and folding-Gray)	1956
• Subranging	1956
• Subranging with error correction	1964
• Pipeline with error correction	1966

Note: Dates are first publications or patent filings

Figure 3.101: High Speed ADC Architecture Timeline

Counting and Integrating ADC Architectures

Although counting-based ADCs are not well suited for high speed applications, they are ideal for high resolution low frequency applications, especially when combined with integrating techniques. The evolution of counting-based ADCs follows the approximate timeline shown in Figure 3.102. Notice that the development of the various counting/integrating ADC architectures runs roughly in parallel with the development of the higher speed architectures (Figure 3.101). By 1957 (with the exception of triple and quad slope), all the fundamental architectures had been proposed in one form or another.

• Reeve's counting ADC	1939
• Charge run -down:	1946
• Ramp run -up	1951
• Tracking	1950
• Voltage -to-frequency converter (VFC)	1952
• Dual Slope	1957
• Triple Slope	1967
• Quad Slope	1973

Note: Dates are first publications or patent filings

Figure 3.102: Counting and Integrating ADC Architecture Timeline

A. H. Reeves' 5-Bit Counting ADC

As previously discussed, the first ADC suitable for PCM applications was the one documented by A. H. Reeves in his comprehensive 1939 PCM patent (Reference 24). A simplified diagram of the ADC is repeated here in Figure 3.103. The early ADCs for PCM typically had 5–7 bits of resolution and sampling rates of 6 kSPS–10 kSPS. Interestingly enough, Reeves' ADC was based on a counting technique, probably because of his general interest in counters—the Eccles-Jordan bistable multivibrator had been invented only a few years earlier. However other architectures such as successive approximation, flash, bit-per-stage, subranging, and pipeline were much more widely used in later PCM applications.

The counting ADC technique (Figure 3.103) basically uses a sampling pulse to take a sample of the analog signal, set an R/S flip-flop, and simultaneously start a controlled ramp voltage. The ramp voltage is compared with the input, and when they are equal, a pulse is generated which resets the R/S flip-flop. The output of the flip-flop is a pulse whose width is proportional to the analog signal at the sampling instant. This pulsewidth modulated (PWM) pulse controls a gated oscillator, and the number of pulses out of the gated oscillator

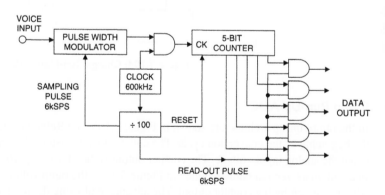

Adapted from: Alec Harley Reeves, "Electric Signaling System," U.S. Patent 2,272,070, Filed November 22, 1939, Issued February 3, 1942

Figure 3.103: A. H. Reeves' 5-bit Counting ADC

represents the quantized value of the analog signal. This pulse train can be easily converted to a binary word by driving a counter. In Reeves' system, a master clock of 600 kHz was used, and a 100:1 divider generated the 6 kHz sampling pulses. The system uses a 5-bit counter, and 31 counts (out of the 100 counts between sampling pulses) therefore represents a full-scale signal. The technique can obviously be extended to higher resolutions.

Charge Run-Down ADC

The charge run-down ADC architecture (see Reference 52) (Figure 3.104) first samples the analog input and stores the voltage on a fixed capacitor. The capacitor is then discharged with a constant current source, and the time required for complete discharge is measured using a counter. Notice that in this approach, the overall accuracy is dependent on the quality and magnitude of the capacitor, the magnitude of the current source, as well as the accuracy of the timebase.

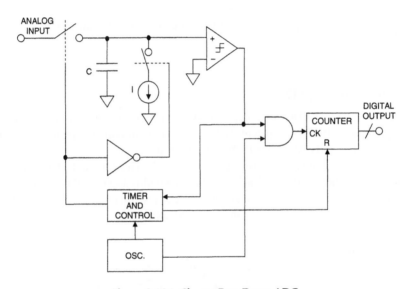

Figure 3.104: Charge Run-Down ADC

Ramp Run-Up ADC

In the ramp run-up architecture shown in Figure 3.105 (see Reference 53), a ramp generator is started at the beginning of the conversion cycle. The counter then measures the time required for the ramp voltage to equal the analog input voltage. The counter output is therefore proportional to the value of the analog input. In an alternate version (shown dotted in Figure 3.105), the ramp voltage generator is replaced by a DAC which is driven by the counter output. The advantage of using the ramp is that the ADC is always mono-tonic, whereas overall monotonicity is determined by the DAC when it is used as a substitute.

The accuracy of the ramp run-up ADC depends on the accuracy of the ramp generator (or the DAC) as well as the oscillator.

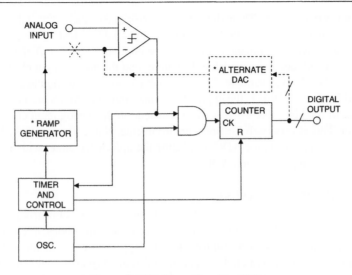

Figure 3.105: Ramp Run-Up ADC

Tracking ADC

The tracking ADC architecture shown in Figure 3.106 (see References 54 and 55) continually compares the input signal with a reconstructed representation of the input signal. The up/down counter is controlled by the comparator output. If the analog input exceeds the DAC output, the counter counts up until they are equal. If the DAC output exceeds the analog input, the counter counts down until they are equal. It is evident that if the analog input changes slowly, the counter will follow, and the digital output will remain close to its correct value. If the analog input suddenly undergoes a large step

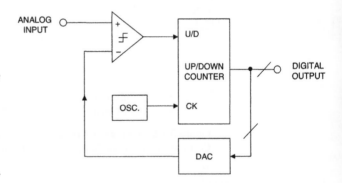

Figure 3.106: Tracking ADC

change, it will be many hundreds or thousands of clock cycles before the output is again valid. The tracking ADC therefore responds quickly to slowly changing signals, but slowly to a quickly changing one.

The simple analysis above ignores the behavior of the ADC when the analog input and DAC output are nearly equal. This will depend on the exact nature of the comparator and counter. If the comparator is a simple one, the DAC output will cycle by 1 LSB from just above the analog input to just below it, and the digital output will, of course, do the same—there will be 1 LSB of flicker. Note that the output in such a case steps every clock cycle, irrespective of the exact value of analog input, and hence always has unity Mark/Space ratio. In other words, there is no possibility of taking a mean value of the digital output and increasing resolution by oversampling.

A more satisfactory, but more complex, arrangement would be to use a window comparator with a window 1–2 LSB wide. When the DAC output is high or low the system behaves as in the previous description, but if the DAC output is within the window, the counter stops. This arrangement eliminates the flicker, provided that the DAC DNL never allows the DAC output to step across the window for 1 LSB change in code.

Tracking ADCs are not very common. Their slow step response makes them unsuitable for many applications, but they do have one asset: their output is *continuously* available. Most ADCs perform conversions: i.e., on receipt of a "start convert" command (which may be internally generated), they perform a conversion and, after a delay, a result becomes available. Providing that the analog input changes slowly, the output of a tracking ADC is always available. This is valuable in synchro-to-digital and resolver to digital converters (SDCs and RDCs), and this is the application where tracking ADCs are most often used. Another valuable characteristic of tracking ADCs is that a fast transient on the analog input causes the output to change only one count. This is very useful in noisy environments. Notice the similarity between a tracking ADC and a successive approximation ADC. Replacing the up/down counter with SAR logic yields the architecture for a successive approximation ADC.

Voltage-to-Frequency Converters (VFCs)

A voltage-to-frequency converter (VFC) is an oscillator whose frequency is linearly proportional to a control voltage. The VFC/counter ADC is monotonic and free of missing codes, integrates noise, and can consume very little power. It is also very useful for telemetry applications, since the VFC, which is small, cheap and low-powered can be mounted on the experimental subject (patient, wild animal, artillery shell, etc.) and communicate with the counter by a telemetry link as shown in Figure 3.107.

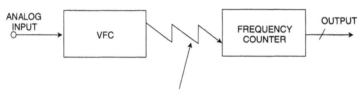

- CONNECTION NEED NOT BE DIRECT
- CIRCUIT IS IDEAL FOR TELEMETRY

Figure 3.107: Voltage-to-Frequency Converter (VFC) and Frequency Counter Make a Low Cost, Versatile, High Resolution ADC

There are two common VFC architectures: the *current- steering multivibrator VFC* and the *charge-balance VFC* (Reference 56). The charge-balanced VFC may be made in *asynchronous* or *synchronous* (clocked) forms. There are many more VFO (variable frequency oscillator) architectures, including the ubiquitous 555 timer, but the key feature of VFCs is linearity—few VFOs are very linear.

The current-steering multivibrator VFC is actually a current-to-frequency converter rather than a VFC, but, as shown in Figure 3.108, practical circuits invariably contain a voltage-to-current converter at the input. The principle of operation is evident: the current discharges the capacitor until a threshold is reached, and when the capacitor terminals are reversed, the half-cycle repeats itself. The waveform across the capacitor is a linear triangular wave, but the waveform on either terminal with respect to ground is the more complex waveform shown.

Practical VFCs of this type have linearities around 14 bits, and comparable stability, although they may be used in ADCs with higher resolutions without missing codes. The performance limits are set by comparator threshold noise, threshold temperature coefficient, and the stability and dielectric absorption (DA) of the capacitor, which is generally a discrete component. The comparator/voltage reference structure shown in the diagram is more of a representation of the function performed than the actual circuit used, which is much more integrated with the switching, and correspondingly harder to analyze.

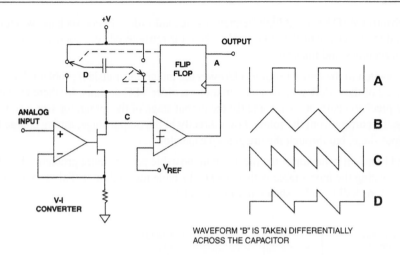

Figure 3.108: A Current-Steering VFC

This type of VFC is simple, inexpensive, and low powered, and most run from a wide range of supply voltages. They are ideally suited for low cost medium accuracy ADC and data telemetry applications.

The charge balance VFC shown in Figure 3.109 is more complex, more demanding in its supply voltage and current requirements, and more accurate. It is capable of 16-18 bit linearity.

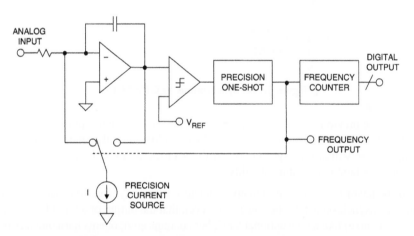

Figure 3.109: Charge Balance Voltage-to-Frequency Converter (VFC)

The integrator capacitor charges from the signal as shown in Figure 3.109. When it passes the comparator threshold, a fixed charge is removed from the capacitor, but the input current continues to flow during the discharge, so no input charge is lost. The fixed charge is defined by the precision current source and the pulse width of the precision monostable. The output pulse rate is thus accurately proportional to the rate at which the integrator charges from the input.

At low frequencies, the limits on the performance of this VFC are set by the stability of the current source and the monostable timing (which depends on the monostable capacitor, among other things). The absolute value and temperature stability of the integration capacitor do not affect the accuracy, although its leakage and dielectric absorption (DA) do. At high frequencies, second-order effects, such as switching transients in the integrator and the precision of the monostable when it is retriggered very soon after the end of a pulse, take their toll on accuracy and linearity.

The changeover switch in the current source addresses the integrator transient problem. By using a change-over switch instead of the on/off switch more common on older VFC designs: (a) there are no on/off transients in the precision current source and (b) the output stage of the integrator sees a constant load—most of the time the current from the source flows directly in the output stage; during charge balance, it still flows in the output stage, but through the integration capacitor.

The stability and transient behavior of the precision monostable present more problems, but the issue may be avoided by replacing the monostable with a clocked bistable multivibrator. This arrangement is known as a *synchronous* VFC or SVFC and is shown in Figure 3.110.

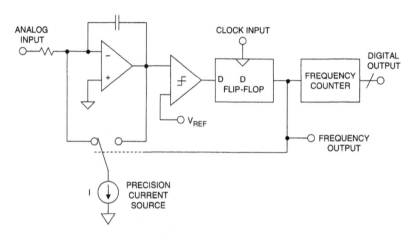

Figure 3.110: Synchronous VFC (SVFC)

The difference from the previous circuit is quite small, but the charge balance pulse length is now defined by two successive edges of the external clock. If this clock has low jitter, the charge will be very accurately defined. The output pulse will also be synchronous with the clock. SVFCs of this type are capable of up to 18-bit linearity and excellent temperature stability.

This synchronous behavior is convenient in many applications, since synchronous data transfer is often easier to handle than asynchronous. It does mean, however, that the output of an SVFC is not a pure tone (plus harmonics, of course) like a conventional VFC, but contains components harmonically related to the clock frequency. The display of an SVFC output on an oscilloscope is especially misleading and is a common cause of confusion—a change of input to a VFC produces a smooth change in the output frequency, but a change to an SVFC produces a change in probability density of output pulses N and N+1 clock cycles after the previous output pulse, which is often misinterpreted as severe jitter and a sign of a faulty device (see Figure 3.111).

Another problem with SVFCs is nonlinearity at output frequencies related to the clock frequency. If we study the transfer characteristic of an SVFC, we find nonlinearities close to sub-harmonics of the clock frequency F_C as shown in Figure 3.112. They can be found at $F_C/3$, $F_C/4$, and $F_C/6$. This is due to stray capacitance on the chip (and in the circuit layout!) and the coupling of the clock signal into the SVFC comparator which causes the device to behave as an injection-locked phase-locked loop (PLL). This problem is intrinsic to SVFCs, but is not often serious: if the circuit card is well laid out, and clock amplitude and dv/dts kept as low as practical, the effect is a discontinuity in the transfer characteristic of less than 8 LSBs (at 18-bit resolution) at $F_C/3$ and $F_C/4$, and less at other sub-harmonics. This is frequently tolerable, since the frequencies where it occurs are known. Of course, if the circuit layout or decoupling is poor, the effect may be much larger, but this is the fault of poor design and not the SVFC itself.

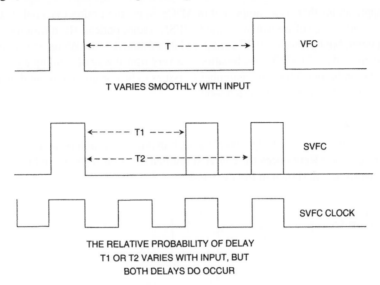

Figure 3.111: VFC and SVFC Waveforms

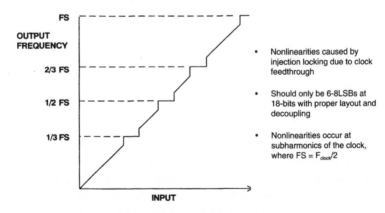

Figure 3.112: SVFC Nonlinearity

It is evident that the SVFC is quantized, while the basic VFC is not. It does NOT follow from this that the counter/VFC ADC has higher resolution (neglecting nonlinearities) than the counter/SVFC ADC, because the clock in the counter also sets a limit to the resolution.

When a VFC has a large input, it runs quickly and (counting for a short time) gives good resolution, but it is hard to get good resolution in a reasonable sample time with a slow-running VFC. In such a case, it may be more practical to measure the period of the VFC output (this does not work for an SVFC), but of course the resolution of this system deteriorates as the input (and the frequency) increases. However, if the counter/timer arrangement is made "smart," it is possible to measure the approximate VFC frequency and the exact period of not one, but N cycles (where the value of N is determined by the approximate frequency), and maintain high resolution over a wide range of inputs. The AD1170 modular ADC released in 1986 is an example of this architecture.

VFCs have more applications than as a component in ADCs. Since their output is a pulse stream, it may easily be sent over a wide range of transmission media (PSN, radio, optical, IR, ultrasonic, etc.). It need not be received by a counter, but by another VFC configured as a frequency-to-voltage converter (FVC). This gives an analog output, and a VFC-FVC combination is a very useful way of sending a precision analog signal across an isolation barrier. There are a number of issues to be considered in building FVCs from VFCs, and these are considered in References 57–60.

Dual Slope/Multislope ADCs

The dual-slope ADC architecture was truly a breakthrough in ADCs for high resolution applications such as digital voltmeters, etc. (see References 61–64). A simplified diagram is shown in Figure 3.113, and the integrator output waveforms are shown in Figure 3.114.

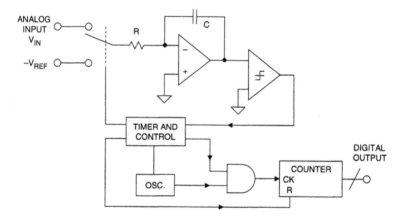

Figure 3.113: Dual Slope ADC

The input signal is applied to an integrator; at the same time a counter is started, counting clock pulses. After a predetermined amount of time (T), a reference voltage having opposite polarity is applied to the integrator. At that instant, the accumulated charge on the integrating capacitor is proportional to the average value of the input over the interval T. The integral of the reference is an opposite-going ramp having a slope of V_{REF}/RC. At the same time, the counter is again counting from zero. When the integrator output reaches zero, the count is stopped, and the analog circuitry is reset. Since the charge gained is proportional to $V_{IN} \times T$, and the equal amount of charge lost is proportional to $V_{REF} \times t_x$, then the number of counts relative to the full scale count is proportional to t_x/T, or V_{IN}/V_{REF}. If the output of the counter is a binary number, it will therefore be a binary representation of the input voltage.

Dual slope integration has many advantages. Conversion accuracy is independent of both the capacitance and the clock frequency, because they affect both the up-slope and the down-slope by the same ratio.

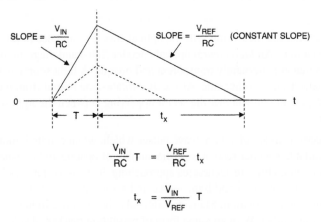

$$\frac{V_{IN}}{RC} T = \frac{V_{REF}}{RC} t_x$$

$$t_x = \frac{V_{IN}}{V_{REF}} T$$

HIGH NORMAL MODE REJECTION AT MULTIPLES OF $\frac{1}{T}$

Figure 3.114: Dual Slope ADC Integrator Output Waveforms

The fixed input signal integration period results in rejection of noise frequencies on the analog input that have periods that are equal to or a submultiple of the integration time T. Proper choice of T can therefore result in excellent rejection of 50 Hz and 60 Hz line ripple as shown in Figure 3.115.

Errors caused by bias currents and the offset voltages of the integrating amplifier and the comparator as well as gain errors can be cancelled by using additional charge/discharge cycles to measure "zero" and "full-scale" and using the results to digitally correct the initial measurement, as in the quad-slope architecture discussed in Reference 65.

The triple-slope architecture (see References 66–68) retains the advantages of the dual slope, but greatly increases the conversion speed at the cost of added complexity. The increase in conversion speed is achieved by accomplishing the reference integration (ramp-down) at two distinct rates: a high-speed rate, and a "vernier" lower speed rate. The counter is likewise divided into two sections, one for the MSBs and one for the LSBs. In a properly designed triple slope converter, a significant increase in speed can be achieved while retaining the inherent linearity, differential linearity, and stability characteristics associated with dual slope ADCs.

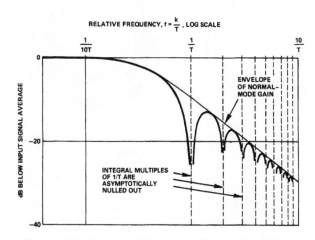

Figure 3.115: Frequency Response of Integrating ADC

Optical Converters

Among the most popular position measuring sensors, optical encoders find use in relatively low reliability and low resolution applications. An *incremental* optical encoder (left-hand diagram in Figure 3.116) is a disc divided into sectors that are alternately transparent and opaque. A light source is positioned on one side of the disc, and a light sensor on the other side. As the disc rotates, the output from the detector switches alternately on and off, depending on whether the sector appearing between the light source and the detector is transparent or opaque.

Thus, the encoder produces a stream of square wave pulses which, when counted, indicate the angular position of the shaft. Available encoder resolutions (the number of opaque and transparent sectors per disc) range from 100 to 65,000, with absolute accuracies approaching 30 arc-seconds (1/43,200 per rotation). Most incremental encoders feature a second light source and sensor at an angle to the main source and sensor, to indicate the direction of rotation. Many encoders also have a third light source and detector to sense a once-per-revolution marker. Without some form of revolution marker, absolute angles are difficult to determine. A potentially serious disadvantage is that incremental encoders require external counters to determine absolute angles within a given rotation. If the power is momentarily shut off, or if the encoder misses a pulse due to noise or a dirty disc, the resulting angular information will be in error.

The *absolute* optical encoder (right-hand diagram in Figure 3.116) overcomes these disadvantages but is more expensive. An absolute optical encoder's disc is divided up into N sectors (N = 5 for example shown), and each sector is further divided radially along its length into opaque and transparent sections, forming a unique N-bit digital word with a maximum count of $2^N - 1$. The digital word formed radially by each sector increments in value from one sector to the next, usually employing Gray code. Binary coding could be used, but can produce large errors if a single bit is incorrectly interpreted by the sensors. Gray code overcomes this defect: the maximum error produced by an error in any single bit of the Gray code is only 1 LSB after the Gray code is converted into binary code. A set of N light sensors responds to the N-bit digital word which corresponds to the disc's absolute angular position. Industrial optical encoders achieve up to 16-bit resolution, with absolute accuracies that approach the resolution (20 arc seconds). Both absolute and incremental optical encoders, however, may suffer damage in harsh industrial environments.

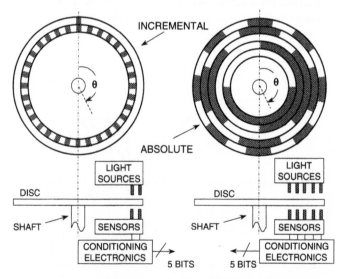

Figure 3.116: Incremental and Absolute Optical Encoders

Resolver-to-Digital Converters (RDCs) and Synchros

Machine-tool and robotics manufacturers have increasingly turned to resolvers and synchros to provide accurate angular and rotational information. These devices excel in demanding factory applications requiring small size, long-term reliability, absolute position measurement, high accuracy, and low noise operation.

A diagram of a typical synchro and resolver is shown in Figure 3.117. Both synchros and resolvers employ single-winding rotors that revolve inside fixed stators. In the case of a simple synchro, the stator has three windings oriented 120° apart and electrically connected in a Y-connection. Resolvers differ from synchros in that their stators have only two windings oriented at 90°.

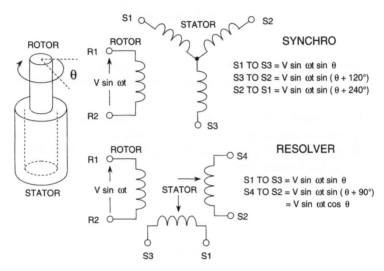

Figure 3.117: Synchros and Resolvers

Because synchros have three stator coils in a 120° orientation, they are more difficult than resolvers to manufacture and are therefore more costly. Today, synchros find decreasing use, except in certain military and avionic retrofit applications.

Modern resolvers, in contrast, are available in a brushless form that employ a transformer to couple the rotor signals from the stator to the rotor. The primary winding of this transformer resides on the stator, and the secondary on the rotor. Other resolvers use more traditional brushes or slip rings to couple the signal into the rotor winding. Brushless resolvers are more rugged than synchros because there are no brushes to break or dislodge, and the life of a brushless resolver is limited only by its bearings. Most resolvers are specified to work over 2 V to 40 V rms and at frequencies from 400 Hz to 10 kHz. Angular accuracies range from 5 arc-minutes to 0.5 arc-minutes. (There are 60 arc-minutes in one degree, and 60 arc-seconds in one arc-minute. Hence, one arc-minute is equal to 0.0167 degrees.)

In operation, synchros and resolvers resemble rotating transformers. The rotor winding is excited by an ac reference voltage, at frequencies up to a few kHz. The magnitude of the voltage induced in any stator winding is proportional to the sine of the angle, θ, between the rotor coil axis and the stator coil axis. In the case of a synchro, the voltage induced across any pair of stator terminals will be the vector sum of the voltages across the two connected coils.

For example, if the rotor of a synchro is excited with a reference voltage, Vsinωt, across its terminals R1 and R2, then the stator's terminal will see voltages in the form:

$$S1 \text{ to } S3 = V \sin\omega t \sin\theta \qquad\qquad Eq. 3.2$$

$$S3 \text{ to } S2 = V \sin\omega t \sin(\theta + 120°) \qquad\qquad Eq. 3.3$$

$$S2 \text{ to } S1 = V \sin\omega t \sin(\theta + 240°) \qquad\qquad Eq. 3.4$$

where θ is the shaft angle.

In the case of a resolver, with a rotor ac reference voltage of Vsinωt, the stator's terminal voltages will be:

$$S1 \text{ to } S3 = V \sin\omega t \sin\theta \qquad\qquad Eq. 3.5$$

$$S4 \text{ to } S2 = V \sin\omega t \sin(\theta + 90°) = V \sin\omega t \cos\theta \qquad\qquad Eq. 3.6$$

It should be noted that the 3-wire synchro output can be easily converted into the resolver-equivalent format using a Scott-T transformer. Therefore, the following signal processing example describes only the resolver configuration.

A typical resolver-to-digital converter (RDC) is shown functionally in Figure 3.118. The two outputs of the resolver are applied to cosine and sine multipliers. These multipliers incorporate sine and cosine lookup tables and function as multiplying digital-to-analog converters. Begin by assuming that the current state of the up/down counter is a digital number representing a trial angle, φ. The converter seeks to adjust the digital angle, φ, continuously to become equal to, and to track θ, the analog angle being measured. The resolver's stator output voltages are written as:

$$V_1 = V \sin\omega t \sin\theta \qquad\qquad Eq. 3.7$$

$$V_2 = V \sin\omega t \cos\theta \qquad\qquad Eq. 3.8$$

where θ is the angle of the resolver's rotor. The digital angle φ is applied to the cosine multiplier, and its cosine is multiplied by V_1 to produce the term:

$$V \sin\omega t \sin\theta \cos\varphi \qquad\qquad Eq. 3.9$$

The digital angle φ is also applied to the sine multiplier and multiplied by V_2 to product the term:

$$V \sin\omega t \cos\theta \sin\varphi \qquad\qquad Eq. 3.10$$

These two signals are subtracted from each other by the error amplifier to yield an ac error signal of the form:

$$V \sin\omega t [\sin\theta \cos\varphi - \cos\theta \sin\varphi] \qquad\qquad Eq. 3.11$$

Using a simple trigonometric identity, this reduces to:

$$V \sin\omega t [\sin(\theta - \varphi)] \qquad\qquad Eq. 3.12$$

The detector synchronously demodulates this ac error signal, using the resolver's rotor voltage as a reference. This results in a dc error signal proportional to $\sin(\theta-\varphi)$.

The dc error signal feeds an integrator, the output of which drives a voltage-controlled-oscillator (VCO). The VCO, in turn, causes the up/down counter to count in the proper direction to cause:

$$\text{Sin } (\theta - \varphi) \rightarrow 0 \qquad\qquad Eq. 3.13$$

When this is achieved,

$$\theta - \varphi \rightarrow 0 \qquad\qquad \text{Eq. 3.14}$$

and therefore

$$\varphi = \theta \qquad\qquad \text{Eq. 3.15}$$

to within one count. Hence, the counter's digital output, φ, represents the angle θ. The latches enable this data to be transferred externally without interrupting the loop's tracking.

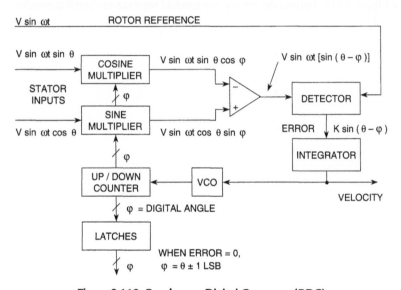

Figure 3.118: Resolver-to-Digital Converter (RDC)

This circuit is equivalent to a so-called type-2 servo loop, because it has, in effect, two integrators. One is the counter, which accumulates pulses; the other is the integrator at the output of the detector. In a type-2 servo loop with a constant rotational velocity input, the output digital word continuously follows, or tracks the input, without needing externally derived convert commands, and with no steady state phase lag between the digital output word and actual shaft angle. An error signal appears only during periods of acceleration or deceleration.

As an added bonus, the tracking RDC provides an analog dc output voltage directly proportional to the shaft's rotational velocity. This is a useful feature if velocity is to be measured or used as a stabilization term in a servo system, and it makes additional tachometers unnecessary.

Since the operation of an RDC depends only on the ratio between input signal amplitudes, attenuation in the lines connecting them to resolvers doesn't substantially affect performance. For similar reasons, these converters are not greatly susceptible to waveform distortion. In fact, they can operate with as much as 10% harmonic distortion on the input signals; some applications actually use square-wave references with little additional error.

Tracking ADCs are therefore ideally suited to RDCs. While other ADC architectures, such as successive approximation, could be used, the tracking converter is the most accurate and efficient for this application.

Because the tracking converter doubly integrates its error signal, the device offers a high degree of noise immunity (12 dB-per-octave roll-off). The net area under any given noise spike produces an error. However, typical inductively coupled noise spikes have equal positive and negative going waveforms. When integrated, this results in a zero net error signal. The resulting noise immunity, combined with the converter's insensitivity to voltage drops, lets the user locate the converter at a considerable distance from the resolver. Noise rejection is further enhanced by the detector's rejection of any signal not at the reference frequency, such as wideband noise.

The AD2S90 is one of a number of integrated RDCs offered by Analog Devices. The general architecture is similar to that of Figure 3.118. Further details on synchro and resolver-to-digital converters can be found in References 69 and 70.

References:
3.2 ADC Architectures

1. Reza Moghimi, "Amplifiers as Comparators," Ask the Applications Engineer 31, Analog Dialogue, Vol. 37-04, Analog Devices, April 2003, www.analog.com.

2. James N. Giles, "High Speed Transistor Difference Amplifier," **U.S. Patent 3,843,934**, filed January 31, 1973, issued October 22, 1974. *(Describes one of the first high-speed ECL comparators, the AM685.)*

3. Christopher W. Mangelsdorf, *A 400 MHz Input Flash Converter with Error Correction*, **IEEE Journal of Solid-State Circuits**, Vol. 25, No. 1, February 1990, pp. 184–191. *(A discussion of the AD770, an 8-bit 200 MSPS flash ADC. The paper describes the comparator metastable state problem and how to optimize the ADC design to minimize its effects.)*

4. Charles E. Woodward, *A Monolithic Voltage-Comparator Array for A/D Converters*, **IEEE Journal of Solid State Circuits**, Vol. SC-10, No. 6, December 1975, pp. 392–399. *(An early paper on a 3-bit flash converter optimized to minimize metastable state errors.)*

5. Paul M. Rainey, "Facimile Telegraph System," **U.S. Patent 1,608,527**, filed July 20, 1921, issued November 30, 1926. *(Although A. H. Reeves is generally credited with the invention of PCM, this patent discloses an electro-mechanical PCM system complete with A/D and D/A converters. The 5-bit electro-mechanical ADC described is probably the first documented flash converter. The patent was largely ignored and forgotten until many years after the various Reeves' patents were issued in 1939–1942.)*

6. R. W. Sears, "Electron Beam Deflection Tube for Pulse Code Modulation," **Bell System Technical Journal**, Vol. 27, pp. 44–57, Jan. 1948. *(Describes an electron-beam deflection tube 7-bit, 100 kSPS flash converter for early experimental PCM work.)*

7. Frank Gray, "Pulse Code Communication," **U.S. Patent 2,632,058**, filed November 13, 1947, issued March 17, 1953. *(Detailed patent on the Gray code and its application to electron beam coders.)*

8. J. O. Edson and H. H. Henning, "Broadband Codecs for an Experimental 224Mb/s PCM Terminal," **Bell System Technical Journal**, Vol. 44, pp. 1887–1940, Nov. 1965. *(Summarizes experiments on ADCs based on the electron tube coder as well as a bit-per-stage Gray code 9-bit solid state ADC. The electron beam coder was 9 bits at 12 MSPS, and represented the fastest of its type at the time.)*

9. R. Staffin and R. D. Lohman, "Signal Amplitude Quantizer," **U.S. Patent 2,869,079**, filed December 19, 1956, issued January 13, 1959. *(Describes flash and subranging conversion using tubes and transistors.)*

10. Goto, et. al., "Esaki Diode High-Speed Logical Circuits," **IRE Transactions on Electronic Computers**, Vol. EC-9, March 1960, pp. 25-29. *(Describes how to use tunnel diodes as logic elements.)*

11. T. Kiyomo, K. Ikeda, and H. Ichiki, "Analog-to-Digital Converter Using an Esaki Diode Stack," **IRE Transactions on Electronic Computers**, Vol. EC-11, December 1962, pp. 791–792. *(Description of a low resolution 3-bit flash ADC using a stack of tunnel diodes.)*

12. H. R. Schindler, "Using the Latest Semiconductor Circuits in a UHF Digital Converter," **Electronics**, August 1963, pp. 37–40. *(Describes a 6-bit 50 MSPS subranging ADC using three 2-bit tunnel diode flash converters.)*

13. J. B. Earnshaw, "Design for a Tunnel Diode-Transistor Store with Nondestructive Read-out of Information," **IEEE Transactions on Electronic Computers**, EC-13, 1964 , pp. 710–722. *(Use of tunnel diodes as memory elements.)*

14. Willard K. Bucklen, "A Monolithic Video A/D Converter," **Digital Video, Vol. 2**, Society of Motion Picture and Television Engineers, March 1979, pp. 34–42. *(Describes the revolutionary TDC1007J 8-bit 20 MSPS video flash converter. Originally introduced at the February 3, 1979 SMPTE Winter Conference in San Francisco, Bucklen accepted an Emmy award for this product in 1988 and was responsible for the initial marketing and applications support for the device.)*

15. J. Peterson, "A Monolithic Video A/D Converter," **IEEE Journal of Solid-State Circuits**, Vol. SC-14, No. 6, December 1979, pp. 932–937. *(Another detailed description of the TRW TDC1007J 8-bit, 20 MSPS flash converter.)*

16. Yukio Akazawa et. al., *A 400 MSPS 8-Bit Flash A/D Converter*, **1987 ISSCC Digest of Technical Papers**, pp. 98–99. *(Describes a monolithic flash converter using Gray decoding.)*

17. A. Matsuzawa *et al.*, *An 8 b 600 MHz Flash A/D Converter with Multistage Duplex-Gray Coding*, **Symposium VLSI Circuits, Digest of Technical Papers**, May 1991, pp. 113–114. *(Describes a monolithic flash converter using Gray decoding.)*

18. Chuck Lane, *A 10-Bit 60 MSPS Flash ADC*, **Proceedings of the 1989 Bipolar Circuits and Technology Meeting**, IEEE Catalog No. 89CH2771-4, September 1989, pp. 44–47. *(Describes an interpolating method for reducing the number of preamps required in a flash converter.)*

19. W. W. Rouse Ball and H. S. M. Coxeter, **Mathematical Recreations and Essays**, Thirteenth Edition, Dover Publications, 1987, pp. 50, 51. *(Describes a mathematical puzzle for measuring unknown weights using the minimum number of weighing operations. The solution proposed in the 1500s is the same basic successive approximation algorithm used today.)*

20. Alec Harley Reeves, "Electric Signaling System," **U.S. Patent 2,272,070**, filed November 22, 1939, issued February 3, 1942. Also **French Patent 852,183** issued 1938, and **British Patent 538,860** issued 1939. *(The ground-breaking patent on PCM. Interestingly enough, the ADC and DAC proposed by Reeves are counting types, and not successive approximation.)*

21. John C. Schelleng, "Code Modulation Communication System," **U.S. Patent 2,453,461**, filed June 19, 1946, issued November 9, 1948. *(An interesting description of a rather cumbersome successive approximation ADC based on vacuum tube technology. This converter was not very practical, but did illustrate the concept. Also in the patent is a description of a corresponding binary DAC.)*

22. W. M. Goodall, "Telephony by Pulse Code Modulation," **Bell System Technical Journal**, Vol. 26, pp. 395–409, July 1947. *(Describes an experimental PCM system using a 5-bit, 8 KSPS successive approximation ADC-based on the subtraction of binary weighted charges from a capacitor to implement the internal subtraction/DAC function. It required 5 internal reference voltages.)*

23. Harold R. Kaiser, et al, "High-Speed Electronic Analogue-to-Digital Converter System," **U.S. Patent 2,784,396**, filed April 2, 1953, issued March 5, 1957. *(One of the first SAR ADCs to use an actual binary-weighted DAC internally.)*

24. B. D. Smith, "Coding by Feedback Methods," **Proceedings of the I. R. E.**, Vol. 41, August 1953, pp. 1053–1058. *(Smith uses an internal DAC and also points out that a nonlinear transfer function can be achieved by using a DAC with nonuniform bit weights, a technique that is widely used in today's voiceband ADCs with built-in companding.)*

25. L.A. Meacham and E. Peterson, "An Experimental Multichannel Pulse Code Modulation System of Toll Quality," **Bell System Technical Journal**, Vol. 27, No. 1, January 1948, pp. 1–43. *(Describes nonlinear diode-based compressors and expanders for generating a nonlinear ADC/DAC transfer function.)*

26. Bernard M. Gordon and Robert P. Talambiras, "Signal Conversion Apparatus," **U.S. Patent 3,108,266**, filed July 22, 1955, issued October 22, 1963. *(Classic patent describing Gordon's 11-bit, 20 kSPS vacuum tube successive approximation ADC done at Epsco. The internal DAC represents the first known use of equal currents switched into an R/2R ladder network.)*

27. Bernard M. Gordon and Evan T. Colton, "Signal Conversion Apparatus," **U.S. Patent 2,997,704**, filed February 24, 1958, issued August 22, 1961. *(Classic patent describes the logic to perform the successive approximation algorithm in a SAR ADC.)*

28. J. R. Gray and S. C. Kitsopoulos, "A Precision Sample-and-Hold Circuit with Subnanosecond Switching," **IEEE Transactions on Circuit Theory**, CT11, September 1964, pp. 389–396. *(One of the first papers on the detailed analysis of a sample-and-hold circuit.)*

29. T. C. Verster, "A Method to Increase the Accuracy of Fast Serial-Parallel Analog-to-Digital Converters," **IEEE Transactions on Electronic Computers**, EC-13, 1964, pp. 471–473. *(One of the first references to the use of error correction in a subranging ADC.)*

30. G. G. Gorbatenko, High-Performance Parallel-Serial Analog to Digital Converter with Error Correction, **IEEE National Convention Record**, New York, March 1966. *(Another early reference to the use of error correction in a subranging ADC.)*

31. D. J. Kinniment, D. Aspinall, and D.B.G. Edwards, "High-Speed Analogue-Digital Converter," **IEE Proceedings**, Vol. 113, pp. 2061–2069, Dec. 1966. *(A 7-bit 9 MSPS three-stage pipelined error corrected converter is described based on recirculating through a 3-bit stage three times. Tunnel [Esaki] diodes are used for the individual comparators. The article also shows a proposed faster pipelined 7-bit architecture using three individual 3-bit stages with error correction, and describes a fast bootstrapped diode-bridge sample-and-hold circuit.)*

32. O. A. Horna, "A 150 Mbps A/D and D/A Conversion System," **Comsat Technical Review**, Vol. 2, No. 1, pp. 52–57, 1972. *(A detailed description and analysis of a subranging ADC with error correction.)*

33. J. L. Fraschilla, R. D. Caveney, and R. M. Harrison, "High Speed Analog-to-Digital Converter," **U.S. Patent 3,597,761**, filed Nov. 14, 1969, issued Aug. 13, 1971. *(Describes an 8-bit, 5 MSPS subranging ADC with switched references to second comparator bank.)*

34. Stephen H. Lewis, Scott Fetterman, George F. Gross, Jr., R. Ramachandran, and T. R. Viswanathan, "A 10 b 20 Msample/s Analog-Digital Converter, **IEEE Journal of Solid-State Circuits**, Vol. 27, No. 3, March 1992, pp. 351–358. *(A detailed description and analysis of an error corrected subranging ADC using 1.5-bit pipelined stages.)*

35. Roy Gosser and Frank Murden, "A 12-Bit 50 MSPS Two-Stage A/D Converter," **1995 ISSCC Digest of Technical Papers**, p. 278. *(A description of the AD9042 error corrected subranging ADC using MagAMP stages for the internal ADCs.)*

36. B. D. Smith, "An Unusual Electronic Analog-Digital Conversion Method," **IRE Transactions on Instrumentation**, June 1956, pp. 155–160. *(Possibly the first published description of the binary-coded and Gray-coded bit-per-stage ADC architectures. Smith mentions similar work partially covered in R. P. Sallen's 1949 thesis at M.I.T.)*

37. N. E. Chasek, "Pulse Code Modulation Encoder," **U.S. Patent 3,035,258**, filed November 14, 1960, issued May 15, 1962. *(An early patent showing a diode-based circuit for realizing the Gray code folding transfer function.)*

38. F. D. Waldhauer, "Analog-to-Digital Converter," **U.S. Patent 3,187,325**, filed July 2, 1962, issued June 1, 1965. *(A classic patent using op amps with diode switches in the feedback loops to implement the Gray code folding transfer function.)*

39. J. O. Edson and H. H. Henning, "Broadband Codecs for an Experimental 224 Mb/s PCM Terminal," **Bell System Technical Journal**, Vol. 44, pp. 1887–1940, Nov. 1965. *(A further description of a 9-bit ADC based on Waldhauer's folding stage).*

40. Udo Fiedler and Dieter Seitzer, "A High-Speed 8-Bit A/D Converter Based on a Gray-Code Multiple Folding Circuit," **IEEE Journal of Solid-State Circuits**, Vol. SC-14, No. 3, June 1979, pp. 547–551. *(An early monolithic folding ADC.)*

41. Rudy J. van de Plassche and Rob E. J. van de Grift, "A High-Speed 7-Bit A/D Converter," **IEEE Journal of Solid-State Circuits**, Vol. SC-14, No. 6, December 1979, pp. 938–943. *(A monolithic folding ADC.)*

42. Rob. E. J. van de Grift and Rudy J. van de Plassche, "A Monolithic 8-Bit Video A/D Converter, **IEEE Journal of Solid State Circuits**, Vol. SC-19, No. 3, June 1984, pp. 374–378. *(A monolithic folding ADC.)*

43. Rob. E. J. van de Grift, Ivo W. J. M. Rutten and Martien van der Veen, "An 8-Bit Video ADC Incorporating Folding and Interpolation Techniques," **IEEE Journal of Solid State Circuits**, Vol. SC-22, No. 6, December 1987, pp. 944–953. *(Another monolithic folding ADC.)*

44. Rudy van de Plassche, **Integrated Analog-to-Digital and Digital-to-Analog Converters**, Kluwer Academic Publishers, 1994, pp. 148–187. *(A good textbook on ADCs and DACs with a section on folding ADCs indicated by the referenced page numbers.)*

45. Carl Moreland, "An 8-Bit 150 MSPS Serial ADC," **1995 ISSCC Digest of Technical Papers**, Vol. 38, p. 272. *(A description of an 8-bit ADC with five folding stages followed by a 3-bit flash converter.)*

46. Carl Moreland, **An Analog-to-Digital Converter Using Serial-Ripple Architecture**, Masters' Thesis, Florida State University College of Engineering, Department of Electrical Engineering, 1995. *(Moreland's early work on folding ADCs.)*

47. Frank Murden, "Analog to Digital Converter Using Complementary Differential Emitter Pairs," **U.S. Patent 5,550,492**, filed December 1, 1994, issued August 27, 1996. *(A description of an ADC based on the MagAMP folding stage.)*

48. Carl W. Moreland, "Analog to Digital Converter Having a Magnitude Amplifier with an Improved Differential Input Amplifier," **U.S. Patent 5,554,943**, filed December 1, 1994, issued September 10, 1996. *(A description of an 8-bit ADC with five folding stages followed by a 3-bit flash converter.)*

49. Frank Murden and Carl W. Moreland, "N-Bit Analog-to-Digital Converter with N-1 Magnitude Amplifiers and N Comparators," **U.S. Patent 5,684,419**, filed December 1, 1994, issued November 4, 1997. *(Another patent on the MagAMP folding architecture applied to an ADC.)*

50. Carl Moreland, Frank Murden, Michael Elliott, Joe Young, Mike Hensley, and Russell Stop, "A 14-Bit 100 Msample/s Subranging ADC, **IEEE Journal of Solid State Circuits**, Vol. 35, No. 12, December 2000, pp. 1791–1798. *(Describes the architecture used in the 14-bit AD6645 ADC.)*

51. Frank Murden and Michael R. Elliott, "Linearizing Structures and Methods for Adjustable-Gain Folding Amplifiers," **U.S. Patent 6,172,636B1**, filed July 13, 1999, issued January 9, 2001. *(Describes methods for trimming the folding amplifiers in an ADC.)*

52. Bernard M. Oliver and Claude E. Shannon, "Communication System Employing Pulse Code Modulation," **U.S. Patent 2,801,281**, filed February 21, 1946, issued July 30, 1957. *(Charge run-down ADC and Shannon-Rack DAC.)*

53. Arthur H. Dickinson, "Device to Manifest an Unknown Voltage as a Numerical Quantity," **U.S. Patent 2,872,670**, filed May 26, 1951, issued February 3, 1959. *(Ramp run-up ADC.)*

54. K. Howard Barney, "Binary Quantizer," **U.S. Patent 2,715,678**, filed May 26, 1950, issued August 16, 1955. *(Tracking ADC.)*

55. Bernard M. Gordon and Robert P. Talambiras, "Information Translating Apparatus and Method," **U.S. Patent 2,989,741**, filed July 22, 1955, issued June 20, 1961. *(Tracking ADC.)*

56. John L. Lindesmith, "Voltage-to-Digital Measuring Circuit," **U.S. Patent 2,835,868**, filed September 16, 1952, issued May 20, 1958. *(Voltage-to-frequency ADC.)*

57. Paul Klonowski, "Analog-to-Digital Conversion Using Voltage-to-Frequency Converters," **Application Note AN-276**, Analog Devices, Inc. *(A good application note on VFCs.)*

58. James M. Bryant, "Voltage-to-Frequency Converters," **Application Note AN-361**, Analog Devices, Inc. *(A good overview of VFCs.)*

59. Walt Jung, "Operation and Applications of the AD654 IC V-F Converter," **Application Note AN-278**, Analog Devices, Inc.

60. Steve Martin, "Using the AD650 Voltage-to-Frequency Converter as a Frequency-to-Voltage Converter," **Application Note AN-279**, Analog Devices, Inc. *(A description of a frequency-to-voltage converter using the popular AD650 VFC.)*

61. Robin N. Anderson and Howard A. Dorey, "Digital Voltmeters," **U.S. Patent 3,267,458**, filed August 20, 1962, issued August 16, 1966. *(Charge balance dual slope voltmeter ADC.)*

62. Richard Olshausen, "Analog-to-Digital Converter," **U.S. Patent 3,281,827**, filed June 27, 1963, issued October 25, 1966. *(Charge balance dual slope ADC.)*

63. Roswell W. Gilbert, "Analog-to-Digital Converter," **U.S. Patent 3,051,939**, filed May 8, 1957, issued August 28, 1962. *(Dual-slope ADC.)*

64. Stephan K. Ammann, "Integrating Analog-to-Digital Converter," **U.S. Patent 3,316,547**, filed July 15, 1964, issued April 25, 1967. *(Dual-slope ADC.)*

65. Ivar Wold, "Integrating Analog-to-Digital Converter Having Digitally-Derived Offset Error Compensation and Bipolar Operation without Zero Discontinuity," **U.S. Patent 3,872,466**, filed July 19, 1973, issued March 18, 1975. *(Quad-slope ADC.)*

66. Hans Bent Aasnaes, "Triple Integrating Ramp Analog-to-Digital Converter," **U.S. Patent 3,577,140**, filed June 27, 1967, issued May 4, 1971. *(Triple-slope ADC.)*

67. Frederick Bondzeit, Lewis J. Neelands, "Multiple Slope Analog-to-Digital Converter," **U.S. Patent 3,564,538**, filed January 29, 1968, issued February 16, 1971. *(Triple-slope ADC.)*

68. Desmond Wheable, "Triple-Slope Analog-to-Digital Converters," **U.S. Patent 3,678,506,** filed October 2, 1968, issued July 18, 1972. *(Triple-slope ADC.)*

69. Dan Sheingold, **Analog-Digital Conversion Handbook**, Prentice-Hall, 1986, ISBN-0-13-032848-0, pp. 441–471. *(This chapter contains an excellent tutorial on optical, synchro, and resolver-to-digital conversion.)*

70. Dennis Fu, "Circuit Applications of the AD2S90 Resolver-to-Digital Converter," **Application Note AN-230,** Analog Devices. *(Applications of the AD2S90 RTD.)*

Sigma-Delta Converters
Walt Kester, James Bryant

Historical Perspective

The sigma-delta (Σ-Δ) ADC architecture had its origins in the early development phases of pulse code modulation (PCM) systems—specifically, those related to transmission techniques called *delta modulation* and *differential PCM*. (An excellent discussion of both the history and concepts of the sigma-delta ADC can be found by Max Hauser in Reference 1.) Delta modulation, like classical PCM, was first invented at the ITT Laboratories in France by the director of the laboratories, E. M. Deloraine, S. Van Mierlo, and B. Derjavitch in 1946 (References 2, 3). The principle was rediscovered several years later, at the Phillips Laboratories in Holland, whose engineers published the first extensive studies of both the single-bit and multibit concepts in 1952 and 1953 (References 4, 5). In 1950, C. C. Cutler of Bell Telephone Labs in the U.S. filed a seminal patent on differential PCM which covered the same essential concepts (Reference 6).

The driving force behind delta modulation and differential PCM was to achieve higher transmission efficiency by transmitting the *changes* (delta) in value between consecutive samples rather than the actual samples themselves.

In *delta modulation*, the analog signal is quantized by a one-bit ADC (a comparator) as shown in Figure 3.119A. The comparator output is converted back to an analog signal with a 1-bit DAC, and subtracted from the input after passing through an integrator. The shape of the analog signal is transmitted as follows: a "1" indicates that a positive excursion has occurred since the last sample, and a "0" indicates that a negative excursion has occurred since the last sample.

If the analog signal remains at a fixed dc level for a period of time, an alternating pattern of "0s" and "1s" is obtained. It should be noted that *differential PCM* (Figure 3.119B) uses exactly the same concept except a multibit ADC is used rather than a comparator to derived the transmitted information.

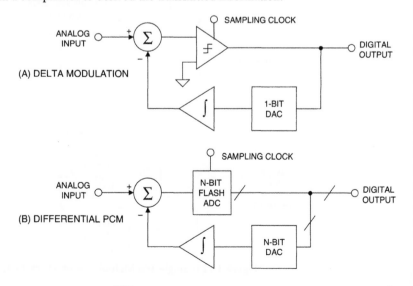

Figure 3.119: Delta Modulation and Differential PCM

Since there is no limit to the number of pulses of the same sign that may occur, delta modulation systems are capable of tracking signals of any amplitude. In theory, there is no peak clipping. However, the theoretical limitation of delta modulation is that the analog signal must not change too rapidly. The problem of slope clipping is shown in Figure 3.120. Here, although each sampling instant indicates a positive excursion, the analog signal is rising too quickly, and the quantizer is unable to keep pace.

Slope clipping can be reduced by increasing the quantum step size or increasing the sampling rate. Differential PCM uses a multibit quantizer to effectively increase the quantum step sizes at the increase of complexity. Tests have shown that in order to obtain the same quality as classical PCM, delta modulation requires very high sampling rates, typically 20× the highest frequency of interest, as opposed to Nyquist rate of 2×.

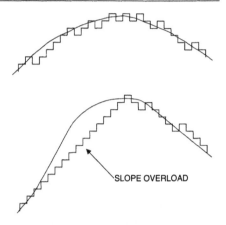

Figure 3.120: Quantization Using Delta Modulation

For these reasons, delta modulation and differential PCM have never achieved any significant degree of popularity; however, a slight modification of the delta modulator leads to the basic sigma-delta architecture, one of the most popular high resolution ADC architectures in use today.

In 1954 C. C. Cutler of Bell Labs filed a very significant patent which introduced the principle of *oversampling* and *noise shaping* with the specific intent of achieving higher resolution (Reference 7). His objective was not to specifically design a Nyquist ADC, but to transmit the oversampled noise-shaped signal without reducing the data rate. Thus Cutler's converter embodied all the concepts in a sigma-delta ADC with the exception of *digital filtering* and *decimation,* which would have been too complex and costly at the time using vacuum tube technology.

The basic single and multibit first-order sigma-delta ADC architecture is shown in Figure 3.121A and 3.121B, respectively. Note that the integrator operates on the error signal, whereas in a delta modulator, the integrator is in the feedback loop. The basic oversampling sigma-delta modulator increases the overall

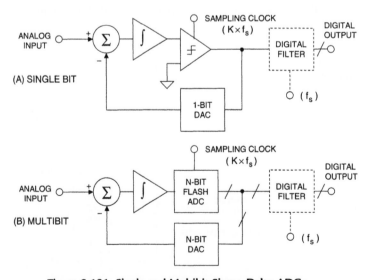

Figure 3.121: Single and Multibit Sigma-Delta ADCs

signal-to-noise ratio at low frequencies by shaping the quantization noise such that most of it occurs outside the bandwidth of interest. The digital filter then removes the noise outside the bandwidth of interest, and the decimator reduces the output data rate back to the Nyquist rate.

Occasional work continued on these concepts over the next several years, including an important patent of C. B. Brahm, filed in 1961, which gave details of the analog design of the loop filter for a second-order multibit noise shaping ADC (Reference 8). Transistor circuits began to replace vacuum tubes over the period, and this opened up many more possibilities for implementation of the architecture.

In 1962, Inose, Yasuda, and Murakami elaborated on the single-bit oversampling noise-shaping architecture proposed by Cutler in 1954 (Reference 9). Their experimental circuits used solid state devices to implement first- and second-order sigma-delta modulators. The 1962 paper was followed by a second paper in 1963 which gave excellent theoretical discussions on oversampling and noise-shaping (Reference 10). These two papers were also the first to use the name *delta-sigma* to describe the architecture. The name *delta-sigma* stuck until the 1970s when AT&T engineers began using the name *sigma-delta*. Since that time, both names have been used; however, sigma-delta may be the more correct of the two (see later discussion in this section by Dan Sheingold on the terminology).

It is interesting to note that all the work described thus far was related to transmitting an oversampled digitized signal directly rather than the implementation of a Nyquist ADC. In 1969 D. J. Goodman at Bell Labs published a paper describing a true Nyquist sigma-delta ADC with a digital filter and a decimator following the modulator (Reference 11). This was the first use of the sigma-delta architecture for the explicit purpose of producing a Nyquist ADC. In 1974 J. C. Candy, also of Bell Labs, described a multibit oversampling sigma-delta ADC with noise shaping, digital filtering, and decimation to achieve a high resolution Nyquist ADC (Reference 12).

The IC sigma-delta ADC offers several advantages over the other architectures, especially for high resolution, low frequency applications. First and foremost, the single-bit sigma-delta ADC is inherently monotonic and requires no laser trimming. The sigma-delta ADC also lends itself to low cost foundry CMOS processes because of the digitally intensive nature of the architecture. Examples of early monolithic sigma-delta ADCs are given in References 13–21. Since that time there have been a constant stream of process and design improvements in the fundamental architecture proposed in the early works cited above. A summary of the key developments relating to sigma-delta is shown in Figure 3.122.

• Delta Modulation	1950
• Differential PCM	1950
• Single and multibit oversampling with noise shaping	1954
• First called Δ-Σ, "delta-sigma"	1962
• Addition of digital filtering and decimation for Nyquist ADC	1969
• Bandpass Sigma-Delta	1988
Note: Dates are first publications or patent filings	

Figure 3.122: Sigma-Delta ADC Architecture Timeline

Sigma-Delta (Σ-Δ) or Delta-Sigma (Δ-Σ)? Editor's Notes from *Analog Dialogue* Vol. 24-2, 1990, by Dan Sheingold

This is not the most earth-shaking of controversies, and many readers may wonder what the fuss is all about—if they wonder at all. The issue is important to both editor and readers because of the need for consistency; we'd like to use the same name for the same thing whenever it appears. But *which* name? In the case of the modulation technique that led to a new oversampling A/D conversion mechanism, we chose *sigma-delta*. Here's why.

Ordinarily, when a new concept is named by its creators, the name sticks; it should not be changed unless it is erroneous or flies in the face of precedent. The seminal paper on this subject was published in 1962 (References 9, 10), and its authors chose the name "delta-sigma modulation," since it was based on *delta* modulation but included an integration (summation, hence Σ).

Delta-sigma was apparently unchallenged until the 1970s, when engineers at AT&T were publishing papers using the term *sigma-delta*. Why? According to Hauser (Reference 1), the precedent had been to name variants of delta modulation with adjectives preceding the word "delta." Since the form of modulation in question is a variant of delta modulation, the sigma, used as an adjective—so the argument went—should precede the delta.

Many engineers who came upon the scene subsequently used whatever term caught their fancy, often without knowing why. It was even possible to find *both* terms used interchangeably in the same paper. As matters stand today, sigma-delta is in widespread use, probably for the majority of citations. Would its adoption be an injustice to the inventors of the technique?

We think not. Like others, we believe that the name delta-sigma is a departure from precedent. Not just in the sense of grammar, but also in relation to the hierarchy of operations. Consider a block diagram for embodying an analog root-mean-square (finding the square root of the mean of a squared signal) computer. First the signal is squared, then it is integrated, and finally it is rooted (see Figure 3.123).

If we were to name the overall function after the causal order of operations, it would have to be called a "square mean root" function. But naming in order of the *hierarchy* of its mathematical operations gives us the familiar—and undisputed—name, *root mean-square*. Consider now a block diagram for taking a difference (delta), and then integrating it (sigma).

Its causal order would give *delta-sigma*, but in functional hierarchy it is *sigma-delta*, since it computes the integral of a difference. We believe that the latter term is correct and follows precedent; and we have adopted it as our standard.

Dan Sheingold, 1990.

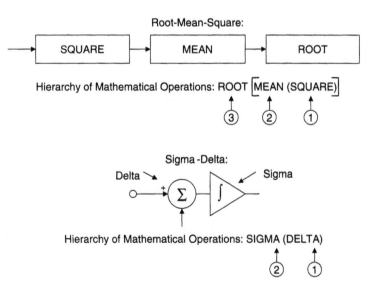

Figure 3.123: Sigma-Delta (Σ-Δ) or Delta-Sigma (Δ-Σ)?

Basics of Sigma-Delta ADCs

Sigma-Delta Analog-Digital Converters (Σ-Δ ADCs) have been known for over thirty years, but only recently has the technology (high-density digital VLSI) existed to manufacture them as inexpensive monolithic integrated circuits. They are now used in many applications where a low cost, low bandwidth, low power, high resolution ADC is required.

There have been innumerable descriptions of the architecture and theory of Σ-Δ ADCs, but most commence with a maze of integrals and deteriorate from there. Some engineers who do not understand the theory of operation of Σ-Δ ADCs are convinced, from study of a typical published article, that it is too complex to comprehend easily.

There is nothing particularly difficult to understand about Σ-Δ ADCs, as long as you avoid the detailed mathematics, and this section has been written in an attempt to clarify the subject. A Σ-Δ ADC contains very simple analog electronics (a comparator, voltage reference, a switch, and one or more integrators and analog summing circuits), and quite complex digital computational circuitry. This circuitry consists of a digital signal processor (DSP) which acts as a filter (generally, but not invariably, a low-pass filter). It is not necessary to know precisely how the filter works to appreciate what it does. To understand how a Σ-Δ ADC works, familiarity with the concepts of *over-sampling, quantization noise shaping, digital filtering,* and *decimation* is required (see Figure 3.124).

```
• Low Cost, High Resolution (to 24 Bits)
• Excellent DNL
• Low Power, but Limited Bandwidth (Voiceband, Audio)
• Key Concepts are Simple, but Math is Complex
    – Oversampling
    – Quantization Noise Shaping
    – Digital Filtering
    – Decimation
• Ideal for Sensor Signal Conditioning
    – High Resolution
    – Self, System, and Auto Calibration Modes
• Wide Applications in Voiceband and Audio Signal Processing
```

Figure 3.124: Sigma-Delta ADCs

Let us consider the technique of over-sampling with an analysis in the frequency domain. Where a dc conversion has a *quantization error* of up to ½ LSB, a sampled data system has *quantization noise*. A perfect classical N-bit sampling ADC has an rms quantization noise of $q/\sqrt{12}$ uniformly distributed within the Nyquist band of dc to $f_s/2$ (where q is the value of an LSB and f_s is the sampling rate) as shown in Figure 3.125A. Therefore, its SNR with a full-scale sinewave input will be (6.02N + 1.76) dB. If the ADC is less than perfect, and its noise is greater than its theoretical minimum quantization noise, then its *effective* resolution will be less than N-bits. Its actual resolution (often known as its Effective Number of Bits or ENOB) will be defined by

$$\text{ENOB} = \frac{\text{SNR} - 1.76 \text{ dB}}{6.02 \text{ dB}}. \qquad \text{Eq. 3.16}$$

If we choose a much higher sampling rate, Kf_s (Figure 3.125B), the rms quantization noise remains $q/\sqrt{12}$, but the noise is now distributed over a wider bandwidth dc to $Kf_s/2$. If we then apply a digital low-pass filter (LPF) to the output, we remove much of the quantization noise, but do not affect the wanted signal—so the ENOB is improved. We have accomplished a high resolution A/D conversion with a low resolution ADC. The factor K is generally referred to as the *oversampling ratio*. It should be noted at this point that oversampling has an added benefit in that it relaxes the requirements on the analog antialiasing filter.

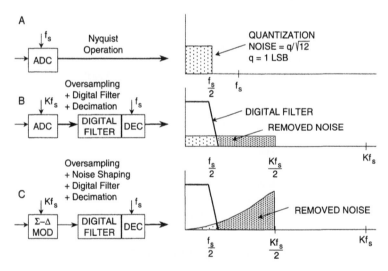

Figure 3.125: Oversampling, Digital Filtering, Noise Shaping, and Decimation

Since the bandwidth is reduced by the digital output filter, the output data rate may be lower than the original sampling rate (Kf_s) and still satisfy the Nyquist criterion. This may be achieved by passing every Mth result to the output and discarding the remainder. The process is known as "decimation" by a factor of M. Despite the origins of the term (*decem* is Latin for ten), M can have any integer value, provided that the output data rate is more than twice the signal bandwidth. Decimation does not cause any loss of information (Figure 3.125B).

If we simply use oversampling to improve resolution, we must oversample by a factor of 2^{2N} to obtain an N-bit increase in resolution. The Σ-Δ converter does not need such a high oversampling ratio because it not only limits the signal pass band, but also shapes the quantization noise so that most of it falls outside this pass band as shown in Figure 3.125C.

If we take a 1-bit ADC (generally known as a comparator), drive it with the output of an integrator, and feed the integrator with an input signal summed with the output of a 1-bit DAC fed from the ADC output, we have a first-order Σ-Δ modulator as shown in Figure 3.126. Add a digital low-pass filter (LPF) and decimator at the digital output, and we have a Σ-Δ ADC—the Σ-Δ modulator shapes the quantization noise so that it lies above the pass band of the digital output filter, and the ENOB is therefore much larger than would otherwise be expected from the over-sampling ratio.

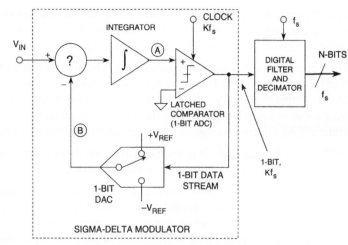

Figure 3.126: First-Order Sigma-Delta ADC

Intuitively, a Σ-Δ ADC operates as follows. Assume a dc input at V_{IN}. The integrator is constantly ramping up or down at node A. The output of the comparator is fed back through a 1-bit DAC to the summing input at node B. The negative feedback loop from the comparator output through the 1-bit DAC back to the summing point will force the average dc voltage at node B to be equal to V_{IN}. This implies that the average DAC output voltage must equal the input voltage V_{IN}. The average DAC output voltage is controlled by the *ones-density* in the 1-bit data stream from the comparator output. As the input signal increases towards $+V_{REF}$, the number of "ones" in the serial bit stream increases, and the number of "zeros" decreases. Similarly, as the signal goes negative towards $-V_{REF}$, the number of "ones" in the serial bit stream decreases, and the number of "zeros" increases. From a very simplistic standpoint, this analysis shows that the average value of the input voltage is contained in the serial bit stream out of the comparator. The digital filter and decimator process the serial bit stream and produce the final output data.

For any given input value in a single sampling interval, the data from the 1-bit ADC is virtually meaningless. Only when a large number of samples are averaged, will a meaningful value result. The sigma-delta modulator is very difficult to analyze in the time domain because of this apparent randomness of the single-bit data output. If the input signal is near positive full-scale, it is clear that there will be more 1s than 0s in the bit stream. Likewise, for signals near negative full-scale, there will be more 0s than 1s in the bit stream. For signals near midscale, there will be approximately an equal number of 1s and 0s. Figure 3.127 shows the output of the integrator for two input conditions. The first is for an input of zero (midscale). To decode the output, pass the output samples through a simple digital low-pass filter that averages every four samples. The output of the filter is 2/4. This value represents bipolar zero. If more samples are averaged, more

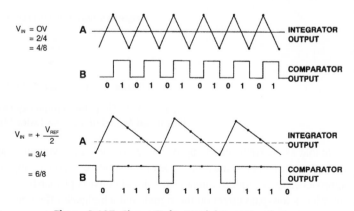

Figure 3.127: Sigma-Delta Modulator Waveforms

237

dynamic range is achieved. For example, averaging four samples gives 2 bits of resolution, while averaging eight samples yields 4/8, or 3 bits of resolution. In the bottom waveform of Figure 3.127, the average obtained for four samples is 3/4, and the average for eight samples is 6/8.

The sigma-delta ADC can also be viewed as a synchronous voltage-to-frequency converter followed by a counter. If the number of 1s in the output data stream is counted over a sufficient number of samples, the counter output will represent the digital value of the input. Obviously, this method of averaging will only work for dc or very slowly changing input signals. In addition, 2^N clock cycles must be counted in order to achieve N-bit effective resolution, thereby severely limiting the effective sampling rate.

Further time-domain analysis is not productive, and the concept of noise shaping is best explained in the frequency domain by considering the simple Σ-Δ modulator model in Figure 3.128.

The integrator in the modulator is represented as an analog low-pass filter with a transfer function equal to $H(f) = 1/f$. This transfer function has an amplitude response that is inversely proportional to the input frequency. The 1-bit quantizer generates quantization noise, Q, which is injected into the output summing block. If we let the input signal be X, and the output Y, the signal coming out of the input summer must be $X - Y$. This is multiplied by the filter transfer function, $1/f$, and the result goes to one input of the output summer. By inspection, we can then write the expression for the output voltage Y as:

$$Y = \frac{1}{f}(X - Y) + Q \qquad\qquad \text{Eq. 3.17}$$

This expression can easily be rearranged and solved for Y in terms of X, f, and Q:

$$Y = \frac{X}{f+1} + \frac{Q \cdot f}{f+1} \qquad\qquad \text{Eq. 3.18}$$

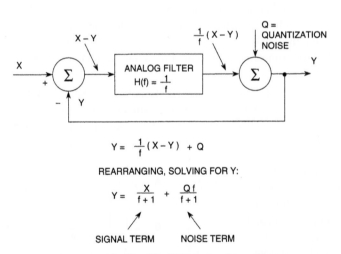

Figure 3.128: Simplified Frequency Domain Linearized Model of a Sigma-Delta Modulator

Note that as the frequency f approaches zero, the output voltage Y approaches X with no noise component. At higher frequencies, the amplitude of the signal component approaches zero, and the noise component approaches Q. At high frequency, the output consists primarily of quantization noise. In essence, the analog filter has a low-pass effect on the signal, and a highpass effect on the quantization noise. Thus the analog filter performs the noise shaping function in the Σ-Δ modulator model.

For a given input frequency, higher order analog filters offer more attenuation. The same is true of Σ-Δ modulators, provided certain precautions are taken.

By using more than one integration and summing stage in the Σ-Δ modulator, we can achieve higher orders of quantization noise shaping and even better ENOB for a given over-sampling ratio as is shown in Figure 3.129 for both a first- and second-order Σ-Δ modulator.

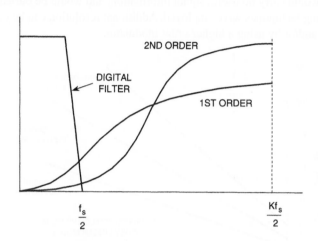

Figure 3.129: Sigma-Delta Modulators Shape Quantization Noise

The block diagram for the second-order Σ-Δ modulator is shown in Figure 3.130. Third-, and higher, order Σ-Δ ADCs were once thought to be potentially unstable at some values of input—recent analyses using *finite* rather than infinite gains in the comparator have shown that this is not necessarily so, but even if instability does start to occur, it is not important, since the DSP in the digital filter and decimator can be made to recognize incipient instability and react to prevent it.

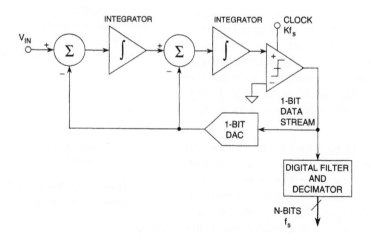

Figure 3.130: Second-Order Sigma-Delta ADC

Figure 3.131 shows the relationship between the order of the Σ-Δ modulator and the amount of over-sampling necessary to achieve a particular SNR. For instance, if the oversampling ratio is 64, an ideal second-order system is capable of providing an SNR of about 80 dB. This implies approximately 13 effective number of bits (ENOB). Although the filtering done by the digital filter and decimator can be done to any degree of precision desirable, it would be pointless to carry more than 13 binary bits to the outside world. Additional bits would carry no useful signal information, and would be buried in the quantization noise unless post-filtering techniques were employed. Additional resolution can be obtained by increasing the oversampling ratio and/or by using a higher-order modulator.

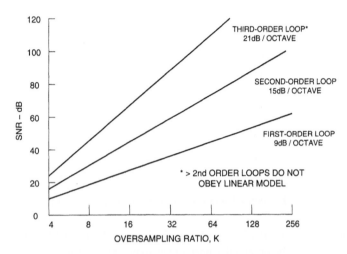

Figure 3.131: SNR versus Oversampling Ratio
for First-, Second-, and Third-Order Loops

Idle Tone Considerations

In our discussion of sigma-delta ADCs up to this point, we have made the assumption that the quantization noise produced by the sigma-delta modulator is random and uncorrelated with the input signal. Unfortunately, this is not entirely the case, especially for the first-order modulator. Consider the case where we are averaging 16 samples of the modulator output in a 4-bit sigma-delta ADC.

Figure 3.132 shows the bit pattern for two input signal conditions: an input signal having the value 8/16, and an input signal having the value 9/16. In the case of the 9/16 signal, the modulator output bit pattern has an extra "1" every 16th output. This will produce energy

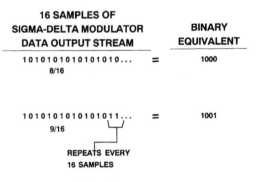

Figure 3.132: Repetitive Bit Pattern in
Sigma-Delta Modulator Output

at $f_s/16$, which translates into an unwanted tone. If the oversampling ratio is less than 16, this tone will fall into the pass band. In audio, the idle tones can be heard just above the noise floor as the input changes from negative to positive fullscale.

Figure 3.133 shows the correlated idling pattern behavior for a first-order sigma-delta modulator, and Figure 3.134 shows the relatively uncorrelated pattern for a second-order modulator. For this reason, virtually all sigma-delta ADCs contain at least a second-order modulator loop, and some use up to fifth-order loops.

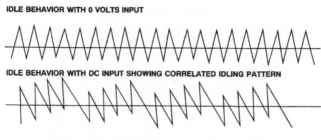

Figure 3.133: Idling Patterns for First-Order Sigma-Delta Modulator (Integrator Output)

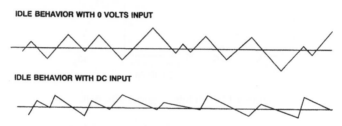

Figure 3.134: Idling Patterns for Second-Order Sigma-Delta Modulator (Integrator Output)

Higher Order Loop Considerations

In order to achieve wide dynamic range, sigma-delta modulator loops greater than second-order are necessary, but present real design challenges. First of all, the simple linear models previously discussed are no longer fully accurate. Loops of order greater than two are generally not guaranteed to be stable under all input conditions. The instability arises because the comparator is a nonlinear element whose effective "gain" varies inversely with the input level. This mechanism for instability causes the following behavior: if the loop is operating normally, and a large signal is applied to the input that overloads the loop, the average gain of the comparator is reduced. The reduction in comparator gain in the linear model causes loop instability. This causes instability even when the signal that caused it is removed. In actual practice, such a circuit would normally oscillate on power-up due to initial conditions caused by turn-on transients. The AD1879 dual audio ADC released in 1994 by Analog Devices used a fifth order loop. Extensive nonlinear stabilization techniques were required in this and similar higher-order loop designs (References 22–26).

Multibit Sigma-Delta Converters

So far we have considered only sigma-delta converters that contain a single-bit ADC (comparator) and a single-bit DAC (switch). The block diagram of Figure 3.135 shows a multibit sigma-delta ADC that uses an n-bit flash ADC and an n-bit DAC. Obviously, this architecture will give a higher dynamic range for a given oversampling ratio and order of loop filter. Stabilization is easier, since second-order loops can generally be used. Idling patterns tend to be more random thereby minimizing tonal effects.

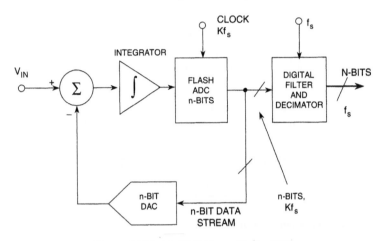

Figure 3.135: Multibit Sigma-Delta ADC

The real disadvantage of this technique is that the linearity depends on the DAC linearity, and thin film laser trimming is required to approach 16-bit performance levels. This makes the multibit architecture extremely impractical to implement on mixed-signal ICs using traditional binary DAC techniques.

However, fully decoded thermometer DACs coupled with proprietary data scrambling techniques as used in a number of Analog Devices' audio ADCs and DACs, including the 24-bit stereo AD1871 (see References 27 and 28) can achieve high SNR and low distortion using the multibit architecture. The multibit data scrambling technique both minimizes idle tones and ensures better differential linearity. A simplified block diagram of the AD1871 ADC is shown in Figure 3.136.

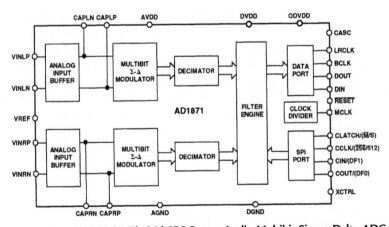

Figure 3.136: AD1871 24-Bit 96 kSPS Stereo Audio Multibit Sigma-Delta ADC

The AD1871's analog Σ-Δ modulator section comprises a second-order multibit implementation using Analog Device's proprietary technology for best performance. As shown in Figure 3.137, the two analog integrator blocks are followed by a flash ADC section that generates the multibit samples.

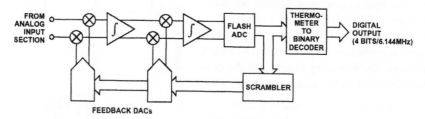

Figure 3.137: Details of the AD1871 Second-Order Modulator and Data Scrambler

The output of the flash ADC, which is thermometer encoded, is decoded to binary for output to the filter sections and is scrambled for feedback to the two integrator stages. The modulator is optimized for operation at a sampling rate of 6.144 MHz (which is $128 \times f_s$ at 48-kHz sampling and $64 \times f_s$ at 96 kHz sampling). The A-weighted dynamic range of the AD1871 is typically 105 dB. Key specifications for the AD1871 are summarized in Figure 3.138.

- Single 5 V Power Supply
- Differential Dual-Channel Analog Inputs
- 16-/20-/24-Bit Word Lengths Supported
- 105 dB (typ) A-Weighted Dynamic Range
- 103 dB (typ) THD+N (–20 dBFS input)
- 0.01 dB Decimator Filter Passband Ripple
- Second-Order, 128-/64-Times Oversampling Multibit Modulator with Data Scrambling
- Less than 350 mW (typ)
- Power-Down Mode
- On-Chip Voltage Reference
- Flexible Serial Output Interface
- 28-Lead SSOP Package

Figure 3.138: AD1871 24-Bit, 96 kSPS Stereo Sigma-Delta ADC Key Specifications

Digital Filter Implications

The digital filter is an integral part of all sigma-delta ADCs—there is no way to remove it. The settling time of this filter affects certain applications especially when using sigma-delta ADCs in multiplexed applications. The output of a multiplexer can present a step function input to an ADC if there are different input voltages on adjacent channels. In fact, the multiplexer output can represent a full-scale step voltage to the sigma-delta ADC when channels are switched. Adequate filter settling time must be allowed, therefore, in such applications. This does not mean that sigma-delta ADCs shouldn't be used in multiplexed applications, just that the settling time of the digital filter must be considered. Some newer sigma-delta ADCs such are actually optimized for use in multiplexed applications.

For example, the group delay through the AD1871 digital filter is 910 μs (sampling at 48 kSPS) and 460 μs (sampling at 96 kSPS)—this represents the time it takes for a step function input to propagate through one-half the number of taps in the digital filter. The total settling time is therefore approximately twice the group delay time. The input oversampling frequency is 6.144 MSPS for both conditions. The frequency response of the digital filter in the AD1871 ADC is shown in Figure 1.139.

In other applications, such as low frequency, high resolution 24-bit measurement sigma-delta ADCs (such as the AD77xx series), other types of digital filters may be used. For instance, the $SINC^3$ response is popular because it has zeros at multiples of the throughput rate. For instance a 10 Hz throughput rate produces zeros at 50 Hz and 60 Hz which aid in ac power line rejection.

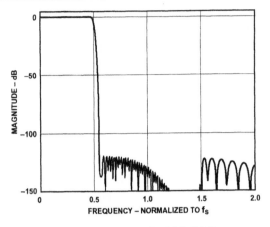

Figure 1.139: AD1871 24-Bit, 96-kSPS Stereo Sigma-Delta ADC Digital Filter Characteristics

Multistage Noise Shaping (MASH) Sigma-Delta Converters

As has been discussed, nonlinear stabilization techniques can be difficult for third order loops or higher. In many cases, the multibit architecture is preferable. An alternative approach to either of these, called multistage noise shaping (MASH), utilizes cascaded stable first-order loops (References 29 and 30). Figure 3.140 shows a block diagram of a three-stage MASH ADC. The output of the first integrator is subtracted from the first DAC output to yield the first stage quantization noise, Q1. Q1 is then quantized by the second stage. The output of the second integrator is subtracted from the second DAC output to yield the second stage quantization noise which is in turn quantized by the third stage.

The output of the first stage is summed with a single digital differentiation of the second stage output and a double differentiation of the third stage output to yield the final output. The result is that the quantization noise Q1 is suppressed by the second stage, and the quantization noise Q2 is suppressed by the third stage yielding the same suppression as a third-order loop. Since this result is obtained using three first-order loops, stable operation is assured.

Figure 3.140: Multistage Noise Shaping Sigma-Delta ADC (MASH)

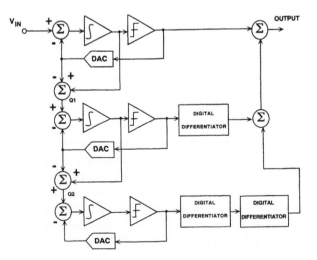

High Resolution Measurement Sigma-Delta ADCs

In order to better understand the capability of sigma-delta measurement ADCs and the power of the technique, a modern example, the AD7730, will be examined in detail. The AD7730 is a member of the AD77xx family and is shown in Figure 3.141. This ADC was specifically designed to interface directly to bridge outputs in weigh scale applications. The device accepts low level signals directly from a bridge and outputs a serial digital word. There are two buffered differential inputs which are multiplexed, buffered, and drive a PGA. The PGA can be programmed for four differential unipolar analog input ranges: 0 V to +10 mV, 0 V to +20 mV, 0 V to +40 mV, and 0 V to +80 mV and four differential bipolar input ranges: ±10 mV, ±20 mV, ±40 mV, and ±80 mV.

The maximum peak-to-peak, or noise-free resolution achievable is 1 in 230,000 counts, or approximately 18 bits. It should be noted that the noise-free resolution is a function of input voltage range, filter cutoff, and output word rate. Noise is greater using the smaller input ranges where the PGA gain must be increased. Higher output word rates and associated higher filter cut-off frequencies will also increase the noise.

The analog inputs are buffered on-chip, allowing relatively high source impedances. Both analog channels are differential, with a common-mode voltage range that comes within 1.2 V of AGND and 0.95 V of AVDD. The reference input is also differential, and the common-mode range is from AGND to AVDD.

The 6-bit DAC is controlled by on-chip registers and can remove TARE (pan weight) values of up to ±80 mV from the analog input signal range. The resolution of the TARE function is 1.25 mV with a 2.5 V reference and 2.5 mV with a 5 V reference.

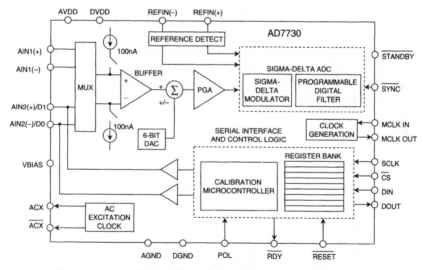

Figure 3.141: AD7730 Sigma-Delta Single-Supply Bridge ADC

The output of the PGA is applied to the Σ-Δ modulator and programmable digital filter. The serial interface can be configured for three-wire operation and is compatible with microcontrollers and digital signal processors. The AD7730 contains self-calibration and system-calibration options and has an offset drift of less than 5 nV/°C and a gain drift of less than 2 ppm/°C. This low offset drift is obtained using a *chop* mode which operates similarly to a chopper-stabilized amplifier.

The oversampling frequency of the AD7730 is 4.9152 MHz, and the output data rate can be set from 50 Hz to 1200 Hz. The clock source can be provided via an external clock or by connecting a crystal oscillator across the MCLK IN and MCLK OUT pins.

The AD7730 can accept input signals from a dc-excited bridge. It can also handle input signals from an ac-excited bridge by using the ac excitation clock signals (ACX and $\overline{ACX}$). These are nonoverlapping clock signals used to synchronize the external switches that drive the bridge. The ACX clocks are demodulated on the AD7730 input.

The AD7730 contains two 100 nA constant current generators, one source current from AVDD to AIN(+) and one sink current from AIN(–) to AGND. The currents are switched to the selected analog input pair under the control of a bit in the Mode Register. These currents can be used in checking that a sensor is still operational before attempting to take measurements on that channel. If the currents are turned on and a full-scale reading is obtained, the sensor has gone open circuit. If the measurement is 0 V, the sensor has gone short circuit. In normal operation, the burnout currents are turned off by setting the proper bit in the Mode Register to 0.

The AD7730 contains an internal programmable digital filter. The filter consists of two sections: a first stage filter, and a second stage filter. The first stage is a sinc^3 low-pass filter. The cutoff frequency and output rate of this first stage filter is programmable. The second stage filter has three modes of operation. In its normal mode, it is a 22-tap FIR filter that processes the output of the first stage filter. When a step change is detected on the analog input, the second stage filter enters a second mode (FASTStep™) where it performs a variable number of averages for some time after the step change, and then the second stage filter switches back to the FIR filter mode. The third option for the second stage filter (SKIP mode) is that it is completely bypassed so the only filtering provided on the AD7730 is the first stage. Both the FASTStep mode and SKIP mode can be enabled or disabled via bits in the control register.

Figure 3.142 shows the full frequency response of the AD7730 when the second stage filter is set for normal FIR operation. This response is with the chop mode enabled and an output word rate of 200 Hz and a clock frequency of 4.9152 MHz. The response is shown from dc to 100 Hz. The rejection at 50 Hz ± 1 Hz and 60 Hz ± 1 Hz is better than 88 dB.

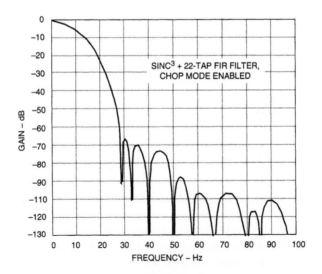

Figure 3.142: AD7730 Digital Filter Response

Figure 3.143 shows the step response of the AD7730 with and without the FASTStep mode enabled. The vertical axis shows the code value and indicates the settling of the output to the input step change. The horizontal axis shows the number of output words required for that settling to occur. The positive input step change occurs at the 5th output. In the normal mode (FASTStep disabled), the output has not reached its final value until the 23rd output word. In FASTStep mode with chopping enabled, the output has settled to the final value by the 7th output word. Between the 7th and the 23rd output, the FASTStep mode produces a settled result, but with additional noise compared to the specified noise

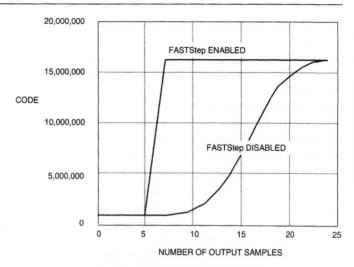

Figure 3.143: AD7730 Digital Filter Settling Time Showing FASTStep Mode

level for normal operating conditions. It starts at a noise level comparable to the SKIP mode, and as the averaging increases ends up at the specified noise level. The complete settling time required for the part to return to the specified noise level is the same for FASTStep mode and normal mode.

The FASTStep mode gives a much earlier indication of where the output channel is going and its new value. This feature is very useful in weigh scale applications to give a much earlier indication of the weight, or in an application scanning multiple channels where the user does not have to wait the full settling time to see if a channel has changed.

Note, however, that the FASTStep mode is not particularly suitable for multiplexed applications because of the excess noise associated with the settling time. For multiplexed applications, the full 23-cycle output word interval should be allowed for settling to a new channel. This points out the fundamental issue of using Σ-Δ ADCs in multiplexed applications. There is no reason why they won't work, provided the internal digital filter is allowed to settle fully after switching channels.

The calibration modes of the AD7730 are given in Figure 3.144. A calibration cycle may be initiated at any time by writing to the appropriate bits of the Mode Register. Calibration removes offset and gain errors from the device.

The AD7730 gives the user access to the on-chip calibration registers allowing an external microprocessor to read the device's calibration coefficients and also to write its own calibration coefficients to the part from prestored values in external E²PROM. This gives the microprocessor much greater control over the AD7730's calibration procedure. It also means that the user can verify that the device has correctly performed its calibration by comparing the coefficients after calibration with prestored values in E²PROM. Since the calibration coefficients are derived by performing a conversion on the input volt-

- Internal Zero -ScaleCalibration
  - 22 Output Cycles (CHP = 0)
  - 24 Output Cycles (CHP = 1)
- Internal Full -Scale Calibration
  - 44 Output Cycles (CHP = 0)
  - 48 Output Cycles (CHP = 1)
- Calibration Programmed via the Mode Register
- Calibration Coefficients Stored in Calibration Registers
- External Microprocessor Can Read or Write to Calibration Coefficient Registers

Figure 3.144: AD7730 Calibration Options

age provided, the accuracy of the calibration can only be as good as the noise level the part provides in the normal mode. To optimize calibration accuracy, it is recommended to calibrate the part at its lowest output rate where the noise level is lowest. The coefficients generated at any output rate will be valid for all selected output update rates. This scheme of calibrating at the lowest output data rate does mean that the duration of the calibration interval is longer.

The AD7730 requires an external voltage reference, however, the power supply may be used as the reference in the ratiometric bridge application shown in Figure 3.145. In this configuration, the bridge output voltage is directly proportional to the bridge drive voltage which is also used to establish the reference

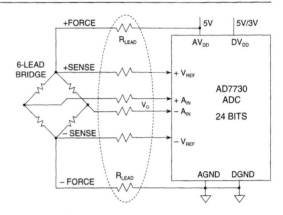

Figure 3.145: AD7730 Bridge Application (Simplified Schematic)

voltages to the AD7730. Variations in the supply voltage will not affect the accuracy. The SENSE outputs of the bridge are used for the AD7730 reference voltages in order to eliminate errors caused by voltage drops in the lead resistances.

Band-Pass Sigma-Delta Converters

The Σ-Δ ADCs we have described so far contain integrators, which are low-pass filters, whose pass band extends from dc. Thus, their quantization noise is pushed up in frequency. At present, most commercially available Σ-Δ ADCs are of this type (although some that are intended for use in audio or telecommunications applications contain band-pass rather than low-pass digital filters to eliminate any system dc offsets). But there is no particular reason why the filters of the Σ-Δ modulator should be LPFs, except that traditionally ADCs have been thought of as being baseband devices, and that integrators are somewhat easier to construct than band-pass filters. If we replace the integrators in a Σ-Δ ADC with band-pass filters (BPFs) as shown in Figure 3.146, the quantization noise is moved up and down in frequency to leave a virtually noise-free region in the pass-band (see References 31, 32, and 33). If the digital filter is then programmed to have its pass-band in this region, we have a Σ-Δ ADC with a band-pass, rather than a low-pass characteristic. Such devices would appear to be useful in direct IF-to-digital conversion, digital radios, ultrasound, and other undersampling applications. However, the modulator and the digital BPF must be designed for the specific set of frequencies required by the system application, thereby somewhat limiting the flexibility of this approach.

In an undersampling application of a band-pass Σ-Δ ADC, the minimum sampling frequency must be at least twice the signal bandwidth, BW. The signal is centered around a carrier frequency, f_c. A typical digital radio application using a 455 kHz center frequency and a signal bandwidth of 10 kHz is described in Reference 32. An oversampling frequency Kf_s = 2 MSPS and an output rate f_s = 20 kSPS yielded a dynamic range of 70 dB within the signal bandwidth.

Another example of a band pass is the AD9870 IF Digitizing Subsystem having a nominal oversampling frequency of 18 MSPS, a center frequency of 2.25 MHz, and a bandwidth of 10 kHz – 150 kHz (see details in Reference 33).

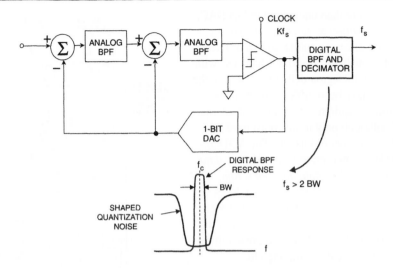

Figure 3.146: Replacing Integrators with Resonators Gives a Band-Pass Sigma-Delta ADC

Sigma-Delta DACs

Sigma-delta DACs operate very similarly to sigma-delta ADCs. However, in a sigma-delta DAC, the noise shaping function is accomplished with a digital modulator rather than an analog one.

A Σ-Δ DAC, unlike the Σ-Δ ADC, is mostly digital (see Figure 3.147A). It consists of an "interpolation filter" (a digital circuit that accepts data at a low rate, inserts zeros at a high rate, and then applies a digital filter algorithm and outputs data at a high rate), a Σ-Δ modulator (which effectively acts as a low-pass filter to the signal but as a high-pass filter to the quantization noise, and converts the resulting data to a high speed bit stream), and a 1-bit DAC whose output switches between equal positive and negative reference voltages. The output is filtered in an external analog LPF. Because of the high oversampling frequency, the complexity of the LPF is much less than the case of traditional Nyquist operation.

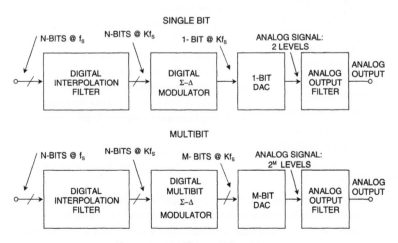

Figure 3.147: Sigma-Delta DACs

It is possible to use more than one bit in the Σ-Δ DAC, and this leads to the *multibit* architecture shown in Figure 3.147B. The concept is similar to that of interpolating DACs previously discussed in Chapter 2, with the addition of the digital sigma-delta modulator. In the past, multibit DACs have been difficult to design because of the accuracy requirement on the n-bit internal DAC (this DAC, although only n-bits, must have the linearity of the final number of bits, N). The AD185x-series of audio DACs, however, use a proprietary *data scrambling* technique (*called data directed scrambling*) which overcomes this problem and produces excellent performance with respect to all audio specifications (see References 27 and 28). For instance, the AD1853 dual 24-bit, 192 kSPS DAC has greater than 104 dB THD + N at a 48 kSPS sampling rate.

One of the newest members of this family is the AD1955 multibit sigma-delta audio DAC shown in Figure 3.148. The AD1955 also uses data directed scrambling, supports a multitude of DVD audio formats and has an extremely flexible serial port. THD + N is typically 110 dB.

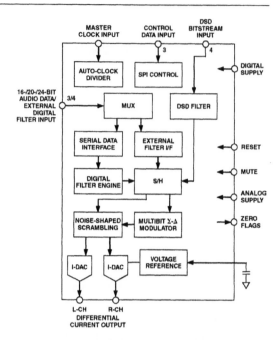

Figure 3.148: AD1955 Multibit Sigma-Delta Audio DAC

Summary

Sigma-delta ADCs and DACs have proliferated into many modern applications including measurement, voiceband, audio, etc. The technique takes full advantage of low cost CMOS processes and therefore makes integration with highly digital functions such as DSPs practical. Resolutions up to 24 bits are currently available, and the requirements on analog antialiasing/anti-imaging filters are greatly relaxed due to oversampling. Modern techniques such as the multibit data scrambled architecture minimize problems with idle tones which plagued early sigma-delta products.

- Inherently Excellent Linearity
- High Resolution Possible (24-Bits)
- Oversampling Relaxes Analog Antialiasing Filter Requirements
- Ideal for CMOS Processes, no Trimming
- No SHA Required
- Added Functionality: On-Chip PGAs, Analog Filters, Autocalibration
- On-Chip Programmable Digital Filters (AD7725: Low-pass, High-pass, Band-pass, Bandstop)
- Upper Sampling Rate Currently Limits Applications to Measurement Voiceband, and Audio, except for Bandpass Sigma-Delta ADCs
- Analog Multiplexer Switching Speed Limited by Internal Filter Settling Time.

Figure 3.149: Sigma-Delta Summary

Many sigma-delta converters offer a high level of user programmability with respect to output data rate, digital filter characteristics, and self-calibration modes. Multichannel sigma-delta ADCs are now available for data acquisition systems, and most users are well educated with respect to the settling time requirements of the internal digital filter in these applications. Figure 3.149 summarizes some final thoughts about sigma-delta converters.

References:
3.3 Sigma-Delta Converters

1. Max W. Hauser, "Principles of Oversampling A/D Conversion," **Journal Audio Engineering Society**, Vol. 39, No. 1/2, January/February 1991, pp. 3–26. *(One of the best tutorials and practical discussions of the sigma-delta ADC architecture and its history.)*

2. E. M. Deloraine, S. Van Mierlo, and B. Derjavitch, "Methode et systéme de transmission par impulsions," **French Patent 932,140**, issued August, 1946. Also **British Patent 627,262**, issued 1949.

3. E. M. Deloraine, S. Van Mierlo, and B. Derjavitch, "Communication System Utilizing Constant Amplitude Pulses of Opposite Polarities," **U.S. Patent 2,629,857**, filed October 8, 1947, issued February 24, 1953.

4. F. de Jager, "Delta Modulation: A Method of PCM Transmission Using the One Unit Code," **Phillips Research Reports**, Vol. 7, 1952, pp. 542–546. *(Additional work done on delta modulation during the same time period.)*

5. H. Van de Weg, "Quantizing Noise of a Single Integration Delta Modulation System with an N-Digit Code," **Phillips Research Reports**, Vol. 8, 1953, pp. 367–385. *(Additional work done on delta modulation during the same time period.)*

6. C. C. Cutler, "Differential Quantization of Communication Signals," **U.S. Patent 2,605,361**, filed June 29, 1950, issued July 29, 1952. *(Recognized as the first patent on differential PCM or delta modulation, although actually first invented in the Paris labs of the International Telephone and Telegraph Corporation by E. M. Deloraine, S. Mierlo, and B. Derjavitch a few years earlier.)*

7. C. C. Cutler, "Transmission Systems Employing Quantization," **U.S. Patent 2,927,962**, filed April 26, 1954, issued March 8, 1960. *(A ground-breaking patent describing oversampling and noise shaping using first and second-order loops to increase effective resolution. The goal was transmission of oversampled noise shaped PCM data without decimation, not a Nyquist-type ADC.)*

8. C. B. Brahm, "Feedback Integrating System," **U.S. Patent 3,192,371**, filed September 14, 1961, issued June 29, 1965. *(Describes a second-order multibit oversampling noise shaping ADC.)*

9. H. Inose, Y. Yasuda, and J. Murakami, "A Telemetering System by Code Modulation: Δ-Σ Modulation,"**IRE Transactions on Space Electronics Telemetry**, Vol. SET-8, September 1962, pp. 204–209. Reprinted in N. S. Jayant, **Waveform Quantization and Coding**, IEEE Press and John Wiley, 1976, ISBN 0-471-01970-4. *(An elaboration on the 1-bit form of Cutler's noise-shaping oversampling concept. This work coined the description of the architecture as 'delta-sigma modulation'.)*

10. H. Inose and Y. Yasuda, "A Unity Bit Coding Method by Negative Feedback," **IEEE Proceedings**, Vol. 51, November 1963, pp. 1524–1535. *(Further discussions on their 1-bit 'delta-sigma' concept.)*

11. D. J. Goodman, "The Application of Delta Modulation of Analog-to-PCM Encoding," **Bell System Technical Journal**, Vol. 48, February 1969, pp. 321–343. Reprinted in N. S. Jayant, **Waveform Quantization and Coding**, IEEE Press and John Wiley, 1976, ISBN 0-471-01970-4. *(The first description of using oversampling and noise shaping techniques followed by digital filtering and decimation to produce a true Nyquist-rate ADC.)*

12. J. C. Candy, "A Use of Limit Cycle Oscillations to Obtain Robust Analog-to-Digital Converters," **IEEE Transactions on Communications**, Vol. COM-22, December 1974, pp. 298–305. *(Describes a multibit oversampling noise shaping ADC with output digital filtering and decimation to interpolate between the quantization levels.)*

13. R. J. van de Plassche, "A Sigma-Delta Modulator as an A/D Converter," **IEEE Transactions on Circuits and Systems**, Vol. CAS-25, July 1978, pp. 510–514.

14. B. A. Wooley and J. L. Henry, "An Integrated Per-Channel PCM Encoder Based on Interpolation," **IEEE Journal of Solid State Circuits**, Vol. SC-14, February 1979, pp. 14–20. *(One of the first all-integrated CMOS sigma-delta ADCs.)*

15. B. A. Wooley et al, "An Integrated Interpolative PCM Decoder," **IEEE Journal of Solid State Circuits**, Vol. SC-14, February 1979, pp. 20–25.

16. J. C. Candy, B. A. Wooley, and O. J. Benjamin, "A Voiceband Codec with Digital Filtering," **IEEE Transactions on Communications**, Vol. COM-29, June 1981, pp. 815–830.

17. J. C. Candy and Gabor C. Temes, **Oversampling Delta-Sigma Data Converters**, IEEE Press, ISBN 0-87942-258-8, 1992.

18. R. Koch, B. Heise, F. Eckbauer, E. Engelhardt, J. Fisher, and F. Parzefall, "A 12-bit Sigma-Delta Analog-to-Digital Converter with a 15 MHz Clock Rate," **IEEE Journal of Solid-State Circuits**, Vol. SC-21, No. 6, December 1986.

19. D. R. Welland, B. P. Del Signore and E. J. Swanson, "A Stereo 16-Bit Delta-Sigma A/D Converter for Digital Audio," **J. Audio Engineering Society**, Vol. 37, No. 6, June 1989, pp. 476–485.

20. B. Boser and Bruce Wooley, "The Design of Sigma-Delta Modulation Analog-to-Digital Converters," **IEEE Journal of Solid-State Circuits**, Vol. 23, No. 6, December 1988, pp. 1298–1308.

21. J. Dattorro, A. Charpentier, D. Andreas, "The Implementation of a One-Stage Multirate 64:1 FIR Decimator for use in One-Bit Sigma-Delta A/D Applications," **AES 7th International Conference**, May 1989.

22. W.L. Lee and C.G. Sodini, "A Topology for Higher-Order Interpolative Coders," **ISCAS PROC.** 1987.

23. P.F. Ferguson, Jr., A. Ganesan and R. W. Adams, "One-Bit Higher Order Sigma-Delta A/D Converters," **ISCAS PROC.** 1990, Vol. 2, pp. 890–893.

24. Wai Laing Lee, **A Novel Higher Order Interpolative Modulator Topology for High Resolution Oversampling A/D Converters**, MIT Masters Thesis, June 1987.

25. R. W. Adams, "Design and Implementation of an Audio 18-Bit Analog-to-Digital Converter Using Oversampling Techniques," **J. Audio Engineering Society**, Vol. 34, March 1986, pp. 153–166.

26. P. Ferguson, Jr., A. Ganesan, R. Adams, et. al., "An 18-Bit 20 kHz Dual Sigma-Delta A/D Converter," **ISSCC Digest of Technical Papers**, February 1991.

27. Robert Adams, Khiem Nguyen, and Karl Sweetland, "A 113 dB SNR Oversampling DAC with Segmented Noise-Shaped Scrambling, "**ISSCC Digest of Technical Papers**, vol. 41, 1998, pp. 62, 63, 413. *(Describes a segmented audio DAC with data scrambling.)*

28. Robert W. Adams and Tom W. Kwan, "Data-directed Scrambler for Multibit Noise-shaping D/A Converters," **U.S. Patent 5,404,142**, filed August 5, 1993, issued April 4, 1995. *(Describes a segmented audio DAC with data scrambling.)*

29. Y. Matsuya, et. al., "A 16-Bit Oversampling A/D Conversion Technology Using Triple-Integration Noise Shaping," **IEEE Journal of Solid-State Circuits**, Vol. SC-22, No. 6, December 1987, pp. 921–929.

30. Y. Matsuya, et. al., "A 17-Bit Oversampling D/A Conversion Technology Using Multistage Noise Shaping," **IEEE Journal of Solid-State Circuits**, Vol. 24, No. 4, August 1989, pp. 969–975.

31. Paul H. Gailus, William J. Turney, and Francis R. Yester, Jr., "Method and Arrangement for a Sigma Delta Converter for Bandpass Signals," **U.S. Patent 4,857,928**, filed January 28, 1988, issued August 15, 1989.

32. S.A. Jantzi, M. Snelgrove, and P.F. Ferguson Jr., "A 4th-Order Bandpass Sigma-Delta Modulator," **IEEE Journal of Solid State Circuits**, Vol. 38, No. 3, March 1993, pp. 282–291.

33. Paul Hendriks, Richard Schreier, Joe DiPilato, "High Performance Narrowband Receiver Design Simplified by IF Digitizing Subsystem in LQFP," **Analog Dialogue**, Vol. 35-3, June-July 2001, available at www.analog.com. *(Describes an IF subsystem with a bandpass sigma-delta ADC having a nominal oversampling frequency of 18 MSPS, a center frequency of 2.25 MHz, and a bandwidth of 10 kHz – 150 kHz.)*

General References on Data Conversion and Related Topics

1. Alfred K. Susskind, **Notes on Analog-Digital Conversion Techniques**, John Wiley, 1957.

2. David F. Hoeschele, Jr., **Analog-to-Digital/Digital-to-Analog Conversion Techniques**, John Wiley and Sons, 1968.

3. K. W. Cattermole, **Principles of Pulse Code Modulation**, American Elsevier Publishing Company, Inc., 1969, New York NY, ISBN 444-19747-8.

4. Hermann Schmid, **Electronic Analog/Digital Conversions**, Van Nostrand Reinhold Co., 1970.

5. Dan Sheingold, **Analog-Digital Conversion Handbook**, First Edition, Analog Devices, 1972.

6. Donald B. Bruck, **Data Conversion Handbook**, Hybrid Systems Corporation, 1974.

7. Eugene R. Hnatek, **A User's Handbook of D/A and A/D Converters**, John Wiley, New York, 1976, ISBN 0-471-40109-9.

8. Nuggehally S. Jayant, **Waveform Quantizing and Coding**, John Wiley-IEEE Press, 1976, ISBN 0-87942-074-X.

9. Dan Sheingold, **Analog-Digital Conversion Notes**, Analog Devices, 1977.

10. C. F. Kurth, editor, **IEEE Transactions on Circuits and Systems Special Issue on Analog/Digital Conversion**, CAS-25, No. 7, July 1978.

11. Daniel J. Dooley, **Data Conversion Integrated Circuits**, John Wiley-IEEE Press, 1980, ISBN 0-471-08155-8.

12. Bernard M. Gordon, **The Analogic Data-Conversion Systems Digest**, Fourth Edition, Analogic Corporation, 1981.

13. Eugene L. Zuch, **Data Acquisition and Conversion Handbook**, Datel-Intersil, 1982.

14. Frank F. E. Owen, **PCM and Digital Transmission Systems**, McGraw-Hill, 1982, ISBN 0-07-047954-2.

15. Dan Sheingold, **Analog-Digital Conversion Handbook**, Analog Devices/Prentice-Hall, 1986, ISBN 0-13-032848-0.

16. Matthew Mahoney, **DSP-Based Testing of Analog and Mixed-Signal Circuits**, IEEE Computer Society Press, 1987, ISBN 0-8186-0785-8.

17. Michael J. Demler, **High-Speed Analog-to-Digital Conversion**, Academic Press, Inc., 1991, ISBN 0-12-209048-9.

18. J. C. Candy and Gabor C. Temes, **Oversampling Delta-Sigma Data Converters**, IEEE Press, ISBN 0-87942-258-8, 1992.

19. David F. Hoeschele, Jr., **Analog-to-Digital and Digital-to-Analog Conversion Techniques**, Second Edition, John Wiley and Sons, 1994, ISBN-0-471-57147-4.

20. Rudy van de Plassche, **Integrated Analog-to-Digital and Digital-to-Analog Converters**, Kluwer Academic Publishers, 1994, ISBN 0-7923-9436-4.

21. David A. Johns and Ken Martin, **Analog Integrated Circuit Design**, John Wiley, 1997, ISBN 0-471-14448-7.

22. Mikael Gustavsson, J. Jacob Wikner, and Nianxiong Nick Tan, **CMOS Data Converters for Communications**, Kluwer Academic Publishers, 2000, ISBN 0-7923-7780-X.

23. R. Jacob Baker, **CMOS Circuit Design Volumes I and II**, John Wiley-IEEE Computer Society, 2002, ISBN 0-4712-7256-6.

24. Rudy van de Plassche, **CMOS Integrated Analog-to-Digital and Digital-to-Analog Converters, Second Edition**, Kluwer Academic Publishers, 2003, ISBN 1-4020-7500-6.

Analog Devices' Seminar Series:

25. Walt Kester, **Practical Analog Design Techniques**, Analog Devices, 1995, ISBN 0-916550-16-8, available for download at www.analog.com.

26. Walt Kester, **High Speed Design Techniques**, Analog Devices, 1996, ISBN 0-916550-17-6, available for download at www.analog.com.

27. Walt Kester, **Practical Design Techniques for Power and Thermal Management**, Analog Devices, 1998, ISBN 0-916550-19-2, available for download at www.analog.com.

28. Walt Kester, **Practical Design Techniques for Sensor Signal Conditioning**, Analog Devices, 1999, ISBN 0-916550-20-6, available for download at www.analog.com.

29. Walt Kester, **Mixed-Signal and DSP Design Techniques**, Analog Devices, 2000, ISBN 0-916550-22-2, available for download at www.analog.com.

30. Walt Kester, **Mixed-Signal and DSP Design Techniques**, Analog Devices and Newnes (An Imprint of Elsevier Science), ISBN 0-75067-611-6, 2003.

31. Walter G. Jung, **Op Amp Applications**, Analog Devices, 2002, ISBN 0-916550-26-5.

CHAPTER 4

Data Converter Process Technology

Data Converter Process Technology

Early Processes
Walt Kester

Vacuum Tube Data Converters

The vacuum tube was the first enabling technology in the development of data converters—starting in the 1920s and continuing well into the late 1950s. As discussed in Chapter 1 of this book, the vacuum tube was invented by Lee De Forest in 1906 (Reference 1). A figure from the patent is shown in Figure 4.1. Vacuum tubes quickly found their way into a variety of electronic equipment, and the Bell Telephone system began using vacuum tube amplifiers in their telephone plants as early as 1914.

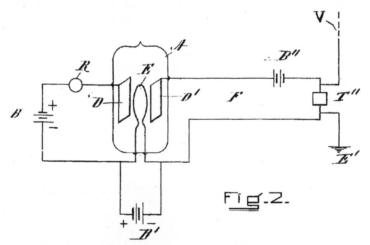

Extracted from: Lee De Forest, "Device for Amplifying Feeble Electrical Currents," U.S. Patent 841,387, Filed October 25, 1906, Issued January 15, 1907

Figure 4.1: The Invention of the Vacuum Tube: 1906

Amplifier development has always been critical to data converter development, starting with these early vacuum tube circuits. A significant contribution was the invention of the feedback amplifier (op amp) by Harold S. Black in 1927 (References 2, 3, 4). Vacuum tube circuit development continued throughout World War II, and many significant contributions came from Bell Labs. For a detailed discussion of the history of op amps, please refer to Walt Jung's book, *Op Amp Applications Handbook* (Reference 5).

257

In the 1920s, 1930s, 1940s, and 1950s, vacuum tubes were the driving force behind practically all electronic circuits. In 1953, George A. Philbrick Researches, Inc., introduced the world's first commercially available op amp, known as the K2-W (Reference 6). A photo and schematic are shown in Figure 4.2.

Pulse code modulation (PCM) was the first major driving force in the development of early data converters, and Alec Hartley Reeves is generally credited for the invention of PCM in 1937. (Reference 7). In his patent, he describes a vacuum tube "counting" ADC and DAC (see Chapter 3 of this book). Data converter development continued at Bell Labs during the 1940s, not only for use in PCM system development, but also in wartime encryption systems.

The development of the digital computer in the late 1940s and early 1950s spurred interest in data analysis, digital process control, etc., and generated more

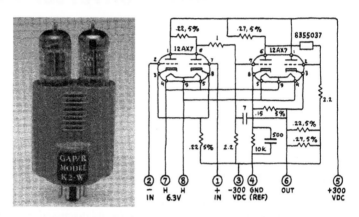

Figure 4.2: The K2-W Op Amp Introduced in 1953
(Courtesy of Dan Sheingold)

commercial interest in data converters. In 1953 Bernard M. Gordon, a pioneer in the field of data conversion, founded a company called Epsco Engineering (now Analogic, Inc.) in his basement in Concord, MA. Gordon had previously worked on the UNIVAC computer, and saw the need for commercial data converters. In 1954 Epsco introduced an 11-bit, 50 kSPS vacuum-tube based SAR ADC called the DATRAC. This converter, shown in Figure 4.3, is generally credited as being the first commercial offering of such a device. The DATRAC was offered in a 19" × 26" × 15" housing, dissipated several hundred watts, and sold for approximately $8000.00.

While the vacuum tube DATRAC was certainly impressive for its time, solid-state devices began to emerge during the 1950s which would eventually revolutionize the entire field of data conversion and electronics in general.

Figure 4.3:
1954 "DATRAC"
11-bit, 50 kSPS
SAR ADC
Designed by
Bernard M. Gordon
at EPSCO

- 19" × 15" × 26"

- 150 lbs

- 500W

- $8,500.00

Courtesy,
Analogic Corporation
8 Centennial Drive
Peabody, MA 01960

www.analogic.com

Solid State, Modular, and Hybrid Data Converters

Although the transistor was invented in 1947 by John Bardeen, Walter Brattain, and William Shockley of Bell Labs (References 8, 9, 10, 11), it took nearly a decade for the technology to find its way into commercial applications. The overall reliability of the devices was partly responsible for this, as the first transistors were germanium, and were limited in terms of leakage currents, general stability, maximum junction temperature, and frequency response.

In May of 1954, Gordon Teal of Texas Instruments developed a grown-junction silicon transistor. These transistors could operate up to 150°C, far higher than germanium. Additional processing refinements were to improve upon the early silicon transistors, and eventually lead a path to the invention of the first integrated circuit in 1958 by Jack Kilby of Texas Instruments (Reference 12).

Kilby's work was paralleled by Robert Noyce at Fairchild, who also developed an IC concept in 1959 (Reference 13). Noyce used interconnecting metal trace layers between transistors and resistors, while Kilby used bond wires. As might be expected from such differences between two key inventions, so closely timed in their origination, there was no instant concensus on the true "IC inventor." Subsequent patent fights between the two inventor's companies persisted into the 1960s. Today, both men are recognized as IC inventors.

In parallel with Noyce's early IC developments, Jean Hoerni (also of Fairchild Semiconductor) had been working on means to protect and stabilize silicon diode and transistor characteristics. Until that time, the junctions of all mesa process devices were essentially left exposed. This was a serious limitation of the mesa process. The mesa process is so-named because the areas surrounding the central base-emitter regions are etched away, thus leaving this area exposed on a plateau, or mesa. In practice, this factor makes a semiconductor so constructed susceptible to contaminants, and as a result, inherently less stable. This was the fatal flaw that Hoerni's invention addressed. Hoerni's solution to the problem was to rearrange the transistor geometry into a flat, or planar surface, thus giving the new process its name (see References 14 and 15). However, the important distinction in terms of device protection is that within the planar process the otherwise exposed regions are left covered with silicon dioxide. This feature reduced the device sensitivity to contaminants; making a much better, more stable transistor or IC. With the arrangement of the device terminals on a planar surface, Hoerni's invention was also directly amenable to the flat metal conducting traces that were intrinsic to Noyce's IC invention. Furthermore, the planar process required no additional process steps in its implementation, so it made the higher performance economical as well. As time has now shown, the development of the planar process was another key semiconductor invention. It is now widely used in production of transistors and ICs.

At a time in the early 1960s shortly after the invention of the planar process, the three key developments had been made as summarized in Figure 4.4. They were the *(silicon) transistor* itself, the *IC*, and the *planar process*. The stage was now set for important solid-state developments in data converters. This was to take place in three stages. First, there would be discrete transistor and modular data converters, second there would be hybrid data converters, and thirdly, the data converter finally became a complete, integral, dedicated IC. Of course, within these developmental stages there were considerable improvements made to device performance. And, as with the vacuum tube/solid-state periods, each stage overlapped the previous and/or the next one to a great extent.

- Invention of the (Germanium) transistor at Bell Labs: John Bardeen, Walter Brattain, and William Shockley in 1947.
- Silicon Transistor: Gordon Teal, Texas Instruments, 1954.
- Birth of the Integrated Circuit:
  - Jack Kilby, Texas Instruments, 1958 (used bond wires for interconnections).
  - Robert Noyce, Fairchild Semiconductor, 1959 (used metallization for interconnections).
- The Planar Process: Jean Hoerni, Fairchild Semiconductor, 1959.

Figure 4.4: Key Solid-State Developments: 1947–1959

The first solid-state data converters utilized discrete transistors, few if any ICs, and required multiple PC boards to implement the analog and digital parts of the conversion process. A typical example was the HS-810, 8-bit, 10 MSPS ADC introduced in 1966 by Computer Labs, Inc. and shown in Figure 4.5. One of the PC boards from the HS-810 is shown in Figure 4.6. (Computer Labs was later acquired by Analog Devices in 1978). The entire converter was built from discrete transistors, resistors, and capacitors, with practically no integrated circuits. The unit was designed to fit in a 19" rack, contained all required power supplies, dissipated over 100 W, and cost over $10,000 at the time of introduction. Data converters such as the HS-810 were primarily used in research applications and in early digital radar receivers.

Figure 4.5: HS-810, 8-bit, 10 MSPS ADC Released by Computer Labs, Inc. in 1966

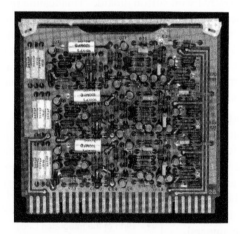

Figure 4.6: Double-Sided PC Board from HS-810 ADC

By the late 1960s and early 1970s, various IC building blocks such as op amps, comparators, and digital logic became available which allowed a considerable reduction in parts count in ADCs and DACs. This led to the *modular* data converter—basically various combinations of ICs, transistors, resistors, capacitors, etc., mounted on a small PC board with pins, and encapsulated in a potted plastic case. The potting compound helped to distribute the heat throughout the module, provided some degree of thermal tracking between critical components, and made it a little more difficult for a competitor to reverse-engineer the circuit design.

A good example of an early converter module was the Analog Devices' industry-standard ADC12QZ, 12-bit, 40 μs SAR ADC introduced in 1972 and shown in Figure 4.7. The ADC12QZ was the first low cost commercial general-purpose 12-bit ADC on the market. The converter used the *quad-switch* ICs in conjunction with precision thin film resistor networks for the internal DAC. (The quad switch AD550 *μDAC* circuits are discussed in more detail in Chapter 3 of this book.)

2" × 4" × 0.4", 1.8W

Figure 4.7: ADC-12QZ General-Purpose 12-Bit, 40 μs SAR ADC Introduced in 1972

Another popular process for data converters that had its origins in the 1970s is the *hybrid*. Hybrid circuits are typically constructed using un-encapsulated die, or "chips," such as ICs, resistors, capacitors, etc., which are bonded to a ceramic substrate with epoxy—in some cases, eutetically bonded. The bond pads on the various chips are connected to pads on the substrate with wire bonds, and interconnections between devices are made with metal paths on the substrate, similar in concept to a PC board. The metal conductor paths are either thick film or thin film, depending upon the process and the manufacturer. For obvious reasons, hybrid technology is often referred to as "chip-and-wire." After assembly, the package is sealed in an inert atmosphere to prevent contamination.

Various technologies are used to construct hybrids, including thick and thin film conductors and resistors, and the devices tend to be rather expensive compared to ICs. The AD572 12-bit, 25 μs ADC released by Analog Devices in 1977 is an excellent example of a hybrid and is shown in Figure 4.8. It is significant that the AD572 was the first 12-bit hybrid ADC circuit to obtain MIL-883B approval.

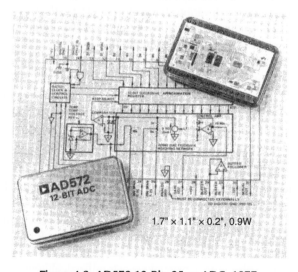

1.7" × 1.1" × 0.2", 0.9W

Figure 4.8: AD572 12-Bit, 25 μs ADC, 1977

The chief motivation behind modules and hybrids was to produce data converters with speed and resolution not achievable with the early IC processes. Hybrid circuit designers could choose from a variety of discrete PNP, NPN, and FET transistors, IC op amps, IC comparators, IC references, IC DACs, IC logic, etc. Coupled with the ability to perform active in-circuit laser trimming of resistors, the hybrid circuits could achieve relatively high levels of performance compared to what was possible in ICs alone. Customers were willing to pay premium prices for the hybrids, because that was the only way to achieve the desired performance. Also, there was usually a period of at least several years before the equivalent function was achievable in completely monolithic form, thereby giving a hybrid a reasonable product life cycle.

Today, however, the situation is reversed—the speed and resolution of modern IC data converters is generally limited by internal process-related parasitics, and these parasitics are much smaller than could ever be achieved in an equivalent hybrid circuit. In other words, it would be impossible to duplicate the performance of most modern IC data converters using conventional hybrid technology. For these reasons, hybrids today serve relatively small niche markets today, such as dc-to-dc and synchro-to-digital converters.

Note the distinction between chip-and-wire hybrids and modern *multichip modules* (MCMs) which basically use surface-mount ICs and other components on small multilayer PC boards to achieve higher levels of functionality than possible in a single IC.

It is also important to distinguish chip-and-wire hybrids and multichip modules from another IC packaging technology—usually referred to as *compound monolithic*—where two die (usually an analog IC and a digital IC) are mounted on a single lead frame, electrically connected with wirebonds, and encapsulated in a plastic IC package.

Calibration Processes

Nearly all data converters require some calibration to ensure overall INL, DNL, gain, and offset errors are within specified limits. For low resolution converters, the accuracy and matching of the various circuit components may be sufficient to ensure this. For high resolution converters (greater than 10 bits or so), methods must generally be provided to accomplish various types of trims. The early rack-mounted and PC board data converters generally used potentiometers and/or selected precision resistors to accomplish the required calibration. In many cases, a precision resistor in the circuit was "padded" with a larger parallel resistor to achieve the desired value.

Modular data converters achieved their accuracy either by using pre-trimmed ICs and precision resistor networks as building blocks, or by manually selecting resistors prior to potting. An interesting trim method was used in the popular DAC-12QZ—the first modular 12-bit DAC which was introduced in 1970. It utilized thick film resistors that were trimmed to the appropriate values by sandblasting.

Because modular data converters had to be calibrated before potting, the effects of the thermal shifts due to potting had to be factored into the actual trim process.

Hybrid circuits generally utilized a variety of types of trimming processes, depending upon the process and the manufacturer. Again, the use of pre-trimmed IC building blocks, such as the AD562 or AD565 IC DAC, minimized substrate-level trimming requirements in such products as the AD572 mentioned previously. Other popular methods included functional laser trimming of thick or thin film resistors on the substrate. These trimmed resistors could be in the form of deposited substrate resistors or resistor networks bonded to the substrate. Both thick and thin film resistor technology was utilized, although thin film resistors generally had better stability.

References:

4.1 Early Processes

1. Lee De Forest, "Device for Amplifying Feeble Electrical Currents," **U.S. Patent 841,387**, filed October 25, 1906, issued January 15, 1907. *(The triode vacuum tube, or 'Audion', the first amplifying device.)*

2. H. S. Black, "Wave Translation System," **U.S. Patent 2,102,671**, filed August 8, 1928, issued December 21, 1937. *(The basis of feedback amplifier systems.)*

3. H. S. Black, "Stabilized Feedback Amplifiers," **Bell System Technical Journal**, Vol. 13, No. 1, January 1934, pp. 1–18. *(A practical summary of feedback amplifier systems.)*

4. Harold S. Black, "Inventing the Negative Feedback Amplifier," **IEEE Spectrum**, December, 1977. *(Inventor's 50th anniversary story on the invention of the feedback amplifier.)*

5. Walter G. Jung, **Op Amp Applications**, Analog Devices, 2002, ISBN 0-916550-26-5.

6. Data Sheet For Model K2-W Operational Amplifier, George A. Philbrick Researches, Inc., Boston, MA, January 1953. See also "40 Years Ago," **Electronic Design**, December 16, 1995, p. 8. *(The George A. Philbrick Research dual triode K2-W, the first commercial vacuum tube op amp.)*

7. Alec Harley Reeves, "Electric Signaling System," **U.S. Patent 2,272,070**, filed November 22, 1939, issued February 3, 1942. Also **French Patent 852,183** issued 1938, and **British Patent 538,860** issued 1939. *(The ground-breaking patent on PCM. Interestingly enough, the ADC and DAC proposed by Reeves are counting types, and not successive approximation.)*

8. Ian M. Ross, "The Foundation of the Silicon Age," **Bell Labs Technical Journal**, Vol. 2, No. 4, Autumn 1997.

9. C. Mark Melliar-Smith et al, "Key Steps to the Integrated Circuit," **Bell Labs Technical Journal**, Vol. 2, No. 4, Autumn 1997.

10. J. Bardeen, W. H. Brattain, "The Transistor, a Semi-Conductor Triode," **Physical Review**, Vol. 74, No. 2, July 15, 1947 pp. 230–231. *(The invention of the germanium transistor.)*

11. W. Shockley, "The Theory of p-n Junctions in Semiconductors and p-n Junction Transistors," **Bell System Technical Journal**, Vol. 28, No. 4, July 1949, pp. 435–489. *(Theory behind the germanium transistor.)*

12. J. S. Kilby, "Invention of the Integrated Circuit," **IRE Transactions on Electron Devices**, Vol. ED- 23, No. 7, July 1976, pp. 648–654. *(Kilby's IC invention at TI.)*

13. Robert N. Noyce, "Semiconductor Device-and-Lead Structure," **U.S. Patent 2,981,877**, filed July 30, 1959, issued April 25, 1961. *(Noyce's IC invention at Fairchild.)*

14. Jean A. Hoerni, "Method of Manufacturing Semiconductor Devices," **U.S. Patent 3,025,589**, filed May 1, 1959, issued March 20, 1962. *(The planar process—a manufacturing means of protecting and stabilizing semiconductors.)*

15. Jean Hoerni, "Planar Silicon Diodes and Transistors," **IRE Transactions on Electron Devices**, Vol. 8, March 1961, p. 168. *(Technical discussion of planar processed devices.)*

Modern Processes
Walt Kester, James Bryant

Bipolar Processes

The basic bipolar IC process of the 1960s was primarily optimized to yield good NPN transistors. However, low beta, low bandwidth PNP transistors were available on the process—the *lateral* PNP and the *substrate* PNP. Clever circuit designers were able to use the PNPs for certain functions such as level shifting and biasing. Bob Widlar of Fairchild Semiconductor Corporation was one of these early pioneers, and designed the first monolithic op amp, the μA702, in 1963. Other op amps followed rapidly, including the μA709 and the industry-standard μA741. Another Widlar design, the μA710/μA711 comparator, was introduced in 1965. These types of linear devices, coupled with the introduction of the 7400-series TTL logic, provided some of the key building blocks for the modular and hybrid data converters of the 1970s. For more details of the history of op amps, please refer to Walt Jung's excellent book, *Op Amp Applications* (Reference 1).

Analog Devices was founded in 1965 by Ray Stata and Matt Lorber, and focused its early efforts on precision modular amplifiers. In 1969, Analog Devices acquired Pastoriza Electronics, then a leader in data conversion products—thereby making a solid commitment to both data acquisition and linear technology. In 1971, Analog Devices acquired a small IC company, Nova Devices of Wilmington, MA, and this later led to many monolithic linear and data converter products.

Thin Film Resistor Processes

There is another process technology which does deserve special mention, since it is crucial to the manufacture of many linear circuits and data converters requiring stable precision resistors and the ability to perform calibrations. This is thin-film resistor technology.

Analog Devices began its efforts to develop thin-film resistor technology in the early 1970s. Much effort has been spent to develop the ability to deposit these stable thin-film resistors on integrated circuit chips, and even more effort to laser trim them at the wafer level. They have temperature coefficients of <20 ppm/°C and matching to within 0.005%. The resistors can be made to match to within 0.01% or better without laser trimming, but to achieve this they must be relatively large—in practice, if resistors must match to better than 0.1% or 0.05%, it is more economical to laser trim them than to design them to meet the specification without trimming.

It is interesting to note that although it is possible to make these resistors very precise (ratiometrically), they usually have quite wide tolerances. The reason is economic—most applications require precision matching and low temperature coefficient but do not actually need very high absolute precision. It is possible to optimize all three, but much less expensive to optimize two out of three—so this is what is usually done.

Many of the precision resistors used in the various data converters are laser trimmed SiCr thin-film resistors, although the new submicron and nonvolatile memory processes make laser trimming unnecessary in many new data converters, where it would have been unavoidable in earlier generations.

In summary, the bipolar process, coupled with thin-film resistors and laser wafer trim technology, led to the proliferation of IC data converters during the 1970s, 1980s, and 1990s. For example, the AD571 was the first complete monolithic 10-bit SAR ADC designed by Paul Brokaw and was introduced in 1978. The AD571 used a bipolar process with integrated-injection-logic (I^2L) as well as thin film laser wafer trimmed resistors. I^2L geometries were made to use a set of diffusions compatible with high performance linear transistors (Reference 2). The AD571 was soon followed by other converters, such as the industry-standard AD574 12-bit ADC in 1980. In addition, the IC converters provided building blocks for high performance modular and hybrid converters during the same period.

Complementary Bipolar (CB) Processes

Although clever IC circuit designers made the best use possible of the poor quality substrate and lateral PNP transistors available on the NPN-based bipolar processes of the 1970s and 1980s, the lack of matching high bandwidth PNP transistors definitely limited circuit design options in linear ICs, especially high speed op amps.

In the mid-1980s, Analog Devices developed the world's first p-epi complementary bipolar (CB) process, and the AD840 series of op amps was introduced starting in 1988. The f_ts of the PNP and NPN transistors in this first-generation 36 V CB process were approximately 700 MHz and 900 MHz, respectively. Since the introduction of the original CB process, several generations of faster CB processes have been developed at Analog Devices designed for even higher speeds with lower breakdowns. Descriptions of the ground-breaking first-generation CB process can be found in References 3 and 4.

Analog Devices' CB processes all have JFETs, allowing high input impedance op amps as well as sample-and-hold amplifiers for data converters. The dielectrically isolated "XFCB" process provided real breakthroughs in speed and distortion performance. Introduced in 1992, this process yields 3 GHz PNPs and 5 GHz matching NPNs. The "XFCB 1.5" process has 5 GHz PNPs and 9 GHz NPNs. A 5 V "XFCB 2" process has 9 GHz PNPs and 16 GHz NPNs.

The XFCB process (and later generations) has been used to produce several notable high-end data converters. For example, the AD9042, designed by Roy Gosser and Frank Murden and introduced in 1995, was the first low distortion 12-bit, 41-MSPS ADC on the market, with greater than 80 dBc SFDR over the Nyquist bandwidth. The AD9042 was followed by several additional XFCB converters, including the AD6645 14-bit, 80/105 MSPS ADC which was introduced in 2001, with 90 dBc SFDR and 75 dB SNR. Both the AD9042 and the AD6645 use laser wafer trimming to achieve their high level of performance.

CMOS Processes

Metal-on-silicon (MOS) devices had their origins in late 1950s and early 1960s in the pursuit of a process tailored for digital devices. The first complementary-metal-oxide-semiconductor (CMOS) devices began to appear in the mid-1960s, and provided both P-channel and N-channel MOS devices on the same process. CMOS offered the potential of much higher packing density and lower power than TTL (bipolar-based) devices, and soon became the IC process of choice for complex VLSI digital devices. The same advances in technology that have enabled cheap, powerful, low power consumption processors with large memory to revolutionize mobile telephony, portable computing and many other fields have also revolutionized data converters.

Data converter designers soon realized the advantages of using CMOS for ADCs and DACs. As discussed in Chapter 7 of this book, CMOS switches make ideal building blocks for data acquisition systems. In addition, CMOS offers the ability to add digital functionality to data converters without incurring significant cost, power, and size penalties.

In 1974, Analog Devices combined its thin-film technology with CMOS to produce the first 10-bit multiplying CMOS DAC, the AD7520, designed by Jim Cecil and Hank Krabbe. In 1976, Analog Devices established a CMOS IC design and manufacturing operation in Limerick, Ireland, and rapidly introduced many more general-purpose CMOS DACs and ADCs starting in the 1970s and continuing to this day.

Although CMOS is capable of efficiently making high density low power logic and can make excellent analog switches, it is not quite as suitable for amplifiers and voltage references as bipolar processes. These considerations caused process technologists to combine bipolar and CMOS processes to achieve both low power high density logic and high accuracy low noise analog circuitry on a single chip. The resulting processes are more complex, and therefore more expensive, than simple bipolar and CMOS processes, but do have better mixed-signal performance. They include BiMOS processes, which are basically bipolar processes to which CMOS structures have been added, and linear compatible CMOS (LC^{2}MOS or LCCMOS), which is basically CMOS with added bipolar capability. Analog Devices' Limerick facility in Ireland began introducing data converters, switches, and multiplexers using its own proprietary LC^{2}MOS process in the mid-1980s.

However, the compromises necessary to combine features mean that neither BiMOS nor LC^{2}MOS offers quite as good performance as its senior parent process does in its own speciality. Thus BiMOS and LC^{2}MOS have not replaced bipolar, complementary bipolar, or CMOS technology, but designers now have four processes to choose from when designing a data converter.

Modern submicron CMOS technology is cheap, fast and low powered. It is also precise—the same techniques that enable submicron features in logic and memory ICs allow us to manufacture matched resistors and capacitors which are smaller, cheaper and more accurately matched without subsequent trimming than has hitherto been possible, and to make switches with lower leakage, lower on-resistance and less stray capacitance. These advances on their own enable the manufacture of smaller, faster, cheaper and lower powered DACs and ADCs and the integration of complex devices which would have been too big to put on a chip a few years ago. The technology brings an additional bonus—logic made with these processes is so small, cheap and low powered that incorporating autocalibration and other computational features to improve data converter performance and accuracy is virtually free.

Cheap, reliable, nonvolatile memory is another recent process innovation that improves the performance of new generation data converters. Gain, offset and even linearity can be adjusted after the chip has been packaged (so packaging stresses will not affect accuracy), at a cost far lower than that of laser trimming. Many data converters trimmed in this way are "locked" before leaving the factory so that the calibration cannot be damaged accidentally—and so the user cannot trim them to his system's requirements. However, the same technology does allow users to store calibration coefficients and similar data. Other converters allow for periodic self-calibrations for gain, offset, and even linearity errors. Various types of "fuse blowing" or "link trimming" techniques are quite often used in the calibration process rather than more expensive thin film laser wafer trimming.

One feature of these new submicron CMOS processes, which is both a benefit and a problem, is that they have low breakdown voltage and must operate on low voltage supplies: 0.6 μm CMOS uses 5 V, and less for the smaller geometry processes (0.35 μm ~ 3.3 V, 0.25 μm ~ 2.5 V, 0.18 μm ~ 2 V, 0.15 μm ~ 1.5 V, and 0.13 μm ~ 1 V). This makes them virtually useless with the traditional precision analog supplies of ±15 V. They will, however, operate accurately and at high speed on the lower supply voltages, making them convenient for low power circuitry. However, this reduces their dynamic range, as their fullscale output is closer (in the case of 3 V single supply circuitry 20 dB closer) to the noise floor.

For high speed data converters, however, the reduced signal amplitude can be an advantage, because it is generally much easier to drive low amplitude signals with low distortion into 50 Ω or 75 Ω loads than larger amplitude signals. The optimum amplitude for the best compromise between noise and distortion generally ranges between 1 V and 2 V peak-to-peak in communications-oriented data converters, although there are some exceptions to this.

A brief summary of data converter processes is given in Figure 4.9.

- NPN-Based bipolar
- NPN-Based bipolar with JFETs and LWT thin-film resistors
- NPN-Based Bipolar with Integrated Injection Logic (I²L) and LWT thin-film resistors
- Complementary Bipolar (CB) with JFETs and LWT thin-film resistors
- Dielectrically Isolated Complementary Bipolar with JFETS and LWT thin-film resistors
- CMOS
- CMOS and LWT thin film resistors
- LC²MOS and BiCMOS with LWT thin-film resistors
- Hybrid (chip-and-wire)
- Multichip Module (MCM)
- GaAs, SiGe

Figure 4.9: Data Converter Processes

Data Converter Processes and Architectures

In the last decade, CMOS has become a dominant process for data converters—replacing more expensive bipolar laser wafer trimmed devices. Submicron CMOS has extremely low parasitic resistance, capacitance, and inductance, and is ideal for a number of data converter architectures, including successive approximation, Σ-Δ, and pipeline. The fine-line lithography techniques associated with submicron processes allow excellent matching between capacitors in a capacitor DAC (a fundamental building block for SAR ADCs). The internal capacitor DACs are then trimmed by adding or subtracting small parallel capacitors using either some form of fuse blowing, link trimming, or autocalibration routine utilizing volatile memory. The addition of analog input multiplexers to form a complete data acquisition system is also relatively easy due to the high quality switches and multiplexers available in CMOS.

CMOS is also the process of choice for all types of Σ-Δ ADCs and DACs, which are also based on switched capacitor circuits. In addition, the Σ-Δ architecture is highly digitally intensive—another reason for utilizing the packing density and low power of CMOS. Statistical matching techniques are popular in the multibit Σ-Δ data converters as a means for higher resolution and dynamic range without the need for trimming.

For high-speed pipelined ADCs, the digital capability of CMOS is ideal to perform the required error correction. Fully differential circuit design techniques, coupled with the high speed switched capacitor capabilities of CMOS, produce excellent performance.

CMOS is also excellent for high speed communications DACs, as exhibited in the Analog Devices' TxDAC family with resolutions up to 16 bits and speeds of several hundred MHz. To illustrate the progression of DAC performance over the last decade, Figure 4.10 shows update rate moving from less than 30 MSPS in 1994 to nearly 1 GSPS in 2004. This is primarily due to the reduction in parasitic capacitance, inductance, and resistance associated with the smaller and smaller submicron processes. Figure 4.11 shows a similar plot for SFDR (10 MHz output signal), which has increased from 50 dBc in 1994 to nearly 90 dBc in 2004.

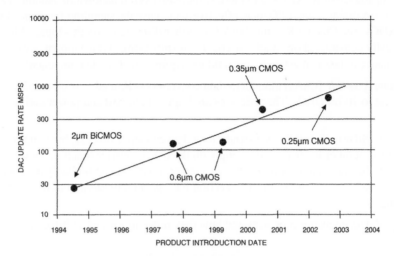

Figure 4.10: High Speed DAC Update Rate Trend

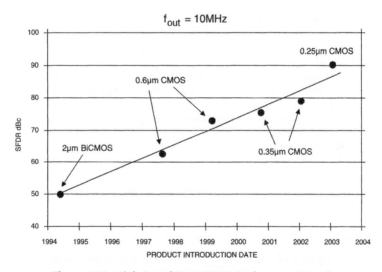

Figure 4.11: High Speed DAC SFDR Performance Trend

The trend in modern data converters is to add much more digital functionality, such as digital filtering, multiplexing, decoding, modulation, etc., and CMOS is the ideal process for this.

As mentioned earlier, analog building blocks such as amplifiers, mixers, and voltage references designed in CMOS cannot achieve the levels of performance attainable in bipolar, hence the need for a process that combines bipolar with CMOS, or BiCMOS. BiCMOS processes are more expensive, but are useful where an ADC with an extremely high performance analog front end is required. Functions such as mixers, sample-and-holds, input buffer amplifiers, and accurate voltage references can be implemented in bipolar, while the digital portion of the data converter is CMOS.

269

Multichip modules offer the flexibility of combining various IC technologies to perform functions otherwise not possible in all-monolithic parts. For instance, high performance RF analog front ends can be tuned to match the input impedance of IC ADCs, and thereby increase overall bandwidth. Another example is the use of digital post processing using FPGAs to effectively increase the sampling frequency by time inter-leaving several ADCs (see Chapter 8 of this book for further discussions on this topic). Modern multichip modules are typically constructed on small low cost PC boards using surface mount components and offer enhanced performance in less real estate than would be required by discrete components.

The role of *Gallium Arsenide* (GaAs) in modern data converters is limited to 6-bit to 8-bit >1 GSPS flash converters and 6-bit to 10-bit DACs. These devices are high in both cost and power and serve small niche markets.

Silicon Germanium (SiGe) offers little as a stand-alone data converter process, but combined with CMOS, could allow the integration of RF front ends along with the data converter function. However, these prod-ucts would probably be very application-specific, as greater flexibility can probably be achieved with the devices in separate packages. (Refer to the next section for a general discussion of the related issue of "smart partitioning.")

A brief summary of data converter processes and how they relate to various architectures is given in Figure 4.12.

- CMOS:
  - Ideal for switched capacitor SAR, Σ-Δ, Pipelined
  - Additional digital functionality
  - Volatile and non-volatile trimming at package level
- BiCMOS
  - Useful if analog front-end requires extremely high performance
  - Amplifiers, mixers, SHAs, highly accurate voltage references
- Calibration processes
  - LWT, fuse blowing, link trimming, volatile and non-volatile memory, autocalibration
- Multichip Module
  - Multiple ADCs and DACs, analog front ends, digital post processing
- GaAs
  - 6-, 8-bit GHz flash ADCs, high power and cost
- SiGe
  - Could be useful combined with CMOS

Figure 4.12: Data Converter Processes and Architectures

Finally, no discussion on data converter processes would be complete without touching upon the issue of packaging. In the last decade, there has been an increase in the demand for small, low cost, high perfor-mance, surface-mount packages suitable for mass production using automated assembly techniques. Today this is possible, primarily because of the lower power and small die size associated with modern submicron processes. Many devices are suitable for packages such as those shown in Figure 4.13, which are repre-sentative of today's trends. Smaller chip-scale-packaged (CSP) devices are available when required, and ball-grid-array (BGA) packages are useful for high pin count, high speed devices.

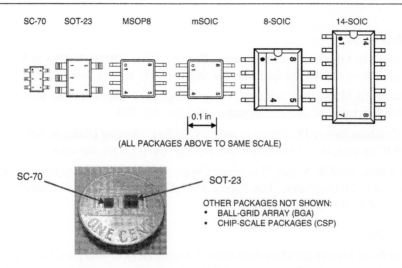

Figure 4.13: Examples of Modern Data Converter Packages

When power dissipation becomes significant, larger packages come equipped with built-in heatsinking "slugs" or "epads," which can be soldered directly to the PC board ground plane to effectively dissipate the heat. The use of high speed serial interfaces is also an important trend in reducing the total package pin count to maintain small package profiles.

References:

4.2 Modern Processes

1. Walter G. Jung, **Op Amp Applications**, Analog Devices, 2003, ISBN 0-916550-26-5. *(The first chapter on op amp history is complete with numerous references to patents, articles, etc.)*

2. A. Paul Brokaw, "A Monolithic 10-Bit A/D Using I²L and LWT Thin-Film Resistors," **IEEE Journal of Solid State Circuits**, Vol. SC-13, December 1978, pp. 736–745.

3. "Op Amps Combine Superb DC Precision and Fast Settling," **Analog Dialogue**, Vol. 22, No. 2, 1988 *(The AD846 IC op amp, the AD840 series, and the high speed CB process used.)*

4. Jerome F. Lapham, Brad W. Scharf, "Integrated Circuit with Complementary Junction-Isolated Transistors and Method of Making Same," **U.S. Patent 4,969,823**, filed May 5, 1988, issued Nov. 13, 1990. *(Design of the ADI CB IC process.)*

Acknowledgments:

Thanks are due to Doug Mercer and Dave Robertson of Analog Devices who provided valuable insights regarding modern data converter IC processes and their relationship to the various trends and architectures. James Bryant provided some of the process-related descriptions, and Walt Jung's *Op Amp Applications* book was the primary source for semiconductor history.

Smart Partitioning
Dave Robertson, Martin Kessler

When Complete Integration Isn't the Optimal Solution

For 30 years, the main path to "smaller, faster, better, cheaper" electronic devices has been through putting more and more of a given system onto a single chip. Large rewards were reaped by those companies that could overcome the various technological barriers to integration, providing more functionality on a single chip. But, as we enter the very deep submicron age, we are approaching some important physical limitations that will change the cost and performance trade-offs that designers have traditionally made.

As we approach the limits of practical reduction in feature size, it will increasingly turn out that a two-chip design will be smaller, faster, better, cheaper than a single, integrated solution. The key in these cases will be selecting the boundary between these chips. Although high levels of integration will continue to be a feature of the most advanced systems, reaching the optimum in cost and performance will no longer be a simple case of steadily increasing integration. Rather, progress will be measured by changes in circuit partitioning that enable system improvements.

There are several examples of this partitioning already evident today. For instance, large amounts of memory are generally cheaper to implement as a separate DRAM chip than to embed into a microprocessor. It is important to note that as integration barriers emerge, we will not step back to the days where the design model was "analog on one chip, digital on another, memory on a third." Chip partitioning will be done along boundaries that optimize the flow of signal information, as well as augment the intellectual property strengths of the chip providers. It will not be a simple case of "dis-integration." Instead, the best systems will reflect carefully considered integration, facilitating a "smart partitioning."

What makes one partitioning "smarter" than another? There are several important factors to consider:

- *Supply Voltage*—Each advance in lithography brings with it a reduced supply voltage. While this generally helps to lower the power in digital circuits, a lower supply voltage can actually cause power dissipation to increase for high performance analog circuits. Lower supply voltages also force the use of smaller signal swings, making it difficult to maintain good signal-to-noise ratios. Many systems will look to implement critical analog functions on technologies that support a higher supply voltage as is being done today in cable modem line drivers.

- *Pin Count*—This still drives package/assembly cost as well as board area, so it is desirable to partition systems in a way that minimizes the number of chip-to-chip interconnections. For example, simple digital-to-analog converter (DAC) functions can still be best integrated onto the digital chip if it allows a single analog output pin rather than a full 12-line digital bus.

- *Interface Bandwidth*—A digital bus running at 500 MHz dissipates more power and generates more EMI than a digital bus running at 5 MHz. Wherever possible, the system should be partitioned across buses running at modest rates; in some cases, the data flow can even be carried as a serial bus, thereby also saving pins. Often, this means putting a large amount of digital processing on an otherwise "analog" chip. Examples of this include decimation filters on analog-to-digital-converters (ADCs), interpolation filters on DACs, and Direct Digital Synthesis with integrated DACs. Low voltage differential signaling (LVDS) can be used for high-speed interfaces (>200 MHz). LVDS provides better signal integrity and lower power dissipation at higher frequencies than a standard CMOS interface but doubles the pin count due to its differential nature.

• *Testability and Yield*—Some levels of integration are technically possible, but a poor choice from a manufacturability perspective. Integrating a finicky function with yield issues onto a very large chip means one is forced to throw the entire large chip away each time the function fails a test-which can be very expensive. It is far more cost effective to segregate the function that is subject to yield fallout.

• *External Components*—When considering integration, it is important to factor in not only the ICs, but also the external passive components (capacitors, inductors, SAW filters, etc.). In many cases, an innovative architecture coupled with smart partitioning can provide significant savings in external components, leading to much smaller form-factor and manufacturing costs. One example is illustrated in Figure 4.14 with ADI's OTHELLO direct-conversion chipset. The multimode

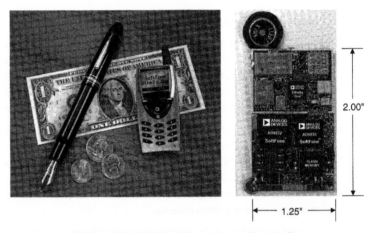

Figure 4.14: Othello Direct Conversion Radio

cellular handset chipset combines circuit innovation with system understanding and smart partitioning to create a breakthrough in form factor, performance and power saving.

• *Flexibility*—For the highest volume applications, a completely optimized solution is generally provided in the form of full-custom ASICs. However, the vast majority of applications never reach the run rate that justifies a fully committed integrated circuit solution. In this case, the designer will look for the highest levels of integration and performance available, often leveraging neighboring high volume applications, and will fill in around these with FPGA or other programmable solutions to customize to the application. Examples include the use of TV tuners in cable modem boxes and the use of cell phone handset chipsets in some low end base stations.

• *Cost*—As CMOS fabrication processes move to finer geometries, digital circuits shrink dramatically and become more cost efficient despite higher silicon wafer costs (Figure 4.15). Analog circuitry however, as illustrated in Figure 4.16, does not shrink as significantly when migrating to finer process geometries. It may in fact even grow in size to maintain performance. At finer process

• As CMOS fab processes move to finer geometries
  – Digital circuits shrink dramatically
  – More features can be packed onto the same size die
  – The cost per unit area of silicon goes up

• Overall, per function cost decreases because circuit sizes shrink faster than silicon costs increase

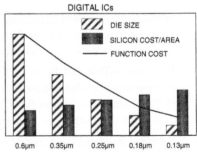

Figure 4.15: Fab Process Geometry Effects on Cost of Digital ICs

geometries, the overall digital per-function cost decreases while the overall analog per-function cost increases. Moreover, chips that integrate digital with analog functions may experience significantly higher yield losses than pure digital chips.

Smart partitioning separates analog and mixed-signal circuits from pure digital circuitry for cost optimization (Figure 4.17). An example of such a partitioning is demonstrated in Figure 4.18 with a set-top box chip-set that can also provide cable-modem functionality. The high density digital ASIC is separated from the analog or mixed-signal components. All ADCs and DACs are integrated with front end digital functions like interpolator, DDS and modulator into a single, mixed-signal front end, the AD9877/AD9879 from Analog Devices.

- As CMOS processes shrink below 0.25μm
  - Analog circuitry does not shrink significantly
  - It may in fact grow to maintain performance
  - Issues of supporting signal dynamic range arise with decreasing supply voltages
- Designing high performance analog circuits becomes harder and takes longer
- Overall, per-function cost increases

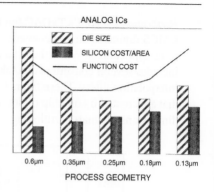

Figure 4.16: Fab Process Geometry Has a Different Effect on Analog Cost

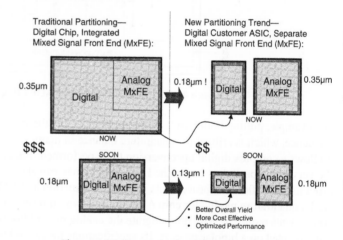

Figure 4.17: Smart Analog/Digital Partitioning

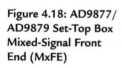
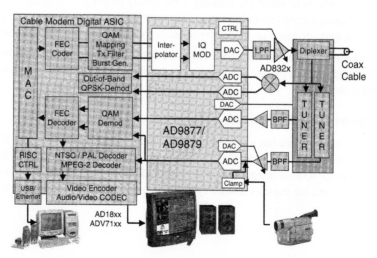

Figure 4.18: AD9877/ AD9879 Set-Top Box Mixed-Signal Front End (MxFE)

- *Performance*—ADI's TxDAC family was launched several years ago, ushering in a new generation of CMOS digital-to-analog converters with exceptional dynamic performance suitable for communications applications. While the product family has three generations of stand-alone converters, it also includes several converters that take advantage of the fine-line CMOS process by integrating digital interpolation filters. These filters take the input data word stream and insert additional sample words that are created by on-chip digital FIR filters. The AD9777 (Figure 4.19) features a few interesting dimensions of smart partitioning.

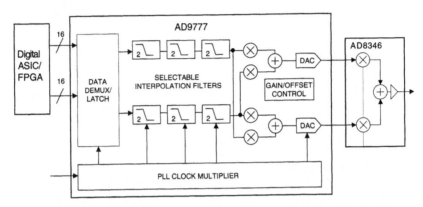

Figure 4.19: AD9777 TxDAC and AD8346 Quadrature Modulator

For example, putting both DACs on the same piece of silicon significantly improves the matching performance, which is critical for quadrature balance in many communications applications. Furthermore, it allows a complex digital upconversion to be performed. Integrating the interpolation filter means that the very high-speed data bus to the DAC (which may run at 400 MHz or more) need never come off-chip, providing significant improvements in power dissipation and EMI. The AD9777 is designed to mate with ADI's quadrature modulator chip AD8346 as a two-chip set, significantly reducing the number of external components required. Why not integrate the analog mixer? For performance and testability reasons, it is implemented on a bipolar process. Its specifications far exceed what is possible in CMOS.

The primary benefit of smart partitioning is the ability to integrate digital functionality onto high performance analog circuits and vice-versa. This frees designers to partition rather than forcing them into a certain arrangement based on the inherent limitations of their chip's functionality.

Combining functionality in high performance analog circuitry and high-performance DSP provides great latitude in partitioning options. This must be combined with a strong system understanding in order to make the wisest choices.

Why Smart Partitioning is Necessary

A single, dominant force has governed the semiconductor universe over the past 25 years: the trend toward ever-higher levels of integration. Gordon Moore of Intel even effectively quantified the slope of this trend, claiming that the level of integration on ICs would double every 18 months. This has become known as Moore's Law, and has been a remarkably accurate predictor of the integration trend for semiconductor circuits.

There have been many critical technology advances that have enabled the industry to keep pace with Moore's prediction. These have included advances to finer and finer lithography, the ability to handle larger and larger wafers, improvements in chemical purity, and reductions in defect density. The benefits of marching down this integration curve have been astounding: exponentially improved processing capability, faster processing speeds, decreasing costs, and reduced size and power consumption.

While the most notable examples of the integration trend have been seen in memory circuits (like DRAM) or in microprocessors, the theme of ever-higher levels of integration pervades virtually every corner of the semiconductor world. The analog world has been no exception. Over the last 30 years, the state of the art in analog has moved from operational amplifiers (op amps) on a single chip, to whole converters, to entire mixed-signal systems that replace 30 to 50 discrete chips.

Originally, analog integrated circuits were implemented on process technologies that differed significantly from those used for digital circuits. During the 1980s and 90s, increasing emphasis was placed on building analog functions on digital processes, allowing the analog and digital circuitry to be integrated onto a signal IC—a "mixed-signal" integrated circuit. This has been highly effective, and mixed-signal ICs are pervasive in today's products, from cell phones to digital still cameras. However, the complexities involved with implementing the analog functions have meant that mixed-signal ICs tend to lag their entirely digital counterparts by at least one lithography generation. While today's state-of-the art digital circuits are being designed on 0.13 μm processes, the most advanced mixed-signal circuits are being done on 0.18 μm or 0.25 μm processes.

What's Changing?

As we enter the new millennium, we are starting to see some changes that could have a significant impact on the seemingly inevitable trend to higher integration. Fundamental laws of physics may ultimately limit the ability to keep shrinking the lithography. We are already seeing some significant increases in costs (refer back to Figure 4.15 and Figure 4.16), and in a few generations, we may reach a point where further feature-size reductions aren't economically practical.

Through most of the 1990s, the analog designers rode the lithography curve that the digital circuits were pushing. However, late in the decade, there was a significant catch: with each lithography shrink below 0.5 μm, the maximum allowable supply voltage also falls. While this was of little consequence to the digital designer (it actually helps to lower power dissipation), it has enormous significance to the analog designer. Shrinking supply voltages force the use of smaller signal voltages, making it even more difficult to preserve the analog signal in the presence of inevitable noise.

Instead of being able to transfer circuit blocks from one lithography generation to the next, each new lithography sends the analog designer back to the drawing board. The consequences can be significant. Instead of mixed-signal lagging digital by one lithography generation, this lag is starting to stretch to two to three generations. Some very high-performance circuits may eventually be impractical (though probably not impossible) on extremely fine line geometries.

As supply voltages on state-of-the-art digital processes continue to drop, specialized processes may become more popular for high-performance mixed-signal circuits. Figure 4.20 illustrates that there are other process technology curves that parallel the digital CMOS curve. These are processes that have been optimized for analog circuits, making different trade-offs more appropriate to the needs of the analog circuit designers.

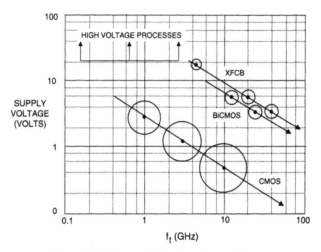

Figure 4.20: Range of Semiconductor Processes

In addition to the difficulties of designing analog circuits in smaller geometries, there is another problem facing "further integration" as the model for future electronic design: process technology has, in many cases, outpaced design and simulation capability. We are now able to integrate larger systems than we are able to effectively simulate, analyze or test.

In the face of these problems, some pundits have predicted that things will go back to the way they were in the mid-to-late 1980s: digital circuits on one chip, analog circuits on another, with different process technologies, simulation tools, and designers used for each. This would essentially constitute a "dis-integration" of the analog from the digital. While some systems may break down this way, in many cases the real answer will be more subtle (and more interesting).

There was an evening panel discussion several years ago at one of the major integrated circuits conferences—the topic was "The Single-chip Cell Phone." The discussion pushed back and forth about the technical challenges associated with combining the different chips in a cellular telephone. At one point, a panelist put a picture of a cell phone circuit board up on the screen. He noted six integrated circuits and 300 passive components. "Stop trying to combine these 6 ICs into one," he said, "and let's do something about all the passive components." The lesson from this story is that integration needs to be used in an intelligent way to reduce the cost, size, and power demands of the overall system, not simply as an exercise in blindly reducing the IC count.

As we approach the practical limits of integration and lithography, intelligence will need to be applied as to where and how to integrate. The key is to find optimal points to break the system into functional blocks. Typically, these will be places that require a minimum information flow across the boundaries between them, allowing the pin counts (and therefore cost and size) of the ICs to be kept low. Consideration should also be given to how the system and ICs will be simulated and analyzed. The partitioning dictated by these factors may or may not correspond to the boundary between analog and digital.

Analog Devices has been working on chipset partitioning for a number of years, and the theme of "smart partitioning" has emerged as one of the most significant factors in optimizing the cost, size, power, and performance for systems that feature both analog and digital circuits.

Figure 4.21 illustrates an example, taken from a real case, where three chips are cheaper than one or two. One of the keys is allowing the majority of the digital functionality to be implemented in the most effective process possible, avoiding the one-to-three generation lag normally associated with attempting to integrate everything. Using higher performance optimized processes for the mixed-signal and analog functions eliminates a large number of external passive/discrete components, thereby significantly reducing system size and cost. The mixed-signal front end AD9860/AD9862 as shown in this application is actually an excellent example for integration of several high speed converters onto a single CMOS chip (see also Figure 4.22).

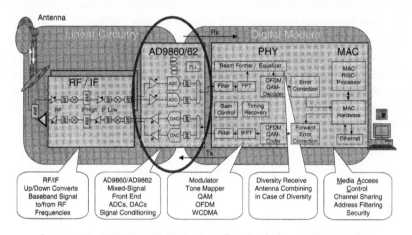

Figure 4.21: AD9860/AD9862: Broadband Wireless OFDM Modem

- Dual 12-Bit, 64MSPS ADCs
- Dual 14-Bit,128MSPS DACs
- Programmable Filters and PGAs
- Versatile Modulation and Clock Generation Circuitry

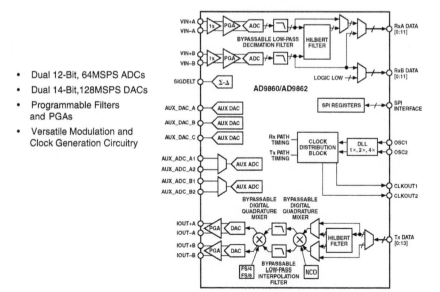

Figure 4.22: AD9860/AD9862 MxFE for Broadband Communications

In that device, putting dual DACs and dual ADCs on the same piece of silicon guarantees matching performance, which is important for applications that require IQ modulation and demodulation. With careful chip design, sufficient isolation between transmit and receive path has been achieved. Variable gain amplifiers as well as auxiliary ADCs and DACs have all been successfully integrated. The CMOS process allowed for embedding complex digital upconversion, interpolation and decimation. This eases digital interface requirements significantly by reducing the data rate between the digital and mixed-signal chip. A lower speed interface draws less power and also improves EMI. Radio components like the mixer, power amplifier, and low noise amplifier are not integrated on this CMOS chip because they achieve better performance when implemented on bipolar processes.

For many in the IC world, the end of Moore's Law seems unthinkable-ever-shrinking lithography has come to be viewed as an inalienable right. Nevertheless, the signs pointing to the end of Moore's law are there for those who will see them, and prudent designers will adapt to the new realities.

It's worth noting that other industries have faced similar technology limitations and are still thriving. For example, the ever-upward trend in aircraft speed was remarkably predictable for forty years, from the Wright brothers to the end of World War II. Yet the sound barrier posed a technology barrier. While it was possible to fly faster than sound, it has not proven economically practical to do so for commercial aircraft. Instead, the aircraft industry has advanced along other dimensions. The electronics industry will do the same. Smart partitioning will be the industry's way forward.

CHAPTER 5

Testing Data Converters

Testing Data Converters

Testing DACs
Walt Kester, Dan Sheingold

Static DAC Testing

The resolution of a DAC refers to the number of unique output voltage levels that the DAC is capable of producing. For example, a DAC with a resolution of 12 bits will be capable of producing 2^{12}—or 4,096—different voltage levels at its output. Similarly, a DAC with a resolution of 16 bits can produce 2^{16}—or 65,536—levels at its output.

Inherent in the specification of resolution, especially for control applications, is the requirement for *monotonicity*. The output of a monotonic DAC always stays the same or increases for an increasing digital code. The quantitative measure of monotonicity is the specification of differential nonlinearity (step size).

The static absolute accuracy of a DAC can be described in terms of three fundamental kinds of errors: *offset errors*, *gain errors*, and *integral nonlinearity*.

Linearity errors are the most important of the three kinds, because in many applications the user can adjust out the offset and gain errors, or compensate for them without difficulty by building end-point auto-calibration into the system design, whereas linearity errors cannot be conveniently or inexpensively nulled out. But before we can understand the nature of linearity errors and how to test for them, the end-point errors must first be established.

There are many methods to measure the static errors of a DAC—the proper choice depends upon the specific objectives of the testing. For instance, an IC manufacturer generally performs production testing on DACs using specialized automatic test equipment. On the other hand, a customer evaluating various DACs for use in a system does not generally have access to sophisticated automatic test equipment and must therefore devise a suitable bench test setup. A basic DAC static test setup is shown in Figure 5.1. This flexible test setup allows the application of various digital codes to the DAC input and uses an accurate digital voltmeter for measuring the DAC output. Computer control can be used to automate the process,

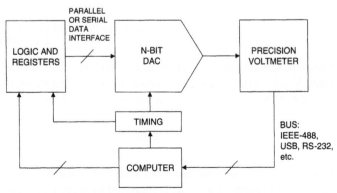

Figure 5.1: Basic Test Setup for Measuring DAC Static Transfer Characteristics

but it should be noted that in many cases DAC static testing can be performed by simply using mechanical switches to apply various codes to the DAC and reading the output with the voltmeter.

Today there are a large number of applications where the static performance of a DAC is rarely of direct concern to the customer—even in the evaluation phase—and ac performance is much more important. DACs used in audio and communications quite often do not even have traditional static specifications listed on the data sheet, and various noise and distortion specifications are of much more interest. However, the traditional static specifications such as differential nonlinearity (DNL) and integral nonlinearity (INL) are most certainly reflected in the ac performance. For instance, low distortion, a key audio and communication requirement, is directly related to low INL. Large INL and DNL errors will increase both the noise and distortion level of a DAC and render it unsuitable for these demanding applications. In fact, one often finds that the static performance of these ac-specified DACs is quite good, even though it is not directly specified.

The following sections on static DAC testing are therefore more oriented to DACs that are used in traditional industrial control, measurement, or instrumentation applications where monotonicity, DNL, INL, gain, and offset are important.

End-Point Errors

The most commonly specified end-point errors associated with DACs are *offset error*, *gain error*, and *bipolar zero error*. Note that bipolar zero error is only associated with bipolar output DACs, whereas offset and gain error is common to both unipolar and bipolar DACs.

Figure 5.2 shows the effects of offset and gain error in a unipolar DAC. Note that in Figure 5.2A, all output points are offset from the ideal (shown as a dotted line) by the same amount. Any such error—either positive or negative—that affects all output points by the same amount is an *offset* error.

The offset error can be measured by applying the all "0"s code to the DAC and measuring the output deviation from 0 volts.

Figure 5.2B shows the effect of *gain* error only. The ideal transfer function has a slope defined by drawing a straight line through the two end points. The slope represents the gain of the transfer function. In nonideal DACs, this slope can differ from the ideal, resulting in a *gain* error—which is usually expressed as a percent because it affects each code by the same percentage. If there is no offset error, gain error is easily determined by applying the all "1"s code to the DAC and measuring its output, designated as V_{111} (assuming a 3-bit DAC). An ideal DAC will measure exactly $V_{FS} - 1$ LSB, so the gain error is computed using the equation:

$$\text{Gain Error}(\%) = 100 \left[\frac{V_{111}}{V_{FS} - 1 \text{ LSB}} - 1 \right] \qquad \text{Eq. 5.1}$$

Figure 5.2C shows the case where there is both offset and gain error. The first step is to measure the offset error, V_{OS}, by applying the all "0"s code and measuring the output. Next, apply the all "1"s code and measure the output V_{111}. The gain error is then calculated using the equation:

$$\text{Gain Error}(\%) = 100 \left[\frac{V_{111} - V_{OS}}{V_{FS} - 1 \text{ LSB}} - 1 \right] \qquad \text{Eq. 5.2}$$

Figure 5.3 shows how gain and offset errors affect the ideal transfer function of a bipolar output DAC. The offset error in Figure 5.3A, V_{OS}, is measured by applying the all "0"s code to the DAC input and measuring the output. Ideally, the DAC should have an output of –FS with all "0"s at its input. The difference between the actual output and –FS is the offset, V_{OS}. In a bipolar DAC it is also common to specify and measure the

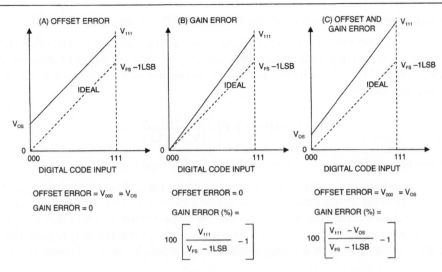

Figure 5.2: Measuring Offset and Gain Error in a Unipolar DAC

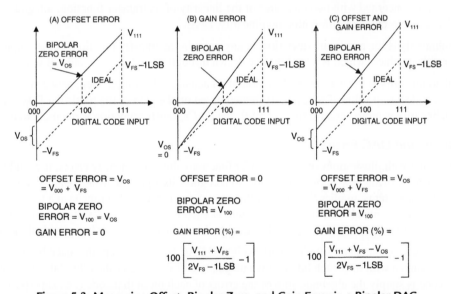

Figure 5.3: Measuring Offset, Bipolar Zero, and Gain Error in a Bipolar DAC

bipolar zero error (or *zero error*) because of its importance in many applications. It is measured by applying the midscale code 100 to the DAC and measuring its output. If there is no gain error, the bipolar zero error is the same as the offset error as shown in Figure 5.3A.

Figure 5.3B shows the case where there is gain error, but no offset error. Notice that the bipolar zero error is affected by the gain error. The DAC output V_{111} is measured by applying the all "1"s code, and the gain error is calculated from the equation:

$$\text{Gain Error}(\%) = 100\left[\frac{V_{111} + V_{FS}}{2V_{FS} - 1\,\text{LSB}} - 1\right]$$

Eq. 5.3

The bipolar zero error is determined by applying 100 to the DAC and measuring its output.

Figure 5.3C shows the case where the bipolar DAC has both gain and offset error. The offset error is determined as above by applying the all "0"s code, measuring the DAC output, and subtracting it from the ideal value, V_{FS}. The all "1"s code is applied to the DAC and the output V_{111} is measured. The gain error is calculated using the equation:

$$\text{Gain Error}(\%) = 100 \left[\frac{V_{111} + V_{FS} - V_{OS}}{2V_{FS} - 1\,\text{LSB}} - 1 \right] \qquad \text{Eq. 5.4}$$

The bipolar zero error is determined by applying 100 to the DAC and measuring its output.

Bipolar zero error in DACs using offset-type coding is a derived, rather than a fundamental quantity, because it is actually the sum of the bipolar offset error, the bipolar gain error, and the MSB linearity error. For this reason, it is important to specify whether this measurement is made before or after offset and gain have been trimmed or taken into account. Because of this error sensitivity, DACs that crucially require small errors at zero are usually unipolar types, with sign-magnitude coding and polarity-switched output amplifiers.

Linearity Errors

In a DAC, we are concerned with two measures of the linearity of its transfer function: integral nonlinearity, INL (or relative accuracy), and differential nonlinearity, DNL.

Integral nonlinearity is the maximum deviation, at any point in the transfer function, of the output voltage level from its ideal value—which is a straight line drawn through the actual zero and full scale of the DAC.

Differential nonlinearity is the maximum deviation of an actual analog output step, between adjacent input codes, from the ideal step value of +1 LSB, calibrated based on the gain of the particular DAC. If the differential nonlinearity is more negative than –1 LSB, the DACs transfer function is nonmonotonic.

Superposition and DAC Errors

Before proceeding with illustrations of DAC transfer functions showing linearity errors, it would be useful to consider the property of *superposition*, and be able to recognize its signature. Mathematically, superposition, a property of linear systems, implies that if the influences of a number of phenomena at a particular point are measured individually, with all other influences at zero as each is asserted, the resulting total, with any number of these influences operating, will always be equal to the arithmetic sum of the individual measurements.

For example, let us assume that a simple binary-weighted DAC is ideal, except that each bit has a small linearity error associated with it. If each bit error is independent of the state of the other bits, then the linearity error at any code is simply the algebraic sum of the errors of each bit in that code (i.e., superposition holds). In addition, by using end-point linearity, we have defined the linearity error at zero and full-scale to be zero. Thus, the sum of all the bit errors must be zero, since all bits are summed to give the all "1"s value.

The bit errors can be either positive or negative; therefore, if their sum is zero, the sum of the positive errors (positive summation) must be equal to the sum of the negative errors (negative summation). These two summations constitute the worst-case integral nonlinearities of the DAC.

Intelligent use of superposition, coupled with a complete understanding of the architecture of the DAC under test, generally allows for a reduction in the number of measurements required to adequately determine DNL and INL. This is significant when one considers that a 16-bit DAC has a total of $2^{16} = 65,536$ possible output levels. Measuring each level individually would be a time-consuming task.

Measuring DAC DNL and INL Using Superposition

One can often determine whether or not superposition holds true by simply examining the architecture of a DAC, and this topic was covered in detail in Chapter 3 of this book (see Section 3.1). Superposition generally holds true for binary weighted DACs, so we will examine this class first. Issues relating to linearity measurements on fully decoded DACs (string DACs), segmented DACs, and sigma-delta DACs will be treated later in this section, as they generally do not obey the rules of superposition.

Figure 5.4 shows the transfer function of a 3-bit DAC where superposition holds. Offset and gain errors have been removed from the data points so that the zero and full-scale errors are zero. This DAC has an error in the

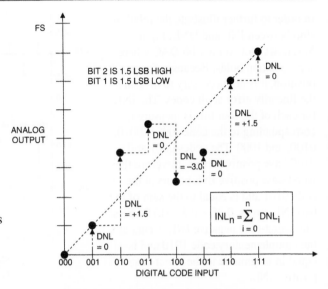

Figure 5.4: Nonideal 3-Bit DAC Transfer Function Where Superposition Holds

first and second bit weights. Bit 1 is 1.5 LSBs low, and Bit 2 is 1.5 LSBs high. The value of the DNL is calculated for each of the eight possible output voltages. The transfer function has a nonmonotonicity at the 100 code with a DNL of –3 LSBs. The INL for any output is the algebraic sum of the DNLs leading up to that particular output. For instance, the INL at the 101 code is equal to $DNL_{001} + DNL_{010} + DLN_{011} + DNL_{100} + DNL_{101} = 0 + 1.5 + 0 - 3 + 0 = -1.5$ LSBs.

The DNL and INL of the 3-bit DAC is plotted in Figure 5.5. The INL has odd symmetry about the midpoint of the transfer function, i.e., the INL of any particular code is equal and opposite in sign from the INL of the complementary code. For instance, the INL at code 010 is +1.5, and the INL at the complementary code 101 is –1.5. In addition, the algebraic sum of all the INLs must equal zero, i.e., the sum of the positive INLs must equal the sum of the negative INLs.

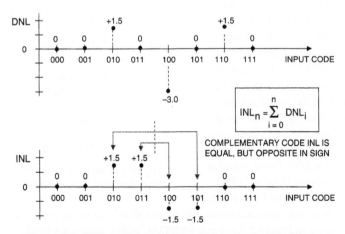

Figure 5.5: Differential and Integral Linearity for Nonideal 3-Bit Binary-Decoded DAC Where Superposition Holds

In order to further illustrate the relationship between INL and DNL, Figure 5.6 shows plots for a 4-bit DAC where superposition holds. Because of superposition, it is not necessary to measure the linearity error at all codes. The INL for each of the four bits is measured, corresponding to the codes 0001, 0010, 0100, and 1000. The codes 0001 and 0100 have positive errors, therefore the worst-case positive INL occurs at the code 0101 and is equal to the sum of the two INL errors, $0.25 + 0.5 = 0.75$ LSBs. The worst-case negative INL occurs at the complementary code 1010 and is equal in magnitude to the worst-case positive INL.

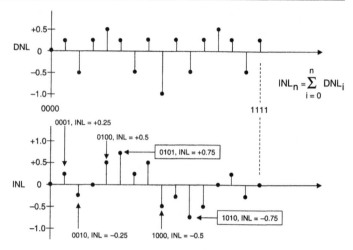

$$INL_n = \sum_{i=0}^{n} DNL_i$$

Figure 5.6: DNL and INL for 4-Bit DAC Where Superposition Holds

Figure 5.7 summarizes the linearity measurements required for a 4-bit binary-weighted DAC where superposition holds. First, the offset and gain error must be removed. The offset corresponds to the all "0"s code output, $V_{OS} = V_{0000}$. Next, measure the all "1"s code output, V_{1111}. The theoretical value of 1 LSB (with gain and offset errors removed) can then be calculated using the equation:

$$1\,LSB = \frac{V_{1111} - V_{0000}}{15} \qquad\qquad \text{Eq. 5.5}$$

The theoretical value of the points being tested is then computed by multiplying the LSB weight (per Eq. 5.5) by the base-10 value of the binary code and adding the offset voltage.

The voltages corresponding to codes 0000, 0001, 0010, 0100, and 1000 are then measured and the linearity error of each bit computed. The all "1"s code, 1111, previously measured should equal the sum of the individual bit voltages: 0000, 0001, 0010, 0100, and 1000. This is a good test to verify that superposition holds. A guideline is that if the algebraic sum of the actual bit voltages is not within ±0.5 LSB of the measured voltage for the all "1"s code, there is enough interaction between the bits to justify more comprehensive all-codes testing. Next, the DNL is measured at each of the major-carry points: 0000 to 0001, 0001 to 0010, 0011 to 0100, and 0111 to 1000. In

CHECK BIT SUM: $V_{1111} \overset{?}{=} V_{0000} + V_{0001} + V_{0010} + V_{0100} + V_{1000}$

$$1\,LSB = \frac{V_{1111} - V_{0000}}{15}$$

$1111 = V_{1111}$

BIT ERRORS

$1000 = V_{1000}$
$0111 = V_{0111}$ $\quad\}$ $DNL4 = V_{1000} - V_{0111} - 1\,LSB$

$0100 = V_{0100}$
$0011 = V_{0011}$ $\quad\}$ $DNL3 = V_{0100} - V_{0011} - 1\,LSB$

$0010 = V_{0010}$
$0001 = V_{0001}$ $\quad\}$ $DNL2 = V_{0010} - V_{0001} - 1\,LSB$

$DNL1 = V_{0001} - V_{0000} - 1\,LSB$

$0000 = V_{0000}$

$V_{OS} = V_{0000}$

Figure 5.7: Major-Carry Bit Tests for 4-Bit Binary-Decoded DAC Where Superposition Holds

a well-behaved DAC where superposition holds, these tests should be sufficient to verify the static performance. The example shown in Figure 5.7 is for a 4-bit DAC which requires a total of eight measurements plus the calculations of the individual errors, although the measurements constitute the major portion of the test time. In the general case of an N-bit DAC, the procedure just discussed requires 2N measurements plus the error calculations.

There is a method to further reduce the number of required measurements to N + 2, but it depends even more heavily upon the validity of superposition. In this method, only the individual bit values and the all "1"s code are measured. For the 4-bit case above, this would correspond to the DAC output for 0000, 0001, 0010, 0100, 1000, and 1111. Superposition should be verified by comparing V_{1111} with $V_{0000} + V_{0001} + V_{0010} + V_{0100} + V_{1000}$. The value of the LSB is computed per Eq. 5.5 above. The theoretical values of the individual bit contributions are then calculated by multiplying the base-10 bit weight by the value of the LSB and adding the offset voltage. The INL values for each bit are then computed. Finally, the DNL values are indirectly calculated from the bit values as follows:

$$DNL1 = V_{0001} - V_{0000} - 1 \text{ LSB}, \hspace{3cm} \text{Eq. 5.6}$$

$$DNL2 = V_{0010} - V_{0001} - 1 \text{ LSB}, \hspace{3cm} \text{Eq. 5.7}$$

$$DNL3 = V_{0100} - (V_{0010} + V_{0001} - V_{0000}) - 1 \text{ LSB}, \hspace{1.5cm} \text{Eq. 5.8}$$

$$DNL4 = V_{1000} - (V_{0100} + V_{0010} + V_{0001} - 2V_{0000}) - 1 \text{ LSB}. \hspace{0.8cm} \text{Eq. 5.9}$$

Here, the voltages corresponding to the 0011 and the 0111 code are computed using superposition rather than measured directly.

Linearity errors in DACs where superposition holds can take many forms. Figure 5.8 shows just two examples for a 3-bit binary-coded DAC. In both examples, the linearity error exhibits odd symmetry about the midpoint of the transfer function. Notice that in Figure 5.8A, the INL is ±1 LSB, and the DAC is nonmonotonic because of the –2 LSB DNL at the 100 code. Figure 5.8B shows a DAC where the INL is ±0.5 LSB and the DNL is –1 LSB at the 100 code, thereby just bordering on nonmonotonicity.

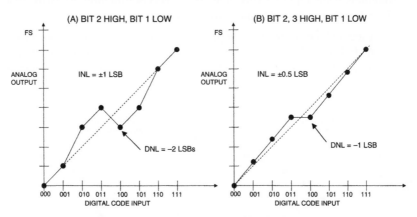

Figure 5.8: Linearity Errors in Binary-Decoded 3-Bit DACs Where Superposition Holds

We are now in a position to make two general statements about INL and DNL and their relationship to each other and to the monotonicity of a DAC. *First, if a DAC has an INL specification of less than ±0.5 LSB, this guarantees that its DNL is no more than ±1 LSB, and that the output is monotonic. On the other hand, just because the DNL error is less than ±1 LSB, and the DAC is monotonic, it cannot be assumed that the INL is less than ±0.5 LSB.*

The second point is illustrated in Figure 5.9 where the worst DNL error is –1 LSB, but the INL is ±0.75 LSBs.

Measuring DAC INL and DNL Where Superposition Does Not Hold

There are several DAC architectures where superposition does not hold, and the techniques described in the previous section may not give a clear picture of the INL and DNL performance. Again, the exact methodology of selecting data points is highly dependent on the architecture of the DAC under test, therefore it is not possible to examine them all but simply to point out some general concepts.

The simple fully-decoded (string DAC) architecture is a good example where the individual bit weights vary depending upon the values of the other bits. Superposition does not hold, and the bit weights are not independent as in the binary-weighted DACs previously discussed.

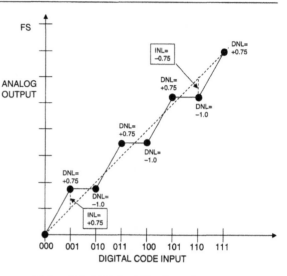

Figure 5.9: DNL Specification of < ±1 LSB Does Not Guarantee INL < ±0.5 LSB

In a string DAC (Kelvin divider), the DNL is primarily determined by the matching of adjacent resistors in the "string," which is generally quite good. However, the INL may follow either a bow- or an s-shaped curve due to gradual changes in the absolute value of the resistors when moving from one end of the string to the other. Figure 5.10 illustrates two possible transfer functions. Figure 5.10A shows a bow-shaped INL function with no trimming, while Figure 5.10B shows the effects of trimming the MSB (described in Chapter 3) where the midscale output is forced to the correct value.

The highest resolution stand-alone string DACs at present are 10-bit digital potentiometers (i.e., AD5231/ AD5235). However, the fully decoded DAC is a common building block in segmented low distortion high resolution DACs as discussed in Chapter 3. In practice, 4-, 5-, or 6-bit fully decoded segments are popular in modern DACs.

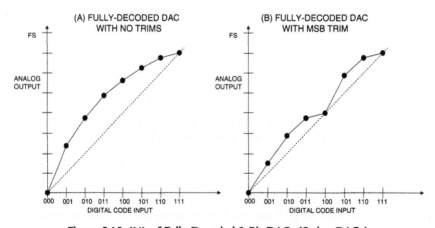

Figure 5.10: INL of Fully-Decoded 3-Bit DACs (String DACs)

The only way to fully characterize a segmented DAC is to perform all-codes testing—a time consuming process, to say the least. However, one can establish a reasonable amount of confidence about segmented DACs by intelligent utilization of knowledge about the particular DAC architecture. As an example, Figure 5.11 shows a simplified segmented DAC consisting of a fully decoded 3-bit MSB DAC and a 4-bit binary-weighted LSB DAC. The logic behind the selection of measurements on the right-hand side of the diagram is as follows.

First, the all "1"s code and all "0"s code are measured so that the weight of the LSB and the offset and can be determined. Next, each of the LSB DAC bits are tested so its DNL and INL can be determined (it is assumed that superposition holds for the LSB DAC, since it is a binary-weighted architecture). This is done with the MSB DAC bits at all "0"s for convenience. The DNL should then be measured at each of the MSB DAC transitions. In addition, the DNL is measured for code above the MSB DAC transitions where the LSB DAC code is 0001. This ensures that the lower and upper range of the LSB DAC is tested on each segment of the MSB DAC. Finally, one INL measurement is made in each of the 7 MSB segments, corresponding to the outputs where the LSB DAC code is 0000.

This technique for selecting test codes can obviously be extended to cover segmented DACs of higher resolution using the same principles in the simple example.

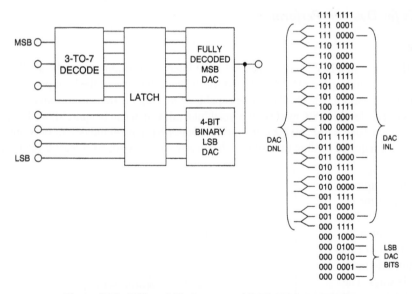

Figure 5.11: 3-Bit × 4-Bit Segmented 7-Bit DAC Test Codes

Much more could be said about testing DAC static linearity, but in light of the majority of today's applications and their emphasis on ac performance, the guidelines put forth in this discussion should be adequate to illustrate the general principles. As we have seen, foremost in importance is a detailed understanding of the particular DAC architecture so that a reasonable test/evaluation plan can be devised which minimizes the actual number of codes tested. This is highly dependent upon whether the DAC is binary-weighted (superposition generally holds), fully-decoded, or segmented. The data sheet for the DAC under test generally will provide adequate information to make this determination.

For the interested reader, References 1 and 2 give more details regarding testing DAC static linearity, including automatic and semi-automatic test methods such as using "reference" DACs as part of the test circuitry to verify the DAC under test. The entire area of ATE testing is much too broad to be treated here and of much more interest to the IC manufacturer than the end user.

Finally, there are several types of DACs which do not generally have static linearity specifications or, if they do, they do not compare very well with more traditional DACs designed for low frequency applications. The first of these are DACs designed for voiceband and audio applications. This type of DAC, although fully specified in terms of ac parameters such as THD, THD + N, etc., generally lacks dc specifications (other than perhaps gain and offset); and generally should not be used in traditional industrial control or instrumentation applications where INL and DNL are critical. However, these DACs almost always use the sigma-delta architecture (either single-bit or multbit with data scrambling) which inherently ensures good DNL performance.

DACs specifically designed for communications applications, such as the TxDAC-family, have extensive frequency-domain specifications; but their static specifications make them less attractive than other more traditional DACs for precision low frequency applications. It is common to see INL and DNL specifications of several LSBs for 14- and 16-bit DACs in this family, with monotonicity guaranteed at the 12-bit level. It should by no means be inferred that these are inferior DACs—it is just that the application requires designs which optimize frequency-domain performance rather than static.

Testing DACs for Dynamic Performance

Settling Time

The precise settling time of a DAC may or may not be of interest, depending upon the application. It is especially important in high speed DACs used in video displays, because of the high pixel rates associated with high resolution monitors. The DAC must be capable of making the transition from all "0"s (black level) to all "1"s (white level) in 5% to 10% of a pixel interval, which can be quite short. For instance, even the relatively common 1024×768, 60 Hz refresh-rate monitor has a pixel interval of only approximately 16 ns. This implies a required settling time of less than a few nanoseconds to at least 8-bit accuracy (for an 8-bit system).

The fundamental definitions of full-scale settling time is repeated in Figure 5.12. The definition is quite similar to that of an op amp. Notice that settling time can be defined in two acceptable ways. The more traditional definition is the amount of time required for the output to settle with the specified error band measured with respect to the 50% point of either the data strobe to the DAC (if it has a parallel register driving the DAC switches) or the time when the input data to the switches changes (if there is no internal register).

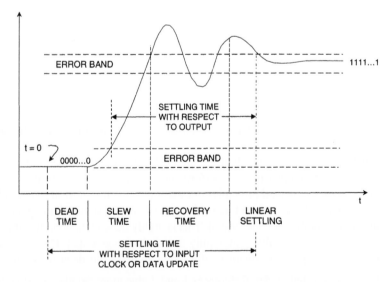

Figure 5.12: DAC Full-Scale Settling Time Measurement

Another equally valid definition is to define the settling time with respect to the time the output leaves the initial error band. This effectively removes the "dead time" from the measurement. In video DAC applications, for instance, settling time with respect to the output is the key specification—the fixed delay (dead time) is of little interest.

The error band is usually defined in terms of an LSB or % full-scale. It is customary, but not mandatory, to define the error band as 1 LSB. However, measuring full-scale settling time to 1 LSB at the 12-bit level (0.025% FS) is possible with care, but measuring it to 1 LSB at the 16-bit level (0.0015% FS) presents a real instrumentation challenge. For this reason, high speed DACs such as the TxDAC family specify 14- and 16-bit settling time to the 12-bit level, 0.025% FS (typically less than 11 ns).

Midscale settling time is of interest, because in a binary-weighted DAC, the transition between the 0111...1 code and the 1000...0 code produces the largest transient. In fact, if there is significant bit skew, the transient amplitude can approach full-scale. Figure 5.13 shows a waveform along with the two acceptable definitions of settling time. As in the case of full-scale settling time, midscale settling time can either be referred to the output or to the latch strobe (or the bit transitions if there is no internal latch).

Figure 5.13: DAC Midscale Settling Time Measurement

Glitch Impulse Area

Glitch impulse area (sometimes incorrectly called glitch energy) is easily estimated from the midscale settling time waveform as shown in Figure 5.14. The areas of the four triangles are used to calculate the net glitch area. Recall that the area of a triangle is one-half the base times the height. If the total positive area equals the total negative glitch area, the net area is zero. The specification given on most data sheets is the net glitch area although, in some cases, the peak area may specified instead.

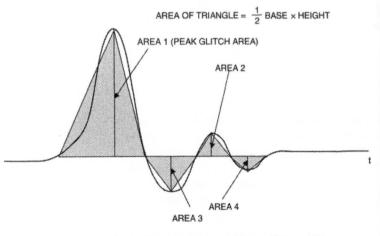

AREA OF TRIANGLE = $\frac{1}{2}$ BASE × HEIGHT

AREA 1 (PEAK GLITCH AREA)

AREA 2

AREA 3

AREA 4

NET GLITCH IMPULSE AREA ≈ AREA 1 + AREA 2 – AREA 3 – AREA 4

Figure 5.14: Glitch Impulse Area Measurement

Oscilloscope Measurement of Settling Time and Glitch Impulse Area

A wideband fast-settling oscilloscope is crucial to settling time measurements. There are several considerations in selecting the proper scope. The required bandwidth can be calculated based on the rise/fall time of the DAC output, for instance, a 1 ns output rise time and fall time corresponds to a bandwidth of $0.35/t_r = 350$ MHz. A scope of at least 500 MHz bandwidth would be required.

Modern digital storage scopes (DSOs) and digital phosphor scopes (DPOs) are popular and offer an excellent solution for performing settling time measurements as well as many other waveform analysis functions (see Reference 3). These scopes offer real-time sampling rates of several GHz and are much less sensitive to overdrive than older analog scopes or traditional sampling scopes. Overdrive is a serious consideration in measuring settling time, because the scope is generally set to maximum sensitivity when measuring a full-scale DAC output change. For instance, measuring 12-bit settling for a 1 V output (20 mA into 50 Ω) requires the resolution of a signal within a 0.25 mV error band riding on the top of a 1 V step function.

From a historical perspective, older analog oscilloscopes were sensitive to overdrive and could not be used to make accurate step function settling time without adding additional circuitry. Quite a bit of work was done during the 1980s on circuits to cancel out portions of the step function using Schottky diodes, current sources, etc. References 4, 5, and 6 are good examples of various circuits which were used during that time to mitigate the oscilloscope overdrive problems.

Even with modern DSOs and DPOs, overdrive should still be checked by changing the scope sensitivity by a known factor and making sure that all portions of the waveform change proportionally. Measuring the midscale settling time can also subject the scope to considerable overdrive if there is a large glitch. The sensitivity of the scope should be sufficient to measure the desired error band. A sensitivity of 1 mV/division allows the measurement of a 0.25 mV error band if care is taken (one major vertical division is usually divided into five smaller ones, corresponding to 0.2 mV/small division). If the DAC has an on-chip op amp, the fullscale output voltage may be larger, perhaps 10 V, and the sensitivity required in the scope is relaxed proportionally.

Although there is a well-known relationship between the rise time and the settling time in a single-pole system, it is inadvisable to extrapolate DAC settling time using rise time alone. There are many higher order nonlinear effects involved in a DAC which dominate the actual settling time, especially for DACs of 12 bits or higher resolution.

Figure 5.15 shows a test setup for measuring settling time. It is generally better to make a direct connection between the DAC output and the 50 Ω scope input and avoid the use of probes. FET probes are notorious for giving misleading settling time results. If probes must be used, compensated passive ones are preferable, but they should be used with care. Skin effect associated with even short lengths of properly terminated coaxial cable can give erroneous settling time results. In making the connection between the DAC and the scope, it is mandatory that a good low impedance ground be maintained. This can

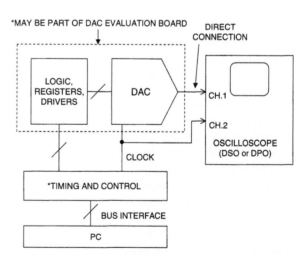

Figure 5.15: Test Setup for Measuring Settling Time and Glitch Impulse Area

be accomplished by soldering the ground of a BNC connector to the ground plane on the DAC test board and using this BNC to connect to the scope's 50 Ω input. A manufacturer's evaluation board can be of great assistance in interfacing to the DAC and should be used if available.

Finally, if the DAC output is specifically designed to drive the virtual ground of an external current-to-voltage converter, and does not have enough compliance to develop a measurable voltage across a load resistor, an external op amp is required and the test circuit measures the settling time of the DAC/op amp combination. In this case, select an op amp that has a settling time which is at least 3 to 5 times smaller than the DAC under test. If the settling time of the op amp is comparable to that of the DAC, the settling time of the DAC can be determined, because the total settling time of the combination is the root-sum-square of the DAC settling time and the op amp settling time. Solving the equation for the DAC settling time yields:

$$DAC\ Settling\ Time = \sqrt{\left(Total\ Settling\ Time\right)^2 - \left(Op\ Amp\ Settling\ Time\right)^2} \qquad Eq.\ 5.10$$

Distortion Measurements

Because so many DAC applications are in communications and frequency analysis systems, practically all modern DACs are now specified in the frequency domain. The basic ac specifications were previously discussed in Chapter 3 of this book and include harmonic distortion, total harmonic distortion (THD), signal-to-noise ratio (SNR), total harmonic distortion plus noise (THD + N), spurious free dynamic range (SFDR), etc. In order to test a DAC for these specifications, a proper digitally-synthesized signal must be generated to drive the DAC (for example, a single or multitone sinewave).

In the early 1970s, when ADC and DAC frequency domain performance became important, "back-to-back" testing was popular, where an ADC and its companion DAC were connected together, and the appropriate analog signal source was selected to drive the ADC. An analog spectrum analyzer was then used to measure the distortion and noise of the DAC output. This approach was logical, because ADCs and DACs were often used in conjunction with a digital signal processor placed between them to perform various functions. Obviously, it was impossible to determine exactly how the total ac errors were divided between the ADC and the DAC. Today, however, ADCs and DACs are used quite independently of one another, so they must be completely tested on their own.

Figure 5.16 shows a typical test setup for measuring the distortion and noise of a DAC. The first consideration, of course, is the generation of the digital signal to drive the DAC. To achieve this, modern arbitrary waveform generators (for example Tektronix AWG2021 with Option 4) or word generators (Tektronix DG2020) allow almost any waveform to be synthesized digitally in software, and are mandatory in serious frequency domain testing of DACs (see Reference 3). In most cases, these generators have standard

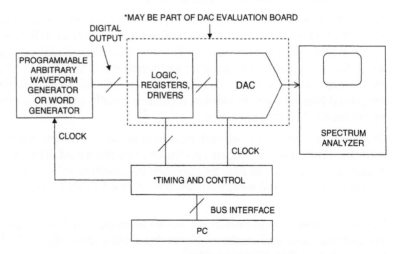

Figure 5.16: Test Setup for Measuring DAC Distortion and Noise

waveforms preprogrammed, such as sinewaves and triangle waves, for example. In many communications applications, however, more complex digital waveforms are required, such as two-tone or multitone sinewaves, QAM, GSM, and CDMA test signals, etc. In many cases, application-specific hardware and software exists for generating these types of signals and can greatly speed up the evaluation process.

Analog Devices and other manufacturers of high performance DACs furnish evaluation boards that greatly simplify interfacing to the test equipment. Because many communications DACs (such as the TxDAC-family) have quite a bit of on-chip control logic, their evaluation boards have interfaces to PCs via the SPI, USB, parallel, or serial ports, as well as Windows®-compatible software to facilitate setting the various DAC options and modes of operation.

Testing DACs that are part of a direct-digital-synthesis (DDS) system is somewhat easier because the DDS portion of the IC acts as the digital signal generator for the DAC. Testing these DACs often requires no more than the manufacturer's evaluation board, a PC, a stable clock source, and a spectrum analyzer. A complete discussion of DDS is included in Chapter 8 of this book and will not be repeated here.

The spectrum analyzer chosen to measure the distortion and noise performance of the DAC should have at least 10 dB more dynamic range than the DAC being tested. The "maximum intermodulation-free range" specification of the spectrum analyzer is an excellent indicator of distortion performance (see Reference 7). However, spectrum analyzer manufacturers may specify distortion performance in other ways. Modern communications DACs such as the TxDAC-series require high performance spectrum analyzers such as the Rohde and Schwarz FSEA30 (Reference 7). As in the case of oscilloscopes, the spectrum analyzer must not be sensitive to overdrive. This can be easily verified by applying a signal corresponding to the full-scale DAC output, measuring the level of the harmonic distortion products, and then attenuating the signal by 6 dB or so and verifying that both the signal and the harmonics drop by the same amount. If the harmonics drop more than the fundamental signal drops, then the analyzer is distorting the signal.

In some cases, an analyzer with less than optimum overdrive performance can still be used by placing a band-stop filter in series with the analyzer input to remove the frequency of the fundamental signal being measured. The analyzer looks only at the remaining distortion products. This technique will generally work satisfactorily, provided the attenuation of the band-stop filter is taken into account when making the distortion measurements. Obviously, a separate band-stop filter is required for each individual output frequency tested, and therefore multitone testing is cumbersome.

Finally, there are a variety of application-specific analyzers for use in communications, video, and audio. In video, the Tektronix VM-700 and VM-5000 series are widely used (Reference 3). In measuring the performance of DACs designed for audio applications, special signal analyzers designed specifically for audio are preferred. The industry standard for audio analyzers is the Audio Precision, System Two (see Reference 8). There are, of course, many other application-specific analyzers available which may be preferred over the general-purpose types. In addition, software is usually available for generating the various digital test signals required for the applications.

Once the proper analyzer is selected, measuring the various distortion and noise-related specifications such as SFDR, THD, SNR, SINAD, etc., is relatively straightforward. Refer back to Chapter 2 of this book for definitions if required. The analyzer resolution bandwidth must be set low enough so that the harmonic products can be resolved above the noise floor. Figure 5.17 shows a typical spectral output where the SFDR is measured.

Figure 5.18 shows how to measure the various harmonic distortion components with a spectrum analyzer. The first nine harmonics are shown. Notice that aliasing causes the 6th, 7th, 8th, 9th, and 10th harmonic to fall back inside the $f_s/2$ Nyquist bandwidth.

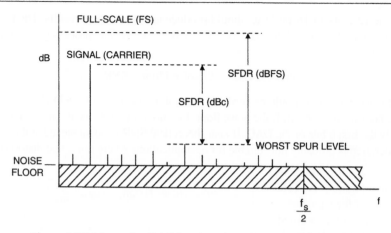

Figure 5.17: Measuring DAC Spurious Free Dynamic Range (SFDR)

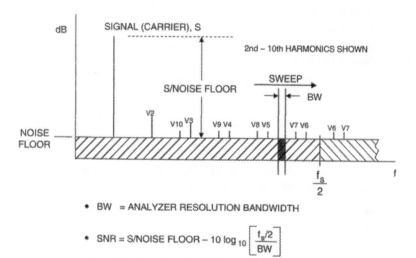

Figure 5.18: Measuring DAC Distortion and SNR with an Analog Spectrum Analyzer

The spectrum analyzer can also be used to measure SNR if the proper correction factors are taken into account. Figure 5.18 shows the analyzer sweep bandwidth, BW, which in most cases will be considerably less than $f_s/2$. First, measure the noise floor level with respect to the signal level at a point in the frequency spectrum which is relatively free of harmonics. This corresponds to the value "S/NOISE FLOOR" in the diagram. The actual SNR over the dc to $f_s/2$ bandwidth is obtained by subtracting the process gain, $10\log_{10}(f_s/2 \times BW)$.

$$SNR = S/NOISE\ FLOOR - 10\log_{10}(f_s/2 \times BW) \qquad \text{Eq. 5.11}$$

In order for this SNR result to be accurate, one must precisely know the analyzer bandwidth. The bandwidth characteristics of the analyzer should be spelled out in the documentation. Also, if there is any signal averaging used in the analyzer, that may affect the net correction factor.

In order to verify the process gain calculation, several LSBs can be disabled—under these conditions, the SNR performance of the DAC should approach ideal. For instance, measuring the 8-bit SNR of a low distortion, low noise 12-, 14-, or 16-bit DAC should produce near theoretical results. The theoretical 8-bit SNR, calculated using the formula SNR = 6.02N + 1.76 dB, is 50 dB. The process gain can then be calculated using the formula:

$$\text{Process Gain} = \text{S/Noise Floor} - \text{SNR} \qquad \text{Eq. 5.12}$$

The accuracy of this measurement should be verified by enabling the 9th bit of the DAC and ensuring that the analyzer noise floor drops by 6 dB. If the noise floor does not drop by 6 dB, the measurement should be repeated using only the first 6 bits of the DAC. If near theoretical SNR is not achieved at the 6-bit level, the DAC under consideration is probably not suitable for ac applications where noise and distortion are important.

The relationship between SINAD, SNR, and THD is shown in Figure 5.19. THD is defined as the ratio of the signal to the root-sum-square (rss) of a specified number harmonics of the fundamental signal. IEEE Std. 1241-2000 (Reference 9) suggests that the first 10 harmonics be included. Various manufacturers may choose to include fewer than 10 harmonics in the calculation. Analog Devices defines THD to be the root-sum-square of the first 6 harmonics (2nd, 3rd, 4th, 5th, and 6th) for example. In practice, the difference in dB between THD measured with 10 versus 6 harmonics is less than a

- $SNR = S/NOISE\ FLOOR - 10\ \log_{10}\left[\dfrac{f_s/2}{BW}\right]$

- $THD = 20\ \log_{10}\sqrt{\left[10^{-V2/20}\right]^2 + \left[10^{-V3/20}\right]^2 + \cdots + \left[10^{-V6/20}\right]^2}$

- $SINAD = 20\ \log_{10}\sqrt{\left[10^{-SNR/20}\right]^2 + \left[10^{-THD/20}\right]^2}$

NOTE: NOISE FLOOR, SNR, THD, SINAD, V2, V3, ... , V6 in units of dBc

Figure 5.19: Calculating S/(N+D) (SINAD) from SNR and THD

few tenths of a dB, unless there is an extreme amount of distortion. The various harmonics, V1 through V6, are measured with respect to the signal level, S, in dBc. They are then converted into a ratio, combined on an rss basis, and converted back into dB to obtain the THD.

The signal-to-noise-and-distortion, SINAD, can then be calculated by combining SNR and THD as a root-sum-square:

$$SINAD = 20\log_{10}\sqrt{\left(10^{-SNR/20}\right)^2 + \left(10^{-THD/20}\right)^2} \qquad \text{Eq. 5.13}$$

One of the most important factors in obtaining accurate distortion measurements is to ensure that the DAC output frequency, f_o is not a subharmonic of the update rate, f_c. If f_c/f_o is an integer, then the quantization error is not random, but is correlated with the output frequency. This causes the quantization noise energy to be concentrated at harmonics of the fundamental output frequency, thereby producing distortion that is an artifact of the sampling process rather than nonlinearity in the DAC. It should be noted that these same artifacts occur in testing ADCs as previously described in Chapter 2 of this book.

To illustrate this point, Figure 5.20 shows simulated results for an ideal 12-bit DAC where the left-hand diagram shows the output frequency spectrum for the case of $f_c/f_o = 32$. Notice that the SFDR is approximately 76 dB. The right-hand spectral output shows the case where the f_c/f_o ratio is a noninteger—the quantization noise is now random—and the SFDR is 92 dB.

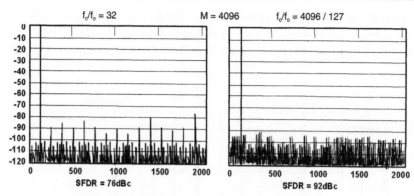

**Figure 5.20: Effect of Ratio of Sampling Clock f_c to
Output Frequency f_o on SFDR for Ideal 12-bit DAC**

Accurately specifying and measuring spectral purity of DACs used in frequency synthesis applications therefore presents a significant challenge to the manufacturer because of the large number of possible combinations of clock and output frequencies as well as output signal amplitudes. Data is traditionally presented in several formats. Figure 5.21 shows two possible spectral outputs of the AD9851 10-bit DDS DAC updated at 180 MSPS. Notice that the two output frequencies (1.1 MHz and 40.1 MHz) are chosen such that they are not a subharmonic of the 180 MSPS clock frequency.

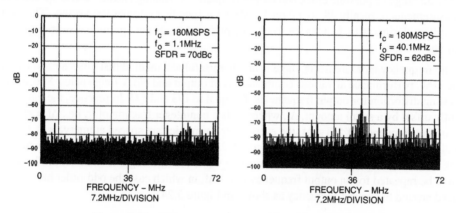

Figure 5.21: AD9851 10-Bit, 180 MSPS DDS Spectral Output

The TxDAC-series of DACs have been specifically optimized for low distortion and noise as required in communications systems. The 14-bit AD9744 is an example of the family, and its single and dual-tone performance for an output frequency of approximately 15 MHz and a clock frequency of 78 MSPS is shown in Figure 5.22.

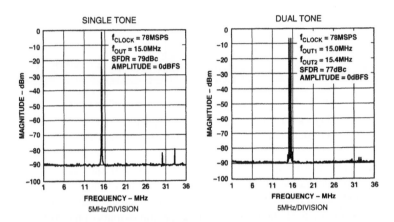

Figure 5.22: AD9744, 14-Bit, 165 MSPS TxDAC Spectral Output

Because of the wide range of possible clock and output frequencies, Analog Devices offers special fast turnaround measurements (typically 48 hours) on TxDACs for specific customer test vectors. This powerful service allows system designers to do advance frequency planning to ensure optimum distortion performance for their application.

In lieu of specific frequency measurements, there is another useful test method that gives a good overall indicator of the DAC performance at various combinations of output and clock frequencies. Specifically, this involves testing distortion for output frequencies, f_o, equal to $f_c/3$ and $f_c/4$. In practice, the output frequency is slightly offset by a small amount, Δf, where Δf is a noninteger fraction of f_c, i.e., $\Delta f = kf_c$, where $k \ll 1$. For an output frequency of $f_c/3 - \Delta f$, the even-order harmonics are spaced at intervals of Δf around the fundamental f_o output frequency as shown in Figure 5.23. The worst even-order harmonic is measured at various clock frequencies up to the maximum allowable while maintaining this same ratio. The same procedure should be repeated for an output frequency $f_c/4 - \Delta f$, in which case the odd-order harmonics are uniformly spaced around the output frequency as shown in Figure 5.24.

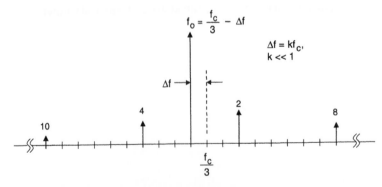

Figure 5.23: Location of Even Harmonics for $f_o = f_c/3 - \Delta f$

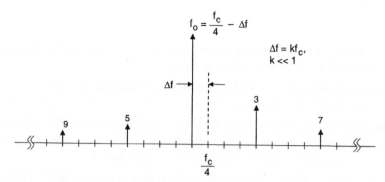

Figure 5.24: Location of Odd Harmonics for $f_o = f_c/4 - \Delta f$

These measurements are relatively easy to make, since once the ratio of f_o to f_c is established by the DDS or digital waveform generator, it is preserved as the clock frequency is changed. Figure 5.25 shows a typical plot of SFDR versus clock frequency for a low distortion DAC with two output frequencies $f_c/3$ and $f_c/4$. In most cases, the $f_c/3$ distortion represents a worst-case condition and is good for comparing various DACs.

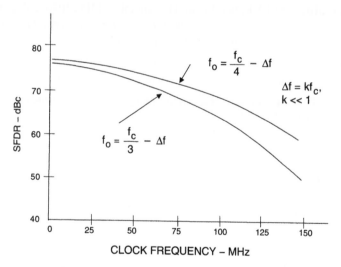

Figure 5.25: Worst Harmonic versus Clock
Frequency for $f_o = f_c/3 - \Delta f$ and $f_o = f_c/4 - \Delta f$

References:

5.1 Testing DACs

1. Jim R. Naylor, "Testing Digital/Analog and Analog/Digital Converters," **IEEE Transactions on Circuits and Systems**, Vol. CAS-25, July 1978, pp. 526–538.

2. Dan Sheingold, **Analog-Digital Conversion Handbook, 3rd Edition**, Analog Devices and Prentice-Hall, 1986, ISBN-0-13-032848-0. *(The defining and classic book on data conversion.)*

3. **Tektronix, Inc.,** 14200 SW Karl Braun Drive, P. O. Box 500, Beaverton, OR 97077, Phone: 800-835-9433, www.tek.com. *(The website contains a wealth of information on oscilloscopes, measurement techniques, probing, etc., as well as complete specifications on products.)*

4. Howard K. Schoenwetter, "High Accuracy Settling Time Measurements," **IEEE Transactions on Instrumentation and Measurement**, Vol. IM-32, No. 1, March 1983, pp. 22–27.

5. James R. Andrews, Barry A. Bell, Norris S. Nahman, and Eugene E. Baldwin, "Reference Waveform Flat Pulse Generator," **IEEE Transactions on Instrumentation and Measurement**, Vol. IM-32, No. 1, March 1983, pp. 27–32.

6. Barry Harvey, "Take the Guesswork out of Settling-Time Measurements," **EDN**, September 19 1985, pp. 177–189.

7. **Rohde & Schwarz, Inc.**, 8661A Robert Fulton Dr.,Columbia, MD 21046-2265, Phone: 410-910-7800, www.rohde-schwarz.com. *(A premier manufacturer of spectrum analyzers, the website contains tutorials on frequency analysis as well as product specifications.)*

8. **Audio Precision**, 5750 S.W. Arctic Drive, Beaverton, Oregon 97005, www.audioprecision.com. *(The recognized industry standard for professional audio measurement equipment.)*

9. **IEEE Std. 1241-2000, IEEE Standard for Terminology and Test Methods for Analog-to-Digital Converters**, IEEE, 2001, ISBN 0-7381-2724-8.

Testing ADCs
Walt Kester

A Brief Historical Overview of Data Converter Specifications and Testing

Although the overall history of data converters has been outlined in Chapter 1 of this book, the evolution of data converter specifications and associated testing methods deserves additional comment here. It is interesting to note how DSP-based tests, in particular, have become nearly universally accepted industry standards for today's ADCs.

The development of data converters, starting in the 1940s and continuing until today, can still be divided roughly into two paths depending on the sampling frequency: low speed (usually associated with higher precision up to 24 bits), and high speed (generally associated with lower precision—but recently extending into the 14- and 16-bit level, so the boundary is becoming less clear). This was certainly true in the 1940s and 1950s, when ADCs and DACs for PCM applications in the Bell System typically required sampling frequencies in the 100 kSPS range at resolutions of 5–9 bits. These converters were generally tested with the ADC and the matching DAC connected in a "back-to-back" fashion (forming a coder-decoder, or *codec*), since the performance of the combination was what determined overall system performance. Analog signals from analog test signal generators drove the ADC, and analog test equipment was used to measure the signal generated by the DAC. As will be discussed later in this section, "back-to-back" testing still has an important role up to 12 bits or so of resolution, particularly in preliminary evaluations of ADC performance.

The first high performance general-purpose commercial data converters became available in the mid-1950s, pioneered by the 11-bit, 50 kSPS DATRAC vacuum tube converter designed by Bernard M. Gordon at Epsco in 1954. Gordon himself was a pioneer is defining the performance of data converters, especially those related to precision applications, and wrote many of the early articles on converter specifications (see References 1, 3, 4, 5, 9, 10, and 20).

The interest in ADCs and DACs increased rapidly in the 1960s, as solid-state data converters as well as mainframe computers became available. Early driving forces were data analysis, instrumentation, PCM, and radar applications.

In ac applications, there was still no way to directly measure the frequency-domain performance of so-called "sampling" ADCs because of the lack of low cost memory and readily available digital computers. Any frequency-domain performance characteristics had to be measured using the "back-to-back" method and required that the reconstruction DAC had better static and dynamic performance than the ADC being tested.

By the mid-1970s, the minicomputer (such as the DEC PDP-series) made frequency domain testing using the fast Fourier transform (FFT) practical for ADC manufacturers (see Reference 13, for example). The IEEE-488 bus (initially the HP-IB) became a convenient way to transfer data from a buffer memory containing the ADC samples to the computer for analysis.

Also during the 1970s, the use of ADCs and DACs in new applications such as digital video (References 17, 18) made application-specific testing a requirement.

The 1980s saw widespread growth in ac testing of ADCs (see References 19, 21, 22, 23, 24, 25, 27). Manufacturers began to standardize on ac specifications such as SNR, SINAD, ENOB, THD, etc., and these became integral parts of all sampling ADC data sheets. These specifications were vital to emerging applications in communications, where wide dynamic range was of utmost importance.

Some of the early pioneering work on ac specifications and testing in the 1980s was done by various IEEE committees involved in preparing ADC/DAC standards for digital video (Reference 28) and waveform recorders (References 22 and 36). The waveform recorder standard (IEEE Std. 1057-1994, R2001) was later expanded to include terminology and test methods for general-purpose ADCs (Reference 37).

In addition to the evolution of ac tests, histogram tests for measuring the static DNL and INL performance of ADCs virtually replaced older methods (see References 21, 22, 24, 25, 27).

By the 1990s, frequency-domain testing of ADCs and DACs became the norm, and readily available FFT software, PCs, and manufacturer's evaluation boards placed it within easy grasp of most customers. Today (2004) nearly all sampling ADCs are fully characterized for ac performance, with the exception of high resolution ADCs designed for measurement applications. Although there are still a few inconsistencies here and there in the industry, most ADC manufacturers have adopted basically the same set of specifications and related terminology as discussed in Chapter 2 of this book.

Static ADC Testing

As seen in Section 5.1, static testing of DACs basically consists of a series of measurements of the output voltage for various digital input codes. Knowledge of the specific DAC architecture and the corresponding error characteristic may allow a reduction in the actual number of individual voltage measurements, however, this only serves to speed up test time, and doesn't change the fundamental test method or concept.

In a DAC, there is one unique output voltage for each digital input code, regardless of DNL or INL errors. An ADC, on the other hand, does not have a unique voltage input corresponding to each output code—there is a small input voltage range equal to 1 LSB in width (for an ideal noiseless ADC) that will produce the same digital output code. This is called the *quantization uncertainty*, and it can be the source of confusion when specifying and measuring ADC static transfer characteristics. Figure 5.26 shows two possible

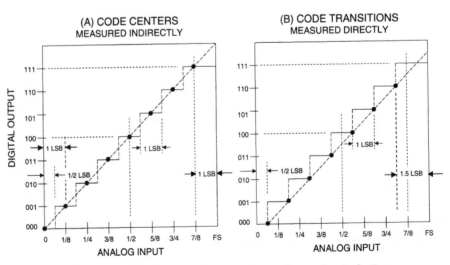

Figure 5.26: Measuring ADC Code Transitions to Determine Code Centers

methods for defining the relationship between the ADC analog input and the digital output code. Method A defines the static transfer characteristic in terms of the *code centers*, however, there is no direct way to measure these points because of the quantization uncertainty. Method B defines the static transfer characteristic in terms of the code transitions, which can be measured directly. All that is required to measure the code transitions is an analog voltage source and DVM connected to the ADC input and a means of observing the digital outputs, such as an LED display. The analog input is varied until the LEDs "flutter" between two codes, and the input voltage is recorded. It should be noted, however, that this method only works well if the ADC (and the input voltage source) has an effective peak-to-peak input-referred noise that is less than 1 LSB. Larger amounts of input-referred noise tend to mask the transitions and make the measurement increasingly difficult.

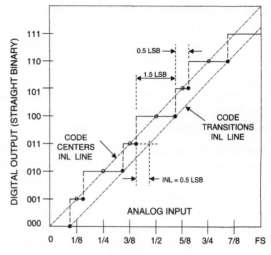

Figure 5.27: Code Transitions Preferable to Code Centers for Measuring ADC DNL and INL

However, assume for the moment that the ADC is relatively noise-free and that the transitions can be easily measured. Once the code transition points are known, the code center can be calculated as the voltage that is halfway between the corresponding code transitions. In fact, the entire ADC transfer function can be defined entirely by the code transitions as shown in Figure 5.26B. Note that the code transition points are shifted ½ LSB to the left of the code centers for the ideal ADC. The advantage of using the code transition method directly is that the DNL for a particular code is simply the difference between the corresponding code transitions.

The code-center method can lead to misleading results as shown in Figure 5.27. Notice that the ADC transfer function has alternating wide and narrow codes, but the line drawn through the endpoints of the code centers indicates perfect INL. On the other hand, the line drawn through the endpoints of the code transitions shows the true INL of ½ LSB.

Figure 5.28 shows a simple test setup for measuring the code transition points of an ADC. The analog input is driven from a precision low noise voltage source, and the ADC digital outputs are observed using an LED display. Displaying the

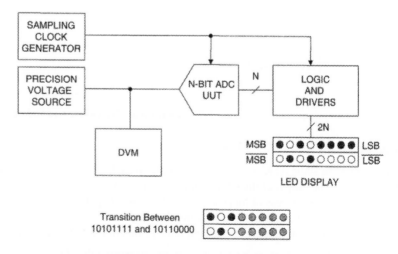

Figure 5.28: Simple Test Setup for Measuring ADC Code Transitions

true and complement of each ADC bit makes determining the precise centers of the code transition points easier simply by adjusting the analog input voltage for equal brightness between the true and complementary LED display for the bits that are "fluttering." Additional logic is obviously required if the output of the ADC is in serial format.

Although quite simple, the manual test setup is very useful with low noise ADCs, especially if only a few code transitions need to be measured, such as when determining offset and gain errors. Figure 5.29 shows the details for measuring ADC gain and offset in terms of the code transitions. The example is for a 3-bit ADC, but is applicable to any resolution.

Notice that for an ideal ADC, the first code transition between 000 and 001 (designated V_{001}) occurs at an analog input voltage of 0.5 LSB and the last transition between codes 110 and 111 (designated V_{111}) occurs at a voltage of FS – 1.5 LSB, following the conventions shown previously in Figure 5.26B. The definitions in Figure 5.29 apply to a unipolar ADC but can easily be modified for a bipolar ADC. In a bipolar ADC (as discussed in Chapter 2 of this book) the convention is that a zero-volt input to the ADC should fall at the code center corresponding to the code 100…0, with adjacent code transitions 0.5 LSB above and below zero volts. As in the case of DACs, the value of the ADC bipolar zero may be of interest and is easily calculated from the corresponding code transition values.

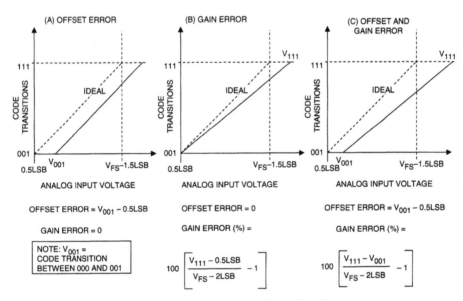

Figure 5.29: Measuring ADC Offset and Gain Error

Back-to-Back Static ADC Testing

Another useful static ADC test method that dates back to the early days of data converters involves connecting a DAC to the output of the ADC and measuring the performance of the ADC using conventional analog test methods. The success of this back-to-back method depends upon the ability to obtain a suitable DAC which has an accuracy significantly greater than the ADC under test. For instance, a 12-bit ADC requires at least a 14-bit accurate DAC, even though only 12 bits of the DAC are actually used.

Figure 5.30 shows a back-to-back test setup for measuring the static performance of the ADC by generating the actual error waveform for the ADC transfer function. The ADC drives an accurate DAC, and the output of the DAC is subtracted from the analog input to the ADC, amplified (if required), and filtered for observation with an oscilloscope. The subtraction function is implemented by complementing the ADC output code before driving the DAC, so that the DAC output is an inverted quantized representation of the ADC input. A simple resistive summer using a potentiometer can then be used to combine the signals. A very low frequency linear ramp is applied to the ADC, and the potentiometer adjusted to null out the effects of gain differences between the ADC and the DAC. A linear ramp input makes the interpretation of the error waveform easier because the horizontal axis (time) can be calibrated directly in LSBs. If desired, the DAC update clock can be a submultiple of the ADC sampling clock to ease the settling time requirements on the DAC and make the error waveform easier to observe. A low-pass filter at the input of the oscilloscope is useful for removing high frequency components that may obscure the actual error waveform.

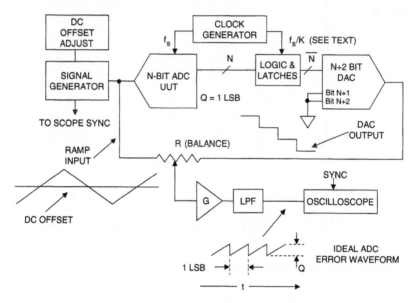

Figure 5.30: ADC/DAC Back-to-Back Static Testing

Figure 5.31A, B, and C show some typical error waveforms that result from various errors in the ADC. Notice that the sawtooth error waveform is composed of a number of "teeth." The width of the tooth corresponds to the code width and can be used to make DNL measurements directly as shown in Figure 5.31A. The dotted vertical lines along the horizontal axis show where the ideal code transitions should occur, and the effects of positive and negative DNL are shown as well as a missing code.

The height of each tooth is determined by the DAC and, if there is a missing code, the height will correspond to 2 LSBs rather than 1 LSB. Figure 5.31B shows a nonmonotonic condition, represented by a temporary reversal in the direction of the error waveform indicated by the arrows.

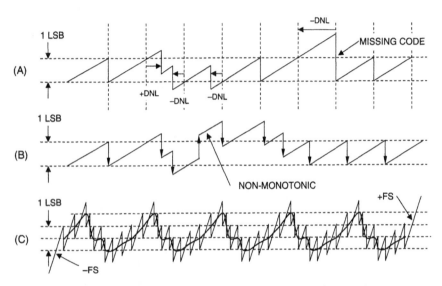

Figure 5.31: ADC Error Waveforms Using "Back-to-Back" Test Setup

When the back-to-back method is used to make static DNL measurements, a low amplitude triangle waveform with a precision dc offset adjustment works best. The dc offset can be slowly adjusted, and the DNL error waveform swept across the entire range of the ADC. If the frequency of the error waveform is too high, the error components cannot be observed. INL measurements, on the other hand, should be made with a full-scale triangle, and the maximum deviation of the error waveform from a line connecting the endpoints can be used to calculate the INL as shown in Figure 5.31C. In this case, it is not necessary to be able to observe each of the individual teeth that make up the overall error waveform. The balance potentiometer should be adjusted to force the endpoints to the same level so that the INL can be measured. The worst-case endpoint INL is approximately +1 LSB in Figure 5.31C.

In summary, the back-to-back method is a useful tool in making static ADC linearity measurements in a simple bench test setup. As the ADC resolution is increased, the frequency of the input signal must be made lower, the amplitude of the error waveform decreases, and the effects of ADC noise and DAC errors become more pronounced.

The technique works best for ADCs of 12 bits of resolution or less, and where the code transition noise (i.e., input-referred noise) is less than a few tenths of an LSB. Measurements on 12-bit ADCs require considerable care, and extending the technique to 14-bit ADCs is quite difficult. Measurements on 16-bit ADCs require at least 18-bit accurate DACs, which are not widely available with traditional dc specifications.

Crossplot Measurements of ADC Linearity

The crossplot method for measuring ADC linearity was developed in the early days of data converters as an easy method for determining integral and differential linearity. This test is suitable for quick evaluations where a high degree of precision is not required. The concept is nearly the same as the back-to-back method previously described, except that only two or three of the ADC LSBs are converted back into analog format. Figure 5.32 shows the basic test setup for the crossplot test.

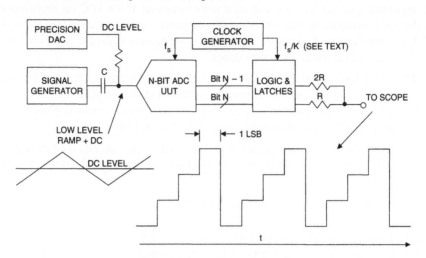

Figure 5.32: ADC Crossplot Test

Two resistors are used for the simple DAC to reconstruct the two LSBs of the ADC output. A precision DAC supplies the dc input to the ADC, and a low-amplitude triangle waveform serves to modulate the dc level. As in the back-to-back test method, the horizontal axis of the oscilloscope can be calibrated in terms of LSBs. The output of the 2-bit DAC is a four-level staircase waveform as shown. The crossplot method was initially developed to test successive approximation ADCs designed around internal binary-weighted DACs.

As previously discussed in this chapter, superposition holds in this type of DAC, and therefore in a SAR ADC based upon such a DAC. All codes need not be checked, only the major carries. In conducting the test, the precision DAC supplies the dc level corresponding to the particular major-carry code transition under test. The horizontal axis of the oscilloscope is calibrated in terms of LSBs. The three waveforms shown in Figure 5.33 illustrate DNL errors and a missing code error. Notice that this method will not accurately detect more than two missing codes above and below the major carry under test. Also, the simplicity and efficiency of the crossplot technique is lost when testing ADCs of architectures other than successive approximation (using binary-weighted internal DACs), as more codes must be tested.

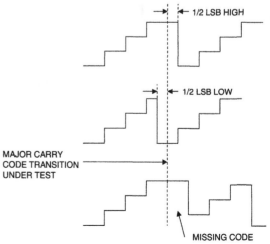

Figure 5.33: Crossplot Waveforms Showing ADC Static Errors Around Major Code Transitions

Servo-Loop Code Transition Test

The servo-loop code transition test setup shown in Figure 5.34 lends itself to automated measurements, either as part of ATE systems or in PC-based controllers. The loop begins with an op amp configured as an integrator. Switches (usually CMOS) at the integrator input are used to select between positive and negative voltage sources. With a constant dc voltage at the integrator input, a linear ramp will be generated which is sampled by the ADC. The output of the ADC goes to the "A" input of a digital comparator. The "B" input to the digital comparator specifies the code transition to be measured. If the ADC output is less than the selected code, the A < B comparator output connects the negative voltage source to the resistor R, and the output of the integrator ramps up. At the point when A ≥ B, the positive voltage source is connected to the resistor R, and the output of the integrator ramps down.

The feedback of the servo loop causes this oscillation to continue, producing a triangle waveform which is centered about the dc level of the code transition. The voltmeter at the integrator output provides an accurate measurement of the code transition voltage. When the loop time constant is properly adjusted, the amplitude of the triangle waveform should ideally be a fraction of an LSB.

This technique can be used to measure the endpoints of the transfer function, i.e., the first and last code transitions, V_1, V_{N-1}. These values can then be used to calculate the nominal LSB value as follows:

$$LSB_{NOM} = \frac{V_{N-1} - V_1}{2^N - 2}$$

Eq. 5.14

The endpoint code transitions and the nominal LSB weight provide the required information to calculate DNL and INL for any desired code transition. Gain and offset errors can also be calculated from the endpoint code transitions.

Hysteresis and/or noise in the ADC can create problems in the servo-loop test method. In some cases, it is beneficial to have control of the integrator time constant, starting out with a short time constant for fast initial response, and ending with a longer one for more accuracy.

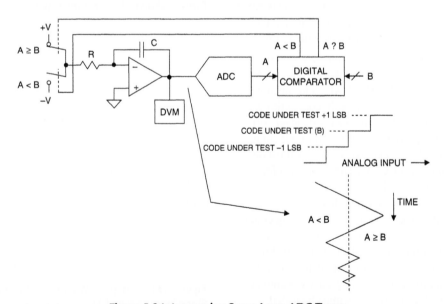

Figure 5.34: Integrating Servo-Loop ADC Tester

Computer-Based Servo-Loop ADC Tester

Figure 5.35 shows a generalized computer-based ADC servo-loop tester suitable for an ATE or PC-based system. This configuration allows the ultimate flexibility by allowing complete software programmability. It also allows the use of averaging techniques to reduce the effects of ADC input-referred noise. It is suitable for more dedicated setups where high volume testing is required, but is most likely not justifiable for simply evaluating ADC performance on the bench.

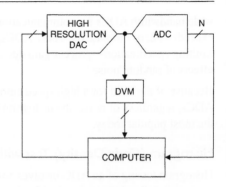

Figure 5.35: Generalized Computer Controlled Servo-Loop ADC Tester

All of the test methods described thus far for determining the code transitions work best when the ADC under test has input-referred noise less than a few tenths of an LSB peak-to-peak, such as shown in Figure 5.36.

The computer-based servo-loop tester of Figure 5.35 is the exception, because it allows for data averaging to remove the effects of excess noise. Some test algorithms which work perfectly for low noise ADCs simply may not converge for high levels of input noise as shown in Figure 5.37.

As ADC technology has evolved over the years, code transition noise has rarely been a problem with 6-, 8-, or 10-bit converters. Even at the 12- and 14-bit level, the effective input bandwidth of early ADCs in the 1970s was low enough so that input-referred noise was still at a level which allowed the back-to-back, crossplot, and integrating servo-loop test methods to work satisfactorily for most converters.

In the 1980s and 1990s, however, the requirements for increasingly higher sampling rate ADCs with associated higher input bandwidth has led to higher input-referred noise, simply because of the fundamental physical laws governing circuit designs such as resistor noise and KT/C noise. Although low input-referred noise 12-, 14-, 16-, and 18-bit ADCs are available today for low sampling rate, low bandwidth applications; many of the

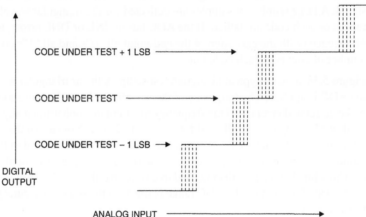

Figure 5.36: ADC Transfer Function with Relatively Low Input-Referred Noise

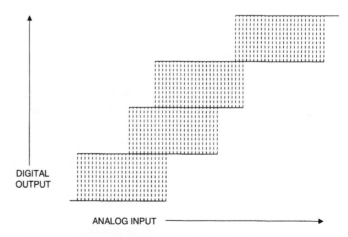

Figure 5.37: ADC Transfer Function with Relatively High Input-Referred Noise

wide bandwidth ADCs for communications applications have peak-to-peak input-referred noise which often exceeds 1, 2, or more LSBs. These ADCs are used in applications where SFDR is actually the performance-limiting specification—process gain due to oversampling and averaging techniques are used to reduce the effects of random noise

Because of the need for a high speed automated test method suitable for measuring DNL and INL of all ADCs, regardless of noise, the histogram (code density) test method (described in the next section) is by far the most popular today.

Histogram (Code Density) Test with Linear Ramp Input

Histogram testing of an ADC involves collecting a large number of digitized samples over a period of time, for a well-defined input signal with a known probability density function. The ADC transfer function is then determined by a statistical analysis of the samples. For example, a linear ramp (in actual practice, a triangular waveform is used) which slightly exceeds both ends of the range of the ADC is a popular histogram test signal. A large number of samples are collected for the triangular waveform input, and the number of occurrences of each code are tallied. If the ADC has no INL or DNL errors, all codes have equal probability of occurrence (with the exception of the end-point all "0"s and all "1"s codes), and there should be the same number of counts in each code bin.

Figure 5.38 shows a typical histogram test setup. A linear triangular waveform which slightly overdrives the ADC is applied. The frequency of the waveform should be low enough such that the ADC does not make ac-related errors, and the frequency must not be subharmonically related to the sampling frequency. A total of M_T samples are collected for codes 1 to $2^N - 2$. Notice that the "overflow" counts that fall in the all "0"s bin (code 0) and the all "1"s bin (code $2^N - 1$) are not included in the M_T total, but still add to the total number of samples required. Therefore, the triangular waveform should be adjusted such that the number of overflow hits is no larger than needed to ensure that the ADC is sufficiently overdriven and that the portion of the waveform within the ADC range is linear to the required accuracy (10% overdrive is reasonable for most ADCs).

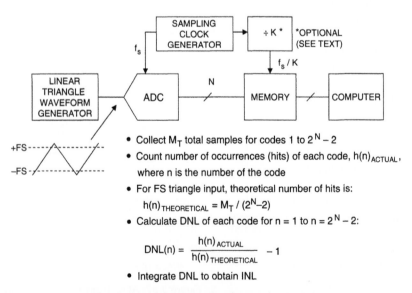

- Collect M_T total samples for codes 1 to $2^N - 2$
- Count number of occurrences (hits) of each code, $h(n)_{ACTUAL}$, where n is the number of the code
- For FS triangle input, theoretical number of hits is:
  $h(n)_{THEORETICAL} = M_T / (2^N - 2)$
- Calculate DNL of each code for n = 1 to n = $2^N - 2$:

$$DNL(n) = \frac{h(n)_{ACTUAL}}{h(n)_{THEORETICAL}} - 1$$

- Integrate DNL to obtain INL

Figure 5.38: Histogram (Code Density) Test Setup

The number of occurrences ("hits"), h(n), in each code bin, n, are then recorded for n = 1 to n = $2^N - 2$. The theoretical number of hits in each bin (assuming perfect INL and DNL) is simply $h(n)_{THEORETICAL} = M_T/(2^N - 2)$. If $h(n)_{ACTUAL}$ is the actual number of hits in a bin, then the DNL of that particular code is given by:

$$DNL(n) = \frac{h(n)_{ACTUAL}}{h(n)_{THEORETICAL}} - 1 \qquad \text{Eq. 5.15}$$

Figure 5.39 shows a typical display of the histogram data. Wide codes, narrow codes, and missing codes are easily spotted in the display. The actual DNL for each code is easily calculated from the histogram data using Eq. 5.15. Once the DNL is calculated, the INL is simply the integral of the DNL as shown in Figure 5.40.

It should be obvious that the histogram test eliminates the effects of input-referred noise by averaging it over all the code bins. The noise and hysteresis associated with each individual code transition is also averaged. Therefore, the histogram test is ideally suited for modern wide bandwidth high precision ADCs and is universally accepted, especially with the proliferation of standard PC-based software and ADC manufacturer's evaluation boards.

Notice, however, that the histogram test alone does not necessarily imply monotonicity in an ADC, i.e., the order in which the codes occur with respect to the input cannot be determined directly. However, a nonmonotonic ADC will also generally have a higher level of distortion, and this condition is easily detected with an FFT analysis of the output data. Since histogram and FFT tests use essentially the same hardware, both are normally part of a comprehensive ADC test plan.

There are several important factors to consider when conducting histogram tests. One of the most important is the number of samples required to accurately measure DNL and INL. Assume a 12-bit, 1 MSPS ADC, and that it is desired to obtain an average of 20 hits in each code bin. With 20 hits per code bin, a DNL resolution of 1/20 = 0.05 LSB is possible. This implies that the input ramp should sweep through all of the $2^{12} = 4,096$ levels, dwelling on each code long enough to produce 20 hits

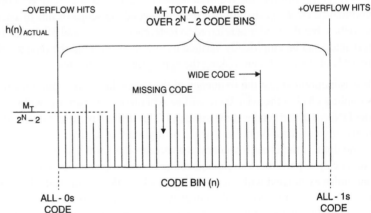

Figure 5.39: Histogram for Linear Ramp Test

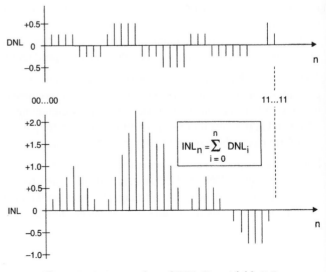

Figure 5.40: Integration of DNL Data Yields INL

313

per level. The total number of samples required is therefore $M_T = 20 \times 4,096 = 81,920$. Since the sampling frequency is 1 MSPS, this implies that the ramp must make a full-scale transition in 82 ms in order to obtain 20 hits per code bin. This assumes no overhead for the overflow samples.

Now assume that the ramp is generated by an ideal 16-bit DAC. This means that the DAC divides each of the 12-bit bins into 16 levels, or $1/16 = 0.06$ LSB. The total uncertainty in measuring the DNL is therefore 0.05 LSB (20 hits per bin) plus 0.06 LSB (16-bit ramp generator DAC), for a measurement uncertainty of 0.11 LSB.

Generating the linear ramp using a DAC becomes impractical for testing ADCs of greater resolution than 12 bits because of the difficulty of designing highly accurate relatively fast settling DACs with greater than 16-bit resolution. In addition, the samples that occur during the DAC settling time must be ignored, thereby further complicating the test. A much more practical method is to use a linear triangular wave function generator with an output frequency that is not a subharmonic of the ADC sampling clock frequency. In this case, determining the required number of samples is a relatively complex statistical problem involving confidence levels, probabilities, etc. It is beyond the scope of this discussion to get into the details of the statistics, but the reader is referred to References 27, 37, and 38 for details. Fortunately, the accuracy of the test setup can be empirically verified by simply examining the repeatability of the measurements on several records of data, and then making the appropriate adjustments.

It was mentioned that the frequency of the triangular wave input must not be a subharmonic of the ADC sampling clock, otherwise there will be repetitive code patterns within each input cycle, which will skew the DNL data and render it meaningless. This artifact is identical to that discussed in Chapter 2 of this book regarding quantization noise. For the histogram test to give accurate results, the quantization noise must be random. If the input frequency is a subharmonic of the sampling frequency, the quantization noise error signal can be periodic, thereby indicating false missing codes and large DNL errors. In most cases, jitter and noise associated with the sampling clock, analog input, and ADC tend to mitigate this effect somewhat in a practical test setup, but the integrity of the histogram measurements should be verified by slightly varying either the sampling clock frequency or the input frequency and making sure the DNL data remains relatively constant.

Finally, the histogram can place constraints on the buffer memory in terms of the speed required to handle the data output from fast ADCs as well as memory size to handle the large number of samples required (possibly several hundred thousand). The speed requirement can be reduced by taking every K^{th} sample from the ADC, since it is not required that the samples be contiguous. This in itself does not reduce the total number of samples required—it simply eases the memory speed requirement at the expense of a longer test time.

The triangular histogram test will give good DNL measurements at 16-bit or higher resolutions if proper precautions are taken as described above. INL measurements, on the other hand, can be no more accurate than the INL of the triangular input waveform. Maintaining suitable waveform INL can be a problem at 12-bit and higher resolutions. In addition, any high frequency noise that may be present on the triangular waveform cannot be removed by filtering, because that will also affect the waveform linearity.

For these reasons, a sinewave rather than a triangular waveform is often used as an input to the ADC when making histogram DNL and INL measurements. Sinewaves can be generated with extremely high linearity and low noise with appropriate filtering. However, unlike the triangular wave, the sinewave input does not yield an equal probability for all codes. It can be shown (see References 21, 24, 27, 32, 36, 37, and 38) that for an N-bit ADC with a full-scale input range equal to $\pm V_{FS}$, and an input sinewave of amplitude A, the probability of occurrence of code n is given by:

$$p(n) = \frac{1}{\pi}\left[\sin^{-1}\left\{\frac{V_{FS}\left(n-2^{N-1}\right)}{A \times 2^N}\right\} - \sin^{-1}\left\{\frac{V_{FS}\left(n-1-2^{N-1}\right)}{A \times 2^N}\right\}\right]$$

Eq. 5.16

This equation is plotted in Figure 5.41. Notice that the probability of occurrence increases at the peaks of the sinewave near $\pm V_{FS}$ because the dv/dt is less (more hits per bin) than at the zero crossing where the dv/dt is the highest (fewer hits per bin).

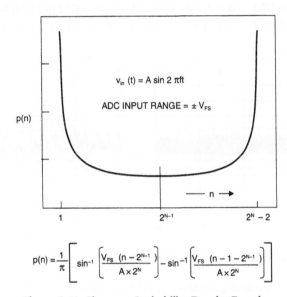

Figure 5.41: Sinewave Probability Density Function

With a sinewave input, the theoretical number of hits for the nth code is given by:

$$h(n)_{THEORETICAL} = p(n)M_T$$

Eq. 5.17

The corresponding DNL error for that code is given by:

$$DNL(n) = \frac{h(n)_{ACTUAL}}{p(n)M_T} - 1$$

Eq. 5.18

Several precautions must be taken in order to obtain accurate results using the sinewave histogram test. As previously discussed for the triangular test waveform, the sinewave frequency must not be a subharmonic of the sampling frequency. The amplitude of the sinewave input, A, should be chosen such that the ADC is slightly overdriven at both ends of its range. The effects of dc offset should then be removed by adjusting the offset of the sinewave such that there are an equal number of hits above and below the midscale point, i.e., the number of hits from code 0 to code $2^{N-1} - 1$ should equal the number of hits from code 2^{N-1} to code $2^N - 1$:

$$\sum_{n=0}^{2^{N-1}-1} h(n) = \sum_{n=2^{N-1}}^{2^N-1} h(n)$$

Eq. 5.19

Finally, the value of A should be estimated using the actual histogram data from the following equation:

$$A_{ESTIMATE} = \cfrac{V_{FS}}{\sin\left[\cfrac{M^T}{M^T + h(0) + h(2^N - 1)} \times \cfrac{\pi}{2}\right]}$$

Eq. 5.20

The estimated value of A predicted by Eq. 5.20 should then be used in Eq. 5.16 for calculating the p(n) values for each code. It is important that these steps be taken to account for the ADC gain and offset in the actual test setup so that accurate values of p(n) can be obtained for the final DNL calculation using Eq. 5.18.

Typical sinewave histogram DNL and INL measurements on the AD9236 12-bit 80 MSPS ADC are shown in Figure 5.42 as an example of how this data typically appears on an ADC data sheet.

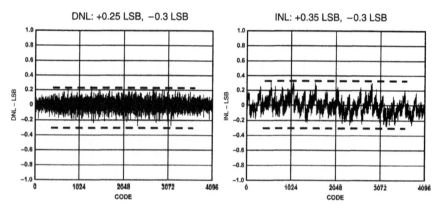

Figure 5.42: Typical Static DNL and INL Histogram Plots for AD9236 12-Bit, 80 MSPS ADC

Another useful application of the histogram test method is measuring the ADC input-referred noise. This is easily accomplished by simply terminating the ADC input with the appropriate resistance, applying a dc input (the actual value is not critical, but a voltage near midscale can be used for convenience), and recording a number of output samples. If the peak-to-peak input-referred ADC noise is less than 1 LSB, the output samples should all correspond to a single code value. If the dc input happens to fall exactly on a code transition, the samples will be divided between two adjacent codes—however, the dc input can be offset by ½ LSB in order to produce a single code at the output. The effect of input-referred noise is to spread the output samples over a number of code bins as shown in Figure 5.43. Assuming the noise is Gaussian, the input-referred noise is simply the standard deviation (σ) of the distribution, and is generally expressed in LSBs rms.

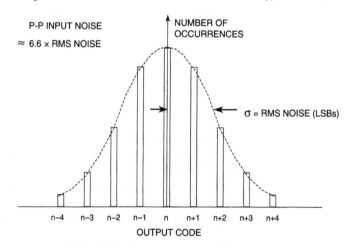

Figure 5.43: Measuring Input-Referred Noise Using "Grounded Input" Histogram

Dynamic ADC Testing

The vast majority of integrated circuit ADCs today are *sampling* ADCs that contain an internal sample-and-hold function of one form or another. In addition to the traditional dc specifications, sampling ADCs are generally fully specified in terms of ac performance characteristics such as SINAD, ENOB, SNR, SFDR, etc. In the early 1970s, the vast majority of ac testing was performed using the back-to-back test method where a high performance DAC was used to reconstruct the ADC output, thereby allowing the use of traditional analog test equipment. In the mid-1970s, DSP-based testing of ADCs began to evolve—today, practically all ADC testing is performed using some type of digital analysis of the ADC output data. While manufacturer-supplied evaluation boards, along with PC-based software packages, have placed these digital techniques within the grasp of most serious ADC users, examination of some of the older methods is still a worthwhile exercise.

This section discusses some of the traditional back-to-back test methods that still have application today in bench test setups, as well as the newer DSP-based methods. Issues relating to signal generation are also discussed where applicable.

Manual "Back-to-Back" Dynamic ADC Testing

A typical "back-to-back" ADC ac test setup is shown in Figure 5.44. As in the case of static testing using this method, the key to its success is finding a DAC that has at least 2 bits better dc and ac performance than the ADC under test. In the 1970s and 1980s this was generally not too much of a problem, because 12-bit fast settling DACs with relatively low distortion were generally available—sometimes an external deglitch-er was required to obtain sufficiently low levels of distortion.

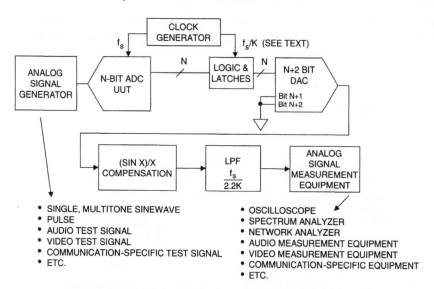

Figure 5.44: "Back-to-Back" Setup for ADC Dynamic Testing

The back-to-back method was widely used with 8- and 10-bit high speed ADCs. Most manufacturer's ADC evaluation boards during the period contained a reconstruction DAC which in most cases was sufficient to perform some basic ac tests. Today, however, if a reconstruction DAC is included on manufacturer's ADC evaluation board, it is generally there to facilitate simple functionality tests—the actual ac evaluation of the ADC should be performed using DSP techniques.

Back-to-back testing is still relevant today in applications where the signal is digitized, processed, and converted back to analog form, such as in audio codecs. In these cases, the back-to-back performance of the ADC/DAC combination is what determines overall system performance, and it is not as important to know exactly how the ac errors are divided between the ADC and the DAC. In these applications, the analog test methods previously described for measuring SNR, SINAD, THD, SFDR can be utilized.

Today, there are many high performance high speed ADCs for which selecting ac-compatible DACs suitable for testing the ADC is extremely difficult. This is especially true at the 12-, 14-, and 16-bit resolution level in communications applications. In many cases, the closest match will be a DAC with complete ac specifications suitable for DDS applications, such as the TxDAC series.

Returning to Figure 5.44, the output of the reconstruction DAC is a series of rectangular pulses whose widths equal the reciprocal of the sampling frequency. As discussed previously in Chapter 2 of this book, the frequency response of this signal follows a (sin x)/x function, therefore system measurements that depend on frequency response must take into account this theoretical roll-off, or include a suitable compensating filter as shown.

In many cases, it is useful to clock the reconstruction DAC at an even submultiple of the ADC sampling frequency (K = 2, 4, 8, etc.) to relax the settling time requirements on the DAC. The output low-pass filter is chosen to have a cut-off frequency of approximately $f_s/2.2K$ so that images are attenuated over the bandwidth of interest.

As previously mentioned, an advantage of the back-to-back test method is that traditional analog test equipment associated with the particular application can be used (see listings in Figure 5.44), and no additional computer hardware or DSP-based software is required.

Two of the most powerful tests for ADC ac linearity using the back-to-back test setup are the *envelope test* and the *beat frequency* test. The envelope test measures the ADC ac performance with a signal which is near $f_s/2$, and the beat frequency test uses a signal near f_s—both tests utilize the same test setup shown in Figure 5.45.

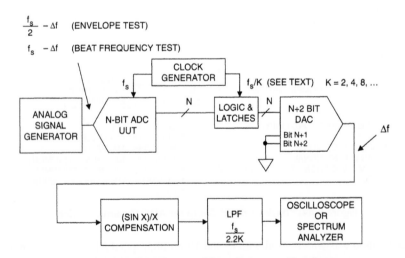

Figure 5.45: Envelope and Beat Frequency Test Setup

Figure 5.46 shows the sampled signal for the envelope test, where the input signal is slightly offset from $f_s/2$ by a small amount, Δf. Notice that the low frequency Δf signal appears in the two envelopes of the sampled signal. If the K factor in Figure 5.45 is set for K = 2, every other ADC sample is clocked into the output reconstruction DAC, yielding the low frequency Δf signal, and removing one of the envelopes. In practice, this reduces the effects of DAC settling time glitches, and the low frequency Δf signal can be easily observed on an oscilloscope for nonlinearities and missing codes, or the distortion measured with a spectrum analyzer. A Δf frequency of a few hundred kilohertz generally performs satisfactorily. Both the sampling clock and the input signal should be derived from stable frequency synthesizers or crystals to prevent phase noise on the low frequency Δf signal.

The *beat frequency* test is essentially the same as the envelope test, except the input signal is placed near the sampling frequency f_s as shown in Figure 5.47. In this case, the low frequency "beat" is obtained directly without the need for dividing the clock to the DAC. However, dividing the clock to the DAC will decrease the sensitivity to settling time glitches as in the envelope test.

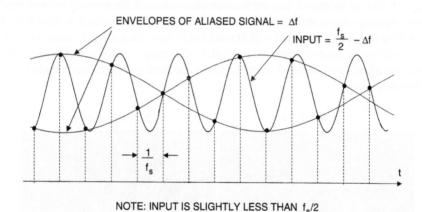

ENVELOPES OF ALIASED SIGNAL = Δf

INPUT = $\dfrac{f_s}{2} - \Delta f$

$\dfrac{1}{f_s}$

NOTE: INPUT IS SLIGHTLY LESS THAN $f_s/2$

TAKE EVERY OTHER SAMPLE TO OBTAIN
SINGLE WAVEFORM AT Δf

Figure 5.46: Envelope Test with Input Frequency Near $f_s/2$

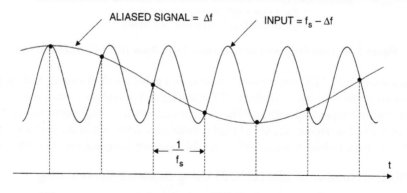

ALIASED SIGNAL = Δf

INPUT = $f_s - \Delta f$

$\dfrac{1}{f_s}$

NOTE: INPUT IS SLIGHTLY LESS THAN f_s

Figure 5.47: Beat Frequency Test with Input Frequency Near f_s

It should be noted that if the DAC clock divider is set for K = 2, the low frequency Δf signal can be observed for inputs that are near any multiple of $f_s/2$, thereby allowing the ADC to be evaluated for use in undersampling applications. The factor K should be at least 2 in order to eliminate the dual envelopes for signals near multiples of $f_s/2$. If the DAC requires more settling time in order to produce a clean display of the low frequency Δf signal, K can be chosen to be 4 or 8.

Gross ac nonlinearities and missing codes can be observed with an oscilloscope as shown in the waveforms of Figure 5.48. These tests were made on an 8-bit flash ADC sampling at 20 MSPS with an input signal slightly offset from 20 MHz. The expanded view on the right dramatically shows the effects of inadequate comparator matching in the flash ADC.

Finally, the back-to-back test setup can be used to easily measure the ADC large-signal input bandwidth. For this test, an oscilloscope is connected directly to the DAC output before the compensation and filtering and synchronized to the sampling clock. The input sinewave is set to a low frequency and adjusted until the DAC indicates the ADC is barely clipping the positive and negative peaks of the input sinewave. The input frequency can then be increased, and the measurement repeated. An increase in input signal level required to cause clipping corresponds to a decrease in the ADC gain; a decrease in input level implies an increase in gain. This test can be extended well beyond the Nyquist frequency, because the DAC response is not critical—it is only being used to detect the point of ADC clipping. Small-signal bandwidth can also be measured, where the input sinewave level is set to activate the same predetermined number of DAC output levels at each test frequency.

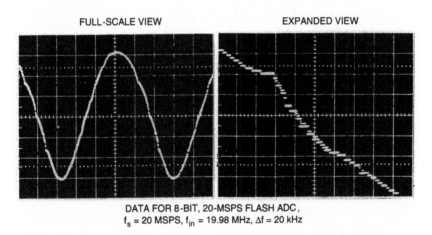

FULL-SCALE VIEW   EXPANDED VIEW

DATA FOR 8-BIT, 20-MSPS FLASH ADC,
f_s = 20 MSPS, f_{in} = 19.98 MHz, Δf = 20 kHz

Figure 5.48: Beat Frequency and Envelope Tests Show ADC AC Nonlinearities

In summary, the back-to-back ADC test method can serve as a quick check of overall ADC ac performance for resolutions of up to about 10 bits. It is also useful in testing end-to-end performance in systems that use a reconstruction DAC in conjunction with an ADC. Serious evaluations of high performance ADCs with 12 bits or more of resolution requires the use of digital techniques described in the next few sections. For this type of ADC, the back-to-back test may still be useful, however, for quick checks for functionality.

Measuring Effective Number of Bits (ENOB) Using Sinewave Curve Fitting

The first DSP-based test to be discussed is the ENOB test using the sinewave curve fitting method. In order for the results to be valid, the input frequency must not be a subharmonic of the sampling frequency. This requirement has been previously discussed in Chapter 2 of this book, as well as the section on DAC testing in this chapter. A number of samples, M (the data record length), are first collected in a buffer memory. A good rule-of-thumb is to make M large enough to contain at least five complete cycles of the input sinewave (Reference 36, p. 28). The data is then read into the computer, and a best-fit sinewave computed. The rms error of the actual sample points referenced to the best-fit sinewave is then used to compute the ENOB. The requirements on the buffer memory can be relaxed using a frequency of f_s/K to clock the memory at the expense of increased test time. The test setup is shown in Figure 5.49.

The algorithm must compute the amplitude, phase, frequency, and offset of the best-fit sinewave (4-parameters). Various algorithms for doing this are available and described in the references (see References 27, 36, and 37). If the input frequency and sampling rate are accurately known, a 3-parameter algorithm (References 36 and 37) should be used (the 4-parameter algorithms may not always converge). Once the best-fit sinewave is known, the actual rms quantization error, Q_A, is computed based on the data points in the data record. This value includes errors due to INL, DNL, missing codes, aperture jitter, noise, etc. The theoretical rms quantization error is the well-known $Q_T = q/\sqrt{12}$, where q is the weight of the LSB. The effective number of bits (ENOB) is then calculated by:

$$ENOB = N - \log_2\left[\frac{Q_A}{Q_T}\right]$$
Eq. 5.21

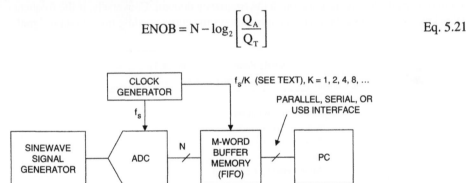

- COLLECT M SAMPLES OF SINEWAVE
- CALCULATE BEST FIT SINEWAVE FROM DATA POINTS
- CALCULATE Q_A = ACTUAL RMS ERROR FROM BEST FIT SINEWAVE
- Q_T = THEORETICAL N-BIT RMS QUANTIZATION ERROR = $q/\sqrt{12}$
- $ENOB = N - \log_2\left[\frac{Q_A}{Q_T}\right]$

Figure 5.49: Sinewave Curve Fit Test Setup for Measuring ADC ENOB

Note that the sinewave curve fitting method gives no information regarding the harmonic distortion content of the error. The FFT spectral analysis technique to be described in the next section must be used if frequency-related performance measurements such as SFDR, THD, etc., are required.

It should be noted that the most popular method today for calculating ENOB makes use of the signal-to-noise and distortion ratio, SINAD, described in Chapter 2 of this book. The SINAD is easily calculated from the FFT output. ENOB is then calculated from full-scale SINAD using the equation,

$$ENOB = \frac{SINAD - 1.76 \text{ dB}}{6.02 \text{ dB}}$$
Eq. 5.22

The two calculations for ENOB for the same ADC under the same conditions should be approximately equal if the input sinewave is full-scale. If the input signal is less than full-scale, Eq. 5.22 must be corrected as follows in order to compare it to the ENOB value predicted by the sinewave curve fit in Eq. 5.21:

$$\text{ENOB} = \frac{\text{SINAD} - 1.76 \text{ dB} + \text{Level of Signal Below FS}}{6.02 \text{ dB}} \qquad \text{Eq. 5.23}$$

FFT Basics

This section covers the basics of FFTs so that their application to ADC testing can be better understood. The emphasis is on the concepts rather than the mathematics. For the interested reader, there are a number of excellent references on the subject, and References 39 and 40 are highly recommended as a starting point. Additional references on FFTs and DSP in general are given at the end of the section (References 45-52).

Fourier analysis forms the basis for much of digital signal processing (see Figure 5.50). Simply stated, the Fourier transform (there are actually several members of this family) allows a time domain signal to be converted into its equivalent representation in the frequency domain. Conversely, if the frequency response of a signal is known, the inverse Fourier transform allows the corresponding time domain signal to be determined.

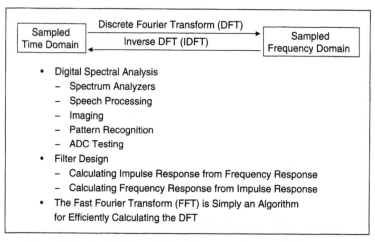

Figure 5.50: Applications of the Discrete Fourier Transform (DFT)

In addition to frequency analysis, these transforms are useful in filter design, since the frequency response of a filter can be obtained by taking the Fourier transform of its impulse response. Conversely, if the frequency response is specified, the required impulse response can be obtained by taking the inverse Fourier transform of the frequency response. Digital filters can be constructed based on their impulse response, because the coefficients of an FIR filter and its impulse response are identical.

The Fourier transform family (*Fourier Transform, Fourier Series, Discrete Time Fourier Series*, and *Discrete Fourier Transform*) is shown in Figure 5.51. These accepted definitions have evolved (not necessarily logically) over the years and depend upon whether the signal is *continuous–aperiodic, continuous–periodic, sampled–aperiodic, or sampled–periodic*. In this context, the term *sampled* is the same as *discrete* (i.e., a *discrete* number of time samples).

The only member of this family which is relevant to digital signal processing is the *Discrete Fourier Transform (DFT)* which operates on a *sampled* time domain signal which is *periodic*. The signal must be periodic

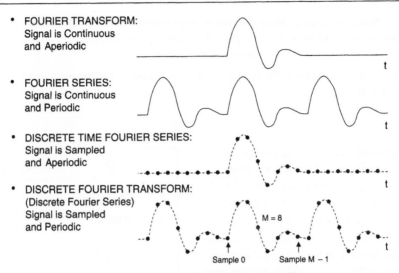

Figure 5.51: Fourier Transform Family

in order to be decomposed into the summation of sinusoids. However, only a finite number of samples (M) are available for inputting into the DFT. This dilemma is overcome by placing an infinite number of groups of the same M samples "end-to-end," thereby forcing mathematical (but not real-world) periodicity as shown in Figure 5.51.

The fundamental analysis equation for obtaining the M-point DFT is as follows:

$$X(k) = \frac{1}{M} \sum_{n=0}^{M-1} x(n) e^{-j2\pi nk/M} = \frac{1}{M} \sum_{n=0}^{M-1} x(n) \left[\cos(2\pi nk/M) - j\sin(2\pi nk/M) \right] \qquad \text{Eq. 5.24}$$

At this point, some terminology clarifications are in order regarding the above equation (also see Figure 5.52). X(k) (capital letter X) represents the DFT frequency output at the kth spectral point, where k ranges from 0 to M–1. The quantity M represents the number of sample points in the DFT data record, and should be a power of 2 (required by FFT routines).

- A Periodic Signal Can be Decomposed into the Sum of Properly Chosen Cosine and Sine Waves (Jean Baptiste Joseph Fourier, 1807)

- The DFT Operates on a Finite Number (M) of Digitized Time Samples, x(n). When These Samples are Repeated and Placed "End-to-End," they Appear Periodic to the Transform.

- The Complex DFT Output Spectrum X(k) is the Result of Correlating the Input Samples with sine and cosine Basis Functions:

$$X(k) = \frac{1}{M} \sum_{n=0}^{M-1} x(n) e^{\frac{-j2\pi nk}{M}} = \frac{1}{M} \sum_{n=0}^{M-1} x(n) \left[\cos\frac{2\pi nk}{M} - j\sin\frac{2\pi nk}{M} \right]$$

$$0 \le k \le M-1$$

Figure 5.52: The Discrete Fourier Transform (DFT)

The quantity x(n) (lower case letter x) represents the nth time sample, where n also ranges from 0 to M – 1. In the general equation, x(n) can be real or complex, however for a single ADC, the time samples have only a real component, and the imaginary component is set to zero.

Notice that the cosine and sine terms in the equation can be expressed in either polar or rectangular coordinates using Euler's equation:

$$e^{j\theta} = \cos\theta + j\sin\theta \qquad \text{Eq. 5.25}$$

The DFT output spectrum can therefore be represented in either polar form (magnitude and phase) or rectangular form (real and imaginary) as shown in Figure 5.53. The conversion between the two forms is straightforward.

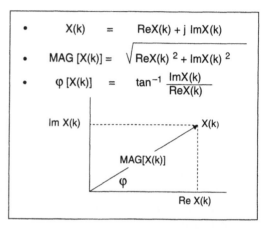

In practice, a Fast Fourier Transform (FFT) is used to compute the DFT. The FFT is simply an algorithm that reduces the required number of mathematical computations. Many FFT algorithms are available, but the most popular is the Radix-2 algorithm. Computing the DFT using Equation 5.24 requires M complex multiply operations and M – 1 complex additions for each of the X(k) terms. For the entire transform, this results in M^2 complex multiplications and M(M – 1) complex additions for a total of $2M^2$ – M complex operations. The FFT was developed to reduce the number of calculations by exploiting the symmetry properties of the DFT to eliminate

Figure 5.53: Converting Real and Imaginary DFT Outputs into Magnitude and Phase

redundant calculations. For example, the Radix-2 algorithm requires only M·log₂M complex operations. For a 1024-point transform, the DFT requires 2,096,128 complex operations compared to only 10,240 for the FFT. To utilize FFT algorithms, M must be an integer power of two.

Regardless of the algorithm, for each time-domain sample, a complex conjugate pair, ReX(k) + jImX(k) will be generated from the FFT. For example, if the time-domain sample size M = 16,384 (2^{14}), the resulting FFT array will contain 16,384 complex samples. In order to generate a frequency domain plot from this data, the magnitude of each complex sample must be calculated using the equation:

$$\text{MAG } X(k) = \sqrt{\text{Re } X(k)^2 + \text{Im } X(k)^2} \qquad \text{Eq. 5.26}$$

If required, the corresponding phase of each point can be calculated using the equation:

$$\varphi\, X(k) = \tan^{-1}\frac{\text{Im } X(k)}{\text{Re } X(k)} \qquad \text{Eq. 5.27}$$

Most FFTs are written to accept complex input data, in which case the FFT output will contain 16,384 magnitude (and phase) values representing frequencies between plus and minus $f_s/2$. If Equation 5.24 is used, the "positive" frequency values occur between FFT outputs k = 0 to k = M/2 (corresponding to the frequency range between dc and $f_s/2$), and the "negative" frequency values occur between FFT outputs k = M/2 and k = M – 1 (corresponding to the frequency range between $f_s/2$ and f_s). Although "complex" ADCs are not available, it is very common to use two ADCs to synchronously sample the I and Q data streams from a quadrature demodulator, in which the FFT input data is complex—however, this is a special case.

In testing a single ADC, the input data to the FFT is real, and the imaginary part of each complex input sample must be set to zero. For real input data, the FFT output samples between k = M/2 and k = M − 1 (the "negative" frequencies) represent an exact mirror image of those between k = 0 and k = M/2, and can be ignored.

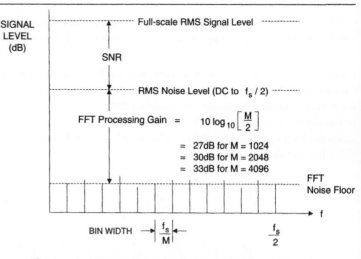

The output of a typical M-point FFT is shown in Figure 5.54. The resolution of the FFT is determined by M, and the frequency *bin width* is f_s/M. The larger M, the more frequency resolution. Figure 5.54 also shows the relationship between the aver-

Figure 5.54: FFT Output Shows Effects of Processing Gain

age noise floor of the FFT and the broadband quantization noise level (quantization noise is approximately uniformly distributed over the bandwidth dc to $f_s/2$). Each time M is doubled, the average noise in the $\Delta f = f_s/M$ frequency bin decreases by 3 dB. Note however, that averaging the results of several individual FFTs does not change the noise floor, but only reduces the variations in the noise components.

As in the previous tests described, the requirements on the buffer memory speed can be relaxed by clocking the memory at a slower rate equal to f_s/K. However, from the perspective of the FFT performed on the actual sampled values, the sampling frequency is now f_s/K rather than the f_s. Input frequencies greater than $f_s/2K$ will appear as aliased signals in the FFT output. This must be kept in mind when interpreting the FFT output spectrum.

In order to obtain spectrally pure results, the FFT data window must contain an exact integral number of sinewave cycles, otherwise spectral leakage will occur. Spectral leakage in FFT processing can best be understood by considering the case of performing an M-point FFT on a pure sinusoidal input. Two conditions will be considered. In Figure 5.55, the ratio between the sampling frequency and the input sinewave frequency is such that precisely an integral number of cycles are contained within the data window (frame, or record). Recall that the DFT assumes that an infinite number of these windows are placed

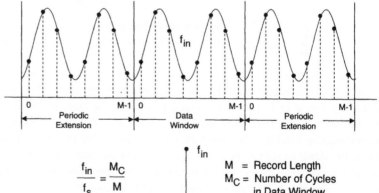

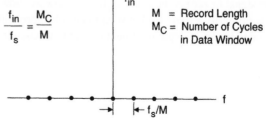

Figure 5.55: FFT of Sinewave Having Integral Number of Cycles in Data Window

end-to-end to form a periodic waveform as shown in the diagram as the periodic extensions. Under these conditions, the waveform appears continuous, and the DFT or FFT output will be a single tone located at the input signal frequency.

Figure 5.56 shows the condition where there are not an integral number of sinewave cycles within the data window. The discontinuities that occur at the endpoints of the data window result in leakage in the frequency domain, because of the sidelobes that are generated. In addition to the sidelobes, the main lobe of the sinewave is smeared over several frequency bins. This process is equivalent to multiplying the input sinewave by a rectangular window pulse which has the familiar sin(x)/x frequency response and associated smearing and sidelobes.

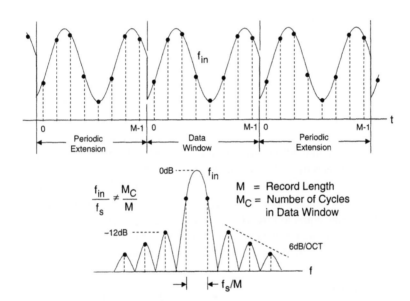

**Figure 5.56: FFT of Sinewave Having
Nonintegral Number of Cycles in Data Window**

Notice that the first sidelobe is only 12 dB below the fundamental, and that the sidelobes roll off at only 6 dB/octave beyond that point. This situation would be unsuitable for most spectral analysis applications. Since in practical FFT spectral analysis applications, the exact input frequencies are unknown, something must be done to minimize these sidelobes. This is accomplished by choosing a window function other than the rectangular window. The input time samples are multiplied by an appropriate window function which brings the signal to zero at the edges of the window as shown in Figure 5.57. The selection of a window function is primarily a trade-off between main-lobe spreading and sidelobe roll-off. Reference 50 is highly recommended for an in-depth treatment of window functions.

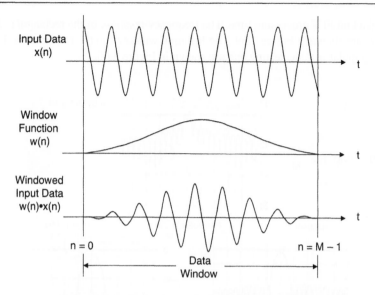

Figure 5.57: Windowing to Reduce Spectral Leakage

The mathematical functions which describe four popular window functions (Hamming, Blackman, Hanning, and Minimum 4-term Blackman-Harris) are shown in Figure 5.58. The computations are straightforward, and the window function data points are usually precalculated and stored in the DSP memory to

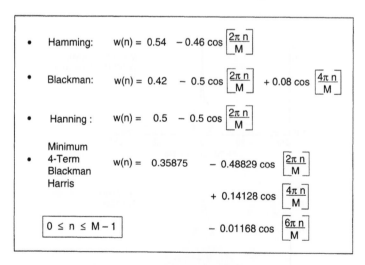

Figure 5.58: Some Window Functions

minimize their impact on FFT processing time. The frequency response of the rectangular, Hamming, and Blackman windows are shown in Figure 5.59. The Hanning window and the Minimum 4-Term Blackman-Harris window are popular in ADC testing and are shown in Figure 5.60.

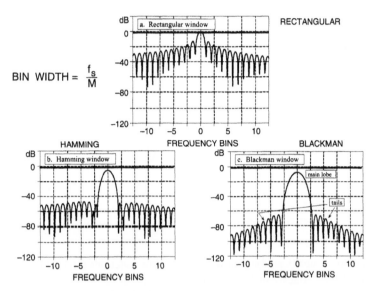

Figure 5.59: Frequency Response of Rectangular, Hamming, and Blackman Windows for M = 256

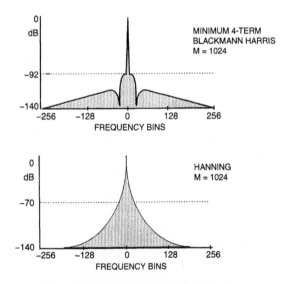

Figure 5.60: Comparison of Two Popular Window Functions Used in ADC Testing

Figure 5.61 shows the trade-off between main-lobe spreading and sidelobe amplitude and roll-off for the popular window functions.

WINDOW FUNCTION	3dB BW (Bins)	6dB BW (Bins)	HIGHEST SIDELOBE (dB)	SIDELOBE ROLLOFF (dB/Octave)
Rectangle	0.89	1.21	−12	6
Hamming	1.3	1.81	− 43	6
Blackman	1.68	2.35	−58	18
Hanning	1.44	2.00	−32	18
Minimum 4-Term Blackman - Harris	1.90	2.72	−92	6

Figure 5.61: Popular Windows and Figures of Merit

FFT Test Setup Configuration and Measurements

The typical FFT test setup shown in Figure 5.62 can be implemented in a number of ways. Maximum utilization of manufacturer's evaluation boards greatly simplifies the test and ensures proper layout of the critical components surrounding the ADC. A well-designed evaluation board should have input buffer amplifiers and/or transformers to drive the ADC, sampling clock conditioning circuits (perhaps even a stable crystal oscillator), voltage references (if required), output data registers, and appropriate input/output connectors. The importance of a low jitter sampling clock source cannot be overemphasized at this point. The effects of sampling clock jitter on the ADC SNR has previously been discussed in Chapter 2, and Chapter 6 of this book has further suggestions regarding low jitter clock sources. Careful attention to grounding, layout, and decoupling is also important, as coupling of the digital outputs into either the sampling clock or the ADC input can degrade SNR, SINAD, and SFDR performance. For these reasons, the use of a manufacturer's evaluation board is highly recommended.

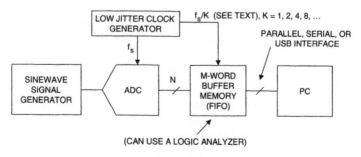

Figure 5.62: FFT Test Setup

It should be noted that there are a variety of popular commercial software packages available today which include suitable FFT routines (References 53, 54, and 55), so there is little need to actually write the FFT itself.

In order to further simplify the evaluation process, Analog Devices offers a High Speed ADC FIFO Evaluation Kit that interfaces directly to a connector on the ADC evaluation board (see Figure 5.63). The FIFO evaluation kit includes a memory board to capture blocks of data from the ADC as well as Windows compatible ('95, '98, 2000, NT) ADC Analyzer™ software. The FIFO board can be connected to the parallel port of a PC through a standard printer cable and used with the ADC Analyzer software to quickly evaluate the performance of the high speed ADC.

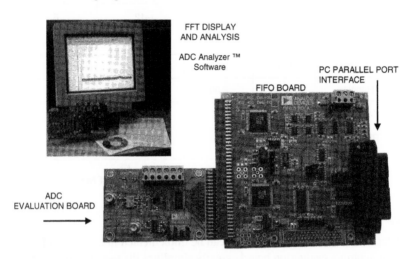

Figure 5.63: Analog Devices' High Speed ADC FIFO Evaluation Kit

The FIFO board contains two 32 K, 16-bit-wide FIFOs, and data can be captured at clock rates up to 133 MSPS on each channel. Memory upgrades are available to increase the size of the FIFO to 64 K, 132 K, or 256 K. Two versions of the FIFO are available—one version is used with dual ADCs or ADCs with demultiplexed digital outputs, and the other version is used with single-channel ADCs. Users can view the FFT output and analyze SNR, SINAD, SFDR, THD, and harmonic distortion information.

After the ADC Analyzer software computes the FFT, there are two ways to evaluate the ADC performance: graphically and computationally. In order to plot the data in a meaningful way, the magnitude data must be converted to decibels (dB). This can be done using the formula:

$$dB = 10\log_{10}\left[\frac{\text{Magnitude}^2}{\text{Magnitude}_{Max}^2}\right] = 20\log_{10}\left[\frac{\text{Magnitude}}{\text{Magnitude}_{Max}}\right] \qquad \text{Eq. 5.28}$$

where *Magnitude* is the individual array elements computed by the FFT, and *Magnitude*$_{Max}$ is the maximum magnitude element in the array.

A typical FFT plot using the FIFO evaluation kit and the ADC Analyzer software is shown in Figure 5.64 for the AD9430 12-bit, 170/210 MSPS ADC. For this test, the sampling rate is 170 MSPS, the input frequency is 10.314 MHz, and the FFT size is M = 16,384.

The key inputs to the ADC Analyzer software configuration file are shown in Figure 5.65. Note that averaging the results of several FFTs does not change the FFT noise floor, it just reduces the variations in the random noise. Most inputs to the file are self-explanatory, but the issue of coherent versus noncoherent sampling deserves further discussion.

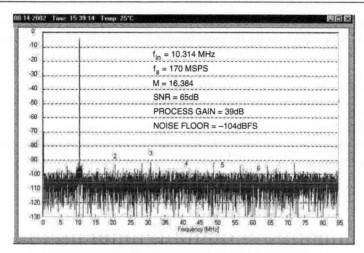

Figure 5.64: Typical FFT Output Display for AD9430 12-Bit, 170/210 MSPS ADC Using FIFO Kit

- Device, Device Number, Number of Bits, Temperature, Default Data Directory
- Number of FFT Samples (16,384 Default, up to 32K), up to 256K with FIFO Upgrades
- Two's Complement or Straight Binary Coding
- Number of FFT Averages (Default 5)
- Sampling Frequency
- Fundamental Leakage (±10 Bins Default for Minimum 4-term Blackman Harris, ±25 Bins for Hanning)
- Harmonic Leakage (±3 Bins Default)
- DC Leakage (6 Bins Default)
- Maximum Number of Harmonics (default – 2nd, 3rd, 4th, 5th, 6th)
- Windowing (Hanning, Min. 4-Term Blackman-Harris-Default, None)
- Utilize Coherent Sampling Frequency Calculator for no windowing
- Power Supply

Figure 5.65: Inputs to ADC Analyzer Configuration File

If coherent sampling is used, the fundamental and its harmonics fall in single frequency bins as shown in Figure 5.66. A strict relationship between the input frequency, f_{in}, and the sampling frequency, f_s, must be observed:

$$\frac{f_{in}}{f_s} = \frac{M_C}{M}$$

Eq. 5.29

where M_C is the integer number of cycles of the input sinewave contained in the data record M. In performing the SINAD calculation, the rss value of all the frequency bins (excluding the dc term) is the value of the noise and distortion energy.

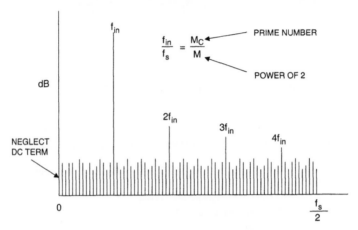

Figure 5.66: FFT Output Signal and Harmonics for Coherent Sampling

In order to prevent repetitive data patterns in the data record and ensure a random quantization noise spectrum, M_C should be a prime number. The input frequency and the sampling clock should be generated from locked frequency synthesizers in order to maintain the exact relationship. The ADC Analyzer software contains a Coherent Sampling Calculator to facilitate the calculation of the input frequency. To use the Coherent Sampling Calculator, the correct sampling frequency must first be entered in the configuration file. Then either the approximate analog input frequency or the number of sinewave cycles is entered. The Calculator then recommends a coherent input frequency and the number of cycles based on the input values.

If noncoherent sampling is used, the energy of the fundamental and its harmonics leaks into adjacent bins as shown in Figure 5.67. As previously discussed, the amount of leakage is dependent upon the particular windowing function used. When calculating the energy of the fundamental signal and its harmonics, the rss value of a number of adjacent bins should be used as shown in Figure 5.67. It is important not to count these bins as noise bins, because this will give an inaccurate SNR calculation.

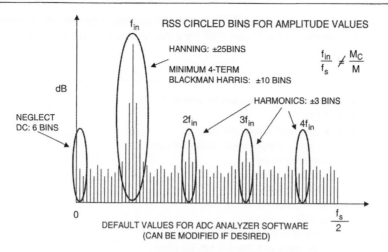

Figure 5.67: FFT Output Signal and Harmonic Leakage for Noncoherent Sampling

The dc component of the FFT output also exhibits leakage with noncoherent sampling, and those bins should also be excluded from the noise calculations to prevent erroneous results. In the ADC Analyzer software, the default signal leakage bins are ±10 bins for the Minimum 4-term Blackman-Harris window and ±25 bins for the Hanning window. The default for the harmonic leakage is ±3 bins, and the dc leakage default is 6 bins. All the leakage values can be modified if required, although the default values work well in most cases.

The SNR and SINAD measurements can be drastically affected if signal leakage is not taken into account. The value of the noise for use in the SNR calculation is obtained by taking the rss value of all noise bins, excluding the leakage bins around dc, the fundamental signal, and the 2nd, 3rd, 4th, 5th, and 6th harmonics. Therefore, including signal or dc leakage bins with amplitudes above the noise floor will give a SINAD measurement that is lower than actual.

The value of noise and distortion for the SINAD calculation is obtained by taking the rss value of all noise and distortion bins excluding the leakage bins around dc and the fundamental signal. Therefore, including signal, harmonic, or dc leakage bins with amplitudes above the noise floor will give an SNR measurement that is lower than actual.

The decision to use coherent or noncoherent testing is largely a matter of preference. While coherent testing eliminates the requirement for windowing, selecting the appropriate frequencies and their ratios can become tedious, especially when multitone tests are required. In addition, locked frequency synthesizers are required to maintain the exact frequency ratios. One can argue that coherent testing is more suitable to a laboratory environment, while noncoherent testing is more like a real-world application, where the exact input frequencies are unknown. In practice, either method will yield approximately the same final results provided the tests are performed correctly.

The ADC Analyzer software not only gives the FFT plot but also calculates the various performance characteristics, and they are summarized in Figure 5.68 for a single-tone test signal and in Figure 5.69 for a two-tone test signal.

- Time domain reconstruction of captured data
- FFT Plot
- Calculated and displayed values
  - Analog and digital power supply voltages
  - Sampling frequency
  - Analog input frequency
  - SNR (relative to signal)
  - SNRFS (relative to full-scale)
  - SINAD
  - Level of fundamental signal (dBFS)
  - Harmonics: 2nd, 3rd, 4th, 5th, 6th (dBc)
  - WoSpur: Worst non-harmonic spur (dBc)
  - THD (rss value of 2nd, 3rd, 4th, 5th, and 6th harmonics), dBc
  - SFDR (dBc)
  - Noise Floor (dBFS)

**Figure 5.68: ADC Analyzer Software
Outputs Single-Tone Input**

- Time domain reconstruction of captured data
- FFT Plot
- Calculated and displayed values
  - Analog and digital power supply voltages
  - Sampling frequency
  - Analog 1: First analog input frequency
  - Analog 2: Second analog input frequency
  - Fundamental 1: Level of first fundamental tone (dBFS)
  - Fundamental 2: Level of second fundamental tone (dBFS)
  - F1 + F2: Sum of the fundamental tones (dBFS)
  - F2 – F1: Difference of the fundamental tones (dBFS)
  - IMD Product Levels at: 2F1 – F2, 2F1 + F2, 2F2 –F1, 2F2 + F1 (dBFS)
  - WoIMD: Worst IMD product (dBc)
  - SFDR (dBc)
  - Noise Floor (dBFS)

**Figure 5.69: ADC Analyzer Software
Outputs Two-Tone Input**

Verifying the FFT Accuracy

In many cases, it is desirable to perform some type of test verification on the FFT software itself, independent of the actual ADC under test. This can be done in a number of ways. A simple method is to simply disable a number of the ADC LSBs and see if the calculated values of SINAD and SNR approach the theoretical numbers for the reduced resolution. The assumption here is that a high performance 12- or 14-bit ADC should approach near theoretical performance at the 8-bit level.. The LSBs can be disabled in hardware or software.

Another method requires the generation of an ideal N-bit digital sinewave, running the data through the FFT, and comparing the results to theoretical. The nth time sample for an ideal N-bit ADC can be written as:

$$v(n) = INT\left[2^{N-1}\sin\left(\frac{2\pi n f_{in}}{f_s}\right)\right]$$

Eq. 5.30

where the *INT* function simply truncates the fractional portion of $v(n)$.

The SINAD and SNR values obtained from the FFT can be compared with theoretical, and the overall dynamic range of the FFT routine can be found by simply increasing N until the SINAD and SNR values no longer increase by 6.02 dB per extra bit of resolution.

Generating Low Distortion Sinewave Inputs

Generating test signals with the spectral purity required to make low distortion high frequency measurements is a challenging task. Although low distortion signal generators are available, they can be quite expensive. For quick evaluations where such equipment is not available, proper filtering can produce a suitable test signal from lower cost generators.

A test setup for generating a low distortion single tone is shown in Figure 5.70. The sinewave oscillator should have low phase noise to prevent elevation of the ADC noise floor. The output of the oscillator is passed through a band-pass (or low-pass) filter, which removes any harmonics present in the oscillator output. The filter may not be required if the distortion of the generator is low enough. The generator distortion should be 10 dB lower than the desired accuracy of the measurement. The 6 dB attenuator isolates the DUT (ADC) from the output of the filter. The impedance at each interface should be maintained at 50 Ω for best performance (75 Ω components can be used, but 50 Ω attenuators and filters are generally more readily available). The termination resistor, R_T, is selected so that the parallel combination of R_T and the input impedance of the DUT is 50 Ω.

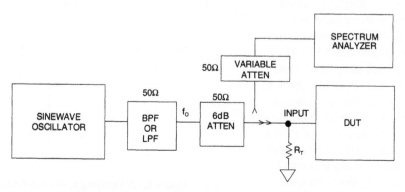

Figure 5.70: Low Distortion Single-Tone Generator

Before performing the actual distortion measurement, the oscillator output should be set to the correct frequency and amplitude. Measure the distortion at the output of the attenuator with the DUT replaced by a 50 Ω termination resistor (generally the 50 Ω input of a spectrum analyzer). Next, replace the 50 Ω load with R_T and the DUT. Measure the distortion at the DUT input a second time. This allows nonlinear DUT loads to be identified. Nonlinear DUT loads (such as flash ADCs with signal-dependent input capacitance, or switched-capacitor CMOS ADCs) can introduce distortion at the DUT input.

Generating two tones suitable for IMD measurements can be very challenging. A low-distortion two-tone generator is shown in Figure 5.71. Two band-pass (or low-pass) filters are required as shown. Harmonic suppression of each filter must be better than the desired measurement accuracy by at least 6 dB. A 6 dB attenuator at the output of each filter serves to isolate the filter outputs from each other and prevent possible cross-modulation. The outputs of the attenuators are combined in a passive 50 Ω combining network, and the combiner drives the DUT. The oscillator outputs are set to the required level, and the IMD of the final output of the combiner is measured. The measurement should be made with a single termination resistor, and again with the DUT connected to identify nonlinear loads.

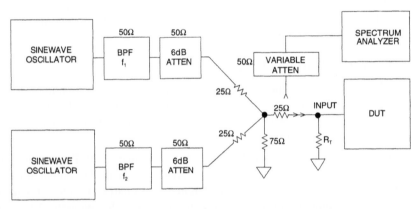

Figure 5.71: Low Distortion Two-Tone Generator

Analog spectrum analyzers should be checked independently for distortion when measuring these types of signals. Most have 50 Ω inputs, therefore an isolation resistor between the device under test (DUT) and the analyzer is required to simulate DUT loads greater than 50 Ω. After adjusting the spectrum analyzer for bandwidth, sweep rate, and sensitivity, check it carefully for input overdrive as shown in Figure 5.72. The simplest method is to use the variable attenuator to introduce a fixed amount of attenuation in the analyzer input path. Both the signal and any harmonics should be attenuated by that same fixed amount (10 dB, for instance) as observed

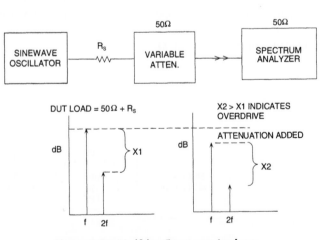

Figure 5.72: Verifying Spectrum Analyzer Sensitivity to Input Overdrive

on the screen of the spectrum analyzer. If the harmonics are attenuated by more than 10 dB, then the input amplifier of the analyzer is introducing distortion, and the sensitivity should be reduced. Many analyzers have an attenuator on the front panel for introducing a known amount of attenuation when checking for overdrive. It should be noted that most high quality modern spectrum analyzers are not generally as sensitive to input overdrive as their older counterparts. However, these simple checks don't take long to perform and are well worth the additional effort to ensure accurate distortion measurements.

If the generator is found to be sensitive to overdrive, a method to minimize the problem is shown in Figure 5.73. The amplitude of the fundamental signal is first measured with the notch filter switched out. The harmonics are measured with the notch filter switched in. The insertion loss of the notch filter, X dB, must be added to the measured level of the harmonics.

Noise Power Ratio (NPR) Testing

As described in Chapter 2 of this book, the Noise Power Ratio (NPR) measure-

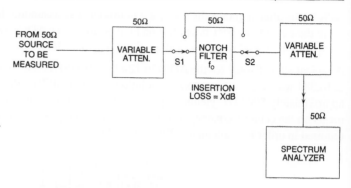

Figure 5.73: Notch Filter Removes the Fundamental Signal to Minimize Analyzer Overdrive

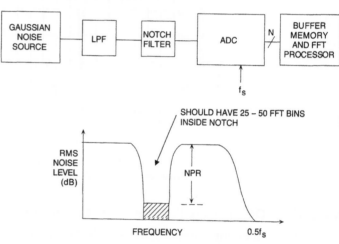

Figure 5.74: Noise Power Ratio (NPR) Test Setup

ment had its origin in the early days of frequency-division-multiplexed (FDM) telephone systems. Multiple FDM channels are simulated with Gaussian noise, and a 4-kHz wide channel is "notched" out of the input signal. The residual signal inside the notch after passing through the transmission system represents clipping noise, thermal noise, and IMD distortion products. The NPR is simply the ratio of the average noise power outside the notch to the average noise power inside the notch, i.e, it is the "depth" of the notch. Because NPR is a function of the signal amplitude, the input signal amplitude is varied until the point of maximum NPR is reached. A simplified test setup for measuring the NPR of an ADC is shown in Figure 5.74.

Conceptually, the test is easily implemented by performing an FFT on the ADC output and analyzing the data. In practice, however, an extremely large buffer memory and FFT is required in order to measure the NPR in a 4 kHz wide voice channel. For example, with a sampling rate of 170 MSPS, a 256 K FFT has a frequency resolution (bin width) of 170 MSPS/256,000 = 664 Hz. This produces only six samples "inside" the notch—not enough to get repeatable readings of the noise level. The solution is to use a wider notch filter, rather than use an impractically large FFT. The notch frequency filter width should be increased such that at least 25 to 50 samples fall inside the notch. Even with the wider notch, several FFT runs should be averaged in order to stabilize the NPR data. Using the wider notch does not invalidate the results by any means, and more closely simulates the wider channels used in modern communications systems. In fact, the

NPR test is often a good substitute for a multitone test, assuming the input tones are not phase-correlated. As in the case of any FFT-based test on high performance ADCs, the sampling clock jitter must be at an acceptably low level so as not to affect the ADC noise floor.

A typical NPR display for the AD9430 12-bit, 170 MSPS ADC is shown in Figure 5.75. The FFT size is 16,384 which gives a frequency resolution of 170 MSPS/16,384 = 10.4 kHz. The notch filter width is approximately 500 kHz, yielding approximately 48 samples inside the notch. Due to the specific requirements on the center frequency, width, and band-stop rejection, custom-made notch filters are generally required in order to implement NPR tests on ADCs.

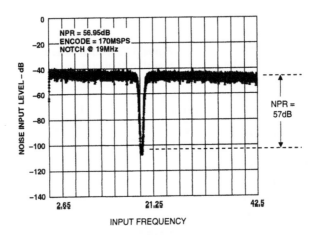

Figure 5.75: AD9430 12-bit, 170 MSPS ADC NPR Measures 57 dB (62.7 dB Theoretical)

Measuring ADC Aperture Jitter Using the Locked-Histogram Test Method

A test setup for measuring aperture jitter using the "locked-histogram" test is shown in Figure 5.76. The ADC input signal and the sampling clock are derived from the same low phase noise clock generator in order to minimize jitter between the two signals. Recall from the discussions in Chapter 2 of this book that the effects of sampling clock jitter are indistinguishable from internal ADC aperture jitter. Ideally, the test is run at the maximum ADC sampling frequency. However, if the input bandwidth of the ADC is not high enough, the test can be run at one-half the maximum sampling rate.

The clock generator drives the sampling clock input of

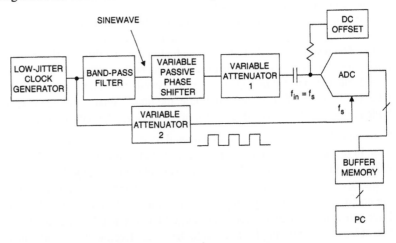

Figure 5.76: Locked-Histogram Test Setup for Measuring ADC Aperture Jitter

the ADC directly through an attenuator. The clock generator also drives a band-pass filter (to convert the squarewave into a sinewave). The output of the band-pass filter passes through a passive variable phase shifter, an attenuator, and is then ac-coupled into the ADC input. The setup is calibrated by setting Attenuator 1 to minimum attenuation and then setting the output level of the clock generator such that the input signal to the ADC is full scale ($2 V_{FS}$ peak-to-peak). The dc offset to the ADC is then adjusted such that the input sinewave is centered around midscale. Attenuator 2 is then adjusted to provide the proper level for the sampling clock input to the ADC.

Next, the attenuation of Attenuator 1 is increased until the peak-to-peak input to the ADC is only exercising a few codes above and below midscale. The variable phase shifter is adjusted such that the midscale code occurs most of the time. Input-referred noise may cause a distribution of codes—if so, the PC software calculates the standard deviation of the distribution, σ_L, in LSBs. This corresponds to the input-referred noise expressed in LSBs. Attenuator 1 is then set for a full-scale input to the ADC. The variable phase shifter is adjusted until the midscale code has the highest probability of occurrence. The standard deviation of the new code distribution, σ_H, now includes the effects of input-referred noise as well as aperture jitter. The noise sources combine on an rss basis:

$$\sigma_H^2 = \sigma_L^2 + \sigma_A^2 \qquad \text{Eq. 5.31}$$

where σ_A is the rms noise (in LSBs) due to aperture jitter.

Equation 5.31 can be solved for σ_A as follows:

$$\sigma_A = \sqrt{\sigma_H^2 - \sigma_L^2} \qquad \text{Eq. 5.32}$$

The full-scale input sinewave is given by $v_{in}(t) = V_{FS}\sin 2\pi f_{in}t$, where the ADC input range is $\pm V_{FS}$. The rate-of-change of the full-scale sinewave at the zero crossing is given by:

$$\left.\frac{dv}{dt}\right|_{max} = V_{FS}\, 2\pi f_{in} \qquad \text{Eq. 5.33}$$

For a slope of $dv/dt|_{max}$, the rms aperture time, t_a, is related to the corresponding rms voltage error, Δv_{rms}, by the equation:

$$t_a = \frac{\Delta v_{rms}}{\left.\dfrac{dv}{dt}\right|_{max}} \qquad \text{Eq. 5.34}$$

The rms noise due to aperture jitter in LSBs, σ_A, can be related to Δv_{rms} by:

$$\Delta v_{rms} = \sigma_A \times \frac{V_{FS}}{2^{N-1}} \qquad \text{Eq. 5.35}$$

Substituting Equation 5.32, 5.35, and 5.33 into Equation 5.34:

$$t_a = \frac{\sqrt{\sigma_H^2 - \sigma_L^2}}{2\pi f_{in} \times 2^{N-1}} \qquad \text{Eq. 5.36}$$

The calculations are summarized in Figure 5.77.

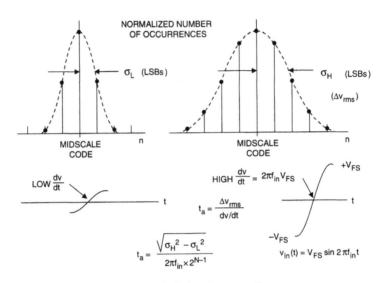

**Figure 5.77: Calculating Aperture Jitter
Based on Locked-Histogram Test**

It should be noted that the test setup can be implemented using a low jitter sinewave signal generator to drive the ADC input and use the sync trigger pulse output of the same generator to drive the sampling clock input of the ADC. However, if aperture jitter measurements of less than 100 ps rms are expected, the jitter between the signal generator output and the trigger output may be of this same magnitude, thereby corrupting the overall measurement.

Measuring Aperture Delay Time

Aperture delay time can be measured with the same test setup used for the locked-histogram aperture jitter test, or the simpler test setup shown in Figure 5.78. The analog input frequency is locked to the sampling clock frequency, and a full-scale sinewave input applied to the ADC input. The delay in the sampling clock signal is adjusted until the PC histogram distribution of codes indicates that the ADC is being sampled at the zero-crossing of the sinewave, corresponding to the midscale code. The aperture delay is simply the difference between the 50% point of the leading edge of the sampling clock and the zero-crossing of the sinewave, measured with a dual-trace oscilloscope.

The aperture delay can be positive or negative as shown in Figure 5.78. The frequency of the sinewave input signal is not critical, but it should be high enough so that the small aperture delay time can be accurately measured. However, the frequency should not exceed the analog input bandwidth of the ADC. For convenience, a frequency of one-half the maximum ADC sampling frequency is a good starting point, and represents a reasonable upper limit.

Measuring ADC Aperture Jitter Using FFTs

The FFT test routine for measuring ADC SNR is an excellent indirect method for measuring aperture jitter. The caveat in this test is that the measurement includes the jitter of the sampling clock generator as well as the ADC internal aperture jitter. Therefore, a generator should be selected with an rms jitter specification that

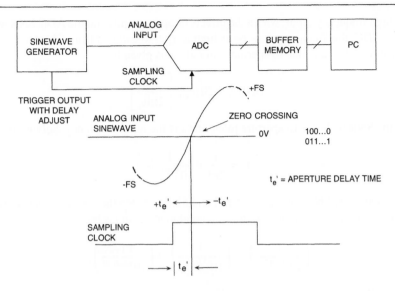

Figure 5.78: Measuring Aperture Delay Time

is several times less than the specified aperture jitter of the ADC under test, since jitter combines on an rss basis. The basic test setup for the aperture jitter test is shown in Figure 5.79 along with the key calculations.

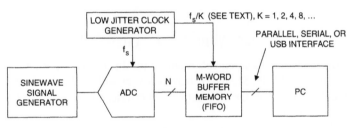

- SNR FOR LOW FREQUENCY FS INPUT = SNRL

- SNR FOR HIGH FREQUENCY FS INPUT = SNRH (FREQUENCY = f)

- $SNRA = 20 \log_{10} \left[\dfrac{1}{2\pi f t_a} \right]$

- $t_a = \dfrac{1}{2\pi f} \sqrt{ \left[10^{-SNRH/20} \right]^2 - \left[10^{-SNRL/20} \right]^2 }$

- INCLUDES JITTER OF CLOCK GENERATOR

Figure 5.79: Measuring Aperture Jitter Based on Degradation in SNR at High Frequencies

There are two SNR measurements required, and both utilize a full-scale input sinewave. The first measurement, SNRL, is made at a relatively low frequency where the noise is primarily the ADC input-referred noise. It should be possible to vary the low input frequency quite a bit and still measure the same SNR value. The sampling frequency is generally set for the maximum allowable. The second measurement, SNRH, is made using a high frequency input, where the effects of aperture jitter on the ADC SNR are noticeable. Depending on the ADC, this frequency may be as high as $f_s/2$. Recall that from Chapter 2, the relationship between the signal-to-noise ratio due to aperture jitter alone is given by:

$$SNRA = 20 \log_{10} \left[\frac{1}{2\pi ft_a} \right]$$

Eq. 5.37

where SNRA is the SNR (dB) due to aperture jitter, and f is the input frequency. Solving for t_a:

$$t_a = \frac{1}{2\pi f} \times \frac{1}{10^{SNRA/20}}$$

Eq. 5.38

The next step is to calculate SNR_A based on SNRH and SNRL. Since the SNRs are in dB, they must first be converted to ratios, and their reciprocals can then be combined on an rss basis:

$$\left(\frac{1}{10^{SNRH/20}} \right)^2 = \left(\frac{1}{10^{SNRL/20}} \right)^2 + \left(\frac{1}{10^{SNRA/20}} \right)^2$$

Eq. 5.39

Rearranging Eq. 5.39:

$$\left(\frac{1}{10^{SNRA/20}} \right) = \sqrt{ \left(\frac{1}{10^{SNRH/20}} \right)^2 - \left(\frac{1}{10^{SNRL/20}} \right)^2 }$$

Eq. 5.40

Substituting Equation 5.40 into Equation 5.38:

$$t_a = \frac{1}{2\pi f} \times \sqrt{ \left(\frac{1}{10^{SNRH/20}} \right)^2 - \left(\frac{1}{10^{SNRL/20}} \right)^2 }$$

Eq. 5.41

It should be emphasized that all the measurements required for this test are SNR and not SINAD. It is extremely important that the 2nd, 3rd, 4th, 5th, and 6th harmonics (as well as the dc components) be removed when making the SNR calculation from the FFT output. Otherwise, the measurement will not give an accurate measure of aperture jitter.

As a final note, measuring rms aperture jitter less than 10 ps rms is extremely difficult, simply because of unwanted jitter which may occur on the input signal or the ADC sampling clock, or layout-induced jitter and noise. Obtaining this level of accuracy requires frequency synthesizers with extremely low jitter as well as detailed attention to layout, signal routing, grounding, and decoupling.

Measuring ADC Analog Bandwidth Using FFTs

Analog input bandwidth can be easily measured by simply observing the FFT output display as the input frequency is swept from low to high frequency. The amplitude of the input signal must be held constant as the frequency is increased, and the relative amplitude of the FFT fundamental output signal is observed. When the amplitude of the fundamental signal in the FFT display drops 3 dB from the initial amplitude, that is defined as the 3 dB bandwidth. The measurement can be made with a full-scale input to obtain the full-power bandwidth (FPBW) or at a low amplitude to obtain the small-signal bandwidth. Notice that this definition of FPBW says nothing about the amount of distortion present in the FFT output at the FPBW frequency, which can be considerable for some ADCs.

In order to include the effects of distortion in the bandwidth measurement, a specification called *effective resolution bandwidth* (ERB) is sometimes used to define the input frequency at which the full-scale SINAD drops by 3 dB, corresponding to the loss of 0.5 ENOB. The ERB is easily measured by the same procedure described above if the calculated SINAD rather than the fundamental signal amplitude is used as the bandwidth criteria.

Settling Time

Settling time can be measured with the test setup shown in Figure 5.80 based around a flat pulse generator and a synchronized adjustable delay clock for the ADC sampling clock input. The flat pulse generator is adjusted to give slightly less than a full-scale step input to the ADC, and the sampling clock delay is adjusted until the output has settled to within 1 LSB of the final value. The delay between the 50% point of the input pulse and the leading edge of the sampling clock can be measured with a dual-channel oscilloscope. If the peak-to-peak input-referred noise of the ADC is less than 1 LSB, a simple LED display can be used to observe the ADC output. If the input-referred noise is more than 1 LSB peak-to-peak, then a histogram may be required to determine the most frequently occurring output code. The sampling frequency does not necessarily need to be set to the maximum value, it can be reduced as required. Note that the aperture delay time must be subtracted from the measured settling time value.

The characteristics of the flat pulse generator must be known precisely, and the distance from its output to

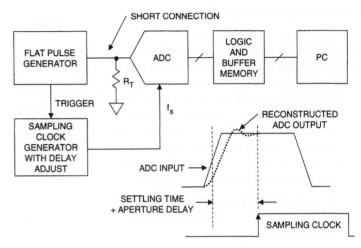

Figure 5.80: Settling Time Test Setup

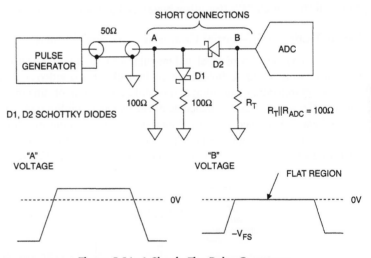

Figure 5.81: A Simple Flat Pulse Generator

the ADC input should be kept to a minimum. A simple flat pulse generator is shown in Figure 5.81, based upon low capacitance Schottky diodes. The input to the network at "A" is adjusted so that the start of the "B" waveform occurs at slightly above the negative full-scale range of the ADC. When the "A" voltage goes positive, diode D2 is reversed biased, and the voltage at "B" is flat, except for a small transient due to the reverse-bias diode capacitance and the small reverse leakage current that flows into the effective 100 Ω termination. The positive portion of the "A" voltage should be no more than required to reverse bias diode D2. Diode D1 ensures that the signal generator is terminated in a net 50 Ω impedance when the voltage at "A" is positive. Another diode can be placed in series with D2 in order to lower the effective coupling capacitance, and the negative portion of the "A" voltage adjusted as required to make up for the voltage drop of the extra diode. This network as shown delivers a ½ full-scale pulse to a bipolar ADC. Key to its success is keeping the diode-resistor network connections extremely short and close to the ADC input.

Overvoltage Recovery Time

The test setup and method for overvoltage recovery is identical to that of the settling time setup, but the voltage at "A" is adjusted so that the start of the waveform is out of the ADC input range as shown in Figure 5.82. The amount of overvoltage is generally specified as a percentage of the ADCs range. For a converter with a 2 V peak-to-peak input range, 50% overvoltage would correspond to 1 V above or below the nominal 2 V range.

As in the transient response test, the aperture delay time must be considered when making the measurement.

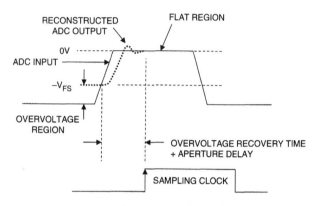

Figure 5.82: Overvoltage Recovery Test Waveform from Flat Pulse Generator

The overvoltage recovery test can verify that the ADC behaves as an ideal limiter for signals outside its nominal range. The ADC should continue to read either all "0"s, or all "1"s (for offset binary coding) while the input signal is outside the range.

Video Testing, Differential Gain and Differential Phase

The development of board-level and modular 8-bit high-speed ADCs in the early 1970s led to widespread interest in using digital techniques in traditionally analog video equipment, such as time base correctors, frame synchronization and storage, standards conversion, on-line monitoring of television signals, digital special effects, and image enhancement (see extensive bibliography in Reference 18). Most of these digital "black boxes" contained an ADC, digital processing of some sort, followed by a reconstruction DAC. Therefore traditional video test equipment could be used to check the performance of the ADC and DAC back-to-back, acting as a *codec* (coder-decoder) as shown in Figure 5.83.

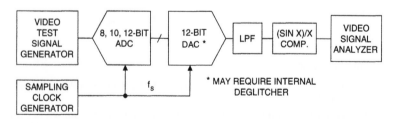

Figure 5.83: Composite Video Test Setup Using Analog Methods

In a composite video signal (see Figure 5.84) the color information is contained in the chrominance (color) signal and which is superimposed on the luminance (black and white) signal. In NTSC the color subcarrier frequency is approximately 3.58 MHz, in PAL it is approximately 4.43 MHz. The amplitude of the chrominance signal determines the color saturation, and the phase of the chrominance signal (with respect to the color burst) determines the actual color.

Differential gain and differential phase are two of the most important specifications in composite video applications. Differential gain is defined as the variation in amplitude (in percent) of a small amplitude sub-carrier signal as it is swept across the video range from black to white. Differential phase is defined as the phase variation (in degrees) of a small amplitude subcarrier signal as it is swept across the video range from black to white. These are important because differential gain errors will distort the degree of color saturation and differential phase errors will cause incorrect hues in the picture (per the definitions above).

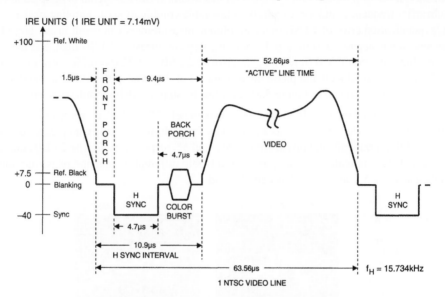

Figure 5.84: NTSC Composite Color Video Line

Some of the popular differential gain and phase video test signals are shown in Figure 5.85. All of them make use of either a staircase or a ramp that is modulated by the color subcarrier. Any of these signals are fine for testing purely analog video systems, but obtaining accurate test results in a system that contains a

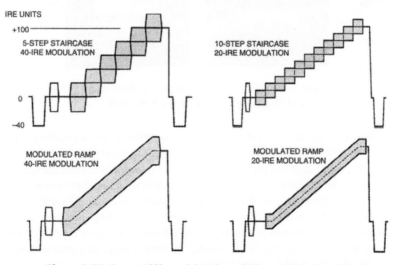

Figure 5.85: Some Differential Gain and Phase Video Test Signals

video codec presents a special challenge. This is because the nature of the test signal causes the quantization noise to give misleading results. The situation is exacerbated because in composite video applications it is very common to operate the sampling clock at exactly four times the color subcarrier frequency (leading to repetitive quantization error patterns as previously discussed).

For instance, consider the 10-step staircase test signal with 20-IRE modulation. This corresponds typically to 30 quantization levels p-p for an 8-bit system with headroom. If the test signal is sampled at exactly four times the subcarrier frequency, and a particularly unfavorable combination of sampling phase and dc level are present, a quantization error of 1 LSB in the amplitude measurement can result in any amplitude measurement on any single step of the staircase. In fact, one step can measure 1 LSB high, and the next 1 LSB low, resulting in a possible differential gain error of 2 LSBs, or 2/30 = 6.7%—even for a perfect 8-bit codec. Early differential gain and phase measurements of this type in the 1970s led to displays on vectorscopes such as those shown in Figure 5.86A for the 10-step, 20-IRE modulated staircase.

Notice that the differential gain and phase error is approximately 7% and 2°, respectively, and is mostly due to the quantization errors. This problem was analyzed by Felix (Reference 43) who predicted a theoretical worst-case 8-bit differential gain and phase error of 8% and 5°, respectively for the 20-IRE unit modulated staircase test signal. Theoretically, a 9-bit system would yield differential gain and phase measurements of approximately 4% and 2.5°, and a 10-bit system 2% and 1.25°.

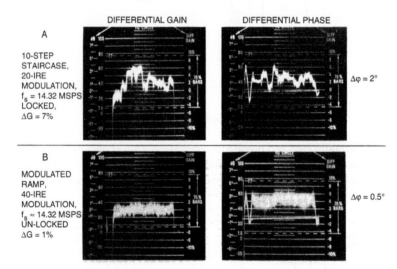

Figure 5.86: Mid-1970s Vectorscope Differential Gain and Phase Measurements for 8-Bit ADC

Several things can be done to help the differential gain and phase measurement problem as recommended in IEEE Standard 746-1984 (Reference 28). Selecting 40-IRE unit modulation immediately divides the theoretical worst-case differential gain and phase numbers by a factor of two, since the quantization error is now a smaller percentage of the modulation. Using the 40-IRE unit modulated ramp test signal rather than the modulated staircase is preferred, because the continuously changing level of a ramp helps to integrate the discontinuities of the quantization levels. Finally, unlocking the sampling frequency from the color subcarrier helps to randomize the quantization levels. Figure 5.86B shows the results of using the 40-IRE unit modulated ramp with an unlocked sampling clock, where the "true" differential gain and phase measurements are now approximately 1% and 0.5°, respectively.

It has also been suggested in an ITU recommendation (Reference 44) that a 30 mV p-p "dither" sinewave at a frequency of 5.162 MHz for NTSC and 6.145 MHz for PAL be summed with the test signal input to the ADC. These frequencies are outside the normal cutoff frequencies of the systems (4.2 MHz for NTSC and 5.0 MHz for PAL). The dither frequencies fall at the second-null points of the respective low-pass filters. Most video signal measuring instruments incorporate various types of filters for easy and accurate measurements. These are usually efficient in removing the dither signals, but if they are not sufficient, external filters must be added.

An all-digital method for differential gain and phase ADC testing is recommended in IEEE Standard 1241-2000 (Reference 37). The test setup is shown in Figure 5.87, and a recommended test signal is shown in Figure 5.88.

The sinewave generator output and the staircase generator output are combined to produce the composite ADC input

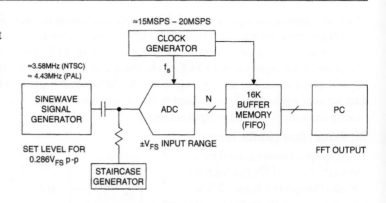

Figure 5.87: Digital Differential Gain and Phase Measurements

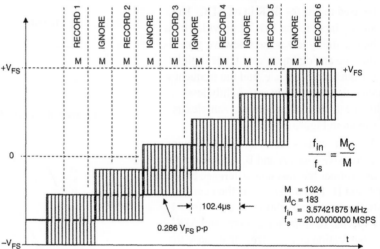

Figure 5.88: Test Signal for Digital Differential Gain and Phase Test

waveform. The sampling frequency, f_s, input frequency, f_{in}, record length, M, and number of cycles within the record, M_C, are all chosen to satisfy the condition for coherent sampling:

$$\frac{f_{in}}{f_s} = \frac{M_C}{M}$$

Eq. 5.42

Notice that 11 records of M = 1024 are taken during the duration of the test waveform, for a total buffer memory requirement of 11,264. Only the records corresponding to the settled waveforms are used in computing the six individual FFTs—the others are ignored, since the data during these intervals is not valid because they occur during the changes of the staircase and the associated settling time. Choosing f_{in}, f_s, M, and M_C is somewhat arbitrary, but the numbers shown in Figure 5.88 approximate video conditions. The starting point is to fix the sampling frequency f_s = 20 MSPS, and the record length M = 1024 (recall M must be a power of 2). M_C should be a prime number, and letting M_C = 183, the resulting input frequency is 3.57421875 MHz, which is close to the NTSC color subcarrier frequency of 3.579545 MHz. The width of each step of the staircase should correspond to 2048 samples taken at 20 MSPS, or 102.4 µs.

The buffer memory is loaded with 11,264 samples relative to the waveform as shown. A total of 11 FFT records are taken, each record containing 1024 points, and the five records corresponding to the waveform transition and settling time intervals are ignored. Only six FFTs are actually computed, and the amplitude and phase of the signals in each FFT can be compared to calculate the differential gain and phase error. Ideally, the phase of each of the six amplitudes should be equal because there is exactly one record of unused data between each of the records used in calculating the FFTs, i.e., the start of each record corresponds to exactly the same point on the input sinewave.

Bit Error Rate (BER) Tests

The concept of errors caused by metastability in ADCs has been discussed at length in Chapter 2. This section concentrates on the test aspects of the resulting bit error rate (BER). The test system shown in Figure 5.89 can be used to test for BER in an ADC. The analog input to the ADC is provided by a high stability low noise sinewave generator. The analog input level is set slightly greater than full-scale, and the frequency such that there is always slightly less than 1-LSB change between samples as shown in Figure 5.90.

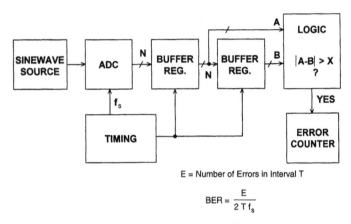

$$E = \text{Number of Errors in Interval } T$$

$$BER = \frac{E}{2\,T\,f_s}$$

Figure 5.89: ADC Bit Error Rate Test Setup

The test set uses series latches to acquire successive codes A and B. A logic circuit determines the absolute difference between A and B. This difference is then compared to the error limit, chosen to allow for expected random noise spikes and ADC quantization errors. Errors that cause the difference to be larger than the limit will increment the counters. The number of errors, E, are counted over a period of time, T. The error rate is then calculated as BER = E/2Tf$_s$. The factor of 2 in the denominator is required because the hardware records a second error when the output returns to the correct code after making the initial error. The error counter is therefore incremented twice for each error.

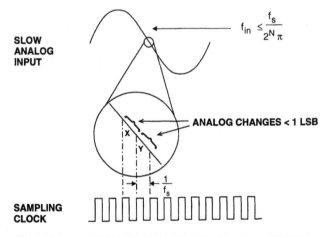

Figure 2.90: ADC Analog Signal for Low Frequency BER Test

It should be noted that the same function can be accomplished in software if the ADC outputs are stored in a memory and analyzed by a computer program.

The input frequency must be carefully chosen such that there is at least one sample taken per code. Assume a full-scale input sinewave having an amplitude of $2^N/2$:

$$v(t) = \frac{2^N}{2} \sin 2\pi ft \qquad\qquad \text{Eq. 5.43}$$

The maximum rate of change of this signal is

$$\left.\frac{dv}{dt}\right]_{max} \leq 2^N \pi f \qquad \text{Eq. 5.44}$$

Letting dv = 1 LSB, dt = 1/f$_s$, and solving for the input frequency:

$$f_{in} \leq \frac{f_s}{2^N \pi} \qquad \text{Eq. 5.45}$$

Choosing an input frequency less than this value will ensure that there is at least one sample per code.

The same test can be conducted at high frequencies by applying an input frequency slightly offset from f$_s$/2 as shown in Figure 5.91. This causes the ADC to slew full-scale between conversions. Every other conversion is compared, and the "beat" frequency is chosen such that there is slightly less than 1 LSB change between alternate samples. The equation for calculating the proper frequency for the high frequency BER test is derived as follows.

Assume an input full-scale sinewave of amplitude $2^N/2$ whose frequency is slightly less than f$_s$/2 by a frequency equal to Δf.

$$v(t)\frac{2^N}{2}\sin\left[2\pi\left(\frac{f_s}{2}-\Delta f\right)t\right] \qquad \text{Eq. 5.46}$$

The maximum rate of change of this signal is

$$\left.\frac{dv}{dt}\right]_{max} \leq 2^N \pi \left(\frac{f_s}{2}-\Delta f\right) \qquad \text{Eq. 5.47}$$

Letting *dv* = 1 LSB and *dt* = 2/f$_s$, and solving for the input frequency Δf:

$$\Delta f \leq \frac{f_s}{2}\left(1-\frac{1}{2\times 2^N \pi}\right) \qquad \text{Eq. 5.48}$$

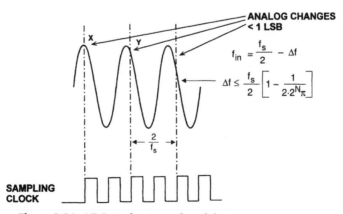

Figure 2.91: ADC Analog Input for High Frequency BER Test

Establishing the BER of a well-behaved ADC is a difficult, time-consuming task—a single unit can sometimes be tested for days without an error. For example, tests on a typical 8-bit flash converter operating at a sampling rate of 75 MSPS yield a BER of approximately 3.7×10^{-12} (1 error per hour) with an error limit of 4 LSBs. Measuring low BER therefore requires long measurement times which increase the probability of power supply transients, noise, etc., causing a false error. Meaningful tests for long periods of time require special attention to EMI/RFI effects (possibly requiring a shielded screen room), isolated power supplies, etc.

Ideally, the BER test requires that each contiguous sample of the ADC output be analyzed, thereby making it difficult to implement using a buffer memory and a PC as in a typical FFT-based ADC test setup. Errors are missed which occur during the time the buffer memory output is being read by the PC. However, this method can be used provided the memory size is large (preferably 256 K), and a large number of data records are taken as shown in Figure 5.92. The "dead time" used by the PC to read the data from the memory and analyze it will add to the total test time, especially if the BER is very low.

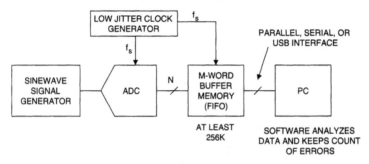

Figure 5.92: Alternate Test Setup for PC-Based BER Test

References:

5.2 Testing ADCs

The following references have been placed in their approximate chronological order so the reader can gain a good historical perspective on the evolution of specifications and testing methods for ADCs. Most of the references are referred to in the text, however some are included simply as additional background material.

1. Bernard M. Gordon, "Definition of Accuracy of Voltage to Digital Converters," **Instruments and Control Systems**, May 1959, p.710.

2. W. M. Gaines, "Terminology for Functional Characteristics of Analog to Digital Converters," **Control Engineering**, February 1961.

3. Bernard M. Gordon, "Designing Sampled Data Systems," **Control Engineering**, April 1961, pp. 127–132.

4. Bernard M. Gordon, "How to Specify Analog-to-Digital Converters," **Electronic Design**, May 10 1961, pp. 36–39.

5. Bernard M. Gordon, "How Much Do Components Limit Converter Performance?" **Electronic Design**, June 21, 1961, pp.52–53.

6. P. Barr, "Influence of Aperture Time and Conversion Rate on the Accuracy of A/D Converters," **Data Systems Engineering**, May 1964, pp. 30–34.

7. A. Van Doren, "Solving Error Problems in Digital Conversion Systems," **Electromechanical Design**, April 1966, pp.44–46.

8. J. Freeman, "Specifying Analog to Digital Converters," **The Electronic Engineer**, June 1968, pp. 44–48.

9. Bernard M. Gordon, "Speaking Out on Analog to Digital Converters," **EEE Magazine**, December 1968.

10. Bernard M. Gordon, "Bernard Gordon of Analogic Speaks Out on What's Wrong with A/D Converter Specs," **EEE Magazine**, February 1969, pp. 54–61.

11. Dan Sheingold, **Analog-Digital Conversion Handbook**, Analog Devices, 1972.

12. Donald B. Bruck, **Data Conversion Handbook**, Hybrid Systems Corporation, 1974.

13. Bill Pratt, "Test A/D Converters Digitally," **Electronic Design**, December 6, 1975.

14. Stuart K. Tewksbury, F.C. Meyer, D.C. Rollenhagen, H.K. Schownwetter, and Thomas Souders, "Terminology Related to the Performance of S/H, A/D, and D/A Circuits," **IEEE Transactions on Circuits and Systems**, Vol. CAS-25, July 1978, pp. 419–426.

15. Eugene L. Zuch, **Data Acquisition and Conversion Handbook**, Datel-Intersil, Inc., 1979, ISBN 0-9602946-0-0.

16. Jim R. Naylor, "Testing Digital/Analog and Analog/Digital Converters," **IEEE Transactions on Circuits and Systems**, Vol. CAS-25, July 1978, pp. 526–538.

17. Walter A. Kester, "Characterizing and Testing A/D and D/A Converters for Color Video Applications," **IEEE Transactions on Circuits and Systems**, Vol. CAS-25, July 1978, pp. 539–550.

18. W. A. Kester, "PCM Signal Codecs for Video Applications," **SMPTE Journal**, Number 88, November 1979, pp. 770–778.

19. Tim Wilhelm, "Test A/D Converters Quickly and Efficiently, **Electronic Design**, October 15, 1981, pp. 193–198.

20. Bernard M. Gordon, **The Analogic Data-Conversion Systems Digest, Fourth Edition**, Analogic Corporation, 1981.

21. Martin Neil and Art Muto, "Tests Unearth A/D Converter's Real-World Performance," **Electronics**, February 24, 1982, pp. 127–132.

22. Bruce E. Peetz, Arthur S. Muto, and Martin Neil, "Measuring Waveform Recorder Performance," **Hewlett-Packard Journal**, Vol. 33, No. 11, November 1982, pp. 21–29.

23. Walter Kester, "Test Video A/D Converters Under Dynamic Conditions," **EDN**, August 18, 1982.

24. Joey Doernberg, Hae-Seung Lee, and David A. Hodges, "Full-Speed Testing of A/D Converters," **IEEE Journal of Solid State Circuits**, Vol. SC-19, No. 6, December 1984, pp. 820–827.

25. T. Michael Souders, Donald R. Flach, and Thick C. Wong, "An Automatic Test Set for the Dynamic Characterization of A/D Converters," **IEEE Transactions on Instrumentation and Measurement**, Vol. IM-32, No. 1, March 1983, pp. 180–186.

26. Dan Sheingold, **Analog-Digital Conversion Handbook, 3rd Edition**, Analog Devices and Prentice-Hall, 1986, ISBN-0-13-032848-0. *(The defining and classic book on data conversion.)*

27. Matthew Mahoney, **DSP-Based Testing of Analog and Mixed-Signal Circuits**, IEEE Computer Society Press, 1987, ISBN 0-8186-0785-8.

28. **IEEE Std. 746-1984, IEEE Standard for Performance Measurements of A/D and D/A Converters for PCM Television Video Circuits**, IEEE, 1984.

29. Walt Kester, "Designer's Guide to Flash ADC Testing Part 1: Flash ADCs Provide the Basis for High-Speed Conversion," **EDN**, January 4, 1990, pp. 101–110.

30. Walt Kester, "Designer's Guide to Flash ADC Testing Part 2: DSP Test Techniques Keep Flash ADCs in Check," **EDN**, January 18, 1990, pp. 133–142.

31. Walt Kester, "Designer's Guide to Flash ADC Testing Part 3: Measure Flash ADC Performance for Trouble-Free Operation," **EDN**, February 1, 1990, pp. 103–114.

32. Michael J. Demler, **High-Speed Analog-to-Digital Conversion**, Academic Press, Inc., 1991, Chapter 6.

33. Walt Kester, "Designer's Guide to Sampling A/D Converters Part 1: Basic Characteristics Distinguish Sampling A/D Converters," **EDN**, September 3, 1992, pp. 135–144.

34. Walt Kester, "Designer's Guide to Sampling A/D Converters Part 2: Peripheral Circuits can Make or Break Sampling-ADC Systems," **EDN**, October 1, 1992, pp. 97–105.

35. Walt Kester, "Designer's Guide to Sampling A/D Converters Part 3: Layout, Grounding, and Filtering Complete Sampling ADC System," **EDN**, October 15, 1992, pp. 127–134.

36. **IEEE Std. 1057-1994 (R2001), IEEE Standard for Digitizing Waveform Recorders**, IEEE, 1994, ISBN 1-55937-488-8.

37. **IEEE Std. 1241-2000, IEEE Standard for Terminology and Test Methods for Analog-to-Digital Converters**, IEEE, 2001, ISBN 0-7381-2724-8.

38. Jerome Blair, "Histogram Measurement of ADC Nonlinearities Using Sine Waves," IEEE Transactions on Instrumentation and Measurement, Vol. 43, No. 3, June 1994, pp. 373–383.

39. Walt Kester, **Mixed-Signal and DSP Design Techniques**, Analog Devices and Newnes (An Imprint of Elsevier Science), ISBN 0-75067-611-6, 2003.

40. Steven W. Smith, **Digital Signal Processing: A Practical Guide for Engineers and Scientists**, Newnes (An Imprint of Elsevier Science), 2002, ISBN 0-75067-444-X.

41. R. Jacob Baker, Harry W. Li, and David E. Boyce, **CMOS Circuit Design, Layout, and Simulation**, Wiley-IEEE Press, 1997, ISBN 0-78033-416-7, Chapter 28, 29.

42. R. Jacob Baker, **CMOS Mixed-Signal Circuit Design**, John Wiley and Sons, 2002, ISBN 0-4712-2754-4.

43. Micheal O. Felix, "Differential Phase and Gain Measurements in Digital Video Signals," **SMPTE Journal**, Vol. 85, February 1976, pp. 76–79.

44. "Measuring Methods for Digital Video Equipment with Analogue Input/Output," **ITU Recommendation ITU-R BT.1204**, 1995, www.itu.int/publications/bookshop.

Additional References on FFTs and DSP

45. C. Britton Rorabaugh, **DSP Primer**, McGraw-Hill, 1999.

46. Richard J. Higgins, **Digital Signal Processing in VLSI**, Prentice-Hall, 1990.

47. V. Oppenheim and R. W. Schafer, **Digital Signal Processing**, Prentice-Hall, 1975.

48. L. R. Rabiner and B. Gold, **Theory and Application of Digital Signal Processing**, Prentice-Hall, 1975.

49. John G. Proakis and Dimitris G. Manolakis, **Introduction to Digital Signal Processing**, MacMillian, 1988.

50. Fredrick J. Harris, "On the Use of Windows for Harmonic Analysis with the Discrete Fourier Transform," **Proc. IEEE**, Vol. 66, No. 1, 1978 pp. 51–83.

51. R. W. Ramirez, **The FFT: Fundamentals and Concepts**, Prentice-Hall, 1985.

52. J. W. Cooley and J. W. Tukey, "An Algorithm for the Machine Computation of Complex Fourier Series," **Mathematics Computation**, Vol. 19, April 1965, pp. 297–301.

FFT and Engineering Data Analysis Software Packages

53. MATLAB®, The MathWorks, Inc., 3 Apple Hill Drive, Natick, MA 01760-2098, Phone: 508-647-7000, www.mathworks.com.

54. LabVIEW™, National Instruments Corporation, 11500 N. Mopac Expressway, Austin, Texas 78759-3504, Phone: 512-683-0100, www.ni.com.

55. Mathcad, Mathsoft Engineering & Education, Inc., 101 Main Street, Cambridge, MA 02142-1521, Phone: 617-444-8000, www.mathsoft.com.

General References on Data Conversion and Related Topics

1. Alfred K. Susskind, **Notes on Analog-Digital Conversion Techniques**, John Wiley, 1957.

2. David F. Hoeschele, Jr., **Analog-to-Digital/Digital-to-Analog Conversion Techniques**, John Wiley and Sons, 1968.

3. K. W. Cattermole, **Principles of Pulse Code Modulation**, American Elsevier Publishing Company, Inc., 1969, New York NY, ISBN 444-19747-8.

4. Hermann Schmid, **Electronic Analog/Digital Conversions**, Van Nostrand Reinhold Co., 1970.

5. Dan Sheingold, **Analog-Digital Conversion Handbook**, First Edition, Analog Devices, 1972.

6. Donald B. Bruck, **Data Conversion Handbook**, Hybrid Systems Corporation, 1974.

7. Eugene R. Hnatek, **A User's Handbook of D/A and A/D Converters**, John Wiley, New York, 1976, ISBN 0-471-40109-9.

8. Nuggehally S. Jayant, **Waveform Quantizing and Coding**, John Wiley-IEEE Press, 1976, ISBN 0-87942-074-X.

9. Dan Sheingold, **Analog-Digital Conversion Notes**, Analog Devices, 1977.

10. C. F. Kurth, editor, **IEEE Transactions on Circuits and Systems Special Issue on Analog/Digital Conversion**, CAS-25, No. 7, July 1978.

11. Daniel J. Dooley, **Data Conversion Integrated Circuits**, John Wiley-IEEE Press, 1980, ISBN 0-471-08155-8.

12. Bernard M. Gordon, **The Analogic Data-Conversion Systems Digest**, Fourth Edition, Analogic Corporation, 1981.

13. Eugene L. Zuch, **Data Acquisition and Conversion Handbook**, Datel-Intersil, 1982.

14. Frank F. E. Owen, **PCM and Digital Transmission Systems**, McGraw-Hill, 1982, ISBN 0-07-047954-2.

15. Dan Sheingold, **Analog-Digital Conversion Handbook**, Analog Devices/Prentice-Hall, 1986, ISBN 0-13-032848-0.

16. Matthew Mahoney, **DSP-Based Testing of Analog and Mixed-Signal Circuits**, IEEE Computer Society Press, 1987, ISBN 0-8186-0785-8.

17. Michael J. Demler, **High-Speed Analog-to-Digital Conversion**, Academic Press, Inc., 1991, ISBN 0-12-209048-9.

18. J. C. Candy and Gabor C. Temes, **Oversampling Delta-Sigma Data Converters**, IEEE Press, ISBN 0-87942-258-8, 1992.

19. David F. Hoeschele, Jr., **Analog-to-Digital and Digital-to-Analog Conversion Techniques**, Second Edition, John Wiley and Sons, 1994, ISBN-0-471-57147-4.

20. Rudy van de Plassche, **Integrated Analog-to-Digital and Digital-to-Analog Converters**, Kluwer Academic Publishers, 1994, ISBN 0-7923-9436-4.

21. David A. Johns and Ken Martin, **Analog Integrated Circuit Design**, John Wiley, 1997, ISBN 0-471-14448-7.

22. Mikael Gustavsson, J. Jacob Wikner, and Nianxiong Nick Tan, **CMOS Data Converters for Communications**, Kluwer Academic Publishers, 2000, ISBN 0-7923-7780-X.

23. R. Jacob Baker, **CMOS Circuit Design Volumes I and II**, John Wiley-IEEE Computer Society, 2002, ISBN 0-4712-7256-6.

24. Rudy van de Plassche, **CMOS Integrated Analog-to-Digital and Digital-to-Analog Converters, Second Edition**, Kluwer Academic Publishers, 2003, ISBN 1-4020-7500-6.

Analog Devices' Seminar Series:

25. Walt Kester, **Practical Analog Design Techniques**, Analog Devices, 1995, ISBN 0-916550-16-8, available for download at www.analog.com.

26. Walt Kester, **High Speed Design Techniques**, Analog Devices, 1996, ISBN 0-916550-17-6, available for download at www.analog.com.

27. Walt Kester, **Practical Design Techniques for Power and Thermal Management**, Analog Devices, 1998, ISBN 0-916550-19-2, available for download at www.analog.com.

28. Walt Kester, **Practical Design Techniques for Sensor Signal Conditioning**, Analog Devices, 1999, ISBN 0-916550-20-6, available for download at www.analog.com.

29. Walt Kester, **Mixed-Signal and DSP Design Techniques**, Analog Devices, 2000, ISBN 0-916550-22-2, available for download at www.analog.com.

30. Walt Kester, **Mixed-Signal and DSP Design Techniques**, Analog Devices and Newnes (An Imprint of Elsevier Science), ISBN 0-75067-611-6, 2003.

31. Walter G. Jung, **Op Amp Applications Handbook**, Newnes (an imprint of Elsevier Science and Technology Books), ISBN 0-7506-7844-5, 2005.

CHAPTER 6

Interfacing to Data Converters

Interfacing to Data Converters

Driving ADC Analog Inputs
Walt Kester

Introduction

Before considering the detailed issues involved in driving ADCs, some general comments about trends in modern data converters are in order. Data converter performance is first and foremost, and maintaining that performance in a system application is extremely important. In low frequency measurement applications (10 Hz bandwidth signals or lower), Σ-Δ ADCs with resolutions up to 24 bits are now quite common. These converters generally have automatic or factory calibration features to maintain required gain and offset accuracy. In higher frequency signal processing, ADCs must have wide dynamic range (low distortion and noise), high sampling frequencies, and generally excellent ac specifications.

In addition to sheer performance, other characteristics such as low power, single supply operation, low cost, and small surface-mount packages also drive the data conversion market. These requirements result in a myriad of application problems because of reduced signal swings, increased sensitivity to noise, etc. As has been mentioned previously in Chapter 3, the analog input to a CMOS ADC is usually connected directly to a switched-capacitor sample-and-hold (SHA), which generates transient currents that must be buffered from the signal source. This can present quite a challenge when selecting a drive amplifier. On the other hand, high performance data converters fabricated on BiCMOS or complementary bipolar processes are more likely to have internal buffering, but generally have higher cost and power consumption than their CMOS counterparts. The general trends in data converters are summarized in Figure 6.1.

- Higher sampling rates, higher resolution, excellent AC performance
- Single supply operation (e.g., 5V, 3V, 2.5V, 1.8V)
- Lower power, shutdown or sleep modes
- Smaller input/output signal swings
- Differential inputs/outputs
- Maximize usage of low cost foundry CMOS processes
- Small surface-mount packages

Figure 6.1: Some General Trends in Data Converters

It should be clear by now that selecting an appropriate drive circuit for a data converter application is highly dependent on the particular converter under consideration. Generalizations are difficult, but some meaningful guidelines can be followed.

To begin with, one shouldn't necessarily assume that a driver amplifier is always required. Some converters have relatively benign inputs and are designed to interface directly to the signal source. In many applications, transformer drive may be preferable. Because there is practically no industry standardization regarding ADC input structures, each ADC must be carefully examined on its own merits before designing the input interface circuitry.

If an amplifier is required, a fundamental requirement is that it not degrade the dc or ac performance of the converter. One might assume that a careful reading of the op amp datasheets would assist in the selection process—simply lay the data converter and the op amp datasheets side by side, and compare each critical performance specification. It is true that this method will provide some degree of success; however, in order to perform an accurate comparison, the op amp must be specified under the exact operating conditions required by the data converter application. Such factors as gain, gain setting resistor values, source impedance, output load, input and output signal amplitude, input and output common-mode (CM) level, power supply voltage, etc., all affect op amp performance to some degree.

It is highly unlikely that even a well-written op amp datasheet will provide an exact match to the operating conditions required in the data converter application. Extrapolation of specified performance to fit the exact operating conditions can give erroneous results. Also, the op amp may be subjected to transient currents from the data converter, and the corresponding effects on op amp performance are rarely found on datasheets.

Converter datasheets themselves can be a good source for recommended op amps and other application circuits. However, this information can become obsolete when newer op amps are introduced after the converter's initial release.

Analog Devices offers a parametric search engine that facilitates part selection (see www.analog.com). For instance, the first search might be for minimum power supply voltage, e.g., 3 V. The next search might be for bandwidth, and further searches on relevant specifications will narrow the selection of op amps even further. While not necessarily suitable for the final selection, this process can narrow the search to a manageable number of amplifiers whose individual datasheets can be retrieved, then reviewed in detail before final selection. Figure 6.2 summarizes the overall selection process.

- Some ADCs (DACs) do not require special input drivers (output buffers)
- The amplifier / transformer should not degrade the performance of the ADC (DAC)
- AC specifications are usually the most important
  - Noise
  - Bandwidth
  - Distortion
  - Settling time from transient currents
- Selection based on op amp data sheet specifications difficult due to varying conditions in actual application circuit with ADC (DAC):
  - Power supply voltages
  - Signal range (differential and common-mode)
  - Loading (static and dynamic)
  - Gain and gain-setting resistor values
- Parametric search engines may be useful
- ADC (DAC) data sheets often recommend op amps but may not include newly released products

---

Figure 6.2: ADC Driver (DAC Buffer) Selection Criteria

Amplifer DC and AC Performance Considerations

As discussed above, the amplifier (if required) should not degrade the performance specifications of the data converter. Today, ac specifications are generally paramount—especially with high speed data converters. Chapter 2 of this book has discussed ADC and DAC specifications in detail, but it is useful to summarize the popular converter dynamic performance specifications in Figure 6.3.

For comparison, the fundamental op amp dc and ac specifications are summarized in Figure 6.4. Although not all op amps will have these specifications, most of them should certainly be listed on the data sheet if the op amp is to be a serious contender for a high performance data converter application.

- Signal-to-Noise-and-Distortion Ratio (SINAD, or S/N +D)
- Effective Number of Bits (ENOB)
- Signal-to-Noise Ratio (SNR)
- Analog Bandwidth (Full-Power, Small-Signal)
- Harmonic Distortion
- Worst Harmonic
- Total Harmonic Distortion (THD)
- Total Harmonic Distortion Plus Noise (THD + N)
- Spurious Free Dynamic Range (SFDR)
- Two-Tone Intermodulation Distortion
- Multi-tone Intermodulation Distortion

**Figure 6.3: Popular Converter
Dynamic Performance Specifications**

- DC
    - Input/Output Signal Range
    - Offset, offset drift
    - Input bias current
    - Open loop gain
    - Integral linearity
    - 1/f noise (voltage and current)
- AC (Highly application dependent!)
    - Wideband noise (voltage and current)
    - Small and Large Signal Bandwidth
    - Harmonic Distortion
    - Total Harmonic Distortion (THD)
    - Total Harmonic Distortion + Noise (THD + N)
    - Spurious Free Dynamic Range (SFDR)
    - Third Order Intermodulation Distortion

**Figure 6.4: Key DC and AC Op Amp
Specifications for ADC/DAC Applications**

Regardless of the importance of the ac specifications, the fundamental dc specifications must not be over-looked—especially in light of the implications of low voltage single-supply operation so popular today. The allowable input and output signal range becomes critically important in single supply applications as illustrated in the fundamental application circuit shown in Figure 6.5.

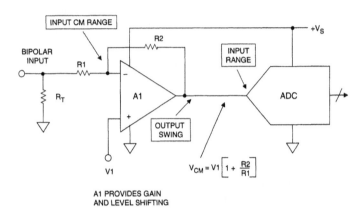

Figure 6.5: Input/Output Signal Swing and Common-Mode Range is Critical in Single-Supply ADC Driver Applications

The circuit of Figure 6.5 shows an op amp as a simple dc-coupled single-supply ADC driver which provides the proper gain and level shifting for the bipolar (ground-referenced) input signal such that it matches the input range of the ADC. Several important points are illustrated in this popular circuit. The first consideration is the input range of the ADC, which in turn determines the output voltage swing requirement of the op amp. There are a number of single-supply CMOS ADCs with inputs that go from 0 V to the positive supply voltage. As will be illustrated shortly, even rail-to-rail output op amps cannot drive the signal completely to each rail. If, however, the ADC input range can be set so that the signal only goes to within a few hundred millivolts of each rail, then a single-supply "almost" rail-to-rail output op amp can often be used.

On the other hand, ADCs fabricated on BiCMOS or complementary bipolar processes typically have fixed input ranges that are usually at least several hundred millivolts from either rail, although many are not centered at the midsupply voltage of $V_S/2$.

Equally important is the input common-mode voltage of the op amp. In the circuit of Figure 6.5, the input common-mode voltage is set by V1, which level shifts the amplifier output to the correct value. Obviously, V1 must lie within the input common-mode voltage range of the op amp in order for the circuit to work properly.

These restrictions can become quite severe when operating the entire circuit on a single low-voltage supply, and therefore a brief discussion of rail-to-rail op amps follows in order to better understand how to properly select the drive amplifier. We will discuss the input and output stage considerations separately.

Rail-Rail Input Stages

Today, there is common demand for op amps with input common-mode voltage that includes *both* supply rails—i.e., *rail-to-rail* common-mode operation. While such a feature is undoubtedly useful in some applications, engineers should recognize that there are still relatively few applications where it is absolutely essential. These applications should be distinguished from the many more applications where a common-mode input range *close* to the supplies, or one that includes *one* supply is necessary, but true input rail-to-rail operation is not.

In many single-supply applications, it is required that the input common-mode voltage range extend to one of the supply rails (usually ground). High side or low side current-sensing applications are typical examples of this. Many amplifiers can handle 0 V common-mode inputs, and they are easily designed using PNP differential pairs (or N-channel JFET pairs or PMOS pairs) as shown in Figure 6.6. The input common-mode range of such an op amp generally extends from about 200 mV below the negative rail ($-V_S$, or ground), to within 1 V to 2 V of the positive rail ($+V_S$).

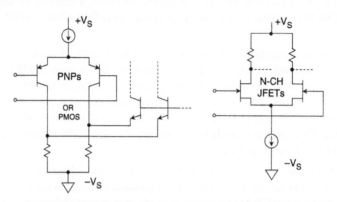

Figure 6.6: PNP, PMOS, or N-Channel JFET Stages Allow Common-Mode Inputs to Include the Negative Rail

An input stage could also be designed with NPN transistors (or P-channel JFET pairs or NMOS pairs), in which case the input common-mode range would include the positive rail, and go to within about 1 V to 2 V of the negative rail. This requirement typically occurs in applications such as high-side current sensing.

A simplified diagram of what has become known as a true rail-to-rail input stage is shown in Figure 6.7. Note that this requires use of *two* long-tailed pairs, one of PNP bipolar transistors Q1-Q2, the other of NPN transistors Q3-Q4. Similar input stages can also be made with CMOS or JFET pairs.

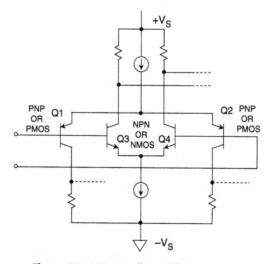

Figure 6.7: A True Rail-to-Rail Input Stage

It should be noted that these two pairs will exhibit *different* offsets and bias currents, so when the applied common-mode voltage changes, the amplifier input offset voltage and input bias current does also. In fact, when both current sources remain active throughout most of the entire input common-mode range, amplifier input offset voltage is the *average* offset voltage of the two pairs. In those designs where the current sources are alternatively switched off at some point along the input common-mode voltage, amplifier input offset voltage is dominated by the PNP pair offset voltage for signals near the negative supply, and by the NPN pair offset voltage for signals near the positive supply. As noted, a true rail-to-rail input stage can also be constructed from CMOS transistors, for example as in the case of the CMOS AD8531/AD8532/AD8534 op amp family.

Amplifier input bias current, a function of transistor current gain, is also a function of the applied input common-mode voltage. The result is relatively poor common-mode rejection (CMR), and a changing common-mode input impedance over the common-mode input voltage range, compared to familiar dual-supply devices. These specifications should be carefully considered when choosing a rail-to-rail input op amp, especially for a noninverting configuration. Input offset voltage, input bias current, and even CMR may be quite good over *part* of the common-mode range, but much worse in the region where operation shifts between the NPN and PNP devices, and vice versa.

True rail-to-rail amplifier input stage designs must transition from one differential pair to the other differential pair, somewhere along the input common-mode voltage range. Some devices like the OP191/ OP291/OP491 family and the OP279 have a common-mode crossover threshold at approximately 1 V below the positive supply (where signals do not often occur). The PNP differential input stage is active from about 200 mV below the negative supply to within about 1 V of the positive supply. Over this common-mode range, amplifier input offset voltage, input bias current, CMR, input noise voltage/current are primarily determined by the characteristics of the PNP differential pair. At the crossover threshold, however, amplifier input offset voltage becomes the average offset voltage of the NPN/PNP pairs and can change rapidly.

Also, as previously noted, amplifier bias currents are dominated by the PNP differential pair over most of the input common-mode range, and change polarity and magnitude at the crossover threshold when the NPN differential pair becomes active.

Op amps like the OP184/OP284/OP484 family, utilize a rail-to-rail input stage design where both NPN and PNP transistor pairs are active throughout most of the entire input common-mode voltage range. With this approach to biasing, there is no common-mode crossover threshold. Amplifier input offset voltage is the average offset voltage of the NPN and the PNP stages, and offset voltage exhibits a smooth transition throughout the entire input common-mode range, due to careful laser trimming of input stage resistors.

In the same manner, through careful input stage current balancing and input transistor design, the OP184 family input bias currents also exhibit a smooth transition throughout the entire common-mode input voltage range. The exception occurs at the very extremes of the input range, where amplifier offset voltages and bias currents increase sharply, due to the slight forward-biasing of parasitic p-n junctions. This occurs for input voltages within approximately 1 V of either supply rail.

When *both* differential pairs are active throughout most of the entire input common-mode range, amplifier transient response is faster through the middle of the common-mode range by as much as a factor of 2 for bipolar input stages and by a factor of $\sqrt{2}$ for JFET input stages. This is due to the higher transconductance of two operating input stages.

The AD8027/AD8028 op amp family (Reference 1) has a pin-selectable crossover threshold that allows the user to choose the crossover point between the PNP/NPN input differential pairs. Depending upon the state of the *select* pin, the threshold can be set for 1.2 V from the positive rail (*select* pin open) or 1.2 V from the negative rail (*select* pin connected to positive supply voltage).

Input stage g_m determines the slew rate and the unity-gain crossover frequency of the amplifier, hence response time degrades slightly at the extremes of the input common-mode range when either the PNP stage (signals approaching the positive supply rail) or the NPN stage (signals approaching the negative supply rail) are forced into cutoff. The thresholds at which the transconductance changes occur are approximately within 1 V of either supply rail, and the behavior is similar to that of the input bias currents.

In light of the many quirks of true rail-to-rail op amp input stages, applications that do require true rail-to-rail inputs should be carefully evaluated, and an amplifier chosen to ensure that its input offset voltage, input bias current, common-mode rejection, and noise (voltage and current) are suitable.

Output Stages

The earliest IC op amp output stages were NPN emitter followers with NPN current sources or resistive pull-downs, as shown in Figure 6.8A and B. Naturally, the slew rates were greater for positive-going than they were for negative-going signals.

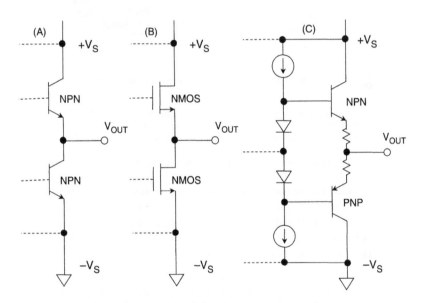

Figure 6.8: Some traditional Op Amp Output Stages

While all modern op amps have push-pull output stages of some sort, many are still asymmetrical, and have a greater slew rate in one direction than the other. Asymmetry tends to introduce distortion on ac signals and generally results from the use of IC processes with faster NPN than PNP transistors. It may also result in an ability of the output to approach one supply more closely than the other in terms of saturation voltage.

In many applications, the output is required to swing to only one rail, usually the negative rail (i.e., ground in single-supply systems). A pull-down resistor to the negative rail will allow the output to approach that rail (provided the load impedance is high enough, or is also grounded to that rail), but only slowly. Using an FET current source instead of a resistor can speed things up, but this adds complexity, as shown in Figure 6.8B.

With modern complementary bipolar (CB) processes, well-matched high speed PNP and NPN transistors are readily available. The complementary emitter follower output stage shown in Figure 6.8C has many advantages, but the most outstanding one is the low output impedance. However, the output voltage of this stage can only swing within about one V_{BE} drop of either rail. Therefore, a usable output voltage range of 1 V to 4 V is typical of such a stage, when operated on a single 5 V supply.

The complementary common-emitter/common-source output stages shown in Figure 6.9A and B allow the op amp output voltage to swing much closer to the rails, but these stages have much higher open-loop output impedance than do the emitter-follower-based stages of Figure 6.8C

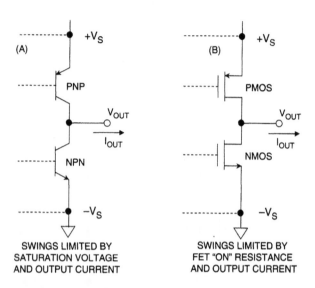

SWINGS LIMITED BY
SATURATION VOLTAGE
AND OUTPUT CURRENT

SWINGS LIMITED BY
FET "ON" RESISTANCE
AND OUTPUT CURRENT

Figure 6.9: "Almost" Rail-To-Rail Output Stages

In practice, however, the amplifier's high open-loop gain and the applied feedback can still produce an application with low output impedance (particularly at frequencies below 10 Hz). What should be carefully evaluated with this type of output stage is the loop gain within the application, with the load in place. Typically, the op amp will be specified for a minimum gain with a load resistance of 10 kΩ (or more). Care should be taken that the application loading doesn't drop lower than the rated load, or gain accuracy may be lost.

It should also be noted that these output stages can cause the op amp to be more sensitive to capacitive loading than the emitter-follower type. Again, this will be noted on the device data sheet, which will indicate a maximum of capacitive loading before overshoot or instability will be noted.

The complementary common emitter output stage using BJTs (Figure 6.9A) cannot swing completely to the rails, but only to within the transistor saturation voltage (V_{CESAT}) of the rails. For small amounts of load current (less than 100 µA), the saturation voltage may be as low as 5 mV to 10 mV, but for higher load currents, the saturation voltage can increase to several hundred mV (for example, 500 mV at 50 mA).

On the other hand, an output stage constructed of CMOS FETs (Figure 6.9B) can provide nearly true rail-to-rail performance, but only under no-load conditions. If the op amp output must source or sink substantial current, the output voltage swing will be reduced by the I × R drop across the FET's internal "on" resistance. Typically this resistance will be on the order of 100 Ω for precision amplifiers, but it can be less than 10 Ω for high current drive CMOS amplifiers.

For the above basic reasons, it should be apparent that there is no such thing as a *true* rail-to-rail output stage, hence the caption of Figure 6.9 ("Almost" Rail-to-Rail Output Stages). The best any op amp output stage can do is an "almost" rail-to-rail swing, when it is lightly loaded.

Gain and Level-Shifting Circuits Using Op Amps

In dc-coupled applications, the drive amplifier must provide the required gain and offset voltage, to match the signal to the input voltage range of the ADC. Figure 6.10 summarizes various op amp gain and level-shifting options. The circuit of Figure 6.10A operates in the noninverting mode, and uses a low impedance reference voltage, V_{REF}, to offset the output. Gain and offset interact according to the equation:

$$V_{OUT} = [1 + (R2/R1)] \times V_{IN} - [(R2/R1) \times V_{REF}] \qquad \text{Eq. 6.1}$$

The circuit in Figure 6.10B operates in the inverting mode, and the signal gain is independent of the offset. The disadvantage of this circuit is that the addition of R3 increases the noise gain, and hence the sensitivity to the op amp input offset voltage and noise. The input/output equation is given by:

$$V_{OUT} = -(R2/R1) \times V_{IN} - (R2/R3) \times V_{REF} \qquad \text{Eq. 6.2}$$

The circuit in Figure 6.10C also operates in the inverting mode, and the offset voltage V_{REF} is applied to the noninverting input without noise gain penalty. This circuit is also attractive for single-supply applications ($V_{REF} > 0$). The input/output equation is given by:

$$V_{OUT} = -(R2/R1) \times V_{IN} + [R4/(R3+R4)][\ 1 + (R2/R1)] \times V_{REF} \qquad \text{Eq. 6.3}$$

Note that the circuit of Figure 6.10A is sensitive to the impedance of V_{REF}, unlike the counterparts in B and C. This is due to the fact that the signal current flows into/from V_{REF}, due to V_{IN} operating the op amp over its common-mode range. In the other two circuits the common-mode voltages are fixed, and no signal current flows in V_{REF}.

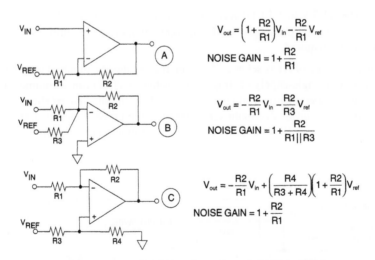

Figure 6.10: Op Amp Gain and Level Shifting Circuits

The circuit of Figure 6.10C is ideally suited to a single-supply level shifter and is identical to the one previously shown in Figure 6.5. It will now be examined further in light of single-supply and common-mode issues. Figure 6.11 shows this type of level shifter driving an ADC with an input range of 1.5 V to 3.5 V. Note that the circuit operates on a single 5 V supply.

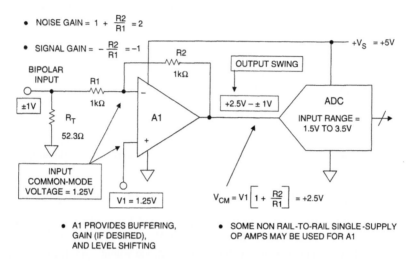

Figure 6.11: Single-Ended Single-Supply DC-Coupled Level Shifter

The input range of the ADC (1.5 V to 3.5 V) determines the output range of the A1 op amp. Since most complementary emitter follower output stages (see Figure 6.8C) will drive to within 1 V of either rail, a rail-to-rail output stage is not required.

The input common-mode voltage of A1 is set at 1.25 V which generates the required output offset of 2.5 V. Note that many nonrail-to-rail single-supply op amps (such as the AD8057) can accommodate this input common-mode voltage when operating on a single 5 V supply. This circuit is an excellent example of where careful analysis of dc voltages is invaluable to the amplifier selection process. However, if we modify the circuit slightly as shown in Figure 6.12, an entirely different set of input/output requirements is placed on the op amp.

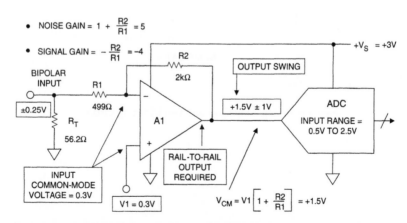

Figure 6.12: Single-Ended Level-Shifter with Gain Requires Rail-to-Rail Op Amp

The input range of the ADC in Figure 6.12 is now 0.5 V to 2.5 V, and the entire circuit must operate on a 3 V power supply. A rail-to-rail output op amp is therefore required for A1 in order to ensure adequate output signal swing. Note that, in addition, the input common-mode voltage of A1 is now 0.3 V in order to set the output common-mode voltage of 1.5 V (noise gain = +5, signal gain = –4). In order to allow an input common-mode voltage of 0.3 V, A1 must have either a PNP or PMOS input stage or a rail-to-rail input stage as previously shown in Figure 6.7.

This simple example serves to illustrate the importance of carefully examining the input/output signal level requirements placed on the op amp by the circuit conditions and the ADC interface. After the amplifier signal level requirements are established, then ac performance should be determined.

Op Amp AC Specifications and Data Converter Requirements

Modern op amps come with what may appear to be a relatively complete set of dc and ac specifications; however, fully specifying an op amp under all possible circuit conditions is almost impossible. For example, Figure 6.13 shows some key specifications taken from the table of specifications on the datasheet for the AD8057/AD8058 high speed, low distortion op amp (Reference 2). Note that the specifications depend on the supply voltage, the signal level, output loading, etc. It should also be emphasized that it is customary to provide only *typical* ac specifications (as opposed to *maximum* and *minimum* values) for most op amps. In addition, we have seen that there are restrictions on the input and output common-mode signal ranges, which are especially important when operating on low voltage dual (or single) supplies.

SPECIFICATION	$V_S = \pm 5V$	$V_S = +5V$
Input Common-Mode Voltage Range	–4.0V to +4.0V	+0.9V to +3.4V
Output Common-Mode Voltage Range	–4.0V to +4.0V	+0.9V to +4.1V
Input Voltage Noise	$7nV/\sqrt{Hz}$	$7nV/\sqrt{Hz}$
Small Signal Bandwidth	325MHz	300MHz
THD @ 5MHz, V_O = 2V p-p, R_L = 1kΩ	– 85dBc	– 75dBc
THD @ 20MHz, V_O = 2V p-p, R_L = 1kΩ	– 62dBc	– 54dBc

Figure 6.13: AD8057/AD8058 Op Amp Key Specifications, G = +1

Most op amp datasheets contain a section that provides supplemental performance data for various other conditions not explicitly specified in the primary specification tables. For instance, Figure 6.14 shows the AD8057/AD8058 distortion as a function of frequency for G = +1 and V_S = ±5 V. Unless it is otherwise specified, the data represented by these curves should be considered typical (it is usually marked as such).

Note however that the data in both Figure 6.14 (and also the following Figure 6.15) is given for a dc load of 150 Ω. This is a load presented to the op amp in the popular application of driving a source and load-terminated 75 Ω cable. Distortion performance is generally better with lighter dc loads, such as 500 Ω to 1000 Ω (more typical of many ADC inputs), and this data may or may not be found on the datasheet.

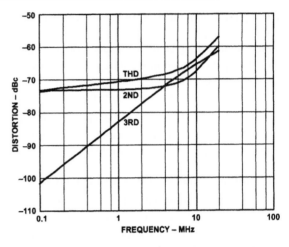

Figure 6.14: AD8057/AD8058 Op Amp Distortion versus Frequency G = +1, R_L = 150 Ω, V_S = ±5 V

On the other hand, Figure 6.15 shows distortion as a function of output signal level for a frequencies of 5 MHz and 20 MHz.

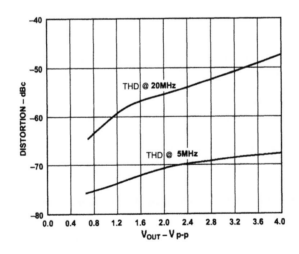

Figure 6.15: AD8057/AD8058 Op Amp Distortion versus Output Voltage G = +1, R_L = 150 Ω, V_S = ±5 V

Whether or not specifications such as those just described are complete enough to select an op amp for an ADC driver application depends upon the ability to match op amp specifications to the actually required ADC operating conditions. In many cases, these comparisons will at least narrow the op amp selection process. The following sections will examine a number of specific driver circuit examples using various types of ADCs, ranging from high resolution measurement to high speed, low distortion applications.

Driving High Resolution Σ-Δ Measurement ADCs

The AD77xx family of ADCs is optimized for high resolution (16–24 bits) low frequency transducer measurement applications. Details of operation for this family can be found in Reference 3, and general characteristics of the family are listed in Figure 6.16.

- Resolution: 16-24 bits
- Input signal bandwidth: <60Hz
- Effective sampling rate: <100Hz
- Designed to interface directly to sensors (< 1 kΩ) such as bridges with no external buffer amplifier (e.g., AD77xx series)
  - On-chip PGA and high resolution ADC eliminates the need for external amplifier
- If buffer is used, it should be precision low noise (especially 1/f noise)
  - OP1177
  - OP177
  - AD797

Figure 6.16: Characteristics of AD77xx-family High Resolution Σ-Δ Measurement ADCs

Some members of this family, such as the AD7730, have a high impedance input buffer which isolates the analog inputs from switching transients generated in the front-end programmable gain amplifier (PGA) and the Σ-Δ modulator. Therefore, no special precautions are required in driving the analog inputs. Other members of the AD77xx family, however, either do not have the input buffer or, if one is included on-chip, it can be switched either in or out under program control. Bypassing the buffer offers a slight improvement in noise performance.

The equivalent input circuit of the AD77xx family without an input buffer is shown below in Figure 6.17. The input switch alternates between the 10 pF sampling capacitor and ground. The 7 kΩ internal resistance, R_{INT}, is the on-resistance of the input multiplexer. The switching frequency is dependent on the frequency of

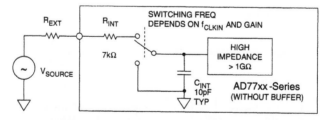

- R_{EXT} Increases C_{INT} Charge Time and May Result in Gain Error

- Charge Time Dependent on the Input Sampling Rate and Internal PGA Gain Setting

- Refer to Specific Data Sheet for Allowable Values of R_{EXT} to Maintain Desired Accuracy

- Some AD77xx-Series ADCs Have Internal Buffering Which Isolates Input from Switching Circuits

Figure 6.17: Driving Unbuffered AD77xx-Series Σ-Δ ADC Inputs

the input clock and also the internal PGA gain. If the converter is working to an accuracy of 20 bits, the 10 pF internal capacitor, C_{INT}, must charge to 20-bit accuracy during the time the switch connects the capacitor to the input. This interval is one-half the period of the switching signal (it has a 50% duty cycle). The input RC time constant due to the 7 kΩ resistor and the 10 pF sampling capacitor is 70 ns. If the charge is to achieve 20-bit accuracy, the capacitor must charge for at least 14 time constants, or 980 ns. Any external resistance in series with the input will increase this time constant.

There are tables on the datasheets for the various AD77xx ADCs, which give the maximum allowable values of R_{EXT} in order to maintain a given level of accuracy. These tables should be consulted if the external source resistance is more than a few kΩ.

Note that for instances where an external op amp buffer is found to be required with this type of converter, guidelines exist for best overall performance. This amplifier should be a precision low noise bipolar-input type, such as the OP1177, OP177, or the AD797.

Driving Single-Ended Input Single-Supply 1.6 V to 3.6 V Successive Approximation ADCs

The need for low power, low supply voltage ADCs in small packages led to the development of the AD7466/AD7467/AD7468 12-/10-/ and 8-bit family of converters (Reference 4). These devices operate on supply voltages from 1.6 V to 3.6 V and utilize a successive approximation architecture that allows sampling rates up to 200 kSPS. The converters are packaged in a 6-lead SOT-23 package and offer this performance at only 0.62 mW with a 3 V supply and 0.12 mW with a 1.6 V supply. An automatic power-down mode reduces the supply current to 8 nA. Data is transferred via a simple serial interface. It is useful to examine these converters in more detail, because they illustrate some of the trade-offs which must be made in designing appropriate interface circuits.

A simplified block diagram of the series is shown in Figure 6.18. As mentioned, the ADC utilizes a standard successive approximation architecture based on a switched capacitor CMOS charge redistribution DAC. The input CMOS switches, SW1 and SW2, comprise the sample-and-hold function, and are shown in the track mode in the diagram. Capacitor C1 represents the equivalent parasitic input capacitance, C_H is the hold capacitor, and R_S is the equivalent on-resistance of SW2. In the track mode, SW1 is connected

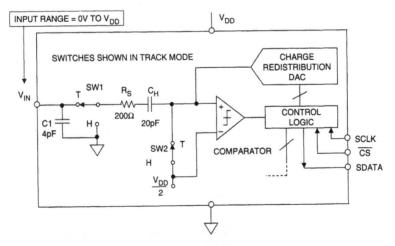

Figure 6.18: Input Circuit of AD7466 1.6 V to 3.6 V,
12-Bit, 200 kSPS SOT-23-6 ADC

to the input, and SW2 is closed. In this condition, the comparator is balanced, and the hold capacitor C_H is charged to the value of the input signal. Assertion of the $\overline{CS}$ (convert start) starts the conversion process: SW2 opens, and SW1 is connected to ground, causing the comparator to become unbalanced. The control logic and the charge redistribution DAC are used to add and subtract fixed amounts of charge from the hold capacitor to bring the comparator back into balance. At the end of the appropriate number of clock pulses, the conversion is complete.

The switching action of CMOS switches SW1 and SW2 places certain requirements on the input drive circuit with respect to the transient currents. In addition, the input signal must charge and discharge C_H in the track mode. In most cases, no input drive amplifier is required provided the source impedance is less than 1 kΩ (although a slight degradation in THD will be observed at input frequencies approaching 100 kHz).

The input voltage range of the AD746x ADC is from 0 V to the supply voltage, and the supply also acts as the reference. If more accuracy or stability is required, the supply voltage can be derived from a voltage reference or an LDO.

Although single-supply rail-to-rail 1.8 V op amps are available (such as the AD8515, AD8517, and AD8631), these op amps will not drive signals completely to either rail due to the saturation voltage of the output transistors (this has previously been discussed in detail). If these are used as drive amplifiers to the AD746x, the usable input range of the ADC will be reduced by an amount that depends not only on this saturation voltage but the amount of additional headroom required at the amplifier output in order to give acceptable distortion performance at the higher input frequencies.

The overall conclusion of this discussion is that low voltage single-supply ADCs such as the AD746x are best driven directly from low impedance sources (< 1 kΩ). If a drive amplifier is required, it must operate on a higher supply voltage in order to utilize the full input range of the ADC.

Driving Single-Supply ADCs with Scaled Inputs

Even with the widespread popularity of single-supply systems, there are still applications where it is desirable for the ADC to process bipolar input signals. This can be handled in a number of ways, but a simple method is to provide an appropriate thin-film resistive divider/level-shifter at the input of the ADC. The AD789x and AD76xx family of single supply SAR ADCs (as well as the AD974, AD976, and AD977) include such a thin-film resistive attenuator and level shifter on the analog input to allow a variety of input range options, both bipolar and unipolar.

A simplified diagram of the input circuit of the AD7890-10 12-bit, 8-channel ADC is shown in Figure 6.19 (Reference 5). This arrangement allows the converter to digitize a ±10 V input while operating on a single +5 V supply.

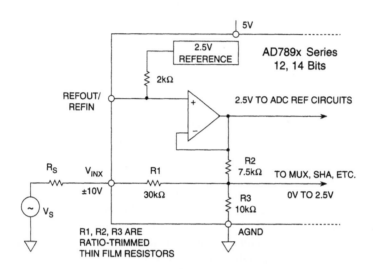

Figure 6.19: Driving Single-Supply Data Acquisition ADCs With Scaled Inputs

Within the ADC, the R1/R2/R3 thin film network provides attenuation and level shifting to convert the ±10 V input to a 0 V to +2.5 V signal that is digitized. This type of input requires no special drive circuitry, because R1 isolates the input from the actual converter circuitry that may generate transient currents due to the conversion process. Nevertheless, the external source resistance R_S should be kept reasonably low, to prevent gain errors caused by the $R_S/R1$ divider.

Driving Differential Input CMOS Switched Capacitor ADCs

CMOS ADCs are quite popular because of their low power, high performance, and low cost. The equivalent input circuit of a typical CMOS ADC using a differential sample-and-hold is shown in Figure 6.20. While the switches are shown in the *track* mode, note that they open/close at the sampling frequency. The 16 pF capacitors represent the effective capacitance of switches S1 and S2, plus the stray input capacitance. The C_S capacitors (4 pF) are the sampling capacitors, and the C_H capacitors are the hold capacitors. Although the input circuit is completely differential, this ADC structure can be driven either single-ended or differentially. Optimum performance, however, is generally obtained using a differential transformer or differential op amp drive.

In the *track* mode, the differential input voltage is applied to the C_S capacitors. When the circuit enters the *hold* mode, the voltage across the sampling capacitors is transferred to the C_H hold capacitors and buffered by the amplifier A (the switches are controlled by the appropriate sampling clock phases). When the SHA returns to the *track* mode, the input source must charge or discharge the voltage stored on C_S to the new input voltage. This action of charging and discharging C_S, averaged over a period of time and for a given f_S sampling frequency, makes the input impedance appear to have a benign resistive component. However, if this action is analyzed within a sampling period ($1/f_S$), the input impedance is dynamic, and certain input drive source precautions should be observed.

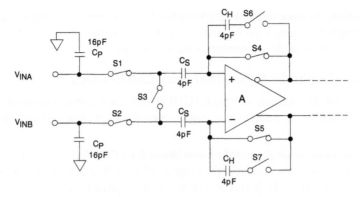

SWITCHES SHOWN IN TRACK MODE

**Figure 6.20: Simplified Input Circuit for a Typical
Switched Capacitor CMOS Sample-and-Hold**

The resistive component of the input impedance can be computed by calculating the average charge that is drawn by C_H from the input drive source. It can be shown that if C_S is allowed to fully charge to the input voltage before switches S1 and S2 are opened that the average current into the input is the same as if there were a resistor equal to $1/(C_S f_S)$ connected between the inputs. Since C_S is only a few picofarads, this resistive component is typically greater than several kΩ for an f_S = 10 MSPS.

Over a sampling period, the SHA's input impedance appears as a dynamic load. When the SHA returns to the track mode, the input source should ideally provide the charging current through the R_{ON} of switches S1 and S2 in an exponential manner. The requirement of exponential charging means that the source impedance should be both low and resistive up to and beyond the sampling frequency.

A differential input CMOS ADC can be driven single-ended with some ac performance degradation. An important consideration in CMOS ADC applications is the input switching transients previously discussed. Typical single-ended transients for a CMOS ADC are shown in Figure 6.21 for the AD9225 12-bit, 25 MSPS ADC. This data was taken driving the ADC with an equivalent 50 Ω source impedance. During the

- Hold-to-Sample Mode Transition – C_S Returned to Source for "recharging." Transient Consists of Linear, Nonlinear, and Common-Mode Components at Sample Rate.
- Sample-to-Hold Mode Transition – Input Signal Sampled when C_S is disconnected from Source.

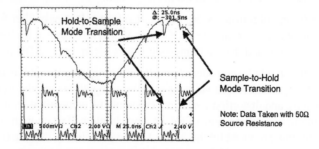

**Figure 6.21: Single-Ended Input Transients for a
Typical CMOS ADC Sampling at 25 MSPS**

sample-to-hold transition, the input signal is sampled when C_S is disconnected from the source. Notice that during the hold-to-sample transition, C_S is reconnected to the source for recharging. The transients consist of linear, nonlinear, and common-mode components at the sample rate.

Single-Ended Drive Circuits for Differential Input CMOS ADCs

A few simple single-ended drive circuits suitable for CMOS ADCs will now be examined. Although differential drive is preferable for best ac performance, single-ended drivers are often adequate in less demanding applications.

Figure 6.22 shows a generalized single-ended op amp driver for a CMOS ADC. In this circuit, series resistor R_S has a dual purpose. Typically chosen in the range of 25 Ω–100 Ω, it limits the peak transient current from the driving op amp. Importantly, it also decouples the driver from the ADC input capacitance (and possible phase margin loss).

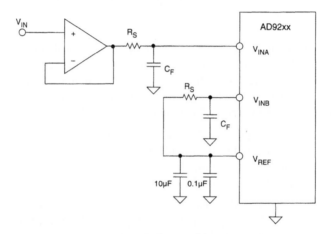

**Figure 6.22: Optimizing a Single-Ended Switched
Capacitor ADC Input Drive Circuit**

Another feature of the circuit are the dual networks of R_S and C_F. Matching both the dc and ac source impedance for the ADC's V_{INA} and V_{INB} inputs ensures symmetrical settling of common-mode transients, for optimum noise and distortion performance. At both inputs, the C_F shunt capacitor also acts as a charge reservoir and steers the common-mode transients to ground.

In addition to the buffering of transients, R_S and C_F also form a low-pass filter for V_{IN}, which limits the output noise of the drive amplifier into the ADC input V_{INA}. The exact values for R_S and C_F are generally optimized within the circuit, and the recommended values given on the ADC datasheet.

Many important factors should be considered in selecting an appropriate drive amplifier. As discussed previously, common-mode input and output voltages must be compatible with the ADC power supply and input range. The op amp noise and distortion performance should be compatible with the expected performance of the ADC. In addition, the settling time of the op amp should be fast enough so that the output can settle from the transient currents produced by the ADC. A good guideline is that the 0.1% settling time of the op amp should be no more than one-half the period of the maximum sampling frequency. The most important factor is simply to consult the ADC data sheet for recommended drive op amps and the associated circuits.

A generalized dc-coupled single-ended op amp driver and level shifter for the AD922x-series of ADCs is shown in Figure 6.23. The values in this circuit are suitable for sampling rates up to about 25 MSPS. This circuit interfaces a ±2 V ground-referenced input signal to the single-supply ADC, and also provides transient current isolation. The ADC input voltage range is 0 V to +4 V, and a dual supply op amp is required, since the ADC minimum input is 0 V.

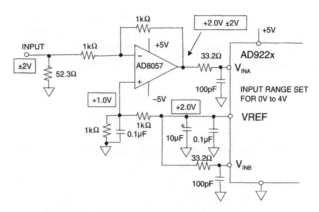

Figure 6.23: Single-Ended DC-Coupled Level Shifter and Driver for the AD922x ADC

The noninverting input of the AD8057 is biased at 1 V, which sets the output common-mode voltage at V_{INA} to 2 V for a bipolar input signal source. Note that the V_{INA} and V_{INB} source impedances are matched for better common-mode transient cancellation. The 100 pF capacitors act as small charge reservoirs for the input transient currents, and also form low-pass noise filters with the 33.2-Ω series resistors.

A similar single-ended level shifter and driver is shown in Figure 6.24; however, this circuit is designed to operate on a single 5 V supply. In this circuit the bipolar ±1 V input signal is interfaced to the input of the

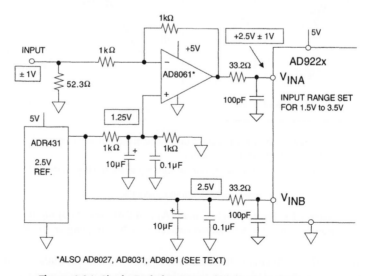

*ALSO AD8027, AD8031, AD8091 (SEE TEXT)

Figure 6.24: Single-Ended DC-Coupled Single-Supply Level Shifter for Driving AD922x ADC

ADC whose range is set for 2 V about a +2.5 V common-mode voltage. The AD8061 rail-to-rail output op amp is used, although others are suitable depending upon bandwidth and distortion requirements (for example, the AD8027, AD8031, or AD8091). The 1.25 V input common-mode voltage for the AD8061 is developed by a voltage divider from the external AD780 2.5 V reference.

Differential Input ADC Drivers

As previously discussed, most high performance ADCs are now being designed with differential inputs. A fully differential ADC design offers the advantages of good common-mode rejection, reduction in second-order distortion products, and simplified factory trimming algorithms. Although most differential input ADCs can be driven single-ended, as previously described, a fully differential driver usually optimizes overall performance.

In the following discussions, it is useful to keep in mind that there are currently two popular IC processes used for high performance ADCs, and each one has certain application implications. Many medium-to-high performance ADCs are fabricated on high density foundry CMOS processes, and these typically use switched capacitor sample-and-hold techniques (previously described) which tend to generate transient currents at the ADC inputs. In many cases, however, ultrahigh performance ADCs are designed on either BiCMOS (bipolar and CMOS devices on the same process) or CB (complementary bipolar) processes. ADCs designed on BiCMOS or CB processes typically provide input buffers as part of a more conventional diode-switched sample-and-hold circuit which minimizes the effects of input transient currents; however, the input range is generally less flexible than in CMOS-based designs.

In order to understand the advantages of common-mode rejection of input transient currents, we will next examine the waveforms at the two inputs of the AD9225 12-bit, 25 MSPS CMOS ADC as shown in Figure 6.25A, designated as V_{INA} and V_{INB}. The balanced source impedance is 50 Ω, and the sampling frequency is set for 25 MSPS. The diagram clearly shows the switching transients due to the internal ADC switched capacitor sample-and-hold. Figure 6-25B shows the difference between the two waveforms, $V_{INA} - V_{INB}$.

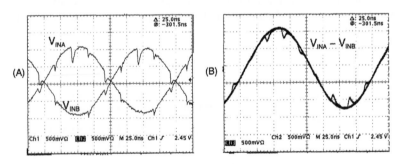

- Differential charge transient is symmetrical around mid-scale and dominated by linear component
- Common-mode transients cancel with equal source impedance

Note: Data Taken with 50Ω Source Resistances

Figure 6.25: Typical Single-Ended (A) and Differential (B) Input Transients of CMOS Switched Capacitor ADC Sampling at 25 MSPS

Note that the resulting differential charge transients are symmetrical about midscale, and that there is a distinct linear component to them. This shows the reduction in the common-mode transients, and also leads to better distortion performance than would be achievable with a single-ended input.

Transformer coupling into a differential input ADC provides excellent common-mode rejection and low distortion, provided performance to dc is not required. Figure 6.26 shows a typical circuit. The transformer is a Mini-Circuits RF transformer, model #ADT4-1WT which has an impedance ratio of 4 (turns ratio of 2). The 3 dB bandwidth of this transformer is 2 MHz to 775 MHz (Reference 6). The schematic assumes that the signal source impedance is 50 Ω. The 1:4 impedance ratio requires the 200 Ω secondary termination for optimum power transfer and low VSWR. The center tap of the transformer secondary winding provides a convenient means of level shifting the input signal to the optimum $V_c/2$ common-mode voltage of the ADC (some ADCs may have a common-mode voltage different than $V_c/2$, so the data sheet should be consulted).

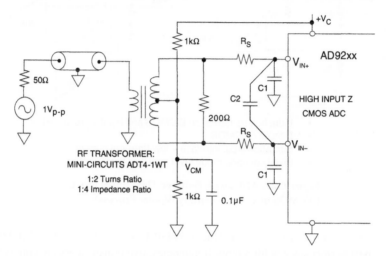

Figure 6.26: Transformer Coupling into a Differential Input CMOS ADC

Transformers with other turns ratios may also be selected to optimize the performance for a given application. For example, a given input signal source or amplifier may realize an improvement in distortion performance at reduced output power levels and signal swings. Hence, selecting a transformer with a higher impedance ratio effectively "steps up" the signal level thus reducing the driving requirements of the signal source.

The network consisting of R_S, C1, and C2 is relatively common when driving CMOS switched capacitor ADC inputs with a transformer. The R_S resistors serve to isolate the transformer secondary winding from the switching transients, and the optimum value (determined empirically) generally ranges from 25 Ω to 100 Ω. The C1 capacitors serve as common-mode charge reservoirs for the switching transients and also provide noise filtering (in conjunction with the R_S resistors). The C1 capacitors should have no greater than 5% tolerance to prevent common-mode to differential signal conversion. If needed, C2 can be added for additional differential filtering. Data sheets for most CMOS ADCs typically recommend optimum values for R_S, C1, and C2 and should be consulted in all cases.

As previously mentioned, BiCMOS or complementary bipolar ADCs typically provide some amount of input buffering, and therefore have lower input transient currents than CMOS converters. Figure 6.27 shows two typical input configurations for buffered BiCMOS or CB ADCs. Although this can simplify the interface, the fixed input common-mode level may limit flexibility. In Figure 6.27A the common-mode voltage is developed with a resistive divider connected between ground and the positive analog supply voltage. In Figure 6.27B, the common-mode voltage is generated by an internal reference voltage.

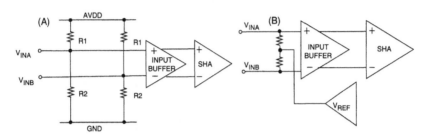

- Input buffers typical on BiCMOS and bipolar processes
- Difficult on CMOS
- Simplified input interface – low transient currents
- Fixed common-mode level may limit flexibility

Figure 6.27: ADCs with Buffered Differential Inputs (BiCMOS or Complementary Bipolar Process)

Figure 6.28 shows a transformer drive circuit for the AD9430 12-bit, 170/210 MSPS BiCMOS ADC (Reference 7). For best performance at high input frequencies, two transformers are connected in series as shown to minimize even-order harmonic distortion. The first transformer converts the single-ended signal to a differential signal—however the grounded input on the primary side degrades the amplitude balance on the secondary winding because of the stray capacitive coupling between the windings. The second

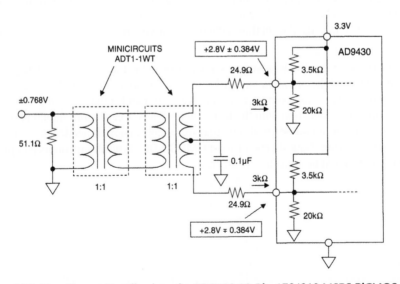

Figure 6.28: Transformer Coupling into the AD9430 12-Bit, 170/210 MSPS BiCMOS ADC

transformer improves the amplitude balance, and thus the harmonic distortion. A wideband transformer, such as the Mini Circuits ADT1-1WT is recommended for these applications. The 3 dB bandwidth of the ADT1-1WT is 0.4 MHz to 800 MHz. Note that the bandwidth through the two transformers is equal to the bandwidth of a single transformer divided by $\sqrt{2}$.

The net impedance seen by the secondary winding of the second transformer is the sum of the ADC input impedance (6 kΩ) and the two 24.9 Ω series resistors, or approximately 6050 Ω. The 51.1 Ω termination resistor in parallel with 6050 Ω yields the desired impedance of approximately 50 Ω.

There is no requirement for input filtering, since the BiCMOS buffered input circuit generates minimal transient currents. The 24.9 Ω series resistors simply buffers the transformer from the small input capacitance of the ADC (~5 pF). The input common-mode voltage is set at 2.8 V by the 3.5 kΩ/20 kΩ resistive divider (when operating on a 3.3 V supply). This serves to illustrate the point made earlier that BiCMOS and complementary bipolar ADCs may not have a common-mode voltage that is exactly mid-supply. In this circuit, the most positive input voltage on either input is 2.8 V + 0.384 V = 3.184 V which is only 116 mV from the 3.3 V supply. The implication therefore is that for low distortion performance in a 3.3 V system, the AD9430 must either be driven from a transformer or from an ac-coupled differential amplifier.

If dc coupling is required, the driving amplifier must operate on a higher supply voltage, because even rail-to-rail output stages will give poor high frequency distortion performance if only 116-mV of headroom is available.

Note that the center tap of the secondary winding of the transformer is decoupled to ground to ensure a balanced drive.

A similar transformer drive circuit for the AD6645 14-bit, 80/105 MSPS (bipolar process) ADC is shown in Figure 6.29 (Reference 8). Note that the input common-mode voltage is developed by the two 500 Ω resistors connected to each input from the internal 2.4 V reference. The differential input resistance of the ADC is therefore 1 kΩ. As in the previous circuit, the 24.9 Ω series resistors isolate the transformer secondary winding from the small input capacitance of the ADC. The net differential impedance seen by the secondary winding of the transformer is therefore 1050 Ω.

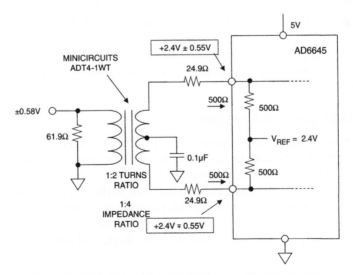

Figure 6.29: Transformer Coupling into the AD6645 14-Bit, 80-/105-MSPS Complementary Bipolar Process ADC

In this circuit, a Mini Circuits ADT4-1WT 1:4 impedance ratio (1:2 turns ratio) transformer is used to match the 1050 Ω differential resistance to the 50 Ω source. The 1050 Ω resistance is 262.5 Ω referred to the primary winding, and the 61.9 Ω termination resistor in parallel with 262.5 Ω is approximately 50 Ω. The 3 dB bandwidth of the transformer is 2 MHz to 775 MHz.

Theoretically, a 1:20 impedance ratio (corresponding to a 1 : 4.47 turns ratio) transformer would perfectly match the AD6645 1000 Ω input to the 50 Ω source and provide a "noise-free" voltage gain of 4.47 (+13 dB). However, this large turns ratio could result in unsatisfactory bandwidth and distortion performance.

To illustrate the effects of utilizing transformers for voltage gain on system noise figure (NF), Figure 6.30 shows the AD6645 sampling a 40 MHz bandwidth signal at 80 MSPS for turns ratios of 1:1 (Figure 6.30A), 1:2 (Figure 6.30B) and 1:4 (Figure 6.30C). (Refer back to Chapter 2 for the basic definitions and calculations of ADC noise figure).

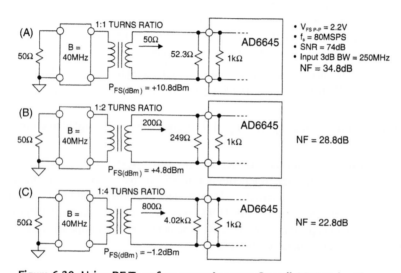

Figure 6.30: Using RF Transformers to Improve Overall ADC Noise Figure

Notice that each time the turns ratio is doubled, the noise figure decreases by 6 dB. In practice, however, empirical data indicates that bandwidth and distortion are compromised when driving the AD6645 with a turns ratio of greater than 1:2.

Driving ADCs with Differential Amplifiers

Certainly for most RF and IF applications, transformer ADC drivers yield the best overall distortion and noise performance, especially if the transformer can be utilized to achieve some amount of "noise free" voltage gain. There are, however, many applications where differential input ADCs cannot be driven with transformers because the frequency response must extend to dc. In these cases, the op amp common-mode input and output voltage, gain, distortion, and noise must be carefully considered in designing dc-coupled drive circuitry. The following two subsections discuss two types of differential op amp drivers: the first is based on utilizing dual op amps, and the second utilizes fully integrated differential amplifiers.

Dual Op Amp Drivers

Figure 6.31 shows how the dual AD8058 op amp can be connected to convert a single-ended bipolar signal to a differential one suitable for driving the AD92xx family of CMOS ADCs. Utilizing a dual op amp provides better gain and phase matching than would be achieved by simply using two single op amps. The input range of the ADC is set for a 2 V p-p differential input signal (1 V p-p on each input), and a common-mode voltage of 2 V. As shown for previous CMOS ADCs, the 100 pF capacitors serve as charge reservoirs for the transient currents, and also act as low-pass noise filters in conjunction with the 33.2 Ω resistors.

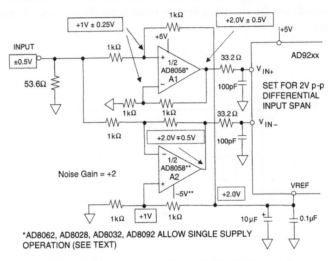

Figure 6.31: Op Amp Single-Ended to Differential DC-Coupled Driver with Level Shifting

The A1 amplifier is configured as a noninverting op amp. The 1 kΩ divider resistors level shift the ±0.5 V input signal to +1 V ±0.25 V at the noninverting input of A1. The output of A1 is therefore +2 V ±0.5 V, because the noninverting gain of A1 is 2.

The A2 op amp inverts the input signal, and the 1 kΩ divider resistors establish a 1 V common-mode voltage on its noninverting input. The output of A2 is therefore +2 V ∓0.5 V.

This circuit provides good matching between the two op amps because they are duals on the same die and are both operated at the same noise gain of 2. However, the input voltage noise of the AD8058 is $20 \text{ nV}/\sqrt{\text{Hz}}$, and this appears as $40 \text{ nV}/\sqrt{\text{Hz}}$ at the output of both A1 and A2, thereby introducing possible SNR degradation in some applications. In the circuit of Figure 6.31, this is somewhat mitigated by the input RC network which not only reduces the input noise, but also absorbs some of the transient currents.

The AD8058 op amp does not have rail-to-rail inputs or outputs, and the following simple analysis shows that the circuit as shown in Figure 6.31 must use dual supplies. The output common-mode voltage of the AD8058 operating on a single 5 V supply is 0.9 V to 3.4 V, which would be acceptable in this circuit, because the required signal swing is only 1.5 V to 2.5 V. However, the input common-mode voltage of the AD8058 operating on a single 5 V supply is specified as 0.9 V to 4.1 V; but the circuit requires that the input common-mode voltage go to 0.75 V, which is outside the allowable range. Therefore, a dual supply is required for the op amp.

If single-supply operation is required, however, there are a number of dual rail-to-rail op amps which should be considered, such as the AD8062, AD8028, AD8032, and the AD8092.

Fully Integrated Differential Amplifier Drivers

A block diagram of the AD813x family of fully differential amplifiers optimized for ADC driving is shown in Figure 6.32 (see Reference 9). Figure 6.32A shows the details of the internal circuit, and Figure 6.32B shows the equivalent circuit. The gain is set by the external resistors R_F and R_G, and the common-mode voltage is set by the voltage on the V_{OCM} pin. The internal common-mode feedback forces the V_{OUT+} and V_{OUT-} outputs to be balanced, i.e., the signals at the two outputs are always equal in amplitude but 180° out of phase per the equation,

$$V_{OCM} = (V_{OUT+} + V_{OUT-})/2 \qquad \text{Eq. 6.4}$$

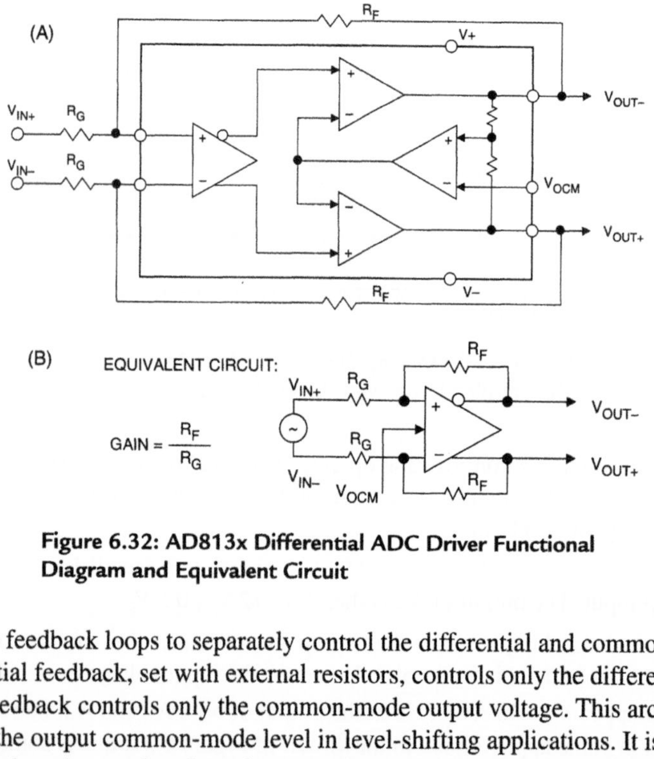

Figure 6.32: AD813x Differential ADC Driver Functional Diagram and Equivalent Circuit

The AD813x uses two feedback loops to separately control the differential and common-mode output voltages. The differential feedback, set with external resistors, controls only the differential output voltage. The common-mode feedback controls only the common-mode output voltage. This architecture makes it easy to arbitrarily set the output common-mode level in level-shifting applications. It is forced, by internal common-mode feedback, to be equal to the voltage applied to the V_{OCM} input, without affecting the differential output voltage. The result is nearly perfectly balanced differential outputs of identical amplitude and exactly 180° apart in phase over a wide frequency range. The circuit can be used with either a differential or a single-ended input, and the voltage gain is equal to the ratio of R_F to R_G.

The circuit can be analyzed using the assumptions and procedures summarized in Figure 6.33. As in the case of op amp circuit dc analysis, one can first make the assumption that the currents into the inverting and noninverting input are zero (i.e., the input impedances are high relative to the values of the feedback resistors). The second assumption is that feedback forces the noninverting and inverting input voltages to be equal. The third assumption is that the output voltages are 180° out of phase and symmetrical about V_{OCM}.

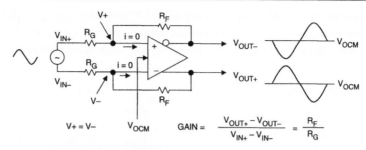

+ and − input currents are zero
+ and − input voltages are equal
Output voltages are 180° out of phase and symmetrical about V_{OCM}
Gain = R_F/R_G

Figure 6.33: Analyzing Voltage Levels in Differential Amplifiers

Even if the external feedback networks (R_F/R_G) are mismatched, the internal common-mode feedback loop will still force the outputs to remain balanced. The amplitudes of the signals at each output will remain equal and 180° out of phase. The input-to-output differential-mode gain will vary proportionately to the feedback mismatch, but the output balance will be unaffected. Ratio matching errors in the external resistors will result in a degradation of the circuit's ability to reject input common-mode signals, much the same as for a four-resistor difference amplifier made from a conventional op amp.

Also, if the dc levels of the input and output common-mode voltages are different, matching errors will result in a small differential-mode output offset voltage. For the G = 1 case with a ground-referenced input signal and the output common-mode level set for 2.5 V, an output offset of as much as 25 mV (1% of the difference in common-mode levels) can result if 1% tolerance resistors are used. Resistors of 1% tolerance will result in a worst case input CMRR of about 40 dB, worst case differential mode output offset of 25 mV due to 2.5 V level-shift, and no significant degradation in output balance error.

The effective input impedance of a circuit, such as the one in Figure 6.33, at V_{IN+} and V_{IN-} will depend on whether the amplifier is being driven by a single-ended or differential signal source. For balanced differential input signals, the input impedance ($R_{IN,dm}$) between the inputs (V_{IN+} and V_{IN-}) is simply:

$$R_{IN,dm} = 2 \times R_G \qquad \text{Eq. 6.5}$$

In the case of a single-ended input signal (for example, if V_{IN-} is grounded, and the input signal is applied to V_{IN+}), the input impedance becomes:

$$R_{IN,dm} = \left(\frac{R_G}{1 - \dfrac{R_F}{2 \times (R_G + R_F)}} \right) \qquad \text{Eq. 6.6}$$

The circuit's input impedance is effectively higher than it would be for a conventional op amp connected as an inverter, because a fraction of the differential output voltage appears at the inputs as a common-mode signal, partially bootstrapping the voltage across the input resistor R_G.

Figure 6.34 shows some of the possible configurations for the AD813x differential amplifier. Figure 6.34A is the standard configuration which utilizes two feedback networks, characterized by feedback factors β1 and β2, respectively. Note that each feedback factor can vary anywhere between 0 and 1.

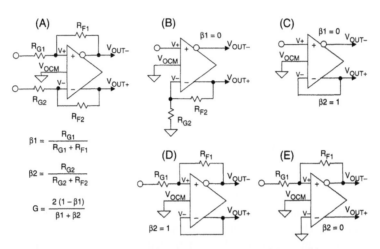

$$\beta 1 = \frac{R_{G1}}{R_{G1} + R_{F1}}$$

$$\beta 2 = \frac{R_{G2}}{R_{G2} + R_{F2}}$$

$$G = \frac{2\,(1 - \beta 1)}{\beta 1 + \beta 2}$$

Figure 6.34: Some Configurations for Differential Amplifiers

Figure 6.34B shows a configuration where there is no feedback from V_{OUT-} to V+, i.e., β1 = 0. In this case, β2 determines the amount of V_{OUT+} that is fed back to V−, and the circuit is similar to a noninverting op amp configuration, except for the presence of the additional complementary output. Therefore, the overall gain is twice that of a noninverting op amp, or $2 \times (1 + R_{F2}/R_{G2})$, or $2 \times (1/\beta 2)$.

Figure 6.34C shows a circuit where β1 = 0 and β2 = 1. This circuit is essentially provides a resistorless gain of 2.

Figure 6.34D shows a circuit where β2 = 1, and β1 is determined by R_{F1} and R_{G1}. The gain of this circuit is always less than 2.

Finally, the circuit of Figure 6.34E has β2 = 0, and is very similar to a conventional inverting op amp, except for the additional complementary output at V_{OUT+}.

The AD813x-series are also well suited to balanced differential line driving as shown in Figure 6.35 where the AD8132 drives a 100-Ω twisted pair cable. The AD8132 is configured as a gain of 2 driver to account for the factor of 2 loss due to the source and load terminated cable. In this configuration, the bandwidth of the AD8132 is approximately 160 MHz.

The line receiver is an AD8130 differential receiver which has a unique architecture called "active feed-back" to achieve approximately 70 dB common-mode rejection at 10 MHz (Reference 10). For a gain of 1, the AD8130 has a 3 dB bandwidth of approximately 270 MHz.

The AD8130 utilizes two identical input transconductance (g_m) stages whose output currents are summed together at a high impedance node and then buffered to the output. The output currents of the two g_m stages must be equal but opposite in sign, therefore, the respective input voltages must also be equal but opposite in sign. The differential input signal is applied to one of the stages (G_{M1}), and negative feedback is applied to the other (G_{M2}) as in a traditional op amp. The gain is equal to 1 + R2/R1. The G_{M1} stage therefore provides a truly balanced input for the terminated twisted pair for the best common-mode rejection. Further details of operation of the AD8130 can be found in Reference 10.

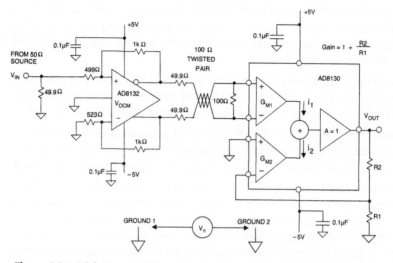

Figure 6.35: High Speed Differential Line Driver, Line Receiver Applications

Driving Differential Input ADCs with Integrated Differential Drivers

The AD8131, AD8132, AD8137, AD8138, and AD8139 differential ADC drivers are ideal replacements for transformer drivers when direct coupling is required. They can also provide the necessary gain and level shifting required to interface a bipolar signal to a high performance ADC input. In addition, the AD8137 has rail-to-rail outputs to simplify interfacing to low voltage differential input ADCs, and the AD8139 (also rail-to-rail output) is optimized for low noise and low distortion for 14- to 16-bit applications.

Figure 6.36 shows an application where the AD8137 differential amplifier is used as a level shifter and driver for the AD7450A 12-bit, 1 MSPS 3 V ADC (Reference 11). The AD7450A has fully differential inputs, and the input range is 4 V p-p differential when an external 2 V reference (ADR390) is applied. This

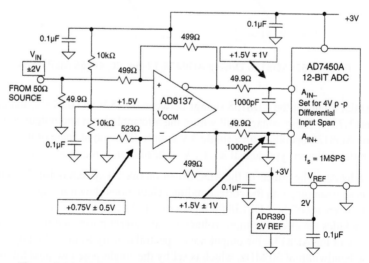

Figure 6.36: AD8137 Driving AD7450A 12-Bit, 1 MSPS, 3 V ADC

implies that the signals at each output of the AD8137 driver must swing between 0.5 V and 2.5 V (out of phase) when operating on a single 3 V supply. The rail-to-rail output structure of the AD8137 will provide this voltage swing with some safety margin. The 1.5 V common-mode voltage for the AD8137 is set by a resistive divider connected to the 3 V supply.

The inputs to the AD8137 must swing between 0.25 V and 1.25 V. This is not a problem, since the input of the AD8137 is a differential PNP pair. The 523 Ω resistor from the inverting input to ground approximately balances the net feedforward resistance seen at the noninverting input (499 Ω + 25 Ω = 524 Ω).

For higher frequency applications, the AD8138 differential amplifier has a 3 dB small-signal bandwidth of 320 MHz (G = +1) and is designed to give low harmonic distortion as an ADC driver. The circuit provides excellent output gain and phase matching, and the balanced structure suppresses even-order harmonics.

Figure 6.37 shows the AD8138 driving the AD9235 12-bit, 20/40/65 MSPS CMOS ADC (see Reference 12). This entire circuit operates on a single 3 V supply. A 1 V p-p bipolar single-ended input signal produces a 1 V p-p differential signal at the output of the AD8138, centered around a common-mode voltage of 1.5 V (mid-supply). The feedback network is chosen to provide a gain of 1, and the 523 Ω resistor from the inverting input to ground approximately balances the net feedforward resistance seen at the noninverting input as in the previous example.

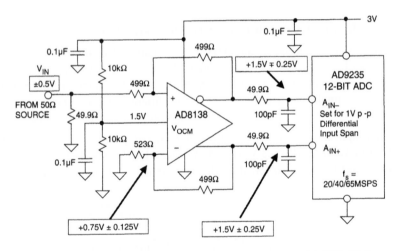

Figure 6.37: AD8138 Driving AD9235 12-Bit, 20/40/65 MSPS ADC

Each of the differential inputs of the AD8138 swings between 0.625 V and 0.875 V, and each output swings between 1.25 V and 1.75 V. These voltages fall within the allowable input and output common-mode voltage range of the AD8138 operating on a single 3 V supply. The output stage of the AD8138 is of the complementary emitter-follower type, and at least 1 V of headroom is required from either supply rail.

It is important to understand the effects of the ADC driver on overall system noise. The circuit of Figure 6.37 will be used as an example, with the corresponding calculations shown in Figure 6.38. The output voltage noise spectral density of the AD8138 for a gain of 1 is 11.6 nV/$\sqrt{\text{Hz}}$ (taken directly from the data sheet). This value includes the effects of input voltage noise, current noise, and resistor noise. To obtain the total rms output noise of the AD8138, the output noise spectral density is multiplied by the square root of the equivalent noise bandwidth of 50 MHz, which is set by the single-pole low-pass filters placed between the differential amplifier outputs and the ADC inputs.

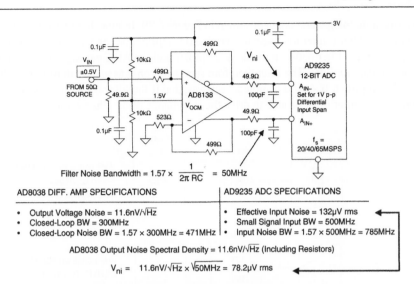

Figure 6.38: Noise Calculations for the AD8138 Differential Op Amp Driving the AD9235 12-Bit, 20/40/65 MSPS ADC

Note that the closed-loop bandwidth of the AD8138 is 300 MHz, and the input bandwidth of the AD9235 is 500 MHz. With no filter, the output noise of the AD8138 would be integrated over the full 300 MHz amplifier closed-loop bandwidth. (In general, with no filtering, the amplifier noise must be integrated over either the amplifier closed-loop bandwidth or the ADC input bandwidth, whichever is less, or the geometric mean if the frequencies are close to each other).

However, the sampling frequency of the ADC is 65 MSPS, thereby implying that signals above 32.5 MHz are not of interest, assuming Nyquist operation (as opposed to undersampling applications where the input signal can be greater than the Nyquist frequency, $f_s/2$). The addition of this simple filter significantly reduces noise effects as will be demonstrated.

The noise at the output of the low-pass filter, V_{ni}, is calculated to be approximately 78.2 µV rms, which is only slightly more than half the effective input noise of the AD9235, 132 µV rms. The effective input noise of the AD9235 is specified as 0.54 LSB rms, which corresponds to $(1 \text{ V}/4096) \times (0.54) = 132 \text{ µV rms}$. Without the filter, the noise from the op amp would be integrated over the full 471 MHz closed-loop noise bandwidth of the AD8138 ($1.57 \times 300 \text{ MHz} = 471 \text{ MHz}$). This would yield a noise of 252 µV rms, compared to 78.2 µV rms obtained with low-pass filtering.

This serves to illustrate the general concept shown in Figure 6.39. In most high speed system applications, a passive antialiasing filter (either low-pass for baseband sampling, or bandpass for undersampling) is required, and placing this filter between the drive amplifier and the ADC can significantly reduce the noise contribution due to the amplifier. The filter therefore serves not only as an antialiasing filter but also as a noise filter for the amplifier. It should be noted, however, that if the filter is placed between the amplifier and the ADC, then the amplifier must be able to drive the impedance of the filter without significant distortion.

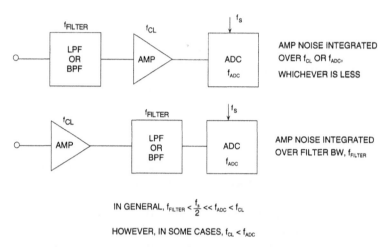

**Figure 6.39: Proper Positioning of the Antialiasing Filter
Will Reduce the Effects of Op Amp Noise**

High speed wide dynamic range ADCs such as the AD6645 14-bit 80/105 MSPS ADC require very low noise, low distortion drivers, and RF transformers generally give optimum performance as previously described. However, there are applications where dc coupling is required, and this places an extremely high burden on the differential driver. Figure 6.40 shows the AD8139 differential driver operating as a dc-coupled level shifter as in the previous examples. The AD8139 has a rail-to-rail output stage, a bandwidth of 370 MHz, and a voltage noise of $2\ nV/\sqrt{Hz}$. SFDR is greater than 88 dBc for a 20 MHz, 2 V p-p output.

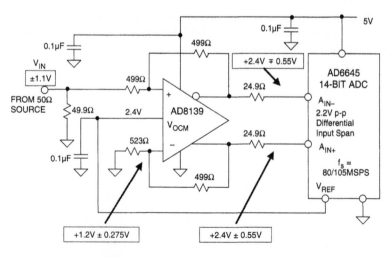

**Figure 6.40: AD8139
Application as DC-Coupled
Driver for the AD6645
14-bit, 80/105 MSPS ADC**

The AD6645 input common-mode voltage is set by its internal reference of 2.4 V, and this voltage is in turn applied to the AD8139 V_{OCM} input pin. The outputs of the AD8139 swing between 1.85 V and 2.95 V, well within the common-mode output range of the amplifier.

Another RF/IF differential amplifier useful for an ac-coupled driver for ADCs such as the AD6645 is the 2.2 GHz AD8351 (Reference 13). A typical application circuit is shown in Figure 6.41.

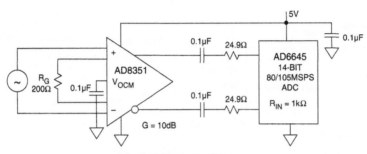

AD8351 KEY FEATURES
- 3dB Bandwidth: 2.2GHz for gain of 12dB
- Slew rate: 13,000V/μs
- Single resistor programmable gain, 0dB to 26dB
- Input noise: 2.7nV/√Hz
- Single supply: 3V to 5.5V
- Adjustable output common-mode voltage

Figure 6.41: AD8351 Low Distortion Differential RF/IF Amplifier Application

The AD8351 sets the standard in high performance, low distortion differential ADC drivers. It is ideal where additional low-noise gain is required ahead of the ADC. Gain is resistor programmable from 0 dB to 26 dB. Output common-mode voltage is set via the V_{OCM} pin. The AD8351 input stage operates at a common-mode voltage of about 2.5 V and is not designed for dc coupling.

Typical performance data as a driver for the AD6645 is shown in Figure 6.42. The data was taken for a sampling rate of 80 MSPS with an input signal of 65 MHz. The undersampled 65-MHz signal appears in the FFT output spectrum at 15 MHz (80 MHz – 65 MHz = 15 MHz). The gain of the AD8351 is set for 10 dB, and the SFDR is 78.2 dBc.

Figure 6.42: AD8351 Differential ADC Driver Performance with AD6645 ADC (G = 10 dB)

SAMPLING RATE:	80MSPS
INPUT FREQ:	65MHz
SNR:	69.1dB
HD2:	−78.5dBc
HD3:	−80.7dBc
THD:	−75.9dBc
SFDR:	78.2dBc

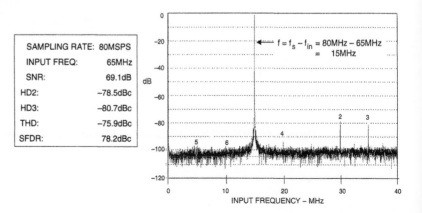

The AD8351 can also be used as an ac-coupled single-ended-to-differential converter when working with single-ended signals as shown in the application circuit of Figure 6.43. The external resistors R_F and R_G are selected per the data sheet recommendations. Even though the differential balance is not perfect under these conditions, the SFDR for a 65 MHz input is reduced by only a few dB relative to the fully differential case shown in Figure 6.42.

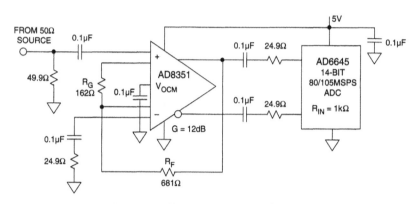

Figure 6.43: Using the AD8351 as a Single-Ended-to-Differential Converter

For low distortion differential 12-bit ADC driver applications where programmable variable gain is required, the AD8370 digitally controlled variable-gain amplifier (VGA) is an excellent choice (Reference 14). Figure 6.44 shows the AD8370 as a single-ended-to-differential converter driving the AD9433 12-bit, 105/125 MSPS BiCMOS ADC. The 3 dB bandwidth of the AD8370 is 700 MHz, and the gain is programmable over two ranges (−11 dB to +17 dB and +6 dB to +34 dB) via a 3-wire serial interface. The AD8370 is designed for use at IF frequencies up to 380 MHz.

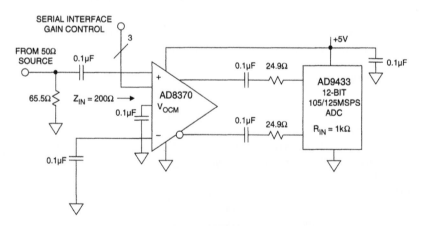

Figure 6.44: AD8370 Variable Gain Amplifier as a Low Distortion ADC Driver

Driving low distortion high performance 16-bit ADCs such as the AD7677 16-bit, 1 MSPS ADC requires special care, especially with respect to noise and linearity. For example, the AD7677 has an INL specification of ±1 LSB, THD of –110 dB at 45 kHz, and 94 dB SINAD @ 45 kHz.

The AD7677 is a CMOS charge redistribution switched capacitor SAR design that operates on a single 5 V supply (Reference 15). Typical power dissipation is only 115 mW when operating at 1 MSPS. The converter is optimized for a differential drive input. Input referred noise is only 0.35 LSB rms, so a low noise drive amplifier is required. The AD8021 200 MHz op amp was especially designed with 16-bit systems in mind (Reference 16). Voltage noise is only $2.1\ nV/\sqrt{Hz}$, and distortion is less than 90 dBc for a 1 MHz output. The AD8021 also has dc precision with 1 mV maximum offset voltage, and 0.5 μV/°C drift. Quiescent current is 7 mA.

The low noise drive circuit in Figure 6.45 shows a single-ended-to-differential conversion using a pair of AD8021 op amps. The output common-mode voltage is set for 1.25 V by applying 1.25 V to the noninverting input of the bottom AD8021. With no input filtering, the output noise of the differential driver must be integrated over the entire 16 MHz input bandwidth of the AD7677. This noise contribution can be reduced to approximately 0.13 LSB rms by the addition of a simple single-pole 4 MHz RC low-pass filter as shown.

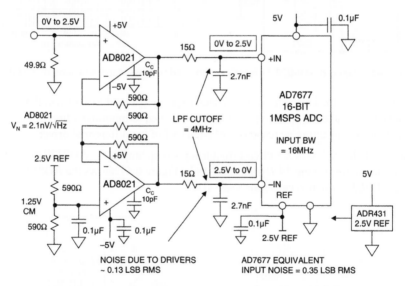

Figure 6.45: A True 16-bit ADC Requires a True 16-Bit Driver

The circuit shown in Figure 6.45 will operate with excellent matching up to several MHz. However, the matching of the outputs can be extended to greater than 100 MHz by individually compensating the two AD8021 op amps as shown in Figure 6.46. The inverting and noninverting bandwidths can be closely matched using this technique, thus minimizing distortion. This circuit illustrates an inverter-follower driver operating at a gain of 2, using individually compensated AD8021s.

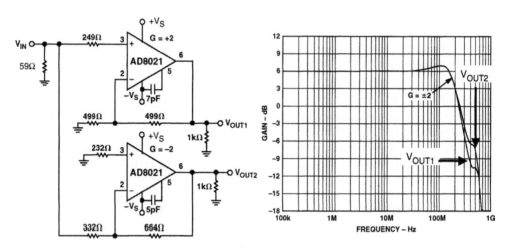

Figure 6.46: Balanced AD8021 Driver Compensated to Give Matched Gains to > 100 MHz

The values of feedback and load resistors were selected to provide a total load of less than 1 kΩ, and the equivalent resistances seen at each op amp's inputs were matched to minimize offset voltage and drift. Figure 6.46 also shows the resulting ac responses of each half of the differential driver.

Rather than using the balanced AD8021 circuit (requiring two op amps), the AD8139 differential amplifier offers another alternative for driving 14-/16-/18-bit ADCs. Figure 6.47 shows the AD8139 driving the 18-bit, 800-kSPS AD7674 switched capacitor SAR ADC.

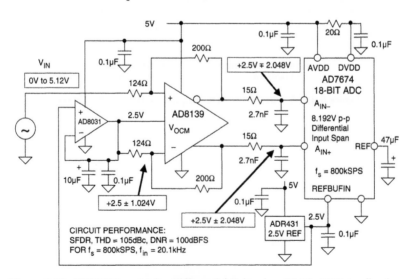

Figure 6.47: AD8139 Low Noise Differential Driver in a 18-Bit ADC Application

Applying a 2.5 V reference to the REFBUFIN pin of the AD7674 generates an internal reference voltage of 4.096 V. The input range of the ADC is then equal to 8.192 V p-p differential.

The circuit scales and level shifts the unipolar 0 V to 5.12 V input voltage to fit the range of the AD7674. The required gain of 1.6 is set by the ratio of the feedback to the feedforward resistor: $200\ \Omega/124\ \Omega = 1.6$. The required common-mode voltage of 2.5 V is developed from the external ADR431 reference which also drives the AD7674 REFBUFIN. This voltage must be buffered by the AD8031 wideband op amp because of the required sink/source current of approximately ± 8.2 mA. The signals at the outputs of the AD8139 must swing from 0.5 V to 4.5 V (out of phase), which is within the range of the AD8139 operating on a single 5 V supply. The signals at the inputs of the AD8139 swing between 1.476 V and 3.524 V, which is well within the allowable input common-mode range when operating on a single 5 V supply.

As in the circuit previously shown in Figure 6.45, the 15 Ω resistors in conjuction with the 2.7 nF capacitors form a 4 MHz low-pass filter to the output noise of the AD8139.

References:
6.1 Driving ADC Analog Inputs

1. Data sheet for AD8027 Low Distortion, High Speed, Rail-to-Rail Input/Output Amplifier, Data sheet for AD8028 Low Distortion, Dual, High Speed, Rail-to-Rail Input/Output Amplifier, www.analog.com.

2. Data sheet for AD8057/AD8058 Low Cost, High Performance Voltage Feedback, 325 MHz Amplifiers, www.analog.com.

3. Chapter 8 of Walt Kester, Editor, Practical Design Techniques for Sensor Signal Conditioning, Analog Devices, 1999, ISBN: 0-916550-20-6. (Also available at www.analog.com.)

4. Data Sheet for AD7466/AD7467/AD7468 1.6 V, Micropower 12-/10-/8-Bit ADCs in 6-Lead SOT-23, www.analog.com.

5. Data Sheet for AD7890 8-Channel, 12-Bit Serial, Data Acquisition System, www.analog.com.

6. Mini-Circuits, P.O. Box 350166, Brooklyn, NY, 11235, 718-934-4500, www.minicircuits.com.

7. Data Sheet for AD9430 12-Bit, 170 MSPS/210 MSPS 3.3 V A/D Converter, www.analog.com.

8. Data Sheet for AD6645 14-Bit, 80 MSPS/105 MSPS A/D Converter, www.analog.com.

9. Data Sheets for AD8131, AD8132, AD8137, AD8138, AD8139 Differential Amplifiers, www.analog.com.

10. Data Sheets for AD830, AD8129, AD8130 Differential Receiver Amplifiers, www.analog.com.

11. Data Sheet for AD7450A/AD7440 Differential Input, 1 MSPS, 12- and 10-Bit ADCs in 8-Lead SOT-23, www.analog.com.

12. Data Sheet for AD9235 12-Bit, 20/40/65 MSPS 3 V A/D Converter, www.analog.com.

13. Data Sheet for AD8351 Low Distortion Differential RF/IF Amplifier, www.analog.com.

14. Data Sheet for AD8370 Digital Control VGA 700MHz Differential Amplifier, www.analog.com.

15. Data Sheet for AD7677 16-Bit, 1 LSB INL, 1 MSPS Differential ADC, www.analog.com.

16. Data Sheet for AD8021 Low Noise, High Speed Amplifier for 16-Bit Systems, www.analog.com.

ADC and DAC Digital Interfaces (and Related Issues)
Walt Kester

Introduction

A discussion of the broad area of data converter digital interfaces, timing, and so forth, can quickly become detailed and very tedious because of the many variations associated with specific products. We will therefore attempt to point out only the highlights in this section. While it is possible to generalize to some degree, the fact is that there is absolutely no substitute for careful study of the particular converter data sheet to clarify key points.

Modern data converters are much more digitally intensive than their predecessors of a few years ago. For example, high resolution Σ-Δ measurement ADCs typically have a number of internal control registers that are used to determine channel selection, set filter bandwidth, throughput rate, PGA gain, etc. These registers must be properly loaded by sending data to them via a serial interface port. This same serial port is often used to read the data out of the ADC at the end of a conversion cycle. Modern high frequency communications converters have also become digitally intensive. For instance, direct digital synthesis (DDS) ICs have internal registers that control the output frequency, amplitude, phase, type of modulation, etc.

Other issues relating to the digital and timing portions of data converters are the condition of logic states immediately after power-on, the effect of pipeline delays, burst mode operation (some will, some won't), minimum sampling frequency, sleep and standby modes, etc. Many of these topics are very similar to those encountered when designing with microprocessors, microcontrollers, and DSPs. However, successful designing with data converters not only requires understanding of digital and timing issues but also diligent attention to the analog design—layout, grounding, decoupling, etc. These hardware design topics are covered in considerable detail in Chapter 9 of this book.

Power-On Initialization of Data Converters

When power is first applied to a simple flip-flop—the fundamental digital storage element—there is generally no way to accurately predict what its output state will be. Without the addition of additional *power-on* circuitry or initialization procedures, the same is true of the many registers contained inside microprocessors, microcontrollers, DSPs, and of course, mixed-signal devices such as ADCs and DACs.

While power-on reset features have been common with microprocessors, microcontrollers, and DSPs, such features are now included in some data converters—especially those that are highly digitally intensive, or where it is critical that signals be at certain levels after power-on.

A good example is a DAC that is used inside an industrial control loop. If the DAC is controlling an actuator, such as a vibration table, one can easily visualize a potential problem if the DAC analog output is full scale at power-on. For this reason, many IC DACs used in industrial applications have internal power-on circuitry that forces the initial digital data into the DAC register (the register that controls the state of the DAC switches) to a known value (generally all 0s or midscale).

A digital potentiometer is another example of a device where the power-on state can be important. For this reason, some digital pots have on-chip nonvolatile memory that stores the desired setting. Other digital pots without nonvolatile memory usually have on-chip circuitry which forces the initial power-on value to either zero or midscale (the actual choice is pin selectable in some cases).

There is less reason to be concerned with the state of ADC outputs on power-on, because one is not generally interested in the ADC output until after a conversion command of some sort is applied. However, pipelined and Σ-Δ ADCs generally do require a number of sample clock cycles before the digital "pipeline" is flushed out and the output data is valid. Again, the data sheet for the device specifies this parameter.

Initialization of Data Converter Internal Control Registers

Modern data converters, especially those offering a high degree of functionality, often utilize internal control registers to set various operational parameters. For instance, the AD77xx family of Σ-Δ ADCs offer programmable throughput rate, filter cutoff frequency, amplifier gain, channel selection, etc. These parameters must be loaded into the ADC after power-on via a serial port. In order to ensure proper operation, these ADCs generally incorporate power-on reset and initialization circuitry which programs a known set of default values into the critical registers upon power-on. This allows the user to start system initialization with the ADC in valid operational state—an invaluable feature when troubleshooting an initial design at the PC board level.

In addition to the power-on reset feature, these types of ADCs generally have a separate reset pin which allows the converter to be put into a known state any time after power is connected. In some cases, the ADCs can also be reset to default conditions under software control.

Highly integrated DACs, Direct Digital Synthesis (DDS) systems, and many other mixed-signal ICs also have initialization features such as power-on reset, default modes, etc. As previously discussed, some have on-chip nonvolatile memory which can be used to store the desired settings. The trend towards more integration and more programmability will ultimately lead to even more devices with on-chip volatile and nonvolatile memory.

Low Power, Sleep, and Standby Modes

In order to conserve power, especially in battery-powered applications, most modern data converters have some type of low power, sleep, or standby mode, where the major portion of the internal circuitry is powered down—usually initiated by the application of a signal to one of the pins, but sometimes under software control via internal control registers. In many applications where the converter is not required to operate continuously, this feature can lead to considerable power savings. Some converters have several reduced-power modes, depending upon the amount of circuitry to be shut down. In some cases, additional power savings can be achieved by disabling some or all of the external clocks.

Sleep-mode power supply current varies widely between devices, and can range from a few μA to tens of mA depending upon the normal-mode power dissipation. Recovery time from the sleep mode, or power-up time, is also a critical specification and can vary widely, depending upon the device, but generally is in the order of a few μs to 100 μs.

During the sleep mode, power is maintained on critical internal mode-controlling registers, etc.; however, the conversion process is usually disabled. If the converter is pipelined or has internal digital filters (such as Σ-Δ ADCs or certain DACs with internal digital filters), a sufficient number of clock cycles must be allowed after power-up to flush out the pipelines before output data is valid.

Single-Shot Mode, Burst Mode, and Minimum Sampling Frequency

This brings up an interesting timing issue with respect to pipelined ADCs. It is not specifically related to the digital interface, but has a direct bearing on the application—the ability (or lack thereof) to operate at very low sampling rates, the burst mode, or the single-shot mode.

Many early successive approximation ADCs, such as the industry-standard AD574, were designed with internal clock generators that were triggered upon the receipt of an external *convert-start* signal. At the end of the conversion cycle, the signal on an output line (labeled *busy, conversion complete, data ready,* etc.) was asserted, indicating that the data was valid and that the conversion was complete. This type of ADC can be utilized in the single-shot mode, burst mode, or operated continuously, with no significant effect on performance. Many modern successive approximation ADCs require that the user supply a continuous high frequency clock (which controls the various steps in the conversion process) as well as the traditional *convert-start* pulse to initiate the actual conversion. The *convert-start* pulse can be synchronous or asynchronous with the high frequency clock in many converters. As long as the user supplies a continuous high frequency clock, these types of ADCs can generally be operated in the single-shot or burst mode.

It should be noted that this is one of the fundamental reasons why SAR ADCs are still so popular in data acquisition, especially multichannel systems where an analog multiplexer drives the ADC. A single convert-start command yields the corresponding data—with no pipeline delay—thereby making it easy to identify the output data corresponding to a particular channel and clock pulse.

On the other hand, pipelined ADCs (see Chapter 3 of this book for detailed descriptions) require a number of sample clocks after power-on before the pipeline is cleared, and valid data appears at the output. In addition, the cascaded internal sample-and-hold amplifiers act as analog delay lines, and they are typically controlled by one or both phases of the actual sampling clock. That is, the "1" state of the sampling clock places some of the SHAs in the *track* mode, and the others in the *hold* mode. The "0" state of the sampling clock reverses the track/hold states of the SHAs. The direct or indirect utilization of the sampling clock phases to control all internal operations reduces chip area, cost, and improves performance by eliminating additional internal clocks which could easily increase overall ADC noise and distortion were they to become noisy because of stray coupling from other parts of the circuit.

However, one can see that as the sampling frequency is decreased, the *hold* time of the SHAs increases proportionally—at some point, the *droop* (caused by leakage current flowing into or out of the hold capacitor) associated with the long hold times will produce large errors in the conversion, thereby rendering the output data invalid. In addition, internal circuits may enter saturation. Therefore, pipelined ADCs often have a *minimum* specified sampling frequency as well as the traditional maximum.

Although most pipelined ADCs cannot be directly operated in the single-shot or burst mode, they can be operated with a continuous sampling clock, and the output data gated to correspond with the desired sampling intervals.

- Power-on Reset and Initialization
  - DACs and Digital Pots
  - ADCs with Internal Control Registers
  - Default Conditions
  - Pipelined ADCs
- Low Power, Sleep, Standby Modes
  - Power Savings
  - Recovery or Power-Up Time
- Single-Shot, Burst Mode, and Minimum Sampling Frequency
  - SAR ADCs
  - Pipelined ADCs
- Get to Know Your Friendly Data Sheet
  - "Getting to know you, getting to know all about you ..." "Anna," from Rogers and Hammerstein's, The King and I

Figure 6.48: Some Important Digital and Timing Interface Issues for Data Converters

Much more could be said about the topics discussed so far, but the reader should at least now be aware of some of the important issues that only a thorough study of the data sheet can clarify. The same can be said about the following section on the digital interface itself.

ADC Digital Output Interfaces

Early ADCs typically provided parallel output data interfaces. As resolutions increased, and microprocessors, microcontrollers, and DSPs became widespread, the serial interface became popular. Today, most 12-bit or greater ADCs that operate at or above 10 MSPS typically have a parallel output data interface, while low frequency high resolution Σ-Δ measurement ADCs almost exclusively utilize a serial interface. In between these two sampling frequency ranges, one finds a wide variety of ADCs—some with parallel, some with serial, and some with options for both parallel and serial output data interfaces.

ADC Serial Output Interfaces

Serial interfaces are typically 3-wire (sometimes 2-wire), and therefore there is a big savings in package pin count and cost versus the parallel interface, especially with high resolution ADCs. It is also very convenient to provide serial outputs on SAR-based and Σ-Δ ADCs since their conversion architecture is essentially serial. If an ADC is operating continuously, the period of the sampling clock must be long enough to transfer all the serial data across the interface at the interface data rate, with some appropriate amount of headroom. For instance, a 16-bit, 1 MSPS sampling ADC requires a serial output data rate of at least 16 MHz, which would not be a problem with most modern microprocessors, microcontrollers, or DSPs.

Most 3-wire serial interfaces associated with ADCs and DACs are compatible with standard serial interfaces such as SPI®, QSPI™, MICROWIRE™, and DSPs. Figure 6.49 shows the timing diagram for a typical serial output converter, the AD7466 12-bit, 200 kSPS ADC which is packaged in a 6-lead SOT-23 package.

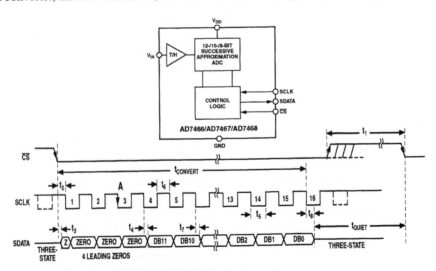

**Figure 6.49: AD7466 12-Bit, 200-kSPS
Serial Output Data Timing Diagram**

The AD7466 is normally in the power-down mode with the $\overline{CS}$ signal high. The part begins to power up on the $\overline{CS}$ falling edge. The falling edge of $\overline{CS}$ puts the track-and-hold into the track mode and takes the bus out of three-state. The conversion is also initiated at this point. On the third SCLK falling edge after the

$\overline{CS}$ falling edge, the part should be fully powered up, as shown in Figure 6.49 at point "A," and the track-and-hold will return to hold. For the AD7466, the SDATA line will go back into three-state, and the part will enter power-down on the 16th SCLK falling edge. If the rising edge of $\overline{CS}$ occurs before 16 SCLKs have elapsed, the conversion will be terminated, the SDATA line will go back into three-state, and the part will enter power-down; otherwise SDATA returns to three-state on the 16th SCLK falling edge, as shown in Figure 6.49. Sixteen serial clock cycles are required to perform the conversion process and to access data from the AD7466.

$\overline{CS}$ going low provides the first leading zero to be read in by the microcontroller or DSP. The remaining data is then clocked out by subsequent SCLK falling edges, beginning with the second leading zero; thus the first clock falling edge on the serial clock has the first leading zero provided and also clocks out the second leading zero. For the AD7466, the final bit in the data transfer is valid on the 16th SCLK falling edge, having been clocked out on the previous (15th) SCLK falling edge. In applications with a slow SCLK, it is possible to read in data on each SCLK rising edge. In such a case, the first falling edge of SCLK after the $\overline{CS}$ falling edge will clock out the second leading zero and could be read in the following rising edge. If the first SCLK edge after the $\overline{CS}$ falling edge is a falling edge, the first leading zero that was clocked out when $\overline{CS}$ went low will be missed unless it is not read on the first SCLK falling edge. The 15th falling edge of SCLK will clock out the last bit and it could be read in the following rising SCLK edge. If the first SCLK edge after $\overline{CS}$ falling edge is a rising edge, $\overline{CS}$ will clock out the first leading zero as before, and it may be read on the SCLK rising edge. The next SCLK falling edge will clock out the second leading zero, and it could be read on the following rising edge.

Looking at higher speed applications, LVDS (low voltage differential signaling) interfaces can be as high as 800 Mbits/s, thereby making serial data transfer practical even for some high speed ADCs. For instance, the AD9289 quad 12-bit, 65 MSPS ADC uses four serial LVDS outputs, each operating at 780 Mbits/s. A functional block diagram of the quad ADC is shown in Figure 6.50 (also see Reference 2).

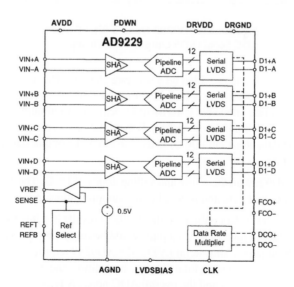

**Figure 6.50: AD9229 Quad 12-Bit,
65 MSPS ADC with Serial LVDS Outputs**

The AD9229 is a quad 12-bit, 65 MSPS ADC converter with an on-chip track-and-hold circuit and is designed for low cost, low power, small size and ease of use. The converter operates up to 65 MSPS conversion rate and is optimized for outstanding dynamic performance where a small package size is critical. The ADC requires a single 3 V power supply and CMOS/TTL sample rate clock for full performance operation. No external reference or driver components are required for many applications. A separate output power supply pin supports LVDS-compatible serial digital output levels. The ADC automatically multiplies the sample rate clock for the appropriate LVDS serial data rate. An MSB trigger is provided to signal a new output byte. Power down is supported, and the ADC consumes less than 3 mW when enabled. A timing diagram is shown in Figure 6.51.

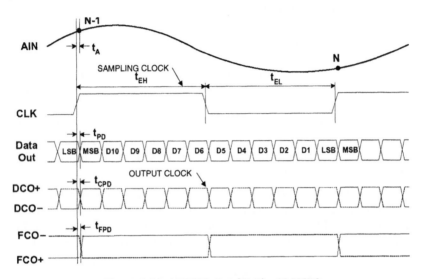

Figure 6.51: AD9289 Quad 8-Bit, 65 MSPS
ADC Serial LVDS Output Data Timing

Data from each ADC is serialized and provided on a separate channel. The data rate for each serial stream is equal to 12 bits times the sample clock rate, with a maximum of 780 MHz (12 bits × 65 MSPS = 780 MHz). The lowest typical conversion rate allowable is 10 MSPS (recall that minimum sampling frequency specifications are characteristic of CMOS pipelined ADCs). Two output clocks are provided to assist in capturing data from the AD9289. The data clock out (DCO) is used to clock the output data and is equal to six times the sampling clock (CLK) rate.

Data is clocked out of the AD9229 on the rising and falling edges of DCO. The MSB clock (FCO) is used to signal the MSB of a new output byte and is equal to the sampling clock rate.

The use of high speed serial LVDS data outputs in the AD9229 results in a huge savings in the pin count, compared with parallel outputs. A total of 48 data pins would be required to provide four individual parallel 12-bit single-ended CMOS outputs. Using serial LVDS, the AD9289 requires only four differential LVDS data outputs, or eight pins, thereby saving a total 40 pins. In addition, the use of LVDS rather than CMOS reduces digital output transient currents and the overall ADC noise. A typical LVDS output driver designed in CMOS is shown in Figure 6.52. Further details regarding the LVDS specification can be found in Chapter 9 of this book.

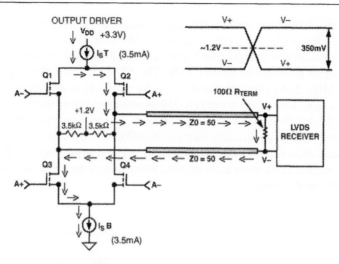

Figure 6.52: LVDS Driver Designed in CMOS

ADC Serial Interface to DSPs

Because of its simplicity and efficiency, the serial interface has become a very popular way to interface ADCs and DACs to DSPs, and real-time operation is possible in many instances. We will consider a typical example of such an interface between a general purpose ADC and a fixed-point DSP.

The AD7853/AD7853L is a 12-bit, 200/100 kSPS ADC that operates on a single 3 V to 5.5 V supply and dissipates only 4.5 mW (3 V supply, AD7853L). After each conversion, the device automatically powers down to 25 μW. The AD7853/AD7853L is based on a successive approximation architecture and uses a charge redistribution (switched capacitor) DAC. A calibration feature removes gain and offset errors. A block diagram of the device is shown in Figure 6.53 (for more details, see Reference 3).

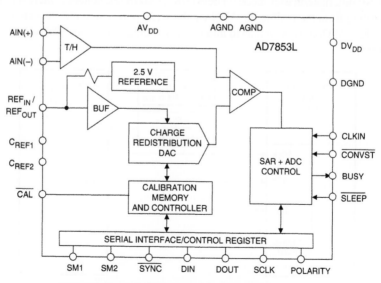

Figure 6.53: AD7853/AD7853L 3 V Single-Supply
12-Bit 200/100 kSPS Serial Output ADC

The AD7853 operates on a 4 MHz maximum external clock frequency. The AD7853L operates on a 1.8 MHz maximum external clock frequency. The timing diagram for AD7853L is shown in Figure 6.54. The AD7853/AD7853L has modes that configure the $\overline{\text{SYNC}}$ and SCLK as inputs or outputs. In the example shown here they are outputs generated by the AD7853L. The AD7853L serial clock operates at a maximum frequency of 1.8 MHz (556 ns period). The data bits are valid 330 ns after the positive-going edges of SCLK. This allows a setup time of approximately 330 ns minimum before the negative-going edges of SCLK, easily meeting the ADSP-2189M 4 ns t_{SCS} requirement. The hold-time after the negative-going edge of SCLK is approximately 226 ns, again easily meeting the ADSP-2189M 7 ns t_{SCH} timing requirement. These simple calculations show that the data and RFS setup and hold requirements of the ADSP-2189M are met with considerable margin. For a much more detailed discussion of the serial interface timing between ADCs, DACs, and DSPs see Reference 5.

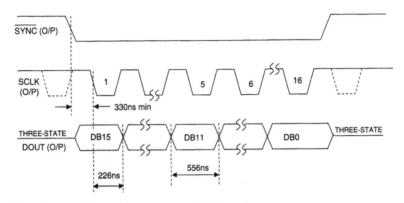

Figure 6.54: AD7853L Serial ADC Output Timing 3 V Supply, SCLK = 1.8 MHz

Figure 6.55 shows the AD7853L interfaced to the ADSP-2189M connected in a mode to transmit data from the ADC to the DSP (alternate/master mode). The AD7853/AD7853L contains internal registers that can be accessed by writing from the DSP to the ADC via the serial port. These registers are used to set various modes in the AD7853/AD7853L as well as to initiate the calibration routines. These connections are not shown in the diagram.

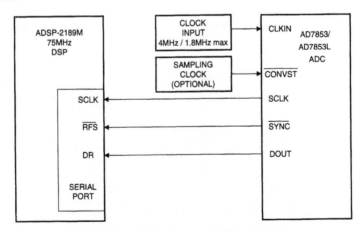

**Figure 6.55: Interfacing the AD7853/AD7853L
Serial Output ADCs to the ADSP-2189M DSP**

ADC Parallel Output Interfaces

Parallel ADC output interfaces are popular, straightforward, and must be used when the product of sampling rate and resolution exceeds the capacity available serial links. For instance, using a maximum LVDS serial data link of 600 Mbits/s requires parallel data transmission for resolutions/sampling rates greater than 8 bits at 75 MSPS, 10 bits at 60 MSPS, 12 bits at 50 MSPS, 14 bits at 43 MSPS, 16 bits at 38 MSPS, etc.

Parallel ADC interface timing is relatively straightforward. At some specified time relative to the assertion of the appropriate edge of the sampling clock, the output data is valid. This time is specified on the data sheet, and may or may not be indicated by a *data ready*, or *data valid* output from the ADC. Also, the data appearing at the output may correspond to a previously applied sampling clock edge due to the pipeline delay of the ADC. In most cases, the output data is valid for an entire sampling clock period (neglecting the rise and fall times). Some parallel output ADCs have a *chip enable* function which allows the data outputs to be connected to a data bus, and the outputs are three-state until the chip enable is asserted by an external DSP, microcontroller, or microprocessor. However, general precautions must be taken when connecting this type of output to a data bus—the most important is to ensure that there is no activity on the bus during the actual ADC conversion interval. Otherwise, bus activity may couple back into the ADC via the stray pin capacitance and corrupt the conversion. In addition, if the capacitive load of the bus is significant, there may be additional ADC digital output transients which can corrupt the conversion.

We will use the AD9430 12-bit, 170/210 MSPS ADC to illustrate the timing associated with a modern high speed parallel output device (Reference 6). An overall block diagram of the AD9430 is shown in Figure 6.56. Notice that this ADC offers two output data options: demultiplexed CMOS outputs on two ports (each at one-half the sampling rate) or differential LVDS outputs at the full sampling rate. There is no penalty in pin count by providing these two options, because demuxed single-ended outputs on two ports require the same number of pins as differential LVDS outputs on a single port.

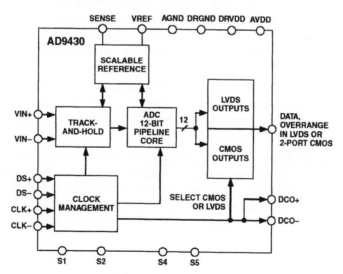

Figure 6.56: AD9430 12-Bit, 170/210 MSPS ADC with LVDS or Demuxed CMOS Output Data Options

Figure 6.57 shows the AD9430 timing when using the LVDS output mode. The AD9430 operates on an LVDS-compatible differential sampling clock which passes through internal *clock management* circuitry that stabilizes the duty cycle and thereby removes the sensitivity of the conversion process to variations in input sampling clock duty cycle.

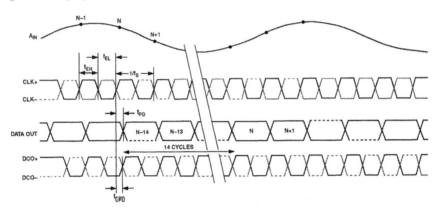

Figure 6.57: AD9430 LVDS Output Data Timing

If the sampling frequency is known, the timing diagram in conjunction with the associated specifications for t_{PD} and t_{CPD} can be used to predict when the output data is valid with respect to either the positive-going edge of the sampling clock (CLK+) or the positive-going or negative-going edge of the data output clock (DCO+).

Because of the high 210 MSPS sampling rate (period = 4.76 ns), it is critical that both the ADC output timing and the receiver input timing be carefully examined so that the receiving register or memory can be clocked when the output data is stable. This "window" is short and, in the case of the AD9430, a *data valid* time of 2 ns minimum is guaranteed. ADC output timing, PC board trace delay and the input register (usually an FPGA) setup and hold time specifications all factor into determining the proper timing for a particular design, and great care must be taken in the analysis to ensure valid data is obtained.

Although best distortion and noise performance is obtained in the LVDS mode, the AD9430 can also be operated in the CMOS data output mode, in which case the output data is demultiplexed and available on two output ports at one-half the overall ADC sampling rate. The timing diagram for the CMOS mode is shown in Figure 6.58. Note that data is available in either interleaved or parallel format, depending upon the option selected.

High speed ADCs such as the AD9430 typically interface to an FPGA or buffer memory. Lower speed parallel output ADCs can interface directly to microcontrollers or DSPs via a standard parallel data bus. A good example is the AD7854/AD7854L 3 V, 12-bit, 200/100 kSPS parallel output ADC (Reference 7). This device uses a successive approximation architecture based on a charge redistribution (switched capacitor) DAC. A calibration mode removes offset and gain errors. A block diagram of this general-purpose converter is shown in Figure 6.59.

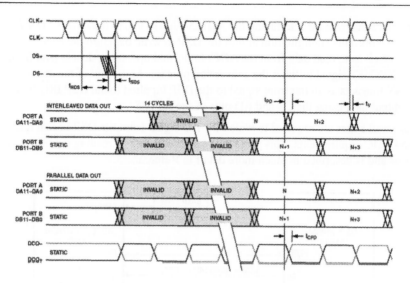

Figure 6.58: AD9430 Demuxed CMOS Output Data Timing

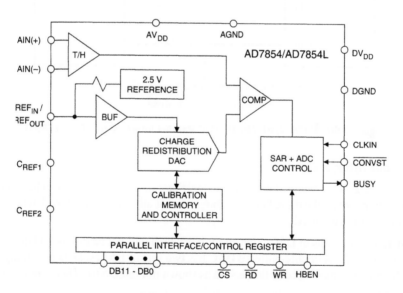

Figure 6.59: AD7854/AD7854L, 3 V Single Supply,
12-bit, 200/100 kSPS Parallel Output ADC

A simplified interface diagram for interfacing the AD7854/AD7854L to the ADSP-2189 75 MHz DSP is shown in Figure 6.60. This configuration allows the DSP to write data into the ADC parallel interface control register as well as to read data from the ADC. In normal operation, data is read from the ADC. The assertion of the $\overline{\text{CONVST}}$ signal initiates the conversion process. At the end of the conversion, the assertion of the ADC BUSY line acts as an interrupt signal to the DSP (applied to the DSP $\overline{\text{IRQ}}$ input). The DSP then reads the ADC output data using the $\overline{\text{CS}}$ and $\overline{\text{RD}}$ pins of the AD7854.

The five software wait states are required to widen the $\overline{\text{RD}}$ signal from the DSP so that it is compatible with the AD7854 ADC requirements. This process is a standard way of reading data from memory-mapped peripheral devices and is described in much more detail in Reference 5.

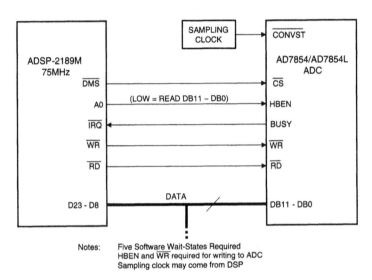

Figure 6.60: AD7854/AD7854L ADC Parallel Interface to ADSP-2189M

DAC Digital Input Interfaces

The earliest monolithic DACs contained little, if any, logic circuitry, and parallel data had to be maintained on the digital input to maintain the digital output. Today almost all DACs have input latches, and data need only be written once, not maintained.

There are innumerable variations in DAC input structures which will not be discussed here, but the majority today are "double-buffered." A double-buffered DAC has two sets of latches. Data is initially latched in the first rank and subsequently transferred to the second as shown in Figure 6.61. There are three reasons why this arrangement is useful.

The first is that it allows data to enter the DAC in many different ways. A DAC without a latch, or with a single latch, must be loaded with all bits at once, in parallel, since otherwise its output during loading may be totally different from what it was or what it is to become. A double-buffered DAC, on the other hand, may be loaded with parallel data, serial data, or with 4-bit or 8-bit words, or whatever, and the output will be unaffected until the new data is completely loaded and the DAC receives its update instruction.

The second feature of this type of input structure is that the output clock can operate at a fixed frequency (the DAC update rate), while the input latch can be loaded asynchronously. This is useful in real-time signal reconstruction applications.

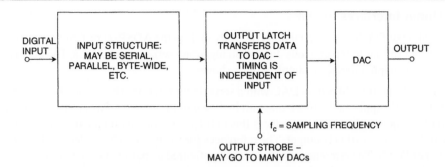

**Figure 6.61: Double-Buffered DAC Permits Complex
Input Structures and Simultaneous Update**

The third convenience of the double-buffered structure is that many DACs may be updated simultaneously: data is loaded into the first rank of each DAC in turn and, when all is ready, the output buffers of all DACs are updated at once. There are many DAC applications where the output of several DACs must change simultaneously, and the double-buffered structure allows this to be done very easily.

Most early monolithic high resolution DACs had parallel or byte-wide data ports and tended to be connected to parallel data buses and address decoders and addressed by microprocessors as if they were very small write-only memories (some DACs are not write-only, but can have their contents read as well—this is convenient for some ATE applications but is not very common). A DAC connected to a data bus is vulnerable to capacitive coupling of logic noise from the bus to the analog output. Many DACs today have serial data structures and are less vulnerable to such noise (since fewer noisy pins are involved), use fewer pins and therefore take less space, and are frequently more convenient for use with modern microprocessors, many of which have serial data ports. Some, but not all, of such serial DACs have data outputs as well as data inputs so that several DACs may be connected in series and data clocked to them all from a single serial port. The arrangement is referred to as "daisy-chaining."

DAC Serial Input Interfaces to DSPs

Interfacing serial input DACs to the serial ports of DSPs such as the ADSP-21xx family is also relatively straightforward and similar to the previous discussion regarding serial output ADCs. The details will not be repeated here, but a simple interface example will be shown.

The AD5322 is a 12-bit, 100 kSPS dual DAC with a serial input interface (Reference 8). It operates on a single 2.5 V to 5.5 V supply, and a block diagram is shown in Figure 6.62. Power dissipation on a 3 V supply is 690 µW. A power-down feature reduces this to 0.15 µW. Total harmonic distortion is greater than 70 dB below full scale for a 10 kHz output. The references for the two DACs are derived from two reference pins (one per DAC). The reference inputs may be configured as buffered or unbuffered inputs. The outputs of both DACs may be updated simultaneously using the asynchronous $\overline{\text{LDAC}}$ input. The device contains a power-on reset circuit that ensures that the DAC outputs power up to 0 V and remain there until a valid write takes place to the device.

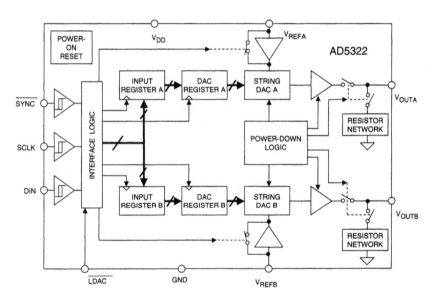

Figure 6.62: AD5322 12-BIT, 100 kSPS Dual Serial DAC

Data is normally input to the AD5322 via the SCLK, DIN, and $\overline{\text{SYNC}}$ pins from the serial port of the DSP. When the $\overline{\text{SYNC}}$ signal goes low, the input shift register is enabled. Data is transferred into the AD5322 on the falling edges of the following 16 clocks. A typical interface between the ADSP-2189M and the AD5322 is shown in Figure 6.63. Notice that the clocks to the AD5322 are generated from the ADSP-2189M clock. It is also possible to generate the SCLK and $\overline{\text{SYNC}}$ signals externally to the AD5322 and use them to drive the ADSP-2189M. The serial interface of the AD5322 is not fast enough to handle the ADSP-2189M maximum master clock frequency. However, the serial interface clocks are programmable and can be set to generate the proper timing for fast or slow DACs.

The input shift register in the AD5322 is 16 bits wide. The 16-bit word consists of four control bits followed by 12 bits of DAC data. The first bit loaded determines whether the data is for DAC A or DAC B. The second bit determines if the reference input will be buffered or unbuffered. The next two bits control the operating modes of the DAC (normal, power-down with 1 kΩ to ground, power-down with 100 kΩ to ground, or power-down with a high impedance output).

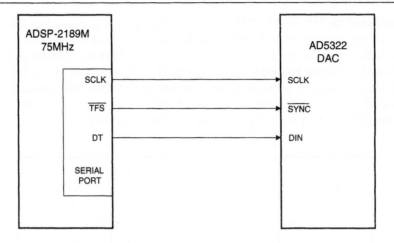

Figure 6.63: AD5322 DAC Serial Interface to ADSP-2189M

DAC Parallel Input Interfaces to DSPs

The AD5340 is a 12-bit 100 kSPS DAC which has a parallel data interface. It operates on a single 2.5 V to 5.5 V supply and dissipates only 345μW (3 V supply). A power-down mode further reduces the power to 0.24 μW. The part incorporates an on-chip output buffer which can drive the output close to both supply rails. The AD5340 allows the choice of a buffered or unbuffered reference input. The device has a power-on reset circuit that ensures that the DAC output powers on at 0 V and remains there until valid data is written to the part. A block diagram is shown in Figure 6.64. The input is double buffered.

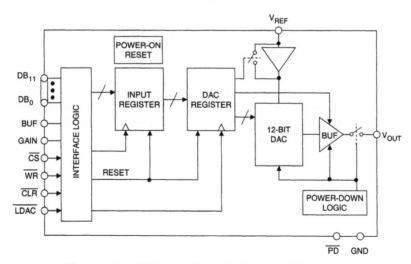

Figure 6.64: AD5340 12-Bit, 100 kSPS Parallel Input DAC

A method for interfacing the AD5340 to a DSP is shown in Figure 6.65. The sampling clock to the DAC updates the internal DAC register via the $\overline{\text{LDAC}}$ input. The sampling clock also generates an interrupt signal to the DSP's $\overline{\text{IRQ}}$ input, thereby requesting a new data word. After the DSP computes the next data word, it puts the word on the data bus and transfers it to the DAC input register via the DAC's $\overline{\text{CS}}$ and $\overline{\text{WR}}$ inputs. Note that this configuration allows real time operation, provided the DSP outputs the new data word before the next sampling clock occurs. The two additional software wait states are required to widen the $\overline{\text{WR}}$ signal from the DSP so that it meets the requirements of the AD5340 DAC.

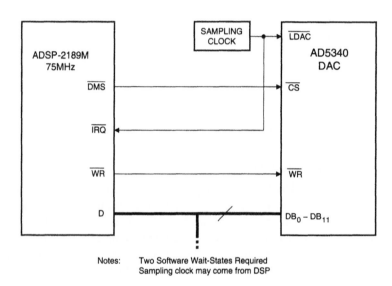

Notes: Two Software Wait-States Required
          Sampling clock may come from DSP

Figure 6.65: AD5340 DAC Parallel Interface to ADSP-2189M

Finally, we will examine a high-speed TxDAC parallel input DAC. The AD9726 is a 16-bit, 600 MSPS DAC that utilizes an LVDS interface to achieve the 600+ MSPS conversion rate (Reference 10). A simplified block diagram is shown in Figure 6.66.

In addition, this device also features unprecedented noise performance of –161 dBm/Hz for output frequencies between 100 MHz and 300 MHz and –169 dBm/Hz at 20 MHz output. This combination of high speed and low noise is ideal for maximizing signal synthesis performance in multicarrier communication systems, as well as in instrumentation and test applications.

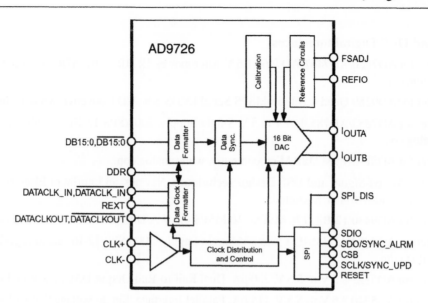

Figure 6.66: AD9726 16-Bit, 600+ MSPS DAC with LVDS Inputs

References:
6.2 ADC And DAC Digital Interfaces

1. Data sheet for AD7466/AD7467/AD7468 1.6 V, Micropower 12-/10-/8-Bit ADCs in 6-Lead SOT-23, www.analog.com.

2. Data sheet for AD9289 Quad 8-Bit, 65-MSPS Serial LVDS 3 V A/D Converter, www.analog.com.

3. Data sheet for AD7853/AD7853L 3 V to 5 V, Single-Supply, 200 kSPS 12-Bit Sampling ADC, www.analog.com.

4. Data sheet for ADSP-2189M DSP Microcomputer, www.analog.com.

5. Walt Kester, **Mixed-Signal and DSP Design Techniques**, Newnes, an Imprint of Elsevier Science, 2003, ISBN-0-75067-611-6, Section 8.

6. Data sheet for AD9430 12-Bit, 170 MSPS/210 MSPS 3.3 V A/D Converter, www.analog.com.

7. Data sheet for AD7854/AD7854L 3 V to 5 V Single-Supply, 200 kSPS 12-Bit Sampling ADC, www.analog.com.

8. Data sheet for AD5322 2.5 V to 5.5 V, 230 µA, Dual Rail-to-Rail Output DAC, www.analog.com.

9. Data sheet for AD5340 2.5 V to 5.5 V, 115 µA, Parallel Interface, Single Voltage Output DAC, www.analog.com.

10. Data sheet for AD9726 16-bit, 600+ MSPS LVDS Input D/A Converter, www.analog.com.

Buffering DAC Analog Outputs
Walt Kester

Introduction

Modern IC DACs provide either voltage or current outputs. Figure 6.67 shows three fundamental configurations, all with the objective of using an op amp for a buffered and/or amplified output voltage.

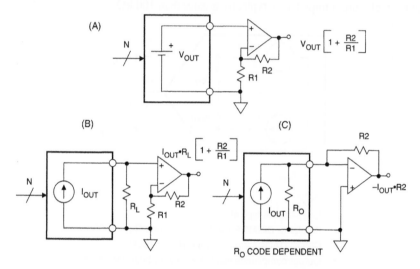

Figure 6.67: Buffering DAC Outputs with Op Amps

Figure 6.67A shows a buffered voltage output DAC. In many cases, the DAC output can be used directly, without additional buffering. If an additional op amp buffer is needed, it is usually configured in a noninverting mode, with gain determined by R1 and R2.

There are two basic methods for dealing with a current output DAC. In Figure 6.67B, a voltage is simply developed across external load resistor, R_L. An external op amp can be used to buffer and/or amplify this voltage if required. Many high speed DACs supply full-scale currents of 20 mA or more, thereby allowing reasonable voltages to be developed across fairly low value load resistors. For instance, fast settling video DACs typically supply nearly 30 mA full-scale current, allowing 1 V to be developed across a source and load terminated 75 Ω coaxial cable (representing a dc load of 37.5 Ω to the DAC output).

A direct method to convert the output current into a voltage is shown in Figure 6.67C. This circuit is usually called a current-to-voltage converter, or I/V. In this circuit, the DAC output drives the inverting input of an op amp, with the output voltage developed across the R2 feedback resistor. In this approach the DAC output always operates at virtual ground (which may give a linearity improvement vis-à-vis Figure 6.67B). Note that an R-2R current-output CMOS DAC must use this configuration, because the output resistance, R_o, is dependent upon the output code (see Chapter 3 of this book on DAC architectures for more details).

The general selection process for an op amp used as a DAC buffer is similar to that of an ADC buffer. The same basic specifications such as dc accuracy, noise, settling time, bandwidth, distortion, etc., apply to DACs as well as ADCs, and the discussion will not be repeated here. Rather, some specific application examples will be shown.

Differential to Single-Ended Conversion Techniques

A general model of a modern current output DAC is shown in Figure 6.68. This model is typical of the AD976x and AD977x TxDAC series (see Reference 1). Current output is more popular than voltage output, especially at audio frequencies and above. If the DAC is fabricated on a bipolar or BiCMOS process, it is likely that the output will sink current, and that the output impedance will be less than 500 Ω (due to the internal R-2R resistive ladder network). On the other hand, a CMOS DAC is more likely to source output current and have a high output impedance, typically greater than 100 kΩ.

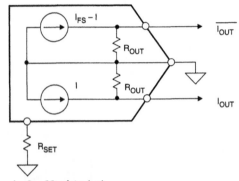

- I_{FS} 2 – 20mA typical
- Bipolar or BiCMOS DACs sink current, R_{OUT} < 500Ω
- CMOS DACs source current, R_{OUT} > 100kΩ
- Output compliance voltage < ±1V for best performance

Figure 6.68: Generalized Model of a High Speed DAC Output such as the AD976x and AD977x Series

Another consideration is the output *compliance voltage*—the maximum voltage swing allowed at the output in order for the DAC to maintain its linearity. This voltage is typically 1 V to 1.5 V, but can vary depending upon the DAC. Best DAC linearity is generally achieved when driving a virtual ground, such as an op amp I/V converter. However, better distortion performance is often achieved when the DAC is allowed to develop a small voltage across a resistive load.

Modern current output DACs usually have differential outputs, to achieve high common-mode rejection and reduce the even-order distortion products. Fullscale output currents in the range of 2 mA to 20 mA are common.

In many applications, it is desirable to convert the differential output of the DAC into a single-ended signal, suitable for driving a coax line. This can be readily achieved with an RF transformer, provided low frequency response is not required. Figure 6.69 shows a typical example of this approach. The high impedance current output of the DAC is terminated differentially with 50 Ω, which defines the source impedance to the transformer as 50 Ω.

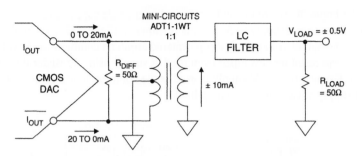

Figure 6.69: Differential Transformer Coupling

The resulting differential voltage drives the primary of a 1:1 RF transformer, to develop a single-ended voltage at the output of the secondary winding. The output of the 50 Ω LC filter is matched with the 50 Ω load resistor R_L, and a final output voltage of 1 V p-p is developed.

The transformer not only serves to convert the differential output into a single-ended signal, but it also isolates the output of the DAC from the reactive load presented by the LC filter, thereby improving overall distortion performance.

An op amp connected as a differential-to-single-ended converter can be used to obtain a single-ended output when frequency response to dc is required. In Figure 6.70 the AD8055 op amp is used to achieve high bandwidth and low distortion (see Reference 2). The current output DAC drives balanced 25 Ω resistive loads, thereby developing an out-of-phase voltage of 0 V to 0.5 V at each output.

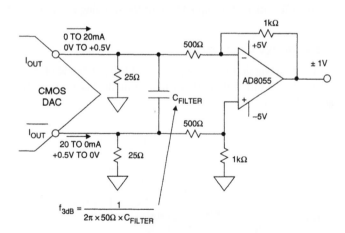

**Figure 6.70: Differential DC-Coupled
Output Using a Dual Supply Op Amp**

The AD8055 is configured for a gain of 2, to develop a final single-ended ground-referenced output voltage of 2 V p-p. Note that because the output signal swings above and below ground, a dual-supply op amp is required.

The C_{FILTER} capacitor forms a differential filter with the equivalent 50 Ω differential output impedance. This filter reduces any slew-induced distortion of the op amp, and the optimum cutoff frequency of the filter is determined empirically to give the best overall distortion performance.

A modified form of the Figure 6.70 circuit can be operated on a single supply, provided the common-mode voltage of the op amp is set to midsupply (2.5 V). This is shown in Figure 6.71, where the AD8061 op amp is used (Reference 3). The output voltage is 2 V p-p centered around a common-mode voltage of 2.5 V. This common-mode voltage can be either developed from the 5 V supply using a resistor divider, or directly from a 2.5 V voltage reference. If the 5 V supply is used as the common-mode voltage, it must be heavily decoupled to prevent supply noise from being amplified.

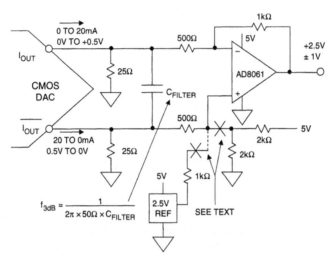

Figure 6.71: Differential DC-Coupled Output Using a Single-Supply Op Amp

Single-Ended Current-to-Voltage Conversion

Single-ended current-to-voltage conversion is easily performed using a single op amp as an I/V converter, as shown in Figure 6.72. The 10 mA full scale DAC current from the AD768 (see Reference 4) develops a 0 to 2 V output voltage across the 200 Ω R_F resistor.

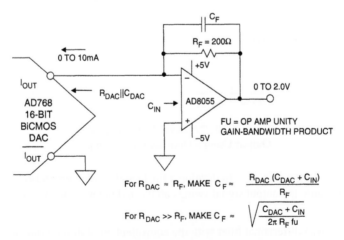

Figure 6.72: Single-Ended I/V Op Amp Interface for Precision 16-Bit AD768 DAC

418

Driving the virtual ground of the AD8055 op amp minimizes any distortion due to nonlinearity in the DAC output impedance. In fact, most high resolution DACs of this type are factory trimmed using an I/V converter.

It should be recalled, however, that using the single-ended output of the DAC in this manner will cause degradation in the common-mode rejection and increased second-order distortion products, compared to a differential operating mode.

The C_F feedback capacitor should be optimized for best pulse response in the circuit. The equations given in the diagram should only be used as guidelines. A much more detailed analysis of this type of circuit is given in Reference 6.

An R-2R based current-output DAC (see Chapter 3 of this book for details of the architecture) has a code-dependent output impedance—therefore, its output must drive the virtual ground of an op amp in order to maintain linearity. The AD5545/AD5555 16-/14-bit DAC is an excellent example of this architecture (Reference 6). A suitable interface circuit is shown in Figure 6.73 where the ADR03 is used as a 2.5 V voltage reference (Reference 7), and the AD8628 chopper-stabilized op amp (Reference 8) is used as an output I/V converter.

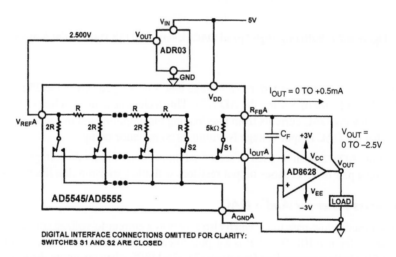

Figure 6.73: AD5545/AD5555 Dual 16-/14-Bit R-2R Current Output DAC Interface

The external 2.5 V references determines the full scale output current, 0.5 mA. Note that a 5 kΩ feedback resistor is included in the DAC, and using it will enhance temperature stability as opposed to using an external resistor. The full scale output voltage from the op amp is therefore –2.5 V. The C_F feedback capacitor compensates for the DAC output capacitance and should be selected to optimize the pulse response, with 20 pF a typical starting point.

Differential Current-to-Differential Voltage Conversion

If a buffered differential voltage output is required from a current output DAC, the AD813x-series of differential amplifiers (Reference 9) can be used as shown in Figure 6.74.

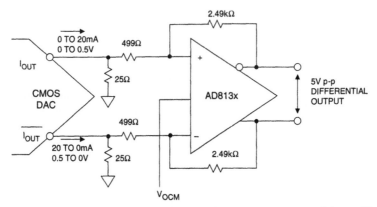

Figure 6.74: Buffering High Speed DACs Using AD813X Differential Amplifier

The DAC output current is first converted into a voltage that is developed across the 25 Ω resistors. The voltage is amplified by a factor of 5 using the AD813x. This technique is used in lieu of a direct I/V conversion to prevent fast slewing DAC currents from overloading the amplifier and introducing distortion. Care must be taken so that the DAC output voltage is within its compliance rating.

The V_{OCM} input on the AD813x can be used to set a final output common-mode voltage within the range of the AD813x. Adding a pair of 75 Ω series output resistors will allow transmission lines to be driven.

An Active Low-Pass Filter for Audio DAC

Figure 6.75 shows an active low-pass filter which also serves as a current-to-voltage converter for the AD1853 Σ-Δ audio DAC (see Reference 10). The filter is a 4-pole filter with a 3 dB cutoff frequency of approximately 75 kHz. Because of the high oversampling frequency (24.576 MSPS when operating the DAC at a 48 kSPS throughput rate), a simple filter is all that is required to remove aliased components above 12 MHz).

The diagram shows a single channel for the dual channel DAC output. U1A and U1B I/V stages form a 1-pole differential filter, while U2 forms a 2-pole multiple-feedback filter that also performs a differential-to-single-ended conversion.

A final fourth passive pole is formed by the 604 Ω resistor and the 2.2 nF capacitor across the output. The OP275 op amp was chosen for operation as U1 and U2 because of its high quality audio characteristics (see Reference 11).

For further details of active filter designs, see Reference 12.

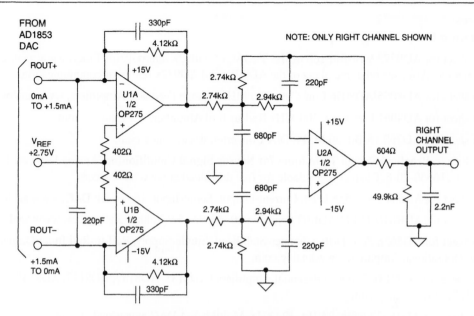

Figure 6.75: A 75 kHz 4-Pole Gaussian Active Filter
for Buffering the Output of the AD1853 Stereo DAC

References:
6.3 Buffering DAC Analog Outputs

1. Data sheet for AD9772A 14-Bit, 160 MSPS TxDAC+® with 2x Interpolation Filter, www.analog.com, for example. Also, see other members of the AD976x and AD977x family of communications DACs.

2. Data sheet for AD8055/AD8056 Low Cost, 300 MHz Voltage Feedback Amplifiers, www.analog.com.

3. Data sheet for AD8061 Low Cost, 300 MHz Rail-to-Rail Amplifier, www.analog.com.

4. Data sheet for AD768 16-Bit, 30 MSPS D/A Converter, www.analog.com.

5. Walt Kester, **Practical Design Techniques for Sensor Signal Conditioning**, Analog Devices, 1999, ISBN-0-916550-20-6, Chapter 5, available for free download at www.analog.com.

6. Data sheet for AD5545/AD5555 Dual, Current-Output, Serial-Input, 16-/14-Bit DAC, www.analog.com.

7. Data sheet for ADR01/ADR02/ADR03 Precision 10 V/5 V/2.5 V Voltage References, www.analog.com.

8. Data sheet for AD8628 Zero-Drift, Chopper-Stabilized, Single-Supply, Rail-to-Rail Input/Output Low Noise Operational Amplifier, www.analog.com.

9. Data sheets for AD813x-Series Differential Amplifiers (AD8131, AD8132, AD8137, AD8138, AD8139), www.analog.com.

10. Data sheet for AD1853 Stereo, 24-Bit, 192 kHz, Multibit Σ-Δ DAC, www.analog.com.

11. Data sheet for OP275 Dual Bipolar/JFET, Audio Operational Amplifier, www.analog.com.

12. Walter G. Jung, **Op Amp Applications Handbook**, Newnes (an imprint of Elsevier Science and Technology Books), ISBN 0-7506-7844-5, 2005, Chapter 5.

Data Converter Voltage References
Walt Kester

In most cases, the accuracy of a data converter is determined by a voltage reference of some sort. An exception to this, of course, is an ADC that operates in a *ratiometric* mode, where both the input signal and input range scale proportionally to the reference. In this specialized case, there is no requirement for an accurate reference, and the power supply is generally adequate. For details on ratiometric operation, see the discussion regarding the AD7730 Σ-Δ ADC in Chapter 3 of this book.

Some ADCs and DACs have internal references, while others do not. Some ADCs use the power supply as a reference. Unfortunately, there is little standardization with respect to ADC/DAC voltage references. In some cases, the dc accuracy of a converter with an internal reference can often be improved by overriding or replacing the internal reference with a more accurate and stable external one. In other cases, the use of an external low noise reference will also increase the noise-free code resolution of a high resolution ADC.

Various ADCs and DACs provide the capability to use external references in lieu of internal ones in various ways. Figure 6.76 shows some of the popular configurations (but certainly not all). Figure 6.76A shows a converter that requires an external reference. It is generally recommended that a suitable decoupling capacitor be added close to the ADC/DAC REF IN pin. The appropriate value is usually specified in the voltage reference data sheet. It is also important that the reference be stable with the required capacitive load (more on this to come).

Figure 6.76B shows a converter that has an internal reference, where the reference is also brought out to a pin on the device. This allows it to be used other places in the circuit, provided the loading does not exceed the rated value. Again, it is important to place the capacitor close to the converter pin. If the internal

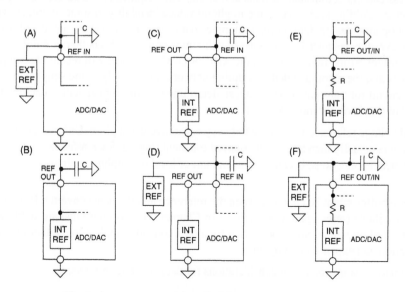

Figure 6.76: Some Popular ADC/DAC Reference Options

reference is pinned out for external use, its accuracy, stability, and temperature coefficient is usually specified on the ADC or DAC data sheet.

If the reference output is to be used other places in the circuit, the data sheet specifications regarding fanout
and loading must be strictly observed. In addition, care must be taken in routing the reference output to
minimize noise pickup. In many cases, a suitable op amp buffer should be used directly at the REF OUT
pin before fanning out to various other parts of the circuit.

Figure 6.76C shows a converter that can use either the internal reference or an external one, but an extra
package pin is required. If the internal reference is used, as in Figure 6.76C, REF OUT is simply externally connected to REF IN, and decoupled if required. If an external reference is used as shown n Figure
6.76D, REF OUT is left floating, and the external reference decoupled and applied to the REF IN pin. This
arrangement is quite flexible for driving similar ADCs or DACs with the same reference in order to obtain
good tracking between the devices.

Figure 6.76E shows an arrangement whereby an external reference can override the internal reference using
a single package pin. The value of the resistor, R, is typically a few kΩ, thereby allowing the low impedance external reference to override the internal one when connected to the REF OUT/IN pin. Figure 6.76F
shows how the external reference is connected to override the internal reference.

The arrangements shown in Figure 6.76 are by no means the only possible configurations for ADC and
DAC references, and the individual data sheets should be consulted in all cases for details regarding
options, fanout, decoupling, etc.

Although the reference element itself can be either a bandgap, buried zener, or XFET (see detailed discussion on voltage references in Chapter 7 of this book), practically all references have some type of output
buffer op amp. The op amp isolates the reference element from the output and also provides drive capability. However, this op amp must obey the general laws relating to op amp stability, and that is what makes
the topic of reference decoupling relevant to the discussion.

Note that a reference input to an ADC or DAC is similar to the analog input of an ADC, in that the internal
conversion process can inject transient currents at that pin. This requires adequate decoupling to stabilize
the reference voltage. Adding such decoupling might introduce instability in some reference types, depending on the output op amp design. Of course, a reference data sheet may not show any details of the output
op amp, which leaves the designer in somewhat of a dilemma concerning whether or not it will be stable
and free from transient errors. In many cases, the ADC or DAC data sheet will recommend appropriate
external references and the recommended decoupling network. Fortunately, some simple lab tests can exercise a reference circuit for transient errors, and also determine stability for capacitive loading. (See Section
7.1 in Chapter 7 of this book for more details.)

A well-designed voltage reference is stable with heavy capacitive decoupling. Unfortunately, some are not
and larger capacitors actually increases the amount of transient ringing. Such references are practically useless in data converter applications, because some amount of local decoupling is almost always required at
the converter.

A suitable op amp buffer might be added between the reference and the data converter. But, there are many
good references available (refer again to Section 7.1 of Chapter 7 in this book) that are stable with an output
capacitor. This type of reference should be chosen for a data converter application, rather than incurring the
further complication and expense of an op amp.

Figure 6.77 summarizes some important considerations for data converter references.

- Data converter accuracy determined by the reference, whether internal or external, but ADC ratiometric operation can eliminate the need for accurate reference
- External references may offer better accuracy and lower noise than internal references
- Bandgap, buried zener, XFET® generally have on-chip output buffer op amp
- Transient loading can cause instability and errors
- External decoupling capacitors may cause oscillation
- Output may require external buffer to source and sink current
- Reference voltage noise may limit system resolution

Figure 6.77: Data Converter Voltage Reference Considerations

Sampling Clock Generation
Walt Kester

Introduction

In Chapter 2 of this book, we derived an extremely important relationship between broadband aperture jitter, t_j, converter SNR, and fullscale sinewave analog frequency, f:

$$SNR = 20 \log_{10} \left[\frac{1}{2\pi f \, t_j} \right]$$

Eq. 6.7

This assumes an ideal ADC (or DAC), where the only error source is jitter. The bandwidth for the SNR measurement is the Nyquist bandwidth, dc to $f_s/2$, where f_s is the sampling rate. Eq. 6.7 also assumes a full-scale sinewave input. The error due to jitter is proportional to the slew rate of the input signal—lower amplitude sinewaves with proportionally lower slew rate yield higher values of SNR (with respect to full scale).

Another interesting case is the theoretical SNR due to jitter for nonsinusoidal signals, in particular those with a Gaussian frequency distribution. Because the average slew rate of this type of signal is less than a fullscale sinewave, the errors due to jitter are smaller. The mathematical treatment of this case is, however, somewhat beyond the scope of the discussion.

It should be noted that t_j in Eq. 6.7 is the combined jitter of the sampling clock, t_{jc}, and the ADC internal aperture jitter, t_{ja}—these terms are not correlated and therefore combine on a root-sum-square (rss) basis:

$$t_j = \sqrt{t_{jc}^2 + t_{ja}^2}$$

Eq. 6.8

In many cases, the sampling clock jitter is several times larger than the ADC aperture jitter, and therefore is the dominate contributor to SNR degradation. For instance, the AD6645 14-bit, 80/105 MSPS ADC has an rms aperture jitter specification of 0.1 ps. Meeting this jitter specification requires a low noise crystal oscillator.

While nothing can be done externally to change the ADC aperture jitter, a number of things can be done to ensure the sampling clock jitter is low enough so that the maximum possible performance is obtained from the ADC.

Figure 6.78 plots Eq. 6.7 and graphically illustrates how SNR is degraded by jitter for various fullscale analog input frequencies (note that we assume t_j includes all jitter sources, including the internal ADC aperture jitter).

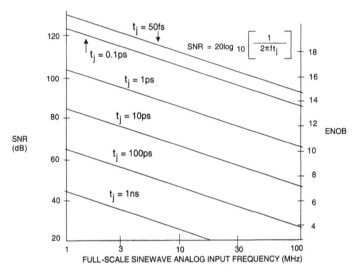

Figure 6.78: Theoretical SNR and ENOB Due to Jitter versus Full-Scale Sinewave Analog Input Frequency

Recall from Chapter 2 of this book that there is a very useful relationship between effective number of bits (ENOB) and the signal-to-noise-plus-distortion ratio (SINAD) given by:

$$ENOB = \frac{SINAD - 1.76 \text{ dB}}{6.02 \text{ dB}}$$

Eq. 6.9

For the purposes of this discussion, assume that the ADC has no distortion, and therefore SINAD = SNR, so Eq. 6.9 becomes:

$$ENOB = \frac{SNR - 1.76 \text{ dB}}{6.02 \text{ dB}}$$

Eq. 6.10

The SNR values on the left-hand vertical axis of Figure 6.78 have been converted into ENOB values on the right-hand vertical axis using Eq. 6.10.

Figure 6.79 shows another plot of Eq. 6.7, where maximum allowable jitter, t_j, is plotted against full scale analog input frequency for various values of ENOB. This plot is useful for determining the jitter requirements on the sampling clock (assuming that it dominates t_j) for various input frequencies and resolutions. For instance, digitization of a full scale 30 MHz input requires less than 0.3 ps rms jitter to maintain 14-bit SNR performance.

If the required sampling clock jitter is selected per the criteria set forth in Figure 6.79, then the SNR due to sampling clock jitter will equal the theoretical SNR of the ADC due to quantization noise.

In order to illustrate the significance of these jitter numbers, consider the typical rms jitter associated with a selection of logic gates shown in Figure 6.80. The values for the 74LS00, 74HCT00, and 74ACT00 were

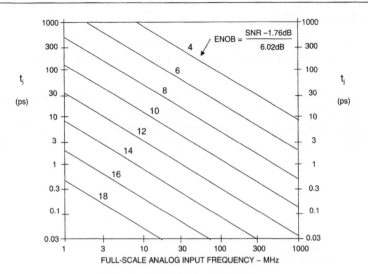

Figure 6.79: Maximum Allowable RMS Jitter versus Fullscale Analog Input Frequency for Various Resolutions (ENOB)

• 74LS00	4.94 ps*
• 74HCT00	2.20 ps*
• 74ACT00	0.99 ps*
• MC100EL16 PECL	0.7 ps**
• NBSG16, Reduced Swing ECL (0.4V)	0.2 ps**
– *Calculated values based on degradation in ADC SNR	
– **Manufacturers' specification	

Figure 6.80: RMS Jitter of Typical Logic Gates

measured with a high performance ADC (aperture jitter less than 0.2 ps rms) using the method described in Chapter 5, where t_j was calculated from FFT-based SNR degradation due to several identical gates connected in series. The jitter due to a single gate was then calculated by dividing by the square root of the total number of series-connected gates. The jitter for the MC100EL16 and NBSG16 was specified by the manufacturer.

Further discussion on aperture jitter in sampled data systems can be found in References 1 and 2 and also in Chapter 2 of this book.

Oscillator Phase Noise and Jitter

The previous analysis centered around broadband jitter, t_j. However, oscillators are most often specified in terms of phase noise. Therefore, the following discussion shows how to approximate the rms jitter based upon the phase noise.

First, a few definitions are in order. Figure 6.81 shows a typical output frequency spectrum of a nonideal oscillator (i.e., one that has jitter in the time domain, corresponding to phase noise in the frequency domain). The spectrum shows the noise power in a 1 Hz bandwidth as a function of frequency. Phase noise is defined as the ratio of the noise in a 1 Hz bandwidth at a specified frequency offset, f_m, to the oscillator signal amplitude at frequency f_0.

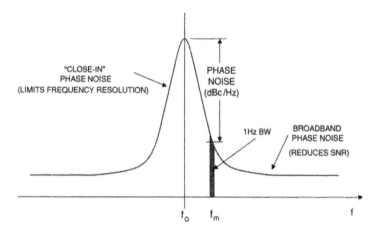

Figure 6.81: Oscillator Power Spectrum Due to Phase Noise

The sampling process is basically a multiplication of the sampling clock and the analog input signal. This is multiplication in the time domain, which is equivalent to convolution in the frequency domain. Therefore, the spectrum of the sampling clock oscillator is convolved with the input and shows up on the FFT output of a pure sinewave input signal (see Figure 6.82). The "close-in" phase noise will "smear" the fundamental signal into a number of frequency bins, thereby reducing the overall spectral resolution. The "broadband" phase noise will cause a degradation in the overall SNR as predicted approximately by Eq. 6.7.

Figure 6.82: Effect of Sampling Clock Phase Noise Ideal Digitized Sinewave

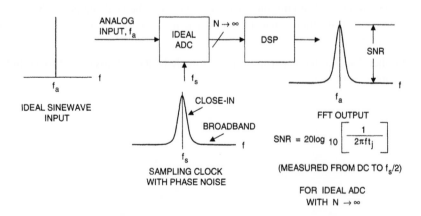

It is customary to characterize an oscillator in terms of its single-sideband phase noise as shown in Figure 6.83, where the phase noise in dBc/Hz is plotted as a function of frequency offset, f_m, with the frequency axis on a log scale. Note the actual curve is approximated by a number of regions, each having a slope of $1/f^x$, where $x = 0$ corresponds to the "white" phase noise region (slope = 0 dB/decade), and $x = 1$ corresponds to the "flicker" phase noise region (slope = –20 dB/decade). There are also regions where $x = 2, 3, 4$, and these regions occur progressively closer to the carrier frequency.

Figure 6.83: Oscillator Phase Noise in dBc/Hz versus Frequency Offset

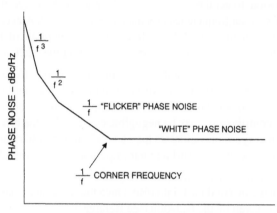

Note that the phase noise curve is somewhat analogous to the input voltage noise spectral density of an amplifier. Like amplifier voltage noise, low 1/f corner frequencies are highly desirable in an oscillator.

We have seen that oscillators are typically specified in terms of phase noise, but in order to relate phase noise to ADC performance, the phase noise must be converted into jitter. In order to make the graph relevant to modern ADC applications, the oscillator frequency (sampling frequency) is chosen to be 100 MHz for discussion purposes, and a typical graph is shown in Figure 6.84. Notice that the phase noise curve is approximated by a number of individual line segments, and the end points of each segment are defined by data points.

Figure 6.84: Calculating Jitter from Phase Noise

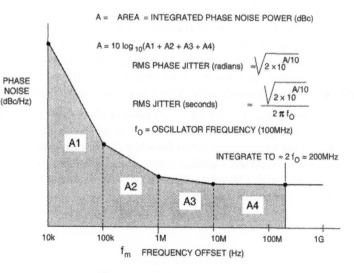

The first step in calculating the equivalent rms jitter is to obtain the integrated phase noise power over the frequency range of interest, i.e., the area of the curve, A. The curve is broken into a number of individual areas (A1, A2, A3, A4), each defined by two data points. Generally speaking, the upper frequency range for the integration should be twice the sampling frequency, assuming there is no filtering between the oscillator and the ADC input. This approximates the bandwidth of the ADC sampling clock input.

Selecting the lower frequency for the integration also requires some judgment. In theory, it should be as low as possible to get the true rms jitter. In practice, however, the oscillator specifications generally will not be given for offset frequencies less than 10 Hz, or so—however, this will certainly give accurate enough results in the calculations. A lower frequency of integration of 100 Hz is also reasonable in most cases, if that specification is available. Otherwise, use either the 1 kHz or 10 kHz data point.

One should also consider that the "close-in" phase noise affects the spectral resolution of the system, while the broadband noise affects the overall system SNR. Probably the wisest approach is to integrate each area separately as explained below and examine the magnitude of the jitter contribution of each area. The low frequency contributions may be negligible compared to the broadband contribution if a crystal oscillator is used. Other types of oscillators may have significant jitter contributions in the low frequency area, and a decision must be made regarding their importance to the overall system frequency resolution.

The integration of each individual area yields individual power ratios. The individual areas are then summed and converted back into dBc. Once the integrated phase noise power is known, the rms phase jitter in radians is given by the equation (see References 3–7 for further details, derivations, etc.),

$$\text{RMS Phase Jitter (radians)} = \sqrt{2 \times 10^{A/10}} \qquad \text{Eq. 6.11}$$

and dividing by $2\pi f_O$ converts the jitter in radians to jitter in seconds:

$$\text{RMS Phase Jitter (radians)} = \frac{\sqrt{2 \times 10^{A/10}}}{2\pi \, f_O} \qquad \text{Eq. 6.12}$$

It should be noted that computer programs and spreadsheets are available online to perform the integration by segments and calculate the rms jitter, thereby greatly simplifying the process (References 8, 9).

Figure 6.85 shows a sample calculation which assumes only broadband phase noise. The broadband phase noise chosen of –150 dBc/Hz represents a reasonably good signal generator specification, so the jitter number obtained represents a practical situation. The phase noise of –150 dBc/Hz (expressed as a ratio) is multiplied by the bandwidth of integration (200 MHz) to obtain the integrated phase noise of –67 dBc. Note that this multiplication is equivalent to adding the quantity $10 \log_{10}[200 \text{ MHz} - 0.01 \text{ MHz}]$ to the phase noise in dBc/Hz. In practice, the lower frequency limit of 0.01 MHz can be dropped from the calculation, as it does not affect the final result significantly. A total rms jitter of approximately 1 ps is obtained using Eq. 6.12.

Crystal oscillators generally offer the lowest possible phase noise and jitter, and some examples are shown for comparison in Figure 6.86. All the oscillators shown have a typical 1/f corner frequency of 20 kHz, and the phase noise therefore represents the white phase noise level. The two Wenzel oscillators are fixed-frequency and represent excellent performance (Reference 9). It is difficult to achieve this level of performance with variable frequency signal generators, as shown by the –150 dBc specification for a relatively high quality generator.

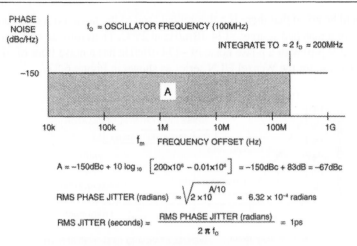

$$A = -150dBc + 10\log_{10}\left[200\times10^6 - 0.01\times10^6\right] = -150dBc + 83dB = -67dBc$$

$$\text{RMS PHASE JITTER (radians)} = \sqrt{2\times10^{A/10}} = 6.32\times10^{-4}\text{ radians}$$

$$\text{RMS JITTER (seconds)} = \frac{\text{RMS PHASE JITTER (radians)}}{2\pi f_0} = 1ps$$

Figure 6.85: Sample Jitter Calculation Assuming Broadband Phase Noise

- Wenzel ULN Series*   −174dBc/Hz @ 10kHz+, ~ $1,500
- Wenzel Sprinter Series,   −165dBc/Hz @ 10kHz+, ~ $350
- High Quality Signal Generator   −150dBc/Hz @ 10kHz+, ~ $10,000

- Thermal noise floor of resistive source in a matched system @ 25°C= −174dBm/Hz
- 0dBm = 1mW = 632mV p-p into 50Ω
- *An oscillator with an output of +13dBm (2.82V p-p) into 50Ω with a phase noise of −174dBc/Hz has a noise floor of +13dBm −174dBc = −161dBm, 13dB above the thermal noise floor

(Wenzel ULN and Sprinter Series Specifications and Pricing Used with Permission of Wenzel Associates)

Figure 6.86: 100 MHz Oscillator Broadband Phase Noise Floor Comparisons (Wenzel ULN and Sprinter Series Specifications and Pricing used with Permission of Wenzel Associates)

At this point, it should be noted that there is a theoretical limit to the noise floor of an oscillator determined by the thermal noise of a matched source: –174 dBm/Hz at 25°C. Therefore, an oscillator with a +13 dBm output into 50 Ω (2.82 V p-p) with a phase noise of –174 dBc/Hz has a noise floor of –174 dBc + 13 dBm = –161 dBm. This is the case for the Wenzel ULN series as shown in Figure 6.87.

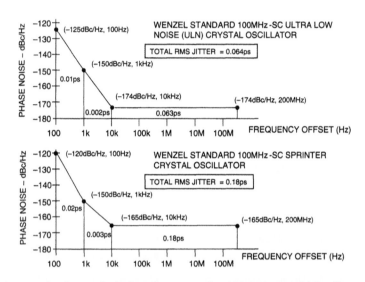

Figure 6.87: Jitter Calculations for Low Noise 100 MHz Crystal Oscillators (Phase Noise Data used with Permission of Wenzel Associates)

Figure 6.87 shows the jitter calculations from the two Wenzel crystal oscillators. In each case, the data points were taken directly for the manufacturer's data sheet. Because of the low 1/f corner frequency, the majority of the jitter is due to the "white" phase noise area. The calculated values of 63 femtoseconds (ULN-Series) and 180 femtoseconds represent extremely low jitter. For informational purposes, the individual jitter contributions of each area have been labeled separately. The total jitter is the root-sum-square of the individual jitter contributors.

In system designs requiring low jitter sampling clocks, the costs of low noise dedicated crystal oscillators is generally prohibitive. An alternative solution is to use a phase-locked-loop (PLL) in conjunction with a voltage-controlled oscillator to "clean up" a noisy system clock as shown in Figure 6.88. There are many good references on PLL design (see References 10–13, for example), and we will not pursue that topic further, other than to state that using a narrow bandwidth loop filter in conjunction with a voltage-controlled crystal oscillator (VCXO) typically gives the lowest phase noise. As shown in Figure 6.88, the PLL tends to reduce the "close-in" phase noise while at the same time, reducing the overall phase noise floor. Further reduction in the white noise floor can be obtained by following the PLL output with an appropriate band-pass filter.

The effect of enclosing a free-running VCO within a PLL is shown in Figure 6.89. Notice that the "close-in" phase noise is reduced significantly by the action of the PLL.

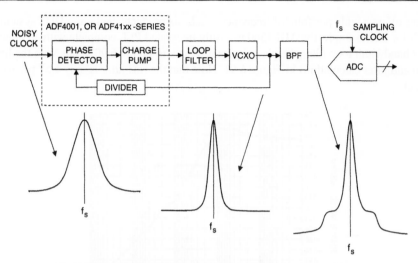

Figure 6.88: Using a Phase-Locked Loop (PLL) and Band-Pass Filter to Condition a Noisy Clock Source

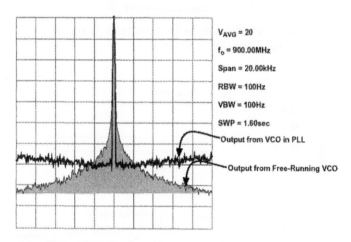

Figure 6.89: Phase Noise for a Free-Running VCO and a PLL-Connected VCO

Analog Devices offers a wide portfolio of frequency synthesis products, including DDS systems, N, and fractional-N PLLs. For example, the ADF4360 is a fully integrated PLL complete with an internal VCO. With a 10 kHz bandwidth loop filter, the phase noise is shown in Figure 6.90, along with the line-segment approximation and jitter calculations in Figure 6.91. Note that the rms jitter is only 1.57 ps, even with a noncrystal VCO.

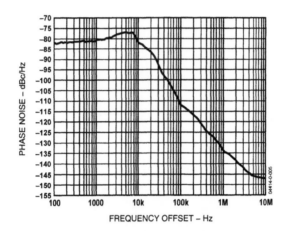

Figure 6.90: Phase Noise for ADF4360 2.25 GHz PLL with Loop Filter BW = 10 kHz

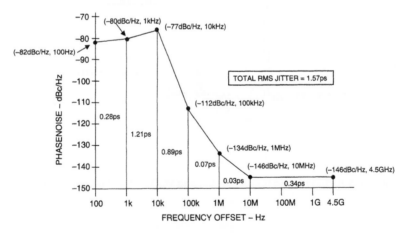

Figure 6.91: Line Segment Approximation to ADF4360 2.25 GHz PLL Phase Noise Showing Jitter

Historically, PLL design relied heavily on textbooks and application notes to assist in the design of the loop filter, etc. Now, with Analog Devices' free downloadable ADIsimPLL software, PLL design is much easier. To start, choose a circuit by entering the desired output frequency range, and select a PLL, VCO, and a crystal reference. Once the loop filter configuration has been selected, the circuit can be analyzed and optimized for phase noise, phase margin, gain, spur levels, lock time, etc., in both the frequency and time domain. The program also performs the rms jitter calculation based on the PLL phase noise, thereby allowing the evaluation of the final PLL output as a sampling clock.

Figure 6.92 summarizes this discussion and should serve as an approximate guideline for selecting the type of sampling clock generator based upon the maximum input frequency and the required resolution in ENOB. The PLL approach with a standard VCO is an excellent one for generating sampling clocks where the rms jitter requirement is approximately 1 ps or greater. However, subpicosecond jitter requires either a VCXO-based PLL or a dedicated low noise crystal oscillator.

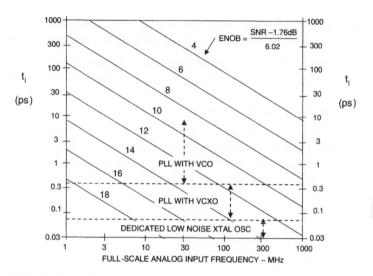

Figure 6.92: Oscillator Requirements versus Resolution and Analog Input Frequency

"Hybrid" Clock Generators

DDSs, mixers, frequency dividers, and frequency doublers can be utilized in conjunction with PLLs to form what is generally referred to as a "hybrid" frequency synthesizer. An excellent tutorial on the subject can be found in Reference 14. A very simple example is shown in Figure 6.93 where a DDS system drives a PLL. The upper output frequency of the DDS system is of course limited by its maximum update rate. The upper frequency of a PLL, on the other hand, is primarily limited by the VCO, which can operate in the GHz range if required.

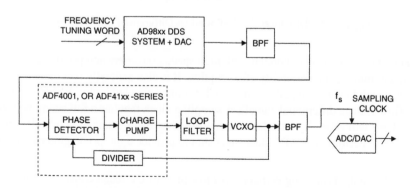

Figure 6.93: A Simple "Hybrid" Sampling Clock Generator

As previously discussed, the phase noise of a PLL can be controlled by the loop filter, the VCXO, and an output filter. DDS systems have phase noise which is produced primarily by the finite resolution of the internal DAC. The system of Figure 6.93 uses the PLL to "clean up" the phase noise produced by the DDS system, thereby generating an output clock which is suitable for high performance ADC/DAC sampling clocks. There are many possible combinations possible if one looks at some of the configurations suggested in Reference 14.

In many less demanding applications, DDS output can, of course, be used directly as a clock generator. Many DDSs have on-chip comparators that facilitate the generation of a square wave output.

Regardless of how it is generated, the overriding requirements on the sampling clock are ultimately dictated by the principles set forth in this section which relate phase noise and timing jitter to SNR.

Driving Differential Sampling Clock Inputs

Data converters that can tolerate sampling clocks with tens of picoseconds or more of jitter can be driven from most any single-ended logic gate. However, for jitter requirements of 10 ps or less, more care must be taken in the selection of an appropriate driver. High performance high speed data converters are almost always designed to accept a differential sampling clock input as shown in Figure 6.94.

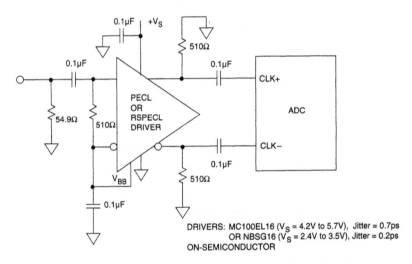

Figure 6.94: Low Jitter Single-Ended-to-Differential Clock Drivers

Differential sampling clock inputs are popular with high speed converters and provide good common-mode rejection, thereby minimizing the possibility of corruption. It is also generally recommended that differential inputs be driven with low level signals such as ECL (emitter-coupled logic), RSECL (reduced signal ECL), or LVDS (low voltage differential signal). A sampling clock that has a full swing between ground and the supply voltage will generally introduce extra noise, thereby degrading the overall converter dynamic performance. A high performance ADC data sheet should provide appropriate guidance for the optimum drive level.

Most oscillator or PLL outputs are single-ended, so a low jitter PECL receiver/driver such as the ON-Semiconductor MC100EL16 or the NBSG16 (Reference 14) is an excellent choice for performing single-ended-to-differential clock conversion. The rms jitter specification is 0.7 ps for the MC100EL16, and 0.2 ps for the NBSG16 (a silicon-germanium device). These parts are basically ECL (Emitter-Coupled-Logic) designs which can be operated on a single positive supply—hence the acronym "PECL" (Positive Emitter Coupled Logic). In almost all cases, the differential sampling clock inputs of the ADC are internally biased at the appropriate dc common-mode level, and the differential driver outputs can simply be ac-coupled to the ADC clock inputs. If the ADC does not have internal biasing, then an external resistor network is required to supply the required bias voltages.

The output voltage swing for the MC100EL16 PECL device is approximately 1 V p-p single-ended (2 V p-p differential), and 0.4 V p-p single-ended (0.8 V p-p differential) for the reduced-swing PECL (RSPECL) NBSG16.

For the ultralow jitter applications, an RF transformer should be used to convert the single-ended oscillator output into a differential signal as shown in Figure 6.95. The back-to-back Schottky diodes limit the differential voltage input swing to about 0.8 V, the 0.1 μF prevents any dc components from causing transformer saturation, and the 100 Ω resistor limits the output current of the drive oscillator. The AD6645 14-bit, 105 MSPS ADC has an aperture jitter specification of 0.1 ps, and the transformer drive circuit in conjunction with a very low noise oscillator will provide optimum performance with this type of low jitter ADC. Some experimentation may be required to determine the amplitude for the input sinewave which gives the best overall SNR.

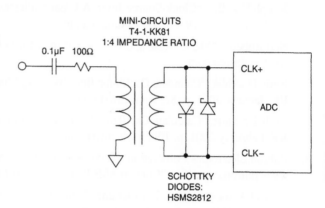

Figure 6.95: Single-Ended-to-Differential Conversion Using RF Transformer

As in the case of ADC analog inputs and DAC analog outputs, there are other possibilities, and the device data sheet must always be consulted for the optimum sampling clock drive recommendations.

Sampling Clock Summary

Earlier in this chapter, we discussed the importance of the drive circuitry for the analog input of an ADC and the analog output buffer for a DAC. Equally important is the ADC or DAC sampling clock. Regarding the sampling clock as simply another "digital" signal is a certain receipt for disaster in a system design.

This section has described the effects of jitter on SNR, assuming that the jitter is solely a combination of the internal ADC aperture jitter and the external sampling clock jitter. However, improper layout, grounding, and decoupling techniques can create additional clock jitter which can drastically degrade dynamic performance, regardless of the specifications of the ADC or sampling clock oscillator.

Routing the sampling clock signal in parallel with noisy digital signals is sure to degrade performance due to stray coupling. In fact, coupling high speed data from parallel output ADCs into the sampling clock not only increases noise, but is likely to create additional harmonic distortion, because the energy contained in the digital output transient currents is signal dependent. For further discussion of these and other critical hardware design techniques, the reader is referred to Chapter 9 of this book.

References:
6.5 Sampling Clock Generation

1. Brad Brannon, "Aperture Uncertainty and ADC System Performance," **Application Note AN-501**, Analog Devices, download at www.analog.com.

2. Bar-Giora Goldberg, "The Effects of Clock Jitter on Data Conversion Devices," **RF Design**, August 2002, pp. 26–32, www.rfdesign.com.

3. Ulrich L. Rohde, **Digital PLL Frequency Synthesizers, Theory and Design**, Prentice-Hall, 1983, ISBN 0-13-214239-2, all of Chapter 2 and pp. 411–418 for computer analysis.

4. Joseph V. Adler, "Clock-Source Jitter: A Clear Understanding Aids Oscillator Selection," **EDN**, February 18, 1999, pp. 79–86, www.ednmag.com.

5. Neil Roberts, "Phase Noise and Jitter – A Primer for Digital Designers," **EEdesign**, July 14, 2003, www.eedesign.com.

6. Boris Drakhlis, "Calculate Oscillator Jitter by using Phase-Noise Analysis Part 1," **Microwaves and RF**, January 2001, p. 82, www.mwrf.com.

7. Boris Drakhlis, "Calculate Oscillator Jitter by using Phase-Noise Analysis Part 2," **Microwaves and RF**, February 2001, p. 109, www.mwrf.com.

8. Raltron Electronics Corporation, 10651 Northwest 19th Street, Miami, Florida 33172, Tel: 305-593-6033, www.raltron.com. (See "Convert SSB Phase Noise to Jitter" under "Engineering Design Tools".)

9. Wenzel Associates, Inc., 2215 Kramer Lane, Austin, Texas 78758, Tel: 512-835-2038, www.wenzel.com. (See "Allan Variance from Phase Noise" under "Spreadsheets".)

10. Mike Curtin and Paul O'Brien, "Phase-Locked Loops for High-Frequency Receivers and Transmitters, Part 1," **Analog Dialogue 33-3**, 1999, www.analog.com.

11. Mike Curtin and Paul O'Brien, "Phase-Locked Loops for High-Frequency Receivers and Transmitters, Part 2," **Analog Dialogue 33-5**, 1999, www.analog.com.

12. R. E. Best, **Phase-Locked Loops: Theory, Design and Applications**, Fourth Edition, McGraw-Hill, 1999, ISBN 0071349030.

13. F. M. Gardner, **Phaselock Techniques**, Second Edition, John Wiley, 1979, ISBN 0471042943.

14. David Crook, "Hybrid Synthesizer Tutorial," **Microwave Journal**, February 2003.

15. ON Semiconductor, 5005 East McDowell Road, Phoenix, AZ 85008, USA, Tel: 602-244-6600, www.onsemi.com.

Data Converter Support Circuits

Data Converter Support Circuits

Voltage References
Walt Jung, Walt Kester, James Bryant

Reference circuits and linear regulators actually have much in common. In fact, the latter could be functionally described as a reference circuit, but with greater current (or power) output. Accordingly, almost all of the specifications of the two circuit types have great commonality (even though the performance of references is usually tighter with regard to drift, accuracy, etc.). This section discusses voltage references, and the next section covers linear regulators, with emphasis on their low dropout operation for highest power efficiency.

Precision Voltage References

Voltage references have a major impact on the performance and accuracy of analog systems. A ±5 mV tolerance on a 5 V reference corresponds to ±0.1% absolute accuracy—only 10 bits. For a 12-bit system, choosing a reference that has a ±1 mV tolerance may be far more cost effective than performing manual calibration, while both high initial accuracy and calibration will be necessary in a system making absolute 16-bit measurements. Note that many systems make *relative* measurements rather than absolute ones, and in such cases the absolute accuracy of the reference is not important, although noise and short-term stability may be. Figure 7.1 summarizes some key points of the reference selection process.

- Tight Tolerance Improves Accuracy, Reduces System Costs

- Temperature Drift Affects Accuracy

- Long-Term Stability, Low Hysteresis Assures Repeatability

- Noise Limits System Resolution

- Dynamic Loading Can Cause Errors

- Power Consumption is Critical to Battery Systems

- Tiny Low Cost Packages Increase Circuit Density

Figure 7.1: Choosing Voltage References for High Performance Systems

Temperature drift or drift due to aging may be an even greater problem than absolute accuracy. The initial error can always be trimmed, but compensating for drift is difficult. Where possible, references should be chosen for temperature coefficient and aging characteristics that preserve adequate accuracy over the operating temperature range and expected lifetime of the system.

Noise in voltage references is often overlooked, but it can be very important in system design. It is generally specified on data sheets, but system designers frequently ignore the specification and assume that voltage references do not contribute to system noise.

There are two dynamic issues that must be considered with voltage references: their behavior at start-up, and their behavior with transient loads. With regard to the first, always bear in mind that voltage references *do not power up instantly* (this is true of references inside ADCs and DACs as well as discrete designs). Thus it is rarely possible to turn on an ADC and reference, whether internal or external, make a reading, and turn off again within a few microseconds, however attractive such a procedure might be in terms of energy saving.

Regarding the second point, a given reference IC may or may not be well suited for pulse-loading conditions, dependent upon the specific architecture. Many references use low power, and therefore low bandwidth, output buffer amplifiers. This makes for poor behavior under fast transient loads, which may degrade the performance of fast ADCs (especially successive approximation and flash ADCs). Suitable decoupling can ease the problem (but some references oscillate with capacitive loads), or an additional external broadband buffer amplifier may be used to drive the node where the transients occur.

References, like almost all other ICs today, are fast migrating to such smaller packages such as SO-8 and MSOP, and the even more tiny SOT-23 and SC-70, enabling much higher circuit densities within a given area of real estate. In addition to the system size reductions these steps bring, there are also tangible reductions in standby power and cost with the smaller and less expense ICs.

Types of Voltage References

In terms of the functionality of their circuit connection, standard reference ICs are often only available in *series*, or *three-terminal* form (V_{IN}, Common, V_{OUT}), and also in positive polarity only. The series types have the potential advantages of lower and more stable quiescent current, standard pretrimmed output voltages, and relatively high output current without accuracy loss. *Shunt*, or *two-terminal* (i.e., diode-like) references are more flexible regarding operating polarity, but they are also more restrictive as to loading. They can in fact eat up excessive power with widely varying resistor-fed voltage inputs. Also, they sometimes come in nonstandard voltages. All of these various factors tend to govern when one functional type is preferred over the other.

Some simple diode-based references are shown in Figure 7.2. In the first of these, a current driven forward biased diode (or diode-connected transistor) produces a voltage, $V_f = V_{REF}$. While the junction drop is somewhat decoupled from the raw supply, it has numerous deficiencies as a reference. Among them are a strong TC of about –0.3%/°C, some sensitivity to loading, and a rather inflexible output voltage: it is only available in 600 mV jumps.

By contrast, these most simple references (as well as all other shunt-type regulators) have a basic advantage, which is the fact that the polarity is readily reversible by flipping connections and reversing the drive current. However, a basic limitation of all shunt regulators is that load current must always be less (usually appreciably less) than the driving current, I_D.

In the second circuit of Figure 7.2, a Zener or avalanche diode is used, and an appreciably higher output voltage realized. While true *Zener* breakdown occurs below 5 V, *avalanche* breakdown occurs at higher voltages and has a positive temperature coefficient. Note that diode reverse breakdown is referred to almost universally today as *Zener*, even though it is usually avalanche breakdown. With a D1 breakdown voltage in the 5 V to 8 V range, the net positive TC is such that it equals the negative TC of forward-biased diode D2, yielding a net TC of 100 ppm/°C or less with proper bias current. Combinations of such carefully chosen

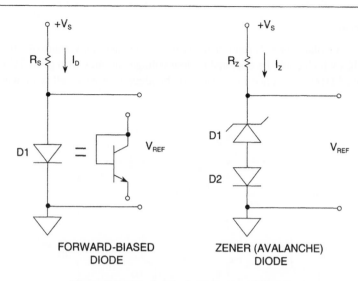

Figure 7.2: Simple Diode Reference Circuits

diodes formed the basis of the early single package "temperature-compensated Zener" references, such as the 1N821-1N829 series.

The temperature-compensated Zener reference is limited in terms of initial accuracy, since the best TC combinations fall at odd voltages, such as the 1N829's 6.2 V. And, the scheme is also limited for loading, since for best TC the diode current must be carefully controlled. Unlike a fundamentally lower voltage (<2 V) reference, Zener-diode-based references must of necessity be driven from voltage sources appreciably higher than 6 V levels, so this precludes operation of Zener references from 5 V system supplies. References based on low TC Zener (avalanche) diodes also tend to be noisy, due to the basic noise of the breakdown mechanism. This has been improved greatly with *monolithic* Zener types, as is described further below.

At this point, we know that a reference circuit can be functionally arranged into either a series or shunt operated form, and the technology within may use either bandgap based or Zener-diode-based circuitry. In practice all permutations of these are available, as well as a third major technology category. The three major reference technologies are now described in more detail.

Bandgap References

The development of low voltage (<5 V) references based on the bandgap voltage of silicon led to the introductions of various ICs which could be operated on low voltage supplies with good TC performance. The first of these was the LM109 (Reference 1), and a basic bandgap reference cell is shown in Figure 7.3.

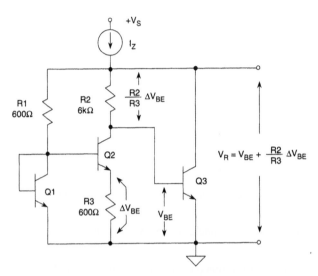

Figure 7.3: Basic Bandgap Reference

This circuit is also called a "ΔV_{BE}" reference because the differing current densities between matched transistors Q1-Q2 produces a ΔV_{BE} across R3. It works by summing the V_{BE} of Q3 with the amplified ΔV_{BE} of Q1-Q2, developed across R2. The ΔV_{BE} and V_{BE} components have opposite polarity TCs; ΔV_{BE} is proportional-to-absolute-temperature (PTAT), while V_{BE} is complementary-to-absolute-temperature (CTAT). The summed output is V_R, and when it is equal to 1.205 V (silicon bandgap voltage), the TC is a minimum.

The bandgap reference technique is attractive in IC designs because of several reasons; among these are the relative simplicity, and the avoidance of Zeners and their noise. However, very important in these days of ever decreasing system power supplies is the fundamental fact that bandgap devices operate at low voltages, i.e., <5 V. Not only are they used for standalone IC references, but are also used within the designs of many other ICs, such as ADCs and DACs.

Buffered forms of 1.2 V two terminal shunt bandgap references, such as the AD589 IC, remain stable under varying load currents. The AD589 (introduced in 1980), a 1.235 V reference, handles 50 μA to 5 mA with an output impedance of 0.6 Ω, and TCs ranging between 10 and 100 ppm/°C. The more recent and functionally similar AD1580, a 1.225 V shunt reference, is in the tiny SOT-23 package and handles the same nominal currents as the AD589, with TCs of 50 and 100 ppm/°C. The ADR510 shunt reference supplies 1.000 V, and the ADR512 supplies 1.200 V.

However, the basic designs of Figure 7.3 suffer from load and current drive sensitivity, plus the fact that the output needs accurate scaling to more useful levels, i.e., 2.5 V, 5 V, etc. The load drive issue is best addressed with the use of a buffer amplifier, which also provides convenient voltage scaling to standard levels.

An improved three-terminal bandgap reference, the AD580 (introduced in 1974) is shown in Figure 7.4. Popularly called the "Brokaw Cell" (see References 2 and 3), this circuit provides on-chip output buffering,

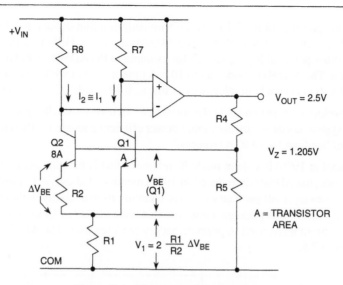

**Figure 7.4: AD580 Precision Bandgap
Reference Uses Brokaw Cell (1974)**

which allows good drive capability and standard output voltage scaling. The AD580 was the first precision bandgap based IC reference, and variants of the topology have influenced further generations of both industry standard references such as the REF01, REF02, and REF03 series, as well as more recent ADI bandgap parts such as the REF19x series, the AD680, AD780, the AD1582-85 series, the ADR38x series, the ADR39x series, and recent SC-70 and SOT-23 offerings of improved versions of the REF01, REF02, and REF03 (designated ADR01, ADR02, and ADR03).

The AD580 has two 8:1 emitter-scaled transistors Q1-Q2 operating at identical collector currents (and thus 1/8 current densities), by virtue of equal load resistors and a closed loop around the buffer op amp. Due to the resultant smaller V_{BE} of the 8× area Q2, R2 in series with Q2 drops the ΔV_{BE} voltage, while R1 (due to the current relationships) drops a PTAT voltage V1:

$$V_1 = 2 \times \frac{R1}{R2} \times \Delta V_{BE} \qquad \text{Eq. 7.1}$$

The bandgap cell reference voltage V_Z appears at the base of Q1, and is the sum of V_{BE} (Q1) and V_1, or 1.205 V, the bandgap voltage:

$$V_Z = V_{BE(Q1)} + V_1 \qquad \text{Eq. 7.2}$$

$$= V_{BE(Q1)} + 2 \times \frac{R1}{R2} \times \Delta V_{BE} \qquad \text{Eq. 7.3}$$

$$= V_{BE(Q1)} + 2 \times \frac{R1}{R2} \times \frac{kT}{q} \times \ln \frac{J1}{J2} \qquad \text{Eq. 7.4}$$

$$= V_{BE(Q1)} + 2 \times \frac{R1}{R2} \times \frac{kT}{q} \times \ln 8 \qquad \text{Eq. 7.5}$$

$$= 1.205V. \qquad \text{Eq. 7.6}$$

Note that J1 = current density in Q1, J2 = current density in Q2, and J1/J2 = 8.

However, because of the presence of the R4/R5 (laser trimmed) thin film divider and the op amp, the actual voltage appearing at V_{OUT} can be scaled higher, in the AD580 case 2.5 V. Following this general principle, V_{OUT} can be raised to other practical levels, such as for example in the AD584, with taps for precise 2.5, 5, 7.5, and 10 V operation. The AD580 provides up to 10 mA output current while operating from supplies between 4.5 V and 30 V. It is available in tolerances as low as 0.4%, with TCs as low as 10 ppm/°C.

Many of the recent developments in bandgap references have focused on smaller package size and cost reduction, to address system needs for smaller, more power efficient and less costly reference ICs. Among these are several recent bandgap-based IC references.

The AD1580 (introduced in 1996) is a shunt mode IC reference that is functionally quite similar to the classic shunt IC reference, the AD589 (introduced in 1980) mentioned above. A key difference is the fact that the AD1580 uses a newer, small geometry process, enabling its availability within the tiny SOT-23 package. The very small size of this package allows use in a wide variety of space limited applications, and the low operating current lends itself to portable battery powered uses. The AD1580 circuit is shown in simplified form in Figure 7.5.

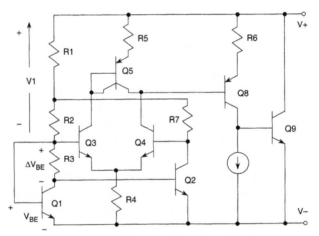

Figure 7.5: AD1580 1.2 V Shunt Type Bandgap Reference has Tiny Size in SOT-23 Footprint

In this circuit, transistors Q1 and Q2 form the bandgap core, and are operated at a current ratio of 5 times, determined by the ratio of R7 to R2. An op amp is formed by the differential pair Q3-Q4, current mirror Q5, and driver/output stage Q8-Q9. In closed loop equilibrium, this amplifier maintains the bottom ends of R2–R7 at the same potential.

As a result of the closed loop control described, a basic ΔV_{BE} voltage is dropped across R3, and a scaled PTAT voltage also appears as V1, which is effectively in series with V_{BE}. The nominal bandgap reference voltage of 1.225 V is then the sum of Q1's V_{BE} and V1. The AD1580 is designed to operate at currents as low as 50 μA, also handling maximum currents as high as 10 mA. It is available in grades with voltage tolerances of ±1 or ±10 mV, and with corresponding TCs of 50 or 100 ppm/°C. Newer members of the Analog Devices' family of shunt regulators are the ADR510 (1.000 V), and the ADR512 (1.200 V).

The ADR520 (2.048 V), ADR525 (2.500 V), ADR530 (3.000 V), ADR540 (4.096 V), ADR545 (4.5 V), and ADR550 (5.0 V) are the latest in the shunt regulator family, with initial accuracies of 0.2%, and available in either SC-70 or SOT-23 packages.

The AD1582-AD1585 series comprises a family of *series* mode IC references, which produce voltage outputs of 2.5, 3.0, 4.096 and 5.0 V. Like the AD1580, the series uses a small geometry process to allow packaging within an SOT-23. The AD1582 series specifications are summarized in Figure 7.6.

- V_{OUT} : 2.500, 3.000, 4.096, & 5.000V
- 2.7V to 12V Supply Range (200mV Headroom)
- Supply Current : 65µA max
- Initial Accuracy: ±0.1% max
- Temperature Coefficient: 50 ppm/°C max
- Noise: 70µV p-p (0.1Hz – 10Hz)
- Noise: 50µV rms (10Hz – 10kHz)
- Long-Term Drift: 100ppm/1khrs
- High Output Current: ±5mA min
- Temperature Range –40°C to +85°C
- Low Cost SOT-23 Package

**Figure 7.6: AD1582/AD1585 2.5 V to 5 V
Series-Type Bandgap Reference Specifications**

The circuit diagram for the series, shown in Figure 7.7, may be recognized as a variant of the basic Brokaw bandgap cell, as described under Figure 7.4. In this case Q1-Q2 form the core, and the overall loop operates to produce the stable reference voltage V_{BG} at the base of Q1. A notable difference here is that the op amp's output stage is designed with push-pull common-emitter stages. This has the effect of requiring an output capacitor for stability, but it also provides the IC with relatively low dropout operation.

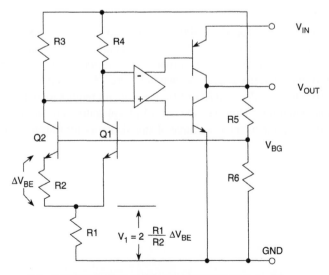

$$V_1 = 2 \frac{R1}{R2} \Delta V_{BE}$$

**Figure 7.7: AD1582/AD1585 2.5 V to 5 V Series-Type
Bandgap References in SOT-23 Footprint**

The low dropout feature means essentially that V_{IN} can be lowered to as close as several hundred mV above the V_{OUT} level without disturbing operation. The push-pull operation also means that this device series can actually both sink and source currents at the output, as opposed to the classic reference operation of sourcing current (only). For the various output voltage ratings, the divider R5-R6 is adjusted for the respective levels.

The AD1582 series is designed to operate with quiescent currents of only 65 μA (maximum), which allows good power efficiency when used in low power systems with varying voltage inputs. The rated output current for the series is 5 mA, and they are available in grades with voltage tolerances of ±0.1 or ±1% of V_{OUT}, with corresponding TCs of 50 or 100 ppm/°C.

Because of stability requirements, devices of the AD1582 series must be used with both an output and input bypass capacitor. Recommended optimum values for these are shown in the hookup diagram of Figure 7.8. For the electrical values noted, it is likely that tantalum chip capacitors will be the smallest in size.

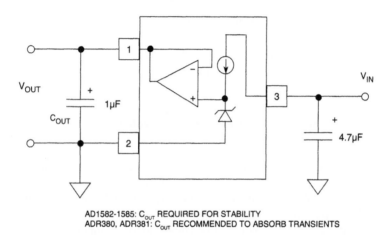

AD1582-1585: C_{OUT} REQUIRED FOR STABILITY
ADR380, ADR381: C_{OUT} RECOMMENDED TO ABSORB TRANSIENTS

Figure 7.8: AD1582/AD1585 Series Connection Diagram

ADR38x and ADR39x series are low dropout (300 mV) bandgap references in SOT-23 packages. Noise is typically 5 μV p-p in the 0.1 Hz to 10 Hz bandwidth. Quiescent current is typically 100 μA, and the ADR39x series have a shutdown pin (shutdown current < 3 μA) as well as a "sense" pin for Kelvin sensing. A connection diagram for the ADR39x series is shown in Figure 7.9, and key specifications for the family are shown in Figure 7.10. The ADR38x and ADR39x-series do not require an output capacitor for stability, regardless of the load conditions. However, at least a 1 μF capacitor is recommended to filter out noise. Larger capacitors may be desirable to act as a source of stored energy for transient loads.

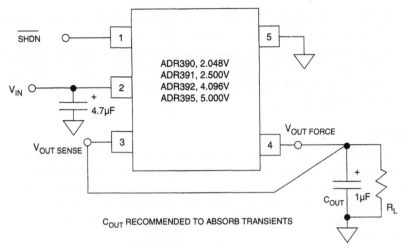

Figure 7.9: ADR390/ADR391/ADR392/ADR395 Connection Diagram

- V_{OUT}: 2.048, 2.500, 4.096, & 5.000V
- 2.3V to 15V Supply Range (300mV Headroom)
- Supply Current : 120µA max
- Initial Accuracy: ±6mV max
- Temperature Coefficient: 25 ppm/°C max
- Noise: 5µV p-p (0.1Hz – 10Hz)
- Long-Term Drift: 50ppm/1khrs
- High Output Current: +5mA min
- Temperature Range –40°C to +85°C
- Shutdown Feature: <3µA max
- Kelvin Sensing (Force and Sense Pins)
- Low Cost SOT-23 (5 pin) Package

Figure 7.10: ADR390/ADR395 2.048 V to 5 V Series-Type Bandgap Reference Specifications

Buried Zener References

In terms of the design approaches used within the reference core, the two most popular basic types of IC references consist of the bandgap and buried Zener units. Bandgaps have been discussed, but Zener-based references warrant some further discussion.

In an IC chip, surface operated diode junction breakdown is prone to crystal imperfections and other contamination, thus Zener diodes formed at the surface are more noisy and less stable than are *buried* (or subsurface) ones. ADI Zener-based IC references employ the much preferred buried Zener. This improves substantially upon the noise and drift of surface-mode operated Zeners (see Reference 4). Buried Zener references offer very low temperature drift, down to the 1–2 ppm/°C (AD588 and AD586), and the lowest noise as a percent of full-scale, i.e., $100 \, nV/\sqrt{Hz}$ or less. On the downside, the operating current of Zener type references is usually relatively high, typically on the order of several mA.

An important general point arises when comparing noise performance of different references. The best way to do this is to compare the ratio of the noise (within a given bandwidth) to the dc output voltage. For example, a 10 V reference with a $100 \, nV/\sqrt{Hz}$ noise density is 6 dB more quiet in relative terms than is a 5 V reference with the same noise level.

XFET References

A third and relatively new category of IC reference core design is based on the properties of junction field effect (JFET) transistors. Somewhat analogous to the bandgap reference for bipolar transistors, the JFET based reference operates a pair of junction field effect transistors with different pinchoff voltages, and amplifies the differential output to produce a stable reference voltage. One of the two JFETs uses an extra ion implantation, giving rise to the name XFET (eXtra implantation junction Field Effect Transistor) for the reference core design.

The basic topology for the XFET reference circuit is shown in Figure 7.11. J1 and J2 are the two JFET transistors, which form the core of the reference. J1 and J2 are driven at the same current level from matched current sources, I1 and I2. To the right, J1 is the JFET with the extra implantation, which causes the difference in the J1-J2 pinchoff voltages to differ by 500 mV. With the pinchoff voltage of two such FETs purposely skewed, a differential voltage will appear between the gates for identical current drive conditions and equal source voltages. This voltage, ΔV_P, is:

$$\Delta V_P = V_{P1} - V_{P2} \qquad \text{Eq. 7.7}$$

where V_{P1} and V_{P2} are the pinchoff voltages of FETs J1 and J2, respectively.

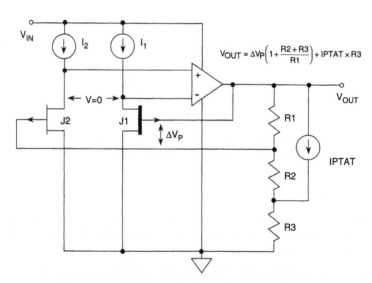

Figure 7.11: ADR290/ADR293 2.048 V to 5 V XFET References Feature High Stability and Low Power

Note that, within this circuit, the voltage ΔV_P exists between the *gates* of the two FETs. We also know that, with the overall feedback loop closed, the op amp axiom of zero input differential voltage will hold the sources of the two JFETs at the same potential. These source voltages are applied as inputs to the op amp, the output of which drives feedback divider R1–R3. As this loop is configured, it stabilizes at an output voltage from the R1-R2 tap which does in fact produce the required ΔV_P between the J1–J2 gates. In essence, the op amp amplifies ΔV_P to produce V_{OUT}, where

$$V_{OUT} = \Delta V_P \left(1 + \frac{R2 + R3}{R1}\right) + \left(I_{PTAT}\right)(R3) \qquad \text{Eq. 7.8}$$

As can be noted, this expression includes the basic output scaling (leftmost portion of the right terms), plus a rightmost temperature dependent term including I_{PTAT}. The I_{PTAT} portion of the expression compensates for a basic negative temperature coefficient of the XFET core, such that the overall net temperature drift of the reference is typically in a range of 3 to 8 ppm/°C.

During manufacture, the R1–R3 scaling resistance values are adjusted to produce the different voltage output options of 2.048 V, 2.5 V, 4.096 V, and 5.0 V for the ADR290, ADR291, ADR292 and ADR293 family (ADR29x). This ADR29x family of series mode references is available in 8-pin packages with a standard footprint. They operate from supplies of V_{OUT} plus 500 mV to 15 V, with a maximum quiescent current of 12 μA, and output currents of up to 5 mA. A summary of specifications for the family appears in Figure 7.12.

The ADR43x series are the second generation of low noise, low drift XFET references. Standard voltage outputs are 2.048 V, 2.500 V, 3.000 V, 4.096 V, and 5.000 V. These devices operate from supplies of V_{OUT} + 1 V to 18 V with quiescent currents of 0.5 mA maximum and output currents of ±10 mA. Temperature drift is 3 ppm/°C maximum. The 0.1 Hz to 10 Hz noise is an incredibly low 1.5 μV p-p. This ADR43x family of series mode references is available in 8 pin packages with a standard footprint. Key specifications for the family are summarized in Figure 7.13.

- V_{OUT}: 2.048V, 2.500V, 4.096V, and 5.000V
- 2.7V to 15V Supply Range (0.5V Headroom)
- Supply Current : 12μA max
- Initial Accuracy: ±0.1%
- Temperature Coefficient: 8 ppm/°C max
- Low Noise: 6μV p-p (0.1 – 10Hz)
- Wideband Noise: 420nV/√Hz @ 1kHz
- Long-Term Drift: 50ppm/1000 hours
- High Output Current: 5mA min
- Temperature Range –40°C to +125°C
- Standard REF02 Pinout
- 8-Lead Narrow Body SOIC, 8-Lead TSSOP

Figure 7.12: ADR290/ADR293 XFET Series Specifications

- V_{OUT}: 2.048V, 2.500V, 4.096V, and 5.000V
- 3V to 18V Supply Range (1V Headroom)
- Supply Current : 500μA max
- Initial Accuracy: ±0.05%
- Temperature Coefficient: 3 ppm/°C max
- Low Noise: 1.75μV p-p (0.1 – 10Hz)
- Wideband Noise: 60nV/√Hz @ 1kHz
- Long-Term Drift: 50ppm/1000 hours
- High Output Current: ±10mA min
- Temperature Range –40°C to +125°C
- 8-Lead MSOP, 8-Lead TSSOP

Figure 7.13: ADR430-ADR439 XFET Series Specifications

The XFET architecture offers performance improvements over bandgap and buried Zener references, particularly for systems where operating current is critical, yet drift and noise performance must still be excellent. XFET noise levels are lower than bandgap-based bipolar references operating at an equivalent current, the temperature drift is low and linear at 3–8 ppm/°C (allowing easier compensation when required), and the series has lower hysteresis than bandgaps. Thermal hysteresis is a low 50 ppm over a –40°C to +125°C range, less that half that of a typical bandgap device. Finally, the long-term stability is excellent, typically only 50 ppm/1000 hours.

Figure 7.14 summarizes the pro and con characteristics of the three reference architectures; bandgap, buried Zener, and XFET.

BANDGAP	BURIED ZENER	XFET
< 5V Supplies	> 5V Supplies	< 5V Supplies
High Noise @ High Power	Low Noise @ High Power	Low Noise @ Low Power
Fair Drift and Long Term Stability	Good Drift and Long Term Stability	Excellent Drift and Long Term Stability
Fair Hysteresis	Fair Hysteresis	Low Hysteresis

Figure 7.14: Characteristics of Reference Architectures

Modern IC references come in a variety of styles, but series operating, fixed output positive types do tend to dominate. These devices can use bandgap-based bipolars, JFETs, or buried Zeners at the device core, all of which has an impact on the part's ultimate performance and application suitability. They may or may not also be low power, low noise, and/or low dropout, and be available within a certain package. Of course, in a given application, any single one of these differentiating factors can drive a choice, thus it behooves the designer to be aware of all the different devices available.

Figure 7.15 shows the standard footprint for such a series type IC positive reference in an 8-pin package. (Note that pin numbers shown refer to the standard pin for that function.) Several details are important. Many references allow optional trimming by connecting an external trim circuit to drive the references' *trim* input pin (5). Some bandgap references also have a high impedance PTAT output (V_{TEMP}) for temperature sensing (Pin 3). The intent here is that no appreciable current be drawn from this pin, but it can be useful for such nonloading types of connections as comparator inputs, to sense temperature thresholds, etc.

All references should use decoupling capacitors on the input pin (2), but the amount of decoupling (if any) placed on the output (Pin 6) depends upon the stability of the reference's output op amp with capacitive load. Simply put, there is no hard and fast rule for capacitive loads here. For example, some three terminal types *require* the output capacitor for stability (i.e., REF19x and AD1582/AD1585 series), while with others it is optional for performance improvement (AD780, REF43, ADR29x, ADR43x, AD38x, AD39x, ADR01, ADR02, ADR03). Even if the output capacitor is optional, it may still be required to supply the

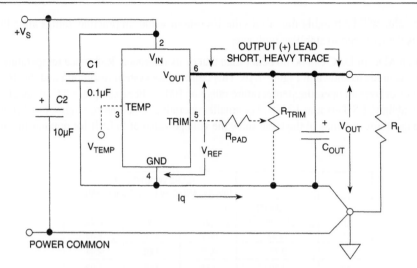

Figure 7.15: Standard Positive Output Three Terminal Reference Hookup (8-Pin DIP Pinout)

energy for transient load currents, as presented by some ADC reference input circuits. The safest rule then is to use the data sheet to verify the specific capacitive loading ground rules for the reference you intend to be used, for the load conditions the circuit presents.

Voltage Reference Specifications

Tolerance

It is usually better to select a reference with the required value and accuracy and to avoid external trimming and scaling if possible. This allows the best TCs to be realized, as tight tolerances and low TCs usually go hand in hand. Tolerances as low as approximately 0.04% can be achieved with the AD586, AD780, REF195, and ADR43x series, while the AD588 is 0.01%. If and when trimming must be used, be sure to use the recommended trim network with no more range than is absolutely necessary. When/if additional external scaling is required, a precision op amp should be used, along with ratio-accurate, low TC tracking thin film resistors.

Drift

The XFET and buried Zener reference families have the best long-term drift and TC performance. The XFET ADR43x-series have TCs as low as 3 ppm/°C. TCs as low as 1–2 ppm/°C are available with the AD586 and AD588 buried Zener references, and the AD780 bandgap reference is almost as good at 3 ppm/°C.

The XFET series achieve long terms drifts of 50 ppm/1000 hours, while the buried Zener types come in at 25 ppm/1000 hours. Note that where a figure is given for long-term drift, it is usually drift expressed in ppm/1000 hours. There are 8766 hours in a year, and many engineers multiply the 1000-hour figure by 8.77 to find the annual drift—this is not correct, and can in fact be quite pessimistic. Long-term drift in precision analog circuits is a "random walk" phenomenon and increases with the *square root* of the elapsed time. (This supposes that drift is due to random micro-effects in the chip and not some over-riding cause such as contamination.) The 1-year figure will therefore be about $\sqrt{8.766} \approx 3$ times the 1000-hour figure, and the ten year value will be roughly nine times the 1000-hour value. In practice, things are a little better even than this, as devices tend to stabilize with age.

The accuracy of an ADC or DAC can be no better than that of its reference. Reference temperature drift affects full-scale accuracy as shown in Figure 7.16. This table shows system resolution and the TC required to maintain ½ LSB error over an operating temperature range of 100°C. For example, a TC of about 1 ppm/°C is required to maintain ½ LSB error at 12 bits. For smaller operating temperature ranges, the drift requirement will be less. The last three columns of the table show the voltage value of ½ LSB for popular full-scale ranges.

BITS	REQUIRED DRIFT (ppm/°C)	½ LSB WEIGHT (mV) 10V, 5V, AND 2.5V FULL-SCALE RANGES		
		10V	5V	2.5V
8	19.53	19.53	9.77	4.88
9	9.77	9.77	4.88	2.44
10	4.88	4.88	2.44	1.22
11	2.44	2.44	1.22	0.61
12	1.22	1.22	0.61	0.31
13	0.61	0.61	0.31	0.15
14	0.31	0.31	0.15	0.08
15	0.15	0.15	0.08	0.04
16	0.08	0.08	0.04	0.02

Figure 7.16: Reference Temperature Drift Requirements for Various System Accuracies (1/2 LSB Criteria, 100°C Span)

Supply Range

IC reference supply voltages range from about 3 V (or less) above rated output, to as high as 30 V (or more) above rated output. Exceptions are devices designed for low dropout, such as the REF19x, AD1582/AD1585, ADR38x, ADR39x series. At low currents, the REF195 can deliver 5 V with an input as low as 5.1 V (100 mV dropout). Note that due to process limits, some references may have more restrictive maximum voltage input ranges, such as the AD1582/AD1585 series (12 V), the ADR29x series (15 V), and the ADR43x series (18 V).

Load Sensitivity

Load sensitivity (or output impedance) is usually specified in µV/mA of load current, or mΩ, or ppm/mA. While figures of 70 ppm/mA or less are quite good (AD780, REF43, REF195, ADR29x, ADR43x), it should be noted that external wiring drops can produce comparable or worse errors at high currents, without care in layout. Load current dependent errors are minimized with short, heavy conductors on the (+) output and on the ground return. For the highest precision, buffer amplifiers and Kelvin sensing circuits (AD588, AD688, ADR39x) are used to ensure accurate voltages at the load.

The output of a buffered reference is the output of an op amp, and therefore the source impedance is a function of frequency. Typical reference output impedance rises at 6 dB/octave from the dc value, and is nominally about 10 Ω at a few hundred kHz. This impedance can be lowered with an external capacitor, provided the op amp within the reference remains stable for such loading.

Line Sensitivity

Line sensitivity (or regulation) is usually specified in μV/V, (or ppm/V) of input change, and is typically 25 ppm/V (–92 dB) in the REF43, REF195, AD680, AD780, ADR29x, ADR39x, and ADR43x. For dc and very low frequencies, such errors are easily masked by noise.

As with op amps, the line sensitivity (or power supply rejection) of references degrades with increasing frequency, typically 30 dB to 50 dB at a few hundred kHz. For this reason, the reference input should be highly decoupled (LF and HF). Line rejection can also be increased with a low dropout pre-regulator, such as one of the ADP3300 series parts.

Figure 7.17 summarizes the major reference specifications along with typical values available.

• Tolerance:		
– AD588	0.01%	
– ADR43x, AD780, REF195	0.04%	
• Drift (TC):		
– AD586, AD588	1–2ppm/°C	
– AD780, ADR42x, ADR43x, ADR01, ADR02, ADR03	3 ppm/°C	
• Drift (long term):		
– ADR29x,ADR42x, ADR43x	50 ppm/1000 hours	
– AD588	25 ppm/1000 hours	
• Supply Range:		
– REF19x, ADR38x, ADR39x, AD158x, AD780	V_{OUT} plus 0.3V – 15V	
• Load Sensitivity	70ppm/mA (350mΩ @ 5V)	
• Line Sensitivity	25ppm/V (–92 dB @ 5V)	

Figure 7.17: Voltage Reference DC Specifications (Typical Values Available)

Noise

Reference noise is not always specified, and when it is, there is not total uniformity on how it is measured. For example, some devices are characterized for peak-to-peak noise in a 0.1 Hz to 10 Hz bandwidth, while others are specified in terms of wideband rms or peak-to-peak noise over a specified bandwidth. The most useful way to specify noise (as with op amps) is a plot of noise voltage spectral density $\left(nV/\sqrt{Hz}\right)$ versus frequency.

Low noise references are important in high resolution systems to prevent loss of accuracy. Since white noise is statistical, a given noise density must be related to an equivalent peak-to-peak noise in the relevant bandwidth. Strictly speaking, the peak-to-peak noise in a gaussian system is infinite (but its probability is infinitesimal). Conventionally, the figure of 6.6 × rms is used to define a practical peak value—statistically, this occurs less than 0.1% of the time. This peak-to-peak value should be less than ½ LSB in order to maintain required accuracy. If peak-to-peak noise is assumed to be six times the rms value, then for an N-bit system, reference voltage full scale V_{REF}, reference noise bandwidth (BW), the required noise voltage spectral density $E_n\left(V/\sqrt{Hz}\right)$ is given by:

$$E_n \leq \frac{V_{REF}}{12 \times 2^N \times \sqrt{BW}} \qquad \text{Eq. 7.9}$$

For a 10 V, 12-bit, 100 kHz system, the noise requirement is a modest $643 \, nV/\sqrt{Hz}$. Figure 7.18 shows that increasing resolution and/or lower full-scale references make noise requirements more stringent. The 100 kHz bandwidth assumption is somewhat arbitrary, but the user may reduce it with external filtering, thereby reducing the noise. Most good IC references have noise spectral densities around $100 \, nV/\sqrt{Hz}$, so additional filtering is obviously required in most high resolution systems, especially those with low values of V_{REF}.

BITS	NOISE DENSITY (nV/√Hz) FOR 10V, 5V, AND 2.5V FULL-SCALE RANGES		
	10V	5V	2.5V
12	643	322	161
13	322	161	80
14	161	80	40
15	80	40	20
16	40	20	10

Figure 7.18: Reference Noise Requirements for Various System Accuracies (1/2 LSB / 100 kHz Criteria)

Some references, for example the AD587 buried Zener type, have a pin designated as the *noise reduction pin* (see data sheet). This pin is connected to a high impedance node preceding the on-chip buffer amplifier. Thus an externally connected capacitor C_N will form a low-pass filter with an internal resistor, to limit the effective noise bandwidth seen at the output. A 1 µF capacitor gives a 3 dB bandwidth of 40 Hz. Note that this method of noise reduction is by no means universal, and other devices may implement noise reduction differently, if at all.

There are also general purpose methods of noise reduction, which can be used to reduce the noise of any reference IC, at any standard voltage level. The reference circuit of Figure 7.19 (References 5 and 6) is one such example. This circuit uses external filtering and a precision low-noise op amp to provide both very low

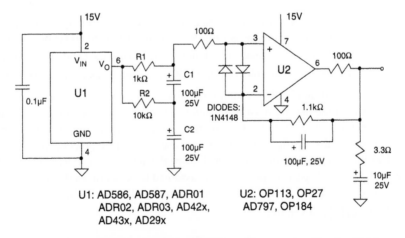

U1: AD586, AD587, ADR01
ADR02, ADR03, AD42x,
AD43x, AD29x

U2: OP113, OP27
AD797, OP184

Figure 7.19: Combining Low-Noise Amplifier with Extensive Filtering Yields Exceptional Reference Noise Performance of (1.5 to 5 nV) /√Hz @ 1 kHz

noise and high dc accuracy. Reference U1 is a 2.5 V, 3.0 V, 5 V, or 10 V reference with a low noise buffered output. The output of U1 is applied to the R1-C1/C2 noise filter to produce a corner frequency of about 1.7 Hz. Electrolytic capacitors usually imply dc leakage errors, but the bootstrap connection of C1 causes its applied bias voltage to be only the relatively small drop across R2. This lowers the leakage current through R1 to acceptable levels. Since the filter attenuation is modest below a few Hertz, the reference noise still affects overall performance at low frequencies (i.e., <10 Hz).

The output of the filter is then buffered by a precision low noise unity-gain follower, such as the OP113EP. With less than ±150 µV offset error and under 1 µV/°C drift, the buffer amplifier's dc performance will not seriously affect the accuracy/drift of most references. For example, an ADR292E for U1 will have a typical drift of 3 ppm/°C, equivalent to 7.5 µV/°C, higher than the buffer amplifier.

Almost any op amp will have a current limit higher than a typical IC reference. Further, even lower noise op amps are available for 5 V to 10 V use. The AD797 offers 1 kHz noise performance less than $2 \ nV/\sqrt{Hz}$ in this circuit, compared to about $5 \ nV/\sqrt{Hz}$ for the OP113. With any amplifier, Kelvin sensing can be used at the load point, a technique which can eliminate I × R related output voltage errors.

Scaled References

A useful approach when a nonstandard reference voltage is required is to simply buffer and scale a basic low voltage reference diode. With this approach, a potential difficulty is getting an amplifier to work well at such low voltages as 3 V. A workhorse solution is the low power reference and scaling buffer shown in Figure 7.20. Here a low current 1.2 V two terminal reference diode is used for D1, which can be either a 1.200 V ADR512, 1.235 V AD589, or the 1.225 V AD1580. Resistor R1 sets the diode current in either case, and is chosen for the diode minimum current requirement at a minimum supply of 2.7 V. Obviously, loading on the unbuffered diode must be minimized at the V_{REF} node.

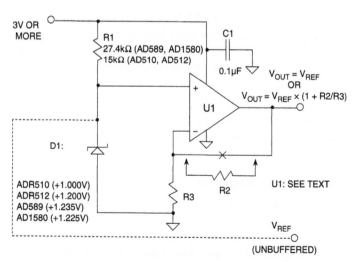

Figure 7.20: Rail-to-Rail Output Op Amps Allow Greatest Flexibility in Low Dropout References

The amplifier U1 both buffers and optionally scales up the nominal 1.0 V or 1.2 V reference, allowing much higher source/sink output currents. Of course, a higher op amp quiescent current is expended in doing this, but this is a basic trade-off of the approach. Quiescent current is amplifier dependent, ranging from 45 µA/channel with the OP196/OP296/OP496 series to 1000 µA–2000 µA/channel with the OP284 and OP279. The former series is most useful for very light loads (<2 mA), while the latter series provide device dependent outputs up to 50 mA. Various devices can be used in the circuit as shown, and their key specs are summarized in Figure 7.21.

DEVICE*	Iq, mA per channel	Vsat (+) V (min @ mA)	Vsat (−) V (max @ mA)	Isc, mA min
OP281/OP481	0.003	4.93 @ 0.05	0.075 @ 0.05	±3.5
OP193/OP293	0.017	4.20 @ 1	0.280 @ 1 (typ)	±8
OP196/OP296/OP496	0.045	4.30 @ 1	0.400 @ 1	±4 (typ)
AD8541/AD42/AD44	0.045	4.97 @ 1	0.025 @ 1	±60
OP777	0.220	4.91 @ 1	0.126 @ 1	±10
AD820/AD822	0.620	4.89 @ 2	0.055 @ 2	±15
OP184/OP284/OP484	1.250**	4.85 @ 2.5	0.125 @ 2.5	±7.5
AD8531/AD32/AD34	1.400	4.90 @ 10	0.100 @ 10	±250

*Typical device specifications @ Vs = 5V, TA = 25°C, unless otherwise noted
**Maximum

Figure 7.21: Op Amps Useful in Low Voltage Rail-Rail References and Regulators

In Figure 7.20, without gain scaling resistors R2-R3, V_{OUT} is simply equal to V_{REF}. With the use of the scaling resistors, V_{OUT} can be set anywhere between a lower limit of V_{REF}, and an upper limit of the positive rail, due to the op amp's rail-rail output swing. Also, note that this buffered reference is inherently low dropout, allowing a 4.5 V (or more) reference output on a 5 V supply, for example. The general expression for V_{OUT} is shown in the figure, where V_{REF} is the reference voltage.

Amplifier standby current can be further reduced below 20 µA, if an amplifier from the OP181/OP281/OP481 or the OP193/OP293/OP493 series is used. This choice will be at some expense of current drive, but can provide very low quiescent current if necessary. All devices shown operate from voltages down to 3 V (except the OP279, which operates at 5 V).

Voltage Reference Pulse Current Response

The response of references to dynamic loads is often a concern, especially in applications such as driving some ADCs and DACs. Fast changes in load current invariably perturb the output, often outside the rated error band. For example, the reference input to a sigma-delta ADC may be the switched capacitor circuit shown in Figure 7.22. The dynamic load causes current spikes in the reference as the capacitor C_{IN} is charged and discharged. As a result, noise may be induced on the ADC reference circuitry.

Although sigma-delta ADCs have an internal digital filter, transients on the reference input can still cause appreciable conversion errors. Thus it is important to maintain a low noise, transient free potential at the ADC's reference input. Be aware that if the reference source impedance is too high, dynamic loading can cause the reference input to shift by more than 5 mV.

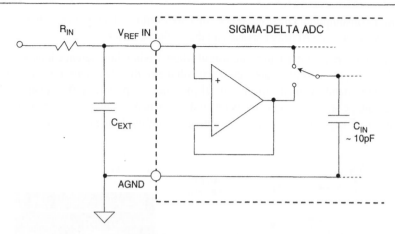

**Figure 7.22: Switched Capacitor Input of Sigma-Delta
ADC Presents a Dynamic Load to the Voltage Reference**

A bypass capacitor on the output of a reference may help it to cope with load transients, but many references are unstable with large capacitive loads. Therefore it is quite important to verify that the device chosen will satisfactorily drive the output capacitance required. In any case, the converter reference inputs should always be decoupled—with at least 0.1 μF, and with an additional 5 μF–50 μF if there is any low frequency ripple on its supply. See Figure 7.15.

Since some references misbehave with transient loads, either by oscillating or by losing accuracy for comparatively long periods, it is advisable to test the pulse response of voltage references which may encounter transient loads. A suitable circuit is shown in Figure 7.23. In a typical voltage reference, a step change of 1 mA produces the transients shown. Both the duration of the transient, and the amplitude of the ringing *increase* when a 0.01 μF capacitor is connected to the reference output.

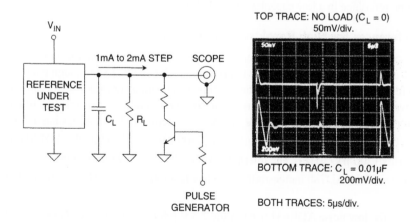

Figure 7.23: Make Sure Reference is Stable with Large Capacitive Loads

Where possible, a reference should be designed to drive large capacitive loads. The AD780 is designed to drive unlimited capacitance without oscillation, it has excellent drift and an accurate output, in addition to relatively low power consumption. Other references that are useful with output capacitors are the REF19x, the AD1582/AD1585 series, the ADR29x series, and the ADR43x series.

As noted above, reference bypass capacitors are useful when driving the reference inputs of successive-approximation ADCs. Figure 7.24 illustrates reference voltage settling behavior immediately following the "Start Convert" command. A small capacitor (0.01 µF) does not provide sufficient charge storage to keep the reference voltage stable during conversion, and errors may result. As shown by the bottom trace, decoupling with a ≥ 1 µF capacitor maintains the reference stability during conversion.

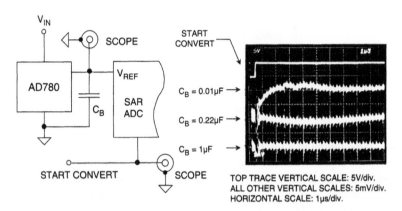

Figure 7.24: Successive-Approximation ADCs Can Present a Dynamic Transient Load to the Reference

Where voltage references are required to drive large capacitances, it is also critically important to realize that their turn-on time will be prolonged. Experiment may be needed to determine the delay before the reference output reaches full accuracy, but it will certainly be much longer than the time specified on the data sheet for the same reference in a low capacitance loaded state.

Low Noise References for High Resolution Converters

High resolution converters (both sigma-delta and high speed) can benefit from recent improvements in IC references, such as lower noise and the ability to drive capacitive loads. Even though many data converters have internal references, the performance of these references is often compromised because of the limitations of the converter process. In such cases, using an external reference rather than the internal one often yields better overall performance. For example, the AD7710 series of 22-bit ADCs has a 2.5 V internal reference with a 0.1 Hz to 10 Hz noise of 8.3 µV rms $\left(2600 \text{ nV}/\sqrt{\text{Hz}}\right)$, while the AD780 reference noise is only 0.67 µV rms $\left(200 \text{ nV}/\sqrt{\text{Hz}}\right)$. The internal noise of the AD7710 series in this bandwidth is about 1.7 µV rms. The use of the AD780 increases the effective resolution of the AD7710 from about 20.5 bits to 21.5 bits.

Figure 7.25 shows the low noise ADR431 used as the 2.5 V reference for the AD77xx series ADCs. Optimally, the use of the ADR433 (3 V output) enhances the dynamic range of the ADC, while lowering overall system noise as described above. In addition, the ADR43x series allow a large decoupling capacitor on its output thereby minimizing conversion errors due to transients.

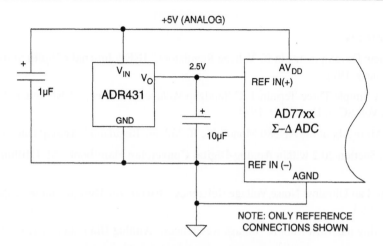

Figure 7.25: The AD431 XFET Reference is Ideal for Driving Precision Sigma-Delta ADCs

There is one possible but yet quite real problem when replacing the internal reference of a converter with a higher precision external one. The converter in question may have been trimmed during manufacture to deliver its specified performance with a relatively inaccurate internal reference. In such a case, using a more accurate external reference with the converter may actually introduce additional gain error. For example, the early AD574 had a guaranteed uncalibrated gain accuracy of 0.125% when using an internal 10 V reference (which itself had a specified accuracy of only ±1%). It is obvious that if such a device, having an internal reference at one end of the specified range, is used with an external reference of exactly 10 V, its gain will be about 1% in error.

References:
7.1 Voltage References

1. Bob Widlar, "New Developments in IC Voltage Regulators," **IEEE Journal of Solid State Circuits**, Vol. SC-6, February, 1971.

2. Paul Brokaw, "A Simple Three-Terminal IC Bandgap Voltage Reference," **IEEE Journal of Solid State Circuits**, Vol. SC-9, December, 1974.

3. Paul Brokaw, "More About the AD580 Monolithic IC Voltage Regulator," **Analog Dialogue**, 9-1, 1975.

4. Dan Sheingold, Section 20.2 within **Analog-Digital Conversion Handbook, 3d. Edition**, Prentice-Hall, 1986.

5. Walt Jung, "Build an Ultralow Noise Voltage Reference," **Electronic Design Analog Applications Issue**, June 24, 1993.

6. Walt Jung, "Getting the Most from IC Voltage References," **Analog Dialogue**, 28-1, 1994, pp. 13–21.

Low Dropout Linear Regulators
Walt Jung

Introduction

Linear IC voltage regulators have long been standard power system building blocks. After an initial introduction in 5 V logic voltage regulator form, they have since expanded into other standard voltage levels spanning from 1.5 V to 24 V, handling output currents from as low as 100 mA (or less) to as high as 5 A (or more). For several good reasons, linear style IC voltage regulators have been valuable system components since the early days. One reason is the relatively low noise characteristic vis-à-vis the switching type of regulator. Others are a low parts count and overall simplicity compared to discrete solutions. But, because of their power losses, these linear regulators have also been known for being relatively inefficient. Early generation devices (of which many are still available) required 2 V or more of unregulated input above the regulated output voltage, making them lossy in power terms.

More recently however, linear IC regulators have been developed with more liberal (i.e., lower) limits on minimum input-output voltage. This voltage, known more commonly as *dropout* voltage, has led to what is termed the *low drop out* regulator, or more popularly, the LDO. Dropout voltage (V_{MIN}) is defined simply as that minimum input-output differential where the regulator undergoes a 2% reduction in output voltage. For example, if a nominal 5.0 V LDO output drops to 4.9 V (–2%) under conditions of an input-output differential of 0.5 V, by this definition the LDO's V_{MIN} is 0.5 V.

As will be shown in this section, dropout voltage is extremely critical to a linear regulator stage's power efficiency. The lower the voltage allowable across a regulator while still maintaining a regulated output, the less power the regulator dissipates as a result. A low regulator dropout voltage is the key to this, as it takes this lower dropout to maintain regulation as the input voltage lowers. In performance terms, the bottom line for LDOs is simply that more useful power is delivered to the load and less heat is generated in the regulator. LDOs are key elements of power systems that must provide stable voltages from batteries, such as portable computers, cellular phones, etc. This is simply because they maintain their regulated output down to lower points on the battery's discharge curve. Or, within classic mains-powered raw dc supplies, LDOs allow lower transformer secondary voltages, reducing system susceptibility to shutdown under brownout conditions as well as allowing cooler operation.

Linear Voltage Regulator Basics

A brief review of three terminal linear IC regulator fundamentals is necessary to understanding the LDO variety. As it turns out, almost all LDOs available today, as well as many of the more general three-terminal regulator types, are *positive leg, series style* regulators. This simply means that they control the regulated voltage output by means of a pass element which is in series with the positive side of the unregulated input.

This is shown more clearly in Figure 7.26, which is a hookup diagram for a hypothetical three-terminal style regulator. To reiterate what was said earlier in the chapter about reference ICs, in terms of their basic functionality, many standard voltage regulator ICs are available in the series three-terminal form as is shown here (V_{IN}, GND or Common, V_{OUT}).

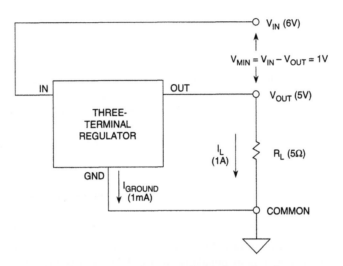

Figure 7.26: A Basic Three-Terminal Voltage Regulator

This diagram also allows some statements to be made about power losses in the regulator. There are two components to power which are dissipated in the regulator, one a function of $V_{IN} - V_{OUT}$ and I_L, plus a second which is a function of V_{IN} and I_{GROUND}. If we call the total power P_D, this then becomes:

$$P_D = (V_{IN} - V_{OUT})(I_L) + (V_{IN})(I_{GROUND})$$

Eq. 7.10

Obviously, the magnitude of the load current and the regulator dropout voltage both greatly influence the power dissipated. However, it is also easy to see that for a given I_L, as the dropout voltage is lowered, the first term of P_D is reduced. With an intermediate dropout voltage rating of 1 V, a 1 A load current will produce 1 W of heat in this regulator, which will require a heat sink for continuous operation. It is this first term of the regulator power which usually predominates, at least for loaded regulator conditions.

The second term, being proportional to I_{GROUND} (typically only 1 mA–2 mA, sometimes even less) usually only becomes significant when the regulator is unloaded, and the regulator's quiescent or standby power then produces a constant drain on the source V_{IN}.

However, it should be noted that in some types of regulators (notably those which have very low β pass devices such as lateral PNP transistors) the I_{GROUND} current under load can actually run quite high. This effect is worst at the onset of regulation, or when the pass device is in saturation, and can be noted by a sudden I_{GROUND} current "spike," where the current jumps upward abruptly from a low level. All LDO regulators using bipolar transistor pass devices which can be saturated (such as PNPs) can show this effect. It is much less severe in PNP regulators using vertical PNPs (since these have a higher intrinsic β) and doesn't exist to any major extent in PMOS LDOs (since PMOS transistors are controlled by voltage level, not current).

In the example shown, the regulator delivers 5 V × 1 A, or 5 W to the load. With a dropout voltage of 1 V, the input power is 6 V times the same 1 A, or 6 W. In terms of power efficiency, this can be calculated as:

$$P_{EFF}(\%) = 100 \times \frac{P_{OUT}}{P_{IN}}$$

Eq. 7.11

where P_{OUT} and P_{IN} are the total output and input powers, respectively.

In these sample calculations, the relatively small portion of power related to I_{GROUND} will be ignored for simplicity, since this power is relatively small. In an actual design, this simplifying step may not be justified.

In the case shown, the efficiency would be 100 × 5/6, or about 83%. But by contrast, if an LDO were to be used with a dropout voltage of 0.1 V instead of 1 V, the input voltage can then be allowed to go as low as 5.1 V. The new efficiency for this condition then becomes 100 × 5/5.1, or 98%. It is obvious that an LDO can potentially greatly enhance the power efficiency of linear voltage regulator systems.

A more detailed look within a typical regulator block diagram reveals a variety of elements, as is shown in Figure 7.27.

In this diagram virtually all of the elements shown can be considered to be fundamentally necessary, the exceptions being the shutdown control and saturation sensor functions (shown dotted). While these are present on many current regulators, the shutdown feature is relatively new as a standard function, and certainly isn't part of standard three-terminal regulators. When present, shutdown control is a logic level controllable input, whereby a digital HIGH (or LO) is defined as regulation active (or vice-versa).

The error output, $\overline{ERR}$, is useful within a system to detect regulator overload, such as saturation of the pass device, thermal overload, etc. The remaining functions shown are always part of an IC power regulator.

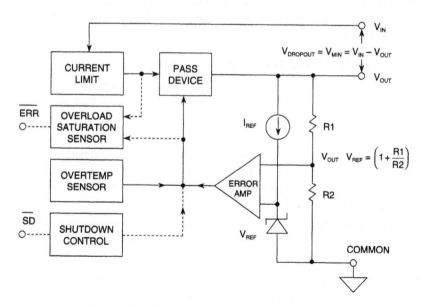

Figure 7.27: Block Diagram of a Voltage Regulator

In operation, a voltage reference block produces a stable voltage V_{REF}, which is almost always a bandgap-based voltage, typically ~1.2 V, which allows output voltages of 3 V or more from supplies as low as 5 V. This voltage is presented to one input of an error amplifier, with the other input connected to the V_{OUT} sensing divider, R1-R2. The error amplifier drives the pass device, which in turn controls the output. The resulting regulated voltage is then simply:

$$V_{OUT} = V_{REF}\left(1 + \frac{R1}{R2}\right)$$

Eq. 7.12

With a typical bandgap reference voltage of 1.2 V, the R1/R2 ratio will be approximately 3/1 for a 5 V output. When standby power is critical, several design steps will be taken. The resistor values of the divider will be high, the error amplifier and pass device driver will be low power, and the reference current I_{REF} will also be low. By these means the regulator's unloaded standby current can be reduced to a mA or less using bipolar technology, and to only a few μA in CMOS parts. In regulators that offer a shutdown mode, the shutdown state standby current will be reduced to a μA or less.

Nearly all regulators will have some means of current limiting and over temperature sensing, to protect the pass device against failure. Current limiting is usually by a series sensing resistor for high current parts, or alternately by a more simple drive current limit to a controlled β pass device (which achieves the same end). For higher voltage circuits, this current limiting may also be combined with voltage limiting, to provide complete load line control for the pass device.

All power regulator devices will also have some means of sensing over temperature, usually by means of a fixed reference voltage and a V_{BE}-based sensor monitoring chip temperature. When the die temperature exceeds a dangerous level (above ~150°C), this can be used to shut down the chip, by removing the drive to the pass device. In some cases an error flag output may be provided to warn of this shutdown (and also loss of regulation from other sources).

Pass Devices and their Associated Trade Offs

The discussion thus far has not treated the pass device in any detail. In practice, this major part of the regulator can actually take on quite a number of alternate forms. Precisely which type of pass device is chosen has a major influence on almost all major regulator performance issues. Most notable among these is the dropout voltage, V_{MIN}.

Figure 7.28a through 7.28e illustrates a number of pass devices that are useful within voltage regulator circuits, shown in simple schematic form. The figure also lists the salient V_{MIN} for the device as it would typically be used, which directly indicates its utility for use in an LDO. Not shown in these various minifigures are the remaining circuits of a regulator.

It is difficult to fully compare all of the devices from their schematic representations, since they differ in so many ways beyond their applicable dropout voltages. For this reason, the chart of Figure 7.29 is useful.

This chart compares the various pass elements in greater detail, allowing easy comparison between the device types, dependent upon which criteria is most important. Note that columns A–E correspond to the schematics of Figure 7.28a–7.28e. Note also that the pro/con comparison items are in *relative* terms, as opposed to a hard specification limit for any particular pass device type.

For example, it can be seen that the all-NPN pass devices of columns A and B have the attributes of a follower circuit, which allows high bandwidth and provides relative immunity to cap loading because of the characteristic low Z_{OUT}. However, neither the single NPN nor the Darlington NPN can achieve low dropout,

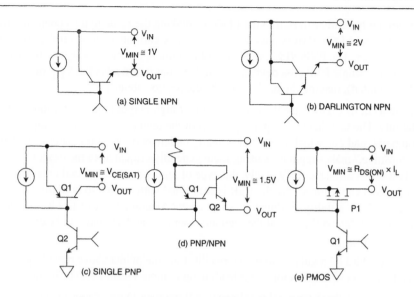

Figure 7.28: Pass Devices Useful in Voltage Regulators

A	B	C	D	E
SINGLE NPN	DARLINGTON NPN	SINGLE PNP	PNP/NPN	PMOS
$V_{MIN} \sim 1V$	$V_{MIN} \sim 2V$	$V_{MIN} \sim 0.1V$	$V_{MIN} \sim 1.5V$	$V_{MIN} \sim R_{DS(ON)} \times I_L$
$I_L < 1A$	$I_L > 1A$	$I_L < 1A$	$I_L > 1A$	$I_L > 1A$
Follower	Follower	Inverter	Inverter	Inverter
Low Z_{OUT}	Low Z_{OUT}	High Z_{OUT}	High Z_{OUT}	High Z_{OUT}
Wide BW	Wide BW	Narrow BW	Narrow BW	Narrow BW
C_L Immune	C_L Immune	C_L Sensitive	C_L Sensitive	C_L Sensitive

Figure 7.29: Pros and Cons of Voltage Regulator Pass Devices

for any load current. This is because the $V_{BE}(s)$ of the pass device appears in series with the input, preventing its saturation, and thus setting a V_{MIN} of about 1 V or 2 V.

By contrast, the inverting mode device connections of both columns C and E do allow the pass device to be effectively saturated, which lowers the associated voltage losses to a minimum. This single factor makes these two pass device types optimum for LDO use, at least in terms of power efficiency.

For currents below 1 A, either a single PNP or a PMOS pass device is most useful for low dropout, and they can both achieve a V_{MIN} of 0.1 V or less at currents of 100 mA. The dropout voltage of a PNP will be highly dependent upon the actual device used and the operating current, with vertical PNP devices being superior for saturation losses, as well as minimizing the I_{GROUND} spike when in saturation. PMOS pass devices offer the potential for the lowest possible V_{MIN}, since the actual dropout voltage will be the product of the device $R_{DS(ON)}$ and I_L. Thus a low $R_{DS(ON)}$ PMOS device can always be chosen to minimize V_{MIN} for a given I_L.

469

PMOS pass devices are typically *external* to the LDO IC, making the IC actually a controller (as opposed to a complete and integral LDO). PMOS pass devices can allow currents up to several amps or more with very low dropout voltages. The PNP/NPN connection of column D is actually a hybrid hookup, intended to boost the current of a single PNP pass device. This it does, but it also adds the V_{BE} of the NPN in series (which cannot be saturated), making the net V_{MIN} of the connection about 1.5 V.

All of the three connections C/D/E have the characteristic of high output impedance, and require an output capacitor for stability. The fact that the output cap is part of the regulator frequency compensation is a most basic application point, and one which needs to be clearly understood by the regulator user. This factor, denoted by "C_L sensitive," makes regulators using them generally critical as to the exact C_L value, as well as its ESR (equivalent series resistance). Typically, this type of regulator must be used only with a specific size as well as type of output capacitor, where the ESR is controlled with respect to both time and temperature to fully guarantee regulator stability. Fortunately, some recent Analog Devices LDO IC circuit developments have eased this burden on the part of the regulator user a great deal, and will be discussed below in further detail.

Some examples of standard IC regulator architectures illustrate the points above regarding pass devices, and allow an appreciation of regulator developments leading up to more recent LDO technologies.

The classic LM309 5 V/1 A three-terminal regulator (see Reference 1) was the originator in a long procession of regulators. This circuit is shown in much simplified form in Figure 7.30, with current limiting and over temperature details omitted. This IC type is still in standard production today, not just in original form, but in family derivatives such as the 7805, 7815 etc., and their various low and medium current alternates. Using a Darlington pass connection for Q18-Q19, the design has never been known for low dropout characteristics (~1.5 V typical), or for low quiescent current (~5 mA). It is, however, relatively immune to instability issues, due to the internal compensation of C1 and the buffering of the emitter follower output. This helps make it easy to apply.

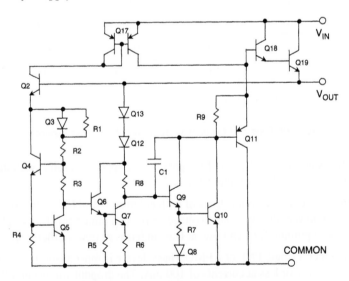

**Figure 7.30: Simplified Schematic of LM309
Fixed 5 V/1 V Three-Terminal Regulator**

The LM109/309 bandgap voltage reference actually used in this circuit consists of a more involved scheme, as opposed to the basic form which was described with Figure 7.3. Resistor R8 drops a PTAT voltage, which drives the Darlington connected error amplifier, Q9-Q10. The negative TC V_{BE}s of Q9-Q10 and Q12-Q13 are summed with this PTAT voltage, and this sum produces a temperature-stable 5 V output voltage. Current buffering of the error amplifier Q10 is provided by PNP Q11, which drives the NPN pass devices.

Later developments in references and three-terminal regulation techniques led to the development of the voltage adjustable regulator. The original IC to employ this concept was the LM317 (see Reference 2), which is shown in simplified schematic form in Figure 7.31. Note that this design does not use the same ΔV_{BE} form of reference as in the LM309. Instead, Q17–Q19, etc. are employed as a form of a Brokaw bandgap reference cell (see Figure 7.4 again, and Reference 3).

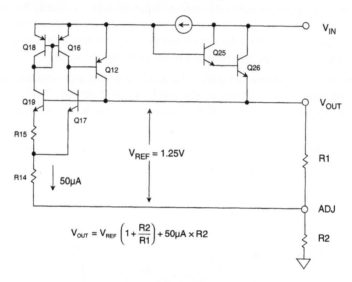

Figure 7.31: Simplified Schematic of LM317 Adjustable Three-terminal Regulator

This adjustable regulator bootstraps the reference cell transistors Q17–19 and the error amplifier transistors Q16–18. The output of the error amplifier drives Darlington pass transistors Q25-Q26, through buffer Q12. The basic reference cell produces a fixed voltage of 1.25 V, which appears between the V_{OUT} and ADJ pins of the IC as shown. External scaling resistors R1 and R2 set up the desired output voltage, which is:

$$V_{OUT} = V_{REF}\left(1+\frac{R2}{R1}\right)+50\mu A \times R2 \qquad \text{Eq. 7.13}$$

As can be noted, the voltage output is a scaling of V_{REF} by R2-R1, plus a small voltage component that is a function of the 50 μA reference cell current. Typically, the R1-R2 values are chosen to draw >5 mA, making the rightmost term relatively small by comparison. The design is internally compensated, and in many applications will not necessarily need an output bypass capacitor.

Like the LM309 fixed voltage regulator, the LM317 series has relatively high dropout voltage, due to the use of Darlington pass transistors. It is also not a low power IC (quiescent current typically 3.5 mA). The strength of this regulator lies in the wide range of user voltage adaptability it allows.

Subsequent variations on the LM317 pass device topology modified the method of output drive, substituting a PNP/NPN cascade for the LM317's Darlington NPN pass devices. This development achieves a lower V_{MIN}, 1.5 V or less (see Reference 4). The modification also allows all of the general voltage programmability of the basic LM317, but at some potential increase in application sensitivity to output capacitance. This sensitivity is brought about by the fundamental requirement for an output capacitor for the IC's frequency compensation, which is a differentiation from the original LM317.

Low Dropout Regulator Architectures

As has been shown thus far, all LDO pass devices have the fundamental characteristics of operating in an inverting mode. This allows the regulator circuit to achieve pass device saturation, and thus low dropout. A byproduct of this mode of operation is that this type of topology will necessarily be more susceptible to stability issues. These basic points give rise to some of the more difficult issues with regard to LDO performance. In fact, these points influence both the design and the application of LDOs to a very large degree, and in the end, determine how they are differentiated in the performance arena.

A traditional LDO architecture is shown in Figure 7.32, and is generally representative of actual parts employing either a PNP pass device as shown, or alternately, a PMOS device. There are both dc and ac design and application issues to be resolved with this architecture, which are now discussed.

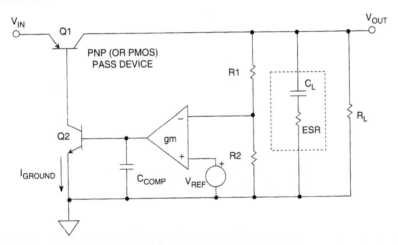

Figure 7.32: Traditional LDO Architecture

In dc terms, perhaps the major issue is the type of pass device used, which influences dropout voltage and ground current. If a lateral PNP device is used for Q1, the β will be low, sometimes only on the order of 10 or so. Since Q1 is driven from the collector of Q2, the relatively high base current demanded by a lateral PNP results in relatively high emitter current in Q2, or a high I_{GROUND}. For a typical lateral PNP based regulator operating with a 5 V/150 mA output, I_{GROUND} will be typically ~18 mA, and can be as high as 40 mA. To compound the problem of high I_{GROUND} in PNP LDOs, there is also the "spike" in I_{GROUND}, as the regulator is operating within its dropout region. Under such conditions, the output voltage is out of tolerance, and the regulation loop forces higher drive to the pass device, in an attempt to maintain loop regulation. This results in a substantial spike upward in I_{GROUND}, which is typically internally limited by the regulator's saturation control circuits.

PMOS pass devices do not demonstrate a similar current spike in I_{GROUND}, since they are voltage controlled. But, while devoid of the I_{GROUND} spike, PMOS pass devices do have some problems of their own. Problem number one is that high quality, low R_{ON}, low threshold PMOS devices generally aren't compatible with many IC processes. This makes the best technical choice for a PMOS pass device an external part, driven from the collector of Q2 in the figure. This introduces the term "LDO controller," where the LDO architecture is completed by an external pass device. While in theory NMOS pass devices would offer lower R_{ON} choice options, they also demand a boosted voltage supply to turn on, making them impractical for a simple LDO. PMOS pass devices are widely available in low both R_{ON} and low threshold forms, with current levels up to several amperes. They offer the potential of the lowest dropout of any device, since dropout can always be lowered by picking a lower R_{ON} part.

The dropout voltage of lateral PNP pass devices is reasonably good, typically around 300 mV at 150 mA, with a maximum of 600 mV. These levels are however considerably bettered in regulators using vertical PNPs, which have a typical β of ~150 at currents of 200 mA. This leads directly to an I_{GROUND} of 1.5 mA at the 200 mA output current. The dropout voltage of vertical PNPs is also an improvement vis-à-vis that of the lateral PNP regulator, and is typically 180 mV at 200 mA, with a maximum of 400 mV.

There are also major ac performance issues to be dealt with in the LDO architecture of Figure. 7.32. This topology has an inherently high output impedance, due to the operation of the PNP pass device in a common-emitter (or common-source with a PMOS device) mode. In either case, this factor causes the regulator to appear as a high source impedance to the load.

The internal compensation capacitor of the regulator, C_{COMP}, forms a fixed frequency pole, in conjunction with the g_m of the error amplifier. In addition, load capacitance C_L forms an output pole, in conjunction with R_L. This particular pole, because it is a second (and sometimes variable) pole of a two-pole system, is the source of a major LDO application problem. The C_L pole can strongly influence the overall frequency response of the regulator, in ways that are both useful as well as detrimental. Depending upon the relative positioning of the two poles in the frequency domain, along with the relative value of the ESR of capacitor C_L, it is quite possible that the stability of the system can be compromised for certain combinations of C_L and ESR. Note that C_L is shown here as a real capacitor, which is actually composed of a pure capacitance plus the series parasitic resistance ESR.

Without a heavy duty exercise into closed-loop stability analysis, it can safely be said that LDOs, like other feedback systems, need to satisfy certain basic stability criteria. One of these is the gain-versus-frequency rate-of-change characteristic in the region approaching the system's unity loop gain crossover point. For the system to be closed loop stable, the phase shift must be less than 180° at the point of unity gain. In practice, a good feedback design needs to have some phase margin, generally 45° or more to allow for various parasitic effects. While a single-pole system is intrinsically stable, two-pole systems are *not* necessarily so—they may in fact be stable, or they may also be unstable. Whether or not they are stable for a given instance is highly dependent upon the specifics of their gain-phase characteristics.

If the two poles of such a system are widely separated in terms of frequency, stability may not be a serious problem. The emitter-follower output of a classic regulator like the LM309 is an example with widely separated pole frequencies, as the very low Z_{OUT} of the NPN follower pushes the output pole due to load capacitance far out in frequency, where it does little harm. The internal compensation capacitance (C1 of Figure 7.30, again) then forms part of a *dominant pole*, which reduces loop gain to below unity at the much higher frequencies where the output pole does occur. Thus stability is not necessarily compromised by load capacitance in this type of regulator.

Figure 7.33 summarizes the various dc and ac design issues of LDOs.

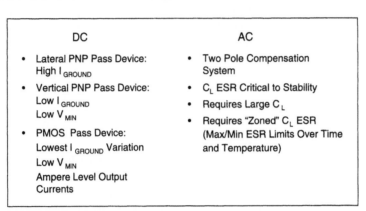

DC	AC
• Lateral PNP Pass Device: High I_{GROUND}	• Two Pole Compensation System
• Vertical PNP Pass Device: Low I_{GROUND} Low V_{MIN}	• C_L ESR Critical to Stability
• PMOS Pass Device: Lowest I_{GROUND} Variation Low V_{MIN} Ampere Level Output Currents	• Requires Large C_L • Requires "Zoned" C_L ESR (Max/Min ESR Limits Over Time and Temperature)

Figure 7.33: DC and AC Design Issues in Low Dropout Regulators

By their nature however, LDOs simply can't afford the luxury of emitter follower outputs, they must instead operate with pass devices capable of saturation. Thus, given the existence of two or more poles (one or more internal and a second formed by external loading) there is the potential for the cumulative gain-phase to add in a less than satisfactory manner. The potential for instability under certain output loading conditions is, for better or worse, a fact-of-life for most LDO topologies.

However, the output capacitor that gives rise to the instability can, for certain circumstances, also be the solution to the same instability. This seemingly paradoxical situation can be appreciated by realizing that almost all practical capacitors are actually as shown in Figure 7.32, a series combination of the capacitance C_L and a parasitic resistance, ESR. While load resistance R_L and C_L do form a pole, C_L and its ESR also form a zero. The effect of the zero is to mitigate the destabilizing effect of C_L for certain conditions.

For example, if the pole and zero in question are appropriately placed in frequency relative to the internal regulator poles, some of the deleterious effects can be made to essentially cancel, leaving little or no problematic instability (see Reference 5).

The basic problem with this setup is simply that the capacitor's ESR, being a parasitic term, is not at all well controlled. As a result, LDOs which depend upon output pole-zero compensation schemes must very carefully limit the capacitor ESR to certain *zones*, such as shown by Figure 7.34.

A zoned ESR chart such as this is meant to guide the user of an LDO in picking an output capacitor which confines ESR to the stable region, i.e., the central zone, for all operating conditions. Note that this generic chart is not intended to portray any specific device, just the general pattern. Unfortunately, capacitor facts of life make such data somewhat limited in terms of the real help it

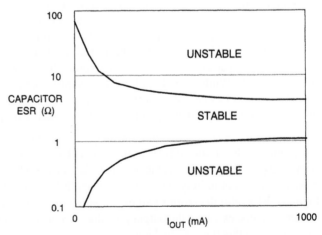

Figure 7.34: Zoned Load Capacitor ESR Can Make an LDO Applications Nightmare

provides. Bearing in mind the requirements of such a zoned chart, it effectively means that general-purpose aluminum electrolytics are prohibited from use, since they deteriorate in terms of ESR at cold temperatures. Very low ESR types such as OS-CON or multilayer ceramic units have ESRs that are too low for use. While they could, in theory, be padded up into the stable zone with external resistance, this would hardly be a practical solution. This leaves tantalum types as the best all around choice for LDO output use. Finally, since a large capacitor value is likely to be used to maximize stability, this effectively means that the solution for an LDO such as Figure 7.32 must use a more expensive and physically large tantalum capacitor. This is not desirable if small size is a major design criteria.

The anyCAP Low Dropout Regulator Family

Some novel modifications to the basic LDO architecture of Figure 7.32 allow major improvements in terms of both dc and ac performance. These developments are shown schematically in Figure 7.35, which is a simplified diagram of the Analog Devices ADP330x, and ADP333x-series LDO regulator family. These regulators are also known as the anyCAP® family, so named for their relative insensitivity to the output capacitor in terms of both size and ESR. They are available in power efficient packages such as the Thermal Coastline (discussed below), in both stand-alone LDO and LDO controller forms, and also in a wide span of output voltage options.

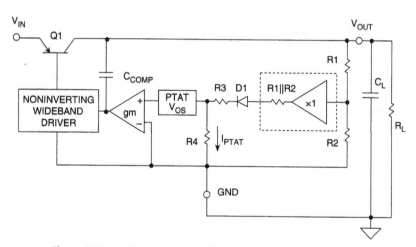

**Figure 7.35: ADP330x and ADP333x anyCAP Topology Features
Improved DC and AC Performance Over Traditional LDOs**

Design Features Related to DC Performance

One of the key differences in the ADP330x/ADP333x series is the use of a high gain vertical PNP pass device, with all of the advantages described above with Figs. 7.32 and 7.33 (also, see Reference 6). This allows the typical dropout voltages for the series to be on the order of 1 mV/mA for currents of 200 mA or less.

It is important to note that the topology of this LDO is distinctly different from that of the generic form in Figure 7.32, as there is no obvious V_{REF} block. The reason for this is the fact that the ADP330x/ADP333x series uses what is termed a "merged" amplifier-reference design. The operation of the integral amplifier and reference scheme illustrated in Figure 7.35 can be described as follows.

In this circuit, V_{REF} is defined as a reference voltage existing at the output of a zero impedance divider of ratio R1/R2. In the figure, this is depicted symbolically by the (dotted) unity gain buffer amplifier fed by R1/R2, which has an output of V_{REF}. This reference voltage feeds into a series connection of (dotted) R1‖R2, then actual components D1, R3, R4, etc.

The error amplifier, shown here as a gm stage, is actually a PNP input differential stage with the two transistors of the pair operated at different current densities, so as to produce a predictable PTAT offset voltage. Although shown here as a separate block V_{OS}, this offset voltage is inherent to a bipolar pair for such operating conditions. The PTAT V_{OS} causes a current I_{PTAT} to flow in R4, which is simply:

$$I_{PTAT} = \frac{V_{OS}}{R4}$$

Eq. 7.14

Note that this current also flows in series connected R4, R3, and the Thevenin resistance of the divider, R1‖R2, so:

$$V_{PTAT} = I_{PTAT}\left(R3 + R4 + R1\|R2\right)$$

Eq. 7.15

The *total* voltage defined as V_{REF} is the sum of two component voltages:

$$V_{REF} = V_{PTAT} + V_{DI}$$

Eq. 7.16

where the I_{PTAT} scaled voltages across R3, R4, and R1‖R2 produce a net PTAT voltage V_{PTAT}, and the diode voltage V_{DI} is a CTAT voltage. As in a standard bandgap reference, the PTAT and CTAT components add up to a temperature stable reference voltage of 1.25 V. In this case, however, the reference voltage is not directly accessible but, instead, exists in the virtual form described above. It acts as it would be seen at the output of a zero impedance divider of a numeric ratio of R1/R2, which is then fed into the R3-D1 series string through a Thevenin resistance of R1‖R2 in series with D1.

With the closed loop regulator at equilibrium, the voltage at the virtual reference node will be:

$$V_{REF} = V_{OUT}\left(\frac{R2}{R1 + R2}\right)$$

Eq. 7.17

With minor rearrangement, this can be put into the standard form to describe the regulator output voltage, as:

$$V_{OUT} = V_{REF}\left(1 + \frac{R1}{R2}\right)$$

Eq. 7.18

In the various devices of the ADP330x/ADP333x series, the R1-R2 divider is adjusted to produce various standard output voltages from 1.5 V to 5.0 V.

As can be noted from this discussion, unlike a conventional reference setup, there is no power wasting reference current such as used in a conventional regulator topology (I_{REF} of Figure 7.27). In fact, the Figure 7.35 regulator behaves as if the entire error amplifier has simply an offset voltage of V_{REF} volts, as seen at the output of a conventional R1-R2 divider.

Design Features Related to AC Performance

While the above-described dc performance enhancements of the ADP330x series are worthwhile, the most dramatic improvements come in areas of ac-related performance. These improvements are in fact the genesis of the anyCAP series name.

Capacitive loading and the potential instability it brings is a major deterrent to easily applying LDOs. While low dropout goals prevent the use of emitter follower type outputs, and so preclude their desirable buffering effect against cap loading, there is an alternative technique of providing load immunity. One method of providing a measure of insusceptibility against variation in a particular amplifier response pole is called *pole splitting* (see Reference 8). It refers to an amplifier compensation method whereby two response poles are shifted in such a way so as to make one a dominant, lower frequency pole. In this manner the secondary pole (which in this case is the C_L related output pole) becomes much less of a major contributor to the net ac response. This has the desirable effect of greatly desensitizing the amplifier to variations in the output pole.

A Basic Pole-Splitting Topology

A basic LDO topology with frequency compensation as modified for pole splitting is shown in Figure 7.36. Here the internal compensation capacitor C_{COMP} is connected as an integrating capacitor, around pass device Q1 (C1 is the pass device input capacitance). While it is true that this step will help immunize the regulator to the C_L related pole, it also has a built in fatal flaw. With C_{COMP} connected directly to the Q1 base as shown, the line rejection characteristics of this setup will be quite poor. In effect, when doing it this way one problem (C_L sensitivity) will be exchanged for another (poor line rejection).

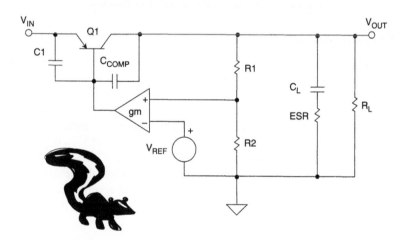

**Figure 7.36: The Solution to CL Sensitivity: Pole Split
Compensation (Wrong Way Example)**

The anyCAP Pole-Splitting Topology

Returning to the anyCAP series topology, (Figure 7.35, again) it can be noted that in this case C_{COMP} is isolated from the pass device's base (and thus input ripple variations), by the wideband noninverting driver. But insofar as frequency compensation is concerned, because of this buffer's isolation, C_{COMP} still functions as a modified pole splitting capacitor (see Reference 9), and it does provide the benefits of a buffered, C_L independent single-pole response. The regulator's frequency response is dominated by the internal compensation, and becomes relatively immune to the value and ESR of load capacitor C_L. Thus the name anyCAP for the series is apt, as the design is tolerant of virtually any output capacitor type.

The benefits of the anyCAP topology are summarized by Figure 7.37. As can be noted, C_L can be as low as 0.47 µF, and it can also be a multilayer ceramic capacitor (MLCC) type. This allows a very small physical size for the entire regulation function, such as when a SOT-23 packaged anyCAP LDO is used, for example the ADP3300 device. Because of the insensitivity to C_L, the designer needn't worry about such things as ESR zones, and can better concentrate on the system aspects of the regulator application.

- Internal C_{COMP} Dominates Response Rolloff

- C_L Can Range from 0.47µF(min) to Infinity

- Low and Ultralow C_LESR is OK

- MLCC Types for C_LWork, is Physically Smallest Solution

- No ESR Exclusion Zones

- Fast Load Transient Response and Good Line Rejection

Figure 7.37: Benefits of anyCAP LDO Topology

The anyCAP LDO series devices

The major specifications of the ADP330x series anyCAP LDO regulators are summarized in Figure. 7.38. The devices include both single and dual output parts, with current capabilities ranging from 50 mA to 200 mA. Rather than separate individual specifications for output tolerance, line and load regulation, plus temperature, the anyCAP series devices are rated simply for a combined total accuracy figure. For the ADP3300, ADP3301, ADP3302, ADP3303, and ADP3307, this accuracy is either 0.8% at 25°C, or 1.4% over the temperature range with the device operating over an input range of V_{OUT} 0.3 V (or 0.5 V), up to 12 V. The ADP3308 and ADP3309 are similarly specified for a total 25°C accuracy of 1.1% and 2.2% over temperature. With total accuracy being covered by one clear specification, the designer can then achieve a higher degree of confidence. It is important to note that this method of specification also includes operation within the regulator dropout range (unlike some LDO parts specified for higher input-output voltage difference conditions).

Part Number	V_{MIN} @ I_L (V, Typ)	I_L (mA)	Accuracy (Total over Temp, %)	Package (All SO -8 are Thermal coastline)	Comment (Singles have NR, SD, ERR; Dual no NR)
ADP3300	0.08	50	1.4	SOT-23-6	Single
ADP3301	0.10	100	1.4	SO-8	Single
ADP3302	0.10	100	1.4	SO-8	Dual
ADP3303	0.18	200	1.4	SO-8	Single
ADP3307	0.13	100	1.4	SOT-23-6	Single
ADP3308	0.08	50	2.2	SOT-23-5	Single
ADP3309	0.12	100	2.2	SOT-23-5	Single

Figure 7.38: anyCAP Series LDO Regulators

The ADP333x series is a newer family of anyCAP LDOs designed for 200 mA and higher output currents with very low quiescent current, I_Q. For instance, the ADP3330 has a typical no-load current of only 35 µA. The ADP333x series are available in thermally enhanced packages, and Figure 7.39 shows the key specifications for the family.

Part Number	V_{MIN} @ I_L (V, Typ)	I_L (mA)	Accuracy (Total over Temp, %)	Package (SOT23-6 are Chip on Lead)	Comment
ADP3330	0.14	200	1.4	SOT-23-6	Single
ADP3331	0.14	200	1.4	SOT-23-6	Single
ADP3333	0.14	300	1.8	MSOP-8	Single
ADP3334	0.20	500	1.8	SO-8	Single
ADP3335	0.20	500	1.8	MSOP-8	Single
ADP3336	0.20	500	1.8	MSOP-8	Single
ADP3338	0.19	1000	1.4	SOT-223	Single
ADP3339	0.23	1500	1.5	SOT-223	Single

Figure 7.39: anyCAP Series Low I_Q LDOs

Functional Diagram and Basic 50 mA LDO Regulator

A functional diagram common to the various devices of the anyCAP series LDO regulators is shown by Figure 7.40. Operation of the various pins and internal functions is discussed below.

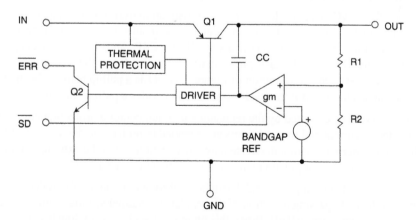

Figure 7.40: anyCAP Series LDO Regulators Functional Diagram

In application, the use of the anyCAP series of LDOs is simple, as shown by a basic 50 mA ADP3300 regulator, in Figure 7.41. This circuit is a general one, to illustrate points common to the entire device series. The ADP3300 is a basic LDO regulator device, designed for fixed output voltage applications while operating from sources over a range of 3 V to 12 V and a temperature range of –40°C to +85°C. The actual ADP3300 device ordered would be specified as ADP3300ART-YY, where the "YY" is a voltage designator suffix such as 2.7, 3, 3.2, 3.3, or 5, for those respective voltages. The "ART" portion of the part number designates the SOT-23 6-lead package. The example circuit shown produces 5.0 V with the use of the ADP3300-5.

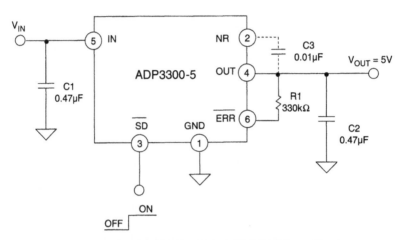

Figure 7.41: A Basic ADP3300 50 mA LDO Regulator Circuit

In operation, the circuit will produce its rated 5 V output for loads of 50 mA or less, and for input voltages above 5.3 V (V_{OUT} + 0.3 V), when the shutdown ($\overline{SD}$) input is in a HIGH state. This can be accomplished either by a logic HIGH control input to the $\overline{SD}$ pin, or by simply tying this pin to V_{IN}. When $\overline{SD}$ is LOW (or tied to ground), the regulator shuts down, and draws a quiescent current of 1 μA or less.

The ADP3300 and other anyCAP series devices maintain regulation over a wide range of load, input voltage and temperature conditions. However, when the regulator is overloaded or entering the dropout region (for example, by a reduction in the input voltage) the open collector $\overline{ERR}$ pin becomes active, by going to a LOW or conducting state. Once set, the $\overline{ERR}$ pin's internal hysteresis keeps the output low, until some margin of operating range is restored. In the circuit of Figure 7.41, R1 is a pullup resistor for the $\overline{ERR}$ output, E_{OUT}. This resistor can be eliminated if the load being driven provides a pullup current.

The $\overline{ERR}$ function can also be activated by the regulator's over temperature protection circuit, which trips at 165°C. These internal current and thermal limits are intended to protect the device against accidental overload conditions. For normal operation, device power dissipation should be externally limited by means of heat sinking, air flow, etc. so that junction temperatures will not exceed 125°C.

A capacitor, C3, connected between pins 2 and 4, can be used for an optional noise reduction (NR) feature. This is accomplished by ac-bypassing a portion of the regulator's internal scaling divider, which has the effect of reducing the output noise ~10 dB. When this option is exercised, only low leakage 10 nF to 100 nF capacitors should be used. Also, input and output capacitors should be changed to 1 μF and 4.7 μF values respectively, for lowest noise and the best overall performance. Note that the noise reduction pin is internally connected to a high impedance node, so connections to it should be carefully done to avoid noise. PC traces and pads connected to this pin should be as short and small as possible.

LDO Regulator Thermal Considerations

To determine a regulator's power dissipation, calculate it as follows:

$$P_D = (V_{IN} - V_{OUT})(I_L) + (V_{IN})(I_{GROUND})$$ Eq. 7.19

where I_L and I_{GROUND} are load and ground current, and V_{IN} and V_{OUT} are the input and output voltages respectively. Assuming I_L= 50 mA, I_{GROUND} = 0.5 mA, V_{IN} = 8 V, and V_{OUT} = 5 V, the device power dissipation is:

$$P_D = (8 - 5)(0.05) + (8)(0.0005) = 0.150 + .004 = 0.154 \text{ W}$$ Eq. 7.20

To determine the regulator's temperature rise, ΔT, calculate it as follows (assume the θ_{JA} of the regulator is 165°C/W):

$$\Delta T = T_J - T_A = P_D \times \theta_{JA} = 0.154W \times 165°C/W = 25.4°C$$ Eq. 7.21

With a maximum junction temperature of 125°C, this yields a calculated maximum safe ambient operating temperature of 125°C – 25.4°C, or just under 100°C. Since this temperature is in excess of the device's rated temperature range of 85°C, the device will then be operated conservatively at an 85°C (or less) maximum ambient temperature.

These general procedures can be used for other devices in the series, substituting the appropriate θ_{JA} for the applicable package, and applying the remaining operating conditions. For reference, a complete tutorial section on thermal management is contained in Chapter 9.

In addition, layout and PCB design can have a significant influence on the power dissipation capabilities of power management ICs. This is due to the fact that the surface-mount packages used with these devices rely heavily on thermally conductive traces or pads, to transfer heat away from the package. Appropriate PC layout techniques should then be used to remove the heat due to device power dissipation. The following general guidelines will be helpful in designing a board layout for lowest thermal resistance in SOT-23 and SO-8 packages:

1. *PC board traces with large cross sectional areas remove more heat. For optimum results, use large area PCB patterns with wide and heavy (2 oz.) copper traces, placed on the uppermost side of the PCB.*

2. *Electrically connect dual V_{IN} and V_{OUT} pins in parallel, as well as to the corresponding V_{IN} and V_{OUT} large area PCB lands.*

3. *In cases where maximum heat dissipation is required, use double-sided copper planes connected with multiple vias.*

4. *Where possible, increase the thermally conducting surface area(s) openly exposed to moving air, so that heat can be removed by convection (or forced air flow, if available).*

5. *Do not use solder mask or silkscreen on the heat dissipating traces, as they increase the net thermal resistance of the mounted IC package.*

A real life example visually illustrates a number of the above points far better than words can do, and is shown in Figure 7.42, a photo of the ADP3300 1.5" square evaluation PCB. The boxed area on the board represents the actual active circuit area.

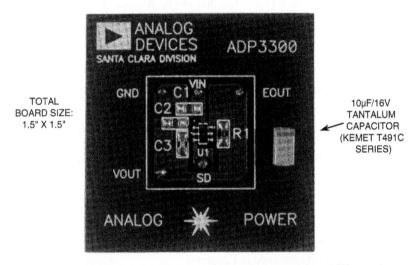

Figure 7.42: ADP3300 Evaluation Board: Capacitor Size Can Make a Difference!

In this figure, a large cross section conductor area can be seen associated with pin 4 and V_{OUT}, the large "U" shaped trace at the lower part within the boxed outline.

Also, the effect of the anyCAP design on capacitor size can be noted from the tiny size of the C1 and C2 0.47 µF input and output capacitors, near the upper left of the boxed area. For comparison purposes, a 10 µF/16 V tantalum capacitor (Kemet T491C series) is also shown outside the box, as it might be used on a more conventional LDO circuit. It is several times the size of output capacitor C2.

Recent developments in packaging have led to much improved thermal performance for power management ICs. The anyCAP LDO regulator family capitalizes on this most effectively, using a thermally improved leadframe as the basis for all eight pin devices. This package is called a "Thermal Coastline" design, and is shown in Figure 7.43. The foundation of the improvement in heat transfer is related to two key parameters

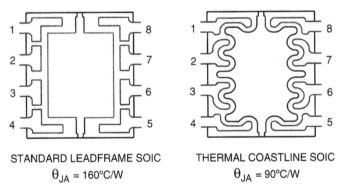

Figure 7.43: anyCAP Series Regulators in SO-8 Use Thermal Coastline Packages

of the leadframe design, distance and width. The payoff comes in the reduced thermal resistance of the leadframe based on the Thermal Coastline, only 90°C/W versus 160°C/W for a standard SO-8 package. The increased dissipation of the Thermal Coastline allows the anyCAP series of SO-8 regulators to support more than one watt of dissipation at 25°C.

Additional insight into how the new leadframe increases heat transfer can be appreciated by Figure. 7.44. In this figure, it can be noted how the spacing of the Thermal Coastline paddle and leads shown on the right is reduced, while the width of the lead ends are increased, versus the standard leadframe, on the left.

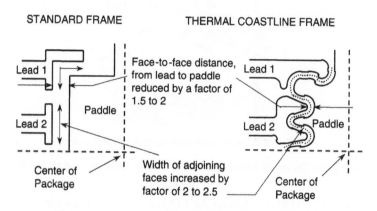

Figure 7.44: Details of Thermal Coastline Package

The ADP3330 and ADP3331 are 200 mA anyCAP LDOs packaged in a 6-lead SOT-23 package that utilizes a proprietary Chip-on-Lead™ packaging technique for thermal enhancement. In a standard SOT-23, the majority of the heat flows out of the ground pin. This new package uses an electrically isolated die attach that allows all pins to contribute to heat conduction. This technique reduces the thermal resistance to 165°C/W on a 4-layer board as compared to >230°C/W for a standard SOT-23 leadframe. Figure 7.45 shows the difference between the standard SOT-23 and the Chip-on-Lead leadframes.

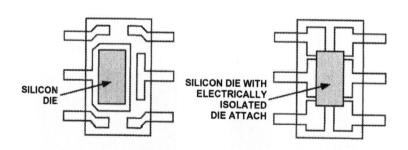

165°C/W vs. >230°C/W for Standard SOT-23

Figure 7.45: Thermally Enhanced Chip-on-Lead SOT-23-6 Package

The ADP3333 (300 mA), ADP3335 (500 mA) and ADP3336 (500 mA) anyCAP LDOs use a patented "paddle-under-lead" package design to ensure the best thermal performance in an MSOP-8 footprint. This package uses an electrically isolated die attach that allows all pins to contribute to heat conduction. This technique reduces the thermal resistance to 110°C/W on a 4-layer board as compared to >160°C/W for a standard MSOP-8 leadframe. Figure 7.46 shows the standard physical construction of the MSOP-8 (left) and the thermally enhanced paddle-under-lead leadframe (right).

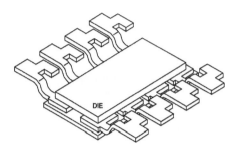

110°C/W vs. >160°C/W for Standard MSOP-8

Figure 7.46: Thermally Enhanced "Paddle-Under-Lead" 8-Lead MSOP Package

The ADP3338 (1 A) and ADP3339 (1.5 A) anyCAP LDOs are packaged in a thermally enhanced SOT-223 package as shown in Figure 7.47. The SOT-223's thermal resistance, θ_{JA}, is determined by the sum of the junction-to-case and the case-to-ambient thermal resistances. The junction-to-case thermal resistance, θ_{JC}, is determined by the package design and specified at 26.8°C/W. However, the case-to-ambient thermal resistance is determined by the printed circuit board design. As shown in Figure 7.47A-C, the amount of copper the ADP3338/ADP3339 is mounted to affects the thermal performance. When mounted to 2 oz. copper with just the minimal pads (Figure 7.47A), the θ_{JA} is 126.6°C/W.

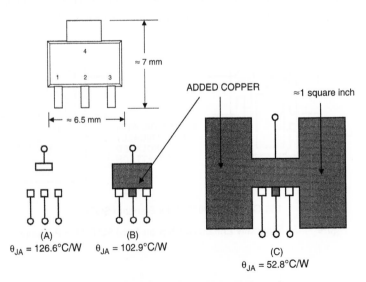

Figure 7.47: Reducing SOT-223 Package θ_{JA}

By adding a small copper pad under the ADP3338 (Figure 7.47B), the θ_{JA} is reduced to 102.9°C/W. Increasing the copper pad to 1 square inch (Figure 7.47C), reduces the θ_{JA} even further to 52.8°C/W. Note that both Pin 2 and Pin 4 (tab) are the LDO output and are internally connected.

LDO Regulator Controllers

To complement the anyCAP series of standalone LDO regulators, there is also the LDO *regulator controller*. The regulator controller IC picks up where the standalone regulator IC is no longer useful in either load current or power dissipation terms, and uses an external PMOS FET for the pass device. The ADP3310 is a basic LDO regulator controller device, designed for fixed output voltage applications while operating from sources over a range of 3.8 V to 15 V and a temperature range of –40°C to +85°C. The actual ADP3310 device ordered would be specified as ADP3310AR-YY, where the "YY" is a voltage designator suffix such as 2.8, 3, 3.3, or 5, for those respective voltages. The "AR" portion of the part number designates the SO-8 Thermal Coastline 8-lead package. A summary of the main features of the ADP3310 device is listed in Figure 7.48.

- Controller drives external PMOS power FETs
  - User FET choice determines I_L and V_{MIN} performance
  - Small, 2-chip regulator solution handles up to 10A
- Advantages compared to integrated solutions
  - High accuracy (1.5%) fixed voltages; 2.8V, 3V, 3.3V, or 5V
  - User flexibility (selection of FET for performance)
  - Small footprint with anyCAP controller and SMD FET
  - Kelvin output sensing possible
  - Integral, low loss current limit sensing for protection

Figure 7.48: anyCAP ADP3310 LDO Regulator Controller Features

Regulator Controller Differences

An obvious basic difference of the regulator controller versus a standalone regulator is the removal of the pass device from the regulator chip. This design step has both advantages and disadvantages. A positive is that the external PMOS pass device can be chosen for the exact size, package, current rating, and power handling most useful to the application. This approach allows the same basic controller IC to be useful for currents of several hundred mA to more than 10 A, simply by choice of the FET. Also, since the regulator controller IC's I_{GROUND} of 800 µA results is very little power dissipation, its thermal drift will be enhanced. On the downside, there are two packages now used to make up the regulator function. And, current limiting (which can be made completely integral to a standalone IC LDO regulator) is now a function that must be split between the regulator controller IC and an external sense resistor. This step also increases the dropout voltage of the LDO regulator controller somewhat, by about 50 mV.

A functional diagram of the ADP3310 regulator controller is shown in Figure 7.49. The basic error amplifier, reference and scaling divider of this circuit are similar to the standalone anyCAP regulator, and will not be described in detail. The regulator controller version does share the same cap load immunity of the standalone versions, and also has a shutdown function, similarly controlled by the EN (enable) pin.

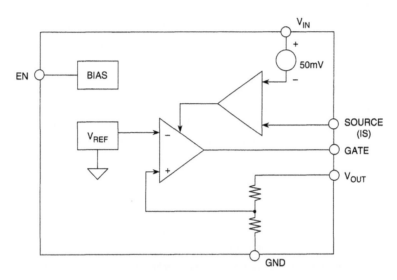

Figure 7.49: Functional Block Diagram of anyCAP Series LDO Regulator Controller

The main differences in the regulator controller IC architecture is the buffered output of the amplifier, which is brought out on the GATE pin, to drive the external PMOS FET. In addition, the current limit sense amplifier has a built in 50 mV threshold voltage, and is designed to compare the voltage between the V_{IN} and IS pins. When this voltage exceeds 50 mV, the current limit sense amplifier takes over control of the loop, by shutting down the error amplifier and limiting output current to the preset level.

A Basic 5 V/1 A LDO Regulator Controller

An LDO regulator controller is easy to use, since a PMOS FET, a resistor and two relatively small capacitors (one at the input, one at the output) is all that is needed to form an LDO regulator. The general configuration is shown by Figure 7.50, an LDO suitable as a 5 V/1 A regulator operating from a V_{IN} of 6 V, using the ADP3310-5 controller IC.

This regulator is stable with virtually any good quality output capacitor used for C_L (as is true with the other anyCAP devices). The actual C_L value required and its associated ESR depends on the g_m and capacitance of the external PMOS device. In general, a 10 µF capacitor at the output is sufficient to ensure stability for load currents up to 10 A. Larger capacitors can also be used, if high output surge currents are present. In such cases, low ESR capacitors such as OS-CON electrolytics are preferred, because they offer lowest ripple on the output. For less demanding requirements, a standard tantalum or aluminum electrolytic can be adequate. When an aluminum electrolytic is used, it should be qualified for adequate performance over temperature. The input capacitor, C_{IN}, is only necessary when the regulator is several inches or more distant from the raw dc filter capacitor. However, since it is a small type, it is usually prudent to use it in most instances, located close to the V_{IN} pin of the regulator.

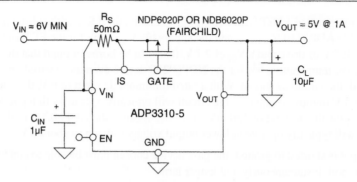

Figure 7.50: A Basic ADP3310 PMOS FET 1 A LDO Regulator Controller Circuit

Selecting the Pass Device

The type and size of the pass transistor are determined by a set of requirements for threshold voltage, input-output voltage differential, load current, power dissipation, and thermal resistance. An actual PMOS pass device selected must satisfy all of these electrical requirements, plus physical and thermal parameters. There are a number of manufacturers offering suitable devices in packages ranging from SO-8 up through TO-220 in size.

To ensure that the maximum available drive from the controller will adequately drive the FET under worst-case conditions of temperature range and manufacturing tolerances, the maximum drive from the controller, $V_{GS(DRIVE)}$, to the pass device must be determined. This voltage is calculated as follows:

$$V_{GS(DRIVE)} = V_{IN} - V_{BE} - \left(I_{L(MAX)}\right)\left(R_S\right)$$

Eq. 7.22

where V_{IN} is the minimum input voltage, $I_{L(MAX)}$ is the maximum load current, R_S the sense resistor, and V_{BE} is a voltage internal to the ADP3310 (~ 0.5 V @ high temp, 0.9 V cold, and 0.7 V at room temp). Note that since $I_{L(MAX)} \times R_S$ will be no more than 75 mV, and V_{BE} at cold temperature $\cong 0.9$ V, this equation can be further simplified to:

$$V_{GS(DRIVE)} \cong V_{IN} - 1V$$

Eq. 7.23

In Figure 7.50 example, $V_{IN} = 6$ V and $V_{OUT} = 5$ V, so $V_{GS(DRIVE)}$ is $6 - 1 = 5$ V

It should be noted that the above two equations apply to FET drive voltages that are *less* than the typical gate-to-source clamp voltage of 8 V (built into the ADP3310, for the purposes of FET protection).

An overall goal of the design is to then select an FET that will have an $R_{DS(ON)}$ sufficiently low so that the resulting dropout voltage will be less than $V_{IN} - V_{OUT}$, which in this case is 1 V. For the NDP6020P used in Figure 7.50 (see Reference 10), this device achieves an $R_{DS(ON)}$ of 70 milliohms (max) with a V_{GS} of 2.7 V, a voltage drive appreciably less than the ADP3310's $V_{GS(DRIVE)}$ of 5 V. The dropout voltage V_{MIN} of this regulator configuration is the sum of two series voltage drops, the FET's drop plus the drop across R_S, or:

$$V_{MIN} = I_{L(MAX)}\left(R_{DS(ON)} + R_S\right)$$

Eq. 7.24

In the design here, the two resistances are roughly comparable to one another, so the net V_{MIN} will be 1 A × (50 + 70 milliohms) = 120 mV.

For a design safety margin, use a FET with a rated V_{GS} at the required R_{DS}, with substantial headroom between the applicable ADP3310 $V_{GS(DRIVE)}$ and the applicable V_{GS} rating for the FET. In the case here, there is ample margin, with 5 V of drive and a V_{GS} of 2.7 V. It should be borne in mind that the FET's V_{GS} and $R_{DS(ON)}$ will change over temperature, but for the NDP6020P device even these variations and a V_{GS} of 4.5 V are still possible with the circuit as shown. With a rated minimum dc input of 6 V, this means that the design is conservative with 5 V output. In practice, the circuit will typically operate with input voltage minimums on the order of V_{OUT} plus the dropout of 120 mV, or ~ 5.12 V. Since the NDP6020P is also a fairly low threshold device, it will typically operate at lower output voltages, down to about 3 V.

In the event the output is shorted to ground, the pass device chosen must be able to conduct the maximum short circuit current, both instantaneously and longer term.

Thermal Design

The maximum allowable thermal resistance between the FET junction and the highest expected ambient temperature must be taken into account, to determine the type of FET package and heat sink used (if any).

Whenever possible to do so reliably, the FET pass device can be directly mounted to the PCB, and the available PCB copper lands used as an effective heat sink. This heat sink philosophy will likely be adequate when the power to be dissipated in the FET is on the order of 1 W–2 W or less. Note that the very nature of an LDO helps this type of design immensely, as the lower voltage drop across the pass device reduces the power to be dissipated. Under normal conditions for example, Q1 of Figure 7.52 dissipates less than 1 W at a current of 1 A, since the drop across the FET is less than 1 V.

To use PCB lands as effective heat sinks with SO-8 and other SMD packages, the pass device manufacturer's recommendations for the lowest θ_{JA} mounting should be followed (see References 11 and 12). In general, these suggestions will likely parallel the five rules noted above, under "LDO regulator thermal considerations" for SO-8 and SOT-23 packaged anyCAP LDOs. For lowest possible thermal resistance, also connect multiple FET pins together, as follows:

Electrically connect multiple FET source and drain pins in parallel, as well as to the corresponding R_S and V_{OUT} large area PCB lands.

Using 2 oz. copper PCB material and one square inch of copper PCB land area as a heatsink, it is possible to achieve a net thermal resistance, θ_{JA}, for mounted SO-8 devices on the order of 60°C/W or less. Such data is available for SO-8 power FETs (see Reference 11). There are also a variety of larger packages with lower thermal resistance than the SO-8, but still useful with surface mount techniques. Examples are the DPAK and D²PAK, etc.

For higher power dissipation applications, corresponding to thermal resistance of 50°C/W or less, a bolt-on external heat sink is required to satisfy the θ_{JA} requirement. Compatible package examples would be the TO-220 family, which is used with the NDP6020P example of Figure 7.52.

Calculating thermal resistance for $V_{IN} = 6.7$ V, $V_{OUT} = 5$ V, and $I_L = 1$ A:

$$\theta_{JA} = \frac{T_J - T_{A(MAX)}}{V_{DS(MAX)} \times I_{L(MAX)}} \qquad \text{Eq. 7.25}$$

where T_J is the pass device junction temperature limit, $T_{A(MAX)}$ is the maximum ambient temperature, $V_{DS(MAX)}$ is the maximum pass device drain-source voltage, and $I_{L(MAX)}$ is the maximum load current.

Inserting some example numbers of 125°C as a max. junction temp for the NDP6020P, a 75°C expected ambient, and the $V_{DS(MAX)}$ and $I_{L(MAX)}$ figures of 1.7 V and 1 A, the required θ_{JA} works out to be (125°C – 75°C) /1.7 = 29.4°C/W. This can be met with a very simple heat sink, which is derived as follows.

The NDP6020P in the TO-220 package has a junction-case thermal resistance, θ_{JC}, of 2°C/W. The required external heatsink's thermal resistance, θ_{CA}, is determined as follows:

$$\theta_{CA} = \theta_{JA} - \theta_{JC} \qquad \text{Eq. 7.26}$$

where θ_{CA} is the required heat sink case-to-ambient thermal resistance, θ_{JA} is the calculated *overall* junction-to-ambient thermal resistance, and θ_{JC} is the pass device junction-to-case thermal resistance, which in this case is 2°C/W typical for TO-220 devices, and NDP6020P.

$$\theta_{CA} = 29.4°C/W - 2°C/W = 27.4°C/W. \qquad \text{Eq. 7.27}$$

For a safety margin, select a heatsink with a θ_{CA} *less* than the results of this calculation. For example, the Aavid TO-220 style clip on heat sink # 576802 has a θ_{CA} of 18.8°C/W, and in fact many others have performance of 25°C/W or less. As an alternative, the NDB6020P D²PAK FET pass device could be used in this same design, with an SMD style heat sink such as the Aavid 573300 series used in conjunction with an internal PCB heat spreader.

Note that many LDO applications like the above will calculate out with very modest heat sink requirements. This is fine, as long as the output is never shorted. With a shorted output, the current goes to the limit level (as much as 1.5 A in this case), while the voltage across the pass device goes to V_{IN} (which could also be at a maximum). In this case, the new pass device dissipation for short circuit conditions becomes 1.5 A × 6.7 V, or 10 W. Continuously supporting this level of power will require the entire heat sink situation to be re-evaluated, as what was adequate for 1.7 W will simply not be adequate for 10 W. In fact, the required heat sink θ_{CA} is about 3°C/W to support the 10 W safely on a continuous basis, which requires a much larger heat sink.

Sensing Resistors for LDO Controllers

Current limiting in the ADP3310 controller is achieved by choosing an appropriate external current sense resistor, R_S, which is connected between the controller's V_{IN} and IS (source) pins. An internally derived 50 mV current limit threshold voltage appears between these pins, to establish a comparison threshold for current limiting. This 50 mV determines the threshold where current limiting begins. For a continuous current limiting, a foldback mode is established, with dissipation controlled by reducing the gate drive. The net effect is that the ultimate current limit level is a factor of 2/3 of maximum. The foldback limiting reduces the power dissipated in the pass transistor substantially.

To choose a sense resistor for a maximum output current I_L, R_S is calculated as follows:

$$R_S = \frac{0.05}{K_F \times I_L} \qquad \text{Eq. 7.28}$$

In this expression, the nominal 50 mV current limit threshold voltage appears in the numerator. In the denominator appears a scaling factor K_F, which can be either 1.0 or 1.5, plus the maximum load current, I_L. For example, if a scaling factor of 1.0 is to be used for a 1 A I_L, the R_S calculation is straightforward, and 50 milliohms is the correct R_S value.

However, to account for uncertainties in the threshold voltage and to provide a more conservative output current margin, a scaling factor of $K_F = 1.5$ can alternately be used. When this approach is used, the same 1 A I_L load conditions will result in a 33 milliohm R_S value. In essence, the use of the 1.5 scaling factor takes into account the foldback scheme's reduction in output current, allowing higher current in the limit mode.

The simplest and least expensive sense resistor for high current applications such as Figure 7.50 is a copper PCB trace controlled in both thickness and width. Both the temperature dependence of copper and the relative size of the trace must be taken into account in the resistor design. The temperature coefficient of resistivity for copper has a positive temperature coefficient of +0.39%/°C. This natural copper TC, in conjunction with the controller's PTAT based current limit threshold voltage, can provide for a current limit characteristic that is simple and effective over temperature.

The table of Figure 7.51 provides resistance data for designing PCB copper traces with various PCB copper thickness (or weight), in ounces of copper per square foot area. To use this information, note that the center column contains a resistance coefficient, which is the conductor resistance in milliohms/inch, divided by the trace width, W. For example, the first entry, for 1/2 ounce copper is 0.983 milliohms/inch/W. So, for a reference trace width of 0.1", the resistance would be 9.83 milliohms/inch. Since these are all linear relationships, everything scales for wider/skinnier traces, or for differing copper weights. As an example, to design a 50 milliohm R_S for the circuit of Figure 7.50 using 1/2 ounce copper, a 2.54" length of a 0.05" wide PCB trace could be used.

Copper Thickness	Resistance Coefficient, milliohms/inch/W (trace width W in inches)	Reference 0.1 inch wide trace, milliohms/inch
1/2 oz/ft²	0.983/W	9.83
1 oz/ft²	0.491/W	4.91
2 oz/ft²	0.246/W	2.46
3 oz/ft²	0.163/W	1.63

**Figure 7.51: Printed Circuit Copper Resistance
Design for LDO Controllers**

To minimize current limit sense voltage errors, the two connections to R_S should be made four-terminal style, as is noted in Figure 7.50. It is not absolutely necessary to actually use four-terminal style resistors, except for the highest current levels. However, as a minimum, the heavy currents flowing in the source circuit of the pass device should not be allowed to flow in the ADP3310 sense pin traces. To minimize such errors, the V_{IN} connection trace to the ADP3310 should connect close to the body of R_S (or the resistor's input sense terminal), and the IS connection trace should also connect close to the resistor body (or the resistor's output sense terminal). Four-terminal wiring is increasingly important for output currents of 1 A or more.

Alternately, an appropriate selected sense resistor such as surface-mount sense devices available from resistor vendors can be used (see Reference 13). Sense resistor R_S may not be needed in all applications, if a current limiting function is provided by the circuit feeding the regulator. For circuits that don't require current limiting, the IS and V_{IN} pins of the ADP3310 must be tied together.

PCB Layout Issues

For best voltage regulation, place the load as close as possible to the controller device's V_{OUT} and GND pins. Where the best regulation is required, the V_{OUT} trace from the ADP3310 and the pass device's drain connection should connect to the positive load terminal via separate traces. This step (Kelvin sensing) will keep the heavy load currents in the pass device's drain out of the feedback sensing path, and thus maximize output accuracy. Similarly, the unregulated input common should connect to the common side of the load via a separate trace from the ADP3310 GND pin.

A 2.8 V/8 A LDO Regulator Controller

With seemingly minor changes to the basic 1 A LDO circuit used in Figure 7.50, an 8 A LDO regulator controller can be configured, as shown in Figure 7.52. This circuit uses an ADP3310-2.8, to produce a 2.8 V output. The sense resistor is dropped to 5 milliohms, which supports currents of up to 10 A (or about 6.7 A, with current limiting active). Four-terminal wiring should be used with the sense resistor to minimize errors.

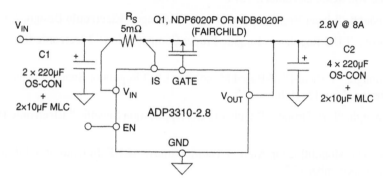

Figure 7.52: A 2.8 V/8 A LDO Regulator Controller

The most significant change over the more generic schematic of Figure 7.50 is the use of multiple, low ESR input and output bypass capacitors. At the output, C2 is a bank of $4 \times 220\ \mu F$ OS-CON type capacitors, in parallel with $2 \times 10\ \mu F$ MLCC chip type capacitors. These are located right at the load point with minimum inductance wiring, plus separate wiring back to the V_{OUT} pin of the ADP3310 and the drain of the pass device. This wiring will maximize the dc output accuracy, while the multiple capacitors will minimize the transient errors at the point-of-load. In addition, multiple bypasses on the regulator input in the form of C1 minimizes the transient errors at the regulator's V_{IN} pin.

Heat sink requirements for the pass device in this application will be governed by the loading and input voltage, and should be calculated by the procedures discussed above.

References:
7.2 Low Dropout Linear Regulators

1. Bob Widlar, "New Developments in IC Voltage Regulators," **IEEE Journal of Solid State Circuits**, Vol. SC-6, February, 1971.

2. Robert C. Dobkin, "3-Terminal Regulator is Adjustable," **National Semiconductor AN-181**, March, 1977.

3. Paul Brokaw, "A Simple Three-Terminal IC Bandgap Voltage Reference," **IEEE Journal of Solid State Circuits**, Vol. SC-9, December, 1974.

4. Frank Goodenough, "Linear Regulator Cuts Dropout Voltage," **Electronic Design**, April 16, 1987.

5. Chester Simpson, "LDO Regulators Require Proper Compensation," **Electronic Design**, November 4, 1996.

6. Frank Goodenough, "Vertical-PNP-Based Monolithic LDO Regulator Sports Advanced Features," **Electronic Design**, May 13, 1996.

7. Frank Goodenough, "Low Dropout Regulators Get Application Specific," **Electronic Design**, May 13, 1996.

8. Jim Solomon, "The Monolithic Op Amp: A Tutorial Study." **IEEE Journal of Solid State Circuits**, Vol. SC-9, No.6, December 1974.

9. Richard J. Reay, Gregory T.A. Kovacs, "An Unconditionally Stable Two-Stage CMOS Amplifier," **IEEE Journal of Solid State Circuits**, Vol. SC-30, No.5, May 1995.

10. NDP6020P/ NDB6020P P-Channel Logic Level Enhancement Mode Field Effect Transistor, Fairchild Semiconductor data sheet, September 1997, www.fairchildsemi.com.

11. Alan Li, et all, "Maximum Power Enhancement Techniques for SO-8 Power MOSFETs," **Fairchild Semiconductor application note AN1029**, April 1996, www.fairchildsemi.com.

12. Rob Blattner, Wharton McDaniel, "Thermal Management in On-Board DC-to-DC Power Conversion," **Temic application note**, www.temic.com.

13. "S" series surface-mount current sensing resistors, KRL/Bantry Components, 160 Bouchard Street, Manchester, NH, 03103-3399, 603-668-3210.

Analog Switches and Multiplexers
Walt Kester

Introduction

Solid-state analog switches and multiplexers have become an essential component in the design of electronic systems that require the ability to control and select a specified transmission path for an analog signal. These devices are used in a wide variety of applications including multichannel data acquisition systems, process control, instrumentation, video systems, etc.

One of the first commercial analog multiplexers is shown in Figure 7.53, the MOSES-8 from the Pastoriza Division of Analog Devices in 1969. This PC board multiplexer consisted of 8 MOSFET switches and 8 switch drivers. The part had a switching time of 100 ns, on-resistance of 500 Ω. Selling price in 1969 was $320. For ±5 V inputs, the multiplexer operated on ±15 V supplies, but for ±10 V inputs, it required not only the +15 V but also a –28 V supply. Today, the ADG725/ADG726/ADG731/ADG732 family offers a 32-channel multiplexer with 4 Ω on-resistance, 20 μA quiescent current, and packaged in 7 mm × 7 mm chip scale (CSP) or thin plastic quad flatpack (TQFP). The price is less than $5.

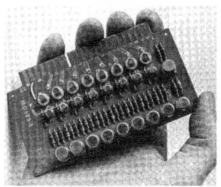

TODAY:

ADG725, ADG726,
ADG731, ADG732:

- 8 Channels
- Switching time: 100 ns
- On Resistance: 500 Ω
- Off Resistance: > 100M Ω
- $320

- 32 Channels
- Switching Time: 30ns
- On Resistance: 4 Ω
- 7 mm^2 CSP or TQFP
- < $5

**Figure 7.53: "MOSES-8" MOSFET Analog Multiplexer
Analog Devices' Pastoriza Division, 1969**

With the development of CMOS processes (yielding good PMOS and NMOS transistors on the same substrate), switches and multiplexers rapidly gravitated to integrated circuit form in the mid-1970s, with product introductions such as the Analog Devices' popular AD7500 series (introduced in 1973). A dielectrically-isolated family of these parts introduced in 1976 allowed input overvoltages of ±25 V (beyond the supply rails) and was insensitive to latch-up.

These early CMOS switches and multiplexers were typically designed to handle signal levels up to ±10 V while operating on ±15 V supplies. In 1979, Analog Devices introduced the popular ADG200 series of switches and multiplexers, and in 1988 the ADG201 series was introduced which was fabricated on a proprietary linear-compatible CMOS process (LC²MOS). These devices allowed input signals to ±15 V when operating on ±15 V supplies.

A large number of switches and multiplexers were introduced in the 1980s and 1990s, with the trend toward lower on-resistance, faster switching, lower supply voltages, lower cost, lower power, and smaller surface-mount packages.

Today, analog switches and multiplexers are available in a wide variety of configurations, options, etc., to suit nearly all applications. On-resistances less than 0.5 Ω, picoampere leakage currents, signal bandwidths greater than 1 GHz, and single 1.8 V supply operation are now possible with modern CMOS technology.

Although CMOS is by far the most popular IC process today for switches and multiplexers, bipolar processes (with JFETs) and complementary bipolar processes (also with JFET capability) are often used for special applications such as video switching and multiplexing where the high performance characteristics required are not attainable with CMOS. Traditional CMOS switches and multiplexers suffer from several disadvantages at video frequencies. Their switching time is generally not fast enough, and they require external buffering in order to drive typical video loads. In addition, the small variation of the CMOS switch on-resistance with signal level (R_{ON} modulation) can introduce unwanted distortion in differential gain and phase. Multiplexers based on complementary bipolar technology offer better solutions at video frequencies—with obvious power and cost increases above CMOS devices.

CMOS Switch Basics

The ideal analog switch has no on-resistance, infinite off-impedance and zero time delay, and can handle large signal and common-mode voltages. Real CMOS analog switches meet none of these criteria, but if we understand the limitations of analog switches, most of these limitations can be overcome.

CMOS switches have an excellent combination of attributes. In its most basic form, the MOSFET transistor is a voltage-controlled resistor. In the "on" state, its resistance can be less than 1 Ω, while in the "off" state, the resistance increases to several hundreds of megohms, with picoampere leakage currents. CMOS technology is compatible with logic circuitry and can be densely packed in an IC. Its fast switching characteristics are well controlled with minimum circuit parasitics.

MOSFET transistors are bilateral. That is, they can switch positive and negative voltages and conduct positive and negative currents with equal ease. A MOSFET transistor has a voltage controlled resistance which varies nonlinearly with signal voltage as shown in Figure 7.54.

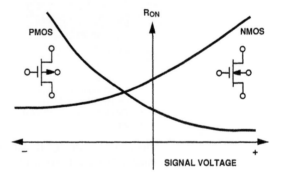

Figure 7.54: MOSFET Switch ON-Resistance versus Signal Voltage

The complementary-MOS process (CMOS) yields good P-channel and N-channel MOSFETs. Connecting the PMOS and NMOS devices in parallel forms the basic bilateral CMOS switch of Figure 7.55. This combination reduces the on-resistance, and also produces a resistance that varies much less with signal voltage.

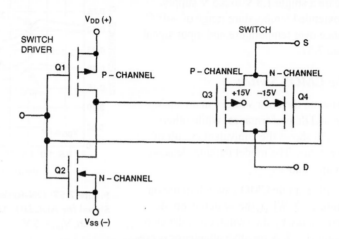

Figure 7.55: Basic CMOS Switch Uses Complementary Pair to Minimize R$_{ON}$ Variation due to Signal Swings

Figure 7.56 shows the on-resistance changing with channel voltage for both N-type and P-type devices. This nonlinear resistance can causes errors in dc accuracy as well as ac distortion. The bilateral CMOS switch solves this problem. On-resistance is minimized, and linearity is also improved. The bottom curve of Figure 7.56 shows the improved flatness of the on-resistance characteristic of the switch.

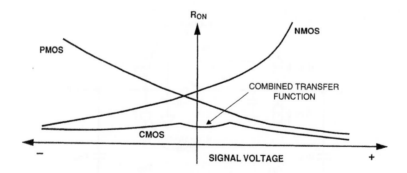

Figure 7.56: CMOS Switch ON-Resistance versus Signal Voltage

The ADG8xx-series of CMOS switches are specifically designed for less than 0.5 Ω on-resistance and are fabricated on a submicron process. These devices can carry currents up to 400 mA, operate on a single 1.8 V to 5.5 V supply, and are rated over an extended temperature range of –40°C to +125°C. On-resistance over temperature and input signal level is shown in Figure 7.57.

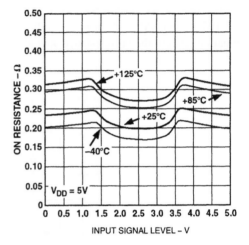

Figure 7.57: ON-Resistance Versus Input Signal for ADG801/ADG802 CMOS Switch, V_{DD} = 5 V

Error Sources in the CMOS Switch

It is important to understand the error sources in an analog switch. Many affect ac and dc performance, while others only affect ac. Figure 7.58 shows the equivalent circuit of two adjacent CMOS switches. The model includes leakage currents and junction capacitances.

DC errors associated with a single CMOS switch in the on state are shown in Figure 7.59. When the switch is on, dc performance is affected mainly by the switch on-resistance (R_{ON}) and leakage current (I_{LKG}). A resistive attenuator is created by the R_G-R_{ON}-R_{LOAD} combination which produces a gain error. The leakage current, I_{LKG}, flows through the equivalent resistance of R_{LOAD} in parallel with the sum of R_G and R_{ON}. Not only can R_{ON} cause gain errors—which can be calibrated using a system gain trim—but its variation with applied signal voltage (R_{ON} modulation) can introduce distortion—for which there is no calibration. Low resistance circuits are more subject to errors due to R_{ON}, while high resistance circuits are affected by leakage currents. Figure 7.59 also gives equations that show how these parameters affect dc performance.

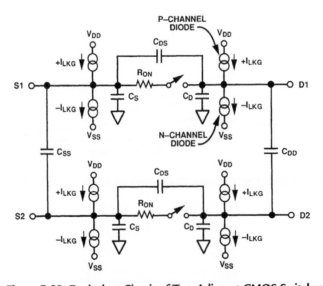

Figure 7.58: Equivalent Circuit of Two Adjacent CMOS Switches

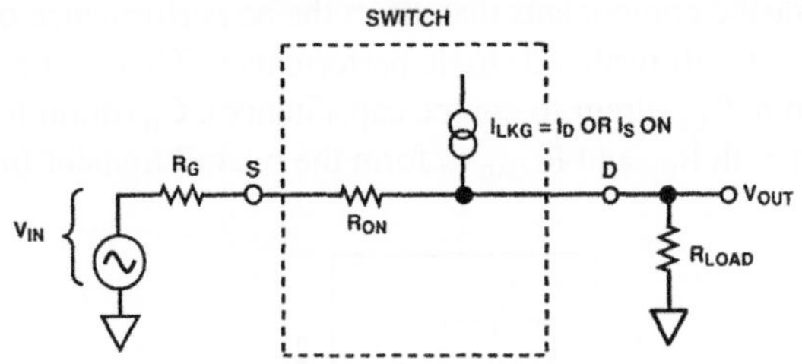

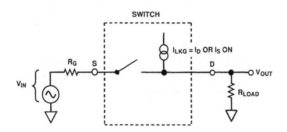

SWITCH

$I_{LKG} = I_D$ OR I_S ON

$$V_{OUT} = V_{IN} \left[\frac{R_{LOAD}}{R_G + R_{ON} + R_{LOAD}} \right] + I_{LKG} \left[\frac{R_{LOAD} (R_{ON} + R_G)}{R_G + R_{ON} + R_{LOAD}} \right]$$

IF $R_G \to 0$,

$$V_{OUT} = V_{IN} \left[\frac{R_{LOAD}}{R_{ON} + R_{LOAD}} \right] + I_{LKG} \left[\frac{R_{LOAD} R_{ON}}{R_{ON} + R_{LOAD}} \right]$$

Figure 7.59: Factors Affecting DC Performance for ON Switch Condition: R_{ON}, R_{LOAD}, and I_{LKG}

When the switch is OFF, leakage current can introduce errors as shown in Figure 7.60. The leakage current flowing through the load resistance develops a corresponding voltage error at the output.

SWITCH

$I_{LKG} = I_D$ OR I_S ON

LEAKAGE CURRENT CREATES ERROR VOLTAGE AT V_{OUT} EQUAL TO:
$V_{OUT} = I_{LKG} \times R_{LOAD}$

Figure 7.60: Factors Affecting DC Performance for OFF Switch Condition: I_{LKG} and R_{LOAD}

Figure 7.61 illustrates the parasitic components that affect the ac performance of CMOS switches. Additional external capacitances will further degrade performance. These capacitances affect feedthrough, crosstalk and system bandwidth. C_{DS} (drain-to-source capacitance), C_D (drain-to-ground capacitance), and C_{LOAD} all work in conjunction with R_{ON} and R_{LOAD} to form the overall transfer function.

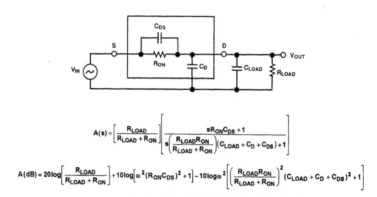

$$A(s) = \left[\frac{R_{LOAD}}{R_{LOAD} + R_{ON}} \right] \left[\frac{sR_{ON}C_{DS} + 1}{s\left(\frac{R_{LOAD}R_{ON}}{R_{LOAD} + R_{ON}} \right)(C_{LOAD} + C_D + C_{DS}) + 1} \right]$$

$$A(dB) = 20\log\left[\frac{R_{LOAD}}{R_{LOAD} + R_{ON}} \right] + 10\log\left[\omega^2(R_{ON}C_{DS})^2 + 1 \right] - 10\log\omega^2\left[\left(\frac{R_{LOAD}R_{ON}}{R_{LOAD} + R_{ON}} \right)^2(C_{LOAD} + C_D + C_{DS})^2 + 1 \right]$$

Figure 7.61: Dynamic Performance Considerations: Transfer Accuracy versus Frequency

In the equivalent circuit, C_{DS} creates a frequency zero in the numerator of the transfer function A(s). This zero usually occurs at high frequencies because the switch on-resistance is small. The bandwidth is also a function of the switch output capacitance in combination with C_{DS} and the load capacitance. This frequency pole appears in the denominator of the equation.

The composite frequency domain transfer function may be rewritten as shown in Figure 7.62 which shows the overall Bode plot for the switch in the on state. In most cases, the pole breakpoint frequency occurs first because of the dominant effect of the output capacitance C_D. Thus, to maximize bandwidth, a switch should have low input and output capacitance and low on-resistance.

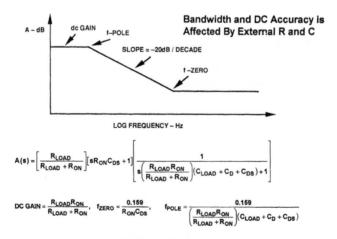

$$A(s) = \left[\frac{R_{LOAD}}{R_{LOAD} + R_{ON}} \right] \left[sR_{ON}C_{DS} + 1 \right] \left[\frac{1}{s\left(\frac{R_{LOAD}R_{ON}}{R_{LOAD} + R_{ON}} \right)(C_{LOAD} + C_D + C_{DS}) + 1} \right]$$

$$DC\ GAIN = \frac{R_{LOAD}R_{ON}}{R_{LOAD} + R_{ON}}, \quad f_{ZERO} = \frac{0.159}{R_{ON}C_{DS}}, \quad f_{POLE} = \frac{0.159}{\left(\frac{R_{LOAD}R_{ON}}{R_{LOAD} + R_{ON}} \right)(C_{LOAD} + C_D + C_{DS})}$$

Figure 7.62: Bode Plot of CMOS Switch Transfer Function in the ON State

The series-pass capacitance, C_{DS}, not only creates a zero in the response in the ON-state, it degrades the feedthrough performance of the switch during its OFF state. When the switch is off, C_{DS} couples the input signal to the output load as shown in Figure 7.63.

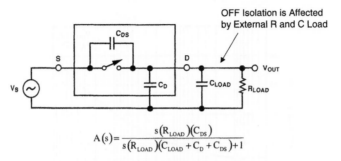

$$A(s) = \frac{s(R_{LOAD})(C_{DS})}{s(R_{LOAD})(C_{LOAD} + C_D + C_{DS}) + 1}$$

Figure 7.63: Dynamic Performance Considerations: Off Isolation

Large values of C_{DS} will produce large values of feedthrough, proportional to the input frequency. Figure 7.64 illustrates the drop in OFF-isolation as a function of frequency. The simplest way to maximize the OFF-isolation is to choose a switch that has as small a C_{DS} as possible.

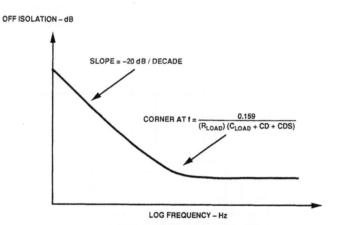

Figure 7.64: Off Isolation versus Frequency

Figure 7.65 shows typical CMOS analog switch OFF-isolation as a function of frequency for the ADG708 8-channel multiplexer. From dc to several kilohertz, the multiplexer has nearly 90 dB isolation. As the frequency increases, an increasing amount of signal reaches the output. However, even at 10 MHz, the switch shown still has nearly 60 dB of isolation.

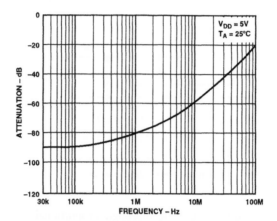

Figure 7.65: OFF-Isolation versus Frequency for ADG708 8-Channel Multiplexer

Another ac parameter that affects system performance is the charge injection that takes place during switching. Figure 7.66 shows the equivalent circuit of the charge injection mechanism.

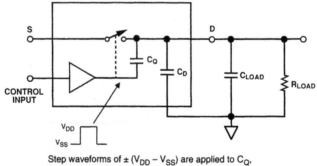

Step waveforms of $\pm (V_{DD} - V_{SS})$ are applied to C_Q, the gate capacitance of the output switches.

Figure 7.66: Dynamic Performance Considerations: Charge Injection Model

When the switch control input is asserted, it causes the control circuit to apply a large voltage change (from V_{DD} to V_{SS}, or vice versa) at the gate of the CMOS switch. This fast change in voltage injects a charge into the switch output through the gate-drain capacitance C_Q. The amount of charge coupled depends on the magnitude of the gate-drain capacitance.

The charge injection introduces a step change in output voltage when switching as shown in Figure 7.67. The change in output voltage, ΔV_{OUT}, is a function of the amount of charge injected, Q_{INJ} (which is in turn a function of the gate-drain capacitance, C_Q) and the load capacitance, C_L.

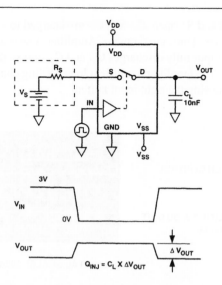

Figure 7.67: Effects of Charge Injection on Output

Another problem caused by switch capacitance is the retained charge when switching channels. This charge can cause transients in the switch output, and Figure 7.68 illustrates the phenomenon.

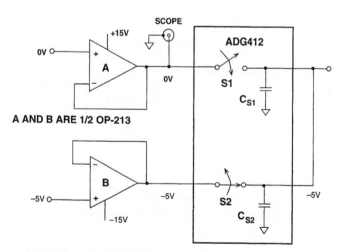

Figure 7.68: Charge Coupling Causes Dynamic Settling Time Transient When Multiplexing Signals

Assume that initially S2 is closed and S1 open. C_{S1} and C_{S2} are charged to –5 V. As S2 opens, the –5 V remains on C_{S1} and C_{S2}, as S1 closes. Thus, the output of Amplifier A sees a –5 V transient. The output will not stabilize until Amplifier A's output fully discharges C_{S1} and C_{S2} and settles to 0 V. The scope photo in Figure 7.69 depicts this transient. The amplifier's transient load settling characteristics will therefore be an important consideration when choosing the right input buffer.

SWITCH CONTROL
5V/div.

AMPLIFIER A OUTPUT
500mV/div.

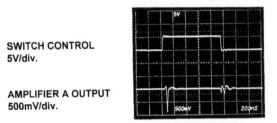

HORIZONTAL SCALE: 200ns/div.

**Figure 7.69: Output of Amplifier Shows Dynamic
Settling Time Transient Due to Charge Coupling**

Crosstalk is related to the capacitances between two switches. This is modeled as the C_{SS} capacitance shown in Figure 7.70.

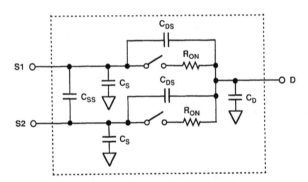

**Figure 7.70: Channel-to-Channel Crosstalk
Equivalent Circuit for Adjacent Switches**

Figure 7.71 shows typical crosstalk performance of the ADG708 8-channel CMOS multiplexer.

Finally, the switch itself has a settling time that must be considered. Figure 7.72 shows the dynamic transfer function. The settling time can be calculated, because the response is a function of the switch and circuit resistances and capacitances. One can assume that this is a single-pole system and calculate the number of time constants required to settle to the desired system accuracy as shown in Figure 7.73.

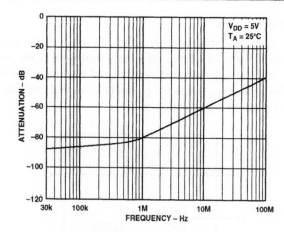

**Figure 7.71: Crosstalk versus Frequency
for ADG708 8-Channel Multiplexer**

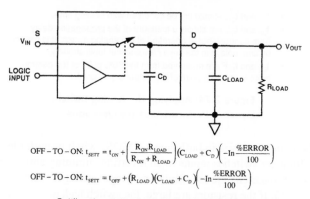

$$\text{OFF} - \text{TO} - \text{ON}: t_{SETT} = t_{ON} + \left(\frac{R_{ON}R_{LOAD}}{R_{ON} + R_{LOAD}}\right)\left(C_{LOAD} + C_D\right)\left(-\ln\frac{\%\text{ERROR}}{100}\right)$$

$$\text{OFF} - \text{TO} - \text{ON}: t_{SETT} = t_{OFF} + \left(R_{LOAD}\right)\left(C_{LOAD} + C_D\right)\left(-\ln\frac{\%\text{ERROR}}{100}\right)$$

Settling time is the time required for the switch output
to settle within a given error band of the final value.

Figure 7.72: Multiplexer Settling Time

RESOLUTION, # OF BITS	LSB (%FS)	# OF TIME CONSTANTS
6	1.563	4.16
8	0.391	5.55
10	0.0977	6.93
12	0.0244	8.32
14	0.0061	9.70
16	0.00153	11.09
18	0.00038	12.48
20	0.000095	13.86
22	0.000024	15.25

**Figure 7.73: Number of Time Constants Required to Settle to
1 LSB Accuracy for a Single-Pole System**

Applying the Analog Switch

Switching time is an important consideration in applying analog switches, but switching time should not be confused with settling time. ON and OFF times are simply a measure of the propagation delay from the control input to the toggling of the switch, and are largely caused by time delays in the drive and level-shift circuits (see Figure 7.74). The t_{ON} and t_{OFF} values are generally measured from the 50% point of the control input leading edge to the 90% point of the output signal level.

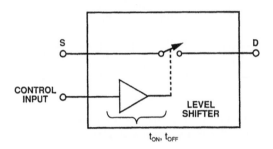

- t_{ON} and t_{OFF} should not be confused with settling time.
- t_{ON} and t_{OFF} are simply a measure of the propagation delay from control input to operation of the analog switch. It is caused by time delays in the drive / level-shifter logic circuitry.
- t_{ON} and $t_{OF}F$ are measured from the 50% point of the control input to the 90% point of the output signal level.

**Figure 7.74: Applying the Analog Switch:
Dynamic Performance Considerations**

We will next consider the issues involved in buffering a CMOS switch or multiplexer output using an op amp. When a CMOS multiplexer switches inputs to an inverting summing amplifier, it should be noted that the on-resistance, and its nonlinear change as a function of input voltage, will cause gain and distortion errors as shown in Figure 7.75. If the resistors are large, the switch leakage current may introduce error. Small resistors minimize leakage current error but increase the error due to the finite value of R_{ON}.

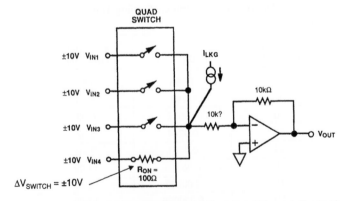

- ΔR_{ON} caused by ΔV_{IN}, degrades linearity of V_{OUT} relative to V_{IN}
- ΔR_{ON} causes overall gain error in V_{OUT} relative to V_{IN}

**Figure 7.75: Applying the Analog Switch: Unity
Gain Inverter with Switched Input**

To minimize the effect of R_{ON} change due to the change in input voltage, it is advisable to put the multiplexing switches at the op amp summing junction as shown in Figure 7.76. This ensures the switches are only modulated with about ±100 mV rather than the full ±10 V—but a separate resistor is required for each input leg.

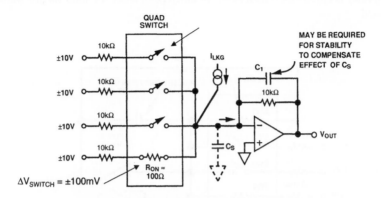

- Switch drives a virtual ground
- Switch sees only ±100mV, not ±10V, minimizes ΔRON

Figure 7.76: Applying the Analog Switch:
Minimizing the Influence of ΔR_{ON}

It is important to know how much parasitic capacitance has been added to the summing junction as a result of adding a multiplexer, because any capacitance added to that node introduces phase shift to the amplifier closed loop response. If the capacitance is too large, the amplifier may become unstable and oscillate. A small capacitance, C_1, across the feedback resistor may be required to stabilize the circuit.

The finite value of R_{ON} can be a significant error source in the circuit shown in Figure 7.77. The gain-setting resistors should be at least 1,000 times larger than the switch on-resistance to guarantee 0.1% gain accuracy. Higher values yield greater accuracy but lower bandwidth and greater sensitivity to leakage and bias current.

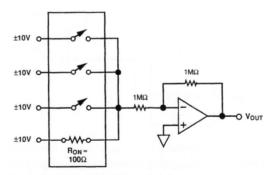

- ΔR_{ON} is small compared to 1MΩ switch load.
- Effect on transfer accuracy is minimized.
- Bias current and leakage current effects are now very important.
- Circuit bandwidth degrades.

Figure 7.77: Applying the Analog Switch: Minimizing Effects
of ΔR_{ON} Using Large Resistor Values

A better method of compensating for R_{ON} is to place one of the switches in series with the feedback resistor of the inverting amplifier as shown in Figure 7.78. It is a safe assumption that the multiple switches, fabricated on a single chip, are well-matched in absolute characteristics and tracking over temperature. Therefore, the amplifier is closed-loop gain stable at unity gain, since the total feedforward and feedback resistors are matched.

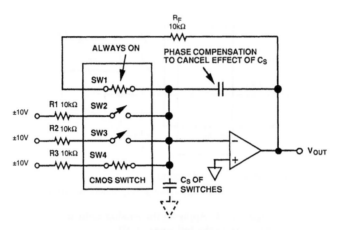

Figure 7.78: Applying the Analog Switch: Using "Dummy" Switch in Feedback to Minimize Gain Error Due to ΔR_{ON}

The best multiplexer design drives the noninverting input of the amplifier as shown in Figure 7.79. The high input impedance of the noninverting input eliminates the errors due to R_{ON}.

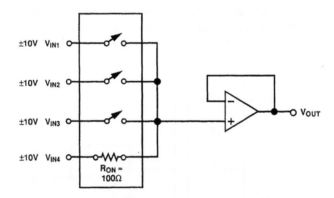

Figure 7.79: Applying the Analog Switch: Minimizing the Influence of ΔR_{ON} Using Noninverting Configuration

CMOS switches and multiplexers are often used with op amps to make programmable gain amplifiers (PGAs). To understand R_{ON}'s effect on their performance, consider Figure 7.80, a poor PGA design. A non-inverting op amp has 4 different gain-set resistors, each grounded by a switch, with an R_{ON} of 100 Ω–500 Ω. Even with R_{ON} as low as 25 Ω, the gain of 16 error would be 2.4%, worse than 8-bit accuracy. R_{ON} also changes over temperature, and from switch-to-switch.

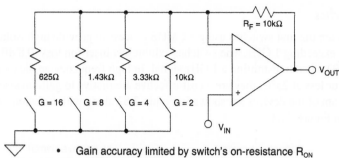

- Gain accuracy limited by switch's on-resistance R_{ON} and R_{ON} modulation
- R_{ON} typically 1 – 500Ω for CMOS or JFET switch
- For R_{ON} = 25Ω, there is a 2.4% gain error for G = 16
- R_{ON} drift over temperature limits accuracy
- Must use very low R_{ON} switches

Figure 7.80: A Poorly Designed PGA Using CMOS Switches

To attempt "fixing" this design, the resistors might be increased, but noise and offset could then be a problem. The only way to improve accuracy with this circuit is to use relays, with virtually no R_{ON}. Only then will the few mΩ of relay R_{ON} be a small error vis-à-vis 625 Ω.

It is much better to use a circuit insensitive to R_{ON}. In Figure 7.81, the switch is placed in series with the inverting input of an op amp. Since the op amp input impedance is very large, the switch R_{ON} is now irrelevant, and gain is now determined solely by the external resistors. Note: R_{ON} may add a small offset error if op amp bias current is high. If this is the case, it can readily be compensated with an equivalent resistance at V_{IN}.

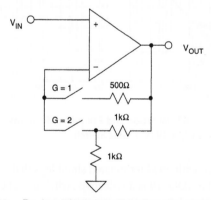

- R_{ON} is not in series with gain setting resistors
- R_{ON} is small compared to input impedance
- Only slight offset errors occur due to bias current flowing through the switches

Figure 7.81: Alternate PGA Configuration Minimizes the Effects of R_{ON}

1 GHz CMOS Switches

The ADG918/ADG919 are the first switches using a CMOS process to provide high isolation and low insertion loss up to and exceeding 1 GHz. The switches exhibit low insertion loss (0.8 dB) and relatively high off isolation (37 dB) when transmitting a 1 GHz signal. In high frequency applications with through-put power of +18 dBm or less at 25°C, they are a cost-effective alternative to gallium arsenide (GaAs) switches. A block diagram of the devices are shown in Figure 7.82 along with isolation and loss versus frequency plots given in Figure 7.83.

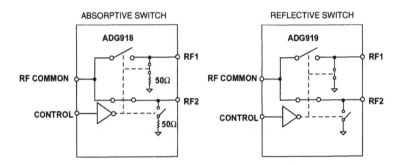

Figure 7.82: 1 GHz CMOS 1.65 V to 2.75 V 2:1 Mux/SPDT Switches

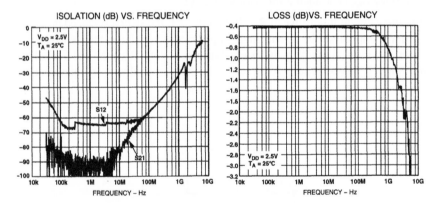

Figure 7.83: Isolation and Frequency Response
of AD918/AD919 1 GHz Switch

The ADG918 is an absorptive switch with 50 Ω terminated shunt legs that allow impedance matching with the application circuit, while the ADG919 is a reflective switch designed for use where the terminations are external to the chip. Both offer low power consumption (<1 μA), tiny packages (8-lead MSOP and 3 mm × 3 mm lead frame chip scale package), single-pin control voltage levels that are CMOS/LVTTL compatible, making the switches ideal for wireless applications and general-purpose RF switching.

Video Switches and Multiplexers

In order to meet stringent specifications of bandwidth flatness, differential gain and phase, and 75 Ω drive capability, high speed complementary bipolar processes are more suitable than CMOS processes for video switches and multiplexers. Traditional CMOS switches and multiplexers suffer from several disadvantages at video frequencies. Their switching time (typically 50 ns or so) is not fast enough for today's applications, and they require external buffering in order to drive typical video loads. In addition, the small variation of the CMOS switch on-resistance with signal level (R_{ON} *modulation*) introduces unwanted distortion in differential gain and phase. Multiplexers based on complementary bipolar technology offer a better solution at video frequencies. The trade-offs, of course, are higher power and cost.

Functional block diagrams of the AD8170/AD8174/AD8180/AD8182 bipolar video multiplexer are shown in Figure 7.84. The AD8183/AD8185 video multiplexer is shown in Figure 7.85. These devices offer a high degree of flexibility and are ideally suited to video applications, with excellent differential gain and phase specifications. Switching time for all devices in the family is 10 ns to 0.1%. The AD8186/AD8187 are single-supply versions of the AD8183/AD8185. Note that these bipolar multiplexers are not bidirectional.

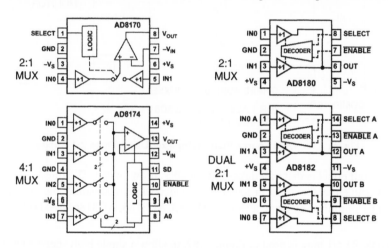

Figure 7.84: AD8170/AD8174/AD8180/AD8182 Bipolar Video Multiplexers

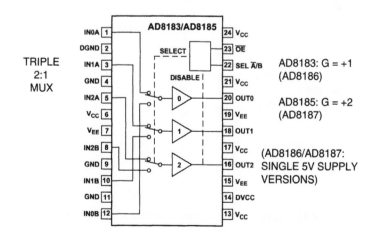

Figure 7.85: AD8183/AD8185 Triple 2:1 Video Multiplexers

The AD8170/AD8174 series of muxes include an on-chip current feedback op amp output buffer whose gain can be set externally. Off channel isolation and crosstalk are typically greater than 80 dB at 5 MHz for the entire family.

Figure 7.86 shows an application circuit for three AD8170 2:1 muxes, where a single RGB monitor is switched between two RGB computer video sources.

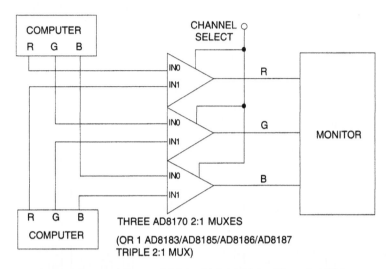

Figure 7.86: Dual Source RGB Multiplexer Using Three 2:1 Muxes

In this setup, the overall effect is that of a three-pole, double-throw switch. The three video sources constitute the three poles, and either the upper or lower of the video sources constitute the two switch states. Note that the circuit can be simplified by using a single AD8183, AD8185, AD8186, or AD8187 triple dual input multiplexer.

The AD8174 or AD8184 4:1 mux is used in Figure 7.87, to allow a single high speed ADC to digitize the RGB outputs of a scanner.

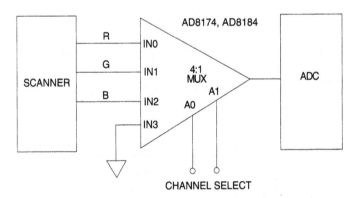

Figure 7.87: Digitizing RGB Signals with One ADC and a 4:1 Mux

The RGB video signals from the scanner are fed in sequence to the ADC, and digitized in sequence, making efficient use of the scanner data with one ADC.

Video Crosspoint Switches

The AD8116 extends the multiplexer concepts to a fully integrated, 16 × 16 buffered video crosspoint switch matrix (Figure 7.88). The 3 dB bandwidth is greater than 200 MHz, and the 0.1 dB gain flatness extends to 60 MHz. Channel switching time is less than 30 ns to 0.1%. Channel-to-channel crosstalk is –70 dB measured at 5 MHz. Differential gain and phase is 0.01% and 0.01° for a 150 Ω load. Total power dissipation is 900 mW on ±5 V.

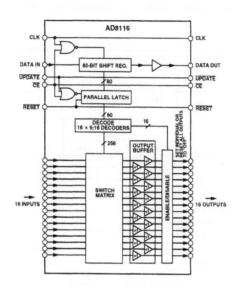

Figure 7.88: AD8116 16 × 16 200 MHz
Buffered Video Crosspoint Switch

The AD8116 includes output buffers that can be put into a high impedance state for paralleling crosspoint stages so that the off channels do not load the output bus. The channel switching is performed via a serial digital control that can accommodate "daisy chaining" of several devices. The AD8116 package is a 128-pin 14 mm × 14 mm LQFP.

Other members of the crosspoint switch family include the AD8108/AD9109 8 × 8 crosspoint switch; the AD8110/AD8111, 260 MHz, 16 × 8, buffered crosspoint switch; the AD8113 audio/video 60 MHz, 16 × 16 crosspoint switch; and the AD8114/AD8115 low cost 225 MHz, 16 × 16, crosspoint switch.

Digital Crosspoint Switches

The AD8152 is a 3.2 Gbps 34 × 34 asynchronous digital crosspoint switch designed for high speed networking (see Figure 7.89). The device operates at data rates up to 3.2 Gbps per port, making it suitable for Sonet/SDH OC-48 with Forward Error Correction (FEC). The AD8152 has digitally programmable current mode outputs that can drive a variety of termination schemes and impedances while maintaining the correct voltage level and minimizing power consumption. The part operates with a supply voltage as low as 2.5 V, with excellent input sensitivity. The control interface is compatible with LVTTL or CMOS/TTL.

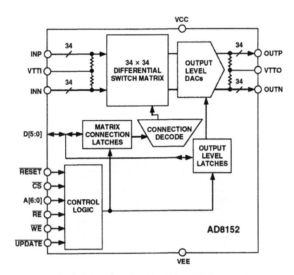

Figure 7.89: AD8152 3.2 Gbps Asynchronous Digital Crosspoint Switch

As the lowest power solution of any comparable crosspoint switch, the AD8152 dissipates less than 2 W at 2.5 V supply with all I/Os active and does not require external heat sinks. The low jitter specification of less than 45 ps makes the AD8152 ideal for high speed networking systems. The AD8152's fully differential signal path reduces jitter and crosstalk while allowing the use of smaller single-ended voltage swings. It is offered in a 256-ball SBGA package that operates over the industrial temperature range of 0°C to 85°C.

Switch and Multiplexer Families from Analog Devices

Selecting the right switch or multiplexer for a particular application can be a difficult task in light of the large number of devices currently offered. Selection guides from Analog Devices can be invaluable in this process. Figure 7.90 summarizes the generic families of CMOS switches and multiplexers, starting with the higher voltage devices and working downward to the newer lower voltage parts. While certainly not all-inclusive, this listing can be useful in getting an overall idea of available choices. Figure 7.91 summarizes the bipolar switch and multiplexer families. The ADG32xx series of Bus Switches are discussed in more detail in Chapter 9.

Parasitic Latchup in CMOS Switches and Muxes

Because multiplexers are often at the front end of a data acquisition system, their inputs generally come from remote locations—hence, they are often subjected to overvoltage conditions. Although this topic is treated in more detail in Chapter 9, an understanding of the problem as it relates to CMOS devices is particularly important. Although this discussion centers around multiplexers, it is germane to nearly all types of CMOS parts.

- ADG2xx, ADG4xx, ADG5xx: ±15V
- ADG508F, ADG509F, ADG528F, ADG438F, ADG439F: Fault-protected ±15V family
- ADG12xx: ±15V, Low R_{ON} (2Ω)
- ADG14xx: ±15V , Low C_{ON} (2pF)
- ADG6xx: Single +5V (some lower) or ±5V
  - 3Ω R_{ON} family
  - 1pC charge injection family
  - 125°C family
- ADG7xx: Single 5V (some as low as 1.8V)
  - Some as low as 2.5Ω R_{ON}
  - Some in CSP
  - 3-5pC charge injection
- ADG8xx: Single +1.8V to +5.5V
  - <0.5Ω R_{ON}
- ADG9xx: Single 1.65V to 2.75V, > 1 GHz RF switches
- ADG3xxx: Bus Switches and Logic Level Shifters

Figure 7.90: CMOS Switches and Multiplexer Families from Analog Devices

---

- Video switches and multiplexers:
  - AD8074, AD8075, AD8170, AD8174, AD8180, AD8182, AD8184, AD8185, AD8186, AD8187
- Video crosspoint switches:
  - AD8108, AD8109, AD8110, AD8111, AD8114, AD8115, AD8116
- Audio and Video crosspoint switch:
  - AD8113
- Digital crosspoint switches:
  - AD8150, AD8151, AD8152, ADSX34

Figure 7.91: High Speed Bipolar Switches and Multiplexers from ADI

Most CMOS analog switches are built using junction-isolated CMOS processes. A cross-sectional view of a single switch cell is shown in Figure 7.92. Parasitic SCR (silicon controlled rectifier) latch-up can occur if the analog switch terminal has voltages more positive than V_{DD} or more negative than V_{SS}. Even a transient situation, such as power-on with an input voltage present, can trigger a parasitic latch-up. If the conduction current is too great (several hundred milliamperes or more), it can damage the switch.

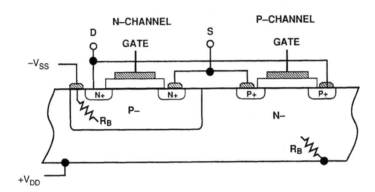

Figure 7.92: Cross-Section of a Junction-Isolation CMOS Switch

The parasitic SCR mechanism is shown in Figure 7.93. SCR action takes place when either terminal of the switch (source or the drain) is either one diode drop more positive than V_{DD} or one diode drop more negative than V_{SS}. In the former case, the V_{DD} terminal becomes the SCR gate input and provides the current to trigger SCR action. In the case where the voltage is more negative than V_{SS}, the V_{SS} terminal becomes the SCR gate input and provides the gate current. In either case, high current will flow between the supplies. The amount of current depends on the collector resistances of the two transistors, which can be fairly small.

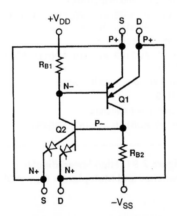

Figure 7.93: Bipolar Transistor Equivalent Circuit
for CMOS Switch Shows Parasitic SCR Latch

In general, to prevent the latch-up condition, the inputs to CMOS devices should never be allowed to be more than 0.3 V above the positive supply or 0.3 V below the negative supply. Note that this restriction also applies when the power supplies are off ($V_{DD} = V_{SS} = 0$ V), and therefore devices can latch up if power is applied to a part when signals are present on the inputs. Manuracturers of CMOS devices invariably place this restriction in the data sheet table of absolute maximum ratings. In addition, the input current under overvoltage conditions should be restricted to 5 mA–30 mA, depending upon the particular device.

In order to prevent this type of SCR latch-up, a series diode can be inserted into the V_{DD} and V_{SS} terminals as shown in Figure 7.94. The diodes block the SCR gate current. Normally the parasitic transistors Q1 and Q2 have low beta (usually less than 10) and require a comparatively large gate current to fire the SCR. The diodes limit the reverse gate current so that the SCR is not triggered.

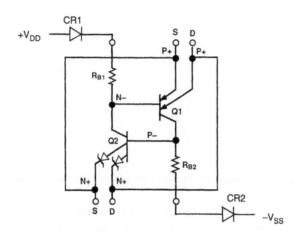

Diodes CR1 and CR2 block base current drive to Q1 and Q2
in the event of overvoltage at S or D.

Figure 7.94: Diode Protection Scheme for CMOS Switch

If diode protection is used, the analog voltage range of the switch will be reduced by one V_{BE} drop at each rail, and this can be inconvenient when using low supply voltages.

As noted, CMOS switches and multiplexers can also be protected from possible overcurrent by inserting a series resistor to limit the current to a safe level as shown in Figure 7.95, generally less than 5 mA–30 mA. Because of the resitive attenuator formed by R_{LOAD} and R_{LIMIT}, this method works only if the switch drives a relatively high impedance load.

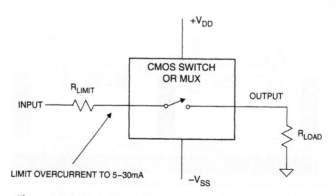

Figure 7.95: Overcurrent Protection Using External Resistor

A common method for input protection is shown in Figure 7.96 where Schottky diodes are connected from the input terminal to each supply voltage as shown. The diodes effectively prevent the inputs from exceeding the supply voltage by more than 0.3 V–0.4 V, thereby preventing latch-up conditions. In addition, if the input voltage exceeds the supply voltage, the input current flows through the external diodes to the supplies, not the device. Schottky diodes can easily handle 50 mA–100 mA of transient current, therefore the R_{LIMIT} resistor can be quite low.

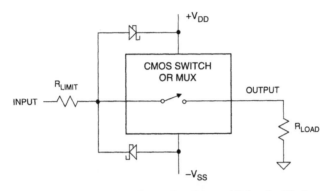

Figure 7.96: Input Protection Using External Schottky Diodes

Most CMOS devices have internal ESD protection diodes connected from the inputs to the supply rails, making the devices less susceptible to latch-up. However, the internal diodes begin conduction at 0.6 V, and have limited current-handling capability, thus adding the external Schottky diodes offers an added degree of protection. However, the effects of the diode leakage and capacitance must be considered.

Note that latch-up protection does not provide overcurrent protection, and vice versa. If both fault conditions can exist in a system, then both protective diodes and resistors should be used.

Analog Devices uses trench-isolation technology to produce its LC²MOS analog switches. The process reduces the latch-up susceptibility of the device, the junction capacitances, increases switching time and leakage current, and extends the analog voltage range to the supply rails.

Figure 7.97 shows the cross-sectional view of the trench-isolated CMOS structure. The buried oxide layer and the side walls completely isolate the substrate from each transistor junction. Therefore, no reverse-biased PN junction is formed. Consequently the bandwidth-reducing capacitances and the possibility of SCR latch up are greatly reduced.

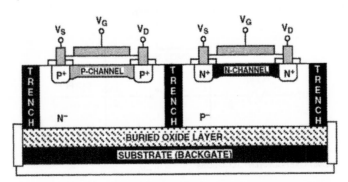

Figure 7.97: Trench-Isolation LC²MOS Structure

The ADG508F, ADG509F, ADG528F, ADG438F, and ADG439F are ±15 V trench-isolated LC^{2}MOS multiplexers which offer "fault protection" for input and output overvoltages between –40 V and +55 V. These devices use a series structure of three MOSFETS in the signal path: an N-channel, followed by a P-channel, followed by an N-channel. In addition, the signal path becomes a high impedance when the power supplies are turned off. This structure offers a high degree of latch-up and overvoltage protection—at the expense of higher R_{ON} (~300 Ω), and more R_{ON} variation with signal level. For more details of this protection method, refer to the individual product data sheets.

Sample-and-Hold Circuits
Walt Kester

Introduction and Historical Perspective

The *sample-and-hold amplifier*, or SHA, is a critical part of most data acquisition systems. It captures an analog signal and holds it during some operation (most commonly analog-digital conversion). The circuitry involved is demanding, and unexpected properties of commonplace components such as capacitors and printed circuit boards may degrade SHA performance.

When the SHA is used with an ADC (either externally or internally), the SHA performance is critical to the overall dynamic performance of the combination, and plays a major role in determining the SFDR, SNR, etc., of the system.

Although today the SHA function has become an integral part of the *sampling* ADC, understanding the fundamental concepts governing its operation is essential to understanding ADC dynamic performance.

When the sample-and-hold is in the sample (or track) mode, the output follows the input with only a small voltage offset. SHAs do exist where the output during the *sample* mode does not follow the input accurately, and the output is only accurate during the *hold* period (such as the AD684, AD781, and AD783). These will not be considered here. Strictly speaking, a sample-and-hold with good tracking performance should be referred to as a *track-and-hold* circuit, but in practice the terms are used interchangeably.

The most common application of a SHA is to maintain the input to an ADC at a constant value during conversion. With many, but not all, types of ADC the input may not change by more than 1 LSB during conversion lest the process be corrupted—this either sets very low input frequency limits on such ADCs, or requires that they be used with a SHA to hold the input during each conversion.

From a historial perspective, it is interesting that the ADC described by A. H. Reeves in his famous PCM patent of 1939 (Reference 1) was a 5-bit 6 kSPS counting ADC where the analog input signal drove a vacuum tube pulsewidth modulator (PWM) directly—the sampling function was incorporated into the PWM. Subsequent work on PCM at Bell Labs led to the use of electron-beam encoder tubes and successive approximation ADCs; and Reference 2 (1948) describes a companion 50 kSPS vacuum tube sample-and-hold circuit based on a pulse transformer drive circuit.

There was increased interest in sample-and-hold circuits for ADCs during the period of the late 1950s and early 1960s as transistors replaced vacuum tubes. One of the first analytical treatments of the errors produced by a solid-state sample-and-hold was published in 1964 by Gray and Kitsopolos of Bell Labs (Reference 3). Edson and Henning of Bell Labs describe the results of experimental work done on a 224 Mbps PCM system, including the 9-bit ADC and a companion 12 MSPS sample-and-hold. References 4, 5, and 6 are representative of work done on sample-and-hold circuits during the 1960s and early 1970s.

In 1969, the newly acquired Pastoriza division of Analog Devices offered one of the first commercial sample-and-holds, the SHA1 and SHA2 as shown in Figure 7.98. The circuits were offered on PC boards, and the SHA1 had an acquisition time of 2 μs to 0.01%, dissipated 0.9 W, and cost approximately $225. The faster SHA2 had an acquisition time of 200 ns to 0.01%, dissipated 1.7 W, and cost approximately $400. They were designed to operate with 12-bit successive approximation ADCs also offered on PC boards.

- Acquisition Time: 2μs to 0.01% (SHA1), 200ns to 0.01% (SHA2)
- Power: 900mW (SHA1), 1.7W (SHA2)
- $225 (SHA1), $400 (SHA2)

**Figure 7.98: "SHA1 and SHA2" Sample-and-Holds
from Analog Devices' Pastoriza Division, 1969**

Modular and hybrid technology quickly made the PC board sample-and-holds obsolete, and the demand for sample-and-holds increased as IC ADCs, such as the industry-standard AD574, came on the market. In the 1970s and into the 1980s, it was quite common for system designers to purchase separate sample-and-holds to drive such ADCs, because process technology did not allow integrating them together onto the same chip. IC SHAs such as the AD582 (4 μs acquisition time to 0.01%), AD583 (6 μs acquisition time to 0.01%), and the AD585 (3 μs acquisition time to 14-bit accuracy) served the lower speed markets of the 1970s and 1980s.

Hybrid SHAs such as the HTS-0025 (25 ns acquisition time to 0.1%), HTC-0300 (200 ns acquisition time to 0.01%), and the AD386 (25 μs acquisition time to 16-bits) served the high speed, high end markets. By 1995, Analog Devices offered approximately 20 sample-and-hold products for various applications, including the following high speed ICs: AD9100/AD9101 (10 ns acquisition time to 0.01%), AD684 (quad 1 μs acquisition time to 0.01%) and the AD783 (250 ns acquisition time to 0.01%).

However, ADC technology was rapidly expanding during the same period, and many ADCs were being offered with internal SHAs (i.e., *sampling* ADCs). This made them easier to specify and certainly easier to use. Integration of the SHA function was made possible by new process developments including high speed complementary bipolar processes and advanced CMOS processes. In fact, the proliferation and popularity of sampling ADCs has been so great that today (2003), one rarely has the need for a separate SHA.

The advantage of a sampling ADC, apart from the obvious ones of smaller size, lower cost, and fewer external components, is that the overall dc and ac performance is fully specified, and the designer need not spend time ensuring that there are no specification, interface, or timing issues involved in combining a discrete ADC and a discrete SHA. This is especially important when one considers dynamic specifications such as SFDR and SNR.

Although the largest applications of SHAs are with ADCs, they are also occasionally used in DAC deglitchers, peak detectors, analog delay circuits, simultaneous sampling systems, and data distribution systems.

Basic SHA Operation

Regardless of the circuit details or type of SHA in question, all such devices have four major components. The input amplifier, energy storage device (capacitor), output buffer, and switching circuits are common to all SHAs as shown in the typical configuration of Figure 7.99.

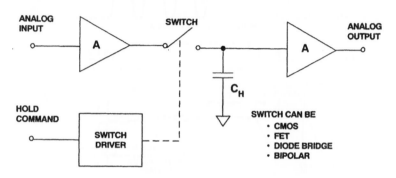

Figure 7.99: Basic Sample-and-Hold Circuit

The energy storage device, the heart of the SHA, is a capacitor. The input amplifier buffers the input by presenting a high impedance to the signal source and providing current gain to charge the hold capacitor. In the *track* mode, the voltage on the hold capacitor follows (or tracks) the input signal (with some delay and bandwidth limiting). In the *hold* mode, the switch is opened, and the capacitor retains the voltage present before it was disconnected from the input buffer. The output buffer offers a high impedance to the hold capacitor to keep the held voltage from discharging prematurely. The switching circuit and its driver form the mechanism by which the SHA is alternately switched between track and hold.

Four groups of specifications describe basic SHA operation: track mode, track-to-hold transition, hold mode, hold-to-track transition. These specifications are summarized in Figure 7.100, and some of the SHA

SAMPLE MODE	SAMPLE-TO-HOLD TRANSITION	HOLD MODE	HOLD-TO-SAMPLE TRANSITION
STATIC: • Offset • Gain Error • Nonlinearity	STATIC: • Pedestal • Pedestal Nonlinearity	STATIC: • Droop • Dielectric • Absorption	
DYNAMIC: • Settling Time • Bandwidth • Slew Rate • Distortion • Noise	DYNAMIC: • Aperture Delay Time • Aperture Jitter • Switching Transient • Settling Time	DYNAMIC: • Feedthrough • Distortion • Noise	DYNAMIC: • Acquisition Time • Switching Transient

Figure 7.100: Sample-and-Hold Specifications

error sources are shown graphically in Figure 7.101. Because there are both dc and ac performance implications for each of the four modes, properly specifying a SHA and understanding its operation in a system is a complex matter.

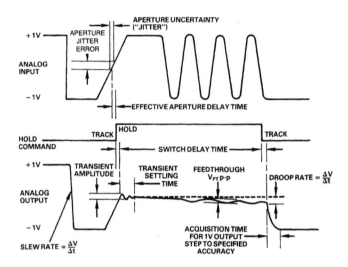

Figure 7.101: Some Sources of Sample-and-Hold Errors

Track Mode Specifications

Since a SHA in the sample (or track) mode is simply an amplifier, both the static and dynamic specifications in this mode are similar to those of any amplifier. (SHAs which have degraded performance in the track mode are generally only specified in the hold mode.) The principle track mode specifications are *offset*, *gain*, *nonlinearity*, *bandwidth*, *slew rate*, *settling time*, *distortion*, and *noise*. However, distortion and noise in the track mode are often of less interest than in the hold mode.

Track-to-Hold Mode Specifications

When the SHA switches from track to hold, there is generally a small amount of charge dumped on the hold capacitor because of nonideal switches. This results in a hold mode dc offset voltage which is called *pedestal* error as shown in Figure 7.102. If the SHA is driving an ADC, the pedestal error appears as a dc

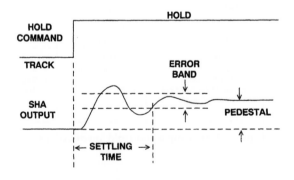

Figure 7.102: Track-to-Hold Mode Pedestal, Transient, and Settling Time Errors

offset voltage that may be removed by performing a system calibration. If the pedestal error is a function of input signal level, the resulting nonlinearity contributes to hold-mode distortion.

Pedestal errors may be reduced by increasing the value of the hold capacitor with a corresponding increase in acquisition time and a reduction in bandwidth and slew rate.

Switching from track to hold produces a transient, and the time required for the SHA output to settle to within a specified error band is called *hold mode settling time*. Occasionally, the peak amplitude of the switching transient is also specified.

Perhaps the most misunderstood and misused SHA specifications are those that include the word *aperture*. The most essential dynamic property of a SHA is its ability to quickly disconnect the hold capacitor from the input buffer amplifier. The short (but nonzero) interval required for this action is called *aperture time*. The various quantities associated with the internal SHA timing are shown in the Figure 7.103.

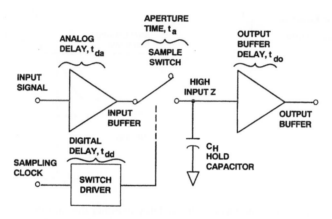

Figure 7.103: SHA Circuit Showing Internal Timing

The actual value of the voltage that is held at the end of this interval is a function of both the input signal and the errors introduced by the switching operation itself. Figure 7.104 shows what happens when the hold command is applied with an input signal of arbitrary slope (for clarity, the sample to hold pedestal and switching transients are ignored). The value that finally gets held is a delayed version of the input signal, averaged over the aperture time of the switch as shown in Figure 7.104. The first-order model assumes that the final value of the voltage on the hold capacitor is approximately equal to the average value of the signal applied to the switch over the interval during which the switch changes from a low to high impedance (t_a).

The model shows that the finite time required for the switch to open (t_a) is equivalent to introducing a small delay in the sampling clock driving the SHA. This delay

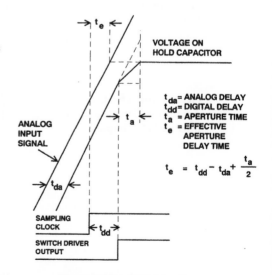

t_{da} = ANALOG DELAY
t_{dd} = DIGITAL DELAY
t_a = APERTURE TIME
t_e = EFFECTIVE APERTURE DELAY TIME

$$t_e = t_{dd} - t_{da} + \frac{t_a}{2}$$

Figure 7.104: SHA Waveforms

is constant and may either be positive or negative. It is called *effective aperture delay time, aperture delay time*, or simply *aperture delay*, (t_e) and is defined as the time difference between the analog propagation delay of the front-end buffer (t_{da}) and the switch digital delay (t_{dd}) plus one-half the aperture time ($t_a/2$). The effective aperture delay time is usually positive, but may be negative if the sum of one-half the aperture time ($t_a/2$) and the switch digital delay (t_{dd}) is less than the propagation delay through the input buffer (t_{da}). The aperture delay specification thus establishes when the input signal is actually sampled with respect to the sampling clock edge.

Aperture delay time can be measured by applying a bipolar sinewave signal to the SHA and adjusting the synchronous sampling clock delay such that the output of the SHA is zero during the hold time. The relative delay between the input sampling clock edge and the actual zero-crossing of the input sinewave is the aperture delay time as shown in Figure 7.105.

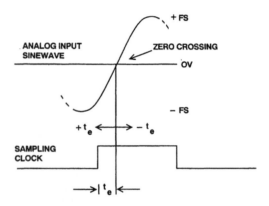

Figure 7.105: Effective Aperture Delay Time

Aperture delay produces no errors, but acts as a fixed delay in either the sampling clock input or the analog input (depending on its sign). If there is sample-to-sample variation in aperture delay (*aperture jitter*), a corresponding voltage error is produced as shown in Figure 7.106. This sample-to-sample variation in the instant the switch opens is called *aperture uncertainty*, or *aperture jitter* and is usually measured in picoseconds rms. The amplitude of the associated output error is related to the rate-of-change of the analog input. For any given value of aperture jitter, the aperture jitter error increases as the input dv/dt increases.

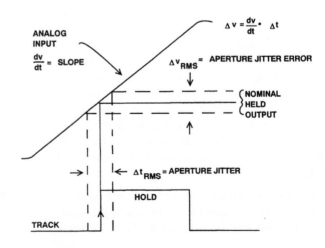

Figure 7.106: Effects of Aperture or Sampling Clock Jitter on SHA Output

Measuring aperture jitter error in a SHA requires a jitter-free sampling clock and analog input signal source, because jitter (or phase noise) on either signal cannot be distinguished from the SHA aperture jitter itself—the effects are the same. In fact, the largest source of timing jitter errors in a system is most often external to the SHA (or the ADC if it is a sampling one) and is caused by noisy or unstable clocks, improper signal routing, and lack of attention to good grounding and decoupling techniques. SHA aperture jitter is generally less than 50 ps rms, and less than 5 ps rms in high speed devices. Details of measuring aperture jitter of an ADC can be found in Chapter 5.

Figure 7.107 shows the effects of total sampling clock jitter on the signal-to-noise ratio (SNR) of a sampled data system. The total rms jitter will be composed of a number of components, the actual SHA aperture jitter often being the least of them.

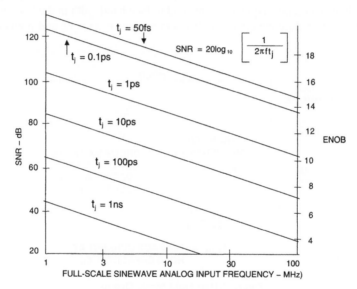

Figure 7.107: Effects of Sampling Clock Jitter on SNR

Hold Mode Specifications

During the hold mode there are errors due to imperfections in the hold capacitor, switch, and output amplifier. If a leakage current flows in or out of the hold capacitor, it will slowly charge or discharge, and its voltage will change. This effect is known as *droop* in the SHA output and is expressed in V/μs. Droop can be caused by leakage across a dirty PC board if an external capacitor is used, or by a leaky capacitor, but is most usually due to leakage current in semiconductor switches and the bias current of the output buffer amplifier. An acceptable value of droop is where the output of a SHA does not change by more than ½ LSB during the conversion time of the ADC it is driving, although this value is highly dependent on the ADC architecture. Where droop is due to leakage current in reversed biased junctions (CMOS switches or FET amplifier gates), it will double for every 10°C increase in chip temperature—which means that it will increase a thousand fold between 25°C and 125°C. Droop can be reduced by increasing the value of the hold capacitor, but this will also increase acquisition time and reduce bandwidth in the track mode. Differential techniques are often used to reduce the effects of droop in modern IC sample-and-hold circuits that are part of the ADC.

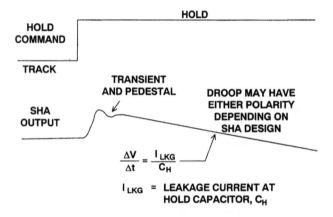

Figure 7.108: Hold Mode Droop

Even quite small leakage currents can cause troublesome droop when SHAs use small hold capacitors. Leakage currents in PCBs may be minimized by the intelligent use of guard rings. A guard ring is a ring of conductor which surrounds a sensitive node and is at the same potential. Since there is no voltage between them, there can be no leakage current flow. In a noninverting application, such as is shown in Figure 7.109, the guard ring must be driven to the correct potential, whereas the guard ring on a virtual ground can be at actual ground potential (Figure 7.110). The surface resistance of PCB material is much lower than its bulk resistance, so guard rings must always be placed on both sides of a PCB—and on multilayer boards, guard rings should be present in all layers.

Hold capacitors for SHAs must have low leakage, but there is another characteristic which is equally important: low *dielectric absorption*. If a capacitor is charged, then discharged, and then left open circuit, it will recover some of its charge as shown in Figure 7.111. The phenomenon is known as *dielectric absorption*, and it can seriously degrade the performance of a SHA, since it causes the remains of a previous sample to contaminate a new one, and may introduce random errors of tens or even hundreds of mV.

Different capacitor materials have differing amounts of dielectric absorption—electrolytic capacitors are dreadful (their leakage is also high), and some high-K ceramic types are bad, while mica, polystyrene and

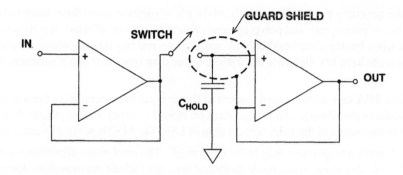

Note: Be Sure a Guard Shield is in Each Layer of the PCB

Figure 7.109: Drive the Guard Shield with the Same Voltage
as the Hold Capacitor to Reduce Board Leakage

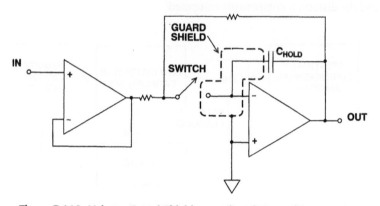

Figure 7.110: Using a Guard Shield on a Virtual Ground SHA Design

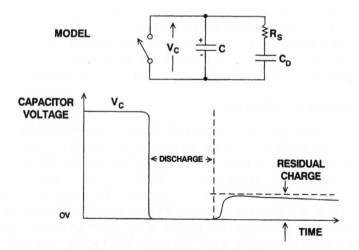

Figure 7.111: Dielectric Absorption

polypropylene are generally good. Unfortunately, dielectric absorption varies from batch to batch, and even occasional batches of polystyrene and polypropylene capacitors may be affected. It is therefore wise to pay 30%–50% extra when buying capacitors for SHA applications and buy devices which are guaranteed by their manufacturers to have low dielectric absorption, rather than types that might generally be expected to have it.

Stray capacity in a SHA may allow a small amount of the ac input to be coupled to the output during hold. This effect is known as *feedthrough* and is dependent on input frequency and amplitude. If the amplitude of the feedthrough to the output of the SHA is more than ½ LSB, the ADC is subject to conversion errors.

In many SHAs, distortion is specified only in the track mode. The *track mode distortion* is often much better than *hold mode distortion*. Track mode distortion does not include nonlinearities due to the switch network, and may not be indicative of the SHA performance when driving an ADC. Modern SHAs, especially high speed ones, specify distortion in both modes. While track mode distortion can be measured using an analog spectrum analyzer, hold mode distortion measurements should be performed using digital techniques as shown in Figure 7.112. A spectrally pure sinewave is applied to the SHA, and a low distortion high speed ADC digitizes the SHA output near the end of the hold time. An FFT analysis is performed on the ADC output, and the distortion components computed.

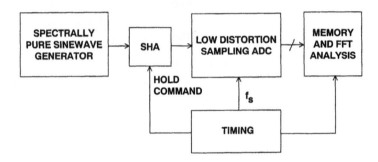

Figure 7.112: Measuring Hold Mode Distortion

SHA *noise* in the track mode is specified and measured like that of an amplifier. Peak-to-peak *hold mode noise* is measured with an oscilloscope and converted to an rms value by dividing by 6.6. Hold mode noise may be given as a spectral density in $nV/\sqrt{Hz}$, or as an rms value over a specified bandwidth. Unless otherwise indicated, the hold mode noise must be combined with the track mode noise to yield the total output noise. Some SHAs specify the total output hold mode noise, in which case the track mode noise is included.

Hold-to-Track Transition Specifications

When the SHA switches from hold to track, it must reacquire the input signal (which may have made a full scale transition during the hold mode). *Acquisition time* is the interval of time required for the SHA to reacquire the signal to the desired accuracy when switching from hold to track. The interval starts at the 50% point of the sampling clock edge, and ends when the SHA output voltage falls within the specified error band (usually 0.1% and 0.01% times are given). Some SHAs also specify acquisition time with respect to the voltage on the hold capacitor, neglecting the delay and settling time of the output buffer. The hold capacitor acquisition time specification is applicable in high speed applications, where the maximum possible time must be allocated for the hold mode. The output buffer settling time must of course be significantly smaller than the hold time.

Acquisition time can be measured directly using modern digital sampling scopes (DSOs) or digital phosphor scopes (DPOs) which are insensitive to large overdrives.

SHA Architectures

As with op amps, there are numerous SHA architectures, and we will examine a few of the most popular ones. The simplest SHA structure is shown in Figure 7.113. The input signal is buffered by an amplifier and applied to the switch. The input buffer may either be open- or closed-loop and may or may not provide gain. The switch can be CMOS, FET, or bipolar (using diodes or transistors) and is controlled by the switch driver circuit. The signal on the hold capacitor is buffered by an output amplifier. This architecture is sometimes referred to as *open-loop* because the switch is not inside a feedback loop. Notice that the entire signal voltage is applied to the switch, therefore it must have excellent common-mode characteristics.

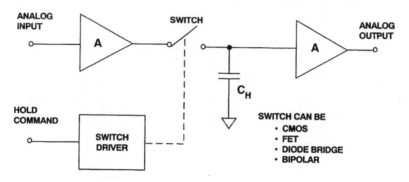

Figure 7.113: Open-Loop SHA Architecture

An implementation of this architecture is shown in Figure 7.114, where a simple diode bridge is used for the switch. In the track mode, current flows through the bridge diodes D1, D2, D3, and D4. For fast slewing input signals, the hold capacitor is charged and discharged with the current, I. Therefore, the maximum slew rate on the hold capacitor is equal to I/C_H. Reversing the bridge drive currents reverse biases the bridge and places the circuit in the hold mode. Bootstrapping the turn-off pulses with the held output signal minimizes common-mode distortion errors and is key to the circuit. The reverse bias bridge voltage is equal to the forward drops of D5 and D6 plus the voltage drops across the series resistors R1 and R2. This circuit

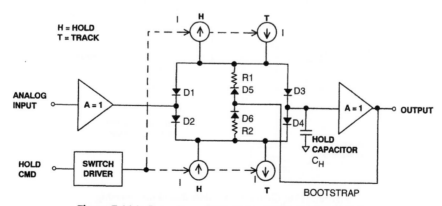

Figure 7.114: Open-Loop SHA Using Diode Bridge Switch

is extremely fast, especially if the input and output buffers are open-loop followers, and the diodes are Schottky ones. The turn-off pulses can be generated with high frequency pulse transformers or with current switches as shown in Figure 7.115. This circuit can be used at any sampling rate, because the diode switching pulses are direct-coupled to the bridge. Variations of this circuit have been used since the mid-1960s in high speed PC board, modular, hybrid, and IC SHAs.

The SHA circuit shown in Figure 7.116 represents a classical *closed-loop* design and is used in many CMOS sampling ADCs. Since the switches always operate at virtual ground, there is no common-mode signal across them.

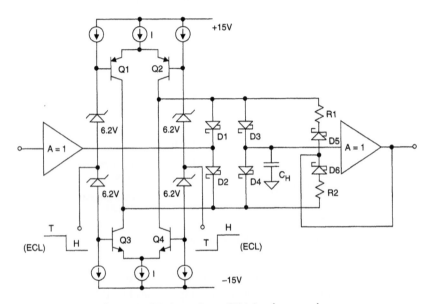

Figure 7.115: Open-Loop SHA Implementation

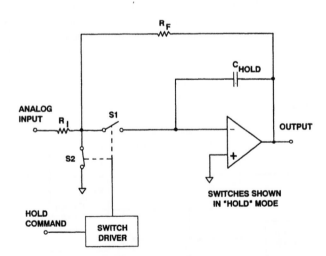

Figure 7.116: Closed-Loop SHA Based on Inverting Integrator Switched at the Summing Point

Switch S2 is required in order to maintain a constant input impedance and prevent the input signal from coupling to the output during the hold time. In the track mode, the transfer characteristic of the SHA is determined by the op amp, and the switches do not introduce dc errors because they are inside the feedback loop. The effects of charge injection can be minimized by using the differential switching techniques shown in Figure 7.117.

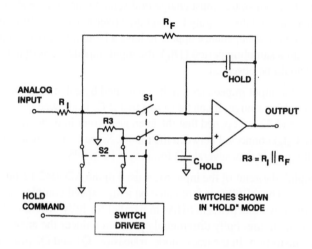

Figure 7.117: Differential Switching Reduces Charge Injection

Internal SHA Circuits for IC ADCs

CMOS ADCs are quite popular because of their low power and low cost. The equivalent input circuit of a typical CMOS ADC using a differential sample-and-hold is shown in Figure 7.118. While the switches are shown in the *track* mode, note that they open/close at the sampling frequency. The 16 pF capacitors represent the effective capacitance of switches S1 and S2, plus the stray input capacitance. The C_S capacitors (4 pF) are the sampling capacitors, and the C_H capacitors are the hold capacitors. Although the input circuit

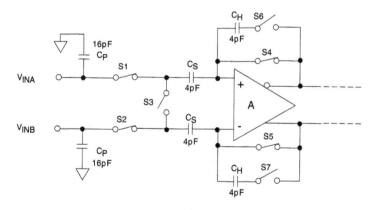

SWITCHES SHOWN IN TRACK MODE

Figure 7.118: Simplified Input Circuit for a Typical Switched Capacitor CMOS Sample-and-Hold

is completely differential, this ADC structure can be driven either single-ended or differentially. Optimum performance, however, is generally obtained using a differential transformer or differential op amp drive.

In the *track* mode, the differential input voltage is applied to the C_S capacitors. When the circuit enters the *hold* mode, the voltage across the sampling capacitors is transferred to the C_H hold capacitors and buffered by the amplifier A (the switches are controlled by the appropriate sampling clock phases). When the SHA returns to the *track* mode, the input source must charge or discharge the voltage stored on C_S to a new input voltage. This action of charging and discharging C_S, averaged over a period of time and for a given sampling frequency f_S, makes the input impedance appear to have a benign resistive component. However, if this action is analyzed within a sampling period ($1/f_S$), the input impedance is dynamic, and certain input drive source precautions should be observed.

The resistive component to the input impedance can be computed by calculating the average charge that is drawn by C_H from the input drive source. It can be shown that if C_S is allowed to fully charge to the input voltage before switches S1 and S2 are opened that the average current into the input is the same as if there were a resistor equal to $1/(C_S f_S)$ connected between the inputs. Since C_S is only a few picofarads, this resistive component is typically greater than several $k\Omega$ for an $f_S = 10$ MSPS.

Figure 7.119 shows a simplified circuit of the input SHA used in the AD9042 12-bit, 41 MSPS ADC introduced in 1995 (Reference 7). The AD9042 is fabricated on a high speed complementary bipolar process, XFCB. The circuit comprises two independent SHAs in parallel for fully differential operation—only one-half the circuit is shown in the figure. Fully differential operation reduces the error due to droop rate and also reduces second-order distortion. In the track mode, transistors Q1 and Q2 provide unity-gain buffering. When the circuit is placed in the hold mode, the base voltage of Q2 is pulled negative until it is clamped by the diode, D1. The on-chip hold capacitor, C_H, is nominally 6 pF. Q3 along with C_F provide output current bootstrapping and reduce the V_{BE} variations of Q2. This reduces third-order signal distortion. Track mode THD is typically –93 dB at 20 MHz. In the time domain, full-scale acquisition time to 12-bit accuracy is 8 ns. In the hold mode, signal-dependent pedestal variations are minimized by the voltage bootstrapping action of Q3 and the A = 1 buffer along with the low feedthrough parasitics of Q2. Hold mode settling time is 5 ns to 12-bit accuracy. Hold-mode THD at a clock rate of 50 MSPS and a 20 MHz input signal is –90 dB.

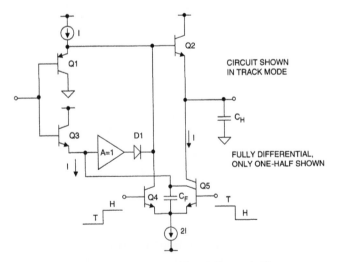

Figure 7.119: SHA Used in AD9042 12-Bit, 41 MSPS ADC Introduced in 1995

Figure 7.120 shows a simplified schematic of one-half of the differential SHA used in the AD6645 14-bit, 105 MSPS ADC recently introduced (Reference 9) gives a complete description of the ADC including the SHA). In the track mode, Q1, Q2, Q3, and Q4 form a complementary emitter follower buffer which drives the hold capacitor, C_H. In the hold mode, the polarity of the bases of Q3 and Q4 is reversed and clamped to a low impedance. This turns off Q1, Q2, Q3, and Q4, and results in double isolation between the signal at the input and the hold capacitor. As previously discussed, the clamping voltages are boostrapped by the held output voltage, thereby minimizing nonlinear effects.

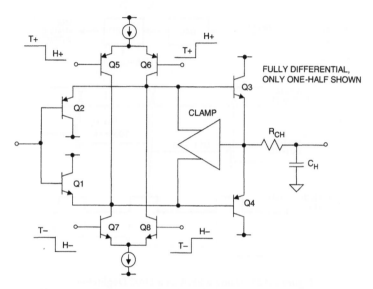

Figure 7.120: SHA Used in AD6645 14-Bit, 105 MSPS ADC

Track mode linearity is largely determined by the V_{BE} modulation of Q3 and Q4 when charging C_H. Hold mode linearity depends on track mode linearity plus nonlinear errors in the track-to-hold transitions caused by imbalances in the switching of the base voltages of Q3 and Q4 and the resulting imbalance in charge injection through their base-emitter junctions as they turn off.

SHA Applications

By far the largest application of SHAs is driving ADCs. Most modern ADCs designed for signal processing are sampling ones and contain an internal SHA optimized for the converter design. Sampling ADCs are completely specified for both dc and ac performance and should be used in lieu of discrete SHA/ADC combinations wherever possible. In a very few selected cases, especially those requiring wide dynamic range and low distortion, there may be advantages to using a discrete combination.

A similar application uses a low distortion SHA to minimize the effects of code-dependent DAC glitches as shown in Figure 7.121. Just prior to latching new data into the DAC, the SHA is put into the hold mode so that the DAC switching glitches are isolated from the output. The switching transients produced by the SHA are not code-dependent, occur at the update frequency, and are easily filterable. This technique may be useful at low frequencies to improve the distortion performance of DACs, but has little value when using high speed low-glitch low distortion DACs designed especially for DDS applications where the update rate is several hundred MHz.

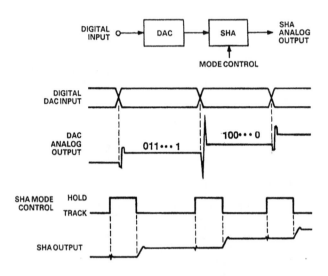

Figure 7.121: Using a SHA as a DAC Deglitcher

Rather than use a single ADC per channel in a simultaneous sampled system, it is often more economical to use multiple SHAs followed by an analog multiplexer and a single ADC (Figure 7.122). Similarly, in data distribution systems multiple SHAs can be used to route the sequential outputs of a single DAC to

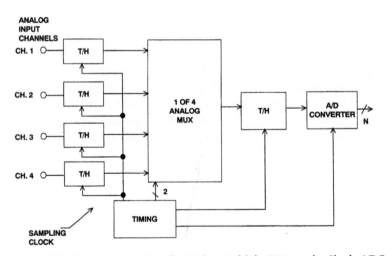

Figure 7.122: Simultaneous Sampling Using Multiple SHAs and a Single ADC

multiple channels as shown in Figure 7.123; although this is not as common, as multiple DACs usually offer a better solution.

A final application for SHAs is shown in Figure 7.124, where SHAs are cascaded to produce analog delay in a sampled data system. SHA 2 is placed in hold just prior to the end of the hold interval for SHA 1. This results in a total pipeline delay greater than the sampling period T. This technique is often used in multistage pipelined subranging ADCs to allow for the conversion delays of successive stages. In pipelined ADCs, a 50% duty cycle sampling clock is common, thereby allowing alternating clock phases to drive each SHA in the pipeline (see Chapter 3 for more details of the pipelined ADC architecture and the use of SHAs for analog delay).

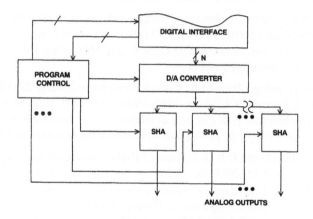

**Figure 7.123: Data Distribution System
Using Multiple SHAs and a Single DAC**

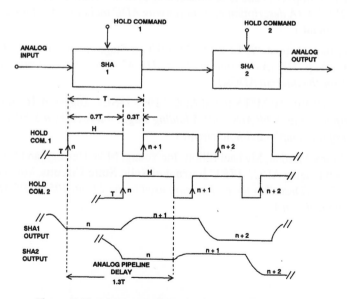

Figure 7.124: SHAs Used for Analog Pipelined Delay

References:
7.4 Sample-and-Hold Circuits

1. Alec Harley Reeves, "Electric Signaling System," **U.S. Patent 2,272,070**, filed November 22, 1939, issued February 3, 1942. Also **French Patent 852,183** issued 1938, and **British Patent 538,860** issued 1939. *(The classic patents on PCM including descriptions of a 5-bit, 6 kSPS vacuum tube ADC and DAC.)*

2. L. A. Meacham and E. Peterson, "An Experimental Multichannel Pulse Code Modulation System of Toll Quality," **Bell System Technical Journal**, Vol 27, No. 1, January 1948, pp. 1–43. *(Describes the culmination of much work leading to this 24-channel experimental PCM system. In addition, the article describes a 50 kSPS vacuum tube sample-and-hold based on a pulse transformer driver.)*

3. J. R. Gray and S. C. Kitsopoulos, "A Precision Sample-and-Hold Circuit with Subnanosecond Switching," **IEEE Transactions on Circuit Theory**, CT11, September 1964, pp. 389–396. *(An excellent description of a solid-state transformer-driven diode bridge SHA, along with a detailed mathematical analysis of the circuit and associated errors.)*

4. J. O. Edson and H. H. Henning, "Broadband Codecs for an Experimental 224Mb/s PCM Terminal," **Bell System Technical Journal**, Vol. 44, pp. 1887–1940, Nov. 1965. *(Summarizes experiments on ADCs based on the electron tube coder as well as a bit-per-stage Gray code 9-bit solid state ADC. The electron beam coder was 9 bits at 12 MSPS, and represented the fastest of its type.)*

5. D. J. Kinniment, D. Aspinall, and D.B.G. Edwards, "High-Speed Analogue-Digital Converter," **IEE Proceedings**, Vol. 113, pp. 2061–2069, Dec. 1966. *(A 7-bit 9 MSPS three-stage pipelined error corrected converter is described based on recirculating through a 3-bit stage three times. Tunnel [Esaki] diodes are used for the individual comparators. The article also shows a proposed faster pipelined 7-bit architecture using three individual 3-bit stages with error correction. The article also describes a fast bootstrapped transformer-driven diode-bridge sample-and-hold circuit.)*

6. O. A. Horna, "A 150 Mbps A/D and D/A Conversion System," **Comsat Technical Review**, Vol. 2, No. 1, pp. 39–72, 1972. *(A description of a subranging ADC including a detailed analysis of the sample-and-hold circuit.)*

7. Roy Gosser and Frank Murden, "A 12-Bit 50 MSPS Two-Stage A/D Converter," **1995 ISSCC Digest of Technical Papers**, p. 278. *(A description of the AD9042 error corrected subranging ADC using MagAMP stages for the internal ADCs.)*

8. Carl Moreland, "An 8-Bit 150 MSPS Serial ADC," **1995 ISSCC Digest of Technical Papers**, Vol. 38, p. 272. *(A description of an 8-bit ADC with 5 folding stages followed by a 3-bit flash converter, including a discussion of the sample-and-hold circuit.)*

9. Carl Moreland, Frank Murden, Michael Elliott, Joe Young, Mike Hensley, and Russell Stop, "A 14-Bit 100 Msample/s Subranging ADC," **IEEE Journal of Solid State Circuits**, Vol. 35, No. 12, December 2000, pp. 1791–1798. *(Describes the architecture used in the 14-bit, 105 MSPS AD6645 ADC and also the sample-and-hold circuit.)*

CHAPTER 8

Data Converter Applications

Data Converter Applications

Precision Measurement and Sensor Conditioning

Introduction

The high resolution Σ-Δ measurement ADC has revolutionized the entire area of precision sensor signal conditioning and data acquisition. Modern Σ-Δ ADCs offer no-missing-code resolutions to 24 bits, and greater than 19 bits of noise-free code resolution. The inclusion of on-chip PGAs coupled with the high resolution virtually eliminates the need for signal conditioning circuitry—the precision sensor can interface directly with the ADC in many cases.

As discussed in detail in Chapter 3 of this book, the Σ-Δ architecture is highly digitally intensive. It is therefore relatively easy to add programmable features and offer greater flexibility in their applications. Throughput rate, digital filter cutoff frequency, PGA gain, channel selection, chopping, and calibration modes are just a few of the possible features. One of the benefits of the on-chip digital filter is that its notches can be programmed to provide excellent 50 Hz/60 Hz power supply rejection. In addition, since the input to a Σ-Δ ADC is highly oversampled, the requirements on the antialiasing filter are not nearly as stringent as in the case of traditional Nyquist-type ADCs. Excellent common-mode rejection is also a result of the extensive utilization of differential analog and reference inputs. An important benefit of Σ-Δ ADCs is that they are typically designed on CMOS processes, and are therefore relatively low cost.

In applying Σ-Δ ADCs, the user must accept the fact that because of the highly digital nature of the devices and the programmability offered, the digital interfaces tend to be more complex than with traditional ADC architectures such as successive approximation, for example. However, manufacturers' evaluation boards and associated development software along with complete data sheets can considerably ease the overall design process.

Some of the architectural benefits and features of the Σ-Δ measurement ADC are summarized in Figure 8.1 and 8.2.

- High Resolution
  - 24 bits no missing codes
  - 22 bits effective resolution (RMS)
  - 19 bits noise-free code resolution (peak-to-peak)
  - On-Chip PGAs
- High Accuracy
  - INL 2ppm of Fullscale ~ 1LSB in 19 bits
  - Gain drift 0.5ppm/°C
- More Digital, Less Analog
  - Programmable Balance between Speed × Resolution
- Oversampling AND Digital Filtering
  - 50/60Hz rejection
  - High oversampling rate simplifies antialiasing filter
- Wide Dynamic Range
- Low Cost

Figure 8.1: Σ-Δ ADC Architecture Benefits

- Analog Input Buffer Options
  - Drives Σ-Δ Modulator, Reduces Dynamic Input Current
- Differential AIN, REFIN
  - Ratiometric Configuration Eliminates Need for Accurate Reference
- Multiplexer
- PGA
- Calibrations
  - Self Calibration, System Calibration, Auto Calibration
- Chopping Options
  - No Offset and Offset Drifts
  - Minimizes Effects of Parasitic Thermocouples

Figure 8.2: Σ-Δ System on Chip Features

Applications of Precision Measurement Σ-Δ ADCs

High resolution measurement Σ-Δ ADCs find applications in many areas, including process control, sensor conditioning, instrumentation, etc. as shown in Figure 8.3. Because of the varied requirements, these ADCs are offered in a variety of configurations and options. For instance, Analog Devices currently (2004) has more than 24 different high resolution Σ-Δ ADC product offerings available. For this reason, it is impossible to cover all applications and products in a section of reasonable length, so we will focus on several representative sensor conditioning examples which will serve to illustrate most of the important application principles.

- Process Control
  - 4–20mA
- Sensors
  - Weigh Scale
  - Pressure
  - Temperature
- Instrumentation
  - Gas Monitoring
  - Portable Instrumentation
  - Medical Instrumentation

WEIGH SCALE

Figure 8.3: Typical Applications of High Resolution Σ-Δ ADCs

Because many sensors such as strain gages, flow meters, pressure sensors, and load cells use resistor-based circuits, we will use the AD7730 ADC as an example in a weigh scale design. A block diagram of the AD7730 is shown in Figure 8.4.

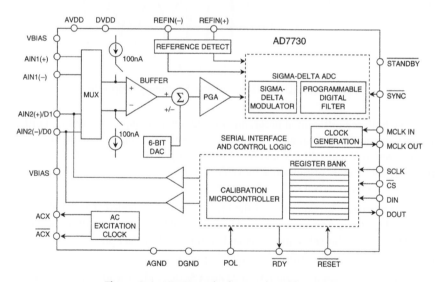

Figure 8.4: AD7730 Single-Supply Bridge ADC

The heart of the AD7730 is the 24-bit Σ-Δ core. The AD7730 is a complete analog front end for weigh-scale and pressure measurement applications. The device accepts low level signals directly from a transducer and outputs a serial digital word. The input signal is applied to a proprietary programmable gain front end based around an analog modulator. The modulator output is processed by a low pass programmable digital filter, allowing adjustment of filter cutoff, output rate and settling time. The response of the internal digital filter is shown in Figure 8.5.

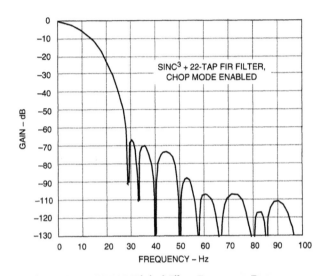

Figure 8.5: AD7730 Digital Filter Frequency Response

The part features two buffered differential programmable gain analog inputs as well as a differential reference input. The part operates from a single 5 V supply. It accepts four unipolar analog input ranges: 0 mV to +10 mV, +20 mV, +40 mV and +80 mV and four bipolar ranges: ±10 mV, ±20 mV, ±40 mV and ±80 mV. The peak-to-peak noise-free code resolution achievable directly from the part is 1 in 230,000 counts. An on-chip 6-bit DAC allows the removal of TARE voltages.

Clock signals for synchronizing ac excitation of the bridge are also provided. The serial interface on the part can be configured for three-wire operation and is compatible with microcontrollers and digital signal processors. The AD7730 contains self-calibration and system calibration options, and features an offset drift of less than 5 nV/°C and a gain drift of less than 2 ppm/°C.

The AD7730 is available in a 24-pin plastic DIP, a 24-lead SOIC and 24-lead TSSOP package. The AD7730L is available in a 24-lead SOIC and 24-lead TSSOP package. Key specifications for the AD7730 are summarized in Figure 8.6. Further details on the operation of the AD7730 can be found in References 1 and 2.

A very powerful *ratiometric* technique that includes Kelvin sensing to minimize errors due to wiring resistance and also eliminates the need for an accurate excitation voltage is shown in Figure 8.7. The AD7730 measurement ADC can be driven from a single supply voltage which is also used to excite the remote bridge. Both the analog input and the reference input to the ADC are high impedance and fully differential. By using the + and – SENSE outputs from the bridge as the differential reference to the ADC, the reference voltage is proportional to the excitation voltage which is also proportional to the bridge output voltage. There is no loss in measurement accuracy if the actual bridge excitation voltage varies.

- Resolution of 80,000 Counts Peak-to-Peak (16.5-Bits) for ± 10mV Fullscale Range
- Chop Mode for Low Offset and Drift
- Offset Drift: 5nV/°C (Chop Mode Enabled)
- Gain Drift: 2ppm/°C
- Line Frequency Common-Mode Rejection: > 150dB
- Two-Channel Programmable Gain Front End
- On-Chip DAC for Offset/TARE Removal
- FASTStep Mode
- AC Excitation Output Drive
- Internal and System Calibration Options
- Single 5V Supply
- Power Dissipation: 65mW, (125mW for 10mV FS Range)
- 24-Lead SOIC and 24-Lead TSSOP Packages

Figure 8.6: AD7730 Key Specifications

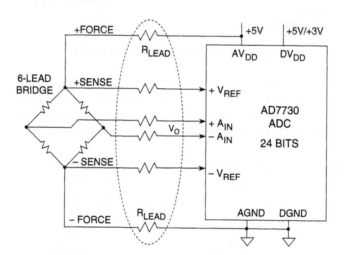

**Figure 8.7: AD7730 Bridge Application Showing
Ratiometric Operation and Kelvin Sensing**

It should be noted that this ratiometric technique can be used in many applications where a sensor output is proportional to its excitation voltage or current, such as a thermistor or RTD.

Weigh Scale Design Analysis Using the AD7730 ADC

We will now proceed with a simple design analysis of a weigh scale based on the AD7730 ADC and a standard load cell. Figure 8.8 shows the overall design objectives for the weigh scale. The key specifications are the fullscale load (2 kg), and the resolution (0.1 g). These specifications primarily determine the basic load cell and ADC requirements.

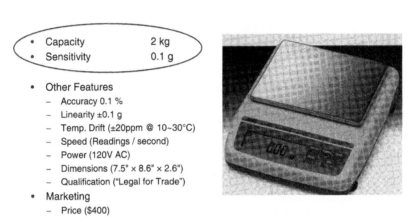

- Capacity   2 kg
- Sensitivity   0.1 g

- Other Features
    - Accuracy 0.1 %
    - Linearity ±0.1 g
    - Temp. Drift (±20ppm @ 10~30°C)
    - Speed (Readings / second)
    - Power (120V AC)
    - Dimensions (7.5" × 8.6" × 2.6")
    - Qualification ("Legal for Trade")
- Marketing
    - Price ($400)

Figure 8.8: Design Example—Weigh Scale

The specifications of a load cell that matches the overall requirements are shown in Figure 8.9. Notice that the load cell is constructed with four individual strain gages connected in a standard bridge configuration. When the load is applied to the beam, R1 and R2 decrease in value, and R3 and R4 increase. This is popularly called the *four-element-varying* bridge configuration and is described in detail in Reference 1, Chapter 2.

The load cell selected has a full-scale load of 2 kg, and an output sensitivity of 2 mV/V. This means that with an excitation voltage of 10 V, the full-scale output voltage is 20 mV. Herein lies the major difficulty in load cell signal conditioning: accurately amplifying and digitizing the low level output signal without

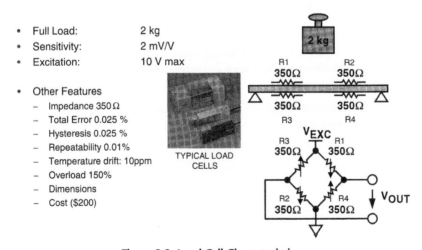

- Full Load: 2 kg
- Sensitivity: 2 mV/V
- Excitation: 10 V max

- Other Features
    - Impedance 350 Ω
    - Total Error 0.025 %
    - Hysteresis 0.025 %
    - Repeatability 0.01%
    - Temperature drift: 10ppm
    - Overload 150%
    - Dimensions
    - Cost ($200)

TYPICAL LOAD CELLS

Figure 8.9: Load Cell Characteristics

corrupting it with noise. The load cell output is analyzed further in Figure 8.10. With the chosen excitation voltage of 5 V, the full-scale bridge output voltage is only 10 mV. Notice that the output is also proportional to (or ratiometric with) the excitation voltage.

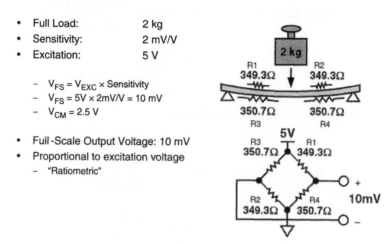

- Full Load: 2 kg
- Sensitivity: 2 mV/V
- Excitation: 5 V

  - $V_{FS} = V_{EXC} \times$ Sensitivity
  - $V_{FS} = 5V \times 2mV/V = 10$ mV
  - $V_{CM} = 2.5$ V

- Full-Scale Output Voltage: 10 mV
- Proportional to excitation voltage
  - "Ratiometric"

Figure 8.10: Determining Full Scale Output of Load Cell with 5 V Excitation

The next step is to determine the resolution requirements of the ADC, and the details are summarized in Figure 8.11. The total number of individual quantization levels (counts) required is equal to the full-scale weight (2 kg) divided by the desired resolution (0.1 g), or 20,000 counts. With a 5 V excitation voltage, the full-scale load cell output voltage is 10 mV for a 2 kg load.

The required noise-free resolution, Vp-p, is therefore given by Vp-p = 10 mV/20,000 = 0.5 μV. This defines the code width, and therefore the peak-to-peak noise must be less than 0.5 μV. The corresponding allowable rms noise is given by V_{RMS} = Vp-p/6.6 = 0.5 μV/6.6 = 0.075 μV rms = 75 nV rms. (The factor 6.6 is used to convert peak-to-peak noise to rms noise, assuming Gaussian noise).

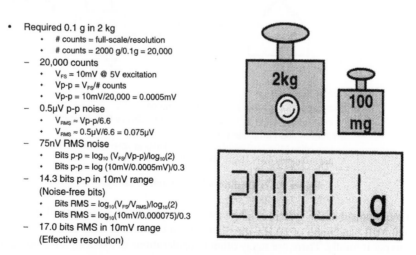

- Required 0.1 g in 2 kg
  - # counts = full-scale/resolution
  - # counts = 2000 g/0.1g = 20,000
  - 20,000 counts
    - V_{FS} = 10mV @ 5V excitation
    - Vp-p = V_{FS}/# counts
    - Vp-p = 10mV/20,000 = 0.0005mV
  - 0.5μV p-p noise
    - $V_{RMS} \approx$ Vp-p/6.6
    - $V_{RMS} \approx$ 0.5μV/6.6 = 0.075μV
  - 75nV RMS noise
    - Bits p-p = $\log_{10}$ (V_{FS}/Vp-p)/$\log_{10}$(2)
    - Bits p-p = log (10mV/0.0005mV)/0.3
  - 14.3 bits p-p in 10mV range
    (Noise-free bits)
    - Bits RMS = $\log_{10}$(V_{FS}/V_{RMS})/$\log_{10}$(2)
    - Bits RMS = $\log_{10}$(10mV/0.000075)/0.3
  - 17.0 bits RMS in 10mV range
    (Effective resolution)

Figure 8.11: Determining Resolution Requirements

The noise-free code resolution of the ADC is calculated as follows:

$$\text{Noise-Free Code Re solution (Bits)} = \frac{\log_{10}\left(\dfrac{V_{FS}}{\text{Vp-p}}\right)}{\log_{10}(2)}$$

$$= \frac{\log_{10}\left(\dfrac{10\text{mV}}{0.5\mu\text{V}}\right)}{\log_{10}(2)} = 14.3 \text{ bits} \qquad \text{Eq. 8.1}$$

The effective resolution of the ADC is calculated as follows:

$$\text{Effective Re solution (Bits)} = \frac{\log_{10}\left(\dfrac{V_{FS}}{\text{Vp-p}/6.6}\right)}{\log_{10}(2)}$$

$$= \frac{\log_{10}\left(\dfrac{10\text{mV}}{0.5\mu\text{V}/6.6}\right)}{0.3} = 17 \text{ bits} \qquad \text{Eq. 8.2}$$

Figure 8.12 shows the traditional sensor conditioning solution to this problem, where an instrumentation amplifier is used to amplify the 10 mV full-scale bridge output signal to 2.5 V, which is compatible with the input of the 14+ bit ADC. This approach requires a low-noise, low drift in amp such as the AD620 precision in amp (Reference 4) which has a 0.1 Hz to 10 Hz peak-to-peak noise of 280 nV, approximately 280 nV ÷ 6.6 = 42 nV rms.

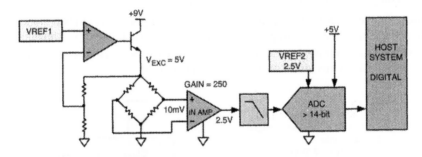

- Complicated design
- Low-pass filter is needed to keep low noise
  - For example, –3dB @ 10Hz, –60dB @ 50Hz (difficult filter design)
- Instrumentation amplifier performance is critical
  - Low noise (AD620: 0.28µV p-p noise in 0.1Hz to 10Hz BW is approximately 42nV RMS), low offset, low gain error

Figure 8.12: Traditional Approach to Design

Another critical requirement of the system is a low-pass filter to remove noise and 50 Hz/60 Hz pickup. Assuming a signal 3 dB bandwidth of 10 Hz, the filter should be down at least 60 dB at 50 Hz—a challenging filter design to put it mildly. There are many other considerations in the design including the stability of the two reference voltages, the VREF1 buffer op amp, etc.

Finally, the ADC presents another serious challenge, requiring 14.3-bit noise-free code performance with a 2.5 V fullscale input signal—implying a 16-bit ADC with no more than approximately 3 LSBs peak-to-peak (0.45 LSBs rms) input-referred noise.

In order to avoid these traditional signal conditioning design problems, the AD7730-based design shown in Figure 8.13 represents a truly elegant solution requiring no instrumentation amplifier, reference, or filter. Note that the bridge interfaces directly with the AD7730 as previously shown in Figure 8.7. The AD7730 input PGA eliminates the need for an external in amp, providing a full-scale input range of 10 mV as a programmable option. Kelvin sensing is used to eliminate errors due to the wiring resistance in the bridge excitation lines. The bridge is driven directly from the 5 V supply, and the sense lines serve as the ADC reference voltage—thereby ensuring fully ratiometric operation as previously described. The need for a complicated filter is also eliminated—simple ceramic capacitor decoupling on each analog and reference input (not shown on the diagram) is sufficient.

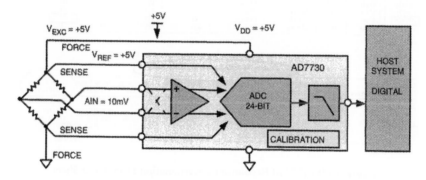

- AD7730 was designed for bridge transducers
  - Chopper, Buffer, PGA, Digital filter, tare DAC, Calibrations, ...
- Fully Ratiometric, changes on $V_{EXC} = V_{REF}$ eliminated
  - Load $\approx V_{OUT}/V_{EXC}$, AD7730 Data $\approx VIN/V_{REF}$, $V_{REF} = V_{EXC}$

Figure 8.13: Design Using AD7730

System performance of the design can be determined by a detailed examination of the AD7730 data sheet, Table I and II, as shown in Figure 8.14. Table I shows the output rms noise in nV as a function of output data rate, digital filter 3 dB frequency, and input range (chopping mode enabled in all cases). An output data rate of 200 Hz yields a filter corner frequency of 7.9 Hz which is reasonable for the application at hand. With an input range of ±10 mV, the output rms noise is 80 nV. This corresponds to a peak-to-peak noise, Vp-p = 80 nV × 6.6 = 528 nV. The number of noise-free counts is obtained as V_{FS}/Vp-p = 10 mV/528 nV = 18,940. The system resolution for a 2-kg load is therefore 2 kg/18,940 = 0.105 g, which is approximately the required specification of 0.1 g.

Table I. Output Noise vs. Input Range and Update Rate (CHP = 1)

Typical Output RMS Noise in nV

Output Data Rate	-3 dB Frequency	SF Word	Settling Time Normal Mode	Settling Time Fast Mode	Input Range = ±80 mV	Input Range = ±40 mV	Input Range = ±20 mV	Input Range = ±10 mV
50 Hz	1.97 Hz	2048	460 ms	60 ms	115	75	55	40
100 Hz	3.95 Hz	1024	230 ms	30 ms	155	105	75	60
150 Hz	5.92 Hz	683	153 ms	20 ms	200	135	95	70
200 Hz*	7.9 Hz	512	115 ms	15 ms	225	145	100	80
400 Hz	15.8 Hz	256	57.5 ms	7.5 ms	335	225	160	110

*Power-On Default

Table II. Peak-to-Peak Resolution vs. Input Range and Update Rate (CHP = 1)

Peak-to-Peak Resolution in Counts (Bits)

Output Data Rate	-3 dB Frequency	SF Word	Settling Time Normal Mode	Settling Time Fast Mode	Input Range = ±80 mV	Input Range = ±40 mV	Input Range = ±20 mV	Input Range = ±10 mV
50 Hz	1.97 Hz	2048	460 ms	60 ms	230k (18)	175k (17.5)	120k (17)	80k (16.5)
100 Hz	3.95 Hz	1024	230 ms	30 ms	170k (17.5)	125k (17)	90k (16.5)	55k (16)
150 Hz	5.92 Hz	683	153 ms	20 ms	130k (17)	100k (16.5)	70k (16)	45k (15.5)
200 Hz*	7.9 Hz	512	115 ms	15 ms	120k (17)	90k (16.5)	65k (16)	40k (15.5)
400 Hz	15.8 Hz	256	57.5 ms	7.5 ms	80k (16.5)	55k (16)	40k (15.5)	30k (15)

*Power-On Default

Figure 8.14: AD7730 Resolution Determination From Data Sheet

Table II can be also used to determine the noise-free code resolution which is 40,000 counts (15.5 noise-free bits) for a ±10 mV input range. This must be divided by a factor of 2 because only one-half the input range is used. Therefore, the actual design will provide approximately 20,000 counts (14.5 noise-free bits), which agrees closely with the previous calculation. The various calculations are summarized in Figure 8.15.

- 80nV RMS noise @ 200Hz
  - VP-P ≈ 6.6 × V_{RMS}
  - VP-P ≈ 6.6 × 80nV = 528nV
  - V_{FS} = 10mV
  - # Counts = V_{FS}/VP-P
  - # Counts = 10mV/0.000528 = 18,940
  - Resolution = full scale/# counts
  - Resolution = 2,000g/18,940 = 0.105g
- 0.105 g Resolution

- 15.5 bits p-p in ±10mV
  (Noise-free bits)
  - V_{FS} = 10mV ~ ½ of 20mV
  - Using only ½ of ADC input range
  - Losing 1 bit
- 14.5 bits p-p in 10mV

- 40,000 counts in ±10mV
  - V_{FS} = 10mV ~ ½ of 20mV
  - Using only ½ of ADC input range
- 20,000 counts in 10mV

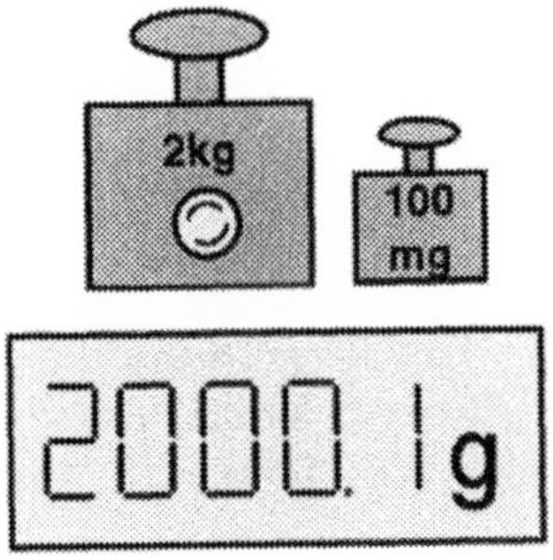

Figure 8.15: AD7730 Resolution @ 200 Hz Data Rate

Note that overall resolution can be increased by dropping back to lower output data rates with correspondingly lower digital filter corner frequencies.

Evaluation of the design is simplified with the AD7730 evaluation board and software as shown in Figure 8.16. The evaluation board can be connected directly to the load cell and the PC. The software allows the various AD7730 options to be varied to evaluate different combinations of data rates, filter frequencies, input ranges, chopping options, etc. Other ADCs in the AD77xx family have similar evaluation boards and software.

A summary of the final weigh scale design and specifications is shown in Figure 8.17.

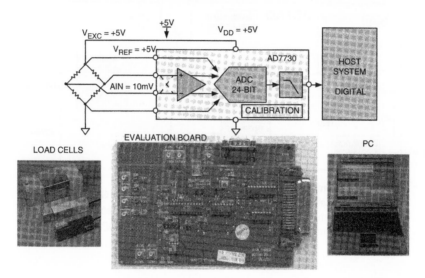

Figure 8.16: Evaluation of Design Using Evaluation Board and Software

	Required	Sensor Load Cell	Circuit AD7730	System Weigh Scale
Capacity	2 kg	2 kg	AIN Range ±10mV	2 kg
Sensitivity	0.1g	2mV / V	Noise 80nV RMS	0.105g

OUTPUT DATA
RATE = 200Hz

Figure 8.17: Final System Performance

Thermocouple Conditioning Using the AD7793

Thermocouples provide accurate temperature measurements over an extremely wide range; however, their relatively small output voltage makes the signal conditioning circuit design difficult. For instance, a Type K thermocouple has a nominal temperature coefficient of 39 µV/°C, so a temperature change of 1000°C produces only a 39 mV output voltage. The thermocouple does not measure temperature directly—its output voltage is proportional to the temperature difference between the actual measuring junction and the "cold" junction where the thermocouple wires are connected to the measuring electronics. (Details of thermocouple operation are described in Reference 1).

Accurate thermocouple measurements therefore require that the temperature of the "cold" junction be measured in some manner to compensate for changes in ambient temperature.

The AD7793 dual channel 24-bit Σ-Δ is ideally suited for direct thermocouple measurements, and a simplified block diagram is shown in Figure 8.18 (Reference 5).

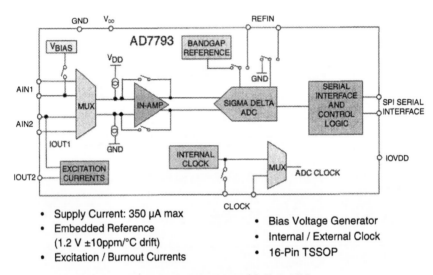

- Supply Current: 350 µA max
- Embedded Reference
  (1.2 V ±10ppm/°C drift)
- Excitation / Burnout Currents

- Bias Voltage Generator
- Internal / External Clock
- 16-Pin TSSOP

Figure 8.18: AD7793 24-bit Σ-Δ ADC

The AD7793 has two differential inputs, an on-chip in amp, reference voltage, bias voltage generator, and burnout/excitation current sources. Single-supply (5 V) power supply current is 350 µA maximum.

A complete solution to a thermocouple measurement design is shown in Figure 8.19. Notice that a thermistor is used to measure the temperature of the "cold" junction via AIN2, and the thermocouple is connected directly to the AIN1 differential input. Note that the internal V_{BIAS} voltage is used to establish the thermocouple common-mode voltage. The R/C filters minimize noise pickup from the remote thermocouple leads, and typical values of 100 Ω and 0.1 µF are reasonable choices.

The AD7793 is first programmed to measure the AIN1 thermocouple voltage using the internal 1.2 V bandgap voltage as a reference. This value is sent to a microcontroller connected to the serial interface. The voltage across the thermistor is established by the IOUT1 excitation current which also flows through a reference resistor, R_{REF}. The voltage developed across R_{REF} drives the auxillary reference input, REFIN. The AD7793 is programmed to use the REFIN reference when measuring the thermistor voltage at AIN2. The thermistor voltage is then sent to the microcontroller which performs the required calculations, including the correction for the temperature of the cold junction, T2. The thermistor is therefore connected in a ratiometric fashion such that variations in IOUT1 do not affect the accuracy of the thermistor measurement. Note that the powerful ratiometric technique will work with any resistive-based sensor including thermistors, bridges, strain gages, and RTDs.

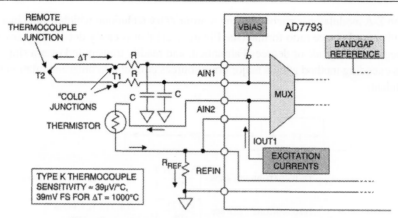

- Bias voltage generator used to generate a common mode voltage for AIN1

- Current source provides current to thermistor for cold junction compensation and ratiometric operation using REFIN

Figure 8.19: Thermocouple Design with Cold Junction Compensation using the AD7793

Direct Digital Temperature Measurements

Temperature sensors with digital outputs have a number of advantages over those with analog outputs, especially in remote applications. Opto-isolators can also be used to provide galvanic isolation between the remote sensor and the measurement system. Although a voltage-to-frequency converter driven by a voltage output temperature sensor accomplishes this function, more sophisticated and more efficient ICs are now available which offer several performance advantages.

The TMP05/TMP06 digital output sensor family includes a voltage reference, V_{PTAT} generator, Σ-Δ ADC, and a clock source (see Figure 8.20). The sensor output is digitized by a first-order Σ-Δ modulator. This converter utilizes time-domain oversampling and a high accuracy comparator to deliver 12 bits of effective accuracy in an extremely compact circuit.

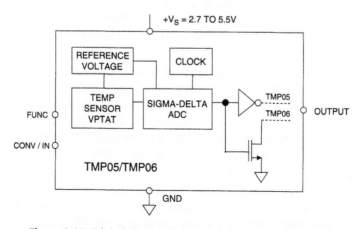

Figure 8.20: Digital Output Temperature Sensors: TMP05/06

The output of the Σ-Δ modulator is encoded using a proprietary technique which results in a serial digital output signal with a mark-space ratio format (see Figure 8.21) that is easily decoded by any microprocessor into either degrees centigrade or degrees Fahrenheit, and readily transmitted over a single wire. Most importantly, this encoding method avoids major error sources common to other modulation techniques, as it is clock-independent.

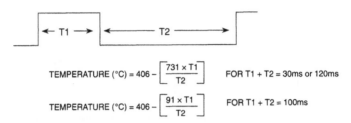

$$\text{TEMPERATURE (°C)} = 406 - \left\lfloor \frac{731 \times T1}{T2} \right\rfloor \quad \text{FOR T1 + T2 = 30ms or 120ms}$$

$$\text{TEMPERATURE (°C)} = 406 - \left\lfloor \frac{91 \times T1}{T2} \right\rfloor \quad \text{FOR T1 + T2 = 100ms}$$

- ±0.5°C Accuracy from 0°C to 70°C
- 0.025°C Resolution
- T1 + T2 = 120ms, 100ms, or 30ms (depending on status of CONV/IN pin
- Specified –40°C to +150°C
- 2.7V to 5.5V supply
- 759µW Power Consumption @ 3.3V, Continuous Mode
- 70µW Power Consumption @ 3.3V, One-Shot Mode (1Hz rate)
- 5-Pin SC-70 or SOT-23 Packages

Figure 8.21: TMP05/TMP06 Output Format

The TMP05/TMP06 output is a stream of digital pulses, and the temperature information is contained in the mark-space ratio per the equations shown in Figure 8.21. The TMP05/TMP06 has three modes of operation. These are *continuously converting*, *daisy chain*, and *one shot*. A three-state FUNC input selects one of the three possible modes. In the one shot mode, the power consumption is reduced to 70 µW at one sample per second.

The CONV/IN input is used to determine the rate with which the TMP05/TMP06 measures temperature in the *continuously converting* and *one shot* mode. In the daisy chain mode, the CONV/IN pin operates as the input to the daisy chain. The daisy chain mode allows multiple TMP05/TMP06s to be connected together and thus allow one input line of the microcontroller to be the sole receiver of all temperature measurements (see Reference 6 for further details).

Popular microcontrollers, such as the 80C51 and 68HC11, have on-chip timers which can easily decode the mark-space ratio of the TMP05/TMP06. A typical interface to the 80C51 is shown in Figure 8.22. Two timers, labeled *Timer 0* and *Timer 1* are 16 bits in length. The 80C51's system clock, divided by twelve, provides the source for the timers. The system clock is normally derived from a crystal oscillator, so timing measurements are quite accurate. Since the sensor's output is ratiometric, the actual clock frequency is not important. This feature is important because the microcontroller's clock frequency is often defined by some external timing constraint, such as the serial baud rate.

Software for the sensor interface is straightforward. The microcontroller simply monitors I/O port P1.0, and starts *Timer 0* on the rising edge of the sensor output. The microcontroller continues to monitor P1.0, stopping *Timer 0* and starting *Timer 1* when the sensor output goes low. When the output returns high, the

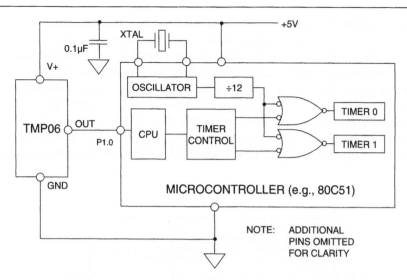

Figure 8.22: Interfacing TMP06 to a Microcontroller

sensor's T1 and T2 times are contained in registers *Timer 0* and *Timer 1*, respectively. Further software routines can then apply the conversion factor shown in the equations above and calculate the temperature.

The TMP05/TMP06 are ideal for monitoring the thermal environment within electronic equipment. For example, the surface mounted package will accurately reflect the thermal conditions that affect nearby integrated circuits.

The TMP05 and TMP06 measure and convert the temperature at the surface of their own semiconductor chip. When they are used to measure the temperature of a nearby heat source, the thermal impedance between the heat source and the sensor must be considered. Often, a thermocouple or other temperature sensor is used to measure the temperature of the source, while the TMP05/TMP06 temperature is monitored by measuring the T1 and T2 pulse widths with a microcontroller. Once the thermal impedance is determined, the temperature of the heat source can be inferred from the TMP05/TMP06 output.

Carrying the integration a step further, we will now look at true temperature-to-digital converters. The basic bandgap reference (see complete discussion in Chapter 6 of this book) has been a building block for ADCs and DACs for many years, and most converters have them integrated on-chip. Inside the bandgap reference circuit, there is invariably a voltage or current that is proportional to absolute temperature (PTAT). There is no fundamental reason why this voltage or current cannot be used to sense the temperature of the IC substrate within the ADC. There is also no fundamental reason why the ADC cannot convert this voltage into a digital output word which represents the chip temperature. In the early days of IC data converters, internal power dissipation was considerable, so an internal temperature sensor would measure a temperature greater than the ambient temperature. Modern low voltage, low power ICs make it quite practical to use such a concept to produce a true temperature-to-digital converter which accurately reflects the ambient or PC board temperature.

This concept has expanded to an entire family of temperature-to-digital converters as well as ADCs with multiplexed inputs, where one input is the on-chip temperature sensor. This is a powerful feature, since modern microprocessor, DSP, and FPGA chips tend to dissipate lots of power, and most require a certain amount of airflow. A simple means of monitoring the PC board temperature is valuable in protecting these critical circuits against damage from excessive temperatures due to fault conditions.

The ADT7301 is a 13-bit digital temperature sensor with a 14th bit as a sign bit (Reference 7). The part contains an on-chip bandgap reference, temperature sensor, a 13-bit ADC, and serial interface logic functions in SOT-23 and MSOP packages. The ADC section consists of a conventional successive-approximation converter based on a switched capacitor DAC architecture. The parts are capable of running on a 2.7 V to 5.5 V power supply. The on-chip temperature sensor allows an accurate measurement of the ambient device temperature to be made. The specified measurement range of the ADT7301 is –40°C to +150°C. It is not recommended to operate the device at temperatures above 125°C for greater than a total of 5% of the projected lifetime of the device. Any exposure beyond this limit will affect device reliability. A simplified block diagram of the ADT7301 is given in Figure 8.23, and key specifications are summarized in Figure 8.24.

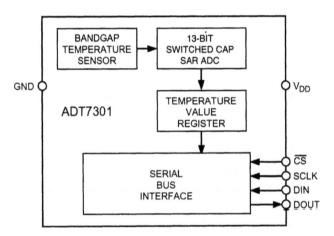

Figure 8.23: ADT7301 13-Bit, ±0.5°C Accurate, Micropower Digital Temperature Sensor

- 13-Bit Temperature-to-Digital Conversion
- –40°C to +150°C Operating Temperature Range
- ±0.5°C Accuracy
- 0.03125°C Temperature Resolution
- 2.7V to 5.5V Supply
- 4.88µW Power Dissipation, for 1 sample/second Conversion Rate
- Serial Interface
- 6-Lead SOT-23 or 8-Lead SOIC Package

Figure 8.24: ADT7301 Key Specifications

The ADT7301 can be used for surface or air-temperature sensing applications. If the device is cemented to a surface with thermally conductive adhesive, the die temperature will be within about 0.1°C of the surface temperature, thanks to the device's low power consumption. Care should be taken to insulate the back and leads of the device from airflow, if the ambient air temperature is different from the surface temperature being measured. The ground pin provides the best thermal path to the die, so the temperature of the die will be close to that of the printed circuit ground track. Care should be taken to ensure that this is in good thermal contact with the surface being measured.

As with any IC, the ADT7301 and its associated wiring and circuits must be kept free from moisture to prevent leakage and corrosion, particularly in cold conditions where condensation is more likely to occur. Water-resistant varnishes and conformal coatings can be used for protection. The small size of the ADT7301 package allows it to be mounted inside sealed metal probes, which provide a safe environment for the device.

Microprocessor Substrate Temperature Sensors

Today's computers require that hardware as well as software operate properly, in spite of the many things that can cause a system crash or lockup. The purpose of hardware monitoring is to monitor the critical items in a computing system and take corrective action should problems occur.

Microprocessor supply voltage and temperature are two critical parameters. If the supply voltage drops below a specified minimum level, further operations should be halted until the voltage returns to acceptable levels. In some cases, it is desirable to reset the microprocessor under "brownout" conditions. It is also common practice to reset the microprocessor on power-up or power-down. Switching to a battery backup may be required if the supply voltage is low.

Under low voltage conditions it is mandatory to inhibit the microprocessor from writing to external CMOS memory by inhibiting the Chip Enable signal to the external memory.

Many microprocessors can be programmed to periodically output a "watchdog" signal. Monitoring this signal gives an indication that the processor and its software are functioning properly and that the processor is not stuck in an endless loop.

The need for hardware monitoring has resulted in a number of ICs, traditionally called "microprocessor supervisory products," that perform some or all of the above functions. These devices range from simple manual reset generators (with debouncing) to complete microcontroller-based monitoring subsystems with on-chip temperature sensors and ADCs. Analog Devices' ADM-family of products is specifically to perform the various microprocessor supervisory functions required in different systems.

CPU temperature is critically important in the Pentium® microprocessors. For this reason, all new Pentium devices have an on-chip substrate PNP transistor that is designed to monitor the actual chip temperature. The collector of the substrate PNP is connected to the substrate, and the base and emitter are brought out on two separate pins of the Pentium.

The ADM1023 Microprocessor Temperature Monitor is specifically designed to process these outputs and convert the voltage into a digital word representing the chip temperature. It is optimized for use with the PentiumIII microprocessor. The simplified analog signal processing portion of the ADM1023 is shown in Figure 8.25.

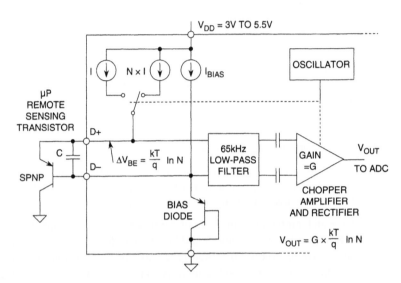

Figure 8.25: ADM1023 Microprocessor Temperature Monitor Input Conditioning Circuits

The technique used to measure the temperature is identical to the "ΔV_{BE}" principle previously discussed in Chapter 7 of this book. Two different currents (I and N·I) are applied to the sensing transistor, and the voltage measured for each. The change in the base-emitter voltage, ΔV_{BE}, is a PTAT voltage and given by the equation:

$$\Delta V_{BE} = \frac{kT}{q} \ln(N) \qquad \text{Eq. 8.3}$$

Figure 8.25 shows the external sensor as a substrate PNP transistor, provided for temperature monitoring in the microprocessor, but it could equally well be a discrete transistor such as a 2N3904 or 2N3906. If a discrete transistor is used, the collector should be connected to the base and not grounded. To prevent ground noise interfering with the measurement, the more negative terminal of the sensor is not referenced to ground, but is biased above ground by an internal diode. If the sensor is operating in a noisy environment, C may be optionally added as a noise filter. Its value is typically 2200 pF, but should be no more than 3000 pF.

To measure ΔV_{BE}, the sensing transistor is switched between operating currents of I and N·I. The resulting waveform is passed through a 65 kHz low-pass filter to remove noise, then to a chopper-stabilized amplifier which performs the function of amplification and synchronous rectification. The resulting dc voltage is proportional to ΔV_{BE} and is digitized by the ADC and stored as an 11-bit word. To further reduce the effects of noise, digital filtering is performed by averaging the results of 16 measurement cycles.

In addition, the ADM1023 contains an on-chip temperature sensor, and its signal conditioning and measurement is performed in the same manner.

One LSB of the ADM1023 corresponds to 0.125°C, and the ADC can theoretically measure from 0°C to 127.875°C. The results of the local and remote temperature measurements are stored in the local and remote temperature value registers, and are compared with limits programmed into the local and remote high and low limit registers as shown in Figure 8.26. An ALERT output signals when the on-chip or remote temperature is out of range. This output can be used as an interrupt, or as an SMBus alert.

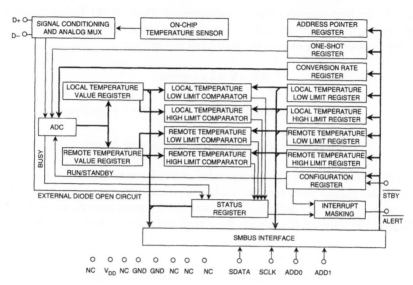

Figure 8.26: ADM1023 Simplified Block Diagram

The limit registers can be programmed, and the device controlled and configured, via the serial System Management Bus (SMBus). The contents of any register can also be read back by the SMBus. Control and configuration functions consist of: switching the device between normal operation and standby mode, masking or enabling the ALERT output, and selecting the conversion rate which can be set from 0.0625 Hz to 8 Hz. Key specifications for the ADM1023 are given in Figure 8.27.

- On-Chip and Remote Microprocessor Sensing
- Offset Registers for System Calibration
- 1°C Accuracy and Resolution on Local Channel
- 0.125°C Resolution/1°C Accuracy on Remote Channel
- Programmable Over/Under Temperature Limits
- Programmable Conversion Rate
- Supports System Management Bus (SMBus) Alert
- 2-Wire SMBus Serial Interface
- 200µA Max. Operating Current (0.25 Conversions/Second)
- 1µA Standby Current
- 3V to 5.5V Supply
- 16-Lead QSOP Package

Figure 8.27: ADM1023 Key Specifications

Applications of ADCs in Power Meters

While electromechanical energy meters have been popular for over 50 years, a solid-state energy meter delivers far more accuracy and flexibility. Just as important, a well designed solid-state meter will have a longer useful life. The ADE775x energy metering ICs are a family products designed to implement this type of meter (References 9, 10, 11).

We must first consider the fundamentals of power measurement (Figure 8.28). Instantaneous ac voltage is given by the expression $v(t) = V \times cos(\omega t)$, and the current (assuming it is in phase with the voltage) by $i(t) = I \times cos(\omega t)$. The *instantaneous power* is the product of $v(t)$ and $i(t)$:

$$p(t) = V \times I \times cos^2(\omega t) \qquad \text{Eq. 8.4}$$

$$\text{Using the trigonometric identity, } 2cos^2(\omega t) = 1 + cos(2\omega t), \qquad \text{Eq. 8.5}$$

$$p(t) = \frac{V \times I}{2}\left[1 + cos\left(2\omega t\right)\right] = \text{Instantaneous Power.} \qquad \text{Eq. 8.6}$$

The *instantaneous real power* is simply the average value of $p(t)$. It can be shown that computing the instantaneous real power in this manner gives accurate results even if the current is not in phase with the voltage (i.e., the power factor is not unity. By definition, the power factor is equal to $cos\theta$, where θ is the phase angle between the voltage and the current). It also gives the correct real power if the waveforms are nonsinusoidal.

- $v(t) = V \times cos(\omega t)$ (Instantaneous Voltage)
- $i(t) = I \times cos(\omega t)$ (Instantaneous Current)
- $p(t) = V \times I \, cos^2(\omega t)$ (Instantaneous Power)
- $p(t) = \dfrac{V \times I}{2}\left[1 + cos(2\omega t)\right]$

Average Value of $p(t)$ = Instantaneous Real Power

Includes Effects of Power Factor and Waveform Distortion

Figure 8.28: Basics of Power Measurements

The ADE7755 implements these calculations, and a block diagram is shown in Figure 8.29. The two ADCs digitize the voltage signals from the current and voltage transducers. These ADCs are 16-bit second-order $\Sigma\text{-}\Delta$ with an input sampling rate of 900 kSPS. This analog input structure greatly simplifies transducer interfacing by providing a wide dynamic range for direct connection to the transducer and also by simplifying the antialiasing filter design. A programmable gain stage in the current channel further facilitates easy transducer interfacing. A high-pass filter in the current channel removes any dc component from the current signal. This eliminates any inaccuracies in the real power calculation due to offsets in the voltage or current signals.

The real power calculation is derived from the instantaneous power signal. The instantaneous power signal is generated by a direct multiplication of the current and voltage signals. In order to extract the real power component (i.e., the dc component), the instantaneous power signal is low-pass filtered. Figure 8.29 illustrates the instantaneous real power signal and shows how the real power information can be extracted by low-pass filtering the instantaneous power signal. This method correctly calculates real power for

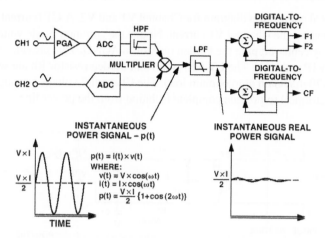

Figure 8.29: ADE7755 Energy Metering IC Signal Processing

nonsinusoidal current and voltage waveforms at all power factors. All signal processing is carried out in the digital domain for superior stability over temperature and time.

The low frequency output of the ADE7755 is generated by accumulating this real power information (see Figure 8.30). This low frequency inherently means a long accumulation time between output pulses. The output frequency is therefore proportional to the average real power. This average real power information can, in turn, be accumulated (e.g., by a counter) to generate real energy information. Because of its high output frequency and shorter integration time, the CF output is proportional to the instantaneous real power. This is useful for system calibration purposes that would take place under steady load conditions.

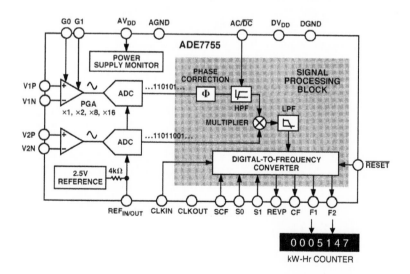

Figure 8.30: ADE7755 Energy Metering IC with Pulse Output

Figure 8.31 shows a typical connection diagram for Channel V1 and V2. A CT (current transformer) is the transducer selected for sensing the Channel V1 current. Notice the common-mode voltage for Channel 1 is AGND and is derived by center tapping the burden resistor to AGND. This provides the complementary analog input signals for V1P and V1N. The CT turns ratio and burden resistor Rb are selected to give a peak differential voltage of ±470 mV/Gain at maximum load. The Channel 2 voltage sensing is accomplished with a PT (potential transformer) to provide complete isolation from the power line.

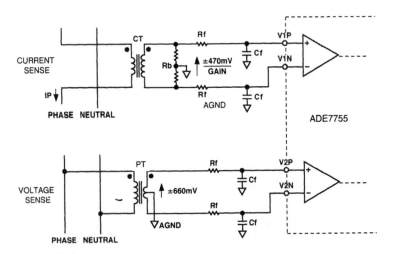

Figure 8.31: Typical Connections for Channel 1 (Current Sense) and Channel 2 (Voltage Sense)

References:
8.1 Precision Measurement And Sensor Conditioning

1. Walt Kester, **Practical Design Techniques for Sensor Signal Conditioning**, Analog Devices, 1999, Chapter 8. Available for download at www.analog.com.

2. Data sheet for AD7730/AD7730L Bridge Transducer ADC, www.analog.com.

3. Walter G. Jung, **Op Amp Applications**, Analog Devices, 2002, ISBN 0-916550-26-5, Chapter 4.

4. Data sheet for AD620 Precision Instrumentation Amplifier, www.analog.com.

5. Data sheet for AD7793 24-bit Dual Sigma-Delta ADC, www.analog.com.

6. Data sheet for TMP05/TMP06 ±0.5°C Accurate PWM Temperature Sensor in 5-Lead SC-70, www.analog.com.

7. Data sheet for ADT7301 13-bit, ±0.5°C Accurate, MicroPower Digital Temperature Sensor, www.analog.com.

8. Data sheet for ADM1023 ACPI Compliant High Accuracy Microprocessor System Temperature Monitor, www.analog.com.

9. Data sheet for ADE7755 Energy Metering IC with Pulse Output, www.analog.com.

10. Paul Daigle, "All-Electronic Power and Energy Meters," **Analog Dialogue**, Volume 33, Number 2, February, 1999, www.analog.com.

11. John Markow, "Microcontroller-Based Energy Metering using the AD7755," **Analog Dialogue**, Volume 33, Number 9, October, 1999, www.analog.com.

Multichannel Data Acquisition Systems
Walt Kester

Data Acquisition System Configurations

There are many applications for data acquisition systems in measurement and process control. All data acquisition applications involve digitizing analog signals for analysis using ADCs. In a measurement application, the ADC is followed by a digital processor that performs the required data analysis. In a process control application, the process controller generates feedback signals that typically must be converted back into analog form using a DAC.

Although a single ADC digitizing a single channel of analog data constitutes a data acquisition system, the term *data acquisition* generally refers to multichannel systems. If there is feedback from the digital processor, DACs may be required to convert the digital responses into analog. This process is often referred to as *data distribution*.

Figure 8.32A shows a data acquisition/distribution process control system where each channel has its own dedicated ADC and DAC. An alternative configuration is shown in Figure 8.32B, where analog multiplexers and demultiplexers are used with a single ADC and DAC. In most cases, especially where there are many channels, this second configuration provides an economical alternative.

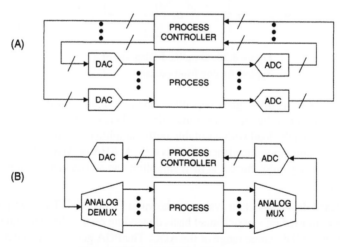

Figure 8.32: Two Approaches to a Multichannel Data Acquisition System

There are many trade-offs involved in designing a data acquisition system. Issues such as filtering, amplification, multiplexing, demultiplexing, sampling frequency, and partitioning must be resolved.

Multiplexing

Multiplexing is a fundamental part of a data acquisition system as shown in Figure 8.32. Multiplexers and switches are examined in more detail in Reference 1, but a fundamental understanding is required to design a data acquisition system—even if the multiplexer is on the same chip as the ADC, which is often the case today.

A simplified diagram of an analog multiplexer is shown in Figure 8.33. The number of input channels typically ranges from 2 to 32, and the devices are generally fabricated on CMOS processes. Most multiplexers have internal channel-address decoding logic and registers, but in a few, these functions must be performed externally. Unused multiplexer inputs *must* be grounded or severe loss of system accuracy may result. The key specifications are *switching time, on-resistance, on-resistance modulation*, and *off-channel isolation (crosstalk)*. For a detailed discussion of the details of analog multiplexers, refer to Chapter 7 of this book.

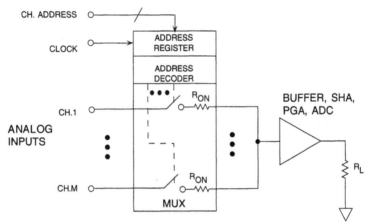

**Figure 8.33: Simplified Diagram of a
Typical Analog Multiplexer**

Multiplexer on-resistance is generally slightly dependent on the signal level (often called R_{on} modulation). This will cause signal distortion if the multiplexer must drive a load resistance, therefore the multiplexer output should be isolated from the load with a suitable buffer amplifier. A separate buffer is not required if the multiplexer drives a high input impedance, such as a PGA, SHA or ADC—but beware, some SHAs and ADCs draw high frequency pulse current at their sampling rate and cannot tolerate being driven by an unbuffered multiplexer.

An M-channel multiplexed data acquisition system is shown in Figure 8.34. The multiplexer output drives a PGA whose gain can be adjusted on a per-channel basis depending on the channel signal level. This ensures that all channels utilize the full dynamic range of the ADC. The PGA gain is changed at the same time as the multiplexer is switched to a new channel. The ADC *Convert Command* is applied after the multiplexer and the PGA have settled to the required accuracy (1 LSB). The maximum sampling frequency (when switching between channels) is limited by the multiplexer switching time t_{mux}, the PGA settling time t_{pga}, and the ADC conversion time t_{conv} as shown in the formula.

In a multiplexed system it is possible to have a positive full-scale signal on one channel and a negative full-scale signal on the other. When the multiplexer switches between these channels its output is a full-scale step voltage. All elements in the signal path must settle to the required accuracy (1 LSB) before the conversion is started. The effect of inadequate settling is dc crosstalk between channels.

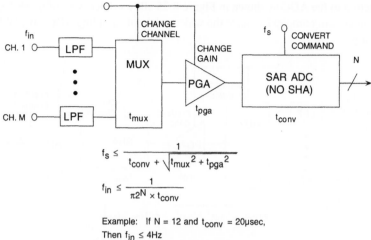

$$f_s \leq \frac{1}{t_{conv} + \sqrt{t_{mux}^2 + t_{pga}^2}}$$

$$f_{in} \leq \frac{1}{\pi 2^N \times t_{conv}}$$

Example: If N = 12 and t_{conv} = 20μsec,
Then $f_{in} \leq$ 4Hz

**Figure 8.34: Multiplexed Data Acquisition
System with PGA and SAR ADC**

The SAR ADC shown in this application has no internal SHA (similar to the industry-standard AD574 series), and therefore the input signal must be held constant (within 1 LSB) during the conversion time in order to prevent encoding errors. This defines the maximum rate-of-change of the input signal:

$$\left. \frac{dv}{dt} \right|_{max} \leq \frac{1LSB}{t_{conv}}$$

Eq. 8.7

The amplitude of a full-scale sinewave input signal is equal to $2^N/2$, or $2^{(N-1)}$, and its maximum rate-of-change is

$$\left. \frac{dv}{dt} \right|_{max} = 2\pi f_{max} \times 2^{N-1} = \pi f_{max} \times 2^N$$

Eq. 8.8

Setting the two equations equal, and solving for f_{max},

$$f_{max} \leq \frac{1}{\pi \times 2^N t_{conv}}.$$

Eq. 8.9

For example, if the ADC conversion time is 20 μsec (corresponding to a maximum sampling rate of slightly less than 50 kSPS, because of overhead), and the resolution is 12 bits, the maximum channel input signal frequency is limited to 4 Hz. This may be adequate if the signals are dc, but the lack of a SHA function severely limits the ability to process dynamic signals.

Adding a SHA function to the ADC as shown in Figure 8.35 allows processing of much faster signals with almost no increase in system complexity, since the wide variety of sampling ADCs available today have the SHA function on-chip.

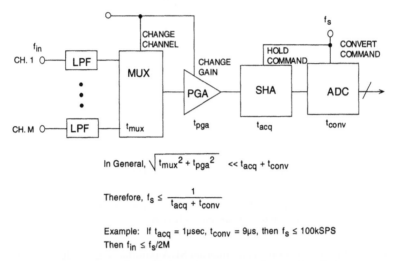

In General, $\sqrt{t_{mux}^2 + t_{pga}^2} \ll t_{acq} + t_{conv}$

Therefore, $f_s \leq \dfrac{1}{t_{acq} + t_{conv}}$

Example: If $t_{acq} = 1\mu sec$, $t_{conv} = 9\mu s$, then $f_s \leq 100kSPS$
Then $f_{in} \leq f_s/2M$

**Figure 8.35: The Addition of a SHA Function to the ADC
Allows Processing of Dynamic Input Signals**

The timing is adjusted such that the multiplexer and the PGA are switched immediately following the acquisition time of the SHA as shown in Figure 8.36. If the combined multiplexer and PGA settling time is less than the ADC conversion time, then the maximum sampling frequency of the system is given by:

$$f_s \leq \frac{1}{t_{acq} + t_{conv}}$$

Eq. 8.10

The per-channel sampling rate is obtained by dividing the ADC sampling rate given in Eq. 8.10 by M.

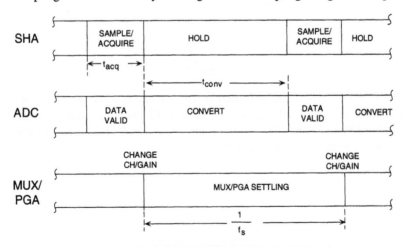

**Figure 8.36: Typical Timing Diagram for Multiplexed
Data Acquisition System Using a SHA**

Filtering Considerations in Data Acquisition Systems

Filtering in data acquisition systems not only prevents aliasing of unwanted signals but also reduces noise by limiting bandwidth. In a multiplexed system, there are basically two places to put filters: in each channel, and at the multiplexer output (see Figure 8.37).

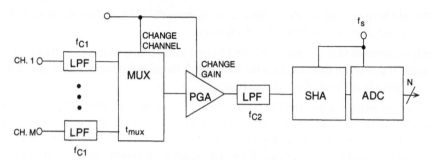

For Sequential Sampling, $f_{C1} < \dfrac{f_s}{2M}$

Figure 8.37: Filtering in a Data Acquisition System

The filter at the input of each channel is used to prevent aliasing of signals that fall outside the Nyquist bandwidth. The per-channel sampling rate (assuming each channel is sampled at the same rate) is f_s/M, and the corresponding Nyquist frequency is $f_s/2M$. The filter should provide sufficient attenuation at $f_s/2M$ to prevent dynamic range limitations due to aliasing.

A second filter can be placed in the signal path between the multiplexer output and the ADC, usually between the PGA and the SHA. The cutoff frequency of this filter must be carefully chosen because of its impact on settling time. In a multiplexed system such as shown in Figure 8.37, there can be a full-scale step voltage change at the multiplexer output when it is switched between channels. This occurs if the signal on one channel is positive full-scale, and the signal on the adjacent channel is negative full-scale. From the timing diagram shown in Figure 8.36, the signal from the filter has essentially the entire conversion period $(1/f_s)$ to settle from the step voltage. The signal should settle to within 1 LSB of the final value in order not to introduce a significant error. The settling time requirement therefore places a lower limit on the filter's cutoff frequency. The single-pole filter settling time required to maintain a given accuracy is shown in Figure 8.38. The settling time requirement is expressed in terms of the filter time constant and also the ratio of the filter cutoff frequency, f_{c2}, to the ADC sampling frequency, f_s.

RESOLUTION, # OF BITS	LSB (%FS)	# OF TIME CONSTANTS	f_{c2}/f_s
6	1.563	4.16	0.67
8	0.391	5.55	0.89
10	0.0977	6.93	1.11
12	0.0244	8.32	1.32
14	0.0061	9.70	1.55
16	0.00153	11.09	1.77
18	0.00038	12.48	2.00
20	0.000095	13.86	2.22
22	0.000024	15.25	2.44

Figure 8.38: Single-Pole Filter Settling Time to Required Accuracy

As an example, assume that the ADC is 12-bits and sampling at 100 kSPS. From the table in Figure 8.38, 8.32 time constants are required for the filter to settle to 12-bit accuracy, and

$$\frac{f_{c2}}{f_s} \geq 1.32, \text{ or } f_{c2} \geq 132 \text{ kSPS} \qquad\qquad \text{Eq. 8.11}$$

While this filter will help prevent wideband noise from entering the SHA, *it does not provide the same function as the antialiasing filters at the input of each channel.*

The above analysis assumes that the multiplexer/PGA combined settling time is significantly less than the filter settling time. If this is not the case, then the filter cutoff frequency must be larger, and in most cases it should be left out entirely in favor of per-channel filters.

We have discussed the importance of the full scale settling time of the multiplexer/PGA/filter combination, but what is equally important is the ability of the ADC to acquire the final value of the step voltage input signal to the required accuracy. Failure of any link in the signal chain to settle will result in dc crosstalk between adjacent channels and loss of accuracy. If the data acquisition system uses a separate SHA and ADC, the key specification to examine is the SHA *acquisition time*, which is usually specified as the amount of time required to acquire a full scale input signal to 0.1% accuracy (10 bits) or 0.01% accuracy (13 bits). In most cases, both 0.1% and 0.01% times are specified. If the SHA acquisition time is not specified for 0.01% accuracy or better, it should not be used in a 12-bit multiplexed application.

If the ADC is a sampling type (with internal SHA), the SHA acquisition time required to achieve a level of accuracy may or may not be specified. Because SHA acquisition time and accuracy are not directly specified for some sampling ADCs, the *transient response* specification should be examined. The transient response of the ADC (settling time to within 1 LSB for a full-scale step input) must be less the $1/f_s$, where f_s is the ADC sampling rate. This often ignored specification may become the weakest link in the signal chain. In some cases neither the SHA acquisition time to specified accuracy nor the transient response specification appear on the data sheet for the particular ADC, in which case it is probably not acceptable for multiplexed applications. Because of the difficulty in measuring and achieving better than 12-bit settling times using discrete components, the accuracy of most multiplexed data acquisition systems made up of discrete components is limited to 14 bits at best. Designing multiplexed systems with greater accuracy is extremely difficult, and using a single ADC per channel should be strongly considered at higher resolutions. The modern alternative, of course, is to use an ADC with an on-chip multiplexer where the overall performance of the combination is specified.

Complete Data Acquisition Systems on a Chip

VLSI mixed-signal CMOS processing allows the integration of large and complex data acquisition circuits on a single chip. Most signal conditioning circuits including multiplexers, PGAs, and SHAs, are now integrated onto the same chip as the ADC. This high level of integration permits data acquisition systems to be specified and tested as a single complex function.

Such functionality relieves the designer of most of the burden of testing and calculating individual component error budgets. The dc and ac characteristics of a complete data acquisition system are specified as a complete function, which removes the necessity of calculating performance from a collection of individual worst case device specifications. A complete monolithic system should achieve a higher performance at much lower cost than would be possible with a system built up from discrete functions. Furthermore, system calibration is easier and in fact many monolithic systems are self calibrating.

With these high levels of integration, it is both easy and inexpensive to make many of the parameters of the device programmable. Parameters that can be programmed include gain, filter cutoff frequency, and even

ADC resolution and conversion time, as well as the obvious digital/MUX functions of input channel selection, output data format, and range selection.

The *data acquisition system on a chip* concept has led to the proliferation of so many ICs, that it would be impossible to discuss all of them in detail. We will, however, discuss a few of the newer devices that are representative of the entire family. All are completely specified in terms of both dc and ac performance, and many come in 8-, 10-, and 12-bit versions.

The AD7908/AD7918/AD7928 are, respectively, 8-bit, 10-bit, and 12-bit, high speed, low power, 8-channel, successive approximation ADCs. The parts operate from a single 2.7 V to 5.25 V power supply and feature throughput rates up to 1 MSPS. The parts contain a low noise, wide bandwidth track-and-hold amplifier that can handle input frequencies in excess of 8 MHz. A block diagram of the AD7908/AD7918/AD7928 is shown in Figure 8.39.

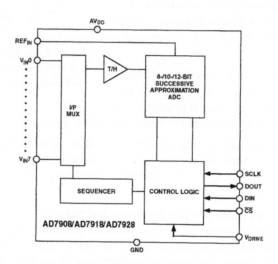

Figure 8.39: AD7908/AD7918/AD7928 8-Channel,
1 MSPS 8-/10-/12-Bit SAR ADCs with Channel Sequencer

The conversion process and data acquisition are controlled using $\overline{\text{CS}}$ (convert start) and the serial clock signal (nominally SCLK = 20 MHz), allowing the device to easily interface with microprocessors or DSPs via a serial interface. The input signal is sampled on the falling edge of $\overline{\text{CS}}$, and conversion is also initiated at this point. There are no pipeline delays associated with the part.

The AD7908/AD7918/AD7928 use advanced design techniques to achieve very low power dissipation at maximum throughput rates. At maximum throughput rates, the AD7908/AD7918/AD7928 consume 2 mA maximum with 3 V supplies; with 5 V supplies, the current consumption is 2.7 mA maximum.

Through the configuration of the Control Register, the analog input range for the part can be selected as 0 V to REF_{IN} or 0 V to $2 \times \text{REF}_{IN}$, with either straight binary or twos complement output coding. The AD7908/AD7918/AD7928 each feature eight single-ended analog inputs with a channel sequencer to allow a preprogrammed selection of channels to be converted sequentially. The conversion time for the AD7908/AD7918/AD7928 is determined by the SCLK frequency, which is also used as the master clock to control the conversion. The maximum throughput rate is 1 MSPS using a serial clock frequency of 20 MHz. The devices are available in a 20-lead TSSOP package.

The AD7938/AD7939 are 12- and 10-bit, high speed, low power, successive-approximation (SAR) ADCs that supply a parallel data output. A simplified block diagram is shown in Figure 8.40. The parts operate from a single 2.7 V to 5.25 V power supply and feature throughput rates up to 1.5 MSPS. The parts contain a low noise, wide bandwidth, differential track/hold amplifier that can handle input frequencies up to 20 MHz.

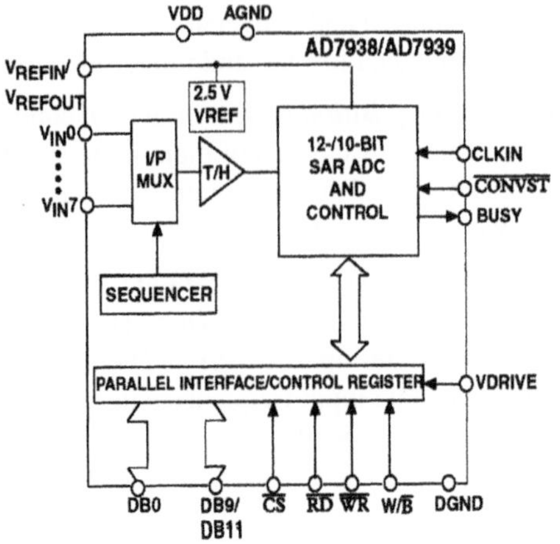

The AD7938/AD7939 feature eight analog input channels with a channel sequencer to allow a preprogrammed selection of channels to be converted sequentially. These parts can operate with either single-ended, fully differential or pseudo-differential analog inputs. The analog input configuration is chosen by setting the relevant bits in the on-chip Control Register.

The conversion process and data acquisition are controlled using standard control inputs allowing easy interfacing to Microprocessors and DSPs. The input signal is sampled on the falling edge of $\overline{\text{CONVST}}$, and the conversion is also initiated at this point.

Figure 8.40: AD7938/AD7939 8-Channel, 1.5 MSPS 12-/10-Bit Parallel Output ADC with Sequencer

The AD7938/AD7939 has an accurate on-chip 2.5 V reference that can be used as the reference source for the analog to digital conversion. Alternatively, this pin can be overridden to provide an external reference in the range 100 mV to 3.5 V. See the AD7938/AD7939 data sheet for performance when using various external reference voltage values.

These parts use advanced design techniques to achieve very low power dissipation at high throughput rates. They also feature flexible power management options. An on-chip Control Register allows the user to set up different operating conditions including analog input range and configuration, output coding, power management, and channel sequencing. The parts are available in a 32-pin LFCSP package.

Multiplexing Inputs to Σ-Δ ADCs

As discussed in Chapter 3, the digital filter is an integral part of a Σ-Δ ADC. When the input to a Σ-Δ ADC changes by a large step, the entire digital filter must fill with the new data before the output becomes valid, which is a slow process. This is why Σ-Δ ADCs are sometimes said to be unsuitable for multichannel multiplexed systems—they are not inherently so, but the time taken to change channels can be inconvenient. Generally speaking, the settling time is on the order of several clock cycles of the output data rate.

However, it is possible to optimize the digital filter and the rest of the Σ-Δ ADC design to yield high throughputs in multiplexed applications. For example, the AD7739 (Reference 3) is an 8-channel input high precision, high throughput Σ-Δ ADC optimized for multiplexed applications. A simplified block diagram is shown in Figure 8.41.

The AD7739 has true 16-bit noise-free code resolution with a total conversion time of 250 μs (4 kHz channel switching), making it ideally suited to high resolution multiplexing applications.

The part can be configured via a simple digital interface, which allows users to balance the noise performance against data throughput up to 15 kHz. The analog front end features eight single-ended or four fully differential input channels with unipolar or bipolar 625 mV, 1.25 V, and 2.5 V input ranges. It accepts a common-mode input voltage from 200 mV above AGND to AVDD – 300 mV. The differential reference input features "No-Reference" detect capability. The ADC also supports per-channel system calibration options.

The digital serial interface can be configured for 3-wire operation and is compatible with microcontrollers and digital signal processors. All interface inputs are Schmitt triggered. The part is specified for operation over the extended industrial temperature range of –40°C to +105°C. Other parts in the AD7739 family are the AD7738, AD7734, and AD7732. The AD7738 is similar to the AD7739 but has higher speed (8.5 kHz channel switching for 16-bit performance) and higher AIN leakage current.

The AD7738 multiplexer output is pinned out externally, allowing the user to implement programmable gain or signal conditioning before being applied to the ADC. The AD7734 ADC features four single-ended input channels with unipolar or true bipolar input ranges to ±10 V while operating from a single 5 V analog

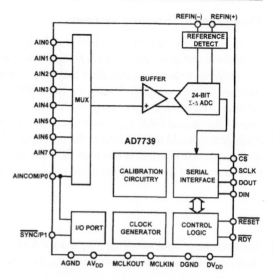

Figure 8.41: AD7739 8-Channel, High Throughput,24-Bit Σ-Δ ADC

supply. The AD7734 accepts an analog input overvoltage to ±16.5 V without degrading the performance of the adjacent channels. The AD7732 is similar to the AD7734, but its analog front end features two fully differential input channels.

The specified conversion time includes one or two settling and sampling periods and a scaling time as shown in Figure 8.42. With chopping enabled, a conversion cycle starts with a settling time of 43 or 44 MCLK cycles (~7.1 µs with a 6.144 MHz MCLK) to allow the circuits following the multiplexer to settle. The Σ-Δ modulator constantly samples the analog signals, and the digital filter processes the digital data stream. The sampling time depends on the channel conversion time register contents. For the example shown, the sampling time is 105.3 µs.

After another settling time of 42 MCLK cycles (~6.8 µs), the sampling time is repeated with a reversed (chopped) analog input signal. Then, during the scaling time of 163 MCLK cycles (~26.5 µs), the two results from the digital filter are averaged, scaled using the calibration registers, and written into the channel data register.

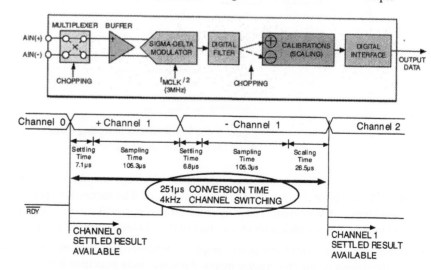

Figure 8.42: AD7739 Timing Showing Channel Switching

Simultaneous Sampling Systems

There are certain applications where it is desirable to sample a number of channels simultaneously such as in-phase and quadrature (I and Q) signal processing. A typical configuration is shown in Figure 8.43. Each channel requires its own filter and SHA. Each SHA is simultaneously placed in the *hold* mode by a common command signal. During the input SHAs' hold time the multiplexer is sequentially switched from channel to channel, and the sampling ADC is used to digitize the signal on each channel. The acquisition time of the second SHA, t_{acq2}, must be considered in determining the maximum ADC sampling rate, f_{s2}. The multiplexer should be switched to the next channel after the single SHA goes into the hold mode. If the multiplexer settling time is less than the ADC conversion time, then the maximum ADC sampling rate f_{s2} is the reciprocal of the sum of the SHA acquisition time and the ADC conversion time.

$$f_{s2} \leq \frac{1}{t_{acq2} + t_{conv}}$$

Eq. 8.12

The maximum input sampling frequency is less than this value divided by M, where M is the number of channels. Additional timing overhead (t_{acq1}) is required for the simultaneous SHAs to acquire the signals.

$$f_{s1} < \frac{1}{t_{acq1} + M\left(t_{conv} + t_{acq2}\right)}$$

Eq. 8.13

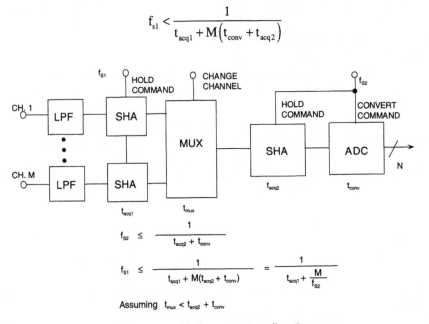

Figure 8.43: Simultaneous Sampling Data Acquisition System Using Sampling ADC

The AD7865 is a fast, low power, four-channel simultaneous sampling 14-bit SAR ADC that operates from a single 5 V supply (Reference 4). The part contains a 2.4 µs successive approximation ADC, four track/hold amplifiers, 2.5 V reference, on-chip clock oscillator, signal conditioning circuitry and a high speed parallel interface. A simplified block diagram of the AD7865 is shown in Figure 8.44.

The input signals on four channels are sampled simultaneously, thus preserving the relative phase information of the signals on the four analog inputs. Aperture delay matching between the sample-and-holds is less than 4 ns. The part accepts analog input ranges of ±10 V, ±5 V, ±2.5 V, 0 V to +2.5 V and 0 V to +5 V. The

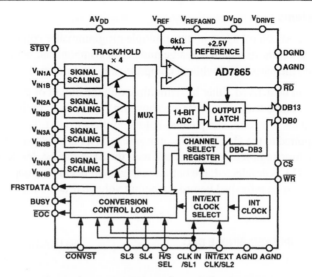

**Figure 8.44: AD7865 4-Channel Simultaneous
Sampling 14-Bit SAR ADC**

part allows any subset of the four channels to be converted in order to maximize the throughput rate on the selected sequence. The channels to be converted can be selected either via hardware (channel select input pins) or via software (programming the channel select register).

A single conversion start signal ($\overline{\text{CONVST}}$) simultaneously places all the track/holds into hold and initiates conversion sequence for the selected channels. The $\overline{\text{EOC}}$ signal indicates the end of each individual conversion in the selected conversion sequence. The BUSY signal indicates the end of the conversion sequence. Data is read from the part via a 14-bit parallel data bus using the standard $\overline{\text{CS}}$ and $\overline{\text{RD}}$ signals. Maximum throughput for a single channel is 350 kSPS. For all four channels the maximum throughput is 100 kSPS. The AD7865 is available in a 44-lead PQFP.

In simultaneous sampling applications using one Σ-Δ ADC per channel, the outputs must be synchronized as shown in Figure 8.45. Although the inputs are sampled at the same instant at a rate Kf_s, the decimated output frequency, f_s, is generally derived internally in each ADC by dividing the input sampling frequency by K (the oversampling rate) as shown in Figure 8.44. The output data must therefore be synchronized by the same clock at a frequency f_s. Most Σ-Δ ADCs provide a SYNC input to allow this synchronization.

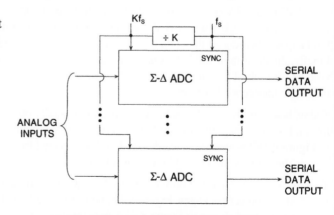

**Figure 8.45: Synchronizing Σ-Δ ADCs in
Simultaneous Sampling Applications**

Data Distribution Systems

In many industrial and process control applications, multiple programmable voltage sources are required. Traditionally, these applications have required a large number of components, but recent product developments have greatly reduced the parts count without compromising performance.

Multiple voltage outputs can be derived either by demultiplexing the output of a single DAC or by employing multiple DACs. These two approaches are shown in Figure 8.46. In the demultiplexed circuit (A), one DAC feeds the inputs of several sample-and-hold amplifiers (SHA). The equivalent digital value for the analog output is applied to the DAC, and the appropriate SHA is selected. After the DAC settling time and SHA acquisition time requirements have been met, the SHA can be deselected and the next channel updated. Once a SHA is deselected, the output voltage will begin to droop at a rate specified for the SHA. Thus, the SHA must be refreshed before the output voltage droop exceeds the required accuracy (typically ½ LSB).

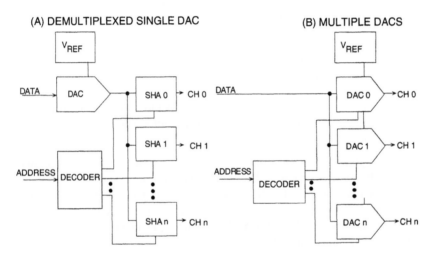

Figure 8.46: Options for Analog Data Distribution

The DAC plus SHA system evolved because, in the past, DACs were more expensive than SHAs. This situation was particularly true for DACs with resolution above 8 bits. In addition, multiple SHAs with on-chip hold capacitors reduced the parts count, printed circuit board area, and cost of demultiplexed DAC systems. Finally, the demultiplexed DAC only required one calibration step, since the same DAC provides the output voltage for each of the output channels. Of course, single-calibration is only valid if the SHA does not introduce unacceptable errors.

Today, however, the DAC plus SHA approach is virtually obsolete because of the availability of high resolution, low cost integrated circuit DACs in duals, quads, octals, etc. The multiple DAC application shown in Figure 8.46B is straightforward. One DAC is provided for each channel, and an address decoder simply selects the appropriate DAC. No refresh is required.

There is a high demand not only for multiple DACs in a single package, but also for single DACs in small low cost low power packages. Figure 8.47 shows two methods for distributing data to several remote locations. The method shown in Figure 8.47A uses multiple DACs to distribute analog data to multiple remote locations. This method requires that the analog signals be protected from noise pickup, and requires the use of shielded cables. If the remote stages are located a long distance from the source, the method of Figure 8.47B is preferred, where the digital data is transmitted over the remote cable link, and individual DACs are used as each of the remote stages.

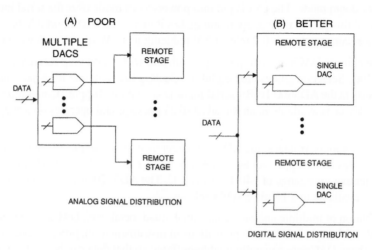

Figure 8.47: Remote, Multichannel Data Distribution

An excellent example of a low cost single DAC is the AD5320 12-bit buffered voltage output DAC (Reference 5). A simplified block diagram is shown in Figure 8.48. AD5320 is one of a family of pin-compatible DACs. The AD5300 is the 8-bit version and the AD5310 is the 10-bit version. The AD5300/AD5310/AD5320 are available in 6-lead SOT-23 packages and 8-lead µSOIC packages.

The AD5320 operates from a single 2.7 V to 5.5 V supply consuming 115 µA at 3 V. Its on-chip precision output amplifier allows rail-to-rail output swing to be achieved. The AD5320 utilizes a versatile 3-wire serial interface that operates at clock rates up to 30 MHz and is compatible with standard SPI, QSPI, MICROWIRE and DSP interface standards. The reference for AD5320 is derived from the power supply inputs and thus gives the widest dynamic output range.

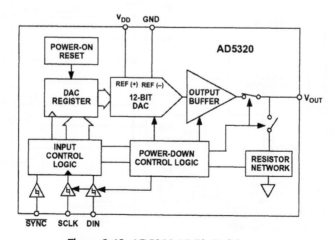

Figure 8.48: AD5320 12-Bit Serial
Input DAC in SOT-23 6-Pin Package

The part incorporates a power-on reset circuit which ensures that the DAC output powers up to zero volts and remains there until a valid write takes place to the device. The part contains a power-down feature that reduces the current consumption of the device to 200 nA at 5 V and provides software-selectable output loads while in power-down mode. The part is put into power-down mode over the serial interface. The low power consumption of this part in normal operation makes it ideally suited to portable battery operated equipment. The power consumption is 0.7 mW at 5 V reducing to 1 µW in power-down mode.

There are many other single DACs in small packages with and without on-chip references. Resolutions range from 8 to 16 bits. Selection guides are helpful in selecting the right one for a particular application. One of the newer parts is the AD5660 16-Bit serial input DAC with a 10-ppm/°C on-chip voltage reference (Reference 6). This device is available in an 8-lead SOT-23 package and operates on a supply voltage of 2.7 V to 5.5 V.

For operation at higher supply voltages, the AD5570 (Reference 7) is a single 16-bit serial input, voltage output DAC that operates from supply voltages of ±12 V up to ±15 V. INL and DNL are accurate to 1 LSB (max) over the full temperature range of –40°C to +125°C. The AD5570 utilizes a versatile 3-wire interface. The AD5570 is available in a 16-pin SSOP package.

For localized distribution of multiple analog signals, dual, quad, octal, etc., DACs are generally much preferred to single DACs. Multiple DACs find applications in instrumentation, process control, ATE, and many other applications. These DACs are generally double-buffered so that data can be loaded via a serial port and then the actual internal parallel DAC register updated either simultaneously or individually. Again, these DACs are available in many resolutions, voltage/current ranges, supply voltage, packages, etc., so a complete discussion of all options is impossible here. We will look at a couple of newer offerings as examples.

The AD5516 consists of 16 12-bit DACs in a single package (Reference 8). A functional block diagram is shown in Figure 8.49. A single reference input pin (REF_IN) is used to provide a 3 V reference for all 16 DACs. To update a DAC's output voltage, the required DAC is addressed via the 3-wire serial interface. Once the serial write is complete, the selected DAC converts the code into an output voltage. The output amplifiers translate the DAC output range to give the appropriate voltage range (±2.5 V, ±5 V, or ±10 V)

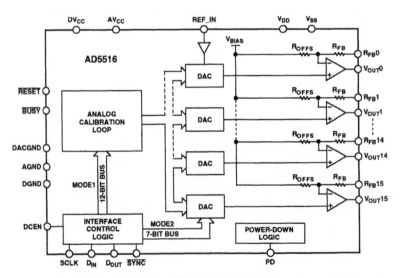

Figure 8.49: AD5516 16-Channel 12-Bit Voltage Output DAC

at output pins $V_{OUT}0$ to $V_{OUT}15$. The AD5516 uses a self-calibrating architecture to achieve 12-bit performance. The calibration routine servos to select the appropriate voltage level on an internal 14-bit resolution. The AD5516 is available a 74-lead CSPBGA package with a body size of 12 mm × 12 mm.

For the maximum available channel count today, the AD5379 contains 40 14-bit DACs in 13mm × 13 mm 108-lead LFBGA package and is ideal for high-end level setting needs in automatic test equipment and in optical networking applications (Reference 9). It has both parallel and 3-wire serial interfaces. A simplified block diagram is shown in Figure 8.50.

The AD5379 has a maximum output voltage span of 17.5 V which corresponds to an output range of −8.75 V to +8.75 V derived from reference voltages of −3.5 V and +5 V. The AD5379 contains a double-buffered parallel interface in which 14 data bits are loaded into one of the input registers under the control of the $\overline{WR}$, $\overline{CS}$ and DAC channel address pins, A0–A7. It also has a 3-wire serial interface compatible with SPI, QSPI, MICROWIRE and DSP interface standards and can handle clock speeds of up to 50 MHz. The DAC outputs are updated on reception of new data into the DAC registers. All the outputs can be updated simultaneously by taking the $\overline{LDAC}$ input low. Each channel has a programmable gain and offset adjust register. Each DAC output is gained and buffered on-chip with respect to an external REFGND input. The DAC outputs can also be switched to REFGND via the $\overline{CLR}$ pin.

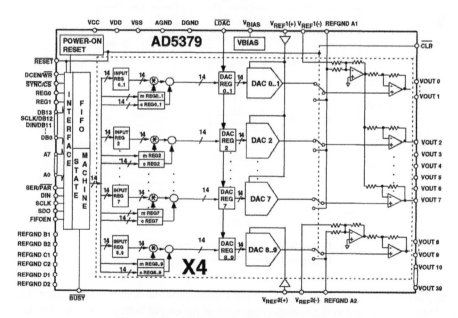

**Figure 8.50: AD5379 40-Channel, 14-Bit,
Parallel and Serial Input, Voltage-Output DAC**

Data Distribution Using an Infinite Sample-and-Hold

An "infinite," or "droopless" sample-and-hold function can be obtained using an ADC and a DAC. For example, the AD5533B 32-channel "infinite sample-and-hold" can be thought of as consisting of an ADC and 32 DACs in a single package (Reference 10). A functional diagram is shown in Figure 8.51. The input voltage V_{IN} is sampled and converted into a digital word. The digital result is loaded into one of the DAC registers and is converted (with gain and offset) into an analog output voltage ($V_{OUT}0$–$V_{OUT}31$). Since the channel output voltage is effectively the output of a DAC, there is no droop associated with it. As long as power to the device is maintained, the output voltage will remain constant until this channel is addressed again.

To update a single channel's output voltage, the required new voltage level is set up on the common input pin, V_{IN}. The desired channel is then addressed via the parallel port or the serial port. When the channel address has been loaded, provided $\overline{TRACK}$ is high, the circuit begins to acquire the correct code to load to the DAC so that the DAC output matches the voltage on V_{IN}. The $\overline{BUSY}$ pin goes low and remains so until the acquisition is complete. The noninverting input to the output buffer is tied to V_{IN} during the acquisition period to avoid spurious outputs while the DAC acquires the correct code. The acquisition is completed in 16 µs max.

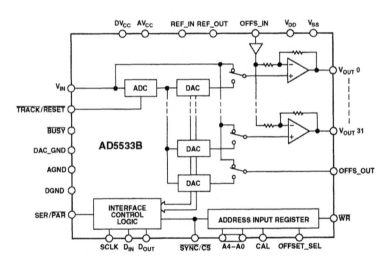

**Figure 8.51: AD5533B 32-Channel
Precision Infinite Sample-and-Hold**

The $\overline{BUSY}$ pin goes high and the updated DAC output assumes control of the output voltage. The output voltage of the DAC is connected to the noninverting input of the output buffer. Since the internal DACs are offset by 70 mV (max) from GND, the minimum V_{IN} in infinite SHA mode is 70 mV. The maximum V_{IN} is 2.96 V, due to the upper dead band of 40 mV (max). On power-on, all the DACs, including the offset channel, are loaded with zeros. Each of the 32 DACs is offset internally by 50 mV (typ) from GND so the outputs $V_{OUT}0$ to $V_{OUT}31$ are 50 mV (typ) on power-on if the OFFS_IN pin is driven directly by the on-board offset channel (OFFS_OUT), i.e., if OFFS_IN = OFFS_OUT = 50 mV => V_{OUT} = (Gain × V_{DAC}) – (Gain – 1) × V_{OFFS_IN} = 50 mV.

The output voltage range is determined by the offset voltage at the OFFS_IN pin and the gain of the output amplifier. It is restricted to a range from V_{SS} + 2 V to V_{DD} – 2 V because of the headroom of the output amplifier.

The AD5533B is operated with AV_{CC} = +5 V ± 5%, DV_{CC} = +2.7 V to +5.25 V, V_{SS} = –4.75 V to –16.5 V, and V_{DD} = +8 V to +16.5 V, and requires a stable 3 V reference on REF_IN as well as an offset voltage on OFFS_IN.

The AD5533B infinite sample-and-hold is ideally suited for use in automatic test equipment. Several ISHAs are required to control pin drivers, comparators, active loads, and signal timing as shown in Figure 8.52. Traditionally, sample-and-hold devices with droop were used in these applications. These required refreshing to prevent the voltage from drifting. The AD5533B has several advantages: no refreshing is required, there is no droop, pedestal error is eliminated, and there is no need for extra filtering to remove glitches. Overall, a higher level of integration is achieved in a smaller area.

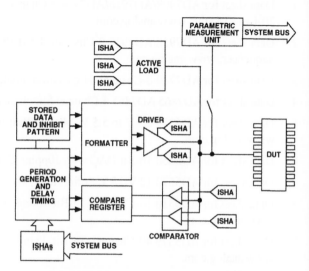

Figure 8.52: Infinite Sample-and-Holds (ISHAs) Used in Automatic Test Equipment Systems

The AD5533B can be used to set up voltage levels on 32 channels as shown in Figure 8.53. An AD780 provides the 3 V reference for the AD5533B, and for the AD5541 16-bit DAC. A simple 3-wire serial interface is used to write to the AD5541. Because the AD5541 has an output resistance of 6.25 kΩ (typ), the time taken to charge/discharge the capacitance at the V_{IN} pin is significant. Thus an AD820 is therefore used to buffer the DAC output. Note that it is important to minimize noise on V_{IN} and REFIN when laying out this circuit.

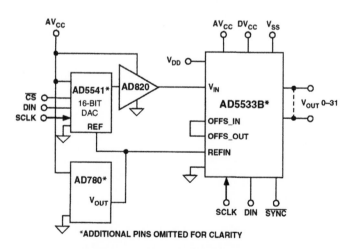

Figure 8.53: AD5533B Infinite Sample-and-Hold Typical Application Circuit

References:

8.2 Multichannel Data Acquisition Systems

1. Data sheet for AD7908/AD7918/AD7928 8-Channel, 1 MSPS 8-/10-/12-Bit ADCs with Sequencer in 20-Lead TSSOP, www.analog.com.

2. Data sheet for AD7938/AD7939 8-Channel, 1.5MSPS, 12- and 10-Bit Parallel Output ADCs with a Sequencer, www.analog.com.

3. Data sheet for AD7739 8-Channel, High Throughput, 24-Bit Σ-Δ ADC, www.analog.com.

4. Data sheet for AD7865 AD7865 4-Channel Simultaneous Sampling 14-Bit SAR ADC, www.analog.com.

5. Data sheet for AD5320 2.7 V to 5.5 V, 140 μA, Rail-to-Rail Output 12-Bit DAC in a SOT-23, www.analog.com.

6. Data sheet for AD5660 16-Bit DAC with 10ppm/°C Max On-Chip Reference, www.analog.com.

7. Data sheet for AD5570 12 V/15 V, Serial Input, Voltage Output, 16-Bit DAC, www.analog.com.

8. Data sheet for AD5516 16-Channel, 12-Bit Voltage-Output DAC with 14-Bit Increment Mode, www.analog.com.

9. Data sheet for AD5379 40-Channel, 14-Bit, Parallel and Serial Input, Voltage-Output DAC, www.analog.com.

10. Data sheet for AD5533B 32-Channel Precision Infinite Sample-and-Hold, www.analog.com.

Digital Potentiometers
Walt Kester, Walt Heinzer

Introduction

Mechanical potentiometers have been used since the earliest days of electronics and provide a convenient method for the adjustment of the output of various sensors, power supplies, or virtually any device that requires some type of calibration. Timing, frequency, contrast, brightness, gain, and offset adjustments are just a few of the possibilities. However, mechanical pots have always suffered from numerous problems including physical size, mechanical wearout, wiper contamination, resistance drift, sensitivity to vibration, temperature, humidity, the need for screwdriver access, layout inflexibility, etc.

Digital potentiometers avoid all the inherent problems associated with mechanical potentiometers and are ideal replacements in new designs where there is either a microcontroller or another digital device to provide the necessary control signals. Manually controlled digital potentiometers are also available for those who do not have any on-board microcontrollers. Unlike mechanical pots, digital pots can be controlled dynamically in active control applications.

The digital potentiometer is based on the CMOS "String DAC" architecture previously described in Chapter 3 of this book, and the basic diagram is shown in Figure 8.54. Note that in the normal string DAC configuration, the A and B terminals are connected between the reference voltage, and the W (wiper) terminal is the DAC output. There is also one more R resistor in the string DAC configuration which connects the A terminal to the reference.

The digital potentiometer configuration essentially makes use of the fact that the CMOS switches' common-mode voltages can be anywhere between the power supplies—the switch selected by the digital input

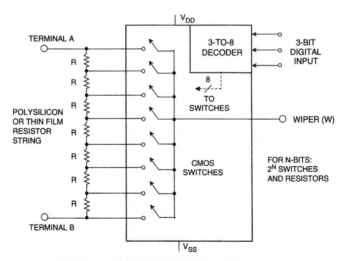

Figure 8.54: 3-Bit CMOS Digital Potentiometer Based on "String DAC" Architecture

simply connects the wiper to the corresponding tap on the resistor string. The relative polarity of A to B can be either positive or negative.

The resistor string represents the end-to-end potentiometer resistance, and the traditional "DAC output" becomes the wiper of the digital potentiometer. The resistors can be either polysilicon (TC ~ 500 ppm/°C) or thin film (TC ~ 35 ppm/°C), depending upon the desired accuracy.

The number of resistors in the string determines the resolution or "step size" of the potentiometer, and ranges from 32 (5 bits) to 1024 (10 bits) at present. The value of the programmable resistors are simply: $R_{WB}(D) = (D/2^N) \cdot R_{AB} + R_W$, and $R_{WA}(D) = [(2^N - D)/2^N] \cdot R_{AB} + R_W$, where R_{WB} is the resistance between W and B terminals, R_{WA} is the resistance between W and A terminals, D is the decimal equivalent of the step value, N is the number of bits, R_{AB} is the nominal resistance, and R_W is the wiper resistance.

The switches are CMOS transmission gates that minimize the on-resistance variations between any given step and the output. The voltages on the A and B terminals can be any value as long as they lie between the power supply voltages V_{DD} and V_{SS}.

Modern Digital Potentiometers in Tiny Packages

Figure 8.55 shows three examples of digital potentiometers that are all offered in small packages. The I²C® serial interface is a very popular one, but digital potentiometers are also available with the SPI, Up/Down Counter, and Manual Increment/Decrement interfaces.

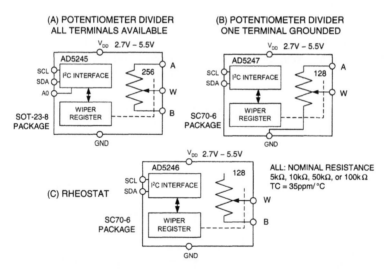

Figure 8.55: Typical Examples of Digital Potentiometers in Tiny Packages

The AD5245 shown in Figure 8.55A is available in an 8-lead SOT-23 package and has 256 positions (8 bits). The A0 pin allows the device to be uniquely identified so that two devices can be placed on the same bus. The thin film resistor string (R_{AB}) is available in 5 kΩ, 10 kΩ, 50 kΩ, or 100 kΩ, and the R_{AB} temperature coefficient is 35 ppm/°C. All three terminals of the potentiometer are available for use. The operating supply voltage can range from 2.7 V to 5.5 V. The power supply current is 8 µA maximum, and an internal command bit is available to shut down the device into a state of zero power consumption. The voltage noise is approximately the thermal noise of R_{AB}. (Recall that the thermal noise of a 1 kΩ resistor at room temperature is approximately $4 \text{ nV}/\sqrt{\text{Hz}}$).

The AD5247 shown in Figure 8.55B is similar to the AD5245, except it has 128 positions (7 bits), the B terminal is grounded, and the part comes in an SC70 6-lead package. The AD5247 does not have the A0 function. Finally, the AD5246 shown in Figure 8.55C is similar to the AD5245, but is connected as a rheostat with the W and B terminals available externally.

In addition to single potentiometers, such as the AD5245, AD5246, and AD5247, digital potentiometers are available as duals, triples, quads, and hex versions. Multiple devices per package offer 1% matching in ganged potentiometer applications as well as reducing PC board real estate requirements. Figure 8.56 summarizes some of the characteristics and features of modern digital potentiometers.

- Resolution (wiper steps): 32 (5 Bits) to 1024 (10 Bits)
- Nominal End-to-End Resistance: 1kΩto 1MΩ
- End-to-End Resistance Temperature Coefficient: 35ppm/°C (Thin Film Resistor String), 500ppm/°C (PolysiliconResistor String)
- Number of Channels: 1, 2, 3, 4, 6
- Interface Data Control: SPI, I^{2}C, Up/Down Counter Input, Increment/Decrement Input
- Terminal Voltage Range: +15V,±15V, +30V, +3V, ±3V, +5V, ±5V
- Memory Options:
  - Volatile (No Memory)
  - Nonvolatile E^{2}MEM
  - One-Time Programmable (OTP) – One Fuse Array
  - Two-Time Programmable – Two Fuse Arrays

Figure 8.56: Characteristics of CMOS Digital Potentiometers

Digital Potentiometers with Nonvolatile Memory

Digital potentiometers, such as the AD5245, AD5246, and AD5247, are used mainly in active control applications, since they do not have nonvolatile memory. Therefore, the setting is lost if power is removed. However, most volatile digital potentiometers have a power-on preset feature that forces the devices to the midscale code when power is applied.

Obviously, there is a demand for digital potentiometers with the ability to retain their setting after power is removed and reapplied. This requires the use of nonvolatile on-chip memory to store the desired setting. The AD5235 is an example of a dual 10-bit digital potentiometer which contains on-chip E²MEM to store the desired settings (Reference 4). A functional block diagram is shown in Figure 8.57.

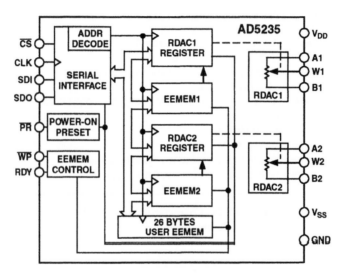

**Figure 8.57: AD5235 Nonvolatile Memory,
Dual 1024-Position Digital Potentiometers**

These devices perform the same electronic adjustment function as a mechanical potentiometer with enhanced resolution, solid state reliability, and superior low temperature coefficient performance. The AD5235's versatile programming via a standard serial interface allows 16 modes of operation and adjustment, including scratch pad programming, memory storing and retrieving, increment/decrement, log taper adjustment, wiper setting readback, and extra user-defined E²MEM. Another key feature of the AD5235 is that the actual resistance tolerance is stored in the E²MEM at 0.1% accuracy. The actual end-to-end resistance can therefore be known, which is valuable for calibration and tolerance matching in precision applications. The new E²MEM family of digital pots (AD5251/AD5252/AD5253/AD5254) also offer such a feature. In the scratch pad programming mode, a specific setting can be programmed directly to the RDAC register, which sets the resistance between terminals W-A and W-B. The RDAC register can also be loaded with a value previously stored in the E²MEM register. The value in the E²MEM can be changed or protected.

When changes are made to the RDAC register, the value of the new setting can be saved into the E²MEM. Thereafter, it will be transferred automatically to the RDAC register during system power on. E²MEM can also be retrieved through direct programming and external preset pin control. The linear step increment and decrement commands cause the setting in the RDAC register to be moved UP or DOWN, one step at a time.

For logarithmic changes in wiper setting, a left/right bit shift command adjusts the level in ±6 dB steps. The AD5235 is available in a thin TSSOP-16 package. All parts are guaranteed to operate over the extended industrial temperature range of –40°C to +85°C.

One-Time Programmable (OTP) Digital Potentiometers

The AD5172/AD5173 are dual channel 256-position, one-time programmable (OTP) digital potentiometers, which employ fuse link technology to achieve the memory retention of resistance setting function (Reference 5). A functional block diagram is shown in Figure 8.58. Note that the AD5172 is configured as a three-terminal potentiometer, while the AD5173 is pinned out as a rheostat. The AD5172/AD5173 is available in 2.5 kΩ, 10 kΩ, 50 kΩ, and 100 kΩ versions. The temperature coefficient of the resistor string is 35 ppm/°C. The power supply voltage can range from 2.7 V to 5.5 V.

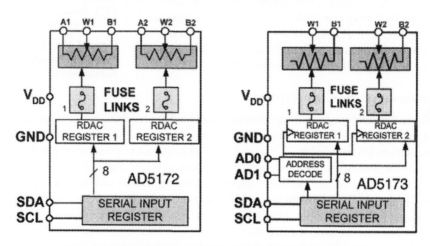

Figure 8.58: AD5172/AD5173 256-Position One-Time Programmable Dual-Channel I²C Digital Potentiometer

OTP is a cost-effective alternative over the E²MEM approach for users who do not need to program the digital potentiometer setting in memory more than once, i.e., "set and forget." These devices perform the same electronic adjustment functions most mechanical trimmers and variable resistors do but offer enhanced resolution, solid-state reliability, and better temperature coefficient performance.

The AD5172/AD5173 are programmed using a 2-wire I²C compatible digital control. They allow unlimited adjustments before permanently setting the resistance value. During the OTP activation, a permanent fuse blown command is sent after the final value is determined; therefore freezing the wiper position at a given setting (analogous to placing epoxy on a mechanical trimmer). Unlike other OTP digital potentiometers in the same family, AD5172/AD5173 have a unique temporary OTP overwriting feature that allows new adjustments if desired, the OTP setting is restored during subsequent power up conditions. This feature allows users to apply the AD5172/AD5173 in active control applications with user-defined presets.

To verify the success of permanent programming, Analog Devices patterned the OTP validation such that the fuse status can be discerned from two validation bits in read mode. For applications that program AD5172/AD5173 in the factories, Analog Devices offers device programming software, which operates across Windows 95 to XP® platforms including Windows NT®. This software application effectively replaces the need for external I²C controllers or host processors and therefore significantly reduces users' development time. An AD5172/AD5173 evaluation kit is available, which include the software, connector, and cable that can be converted for factory programming applications. The AD5172/AD5173 are available in a MSOP-10 package. All parts are guaranteed to operate over the automotive temperature range of −40°C to +125°C. Besides their unique OTP features, the AD5172/AD5173 lend themselves well to other general-purpose digital potentiometer applications due to their programmable preset, superior temperature stability, and small form factor.

The AD5170 (Reference 6) is a two-time programmable 8-bit digital potentiometer, and a functional diagram is shown in Figure 8.59. Note that a second fuse array is provided to allow "second chance" programmability. Like the AD5172/AD5173, there is unlimited programmability before making the permanent setting. The electrical characteristics of the AD5170 are similar to the AD5172/AD5173.

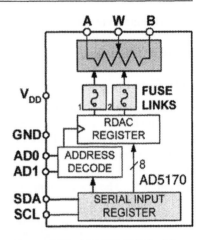

Figure 8.59: AD5170 256-Position Two-Time Programmable I²C Digital Potentiometer

Digital Potentiometer AC Considerations

Digital potentiometers can be used in ac applications, provided the bandwidth limitations created by the internal capacitance are considered. Figure 8.60 shows an ac model of a digital potentiometer, where the capacitances are modeled as C_A, C_B, and C_W. The bandwidth of the digital pot is configuration dependent. It is also dynamic because of the variable resistance. For example, if A terminal is the input, B terminal is grounded, and W terminal is the output; then the bandwidth can be approximated by $BW = 1/[2\pi(R_{WB}\|R_{WA}) \times C_W]$. The lowest bandwidth occurs at midscale, where the equivalent resistance is at its maximum in this configuration. The typical values for the AD5245 are shown as well as the corresponding bandwidths for the various resistance options measured at midscale. This simple model can be used in SPICE simulations to predict circuit performance, such as when the digital potentiometer is used as a part of the feedback network of an op amp. The other issue to consider when placing digital

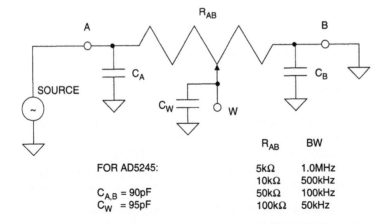

FOR AD5245:	R_{AB}	BW
	5kΩ	1.0MHz
	10kΩ	500kHz
$C_{A,B}$ = 90pF	50kΩ	100kHz
C_W = 95pF	100kΩ	50kHz

BW MEASURED FROM A TO W WITH B GROUNDED, MIDSCALE CODE, DRIVEN FROM A LOW IMPEDANCE SOURCE

Figure 8.60: Digital Potentiometer Bandwidth Model

potentiometers directly in the signal path is their slightly nonlinear resistance as a function of applied voltage. This effect leads to a small amount of distortion. For example, the AD5245 has a THD of 0.05% when a 1 V rms, 1 kHz signal is applied to the configuration described above at midscale. References 10, 11, 14, 15, and 17 show excellent examples of the application of digital potentiometers in ac applications.

Application Examples

Like op amps, digital pots are the building blocks of many electronic circuits. Because they are digitally controlled, digital pots can be used in active control applications, in addition to basic trimming or calibration applications. For example, digital pots can be used in programmable power supplies as shown in Figure 8.61A. Typical adjustable low dropout voltage regulators (such as the anyCAP series) have a FB pin, where applying a resistor divider yields a variable output voltage. As shown, R1 and R2 are the feedback and input resistors, respectively. The FB circuit has an internal noninverting amplifier that gains up a 1.2 V bandgap reference to the desired output voltage.

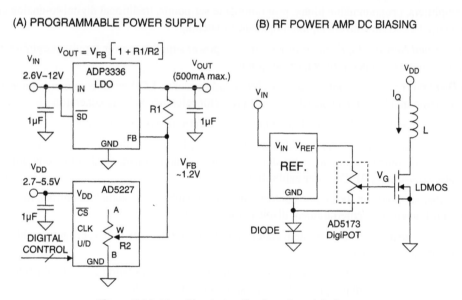

Figure 8.61: Two Circuit Applications for Digital Pots

Similarly, electronic equipment makers use digital potentiometers in power supplies by adjusting the supplies to the tolerances that cover all supply voltage conditions during reliability testing. This voltage-margining approach accelerates the burn-in process, and therefore reduces the system time-to-market.

Because of the optimized cost/performance benefits, digital pots have been gaining popularity in replacing traditional DACs in many applications. For example, in wireless basestations, the optimum threshold voltages of the RF power amplifiers vary widely in production. Such variation affects the transmitted signal linearity and power efficiency. Too much power delivered from a poorly regulated amplifier can also interfere with neighboring cells within the wireless network. Although DACs are widely used in biasing RF power amplifiers, many users find digital pots to be more suitable in such applications because of the availability of nonvolatile memory, which simplifies the designs. As shown in Figure 8.61B, the one-time-programmable digital pot is used to calibrate the dc bias point of the RF power amplifier, and the calibration is programmed by factory software without the need for any external controllers. Note that the diode is added to the circuit to compensate for the amplifier's temperature coefficient.

Summary

Digital potentiometers offer many obvious advantages over mechanical potentiometers and trimpots®, and therefore they have become widely accepted in modern systems. Their reliability, flexibility, and ease of use makes them popular replacements for the traditional potentiometer. Digital pots can also be used as programmable building blocks in many active control applications.

There are virtually endless applications for digital potentiometers in modern electronic systems—one only has to consider the many traditional applications for mechanical pots and trimpots as a starting point. References 7–18 should be consulted for more ideas on how these devices can enhance a design. A few applications are summarized below:

- *General-Purpose Applications:* sensor calibration, system gain and offset adjustments, programmable gain amplifiers, programmable filters, programmable set-points, traditional digital-to-analog converters, voltage-to-current converters, line impedance matching.

- *Computer and Network Equipment:* programmable power supplies, power supply margining, battery charger set-points, temperature control set-points.

- *LCD Displays:* backlight, contrast, and brightness adjustments, LCD panel common voltage adjustment, programmable gamma correction, LCD projector reference voltage generator.

- *Consumer Applications:* PDA backlight adjustment, electronic volume controls.

- *RF Communications:* RF power amplifier biasing, DDS/PLL amplitude adjustment, VCXO frequency tuning, varactor diode biasing, log amp slope and intercept adjustment, quadrature demodulator gain and phase adjustment, RFID reader calibration.

- *Automotive:* set-points in the engine control unit, sensor calibrations, actuator controls, instrumentation control, navigation/entertainment display adjustments.

- *Industrial and Instrumentation:* system calibration, floating reference DACs, programmable 4 mA to 20 mA current transmitters.

- *Optical Communications:* laser bias current adjustments, laser modulation current adjustments, optical receiver signal conditioning, optical attenuators, wavelength controllers.

References:
8.3 Digital Potentiometers

1. Data sheet for AD5245 256-Position I²C Compatible Digital Potentiometer, www.analog.com.

2. Data sheet for AD5247 128-Position I²C CompatibleDigital Potentiometer, www.analog.com.

3. Data sheet for AD5246 128-Position I²C CompatibleDigital Resistor, www.analog.com.

4. Data sheet for AD5235 Nonvolatile Memory, Dual 1024-Position Digital Potentiometers, www.analog.com.

5. Data sheet for AD5172 256-Position One-Time Programmable Dual-Channel I²C Digital Potentiometer, www.analog.com.

6. Data sheet for AD5170 256-Position Two-Time Programmable I²C Digital Potentiometer, www.analog.com.

7. Walt Heinzer, "Design Circuits with Digitally Controllable Variable Resistors," Analog Dialogue, Vol. 29, No. 1, 1995, www.analog.com.

8. Hank Zumbahlen, "Tack a Log Taper onto a Digital Potentiometer," **EDN**, January 20, 2000.

9. Mary McCarthy, "Digital Potentiometers Vary Amplitude In DDS Devices," **Electronic Design**, Ideas for Design, May 29, 2000.

10. Alan Li, "Versatile Programmable Amplifiers Use Digital Potentiometers with Nonvolatile Memory," **Analog Dialogue**, Vol. 35, No. 3, June-July, 2001.

11. Reza Moghimi, "Difference Amplifier Uses Digital Potentiometers ," **EDN**, May 30, 2002.

12. Mark Malaeb, "Single-Chip Digitally Controlled Data-Acquisition as Core of Reliable DWDM Communication Systems," **Analog Dialogue**, Vol. 36, No. 5, September-October, 2002, www.analog.com.

13. Peter Khairolomour, "Rotary Encoder Mates with Digital Potentiometer," **EDN**, Design Idea, March 6, 2003.

14. Alan Li, "Versatile Programmable Amplifiers Using Digital Potentiometers with Nonvolatile Memory," **Application Note AN-579**, www.analog.com.

15. Alan Li, "Programmable Oscillator Uses Digital Potentiometers," **Application Note AN-580**, www.analog.com.

16. Alan Li, "Resolution Enhancements of Digital Potentiometers with Multiple Devices," **Application Note AN-582**, www.analog.com.

17. Alan Li, "AD5232 Programmable Oscillator Using Digital Potentiometers, " **Application Note AN-585**, www.analog.com.

18. Alan Li, " ADN2850 Evaluation Kit User Manual," **Application Note AN-628**, Analog Devices, www.analog.com.

Digital Audio
Walt Kester

Introduction

Voiceband digital audio had its beginnings in the early days of the PCM system development initiated by the 1937 patent filing of A. H. Reeves of the International Telephone and Telegraph Corporation (Reference 1). During the early 1940s, Bell Telephone Laboratories continued PCM work relating to speech encryption systems and, after the war, turned their attention to commercial PCM transmission. An experimental 24-channel PCM system was developed, and the results summarized in 1948 in Reference 2 by L. A. Meacham and E. Peterson. Some of the significant developments of this work were the successive approximation ADC, Shannon-Rack decoder (DAC), and the logarithmic companding/expanding of voiceband signals.

In order to minimize the number of bits per second and still maintain the required dynamic range for voiceband, the early system utilized 7-bit ADCs and DACs with a logarithmic compressor ahead of the ADC and a logarithmic expander after the DAC.

With the solid-state devices available in the mid 1950s, Bell Labs developed the T-1 carrier system which was prototyped in the late 1950s and put into service in the 1960s. The standard sampling rate for voiceband signals was established at 8 kSPS, and the initial system used 7-bit logarithmically encoded ADCs and DACs. Later systems used 8-bit "segmented" ADCs and DACs (see Chapter 3 of this book for a description of the architecture).

Modern wireless systems, such as cellular telephone, use higher resolution *linear* Σ-Δ ADCs and DACs, rather than the *logarithmic* ones used in the early systems. A simplified comparison between the standard 8-bit companded system and the modern cell phone handset is shown in Figure 8.62. Modern cellular transmission systems make use of sophisticated DSP-based speech compression algorithms in order to reduce the overall data rate to acceptable levels, rather than limiting the resolution of the converters. In most cases the ADC and DAC (codec) are integrated into a chip which performs many other digital functions in a handset. It should be noted that if needed, the modern linear codecs can be made backward compatible with the logarithmically encoded 8-bit systems by simply using on-chip lookup tables to convert the high resolution linear data into the 8-bit logarithmic data.

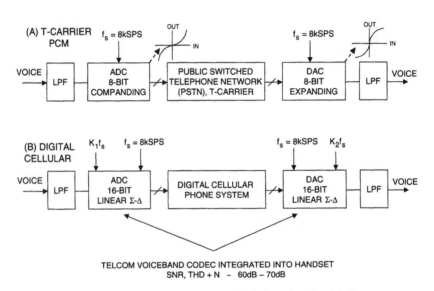

TELCOM VOICEBAND CODEC INTEGRATED INTO HANDSET
SNR, THD + N ~ 60dB – 70dB

Figure 8.62: Voiceband Telcom Digital Audio Simplified

In the 1970s and 1980s, the field of digital audio rapidly expanded to include much more than voiceband for PCM systems. A driving force behind this expansion was the increased availability of low cost high resolution ADCs and DACs with sufficient dynamic range and sampling rates.

In the consumer electronics industry, the compact disk (CD) player has proliferated into nearly every household. Today, high-end DVD audio players give increased levels of performance in home theater systems.

Since the beginning of the 1980s, digital audio equipment has steadily been replacing analog equipment in broadcast and production systems. Some of this digital equipment has analog inputs and outputs and is designed to replace an analog device and operate in an analog environment (i.e., a digital "black box"). However, the trend in broadcast and production is toward the all-digital studio, in which all aspects of recording, processing, and transmission take place in the digital domain. To this end, the AES (Audio Engineering Society) and EBU (European Broadcast Union) have developed standards to facilitate this future transition.

Sampling Rate and THD + N Requirements for Digital Audio

The key audio specifications are total harmonic distortion plus noise (THD + N), dynamic range (DNR), and signal-to-noise ratio (SNR). THD + N is more correctly expressed as (THD + N)/S the ratio of the rms sum of all spectral components in the pass band (20 Hz to 20 kHz), excluding the fundamental, to the rms value of the fundamental input signal. The ratio can be expressed in % or dB. THD + N is a negative number when expressed in dB, but often simply expressed as a positive number, with the minus sign assumed.

Dynamic range (DNR) is the ratio of a fullscale input signal to the integrated noise in the pass band (20 Hz to 20 kHz), expressed in dB. It is measured with a –60 dB input signal and is equal to [S/(THD+N)] + 60dB. The noise level therefore basically establishes the dynamic range. It is specified with and without an A-Weight filter applied.

Signal-to-noise ratio (SNR) is the ratio of the full-scale input signal level to the integrated noise in the pass band (20 Hz to 20 kHz) with no input signal applied, expressed in dB (a positive number). In many cases, the SNR and DNR specifications are approximately equal. Note that this definition of SNR for audio is slightly different than the SNR for standard ADCs and DACs defined in Chapter 2, where it is defined as the ratio of the rms signal to rms value of all other components excluding the harmonics of the fundamental. These definitions are summarized in Figure 8.63.

- Total Harmonic Distortion Plus Noise (THD + N):
  - More correctly: S/(THD + N), the ratio of the rms value of the fundamental input signal to the rms sum of all other spectral components in the passband (20Hz to 20kHz), expressed in % or dB. This is a negative number in dB, but often simply expressed as a positive number, with the minus sign assumed.
- Dynamic Range (DNR):
  - The ratio of a full-scale input signal to the integrated noise in the passband (20Hz to 20kHz), expressed in dB. It is measured with a –60dB input signal and is equal to [S/(THD+N)] + 60dB, so the noise level establishes the dynamic range. It is specified with and without an A-Weight filter applied.
- Signal-to-Noise Ratio (SNR):
  - The ratio of the full-scale input signal level to the integrated noise in the pass band (20Hz to 20kHz) with no input signal applied, expressed in dB. (Approximately equal to DNR in many cases)

Figure 8.63: Key Audio Specifications

Figure 8.64 lists a few of the popular applications of digital audio and some typical THD + N and sample rate requirements. For the purposes of this discussion, THD + N numbers are given as positive numbers in dB below the signal level. In actuality, when a small percentage, for example 0.001% is converted into dB, it is a negative number as previously mentioned above.

	THD + N (dB)	Standard Sample Rates (kSPS)
Telcom	60 – 70	8
FM Stereo	60 – 70	32
Speech Analysis, etc.	70 – 80	8 – 48
Computer Audio	80 – 90	48
Stereo CD, DAT, etc.	> 100	44.1, 48, 88.2, 96
DVD Audio	> 100	48, 96, 192

Figure 8.64: Digital Audio THD + N and Sample Rate Requirements

In most cases, the sampling frequency is chosen to be slightly above twice the highest frequency of interest, however, the actual numbers deserve some further discussion.

The early requirement for the standard PCM T-carrier sampling rate of 8 kSPS has already been discussed. Voiceband audio occupies an approximate bandwidth of 3.5 kHz, and requires an SNR of only 60 dB to 70 dB. Although 16-bit Σ-Δ codecs are used today for the convenience of DSPs, only approximately 11 bits of actual dynamic range is required.

FM stereo has a higher bandwidth (typically 15 kHz), and a sampling rate of 32 kSPS was therefore chosen for digital transmission over land lines connecting the studio to FM transmitter. A minimum THD + N of 60 dB to 70 dB is sufficient for these applications.

Speech analysis and speech processing systems require a THD + N of at least 70 dB to 80 dB, and sampling rates up to 48 kSPS are used, depending upon the application.

The professional audio bandwidth extends from 20 Hz to 20 kHz. Therefore, a minimum sampling frequency of 40 kSPS is required—in practice, 44.1 kSPS (audio CD standard) is the lowest used.

High quality computer audio sound cards need 80 dB to 90 dB THD + N, and 48 kSPS has been adopted as the standard sampling frequency.

High-end stereo CD players, DVD audio players, and studio recording systems require greater than 100 dB of THD + N and DNR, and therefore place the most critical requirements on the ADC and DAC. Regarding sampling frequency, 44.1 kSPS was chosen as the compact standard sampling frequency. It was selected to allow the use of NTSC or PAL "U-Matic" videotape recorders (VTRs) fitted with a PCM adapter to record and play back digital audio signals transformed into "pseudo-video" waveforms. Later, these VTRs were used to master compact disks (CDs), and 44.1 kSPS became a defacto standard that is also used in some digital audio tape (DAT) play-back-only applications.

For studio and broadcast applications, 48 kSPS sampling has become the industry standard for digital audio recording. This frequency was adopted for the following reason. In a digital television environment, the digital audio reference signal must be locked to the video reference signal to avoid drift in the relationship between audio and video signals and allow click-free audio and video switching. A sampling rate of 48 kSPS was chosen for both NTSC and PAL to facilitate the conversion between the two standards and maintain the proper phase relationship between video and audio in both systems (the details of this are explained in Reference 1). In these applications, THD + N requirements are generally greater than 100 dB.

Audio for digital video disk (DVD) utilizes sampling frequencies of 48 kSPS when video and audio are both present. Although most humans cannot hear frequencies above 20 kHz, tests have shown that harmonics and the effects of room acoustics enable some audio above 20 kHz to be heard, or at least felt. Therefore, for high-end DVD audio-only applications, sampling rates of 96 kSPS and 192 kSPS can be used. Higher DVD sampling frequencies and resolution enhance the audio quality over CDs in stereo playback (i.e., 96 kSPS/24-bit versus 44.1 kSPS/16-bit for CDs). THD + N and DNR requirements for DVD audio are typically greater than 100 dB.

Pulse Code Modulation (PCM) is the basic form of digital audio used on most CDs as well as to master virtually all digital recording. PCM encoding on DVD-video products can use up to a 96 kSPS sampling frequency and a 24-bit sample word. Producers of DVD products rely heavily on audio and video compression to fit their source material within the space and bandwidth of the DVD medium. By slightly compressing the audio, DVD producers can make space available for video and other added features without sacrificing perceptible audio performance and quality. The audio compression formats are used only to remove redundant data. The end user does not detect the difference, because the removed data is masked by other sounds. Since the incoming audio stream is altered by the compression, all of the original PCM data cannot be recovered on playback—these formats are referred to as lossy compression. This can result in audio tracks as small as 1/15th the size of the uncompressed PCM master for Dolby Digital compression.

Overall Trends in Digital Audio ADCs and DACs

Digital audio systems place high performance demands on ADCs and DACs because of the wide dynamic range requirements. Early ADCs for digital audio in the 1970s typically utilized either the successive approximation or subranging architectures (see Chapter 3 for architecture descriptions). The resolution was generally 16 bits, and the maximum sampling frequency about 50 kSPS. The ADCs were modules or hybrids, and the DACs were ICs with an input serial-to-parallel converter followed by a traditional parallel binarily-weighted DAC. The DACs generally used thin-film laser trimmed resistors to achieve the required accuracy. The ADCs were relatively costly and primarily used in the recording studios, while the DACs were used in high volume in consumer CD players.

Early CD players used techniques similar to the simplified diagram in Figure 8.65A. The output anti-imaging filter presents a fundamental problem with this approach, especially when audio requirements are considered. Theoretically, a 16-bit parallel DAC updated at 44.1 kSPS could be used at the output, however, because the audio bandwidth extends to 20 kHz, the transition region of the filter is narrow. For example, with a 44.1 kSPS update rate, the transition region is from 20 kHz to 24 kHz, which is $\log_2(24 \div 20) = \log_2(1.2) = 0.263$ octaves. For 60 dB stopband attenuation in 0.263 octaves, the required slope of the filter in the transition region is $60 \div 0.263 = 228$ dB/octave. Assuming 6 dB/octave per pole, a 38-pole filter would be required. This is obviously a difficult and expensive filter, especially if linear phase is desired (as is the case in audio applications). Another problem is the "sin(x)/x" roll-off caused by the DAC reconstruction process. With no compensation, the output is attenuated by approximately 4 dB at the Nyquist frequency of 22.05 MHz (see Chapter 2 of this book for details).

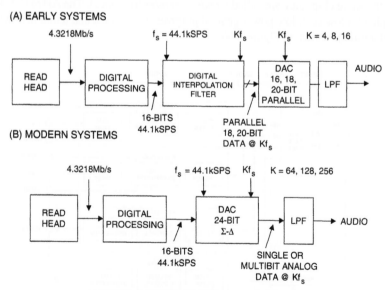

Figure 8.65: Stereo CD Digital Audio Simplified (One Channel Shown)

Therefore, in order to simplify the anti-imaging filter requirement, oversampling techniques are used as shown. The 16-bit, 44.1 kSPS data is passed through a digital interpolation filter which creates extra data points, and produces output data at a multiple K times the original sampling rate of 44.1 kSPS. A side benefit from oversampling is the 3 dB increase in SNR that occurs each time the output sample rate is doubled. Oversampling rates of 4, 8, and 16 were popular in early CD players. With K = 16, for example, the effective output update rate is now $16 \times 44.1 = 705.6$ kSPS. The transition region of the filter now extends from 20 kHz to 685.6 kHz, which is equal to $\log_2(685.6 \div 20) = \log_2(34.3) = 5.1$ octaves. The required slope of the filter in the transition region is now $60 \div 5.1 = 12$ dB/octave, and a 2- or 3-pole filter is sufficient. The high oversampling rate also minimizes the signal attenuation due to the sin(x)/x roll-off previously mentioned.

The Σ-Δ ADC uses a highly oversampled input analog modulator followed by a digital filter and decimator, as described in detail in Chapter 3 of this book. The Σ-Δ DAC uses a digital modulator to produce a single (or multibit) analog output at a highly oversampled rate. The oversampling feature greatly eases the requirements on both the ADC input antialiasing filter and the DAC output anti-imaging filter. There are many other advantages of the Σ-Δ architecture which make it the ideal choice for audio ADCs and DACs. Figure 8.65B shows a modern CD player using a Σ-Δ DAC, where oversampling rates of 64, 128, and 256 are quite commonly used.

Today, Σ-Δ ADCs and DACs dominate the digital audio market because of their high dynamic range, oversampling architecture, low power, and relatively low cost. Because they are typically produced on CMOS processes, the addition of many audio-specific digital features is relatively easy. The following sections examine a few of the many digital audio offerings available from Analog Devices.

Voiceband Codecs

There are many applications where both an ADC and a compatible DAC are required, such as in voice and audio processors, digital video camcorders, cell phone handsets, PC sound cards, etc. When the ADC and DAC are both on the same chip, they are called a *coder-decoder* (or codec). The AD74122 (Reference 4) is a typical example of a low cost, low power general purpose stereo voice and audio bandwidth codec. A simplified block diagram is shown in Figure 8.66.

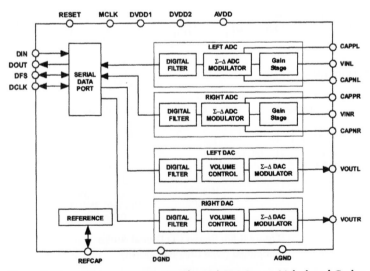

Figure 8.66: AD74122 16-/20-/24-Bit, 48 kSPS Stereo Voiceband Codec

The AD74122 is a 2.5 V Σ-Δ stereo audio codec with 3.3 V tolerant digital interface. It supports sampling rates from 8 kSPS to 48 kSPS and provides 16-/20-/24-bit word lengths. The architecture uses multibit modulators and data directed scrambling DACs for reduced idle tones and noise floor (see Chapter 3 of this book).

The ADC THD + N in the 20 Hz to 20 kHz bandwidth is 67 dB and the dynamic range (DNR) is 85 dB. The DAC THD + N is 88 dB (measured with a sampling rate of 48 kSPS), and the dynamic range is 93 dB. The device has digitally programmable input/output gain, on-chip volume controls per output channel, software controllable clickless mute, and contains an on-chip reference. The serial interface is compatible with popular DSPs. The AD74122 is packaged in a 20-lead TSSOP package.

High Performance Audio ADCs and DACs in Separate Packages

For studio quality digital audio recording, the THD + N and SNR of an ADC should be greater than 100 dB. Early ADCs for digital audio were expensive modules or hybrids, now of course, they are ICs based on the Σ-Δ architecture. The AD1871 stereo 24-bit, 96 kSPS Σ-Δ ADC is an excellent example of an ADC suitable for the exacting requirements of professional audio recording (Reference 5). It uses multibit modulators and data scrambling techniques to yield 103 dB THD + N, and 105 dB SNR/DNR. The device operates at an input oversampling frequency of 6.144 MSPS for an output sampling rate of 48 kSPS (K = 128). It has an SPI-compatible serial port, on-chip reference, and is housed in a 28-lead SSOP package. A functional diagram is shown in Figure 8.67.

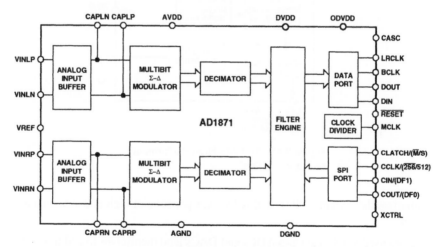

**Figure 8.67: AD1871 Stereo Audio,
24-Bit, 96 kSPS Multibit Σ-Δ ADC**

Looking at DACs, the AD1955 (Reference 6) represents the high-end of digital audio performance. It operates on a 5 V power supply and 16-/18-/20-/24-bit data at up to a 192-kSPS sample rate. It supports the SACD (super audio compact disk, a Phillips standard) DSD (direct stream digital) bit stream, and supports a wide range of PCM sample rates including 32 kSPS, 44.1 kSPS, 48 kSPS, 88.2 kSPS, 96 kSPS, and 192 kSPS. The AD1955 uses a multibit Σ-Δ modulator with "Perfect Differential Linearity Restoration" and data directed scrambling for reduced idle tones and noise floor. The output is a differential current of 8.64 mA p-p.

The AD1955 has 120 dB SNR/DNR at a 48 kSPS sample rate (A-Weighted stereo) and 110 dB THD + N. The digital filter has 110 dB stopband attenuation with ±0.0002 dB pass band ripple. The device has hardware and software controllable clickless mute, serial (SPI) interface, and is housed in a 28-lead SSOP plastic package. A functional diagram is shown in Figure 8.68.

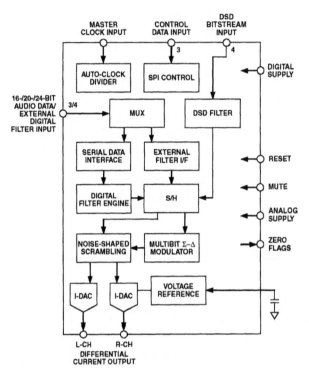

Figure 8.68: AD1955 High Performance Multibit Σ-Δ DAC

Figure 8.69 summarizes the current (2004) Analog Devices portfolio of high-end stereo audio DACs in individual packages.

The CMOS processes used to fabricate Σ-Δ ADCs and DACs lend themselves to additional digital functionality with only moderate increases in chip real estate and power. The SigmaDSP™ family of audio DACs incorporate DSPs which allow speaker equalization, dual-band compression/limiting, delay compensation, and image enhancement. These algorithms can be used to compensate for real-world limitations of speakers, amplifiers, and listening environments, resulting in a dramatic improvement in perceived audio quality.

The AD1953 SigmaDSP 3-channel, 26-bit signal processing DAC (Reference 7) accepts data at sample rates up to 48 kSPS. The signal processing used in the AD1953 is comparable to that found in high-end studio equipment. Most of the processing is done in full 48-bit double-precision mode, resulting in very good low level signal performance and the absence of limit cycles or idle tones. The compressor/limiter uses a sophisticated two-band algorithm often found in high-end broadcast compressors. A simplified block diagram is shown in Figure 8.70.

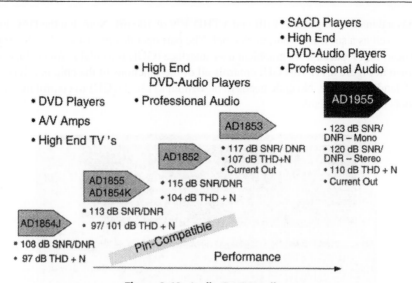

Figure 8.69: Audio DAC Family

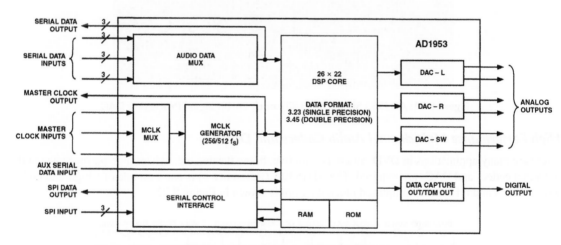

Figure 8.70: AD1953 SigmaDSP 3-Channel, 26-Bit Signal Processing DAC

The AD1953 has a dynamic range of 112 dB and a THD + N of 100 dB. Note that the DAC has left and right channels as well as a subwoofer output channel. The part operates from a single 5 V supply and is housed in a 48-lead LQFP package. A graphical user interface (GUI) is available for evaluation of the AD1953 as shown in Figure 8.71. This GUI controls all of the functions of the chip in a very straightforward and user-friendly interface. No code needs to be written to use the GUI to control the device. Software development tools are also available.

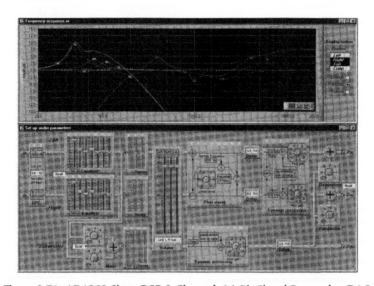

Figure 8.71: AD1953 SigmaDSP 3-Channel, 26-Bit Signal Processing DAC

High Performance Multichannel Audio Codecs and DACs

There are many applications in DVD, audio, automotive, home theater, etc., where high performance multichannel codecs and DACs are required. The AD1839A (Reference 8) is one example of a high-end codec with 2 ADCs and 6 DACs. A simplified block diagram is shown in Figure 8.72.

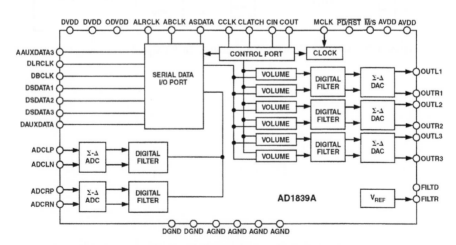

Figure 8.72: AD1839A, Two ADC, Six DAC, 96 kSPS 24-Bit Σ-Δ Codecs

The ADCs have 97 dB THD + N and 105 dB SNR/DNR. The DACs have 92 dB THD + N and 108 dB SNR/DNR. The devices operate on 5 V with 3.3 V tolerant digital interfaces. The maximum sampling rate is 96 kSPS (192 kSPS available on 1 DAC), and 16-/20-/24-bit word lengths are supported. The Σ-Δ modulators are multibit and utilize data directed scrambling. The AD1839A is housed in a 52-lead MQFP package.

A summary of the multichannel audio family of codecs and DACs is shown in Figure 8.73.

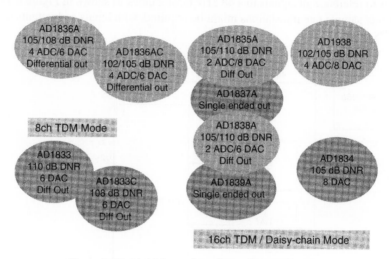

Figure 8.73: Multichannel Audio Codecs and DACs

There are many applications for audio codec products in computer sound cards, and these require compatibility with the AC'97 specification. The Analog Devices' AD1985 SoundMAX® Codec (Reference 9) is a good example of a codec which is compliant with the latest revision of AC'97.

Sample Rate Converters

From an earlier discussion in this section, we saw that there are a variety of standard sampling frequencies associated with digital audio signal processing: 8 kSPS, 32 kSPS, 44.1 kSPS, 48 kSPS, 88.2 kSPS, 96 kSPS, and 192 kSPS, etc. A typical audio studio generally has a common mixing console through which all signals must pass—analog audio and digital audio. These signals must be synchronized, and the most common method is to reference all signals to a 48 kHz master clock as shown in Figure 8.74. It should be noted that many other sample rate translations might be required, with inputs and outputs ranging from 8 kSPS to 192 kSPS.

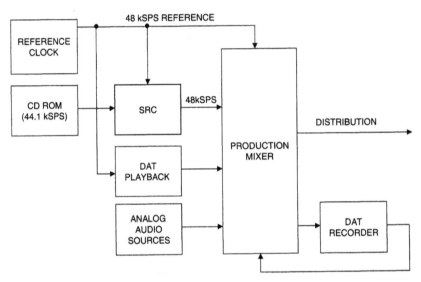

Figure 8.74: The Need for Sample Rate Converters (SRCs)

In order for this to be practical, there must be an easy method for seamlessly translating digital signals from one sampling frequency to another. For instance, the 44.1 kSPS CD player output must be translated into 48 kSPS to interface with the mixer. This requires more than simply changing the sampling frequency—a completely new set of data samples must be generated where these translations occur. Figure 8.75 shows one way to illustrate the concept of a sample rate converter (SRC).

The digital input data is first passed through a DAC updated at the input sampling frequency. The analog output of the DAC is resampled by an ADC which operates at the output sampling frequency. In practice, however, the SRC is an entirely digital device. The process is conceptually one of upsampling the input data followed by zero-stuffing, digital interpolation to generate new sampled data, digital filtering, and finally downsampling to the desired output sampling frequency. The conversion of a 5 kHz sinewave from a sampling frequency of 44.1 kSPS to 48 kSPS using this technique is shown graphically in Figure 8.76.

The AD1896 (Reference 10) is a 24-bit, high performance, single-chip, second generation asynchronous sample rate converter. Based on Analog Devices experience with its first asynchronous sample rate converter, the AD1890, the AD1896 offers improved performance and additional features. This improved performance includes a THD + N range of 117 dB to 133 dB, depending on the sample rate and input frequency, 142 dB (A-Weighted) dynamic range, up to 192 kSPS sampling frequencies for both input and output sample rates, improved jitter rejection, and 1:8 upsampling and 7.75:1 downsampling ratios.

- A SRC is a FULLY DIGITAL ENGINE
- However, one way of thinking about it is:
  - it reconstructs the signal, just as a DAC would
  - it resamples the signal, just as an ADC would

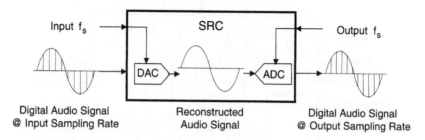

Figure 8.75: The Concept of a Sample Rate Converter

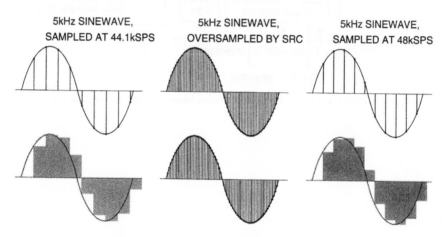

Figure 8.76: Conversion of a 5 kHz Sinewave
from 44.1 kSPS to 48 kSPS Sample Rate

Additional features include more serial formats, a bypass mode, better interfacing to digital signal processors, and a matched-phase mode. The AD1896 has a 3-wire interface for the serial input and output ports that supports left-justified, I²S, and right-justified (16-, 18-, 20-, 24-bit) modes. Additionally, the serial output port supports TDM mode for daisy-chaining multiple AD1896s to a digital signal processor. The serial output data is dithered down to 20, 18, or 16 bits when 20-, 18-, or 16-bit output data is selected. The AD1896 sample rate converts the data from the serial input port to the sample rate of the serial output port. The sample rate at the serial input port can be asynchronous with respect to the output sample rate of the output serial port. The master clock to the AD1896, MCLK, can be asynchronous to both the serial input and output ports.

The AD1896 operates on 3.3 V to 5 V input supplies and 3.3 V core voltages. The part is housed in a 28-lead SSOP package. A functional diagram is shown in Figure 8.77.

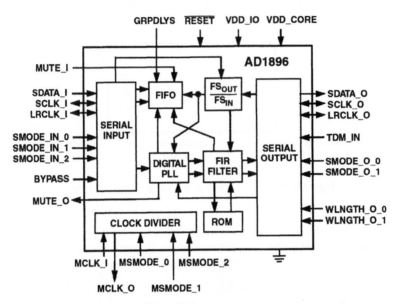

Figure 8.77: AD1896 192 kHz Stereo Asynchronous Sample Rate Converter

References:
8.4 Digital Audio

1. Alec Harley Reeves, "Electric Signaling System," **U.S. Patent 2,272,070**, filed November 22, 1939, issued February 3, 1942. Also **French Patent 852,183** filed 1937, issued 1938, and **British Patent 538,860** issued 1939.

2. L. A. Meacham and E. Peterson, "An Experimental Multichannel Pulse Code Modulation System of Toll Quality," **Bell System Technical Journal**, Vol 27, No. 1, January 1948, pp. 1–43.

3. Michael Robin and Michel Poulin, **Digital Television Fundamentals**, Second Edition, McGraw-Hill, 2000, ISBN 0-07-135581-2, Chapter 6.

4. Data sheet for AD74122 16-/20-/24-Bit, 48-kSPS Stereo Voiceband/Audio Codec, www.analog.com.

5. Data sheet for AD1871 Stereo Audio, 24-Bit, 96 kSPS, Multibit Σ-Δ ADC, www.analog.com.

6. Data sheet for AD1955 High Performance Multibit Σ-Δ DAC, www.analog.com.

7. Data sheet for AD1953 SigmaDSP 3-Channel, 26-Bit Signal Processing DAC, www.analog.com.

8. Data sheet for AD1839A Two ADC, Six DAC, 96 kSPS 24-Bit Σ-Δ Codecs, www.analog.com.

9. Data sheet for AD1985 AC '97 SoundMAX Codec, www.analog.com.

10. Data sheet for AD1896 192 kSPS Stereo Asynchronous Sample Rate Converter, www.analog.com.

11. Ken C. Pohlmann, **Principles of Digital Audio**, Fourth Edition, McGraw Hill, 2000, ISBN 0-07-134819-0.

Digital Video and Display Electronics
Walt Kester

Digital Video

Introduction

Before discussing some video applications for data converters, we will review some basics regarding video signals and specifications. The standard video format is the specification of how the video signal looks from an electrical point of view. Light strikes the surface of an image sensing device within the camera, producing a voltage level corresponding to the amount of light hitting a particular spatial region of the surface. This information is then placed into the standard format and sequenced out of the camera. Along with the actual light and color information, synchronization pulses are added to the signal to allow the receiving device—a television monitor, for instance—to identify where the sequence is in the frame data.

A standard video format image is read out on a line-by-line basis from left to right, top to bottom. A technique called *interlacing* refers to the reading of all even numbered lines, top to bottom, followed by all odd lines as shown in Figure 8.78.

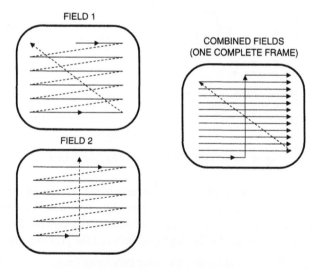

Figure 8.78: Standard Broadcast Television Interlace Format

The standard television picture *frame* is thus divided into even and odd *fields*. Interlacing is used to produce an apparent update of the entire frame in half the time that a full update actually occurs. This results in a television image with less apparent flicker. Typical broadcast television frame update rates are 30 Hz and 25 Hz, depending upon the line frequency. It should be noted that interlacing is generally not used in graphics display systems where the refresh rate is usually greater (typically 60 Hz).

The original black and white, or *monochrome*, television specification in the USA is the EIA RS-170 (replaced by SMPTE 170M) specification that prescribes all timing and voltage level requirements for standard commercial broadcast video signals. The standard American specification for color signals, NTSC, modifies RS-170 to work with color signals by adding color information to the signal which otherwise contains only brightness information.

A video signal comprises a series of analog television lines. Each line is separated from the next by a synchronization pulse called the *horizontal sync*. The fields of the picture are separated by a longer synchronization pulse, called the *vertical sync*. In the case of a monitor receiving the signal, its electron beam scans the face of the display tube with the brightness of the beam controlled by the amplitude of the video signal. A single line of an NTSC color video signal is shown in Figure 8.79.

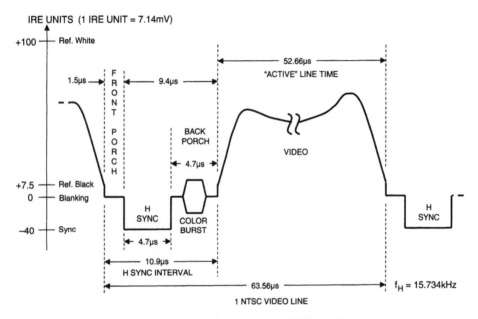

Figure 8.79: NTSC Composite Color Video Line

Whenever a horizontal sync pulse is detected, the beam is reset to the left side of the screen and moved down to the next line position. A vertical sync pulse, indicated by a horizontal sync pulse of longer duration, resets the beam to the top left point of the screen to a line centered between the first two lines of the previous scan. This allows the current field to be displayed between the previous one.

In the NTSC system (used in the U.S. and Japan), the color subcarrier frequency is 3.58 MHz. The PAL system (used in the U.K. and Germany) and SECAM system (used in France) use a 4.43 MHz color subcarrier.

In terms of their key frequency differences, a comparison between the NTSC system and the PAL system are given in Figure 8.80.

Digital Video Formats

Digital video had its beginnings in the early 1970s when 8-bit ADCs with sampling frequencies of 15 MSPS to 20 MSPS became available. Subjective tests, such as those conducted by A. A. Goldberg

	NTSC	PAL
Horizontal Lines	525	625
Color Subcarrier Frequency	3.58MHz	4.43MHz
Frame Frequency	30Hz	25Hz
Field Frequency	60Hz	50Hz
Horizontal Sync Frequency	15.734kHz	15.625kHz

Figure 8.80: NTSC and PAL Signal Characteristics

(Reference 1) showed that 8-bit resolution was sufficient for digitizing the composite video signal at sampling frequencies of three or four times the NTSC color subcarrier frequency (3.58 MHz).

Digital techniques were first applied to "video black boxes" which replaced functions previously implemented using analog techniques. These early digital black boxes had an analog video input and an analog video output, and replaced analog-based equipment such as time-base correctors, frame stores, standards converters, etc. (Reference 2, 3, 4). A typical black box is shown in Figure 8.81.

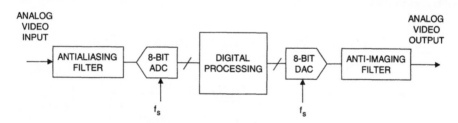

COMPOSITE VIDEO SIGNAL CHARACTERISTICS

	BANDWIDTH	SUBCARRIER, f_{SC}	$f_s = 3\,f_{SC}$	$f_s = 4\,f_{SC}$
NTSC	4.2 MHz	3.58 MHz	10.74 MSPS	14.32 MSPS
PAL	5.0 MHz	4.43 MHz	13.29 MSPS	17.72 MSPS

Figure 8.81: Digital Video "Black Box"

As previously mentioned, the required resolution was determined to be 8 bits early in the 1970s, however, 9-bit and 10-bit resolution eventually became popular as higher resolution low cost ADCs and DACs became available. Some initial black boxes used sampling frequencies of three times the subcarrier frequency—simply due to the lack of faster ADCs, but ultimately four times the subcarrier frequency became the industry standard.

The early ADCs used in these digital block boxes were modular devices, however in 1979, the first commercial 8-bit monolithic flash converter was introduced (Reference 5), and within a few years was soon followed by many others from a variety of IC manufacturers. The availability of low cost IC ADCs played a large role in the growth of digital video.

Digital videotape recorders (VTRs) emerged in the 1980s, based on CCIR recommendations. More digital black boxes, such as digital effects generators, graphic systems, and still stores proliferated, these devices operating in a variety of noncorrelated and incompatible standards. Digital connections between the black boxes were difficult or impossible, and the majority were connected with other equipment using analog input and output ports. As a matter of fact, this was one reason why 9-bit and 10-bit ADCs became popular—the additional resolution reduced the cumulative effects of quantization noise in cascaded devices.

In the 1980s, the Society of Motion Picture and Television Engineers (SMPTE) developed a digital standard (SMPTE 244M, Reference 6) which defined the characteristics of $4f_{SC}$ sampled NTSC composite digital signals as well as the characteristics of a bit-parallel digital interface which allowed up to 10-bit samples. The digital interface consisted of 10 differential ECL-compatible data signals, 1 differential ECL-compatible clock signal, two system grounds, and one chassis ground, for a total of 25 pins. Also in the 1980s, an IEEE standard (Reference 7) was developed which defined test methods for measuring the performance of ADCs and DACs used in composite digital television applications. Later, digital systems using $4f_{SC}$ NTSC composite digital signals adopted a high speed bit-serial interface, with a data rate of 143 Mb/s (defined in SMPTE 259M, Reference 10).

Even before the finalization of the $4f_{SC}$ composite digital standard, work was progressing on digital *component* systems, which offer numerous advantages over the composite digital systems. To understand the differences and advantages, Figure 8.82 shows a generalized block diagram of how the composite broadcast video signal is constructed.

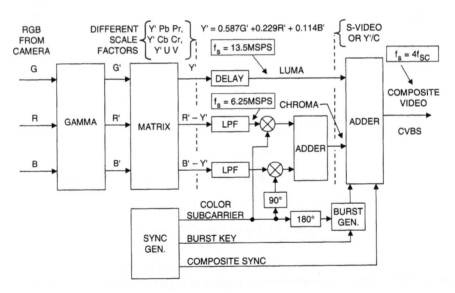

Figure 8.82: Model for Generating the Composite Video Signal from RGB Components

The native RGB signals from the color camera are first passed through a nonlinear *gamma* unit which compensates for the inherent nonlinearity in the receiving CRT. The R'G'B' outputs of the gamma unit then pass through a resistive *matrix* which generates a high-bandwidth *luma* signal (often incorrectly called *luminance*) and two reduced-bandwidth color difference signals. The luma signal, Y', is formed using the relationship Y' = 0.587G' + 0.229R' + 0.114B'. In addition, two *color difference* signals, designated

R' – Y' and B' – Y' are formed. The color subcarrier is then used to modulate the color difference signals in quadrature, and they are summed to form the *chroma* signal (often incorrectly called *chrominance*). The color burst and composite sync signals are then combined with the luma and chroma signals to form the *composite* video signal, designated CVBS (composite video with burst and sync)—the composite signal is ultimately broadcast.

A reverse process occurs in the television receiver, where the composite signal is decomposed into the various components and finally into an RGB signal which ultimately drives the three color inputs to the CRT.

Note that each step in the construction of the composite video signal after the output of the resistive matrix has the potential of introducing artifacts in the signal. For this reason, engineers working in digital video soon realized that it would be advantageous to keep the digital video signal as close to the native R'G'B' format as possible. The first so-called *component* analog video standard developed was designated as *Y'PbPr* (note that the prime notation has been dropped from most modern nomenclature). The corresponding digital standard is designated *Y'CbCr*. Digital Y'CbCr component video is specified in References 8, 9, and 10.

Another analog component standard is designated as Y'UV and is similar to Y'PbPr with different scaling factors for the color difference signals.

The final popular analog component standard to be discussed is the so-called *S-Video*, or simply *Y'/C*. This is a two-component analog system and is often used in high-end VCRs, DVDs, and TV receivers and monitors.

The various analog digital video component standards are summarized in Figure 8.83, and the back panel connections for each are shown in Figure 8.84 for a typical high-end video receiver.

The digital component video standards (References 8, 9, and 10) call for a sampling frequency of 13.5 MSPS for the Y' luma signal and 6.25 MSPS for each of the two color difference signals, Pb and Pr.

- Y'PbPr
  - In component analogvideo, B' – Y' and B' – Y' scaled to form color difference signals Pb and Pr.
- Y'CbCr
  - In component digitalvideo, B' – Y' and B' – Y' scaled to form Cb and Cr components.
- Y'UV
  - In NTSC or PAL, B' – Y' and B' – Y' scaled to form U and V analog components. U and V are lowpass filtered, and combined into a modulated chromacomponent, C. Luma (Y) is summed with chroma to form composite NTSC or PAL signal.
- S-Video, or Y'/C
  - Analog two-component system based on luma (Y') and chroma (C) signals.
- Composite, or CVBS
  - Composite video signal with burst and sync.

Figure 8.83: Analog and Digital Video Component and Composite Standards

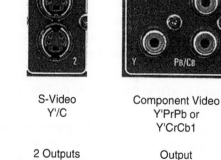

CVBS Composite Video with Burst and Sync	S-Video Y'/C	Component Video Y'PrPb or Y'CrCb1
2 Outputs	2 Outputs	Output

Figure 8.84: What the Analog Connectors Look Like on a High-End Receiver

This is often referred to as "4:2:2 sampling." The luma sampling frequency of 13.5 MSPS was selected to allow an integer number of sample periods in the line periods in both NTSC and PAL standards.

The 13.5 MSPS standard is sufficient for the luma signal whose bandwidth extends to 5.75 MHz. The color difference signal bandwidths extend to 2.75 MHz, and the 6.25 MSPS is also adequate.

It should be noted that if the R'B'G' signals were sampled directly, each would require a sampling frequency of 13.5 MSPS, and this is referred to as "4:4:4 sampling."

Serial Data Interfaces

Because of the large number of digital interconnection lines required with parallel interfaces, the serial digital interface (SDI) has mostly replaced the early parallel interfaces. The current serial interface standard is SMPTE 259M (Reference 10) defines 10-bit serial interfaces for $4f_{sc}$ NTSC at about 143 Mb/s, $4f_{sc}$ PAL at about 177 Mb/s, ITU-R BT601 4:2:2 component video at 270 Mb/s, ITU-R BT601 4:2:2 component video sampled at 18 MSPS (to achieve 16:9 aspect ratio) at 360 Mb/s.

High definition TV (HDTV) standards with a 16:9 aspect ratio are defined in ITU-R BT709 (Reference 12) and SMPTE 292M (Reference 13). A sampling frequency of 74.25 MSPS is used with 4:2:2 component sampling, for a serial bit rate of 1.485 Gb/s. There are various other HDTV scanning standards which are defined in SMPTE 296M and SMPTE 274M.

The high data rates associated with digital television require the use of data compression (such as MPEG) for broadcast transmission, but within the studio the signals are typically transmitted in serial uncompressed format on either coaxial or fiber optic cable.

The field of digital television is somewhat complicated because of the large number of acronyms and standards. For further information, the reader should consult References 14 and 15.

Digital Video ADCs and DACs: Decoders, and Encoders

In the world of digital video, a few definitions are in order. The acronym SDTV simply refers to standard definition TV (as opposed to HDTV, high definition TV).

An SDTV *decoder* converts *analog* composite video (CVBS), S-Video (Y'/C), Y'UV, or Y'PbPr signals into a digital video steam in the form of a Y'CbCr digital stream per ITU-R BT.656 4:2:2 component video compatible with NTSC, PAL B/D/G/H/I/ PAL M, or PAL N. An ADC function is implicit in the definition of the video decoder but, traditionally, the term decoder is more generally used to define the DAC function.

A digital video *encoder* converters digital component video (ITU-R BT.601 4:2:2, for example) into a standard composite analog baseband signal compatible with NTSC, PAL B/D/G/H/I, PAL M, or PAL N. In addition to the composite output signal, there is often the facility to output S-Video (Y'/C), RGB, Y'PbPr, or Y'UV component analog video.

In contrast to the digital video terminology, in ADC and DAC terminology, the terms *encoder* and *decoder* are used to refer to the ADC and the DAC function, respectively, and the combination is called a *codec* (coder-decoder).

The reason for this is that video engineers consider a composite color signal to have the chroma *encoded* on top of the luma signal. The video *decoder* (with the ADCs) *decodes* (separates) the chroma and luma signal and is referred to as the *decoder*. On the other hand, the video *encoder encodes* the chroma and the luma back into the composite signal.

The first ADCs and DACs used in digital television in the 1970s were modular devices, and were typically 8 bits with $4f_{sc}$ sampling, requiring an upper sampling rate of 17.72 MSPS for PAL. Many of the digital video black boxes of the 1980s ultimately went to 9- and even 10-bit resolution when IC ADCs became available. The analog conditioning circuitry, such as clamping, dc restoration, and filtering was typically separate from the ADC function itself.

In the 1990s, many low cost CMOS ADCs and DACs became available with resolutions up to 12 bits, and sampling rates of greater than 20 MSPS, thereby solving the basic data conversion problem and paving the way for higher levels of mixed-signal integration. Modern video decoders and encoders are therefore highly integrated, with on-chip analog signal conditioning and digital signal processing. Video decoders and encoders may operate at typical sampling frequencies of $4f_{SC}$, 13.5 MSPS, 27 MSPS, 54 MSPS, 108 MSPS, 216 MSPS, or 74.25 MSPS.

The ADV7183A 10-bit video decoder (Reference 16) is a good example of a modern highly integrated video signal processor, and a simplified block diagram is shown in Figure 8.85.

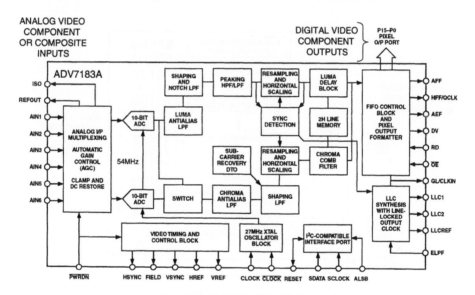

Figure 8.85: ADV7183A 10-Bit Video Decoder

The ADV7183A is an integrated video *decoder* that automatically detects and converts a standard definition analog baseband television signal (SDTV) compatible with worldwide standards NTSC , PAL, or SECAM into 4:2:2 component video data compatible with 16-/8-bit ITU-R BT.601/ITU-R BT.656. The advanced and highly flexible digital output interface enables high performance video decoding and conversion in both frame-buffer-based and line-locked clock-based systems. This makes the device ideally suited for a broad range of applications with diverse analog video characteristics, including tape-based sources, broadcast sources, security/surveillance cameras, and professional systems.

The internal 10-bit accurate A/D conversion provides professional quality SNR performance. This allows true 8-bit resolution in the 8-bit output mode. The analog input channels accept standard composite (CVBS), S-video, and component Y'PrPb video signals in an extensive number of combinations. AGC and clamp restore circuitry allow an input video signal peak-to-peak range of 0.5 V up to 2 V. Alternatively, these can be bypassed for manual settings.

The fixed 54 MHz (4 × 13.5 MSPS) clocking of the ADCs and data path for all modes allows very precise and accurate sampling and digital filtering. The line-locked clock output allows the output data rate, timing signals, and output clock signals to be synchronous, asynchronous, or line-locked even with ±5% line length variation. The output control signals allow glueless interface connection in almost any application. The ADV7183A modes are set up over a 2-wire serial bidirectional port (I²C-compatible). The ADV7183A is fabricated in a +3.3 V CMOS process. Its monolithic CMOS construction ensures greater functionality with lower power dissipation. The ADV7183A is packaged in a small 80-pin LQFP package.

The ADV7310 (Reference 17) is a second generation 12-bit video *encoder* featuring high performance DACs using oversampled Noise Shaped Video (NSV™) techniques to achieve better levels of performance for mid-range consumer electronics and professional video solutions. The maximum sampling rate is 216 MSPS which enables up to 16× oversampling. This means better figures for differential gain, phase and signal to noise ratio. A simplified block diagram of the ADV7310 is shown in Figure 8.86.

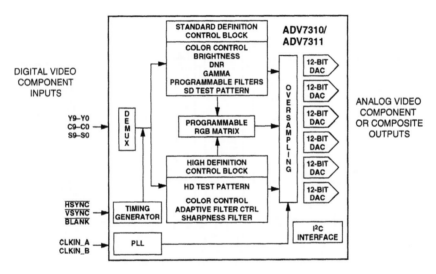

**Figure 8.86: ADV7310 Multiformat
216-MSPS Video Encoder**

The ADV7310 accepts a variety of standard digital video component input formats, including HDTV and SDTV. Outputs are HDTV, SDTV component and composite, S-Video, etc. Six 12-bit NSV precision video DACs provide multiple video outputs and allows 2×, 4×, 8× and 16× oversampling for high definition, progressive scan and standard definition video. A DAC adjust feature allows fine tuning of output levels on analog interfaces internal to many items of equipment, especially TVs, where EMI compliance requirements can be tough to meet.

The ADV7310 has a standard 2-wire serial I²C-compatible interface, and is housed in a 64-lead LQFP package.

Specifications for Video Decoders and Encoders

Video decoders and encoders are generally specified in terms of video performance, in addition to some of the traditional ADC/DAC specifications. Figure 8.87 lists the typical video specifications, and their definitions can be found in Reference 7, 14, and 15.

Note that the differential gain and differential phase specifications (defined in Chapters 2 and 5 of this book) apply only to the composite video signal and not the component video signals.

Display Electronics

Manufacturers of computer graphics displays realized that the resolution provided by standard NTSC and PAL video monitors was not sufficient for serious computer users because of the close viewing distance required. They also realized that the best performance could be obtained by utilizing the native RGB component video rather than composite signals.

For these reasons, a variety of formats for resolution/ scanning standards have evolved, generally defined by the somewhat obsolete EIA RS-343A standard. Unlike broadcast video, the horizontal and vertical resolution as well as the refresh rate in a graphics display system can vary widely depending upon the desired performance.

- Resolution, Sampling Rate, Linearity, Bandwidth
- Differential Gain (CVBS)
- Differential Phase (CVBS)
- SNR
- Chroma-Specific (Component Video)
  - Hue Accuracy
  - Color Saturation Accuracy
  - Color Gain Control Range
  - Analog Color Gain Range
  - Digital Color Gain Range
  - Chroma Amplitude Error
  - Chroma Phase Error
  - Chroma /Luma Intermodulation
- Luma-Specific (Component Video)
  - Luma Brightness Accuracy
  - Luma Contrast Accuracy

Figure 8.87: Video Decoder and Encoder Specifications

The resolution in such a system is defined in terms of *pixels* based on the number of horizontal lines and the number of pixels in each line. For instance, a 640 × 480 monitor has 480 horizontal lines, and each horizontal line is divided into 640 pixels. So a single frame would contain 307,200 pixels. In a color system, each pixel requires RGB intensity data. This data is generally stored as 8- or 10-bit words in a memory. Refresh rates generally vary from 60 Hz to 85 Hz. Most modern raster scan computer graphics monitors are "multisync," i.e, they will automatically synchronize to a variety of refresh rates and resolutions.

Figure 8.88 shows some typical resolutions and pixel rates for common display systems, assuming a 75 Hz, noninterlaced refresh rate. Standard computer graphics monitors, like television monitors, use a display technique known as *raster scan*. This technique writes information to the screen line by line, left to right, top to bottom, as has been previously discussed. The monitor must receive a great deal of information to

NOTATION	RESOLUTION	REFRESH RATE	PIXEL RATE
VGA	640 × 480	75Hz	30MHz
SVGA	800 × 600	75Hz	47MHz
XGA	1024 × 768	75Hz	83MHz
SXGA	1280 × 1024	75Hz	138MHz
UXGA	1600 × 1200	75Hz	202MHz
QXGA	2048 × 1536	75Hz	330MHz

Pixel Rate ≈ Vertical Resolution × Horizontal Resolution × Refresh Rate × 1.4
Refresh rates can be from 60Hz to 100Hz

Figure 8.88: Typical Graphics Resolution and Pixel Rates for 75 Hz Noninterlaced Refresh Rate

display a complete picture. Not only must the intensity information for each pixel be present in the signal but information must be provided to determine when a new line needs to start (HSYNC) and when a new picture frame should start (VSYNC).

The pixel clock frequency gives a good idea of the settling time and bandwidth requirements for any analog component, such as the DAC, which is placed in the path of the RGB signals. The pixel clock frequency can be estimated by finding the product of the horizontal resolution times the vertical resolution times the refresh rate. An additional 40% should be added, called the retrace factor, to allow for overhead.

There are several system architectures which may be used to build a graphics display system. The most general approach is illustrated in Figure 8.89. It consists of a host microprocessor, a graphics controller, a video frame buffer, three color memory banks (lookup tables), one for each of the primary colors red, green, and blue, (only one for monochrome systems). The microprocessor provides the image information to the graphics controller, frame buffer, and to the color lookup tables. This information typically includes position and color information. The graphics controller is responsible for interpreting this information and adding the required output signals such as sync, blanking, and memory management signals. The high speed frame buffer provides pixel address information to the lookup tables at the pixel rate.

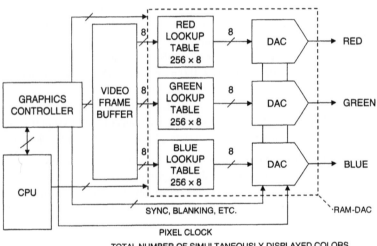

TOTAL NUMBER OF SIMULTANEOUSLY DISPLAYED COLORS
= $2^8 \times 2^8 \times 2^8$ = 16.8 MILLION COLORS

Figure 8.89: Simplified Graphics Control System for Generating RGB Signals

The R, G, and B memories are therefore lookup tables that hold the intensity information for each pixel for one frame. The DACs use the words in the memory and information from the memory controller to write the pixel information to the monitor. This system, when used with 8 bits for each DAC, is known as a 24-bit "true color" system. A total of 16.8 million addressable colors can be displayed simultaneously. The lookup tables and the DACs are generally integrated into an IC called a video "RAM-DAC," thereby minimizing the memory requirements of the graphics controller. Once the data for a frame is loaded into the RAM-DAC, no more data is required from the graphics controller unless there is a change in the pixel content.

In an effort to reduce system costs while maintaining flexibility, an alternative configuration shown in Figure 8.90 was developed, called a "pseudo-color" 8-bit graphics system. In this configuration, only 256 individual colors out of the total of 16.8 million can be displayed simultaneously, which is often acceptable in low-end applications. Today, most graphics controllers are capable of providing 8-bit ("256 color"), 16-bit ("high color"), and 24-bit ("true color") outputs.

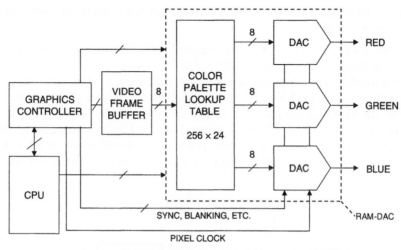

TOTAL NUMBER OF SIMULTANEOUSLY DISPLAYED COLORS
$= 2^8 = 256$ COLORS

Figure 8.90: "Pseudo-Color" 8-Bit RGB Graphics System

Video DACs have some features that distinguish them from traditional high speed DACs. Figure 8.91 shows the output waveform and levels for the green output which contains the sync information. Notice that the current output video DAC is terminated by a 75 Ω resistor to ground (source termination), and the end of the 75 Ω cable is terminated with another 75 Ω resistor (load termination). The net dc load on the output of the DAC is therefore 37.5 Ω. In order to develop a full-scale 1 V output, a current output of 26.67 mA is required for the green output DAC. In a graphics system, 1 IRE unit corresponds to approximately 7 mV.

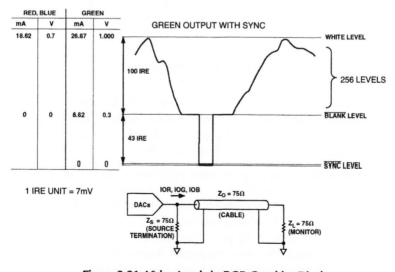

Figure 8.91: Video Levels in RGB Graphics Displays

617

Note that all 8 bits (256 levels) of the DAC are devoted to the active video region between the blanking level and the white level. The 300 mV sync level is generated by a separate current switch in the green DAC. A separate input is also supplied which generates the blanking level. In some systems, sync and blanking is applied to all three color signals, however it is very common practice to apply the sync signal to the green only (sync-on-green, or SOG).

Looking at high-end video DACs, the ADV7125 is a triple 330 MSPS DAC on a single monolithic chip (Reference 18). It consists of three high speed, 8-bit video DACs with complementary current outputs, a standard TTL input interface, and a high impedance analog current output. The ADV7125 has three separate 8-bit-wide input ports. A single 5 V or 3.3 V power supply and clock are all that are required to make the part functional. The ADV7125 has additional video control signals for composite SYNC and BLANK, as well as a power-save mode. The ADV7125 is fabricated in a 5 V CMOS process. Its monolithic CMOS construction ensures greater functionality with lower power dissipation (250 mW @ 3.3 V, 330 MSPS update). The ADV7125 is available in a 48-lead LQFP package. A simplified functional diagram is shown in Figure 8.92.

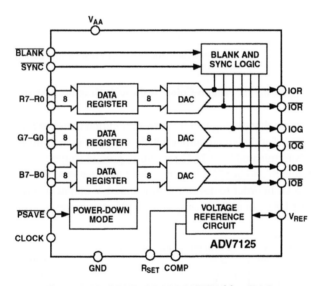

Figure 8.92: ADV7125 330 MSPS Video DAC

Today, there are a variety of video RAM-DACs which include the color lookup tables as well as the DACs, plus additional features such as overlay palettes and various mode controls. These digital features are integrated with the video DACs to provide a high degree of functionality. Figure 8.93 shows the ADV7160/ADV7162 true color, 220 MSPS video RAM-DAC (Reference 19).

The ADV7160/ADV7162 has a 96-bit fully programmable pixel port for support of up to 220 MSPS 1600 × 1280 screen resolution at an 85 Hz refresh rate. The lookup tables are 10 bits, thereby allowing on-chip gamma correction. The device has a fully programmable on-board PLL, and a standard microprocessor I/O interface. The overlay palettes allow the addition of cursors, pull-down menus, grids, pointers, etc., without additional hardware or software overhead. By providing these limited depth overlay palettes, the system software, which is generally responsible for cursor and pointer control, menus, etc., can efficiently control these graphics without altering the main graphic image which is controlled by the application software.

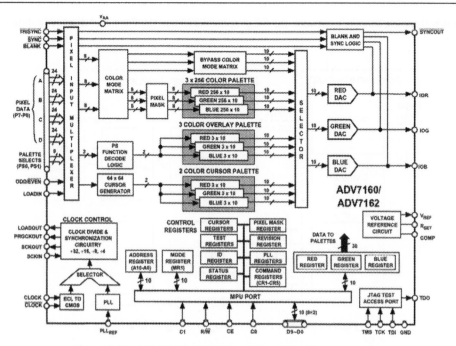

Figure 8.93: ADV7160/ADV7162 220 MSPS Video RAM-DAC

The ADV7160/ADV7162 operates on a single 5 V supply and is housed in either a 160-lead thermally enhanced QPF PQUAD (ADV7160), or a 160-lead plastic quad flatpack QFP (ADV7162).

Flat Panel Display Electronics

The popularity of flat panel LCD-based displays has steadily increased over the last few years, and they are rapidly replacing CRT-based monitors in desktop computer systems. In addition, LCD projectors have virtually replaced 35-mm slide and overhead projectors as a means of delivering presentation material.

The graphics card in a typical desktop computer system converts the digital pixel data to an analog RGB signal for driving an external monitor. In a laptop computer the built-in LCD display is generally driven directly with the digital data, and it is also converted to analog RGB video using video DACs, where it is available on an output connector for driving an external monitor or projector.

The analog RGB interface to the CRT is the primary workhorse in the display of computer-generated graphics data. A large legacy of PC-graphics adapters currently exist that use RAM-DACs to convert digital graphics data to analog RGB signals. The new flat panel displays must therefore be able to interface with this conventional technology to achieve market penetration and fast acceptance (References 20 and 21).

In an effort to establish an industry-wide standard for the next-generation flat panel displays, the Digital Display Working Group (DDWG) developed the Digital Video Interface (DVI 1.0) specification (Reference 22). This specification describes how designers should implement the analog and digital interfaces. Analog timing is described in the Video Electronics Standards Association (VESA) standard for monitors, and the digital interface uses Transistion Minimized Differential Signaling (TMDS) format.

The generalized analog interface between the PC graphics card and the flat panel display is shown in Figure 8.94.

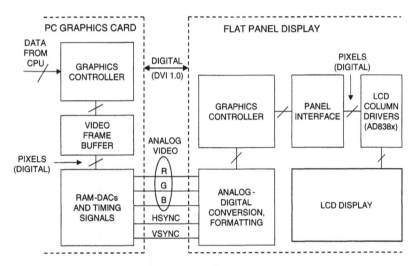

Figure 8.94: Flat Panel Analog and Digital Interfaces

For current flat panel displays, ICs such as the AD9888 (Reference 23) digitize the analog RGB data, generate a pixel clock from the HSYNC, and provide other functions necessary to format the pixel data which ultimately drives the columns of LCD display. A functional diagram of the AD9888 is shown in Figure 8.95.

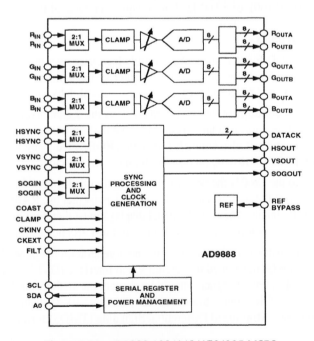

Figure 8.95: AD9888 100/140/170/205 MSPS
Analog Flat Panel Interface

The AD9888 is a complete 8-bit, 205 MSPS monolithic analog interface optimized for capturing RGB graphics signals from personal computers and workstations. Its 205 MSPS encode rate capability and full-power analog bandwidth of 500 MHz supports resolutions up to UXGA (1600 × 1200 @ 75 Hz). For ease of design and to minimize cost, the AD9888 is a fully integrated interface solution for flat panel displays. The AD9888 includes an analog interface with a 205 MSPS triple ADC with internal 1.25 V reference, PLL to generate a pixel clock from HSYNC and COAST, midscale clamping, and programmable gain, offset, and clamp control. The user provides only a 3.3 V power supply, analog input, and HSYNC and COAST signals. Three-state CMOS outputs may be powered from 2.5 V to 3.3 V.

The AD9888's on-chip PLL generates a pixel clock from HSYNC and COAST inputs. Pixel clock output frequencies range from 10 MHz to 205 MHz. PLL clock jitter is typically less than 450 ps p-p at 205 MSPS. When the COAST signal is presented, the PLL maintains its output frequency in the absence of HSYNC. A sampling phase adjustment is provided. Data, HSYNC, and clock output phase relationships are maintained. The PLL can be disabled and an external clock input can be provided as the pixel clock.

The AD9888 also offers full sync processing for composite sync and sync-on-green applications. A clamp signal is generated internally or may be provided by the user through the CLAMP input pin. This interface is fully programmable via a 2-wire serial interface. Fabricated in an advanced CMOS process, the AD9888 is provided in a space-saving 128-lead MQFP surface-mount plastic package and is specified over the 0°C to 70°C temperature range.

The AD9887A (Reference 24) offers designers the flexibility of an analog interface and digital visual interface (DVI) receiver integrated on a single chip. Also included is support for High Bandwidth Digital Content Protection (HDCP). The AD9887A is a complete 8-bit 170 MSPS monolithic analog interface optimized for capturing RGB graphics signals from personal computers and workstations. Its 170 MSPS sampling rate capability and full-power analog bandwidth of 330 MHz supports resolutions up to UXGA (1600 × 1200 at 60 Hz). The analog interface includes a 170-MHz triple ADC with internal 1.25 V reference, a phase-locked loop (PLL), and programmable gain, offset, and clamp control. The user provides only a 3.3 V power supply, analog input, and HSYNC. Three-state CMOS outputs may be powered from 2.5 V to 3.3 V. The AD9887A's on-chip PLL generates a pixel clock from HSYNC. Pixel clock output frequencies range from 12 MHz to 170 MHz. PLL clock jitter is typically 500 ps p-p at 170 MSPS. The AD9887A also offers full sync processing for composite sync and sync-on-green (SOG) applications. A functional diagram of the AD9887A is shown in Figure 8.96.

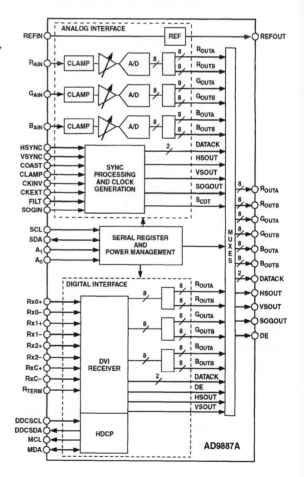

Figure 8.96: AD9887A Dual Flat Panel Interface

The AD9887A contains a DVI 1.0 compatible receiver and supports display resolutions up to UXGA (1600 × 1200 at 60 Hz). The receiver operates with true color (24-bit) panels in 1 or 2 pixel(s)/clock mode and features an intrapair skew tolerance of up to one full clock cycle. With the inclusion of HDCP, displays may now receive encrypted video content. The AD9887A allows for authentication of a video receiver, decryption of encoded data at the receiver, and renewability of that authentication during transmission as specified by the HDCP V1.0 protocol. Fabricated in an advanced CMOS process, the AD9887A is provided in a 160-lead MQFP surface-mount plastic package and is specified over the 0°C to 70°C temperature range.

CCD Imaging Electronics

The *charge-coupled-device* (CCD) and *contact-image-sensor* (CIS) are widely used in consumer imaging systems such as scanners and digital cameras. A generic block diagram of an imaging system is shown in Figure 8.97. The imaging sensor (CCD, CMOS, or CIS) is exposed to the image or picture much like film is exposed in a camera. After exposure, the output of the sensor undergoes some analog signal processing and then is digitized by an ADC. The bulk of the actual image processing is performed using fast digital signal processors. At this point, the image can be manipulated in the digital domain to perform such functions as contrast or color enhancement/correction, etc.

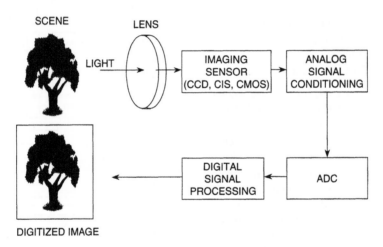

**Figure 8.97: Generic Imaging System for
Scanners or Digital Cameras**

The building blocks of a CCD are the individual light sensing elements called pixels (see Figure 8.98). A single pixel consists of a photo sensitive element, such as a photodiode or photocapacitor, which outputs a charge (electrons) proportional to the light (photons) that it is exposed to. The charge is accumulated during the exposure or integration time, and then the charge is transferred to the CCD shift register to be sent to the output of the device. The amount of accumulated charge will depend on the light level, the integration time, and the quantum efficiency of the photo sensitive element. A small amount of charge will accumulate even without light present; this is called dark signal or dark current and must be compensated for during the signal processing.

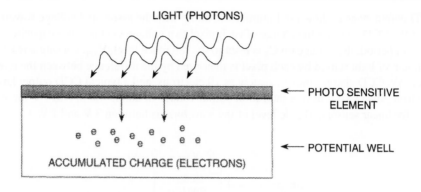

ONE PHOTOSITE OR "PIXEL"

Figure 8.98: Light Sensing Element

The pixels can be arranged in a linear or area configuration as shown in Figure 8.99. Clock signals transfer the charge from the pixels into the analog shift registers, and then more clocks are applied to shift the individual pixel charges to the output stage of the CCD. Scanners generally use the linear configuration, while digital cameras use the area configuration. The analog shift register typically operates at pixel frequencies between 1 MHz and 10 MHz for linear sensors, and 5 MHz to 25 MHz for area sensors.

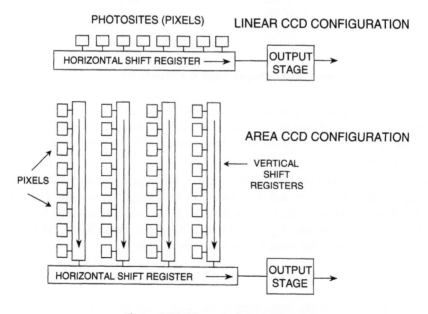

Figure 8.99: Linear and Area CCD Arrays

A typical CCD output stage is shown in Figure 8.100 along with the associated voltage waveforms. The output stage of the CCD converts the charge of each pixel to a voltage via the sense capacitor, C_S. At the start of each pixel period, the voltage on C_S is reset to the reference level, V_{REF}, causing a reset glitch to occur. The amount of light sensed by each pixel is measured by the difference between the reference and the video level, ΔV. CCD charges may be as low as 10 electrons, and a typical CCD output has a sensitivity of 0.6 µV/electron. Most CCDs have a saturation output voltage of about 500 mV to 1 V for area sensors and 2 V to 4 V for linear sensors. The dc level of the waveform is between 3 V and 7 V.

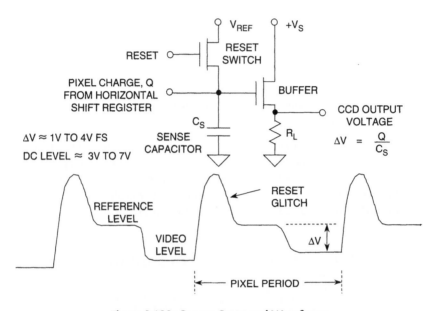

Figure 8.100: Output Stage and Waveforms

CCD processes generally have limited capability to perform on-chip signal conditioning. Therefore the CCD output is generally processed by external conditioning circuits. The nature of the CCD output requires that it be clamped before being digitized by the ADC. In addition, offset and gain functions are generally part of the analog signal processing.

CCD output voltages are small and quite often buried in noise. The largest source of noise is the thermal noise in the resistance of the FET reset switch. This noise may have a typical value of 100 to 300 electrons rms (approximately 60 to 180 mV rms). This noise, called "kT/C" noise, is illustrated in Figure 8.101. During the reset interval, the storage capacitor C_S is connected to V_{REF} via a CMOS switch. The on-resistance of the switch (R_{ON}) produces thermal noise given by the well known equation:

$$\text{Thermal Noise} = \sqrt{4kT \times BW \times R_{ON}} \qquad \text{Eq. 8.14}$$

The noise occurs over a finite bandwidth determined by the $R_{ON} C_S$ time constant. This bandwidth is then converted into equivalent noise bandwidth by multiplying the single-pole bandwidth by $\pi/2$ (1.57):

$$\text{Noise BW} = \frac{\pi}{2}\left[\frac{1}{2\pi R_{ON} C_S}\right] = \frac{1}{4 R_{ON} C_S} \qquad \text{Eq. 8.15}$$

Substituting into the formula for the thermal noise, note that the R_{ON} factor cancels, and the final expression for the thermal noise becomes:

$$\text{Thermal Noise} = \sqrt{\frac{kT}{C}}$$

Eq. 8.16

This is somewhat intuitive, because smaller values of R_{ON} decrease the thermal noise but increase the noise bandwidth, so only the capacitor value determines the noise.

Note that when the reset switch opens, the kT/C noise is stored on C_S and remains constant until the next reset interval. It therefore occurs as a *sample-to-sample* variation in the CCD output level and is common to both the reset level and the video level for a given pixel period.

V_{REF}

RESET SWITCH

R_{ON}

Q

C_S

$$\text{THERMAL NOISE} = \sqrt{4kT \times BW \times R_{ON}}$$

$$\text{NOISE BW} = \frac{\pi}{2}\left[\frac{1}{2\pi R_{ON}C_S}\right] = \frac{1}{4\,R_{ON}C_S}$$

$$\text{THERMAL NOISE} = \sqrt{\frac{kT}{C_S}}$$

SAME VALUE PRESENT DURING
REFERENCE AND VIDEO LEVELS
WHILE RESET SWITCH IS OPEN

Figure 8.101: kT/C Noise

A technique called *correlated double sampling* (CDS) is often used to reduce the effect of this noise. Figure 8.102 shows one circuit implementation of the CDS scheme, though many other implementations exist. The CCD output drives both SHAs. At the end of the reset interval, SHA1 holds the reset voltage level plus the kT/C noise. At the end of the video interval, SHA2 holds the video level plus the kT/C noise. The SHA outputs are applied to a difference amplifier which subtracts one from the other. In this scheme, there is only a

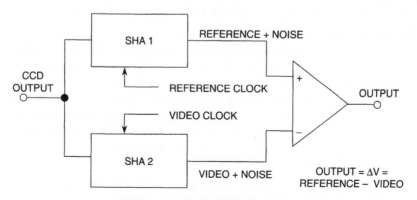

CCD OUTPUT

SHA 1

REFERENCE + NOISE

REFERENCE CLOCK

VIDEO CLOCK

SHA 2

VIDEO + NOISE

OUTPUT

$\text{OUTPUT} = \Delta V =$
$\text{REFERENCE} - \text{VIDEO}$

Figure 8.102: Correlated Double Sampling (CDS)

short interval during which both SHA outputs are stable, and their difference represents ΔV, so the difference amplifier must settle quickly. Note that the final output is simply the difference between the reference level and the video level, ΔV, and that the kT/C noise is removed.

Contact Image Sensors (CIS) are linear sensors often used in facsimile machines and low-end document scanners instead of CCDs. Although a CIS does not offer the same potential image quality as a CCD, it does offer lower cost and a more simplified optical path. The output of a CIS is similar to the CCD output except that it is referenced to or near ground (see Figure 8.103), eliminating the need for a clamping function. Furthermore, the CIS output does not contain correlated reset noise within each pixel period, eliminating the need for a CDS function. Typical CIS output voltages range from a few hundred mV to about 1V fullscale. Note that although a clamp and CDS is not required, the CIS waveform must be sampled by a sample-and-hold before digitization.

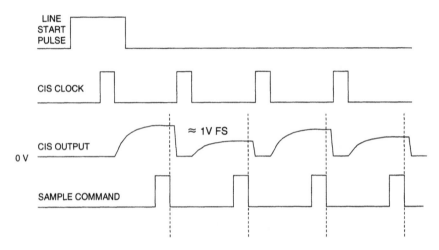

Figure 8.103: Contact Image Sensor (CIS) Waveforms

Analog Devices offers several *analog-front-end* (AFE) integrated solutions for the scanner, digital camera, and camcorder markets. They all comprise the signal processing steps described above. Advances in process technology and circuit topologies have made this level of integration possible in foundry CMOS without sacrificing performance. By combining successful ADC architectures with high performance CMOS analog circuitry, it is possible to design complete low cost CCD/CIS signal processing ICs.

The AD9898 (Reference 28) is a highly integrated CCD signal processor for digital still camera and digital video camera applications. A simplified block diagram is shown in Figure 8.104. It includes a complete analog front end with 10-bit A/D conversion combined with a full function programmable timing generator. A precision timing core allows adjustment of high speed clocks with 1 ns resolution at 20 MSPS operation. The AD9898 is specified at pixel rates as high as 20 MHz. The analog front end includes black level clamping, CDS, VGA, and a 10-bit A/D converter. The timing generator provides all the necessary CCD clocks: RG, H-clocks, V-clocks, sensor gate pulses, substrate clock, and substrate bias pulse. Operation is programmed using a 3-wire serial interface. Packaged in a space saving 48-lead LFCSP, the AD9898 is specified over an operating temperature range of –20°C to +85°C.

Three-channel CCD analog front-ends available from Analog Devices include the AD9816 (12-bit), AD9822 (14-bit), AD9814 (14-bit), and the AD9826 (16-bit) processors.

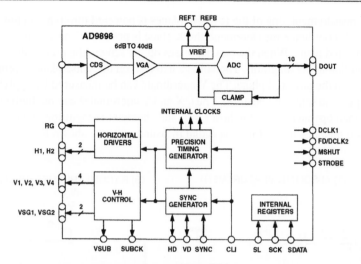

Figure 8.104: AD9898 CCD Signal Processor with
Precision Timing™ Generator

Touchscreen Digitizers

Touchscreens have become widespread in hand-held PDAs (Personal Digital Assistants) and other computer products. The majority of PDA makers use a 4-wire resistive element as the touchscreen due to its low cost and simplicity. For the touchscreen to interface with the host processor, analog waveforms from the screen must first be converted to digital data (Reference 29, 30).

The user enters data on the screen with a stylus. An ADC converts this analog information to digital data that the host microprocessor uses to determine the stylus's position on the screen. There are a number of inherent problems associated with this application that must be overcome by the ADC.

The touchscreen is usually constructed from two layers of transparent resistive material, in most cases indium tin oxide or other resistive polyester material, with silver ink used for electrodes. The resistance of each layer can vary between vendors, but typically ranges from 100 Ω to 900 Ω. The two layers are placed on top of each other on an insulating layer of glass as shown in Figure 8.105.

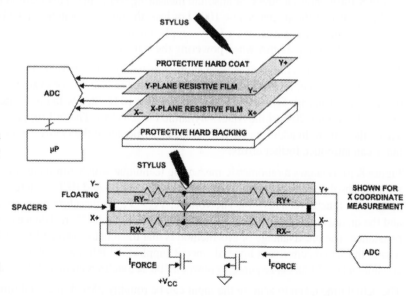

Figure 8.105: 4-Wire Resistive Touchscreen ADC Interface

During coordinate measurement, one of the resistive planes is powered through on-chip switches on the controller ADC. For X coordinate measurement, the X plane is powered. The Y plane senses where the pen is located on the powered plane. When the pen depresses on the screen, the planes short at this location (shown as the dotted line in the diagram). The voltage detected on the sense plane is proportional to the location of the touch on the powered plane. The Y coordinate can be measured by applying power to the Y plane, using the X plane to sense the position. Thus, X and Y coordinates can be digitized from the screen. The digital code is then operated on by the host CPU, and character recognition and position information can be achieved. Two methods for making the actual measurements are shown in Figure 8.106.

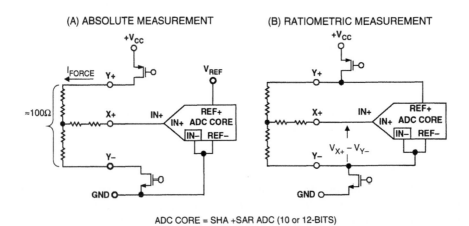

ADC CORE = SHA +SAR ADC (10 or 12-BITS)

Figure 8.106: Absolute and Ratiometric Measurements of Touchscreen Voltages

Figure 8.106A shows a direct, or absolute measuring technique. This method has several problems. Because the impedance of the screen can be 100 Ω or less, the on-chip switches must be carefully designed. For example, assume the system has a 3.3 V supply and a 100 Ω touchscreen. The switches must be capable of sourcing and sinking 33 mA when powering the screen.

The on-chip switches themselves pose another difficulty. They have an inherent ON-resistance, which when powering the screen, results in a voltage drop across the switch. For example, with the 100 Ω screen, if the on-chip switches have ON-resistance values around 10 Ω, then with one switch to the power supply and another to ground, 20% of the ADC's dynamic range is lost. The full supply voltage can never be developed across the screen. In addition, the temperature coefficient of the ON-resistance and the touchscreen resistance can introduce further errors.

Figure 8.106B shows a ratiometric measuring technique which eliminates most of the errors associated with the absolute measurement. In this method, the REF+ and REF– voltages are taken directly across the ends of the touchscreen resistor. The voltage, $V_{X+} - V_{Y-}$, is therefore proportional to the reference voltage, and the digital code is not affected by changes in the switch ON-resistance or the end-to-end touchscreen resistance. The disadvantage of this method is that the touchscreen end-to-end resistance must be powered during the actual conversion interval since it supplies the reference voltage to the ADC. The power can be considerable—33 mA is required to power a 100 Ω touchscreen on a 3.3 V supply, i.e., 109 mW.

The actual time taken to acquire the input can be roughly 25% of the total time taken to acquire a sample (1.5 μs for AD7873) and convert a sample (6 μs for AD7873) by the successive approximation ADC.

In effect, the screen need not be powered when the converter is performing the actual analog-to-digital conversion if the absolute technique is used. It only needs to be powered during the SHA acquisition time. However, in the ratiometric method, the screen must be powered throughout the entire conversion process, as it provides the reference voltage for the ADC.

The AD7873 touchscreen digitizer (Reference 31) has a 12-bit successive approximation ADC with a synchronous serial interface and low ON-resistance switches for driving touch screens (Figure 8.107). The AD7873 operates from a single 2.2 V to 5.25 V power supply and features throughput rates greater than 125 kSPS. The AD7873 features direct battery measurement, temperature measurement, and touch-pressure measurement. The AD7873 also has an on-board reference of 2.5 V which can be used for the auxiliary input, battery monitor, and temperature measurement modes. When not in use, the internal reference can be shut down to conserve power. An external reference can also be applied and can be varied from 1 V to V_{CC}, while the analog input range is from 0 V to V_{REF}.

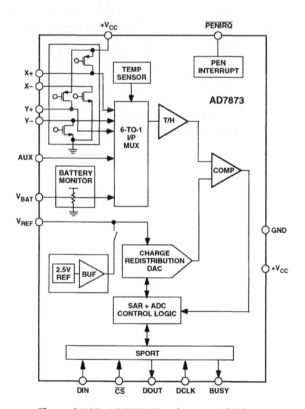

Figure 8.107: AD7873 Touchscreen Digitizer

The device includes a shutdown mode that reduces the current consumption to less than 1 µA. The AD7873 features on-board switches. This, coupled with low power and high-speed operation, makes the device ideal for battery-powered systems such as personal digital assistants with resistive touch screens and other portable equipment. The part is available in a 16-lead 0.15" Quarter Size Outline (QSOP) package, a 16-lead Thin Shrink Small Outline (TSSOP) package, and a 16-lead Lead Frame Chip Scale (LFCSP) package.

The analog input to the ADC is provided via an on-chip multiplexer. This analog input may be any one of the X, Y, and Z panel coordinates, battery voltage, or chip temperature. The multiplexer is configured with low-resistance switches that allow an unselected ADC input channel to provide power and an accompanying pin to provide ground for an external device. For some measurements the ON-resistance of the switches may present a source of error. However, with a ratiometric input to the converter this error can be negated as previously described. A typical application circuit is shown in Figure 8.108.

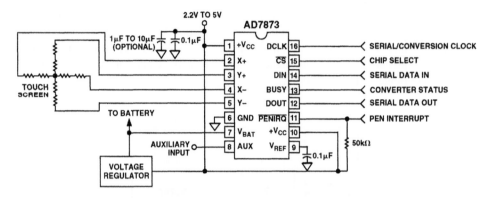

**Figure 8.108: AD7873 Touchscreen
Digitizer Application Circuit**

References:
8.5 Digital Video And Display Electronics

1. A. Goldberg, "PCM-Encoded NTSC Color Television Subjective Tests," **SMPTE Journal**, Vol. 82, pp. 649–654, Aug. 1973.

2. David E. Acker and Richard H. McLean, "Digital Time-Base Correction for Video Signal Processing," **SMPTE Journal**, Vol. 85, pp. 146–150, March 1976.

3. Walter A. Kester, "Characterizing and Testing A/D and D/A Converters for Color Video Applications," **IEEE Transactions on Circuits and Systems**, Vol. CAS-25, July 1978, pp. 539–550.

4. W. A. Kester, "PCM Signal Codecs for Video Applications," **SMPTE Journal**, Number 88, November 1979, pp. 770–778.

5. W. K. Bucklen, "A Monolithic Video A/D Converter," **Digital Video**, Vol. 2, Society of Motion Picture and Television Engineers, pp. 34–42, March 1979. (Describes the revolutionary TDC1007J. Originally introduced at the 3 Feb. 1979 SMPTE Winter Conference in San Francisco). Bill Bucklen actually accepted an Emmy award in 1988 for this product and was responsible for marketing it to the world).

6. SMPTE 244M, "System M/NTSC Composite Video Signals-Bit Parallel Digital Interface," www.smpte.org.

7. IEEE Std. 746-1984, "IEEE Standard for Performance Measurements of A/D and D/A Converters for PCM Television Video Circuits," IEEE, 1984.

8. ITU-Recommendation BT.601, "Universal Sampling Specification for SDTV and HDTV Broadcast Video," www.itu.int.

9. SMPTE 125M, "Component Video Signal 4:2:2 Bit-Parallel Digital Interface," www.smpte.org.

10. SMPTE 259M, "10-Bit, 4:2:2 Component and $4f_{sc}$ Composite Digital Signals—Serial Digital Interface," www.smpte.org.

11. ITU-Recommendation BT.656-4, "Interfaces for Digital Component Video signals in 525-Line and 625-Line Television Systems Operating at the 4:2:2 Level of Recommendation ITU-R BT.601 (Part A)," www.itu.int.

12. ITU-Recommendation BT.709-5, "Parameter Values for the HDTV Standards for Production and International Programme Exchange," www.itu.int. Also, see SMPTE 296M, SMPTE 274M, www.smpte.org.

13. SMPTE 292M, "Bit-Serial Digital Interface for High-Definition Television Systems, www.smpte.org.

14. Charles Poynton, **Digital Video and HDTV Algorithms and Interfaces**, Morgan Kaufmann Publishers, 2003, ISBN 1-55860-792-7.

15. Michael Robin and Michel Poulin, **Digital Television Fundamentals**, Second Edition, McGraw-Hill, 2000, ISBN 0-07-135581-2, Chapter 6.

16. Data sheet for ADV7183A 10-Bit NTSC/PAL/SECAM Video Decoder, www.analog.com.

17. Data sheet for ADV7310 Multiformat 216 MHz Video Encoder with Six NST™ 12-Bit DACs, www.analog.com.

18. Data sheet for ADV7125 CMOS, 330 MHz Triple 8-Bit High Speed Video DAC, www.analog.com.

19. Data sheet for ADV7160/ADV7162 96-Bit, 220 MHz True-Color Video RAM-DAC, www.analog.com.

20. George Diniz and Tim Stroud, "Bringing Displays into the Digital Future," **EDN**, April 26, 2001, pp. 105–114.

21. Doug Bartow, "Smart Integration in Flat Panel Displays," **Information Display**, Vol. 15, October 1999.

22. Digital Display Working Group, Digital Video Interface Specification, DVI 1.0, www.ddwg.org.

23. Data sheet for AD9888 100/140/170/205 MSPS Analog Flat Panel Interface, www.analog.com.

24. Data sheet for AD9887A Dual Interface for Flat Panel Displays, www.analog.com.

25. Erik Barnes, "High Integration Simplifies Signal Processing for CCDs," **Electronic Design**, February 23, 1998, pp. 81–88.

26. Erik Barnes, "Integrated Front Ends for CCD Signal Processing, **Analog Dialogue,** 32-1, Analog Devices, 1998, www.analog.com.

27. Kevin Buckley, "Selecting an Analog Front End for Imaging Applications," **Analog Dialogue**, 34-6, Analog Devices, October, 2000, www.analog.com.

28. Data sheet for AD9898 CCD Signal Processor with Precision Timing™ Generator, www.analog.com.

29. Paul Kearney, "Configure Your ADC Correctly to Interface with the Display," **Portable Design**, July, 2003.

30. Paul Kearney, "The PDA Challenge-Met by the AD7873 Resistive-Touch-Screen Controller ADC, Analog Dialogue, **Analog Dialogue**, 35-04, Analog Devices, 2001, www.analog.com.

31. Data sheet for AD7873 Touchscreen Digitizer, www.analog.com.

Software Radio and IF Sampling
Walt Kester

Introduction

The term *software radio* had its origins in military intelligence receivers of the late 1980s and the early 1990s (References 1 and 2). Since then, the concept has been widely implemented commercially, especially in cellular radio applications (References 3–18).

A software radio receiver uses an ADC to digitize the analog signal in the receiver as close to the antenna as practical, generally at an intermediate frequency (IF). Hence the term, *IF sampling* came into being. Once digitized, the signals are filtered, demodulated, and separated into individual channels using specialized DSPs called *receive signal processors* (RSPs). Similarly, a software radio transmitter performs coding, modulation, etc., in the digital domain—and near the final output IF stage, a DAC is used to convert the signal back to an analog format for transmission. The DSP which precedes the DAC is referred to as the transmit signal processor (TSP). A very simplified generic software radio receiver and transmitter are shown in Figure 8.109.

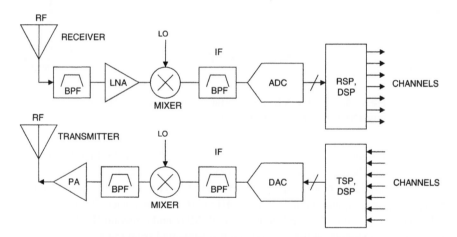

Figure 8.109: Generic IF Sampling Software Radio Receiver and Transmitter

Ideally, the software radio eliminates quite a bit of expensive analog signal processing circuitry and performs these functions in low cost DSPs. The software radio also allows the same hardware to handle various wireless air standards by making changes to the various DSP programs.

Wideband IF sampling places high demands on the ADCs and DACs in terms of SNR and SFDR, as has previously been discussed in Chapter 2. However, converter technology has progressed to the point that software radio is practical for most of the popular wireless air standards. For high volume applications, such as cellular telephone basestations and handsets, software radio has become a reality and a necessity.

Evolution of Software Radio

In order to understand the evolution of software radio, consider the analog superheterodyne receiver invented in 1917 by Major Edwin H. Armstrong (see Figure 8.110). This architecture represented a significant improvement over single-stage direct conversion (homodyne) receivers which had previously been constructed using tuned RF amplifiers, a single detector, and an audio gain stage. A significant advantage of the superhetrodyne receiver is that it is much easier and more economical to have the gain and selectivity of a receiver at fixed intermediate frequencies (IF) than to have the gain and frequency-selective circuits "tune" over a band of frequencies.

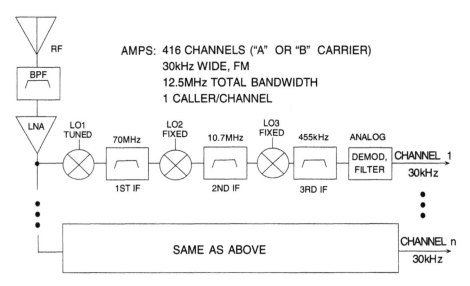

Figure 8.110: U.S. Advanced Mobile Phone Service (AMPS) Superheterodyne Analog Receiver

The frequencies shown in Figure 8.110 correspond to the AMPS (Advanced Mobile Phone Service) analog cellular phone system currently used in the U.S., but quickly being phased out in favor of other digital standards. The receiver is designed for AMPS signals at 900 MHz RF. The signal bandwidth for the "A" or "B" carriers serving a particular geographical area is 12.5 MHz (416 channels, each 30 kHz wide). The receiver shown uses triple conversion, with a first IF frequency of 70 MHz and a second IF of 10.7 MHz, and a third IF of 455 kHz. The image frequency at the receiver input is separated from the RF carrier frequency by an amount equal to twice the first IF frequency (illustrating the point that using relatively high first IF frequencies makes the design of the image rejection filter easier).

The output of the third IF stage is demodulated using analog techniques (discriminators, envelope detectors, synchronous detectors, etc.). In the case of AMPS, the modulation is FM. An important point to notice about the above scheme is that there is *one receiver required per channel*, and only the antenna, prefilter, and LNA can be shared.

It should be noted that in order to make the receiver diagrams more manageable, the interstage amplifiers are not shown. They are, however, an important part of the receiver, and the reader should be aware that they must be present.

Receiver design is a complicated art, and many trade-offs can be made between IF frequencies, single-conversion versus double-conversion or triple conversion, filter cost and complexity at each stage in the receiver, demodulation schemes, etc. There are many excellent references on the subject, and the purpose of this section is only to acquaint the design engineer with some of the emerging architectures, especially in the application of ADCs and DACs in the design of advanced communications receivers.

A Receiver Using Digital Processing at Baseband

With the availability of high performance high speed ADCs and DSPs, it is now becoming common practice to use digital techniques in at least part of the receive and transmit path, and various chipsets are available from Analog Devices to perform these functions for GSM and the other cellular standards. This is illustrated in Figure 8.111, where the output of the last IF stage is converted into a baseband in-phase (I) and quadrature (Q) signal using a quadrature demodulator. The I and Q signals are then digitized by a dual ADC. The RSPs/DSPs then perform the additional signal processing. The signal can then be converted into analog format using a DAC, or it can be processed, mixed with other signals, upconverted, and retransmitted.

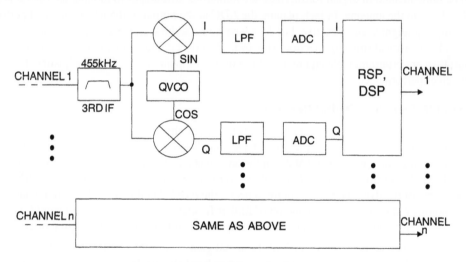

**Figure 8.111: Digital Receiver Using
Baseband Sampling and Digital Processing**

At this point, we should make it clear that *a digital receiver is not the same thing as digital modulation.* In fact, a digital receiver will do an excellent job of receiving an analog signal such as AM or FM. Digital receivers can be used to receive any type of modulation standard including analog (AM, FM) or digital (QPSK, QAM, FSK, GMSK, etc.). Furthermore, since the core of a digital radio is its digital signal processor (RSP/DSP), the same receiver can be used for both analog and digitally modulated signals (simultaneously if necessary), assuming that the RF and IF hardware in front of the RSP/DSP is properly designed. Since it is software that determines the characteristics of the radio, changing the software changes the radio. For this reason, digital receivers are often referred to as *software radios*.

The fact that a radio is software programmable offers many benefits. A radio manufacturer can design a generic radio in hardware. As air interface standards change (as from AMPS to IS-136, or IS-95), the manufacturer is able to make timely design changes to the radio by reprogramming the RSP/TSP/DSP. From a user or service-provider's point of view, the software radio can be upgraded by loading the new software at a small cost, while retaining all of the initial hardware investment. Additionally, the receiver can be tailored for custom applications at very low cost, since only software costs are involved.

A digital receiver performs the same function as an analog one with one difference; some of the analog functions have been replaced with their digital equivalent. The main difference between Figure 8.110 and Figure 8.111 is that the FM discriminator in the analog radio has been replaced with two ADCs and a RSP/DSP. While this is a very simple example, it shows the fundamental beginnings of a digital, or *software* radio.

An added benefit of using digital techniques is that some of the filtering in the radio is now performed digitally. This eliminates the requirement of tight tolerances and matching for frequency-sensitive components such as inductors and capacitors. In addition, since filtering is performed within the RSP/DSP, the filter characteristics can be implemented in software instead of costly and sensitive SAW, ceramic, or crystal filters. In fact, many filters that could never be implemented in a strictly analog receiver can be digitally synthesized.

This simple example is only the beginning. With current technology, much more of the receiver and transmitter can be implemented in digital form. There are numerous advantages to moving the digital portion of the radio closer to the antenna. In fact, placing the ADC at the output of the RF section and performing direct RF sampling might seem attractive, but does have some serious drawbacks, particularly in terms of selectivity and out-of-band (image) rejection. However, the concept makes clear one key advantage of software radios: they are programmable and require little or no component selection or adjustments to attain the required receiver performance.

Narrowband IF-Sampling Digital Receivers

A reasonable compromise in many digital receivers is to convert the signal to digital form at the output of the first or the second IF stage. This allows for out-of-band signals to be filtered before reaching the ADC. It also allows for some automatic gain control (AGC) in the analog stage ahead of the ADC to reduce the possibility of in-band signals overdriving the ADC and allows for maximum signal gain prior to the A/D conversion. This relieves some of the dynamic range requirements on the ADC. Additionally, IF sampling and digital receiver technology reduce costs by elimination of further IF stages (mixers, filters, and amplifiers) and adds flexibility by the replacement of fixed analog filter components with programmable digital ones.

In analyzing an analog receiver design, much of the signal gain is after the first IF stage. This prevents front-end overdrive due to out-of-band signals or strong in-band signals. However, in an IF sampling digital receiver, all of the gain is in the front end, and great care must be taken to prevent in-band and out-of-band signals from saturating the ADC, which results in excessive distortion. Therefore, a method of attenuation must be provided when large in-band signals occur. While additional signal gain can be obtained digitally after the ADC, there are certain restrictions. Gain provided in the analog domain improves the SNR of the signal and only reduces the performance to the degree that the noise figure (NF) degrades noise performance.

Figure 8.112 shows a detailed IF sampling digital receiver for the GSM/EDGE (900 MHz) system. The receiver has RF gain, automatic gain control (AGC), a high performance ADC, digital receive signal processor (RSP), and a DSP.

The heart of the system is the 12-bit 26 MSPS ADC with AGC, the RSSI (Received Signal Strength Indicator), and the RSP. Various chipsets are available which perform these functions for GSM/EDGE (see AD6600, AD6650, AD6620, AD6624, and AD6634) and for WCDMA (see AD6634, and AD6652).

GSM/EDGE (Europe) and IS-136 (United States) are similar multicarrier time-division-multiplexed-access (TDMA) systems, while IS-95, IS-95B, WCDMA, and CDMA2000 are spread spectrum code-division-multiple-access systems (CDMA). The channel bandwidth for CDMA systems is either 1.25 MHz for

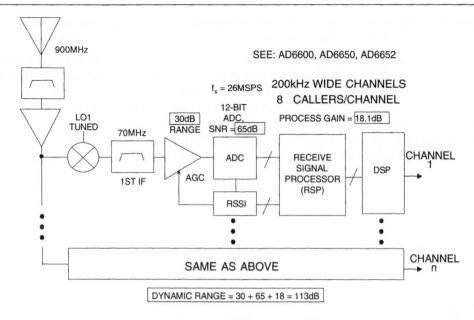

Figure 8.112: Narrowband IF Sampling GSM/EDGE Digital Receiver

IS-95, IS-95B, and CDMA2000, or 5 MHz for WCDMA. There will be more details on these air standards later in this section.

The GSM/EDGE 900 MHz air standard is one of the most stringent with respect to ADC dynamic range, and therefore a narrowband receiver design is most often implemented. In the system shown in Figure 8.112, the total dynamic range is 113 dB comprised by the AGC loop (30 dB), ADC SNR (65 dB), and the process gain (18.1 dB).

The bandwidth of a single GSM channel is 200 kHz, and each channel can handle up to eight simultaneous callers. A typical basestation may be required to handle 50 to 60 simultaneous callers, thereby requiring eight separate signal processing channels.

Figure 8.113 shows the IF frequency of 71.5 MHz centered in the 6th Nyquist zone sampled at a frequency of 26 MSPS. The RSP reverses the frequency sense of the signal when it is translated to the first Nyquist zone as shown.

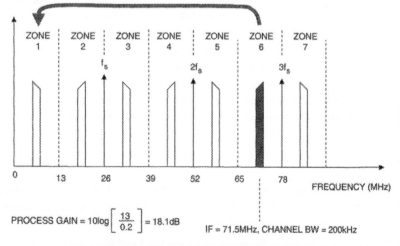

Figure 8.113: Narrowband GSM Receiver Bandpass Sampling of a 200 kHz Channel at 26 MSPS

We now have a 200 kHz baseband signal (generated by undersampling) which is being sampled at 26 MSPS as shown in Figure 8.114A. The RSP then translates the signal to baseband as shown in Figure 8.114B.

The signal is then passed through a digital filter in the RSP which removes all frequency components above 200 kHz, including the quantization noise which falls in the region between 200 kHz and 13 MHz (the Nyquist frequency) as shown in Figure 8.114C. The resultant increase in SNR is 18.1 dB (processing gain). There is no information contained in the signal above 200 kHz, and the output data rate can be reduced (decimated) from 26 MSPS to 541.7 kSPS, a data rate the DSP can handle, as shown in Figure 8.114D. The data corresponding to the 200 kHz channel is transmitted to the DSP over a simple 3-wire serial interface. The DSP then performs such functions as channel equalization, decoding, and spectral shaping.

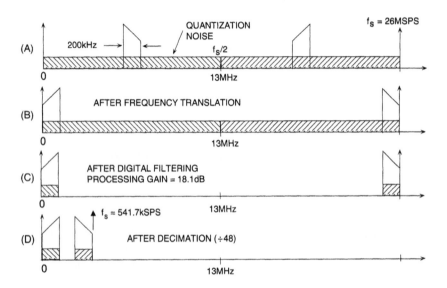

**Figure 8.114: Digital Filtering and Decimation
of the 200 kHz GSM Channel**

The concept of *processing gain* is common to all communications systems, analog or digital, and was discussed in Chapter 2 of this book. In a sampling system, the quantization noise produced by the ADC is spread over the entire Nyquist bandwidth which extends from dc to $f_s/2$. If the signal bandwidth, BW, is less than $f_s/2$, digital filtering can remove the noise components outside this bandwidth, thereby increasing the effective SNR. The processing gain in a sampling system can be calculated from the formula:

$$\text{Processing Gain} = 10\log\left(\frac{f_s}{2\times\text{BW}}\right) \qquad \text{Eq. 8.17}$$

The SNR (noise measured over $f_s/2$ bandwidth) of the ADC at the bandwidth of the signal should be used to compute the actual narrowband SNR by adding the processing gain determined by the above equation. If the ADC is an ideal N-bit converter, then its SNR (measured over the Nyquist bandwidth) is 6.02N + 1.76 dB.

Notice that as shown in the previous narrowband receiver example, there can be processing gain even if the original signal is an undersampled one. The only requirement is that the signal bandwidth be less than $f_s/2$, and that the noise outside the signal bandwidth be removed with a digital filter.

Wideband IF-Sampling Digital Receivers

Thus far, we have avoided a detailed discussion of *narrowband* versus *wideband* digital receivers. A digital receiver can be either, but more detailed definitions are important at this point. By *narrowband*, we mean that sufficient prefiltering has been done such that all undesired signals have been eliminated and that only the signal of interest is presented to the ADC input. This is the case for the GSM/EDGE basestation example previously discussed.

Wideband simply means that a number of channels are presented to the input of the ADC, and further filtering, tuning, and processing is performed digitally. Usually, a wideband receiver is designed to receive an entire band of cellular or other similar wireless services. In fact, one wideband digital receiver can be used to receive *all* channels within the band simultaneously, allowing almost all of the analog hardware (including the ADC) to be shared among all channels as shown in Figure 8.115, which compares the narrowband and the wideband approaches.

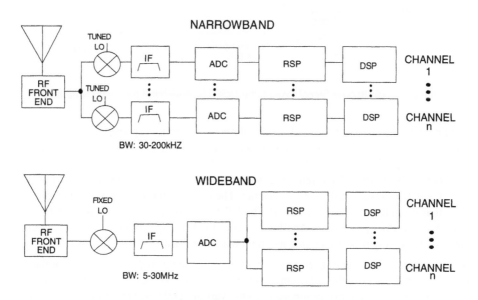

Figure 8.115: Narrowband versus Wideband Digital Receiver

Note that in the narrowband digital radio, there is one front-end LO and mixer required per channel to provide individual channel tuning. In the wideband digital radio, however, the first LO frequency is fixed, and the "tuning" is done in the RSP circuits following the ADC.

A typical wideband digital receiver may process a 5 MHz to 30 MHz band of signals simultaneously. This approach is frequently called *block conversion*. In the wideband digital receiver, the variable local oscillator in the narrowband receiver has been replaced with a fixed oscillator, so tuning must be accomplished digitally. Tuning is performed using a digital down converter (DDC) and digital filter called a *channelizer*. The term channelizer is used because the purpose of these chips is to select one channel out of the many within the broadband spectrum actually present in the ADC output. A typical RSP is shown in Figure 8.116.

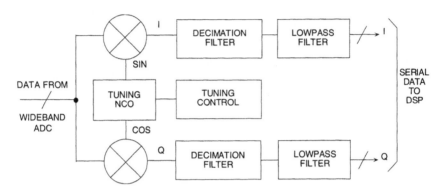

Figure 8.116: Receive Signal Processing (RSP) in Wideband Receiver (Simplified)

It consists of an NCO (Numerically Controlled Oscillator) with tuning capability, dual mixer, and matched digital filters. These are the same functions that would be required in an analog receiver, but implemented in digital form. The digital output from the channelizer is the demodulated signal in I and Q format, and all other signals have been filtered and removed. Since the channelizer output consists of one selected RF channel, one channelizer is required for each channel. The channelizer also serves to decimate the output data rate such that it can be processed by a DSP. The DSP extracts the signal information from the I and Q data and performs further processing. Another effect of the filtering provided by the channelizer is to increase the SNR by adding processing gain as previously described.

The design of a complete wideband receiver is a major project and is highly dependent on the particular air standard. Figure 8.117 shows the approximate evolution of the wireless air standards starting with the first generation (1G) analog systems, progressing to the various TDMA/FDM (time-division-multiple-access, frequency-division-multiplex) and the CDMA (code-division-multiple-access) digital systems of the second-generation (2G), followed by an intermediate generation referred to as 2.5G, and through the projected 3G standards of the future. Details of this evolution can be found in Reference 16.

From a spectral standpoint, there are basically two types of digital air standards. The TDMA/FDM standards use various time slots to multiplex the data on the different channels, and the resulting carriers are then multiplexed in frequency (FDM), with a spacing between channels of either 30 kHz or 200 kHz depending upon the air standard. The total bandwidth allocation per provider for these systems can range from 5 MHz to 15 MHz.

The second class of standards are the ones which use code-division-multiple-access (CDMA) techniques, sometimes referred to as *spread-spectrum*. In these systems, a pseudo-random number sequence modulates the channel data frequency, and the receiver uses an identical sequence to recover the channel data. The combination of multiple channels appears approximately as random noise spread over a bandwidth of either 1.25 MHz or 5 MHz depending on the air standard. Bandwidth allocations per provider of 5 MHz to 20 MHz are typical for these systems.

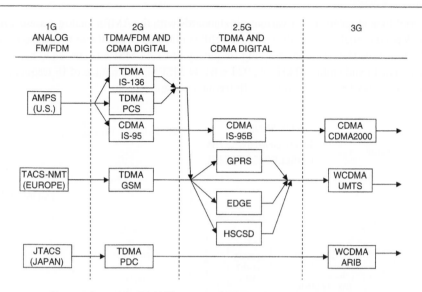

Figure 8.117: 1st Generation to 3rd Generation Wireless Evolution

The wireless spectrum is very crowded and contains many large signals which cause "blockers" from one band to interfere with desired signals in another. Regardless of the air standard, there are many signals that can interfere with the desired carriers. Figure 8.118 shows a typical RF spectrum where numerous narrowband signals surround the two CDMA2000 carriers located midband. The receiver must tolerate all of the narrowband signals while still maintaining the required sensitivity as defined by the particular air standard.

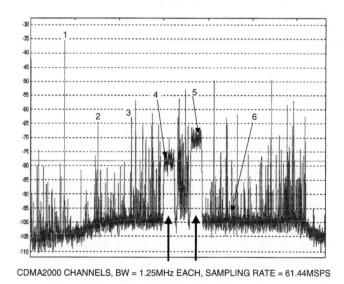

CDMA2000 CHANNELS, BW = 1.25MHz EACH, SAMPLING RATE = 61.44MSPS

Figure 8.118: Typical RF Spectrum of a
Multicarrier CDMA2000 Receiver

We begin our brief look at receivers for various air standards with the AMPS analog system which is ideally suited to the wideband digital receiver design. A simplified diagram of a suitable wideband digital receiver is shown in Figure 8.119. The AD6645 sampling frequency of 61.44 MSPS is chosen to be a power-of-two multiple of the channel bandwidth (30 kHz × 2024 = 61.44 MSPS). The choice of IF frequency is flexible, and a second IF stage may be required if lower IF frequencies are chosen.

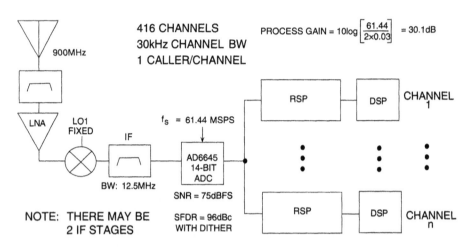

Figure 8.119: AMPS Wideband Digital Receiver

With a sampling frequency of 61.44 MSPS, the 12.5 MHz bandwidth signal can be positioned in the first Nyquist zone (dc to 30.72 MHz) with an IF frequency of 15.36 MHz, or in the second Nyquist zone (30.72 MHz to 61.44 MHz) with an IF frequency of 46.08 MHz.

The receive signal processors (RSPs) provide the receiver tuning and demodulate the signal into the I and Q components. The output data rate to the DSPs after decimation is approximately 60 kSPS. The processing gain incurred for a sampling frequency of 61.44 MSPS is calculated as follows:

$$\text{Processing Gain} = 10\log\left(\frac{61.44}{2 \times 0.03}\right) = 30.1 \text{ dB} \qquad \text{Eq. 8.18}$$

The SNR of the AD6645 over the Nyquist bandwidth is 75 dB, and when the process gain of 30.1 dB is added, the SNR in the 30-kHz bandwidth is 75 + 30.1 = 105.1 dB.

The SFDR of the AD6645 is greater than 96 dBc for signals several dB below full scale (with dither added). The following analysis shows that these values are more than adequate to meet the minimum AMPS requirements for sensitivity of –116 dBm with a blocker level of –26 dBm.

The simplified AMPS receiver analysis for spurious requirements begins as shown in Figure 8.120.

The maximum blocker level is –26 dBm (there is no actual specification for this—it was determined by actual measurements at Analog Devices). The minimum detectable signal (sensitivity) is specified as –116 dBm. Approximately 6 dB carrier-to-interferer ratio (C/I) is required to prevent the interferer from overtaking the desired signal. Therefore, for an input signal of –26 dBm, the worst allowable spurious signal must be at –122 dBm. This implies a minimum SFDR requirement of –26 dBm –(–122 dBm) = 96 dBc. The AD6645 14-bit ADC will meet this requirement with dither added (see discussion on dither later in this section) for a signal 5 dB below the ADC full-scale input of +5 dBm. In practice, dither is not generally

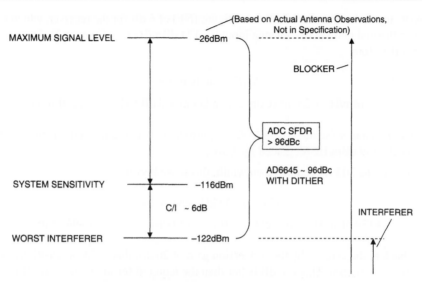

Figure 8.120: AMPS Spurious Requirements

used because there is additional margin in the design because the blocker level of –26 dBm (measured, not specified) is a very conservative number, and in practice can be reduced by several dB without affecting overall system performance.

The simplified AMPS receiver requirement for sensitivity (relating to ADC SNR) is shown in Figure 8.121.

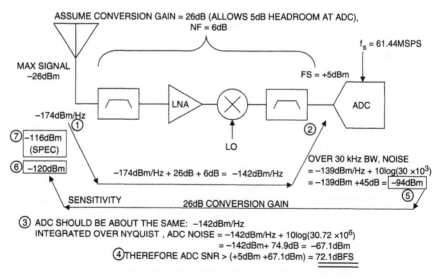

Figure 8.121: AMPS Receiver Sensitivity Requirements

The full-scale input of the ADC is +5 dBm (2.2 V p-p into 200 Ω, matched to 50 Ω with a 1:4 impedance-ratio RF transformer). Allowing 5 dB headroom at the ADC input, this requires a gain from the antenna to the ADC (conversion gain) of 26 dB, which causes the –26 dBm signal at the antenna to appear as a 0 dBm signal at the ADC input. Assume an overall noise figure (NF) of 6 dB for the receiver, which represents a good design. The thermal noise at the antenna input is –174 dBm/Hz (see Chapter 2). The noise reflected to the ADC input is therefore

$$\text{ADC Input Noise} =$$
$$-174\text{dBm/Hz} + 26 \text{ dB (Conversion Gain)} + 6 \text{ dB (NF)} = -142 \text{ dBm/Hz} \qquad \text{Eq. 8.19}$$

Now, assume that the input noise due to the ADC is approximately the same, and this places the total noise at the ADC input at –139 dBm/Hz (degraded by 3 dB).

When integrated over the 30 kHz channel bandwidth, this noise becomes:

$$\text{Noise in 30 kHz channel} =$$
$$-139 \text{ dBm/Hz} + 10\log(30 \times 10^3) = -139 \text{ dBm} + 45 \text{ dB} = -94 \text{ dBm} \qquad \text{Eq. 8.20}$$

When reflected back to the antenna by the conversion gain of 26 dB, this yields a sensitivity of –94 dBm – 26 dB = –120 dBm. This is 4 dB better than the required sensitivity of –116 dBm.

This assumes the ADC can meet the –142 dBm/Hz noise specification. This noise must be integrated over the Nyquist bandwidth (30.72 MHz) to obtain the data sheet number.

$$\text{ADC Noise}_{\text{Nyquist}} =$$
$$-142 \text{ dBm/Hz} + 10\log(30.72 \times 10^6) = -142 \text{ dBm} + 74.9 \text{ dB} = -67.1 \text{ dBm} \qquad \text{Eq. 8.21}$$

Since the full scale ADC input is +5 dBm, the minimum SNR requirement for the ADC is

$$\text{ADC SNR}_{\text{Nyquist}} = +5 \text{ dBm} -(-67.1 \text{ dBm}) = 72.1 \text{ dBFS} \qquad \text{Eq. 8.22}$$

The SNR specification for the AD6645 is 75 dBFS for a 15 MHz input signal, and it therefore meets the sensitivity requirement with nearly 3 dB margin.

The various steps in this analysis are numbered inside the circles in Figure 8.121 to make them easier to follow.

Needless to say, there are a number of other ways to approach this receiver design analysis, and many trade-offs can be made between the various parameters, but the simple method and numbers used above serve to illustrate the process, especially as it relates to the approximate ADC requirements.

Figure 8.122 shows the system requirements for the GSM 900MHz system. This is the most strenuous standard, especially from the standpoint of the ADC. A similar analysis can be used to determine the approximate ADC SFDR requirement. In this case, a C/I ratio of 15 dB is required. The resulting SFDR requirement of 106 dBc cannot be met with current ADCs, therefore narrowband receivers are most often used in this application.

Figure 8.123 shows the sensitivity analysis for the GSM 900 MHz system. The analysis proceeds along the same lines as in the previous AMPS sensitivity analysis, and the resulting ADC SNR requirement is approximately 85 dBFS which cannot be met with current ADCs—a narrowband approach must therefore be used.

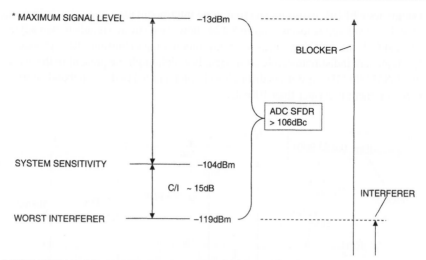

Figure 8.122: GSM 900 MHz Spurious Requirements

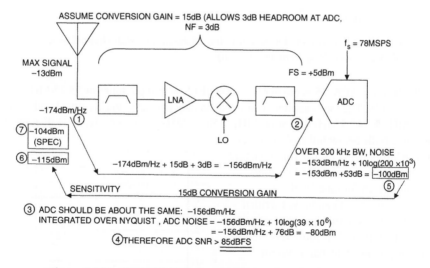

Figure 8.123: GSM 900 MHz Sensitivity Requirements (ADC SNR)

Because of the stringent ADC requirements for wideband GSM 900 MHz, the current receivers for this system are typically single-carrier narrowband types as previously discussed (see Figure 8.112 – Figure 8.114).

It should be noted, however, that the GSM-1800 MHz/1900 MHz (as well as PCS in the U.S.) maximum signal level requirement is reduced to –23 dBm, rather than –13 dBm (the sensitivity requirement is still –104 dBm), and a similar analysis shows an SFDR requirement of 93 dBc and an SNR of 75 dB which are obtainable with modern ADCs such as the AD6645.

In addition to single-tone SFDR, two-tone and multitone intermodulation distortion is important in an ADC for wideband receiver applications. Figure 8.124 shows two strong signals in two adjacent channels at frequencies f_1 and f_2. If the ADC has third-order intermodulation distortion, these products will fall at $2f_2 - f_1$ and $2f_1 - f_2$ and are indistinguishable from signals which might be present in these channels. This is one reason the GSM 900 MHz system is difficult to implement using the wideband approach, since the dynamic range requirement is greater than 100 dBc.

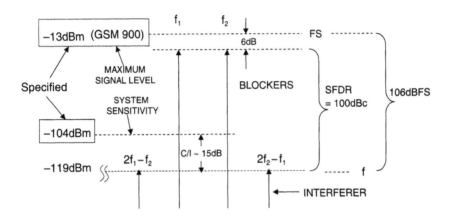

NOTE: MAXIMUM SIGNAL LEVEL FOR GSM 1800/1900 SPEC = –23dBm

Figure 8.124: Two-Tone Intermodulation Distortion in Multichannel System (GSM 900 MHz Requirements Shown)

The two-tone SFDR of the AD6645 is greater than 103 dBFS with input tones at 55.25 MHz and 56.25 MHz as shown in Figure 8.125. The tones are undersampled, so they appear in the Nyquist bandwidth at 80 MHz – 55.25 MHz = 24.75 MHz and at 80 MHz – 56.25 MHz = 23.75 MHz. Note than the amplitude of each tone must be 6 dB below full scale in order to prevent the ADC from being overdriven. It should be noted however that the actual GSM two-tone IMD specification is given for tone levels of –43 dBm. However, this specification was written with single-carrier systems in mind, so the test with tone levels 6 dB below fullscale is more representative of the requirements in a wideband system.

The requirements for multicarrier CDMA system performance are slightly different from TDMA/FDM systems because of the different architecture. In a CDMA receiver, the information to be transmitted is combined with a pseudorandom number (PN) spreading sequence that has a much wider bandwidth, using a function similar to a mixer. This has the effect of spreading the desired information over the wider bandwidth of the spreading signal as shown in Figure 8.126A and B.

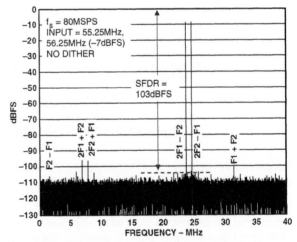

Figure 8.125: AD6645 Two-Tone Intermodulation Performance

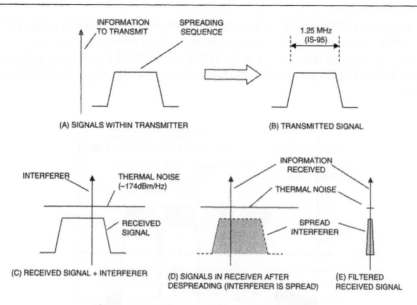

Figure 8.126: Signals Within a CDMA System

In the receiver, the same PN sequence is correlated with the incoming signal. The correlation process has the effect of "gathering" the energy of the desired transmission into the original information bandwidth, allowing it to be detected and further processed. At the same time, any energy, including interferers that do not correlate to the PN sequence, become spread over the wider bandwidth of the PN sequence as shown in Figure 8.126C and D.

Since the information bandwidth is now much narrower than the interfering energy, a low-pass filter can be used to remove all of the interfering energy, except the small amount that appears in the information bandwidth. This energy typically appears as Gaussian noise.

Figures 8.126D and E show the two components to the noise. The thermal noise present in the receiver is one component. The source of this is available atmospheric noise plus the active noise of the receiver and transmitter. In addition to this is the band-limited noise generated by spreading the interferer while the main signal is being despread. Since the receiver does not care about the source of the noise, the effective noise is the root-sum-square of these two.

This information can be used to determine the performance requirements for a 3G receiver, or any other receiver used for spread spectrum reception. Unlike GSM and other narrowband standards, spurious effects are not usually directly specified when it comes to "co-channel" interference, but they may be determined by carefully studying the operations in conjunction with the given standard specifications.

From this it is possible to determine the required performance from an ADC and the rest of the signal chain. (A detailed analysis of the requirements for an IS-95 CDMA system can be found in Reference 18.)

A digitized undersampled FFT output for a 4-carrier WCDMA system is shown in Figure 8.127. The channel spacing is 5 MHz, and the total bandwidth required for the four carriers is approximately 20 MHz. The AD6645 operates at a sampling frequency of 61.44 MSPS. The WCDMA carriers are shifted from a center frequency of 46.08 MHz (2nd Nyquist zone) to the baseband center frequency of 15.36 MHz by the process of undersampling.

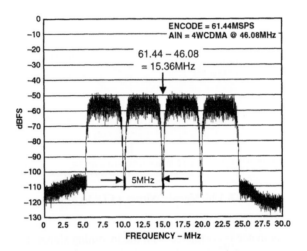

Figure 8.127: AD6645 Sampling at 61.44 MSPS with
Four WCDMA Inputs Centered at 46.08 MHz

Figure 8.128 illustrates an entire 25 MHz bandwith multicarrier signal centered at an IF frequency of 48.75 MHz digitized at 65 MSPS. The AD6645 digitizes signals in the second Nyquist zone with nearly the same dynamic performance as would be obtained if the signal were in the first Nyquist zone.

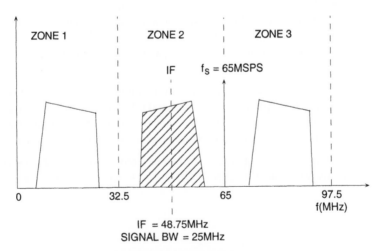

Figure 8.128: Sampling a 25 MHz BW Signal Using AD6645:
IF Frequency = 48.75 MHz, f_s = 65 MSPS

Figure 8.129 summarizes most of the current (2004) air standards and the approximate ADC requirements based on the individual standard specifications for maximum signal level, minimum signal level, etc. Notice that ADCs are currently available which will meet all the standards except for the GSM 900 MHz systems, previously discussed.

	MULTIPLE ACCESS METHOD	CHANNEL SPACING (BW)	TYPICAL TOTAL BW	ADC SAMPLING RATE (TYP.)	ADC SFDR	ADC SNR
AMPS	FDMA	30kHz	12.5MHz	61.44MSPS	96dBc	72dBFS
IS-136	TDMA/FDM	30kHz	5–15MHz	61–92MSPS	88dBc	68dBFS
GSM 900 MHz	TDMA/FDM	200kHz	5–15MHz	61–92MSPS	106dBc	85dBFS
GSM 1800/1900MHz, PCS	TDMA/FDM	200kHz	5–15MHz	61–92MSPS	93dBc	75dBFS
IS-95	CDMA	1.25MHz	5–15MHz	61–92MSPS	83dBc	74dBFS
CDMA2000	CDMA	1.25MHz	5–15MHz	61–92MSPS	79dBc	74dBFS
WCDMA (UMTS)	CDMA	5MHz	5–20MHz	61–92MSPS	79dBc	69dBFS

Figure 8.129: Approximate Wideband ADC Requirements for Popular Wireless Air Interface Standards

Increasing ADC Dynamic Range Using Dither

There are two fundamental limitations to maximizing SFDR in a high speed ADC. The first is the distortion produced by the front-end amplifier and the sample-and-hold circuit. The second is that produced by nonlinearity in the actual transfer function of the encoder portion of the ADC. The key to high SFDR is to minimize the nonlinearity of each.

Nothing can be done externally to the ADC to significantly reduce the inherent distortion caused by the ADC front end. However, the nonlinearity in the ADC encoder transfer function can be reduced by the proper use of dither (external noise that is summed with the analog input signal to the ADC).

Dithering improves ADC SFDR under certain conditions (References 20–23). For example, even in a perfect ADC, there is some correlation between the quantization noise and the input signal. This can reduce the SFDR of the ADC, especially if the input signal is an exact submultiple of the sampling frequency. Summing broadband noise (about ½ LSB rms in amplitude) with the input signal tends to randomize the quantization noise and minimize this effect (see Figure 8.130A). In most systems, however, there is enough noise riding on top of the signal so that adding additional dither noise is not required. Increasing the wideband rms noise level beyond an LSB will proportionally reduce the ADC SNR.

Other schemes have been developed using larger amounts of dither noise to randomize the transfer function of the ADC. Figure 8.130B also shows a dither noise source comprised of a pseudo-random number generator which drives a DAC. This signal is subtracted from the ADC input signal and then digitally added to the ADC output, thereby causing no significant degradation in SNR. An inherent disadvantage of this technique is that the allowable input signal swing is reduced as the amplitude of the dither signal is increased. This reduction in signal amplitude is required to prevent overdriving the ADC. It should be noted that this scheme does not significantly improve distortion created by the front-end of the ADC, only that produced by the nonlinearity of the ADC encoder transfer function.

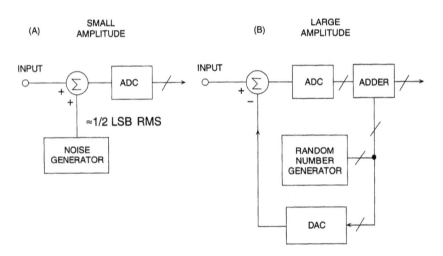

Figure 8.130: Using Dither to Randomize ADC Transfer Function

Another method that is easier to implement, especially in wideband receivers, is to inject a narrowband dither signal *outside the signal band of interest* as shown in Figure 8.131. Usually, there are no signal components located in the frequency range near dc, so this low frequency region is often used for such a dither signal. Another possible location for the dither signal is slightly below $f_s/2$. Because the dither signal occupies only a small bandwidth relative to the signal bandwidth, there is no significant degradation in SNR, as would occur if the dither was broadband.

A subranging pipelined ADC, such as the AD6645 (see Figure 8.132), has small differential nonlinearity errors that occur at specific regions across the ADC range. The AD6645 uses a 5-bit ADC (ADC1) followed by a 5-bit ADC2 and a 6-bit ADC3. The only significant DNL errors occur at the ADC1 transition points— the second and third stage ADC DNL errors are minimal. There are $2^5 = 32$ decision points associated with ADC1, and they occur every 68.75 mV ($2^9 = 512$ LSBs) for a 2.2 V full-scale input range. Figure 8.133 shows a greatly exaggerated representation of these nonlinearities.

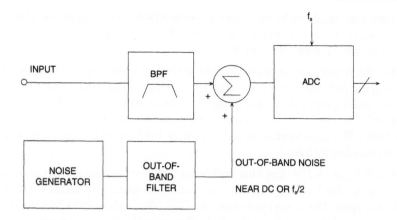

Figure 8.131: Injecting Out-of-Band Dither to Improve ADC SFDR

$2^5 = 32$ ADC 1 TRANSITIONS

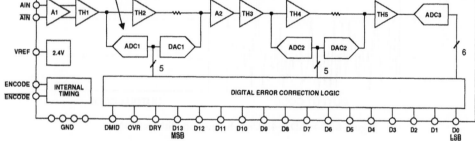

Figure 8.132: AD6645 Subranging Point DNL Errors (Exaggerated)

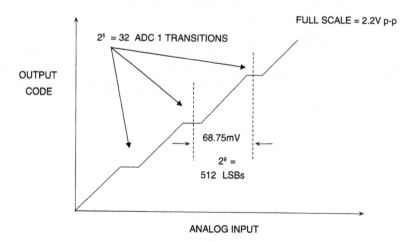

Figure 8.133: AD6645 Undithered and Dithered DNL

The distortion components produced by the front end of the AD6645 up to about 200 MHz analog input are negligible compared to those produced by the encoder. That is, *the static nonlinearity of the AD6645 transfer function* is the chief limitation to SFDR.

The goal is to select the proper amount of out-of-band dither so that the effect of these small DNL errors is *randomized* across the ADC input range, thereby reducing the average DNL error. Experimentally, it was determined that making the peak-to-peak dither noise cover about two ADC1 transitions gives the best improvement in DNL. The DNL is not significantly improved with higher levels of noise. Two ADC1 transitions cover 1024 LSBs peak-to-peak, or approximately 155 LSBs rms (peak-to-peak gaussian noise is converted to rms by dividing by 6.6).

The first plot shown in Figure 8.134 shows the undithered DNL over a small portion of the input signal range. The horizontal axis has been expanded to show two of the subranging points which are spaced 68.75 mV (512 LSBs) apart. The second plot shows the DNL after adding 155 LSBs rms dither. This amount of dither corresponds to approximately –20.6 dBm. Note the dramatic improvement in the DNL.

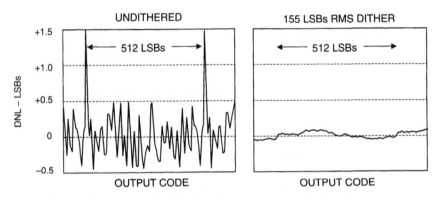

Figure 8.134: AD6645 Undithered and Dithered DNL

Dither noise can be generated in a number of ways. Noise diodes can be used, but simply amplifying the input voltage noise of a wideband bipolar op amp provides a more economical solution. This approach has been described in detail (References 21–23) and will not be repeated here.

The dramatic improvement in SFDR obtained with out-of-band dither is shown in Figure 8.135 using a deep (1,048,576-point) FFT, where the AD6645 is sampling a –35 dBm, 30.5 MHz signal at 80 MSPS. Note that the SFDR without dither is approximately 92 dBFS compared to 108 dBFS with dither, representing a 16 dB improvement. Figure 8.136 shows undithered and dithered SFDR as a function of input signal level and again shows the dramatic improvement.

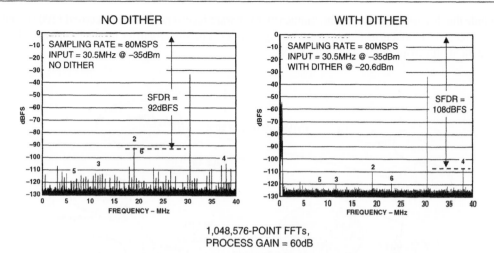

1,048,576-POINT FFTs,
PROCESS GAIN = 60dB

Figure 8.135: AD6645 Undithered and Dithered SFDR FFT Plot

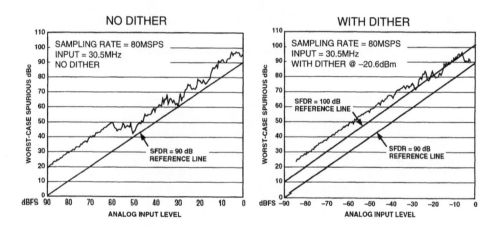

Figure 8.136: AD6645 Undithered and Dithered SFDR

We conclude the discussion of single and multicarrier software receivers with a few current (2004) road-maps of the receiver products available from Analog Devices. The single-carrier family is shown in Figure 8.137, and the multicarrier family in Figure 8.138.

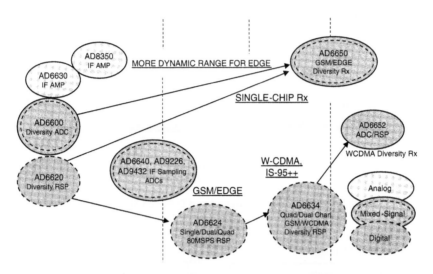

Figure 8.137: Summary: Single Carrier Receivers

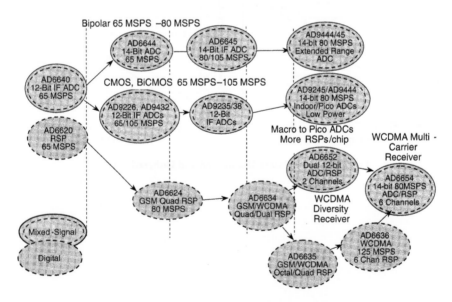

Figure 8.138: Summary: Multicarrier Receivers

Wideband Radio Transmitter Considerations

Many of the same concepts discussed in the previous wideband receiver sections apply to wideband transmitters as well. Two basic transmit architectures are shown in 8.139. In quadrature-based modulation schemes, such as QPSK and QAM, mixers are used to mix the in-phase (I) and quadrature (Q-90 degree out of phase) signals into a composite single-sideband signal for transmission. Figure 8.139A demonstrates a baseband transmit architecture that performs an analog mix of the I and Q. In this example, two DACs are required per transmit channel. This is the traditional architecture used in single-carrier systems. Even at the low output frequencies used in many baseband applications, the TxDAC family are the best choice because all family members combine (1) high SFDR at low output frequencies; (2) low power consumption, single-supply operation to enhance system power efficiency; (3) lower overall cost by oversampling the signal (interpolation) to reduce the DACs' in-band aliased images, thus easing the complexity of the analog band-pass filter; and (4) the variety of resolutions offered in the same pinout allows ultimate cost/performance trade-offs. For example, in many of the TxDAC beta-site applications, users started with one resolution model and later designed in either a higher or lower resolution device based on actual system performance. Details of the TxDAC family can be found in References 24 and 25.

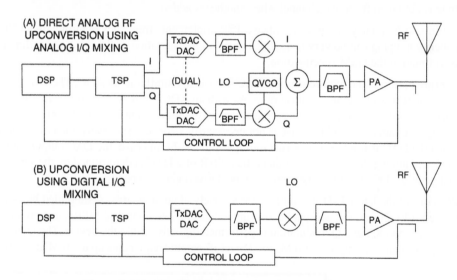

Figure 8.139: Simplified Wireless Transmitter Architectures

The system architecture in Figure 8.139B uses digital mixing of I and Q signals within the transmit signal processor (TSP) and sends the modulated signal directly to a single DAC. In this case, the bandwidth requirements of the DAC are more stringent. This approach is best for multicarrier systems. Current TxDACs can receive data at up to 160 MSPS. With digital modulation, intermediate frequencies (IFs) up to 70 MHz can be generated using TxDAC chips. Here, too, high SFDR, low price, low power, and family pin-compatibility are desirable (required) attributes. If multiple digital I and Q modulators are fed into the single DAC depicted in Figure 8.139B, the system becomes a wideband multicarrier transmit architecture, for which the superior multitone performance of the TxDAC family of products is a major performance attribute.

The transmit signal processor is a numeric post-processor for the DSP. The purpose of the TSP is to replace the first local oscillator, quadrature modulator, channel filtering and data interpolation. Like the RSP in the receiver, the TSP sets the transmitter apart from traditional designs because all channel characteristics are now programmable. This includes data rate, channel bandwidth and channel shape. Since modulation, channel filtering and other aspects of the modulation are done digitally, the filters will always perform exactly alike across all boards, unlike analog solutions that always have tolerances.

Several specifications are important when selecting a TSP. First, the device must be capable of generating data at the rates required to preserve the Nyquist bandwidth over the spectrum of interest. As with the ADC's sample rate, the sample rate of the DAC determines how much spectrum can be faithfully generated. Therefore, the TSP must be capable of generating data at least twice as fast as the band of interest and preferably three times faster as reasoned earlier for antialiasing filter response.

Similar to RSPs, the bus widths are also important, yet for different reasons. In the transmit direction, there are two different issues. If the TSP is used in a single-channel mode, the issue is simply quantization and thermal noise. It is usually not desirable to transmit excess in-band or out-of-band noise, since this wastes valuable transmitter efficiency and causes interference. In a multicarrier application, the concern is slightly different. Here, many channels would be digitally summed before reconstruction with a D/A converter. Therefore, each time the number of channels is doubled, an additional bit should be added so that the dynamic range is not taken from one channel when another is added.

Finally, the ability to frequency hop is vital. Since a TSP implements frequency control with an NCO and a mixer, frequency hopping can be very fast, allowing the implementation of the most demanding hopping applications as found in the GSM specification.

When considering performance requirements, a DAC is basically similar to an ADC. Therefore, the first specification of interest is the signal-to-noise ratio. As with an ADC, SNR is primarily determined by quantization and thermal noise. If either is too large, then the noise figure of the DAC will begin to contribute to the overall signal chain noise. While noise is not necessarily a concern spectrally, the issue does become important when the DAC is used to reconstruct multiple signals. In this case, the DAC output signal swing ("power") is shared among the carriers. The theoretical SNR of a DAC is determined by the same set of equations that govern ADC, and the noise figure can be derived given a specific SNR.

The AD9786 is one of the latest TxDACs suitable for a variety of air standards. A simplified block diagram is shown in Figure 8.140. The device accepts I and Q input data at a rate up to 160 MSPS, and provides on-chip interpolation of 2, 4, and 8. I and Q modulation is performed digitally within the device. The interpolated output sampling rate can be as high as 400 MSPS. Direct IF output frequencies up to 70 MHz are possible.

Figure 8.140: AD9786 16-Bit, 160-MSPS TxDAC+ with 2×/4×/8× Interpolation and Signal Processing

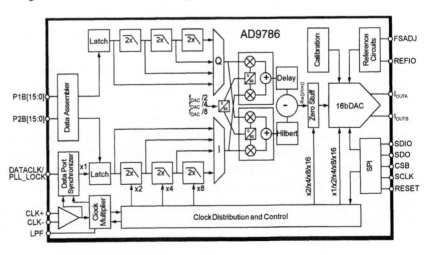

As explained in Chapter 2, oversampling by interpolation relaxes the requirements on the anti-imaging output filter as well as reduces the effects of "sin x/x" roll-off.

The AD9786 has a noise floor of –163 dBm/Hz up to 100 MHz. IMD performance to 300 MHz is less than –80 dBc, and 10 MHz SFDR is 90 dBc. These and other key specifications are summarized in Figure 8.141. Overall performance is more than sufficient to meet the exacting transmitter requirements of all multicarrier air standards, including GSM and WCDMA. A summary of the TxDAC family is shown in Figure 8.142.

- Targeted at the most demanding Multicarrier Macro GSM/WCDMA basestation applications
- 16-/14-/12-Bit resolution with up to 400MSPS DAC Update Rate
- Selectable 2×/4×/8×High Performance Interpolation Filters 160MSPS Data Rate
- Direct IF Transmission Frequencies 70MHz and Higher
- Twos Complement/Straight Binary Selectable Data Format
- LVTTL/CMOS Compatible Inputs
- Programmable via SPI Port
- Noise Floor Performance: –163dBm/Hz out to 100MHz
- IMD to 300MHz: < –80dBc
- SFDR @ 10MHz: 90dBc
- Versatile Clock Interface
- Power Dissipation: ~800mW, Single Supply (+2.5 V / +3 V)
- 80-Pin LQFP Package

Figure 8.141: AD978x 16-Bit Interpolating TxDAC+ Family Key Specifications

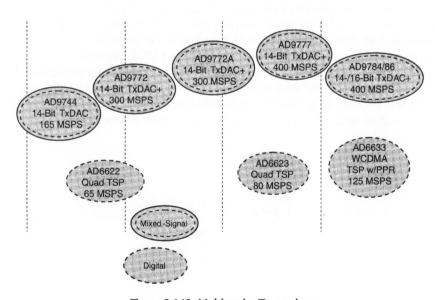

Figure 8.142: Multicarrier Transmitters

Figure 8.143 shows a summary of the entire Analog Devices' receiver and transmitter "Softcell" family.

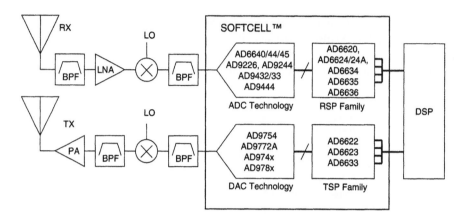

Figure 8.143: Multicarrier Transceiver Summary

For applications requiring analog I/Q modulation, the AD8349 is a silicon monolithic RF IC quadrature modulator, designed for use from 0.8 GHz to 2.7 GHz. Its excellent phase accuracy and amplitude balance enable high performance direct RF modulation. A functional diagram is shown in Figure 8.144. The differential LO signal first passes through a polyphase phase splitter. The I- and Q-channel outputs of the phase splitter are buffered to drive the LO inputs of two Gilbert cell mixers. Two differential V-to-I converters connected to the I- and Q-channel baseband inputs provide the tail currents for the mixers. The outputs of the two mixers are summed together by a differential buffer to drive 50 Ω loads. The device also features an output disable function. The AD8349 can be used as a direct-to-RF transmit modulator in digital communication systems such as GSM, CDMA, WCDMA basestations and QPSK or QAM broadband wireless access transmitters. It can also be used as the IF modulator within LMDS transmitters. Additionally, this quadrature modulator can be used with direct digital synthesizers in hybrid phase-locked loops to generate signals over a wide frequency range with millihertz resolution. The AD8349 is supplied in a 16-lead exposed-paddle TSSOP package. Its performance is specified over a –40°C to +85°C temperature range. This device is fabricated on Analog Devices' advanced complementary silicon bipolar process.

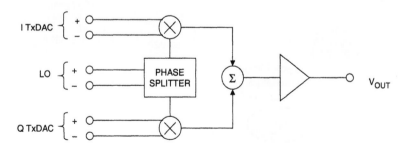

**Figure 8.144: AD8349 800 MHz to
2.7 GHz Quadrature Modulator**

Cellular Telephone Handsets

One of the fastest growing and rapidly changing high volume applications of digital radio is the cellular telephone handset. Each new generation of handsets has a lower components count, lower power, and more features than the previous models. Because of the different air standards, multimode and multiband operation is required. In order to give an overview of the cellular telephone handset, we will limit the discussion to GSM—with the additional understanding that the product examples shown do not necessarily represent the latest generation Analog Devices product offerings due to proprietary considerations.

Figure 8.145 shows a simplified block diagram of the GSM Digital Cellular Telephone System. The *speech encoder and decoder* and *discontinuous transmission* function will be described in detail. Up conversion and downconversion portions of the system contain mixed-signal functions and will be described later. Similar functions are performed digitally such as equalization, convolutional coding, Viterbi decoding, modulation and demodulation.

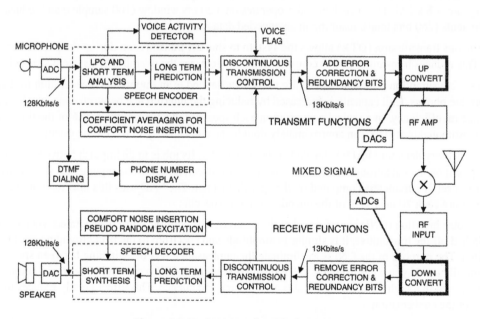

Figure 8.145: GSM Handset Block Diagram

The standard for encoding voice signals has been set in the T-Carrier digital transmission system. In this system, speech is logarithmically encoded to 8 bits at a sampling rate of 8 kSPS. The logarithmic encoding and decoding to 8 bits is equivalent to linear encoding and decoding to 13 bits of resolution. This produces a bit-rate of 104 kb/s. In most handsets, a 16-bit Σ-Δ ADC is used, so the effective bit-rate is 128 kb/s. The Speech Encoder portion of the GSM system compresses the speech signal to 13 kb/s, and the decoder expands the compressed signal at the receiver. The speech encoder is based on an enhanced version of linear predictive coding (LPC). The LPC algorithm uses a model of the human vocal tract that represents the throat as a series of concentric cylinders of various diameters. An excitation (breath) is forced into the cylinders. This model can be mathematically represented by a series of simultaneous equations that describe the cylinders.

The excitation signal is passed through the cylinders, producing an output signal. In the human body, the excitation signal is air moving over the vocal cords or through a constriction in the vocal tract. In a digital system, the excitation signal is a series of pulses for vocal excitation, or noise for a constriction. The signal is input to a digital lattice filter. Each filter coefficient represents the size of a cylinder.

An LPC system is characterized by the number of cylinders it uses in the model. Eight cylinders are used in the GSM system, and eight reflection coefficients must be generated.

Early LPC systems worked well enough to understand the encoded speech, but often the quality was too poor to recognize the voice of the speaker. The GSM LPC system employs two advanced techniques that improve the quality of the encoded speech. These techniques are *regular pulse excitation* (RPE) and *long term prediction* (LTP). When these techniques are used, the resulting quality of encoded speech is nearly equal to that of logarithmic pulse code modulation (companded PCM as in the T-Carrier system).

The actual input to the speech encoder is a series of 16-bit samples of uniform PCM speech data. The sampling rate is 8 kSPS. The speech encoder operates on a 20 ms window (160 samples) and reduces it to 76 coefficients (260 bits total), resulting in an encoded data rate of 13 kb/s.

Discontinuous transmission (DTX) allows the system to shut off transmission during the pauses between words. This reduces transmitter power consumption and increases the overall GSM system's capacity.

Low power consumption prolongs battery life in the handset and is an important consideration for hand-held portable phones. Call capacity is increased by reducing the interference between channels, leading to better spectral efficiency. In a typical conversation each speaker talks for less than 40% of the time, and it has been estimated that DTX can approximately double the call capacity of the radio system.

The voice activity detector (VAD) is located at the transmitter. Its job is to distinguish between speech superimposed on the background noise and noise with no speech present. The input to the voice activity detector is a set of parameters computed by the speech encoder. The VAD uses this information to decide whether or not each 20 ms frame of the encoder contains speech.

Comfort noise insertion (CNI) is performed at the receiver. The comfort noise is generated when the DTX has switched off the transmitter; it is similar in amplitude and spectrum to the background noise at the transmitter. The purpose of the CNI is to eliminate the unpleasant effect of switching between speech with noise, and silence. If you were listening to a transmission without CNI, you would hear rapid alternating between speech in a high noise background (i.e., in a car), and silence. This effect greatly reduces the intelligibility of the conversation.

When DTX is in operation, each burst of speech is transmitted followed by a *silence descriptor* (SID) frame before the transmission is switched off. The SID serves as an end of speech marker for the receive side. It contains characteristic parameters of the background noise at the transmitter, such as spectrum information derived through the use of linear predictive coding.

The SID frame is used by the receiver's comfort noise generator to obtain a digital filter which, when excited by pseudo-random noise, will produce noise similar to the background noise at the transmitter. This comfort noise is inserted into the gaps between received speech bursts. The comfort noise characteristics are updated at regular intervals by the transmission of SID frames during speech pauses.

Redundant bits are then added by the processor for error detection and correction at the receiver, increasing the final encoded bit rate to 22.8 kb/s. The bits within one window, and their redundant bits, are interleaved and spread across several windows for robustness.

The Role of ADCs and DACs in Cellular Telephone Handsets
Doug Grant

The cell phone handset uses quite a bit of ADC and DAC technology. Starting with the audio section, we find a high performance voiceband codec. Unlike the companded voice codecs used in the public switched telephone network, the voice codecs used in cellular handsets are linear-coded and higher resolution, typically 16 bits. Linear coding is preferred, because all cellular systems use DSP compression algorithms to reduce the bit rate to be transmitted, and the math is simpler when linear coding is used. Furthermore, less information is lost in the mathematical operations with linear coding, and this SNR is better than typical companded voiceband codecs.

The voiceband ADCs in cell phones are all Σ-Δ types, and include digital filters compliant with the bandwidth and stopband-rejection specs dictated by the applicable standard. In GSM, these converters provide 16-bit resolution, 8 kSPS sample rates, and 60 dB to70 dB signal-to-noise ratio in the voice band. The ADC section also includes analog interfaces to accommodate a variety of microphone types, with dc bias for electret types, single-ended and differential inputs, programmable gain, switch hook detection, etc., as well as other sources such as built-in FM radio or MP3 decoders. The DAC section includes audio output driver amplifiers suitable for speakers, earpieces, and headphones of various types and impedances, as well as provision for mixing multiple audio sources to an output device. And of course, they are optimized for low voltage and low power operation, with efficient power-up and power-down sequencing to save battery life.

Some advanced handsets now include higher performance DACs to enable playback of ringing and game tones, MP3 audio clips, and even full streaming audio content. These include all the usual features of multistandard audio playback converters, such as sample-rate conversion, but again with the constraints of low voltage, low current drain, and efficient power-up/power-down sequencing.

Converters also play an important part in the radio and baseband signal chain. Most cellular handsets down-convert the modulated RF signal to quadrature (I/Q) baseband components. In order to process these signals, dual Σ-Δ A/D converters are generally used, with integrated digital channel selection filters matched to the transmitted waveform for maximum transfer of signal energy. On the transmit side, most systems calculate the quadrature components of the waveform representing the bit stream to be transmitted, and load the waveforms in a burst RAM prior to transmission. At the appointed time, the RAM contents are clocked into a pair of DACs which modulate an intermediate-frequency carrier which is then upconverted to the appropriate RF carrier frequency, or in some implementations, the DACs modulate the carrier directly. The converter requirements in such a system are dictated by the trade-off of analog and digital filtering used in the system, signal bandwidth, dc offsets in the receive path before the ADCs, and the required signal-to-noise ratio to support the bit-error-rate needed for the system. In a typical GSM/GPRS/EDGE handset, the ADCs are on the order of 16-bit resolution with 65 dB to 75 dB dynamic range and sample rates equal to the symbol rate (270.833 kSPS). And of course, these system-specific parameters are in addition to the general requirements in a handset for low voltage, low current operation, with efficient control of power-up and power-down sequencing.

Cellular handsets also include several additional converters of varying resolutions and speeds for the monitoring and control of handset functions. Some of these functions include battery status and charge control, battery and power amplifier (PA) temperature monitoring, receive-path gain and offset control, transmit burst power ramp-up/down, automatic frequency control, and display brightness control. Most of these functions only require converters with relatively low bandwidth and low-to-moderate (10- to 14-bit) resolution.

SoftFone® and Othello Radio Chipsets from Analog Devices

(The following descriptions of the Othello Radio chipsets do not reflect the latest generation Analog Devices products. Details of the latest generation designs are available from Analog Devices under nondisclosure agreeement.)

Analog Devices offers a chipset which comprises the majority of a GSM handset. The SoftFone chipset performs the baseband and DSP functions, while the Othello radio chipset handles the RF functions as shown in Figure 8.146.

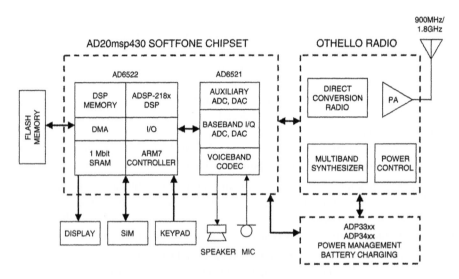

**Figure 8.146: Othello Radio and SoftFone Chipsets
Make Complete GSM/DCS Handset**

Because of the frequency allocations in GSM countries (other than the U.S.), most GSM handsets must be dual-band: capable of handling both GSM and DCS frequencies. The SoftFone and Othello chipsets supply the main functions necessary for implementing dual- or triple-band radios for GSM cellular phones. The AD20msp430 SoftFone chipset comprises the baseband portion of the GSM handset. The AD20msp430 baseband processing chipset uses a combination of GSM system knowledge and advanced analog and digital signal processing technology to provide a new benchmark in GSM/GPRS terminal design. The SoftFone architecture is entirely RAM-based. The software is loaded from FLASH memory and is executed from the on-chip RAM. This allows fast development cycles, since no ROM-code turns are required. Furthermore, the handset software can be updated in the field to enable new features. Combined with the Analog Devices Othello RF chipset, a complete multiband handset design contains less than 200 components, fits in a 20 cm^2 single sided PCB layout, and has a total bill-of-materials cost 20–30% lower than previous solutions.

The AD20msp430 chipset is comprised of two chips, the AD6522 DSP-based baseband processor and the AD6521 voiceband/baseband mixed-signal codec. Together with the Othello radio, the AD20msp430 allows a significant reduction in the component count and bill-of-materials (BOM) cost of GSM voice handsets and data terminals. The software and hardware foundations of the AD20msp430 chipset enjoy a long history of successful integration into GSM handsets. This is Analog Devices' 4th generation of GSM chipsets, each of which has passed numerous type approvals and network operator approvals in OEM handsets. In each generation, additional features have been added, while cost and power have been reduced. Numerous power-saving features have been included in the AD20msp430 chipset to reduce the total power consumption. A programmable state machine allows events to be controlled with a resolution of one-quarter of a bit period. The AD20msp430 chipset uses the SoftFone architecture, where all software resides in RAM or FLASH memory. Since ROM is not used, development time is reduced and additional features can be field-installed easily.

There are two processors in the AD20msp430 chipset. The DSP processor is the ADSP-218x core, proven in previous generations of GSM chipsets, and operated at 65 MIPS in the AD20msp430. This DSP performs the voiceband and channel coding functions previously discussed. The AD6521 voiceband/baseband codec chip contains all analog and mixed-signal functions. These include the I/Q channel ADCs and DACs, high performance multichannel voiceband codec, and several auxiliary ADCs and DACs for AGC (automatic gain control), AFC (automatic frequency control), and power-amplifier ramp control. The microcontroller is an ARM7 TDMI, running at 39 MIPS. The ARM7 handles the protocol stack and the man-machine interface functions. Both processors are field-proven in digital wireless applications. A simplified block diagram of the AD6521 baseband/voiceband codec is shown in Figure 8.147.

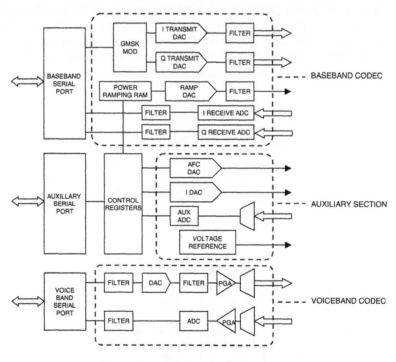

**Figure 8.147: AD6521 Baseband/Voiceband
Codec Simplified Block Diagram**

The AD20msp430 chipset is fully supported by a suite of development tools and software. The development tools allow easy customization of the DSP and/or ARM controller software to allow handset and terminal manufacturers to optimize the feature set and user interface of the end equipment. Software is available for all layers, including both voice and data applications, and is updated as new features become available. The system DMA and interrupt controllers are designed to allow easy upgrades to future generations of

DSP and controller cores. The display interface can be used with either parallel or serial-interface displays. System development can be shortened by the use of the debugging features in the AD20msp430. Most critical signals can be routed under software control to the Universal System Connector. This allows system debugging to take place in the final form factor. In addition, the architecture includes high speed logger and address trace functions in the DSP and single-wire trace/debug in the ARM controller.

Analog Devices' Othello direct-conversion radio eliminates intermediate-frequency (IF) stages, permits the mobile electronics industry to reduce the size and cost of radio sections, and enables flexible, multistandard, multimode operation. The radio includes a Zero-IF Transceiver and a Multiband Synthesizer.

Othello contains the main functions necessary for both a direct-conversion receiver and a direct VCO transmitter, known as the Virtual-IF™ transmitter. It also includes the local-oscillator generation block and a complete on-chip regulator that supplies power to all active circuitry for the radio. Also included is a fractional-N synthesizer that features extremely fast lock times to enable advanced data services over cellular telephones—such as high speed circuit-switched data (HSCSD) and general packet radio services (GPRS). Most digital cellular phones today include at least one "downconversion" in their signal chain. This frequency conversion shifts the desired signal from the allocated RF band for the standard (say, at 900 MHz) to some lower intermediate frequency (IF), where channel selection is performed with a narrow channel-select filter (usually a surface acoustic-wave (SAW) or a ceramic type). The now-filtered signal is then further downconverted to either a second IF or directly to baseband, where it is digitized and demodulated in a digital signal processor (DSP). Figure 8.148 shows the comparison between this *superheterodyne* architecture and the *superhomodyne*™ architecture of the Othello radio receiver.

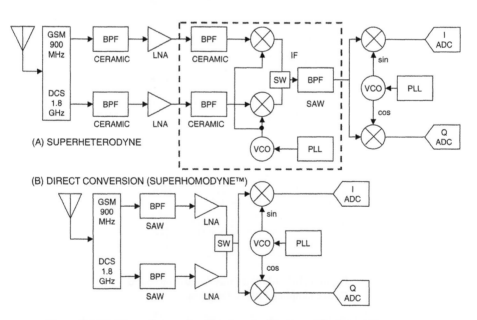

Figure 8.148: Direct Conversion Receiver Architecture Eliminates Components

The idea of using direct-conversion for receivers has long been of interest in RF design. The reason is obvious: in consumer equipment conversion stages add cost, bulk, and weight. Each conversion stage requires a local oscillator, (often including a frequency synthesizer to lock the LO onto a given frequency), a mixer, a filter, and (possibly) an amplifier. No wonder, then, that direct conversion receivers are attractive. All intermediate stages are eliminated, reducing the cost, volume, and weight of the receiver.

The Othello radio reduces the component count even more by integrating the front-end GSM low-noise amplifier (LNA). This eliminates an RF filter (the "image" filter) that is necessary to eliminate the image, or unwanted mixing product of a mixer and the off chip LNA. This stage, normally implemented with a discrete transistor, plus biasing and matching networks, accounts for a total of about 12 components. Integrating the LNA saves a total of about 15 to 17 components, depending on the amount of matching called for by the (now-eliminated) filter.

A simplified functional block diagram of the Othello dual band GSM radio's architecture is shown in Figure 8.149. The receive section is at the top of the figure. From the antenna connector, the desired signal enters the transmit/receive switch and exits on the appropriate path, either 925 MHz–960 MHz for the GSM band or 1805 MHz–1880 MHz for DCS. The signal then passes through an RF band filter (a so-called "roofing filter") that serves to pass the entire desired frequency band while attenuating all other out-of-band frequencies (blockers-including frequencies in the transmission band) to prevent them from saturating the active components in the radio front end. The roofing filter is followed by the low noise amplifier (LNA). This is the first gain element in the system, effectively reducing the contribution of all following stages to system noise. After the LNA, the direct-conversion mixer translates the desired signal from radio frequency (RF) all the way to baseband by multiplying the desired signal with a local oscillator (LO) output at the same frequency.

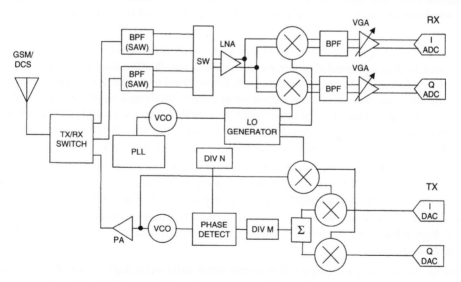

**Figure 8.149: Superhomodyne Direct Conversion
Dual-Band Transceiver Using Othello**

The output of the mixer stage is then sent in quadrature (I and Q channels) to the variable-gain baseband amplifier stage. The VGA also provides some filtering of adjacent channels, and attenuation of in-band blockers. These blocking signals are other GSM channels that are some distance from the desired channel, say 3 MHz and beyond. The baseband amplifiers filter these signals so that they will not saturate the Receive ADCs. After the amplifier stage, the desired signal is digitized by the Receive ADCs.

The Transmit section begins on the right, at the multiplexed I and Q inputs/outputs. Because the GSM system is a time division duplex (TDD) system, the transmitter and receiver are never on at the same time. The Othello radio architecture takes advantage of this fact to save four pins on the transceiver IC's package. The quadrature transmit signals enter the transmitter through the multiplexed I/Os. These I and Q signals are then modulated onto a carrier at an intermediate frequency greater than 100 MHz.

The output of the modulator goes to a phase-frequency detector (PFD), where it is compared to a reference frequency that is generated from the external channel selecting LO. The output of the PFD is a charge pump, operating at above 100 MHz, whose output is filtered by a fairly wide (1 MHz) loop filter. The output of the loop filter drives the tuning port of a voltage-controlled oscillator (VCO), with frequency ranges that cover the GSM and DCS transmit bands.

The output of the transmit VCO is sent to two places. The main path is to the transmit power amplifier (PA), which amplifies the transmit signal from about +3 dBm to +35 dBm, sending it to the transmit/receive switch and low-pass filter (which attenuates power-amplifier harmonics). The power amplifiers are dual band, with a simple CMOS control voltage for the band switch. The VCO output also goes to the transmit feedback mixer by means of a coupler, which is either a printed circuit, built with discrete inductors and capacitors, or a monolithic (normally ceramic) coupling device. The feedback mixer downconverts the transmit signal to the transmit IF, and uses it as the local oscillator signal for the transmit modulator. This type of modulator has several names, but the most descriptive is probably "translation loop." The translation loop modulator takes advantage of one key aspect of the GSM standard: the modulation scheme is Gaussian-filtered minimum-shift keying (GMSK). This type of modulation does not affect the envelope amplitude, which means that a power amplifier can be saturated and still not distort the GMSK signal sent through it.

GMSK can be generated in several different ways. In another European standard (for cordless telephones), GMSK is created by directly modulating a free-running VCO with the Gaussian filtered data stream. In GSM, the method of choice has been quadrature modulation. Quadrature modulation creates accurate phase GMSK, but imperfections in the modulator circuit (or upconversion stages) can produce envelope fluctuations, which can in turn degrade the phase trajectory when amplified by a saturated power amplifier. To avoid such degradations, GSM phone makers have been forced to use amplifiers with somewhat higher linearity, at the cost of reduced efficiency and talk time per battery charge cycle.

The translation loop modulator combines the advantages of directly modulating the VCO and the inherently more accurate quadrature modulation. In effect, the scheme creates a phase locked loop (PLL), comprising the modulator, the LO signal, and the VCO output and feedback mixer. The result is a directly modulated VCO output with a perfectly constant envelope and almost perfect phase trajectory. Phase trajectory errors as low as 1.5 degrees have been measured in Othello, using a signal generator as the LO signal to provide a reference for the loop.

Because Othello radios can be so compact, they enable GSM radio technology to be incorporated in many products from which it has been excluded, such as very compact phones or PCMCIA cards. However, the real power of direct conversion will be seen when versatile third-generation phones are designed to handle multiple standards. With direct-conversion, hardware channel-selection filters will be unnecessary, because channel selection is performed in the digital signal-processing section, which can be programmed to handle multiple standards. Contrast this with the superheterodyne architecture, where multiple radio circuits are required to handle the different standards (because each will require different channel-selection filters), and all the circuits will have to be crowded into a small space. With direct conversion, the same radio chain could in concept be used for several different standards, bandwidths, and modulation types. Thus, Web browsing and voice services could, in concept, occur over the GSM network using the same radio in the handset.

Time-Interleaved IF Sampling ADCs with Digital Post-Processors
Mark Looney

The material in this section was extracted from Mark Looney's Analog Dialogue article, Reference 35.

Time interleaving of multiple analog-to-digital converters by multiplexing the outputs of (for example) a pair of converters at a doubled sampling rate is by now a mature concept—first introduced by Black and Hodges in 1980 (Reference 26, 27). While designing a 7-bit, 4 MHz ADC, they determined that a time-interleaved solution would require less die area than a comparable 2^N comparator flash converter design. This new concept proved of great value in their design, but space-saving was not its only benefit. Time interleaving of ADCs offers a conceptually simple method for multiplying the sample rate of existing high-performing ADCs, such as the 14-bit, 105 MSPS AD6645 and the 12-bit, 210 MSPS AD9430. In many different applications, this concept has been leveraged to benefit systems that require very high sample rate analog-to-digital conversion.

While the speed and resolution of standard ADC products have advanced well beyond 4 MSPS and 7 bits, time-interleaved ADC systems (for good reasons) have not advanced far beyond 8-bit resolution. Nevertheless, at 8-bit performance levels, this concept has been widely adopted in the test and measurement industry, particularly for wideband digital oscilloscopes. That it continues to make an impact in this market is evidenced by the 20 GSPS, 8-bit ADC that was recently developed by Agilent Labs (Reference 28) and adopted by the Agilent Technologies Infiniium™ oscilloscope family (Reference 29). Indeed, time-interleaved ADC systems thrive at the 8-bit level, but they continue to fall short in applications that require the combination of high resolution, wide bandwidth, and wide dynamic range.

The primary limiting factor in time-interleaved ADC systems at 12- and 14-bit levels is the requirement that the channels be matched. An 8-bit system that provides a dynamic range of 50 dB can tolerate a gain mismatch of 0.25% and a clock-skew error of 5 ps. This level of accuracy can be achieved by traditional methods, such as matching physical channel layouts, using common ADC reference voltages, prescreening devices, and active analog trimming, but at higher resolutions the requirements are much tighter. Until now, devices employing more innovative matching techniques have not been commercially available.

This discussion will outline in detail the matching requirements for 12- and 14-bit time-interleaved ADC systems, discuss the idea of advanced digital post-processing techniques as an enabling technology, and introduce a device employing the most promising solution to date, Advanced Filter Bank (AFB™), from V Corp Technologies, Inc. (References 31 and 31).

Time-interleaving ADC systems employ the concept of running M ADCs at a sample rate that is 1/M of the overall system sample rate. Each channel is clocked at a phase that enables the system as a whole to sample at equally spaced increments of time, creating the seamless image of a single ADC sampling at full speed. Figure 8.150 illustrates the block-and-timing diagrams of a typical four-channel, time-interleaved ADC system. Each of the four ADC channels runs at one-fourth the system's sample rate, spaced at 90° intervals. The final output data stream is created by interleaving all of the individual channel data outputs in the proper sequence (e.g., 1, 2, 3, 4, 1, 2, etc.). In a two-converter example, both ADC channels are clocked at one-half of the overall system's sample rate, and they are 180° out of phase with one another.

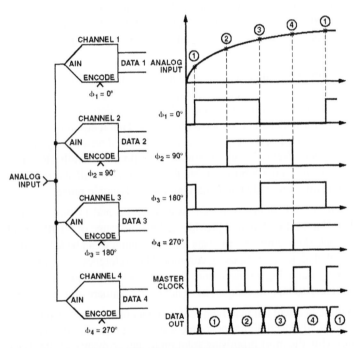

Figure 8.150: Time-Interleaved ADCs

For simplicity, this discussion focuses primarily on two-converter systems, but the concepts can be extended to four-converter (Reference 35). A two-converter interleaved system is shown in Figure 8.151 where two 12-bit, 200 MSPS ADCs are interleaved to produce an effective sampling rate of 400 MSPS.

As mentioned, channel-to-channel matching has a direct impact on the dynamic range performance of a time-interleaved ADC system. Mismatches between the ADC channels result in dynamic range degradation that—in an FFT plot—show up as spurious frequency components called *image spurs* and *offset spurs*. The image spur(s) associated with time-interleaved ADC systems are a direct result of gain and phase mismatches between the ADC channels. The gain and phase errors produce error functions that are orthogonal to one another. Both contribute to the image-spur energy at the same frequency location(s). The offset spur is generated by offset differences between the ADC channels. Unlike the image spur(s), the offset spurs are not dependent on the input signal. For a given offset mismatch, the offset spur(s) will always be at the same level. Extensive studies of the behavior of these spurs have resulted in several mathematical methods for characterizing the relationship between channel matching errors and dynamic range performance (References 32 and 33).

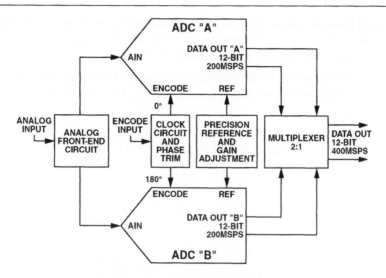

Figure 8.151: Two-Converter Time-Interleaved 12-Bit 400 MSPS ADC

While these methods are thorough and very useful, the "error voltage" approach used here provides a simple method for understanding the relationship without requiring a deep study of complex mathematical derivations. This approach is based on the same philosophy used in Analog Devices Application Note AN-501 (Reference 34) to establish the relationship between aperture jitter and signal-to-noise (SNR) degradation in ADCs. The error voltage is defined as the difference between the "expected" sample voltage and the "actual" sample voltage. These differences are a result of a large subset of errors that fall into three basic categories: gain, phase, and offset mismatches.

In a two-converter interleaved system, the error voltages generated by gain and phase mismatches result in an image spur that is located at Nyquist minus the analog input frequency. The offset mismatch generates an error voltage that results in an offset spur that is located at Nyquist. Since the offset spur is located at the edge of the Nyquist band, designers of two-channel systems can typically plan their system frequency around it, and focus their efforts on gain-and-phase matching. Figure 8.152 displays a typical FFT plot for a two-channel system showing these errors.

In a four-converter interleaving system, there are three image spurs and two offset spurs. The image spurs,

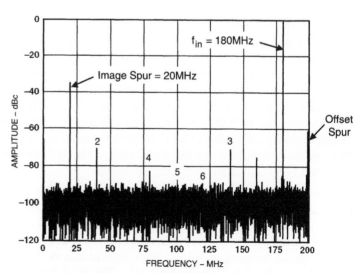

Figure 8.152: Two-Converter Interleaved FFT Plot, f_s = 400 MSPS, f_{in} = 180 MHz

generated by gain and phase mismatches between the ADC channels, are located at (1) Nyquist minus the analog input frequency and (2) one-half Nyquist plus or minus the analog input frequency. The offset spurs are located at Nyquist and at one-half of Nyquist (middle of the band).

Once the error voltages from each of the three mismatch groups are known, the following equations can be used to calculate the image and offset spurs (IS_{gain}, IS_{phase}, IS_{total}, OS_{offset}) in a single-tone, two-converter system:

$$IS_{gain(dB)} = 20\log\left(IS_{gain}\right) = 20\log\left(\frac{G_e}{2}\right) \qquad \text{Eq. 8.23}$$

where

$$G_e = \text{gain error ratio} = \left|1 - \frac{V_{FSA}}{V_{FSB}}\right| \qquad \text{Eq. 8.24}$$

$$IS_{phase(dB)} = 20\log\left(IS_{phase}\right) = 20\log\left(\frac{\theta_{ep}}{2}\right) \qquad \text{Eq. 8.25}$$

$$\text{where } \theta_{ep} = \omega_a \,\Delta t_e \text{ (radians)}$$

$$\omega_a = \text{analog input frequency}$$

$$\Delta t_e = \text{clock skew error}$$

$$IS_{total(dB)} = 20\log\sqrt{\left(IS_{gain}\right)^2 + \left(IS_{phase}\right)^2} \qquad \text{Eq. 8.26}$$

$$OS_{offset(dB)} = 20\log\left(\frac{\text{Offset}}{2 \times \text{Total Codes}}\right) \qquad \text{Eq. 8.27}$$

where Offset = channel-to-channel offset (codes).

As noted earlier, the gain and phase errors generate error functions that are orthogonal (Reference 32), requiring a "root-sum-square" combination of their individual contributions to the image spur. Using these equations, an error budget can be developed to determine what level of matching will be required to maintain a given dynamic range requirement. For example, a 12-bit dynamic range requirement of 74 dBc at an input frequency of 180 MHz would require gain matching better than 0.02% and aperture delay matching better than 300 fs. If the gain can be perfectly matched, the aperture delay matching can be "relaxed" to approximately 350 fs. Figure 8.153 provides the matching requirements for several different cases to illustrate the extreme precision required to make a classical time-interleaved A/D conversion system work at 12- and 14-bit resolutions over wide bandwidths.

The traditional, 2-channel time-interleaved ADC shown in Figure 8.151 achieves the first level of matching by reducing the physical and electrical differences between the channels. For example, gain matching is typically controlled by the use of common reference voltages and carefully matched physical layouts. Phase matching is achieved by manually tuning the electrical length of the clock (or analog input) paths and/or through special trimming techniques that control an electrical characteristic of the clock distribution circuit (rise/fall times, bias levels, trigger level, etc.). The offset matching depends on the offset performance of the individual ADCs.

PERFORMANCE REQUIREMENT AT 180 MHz	SFDR (dBc)	GAIN MATCHING (%)	APERTURE TIME MATCHING (fs)
12 Bits	74	0.04	0
12 Bits	74	0	350
12 Bits	74	0.02	300
14 Bits	86	0.01	0
14 Bits	86	0	88
14 Bits	86	0.005	77

Figure 8.153: Time-Interleaved ADC Matching Requirements

Many of these matching approaches are based on careful analog design and trim techniques. While there have been an abundance of excellent ideas to address these tough matching requirements, many of them require additional circuits that add error sources of their own—defeating the original purpose of achieving precise gain and phase matching. An example of such an idea would be setting the rise and fall times of the two different clock signals. Any circuit that could provide this level of control would be subjected to increased influence of power supply voltage—and temperature—on each channel's phase behavior.

Advanced Digital Post Processing

The development of new digital signal processing techniques, along with the advances in inexpensive, high speed, configurable digital hardware platforms (DSPs, FPGAs, CPLDs, ASICs, etc.), has opened the way for breakthroughs in time-interleaving ADC performance. Digital post-processing approaches have several advantages over classical analog matching techniques. They are flexible in their implementation and can be designed for precision well beyond the ADC resolutions of interest. A conceptual view of how digital signal processing techniques can impact time-interleaved system architectures can be found in Figure 8.154.

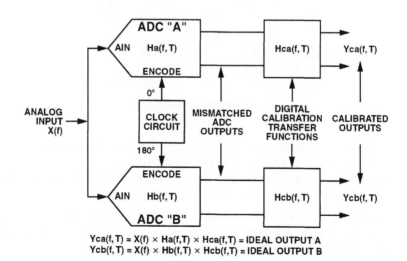

$$Yca(f, T) = X(f) \times Ha(f,T) \times Hca(f,T) = \text{IDEAL OUTPUT A}$$
$$Ycb(f, T) = X(f) \times Hb(f,T) \times Hcb(f,T) = \text{IDEAL OUTPUT B}$$

Figure 8.154: Example of Digital Post-Processing Architecture

This concept employs a set of digital calibration transfer functions that process each ADC's output data, creating a new set of "calibrated outputs." These digital calibration transfer functions can be implemented using a variety of digital filter configurations (FIR, IIR, etc.). They can be as simple as trimming the gain of one channel or as complicated as trimming the gain, phase, and offset of each channel over wide bandwidths and temperature ranges.

Wide bandwidth and temperature matching presents the greatest opportunity—and challenge—for using digital post-processing techniques to improve the performance of time-interleaving ADC systems. The mathematical derivations required for designing the digital calibration transfer functions for multiple ADC channels over wide bandwidths and temperature ranges are extremely complex and not readily available. However, a great deal of academic work has been invested in this area, creating a number of interesting solutions. One of these solutions, known as Advanced Filter Bank (AFB), stands out in its ability to provide a platform for a significant breakthrough.

Advanced Filter Bank (AFB)

AFB is one of the first commercially available digital post- processing technologies to make a significant impact on the performance of time-interleaving ADC systems. By providing precise channel-to-channel gain, phase, and offset matching over wide bandwidths and temperature ranges, AFB is well-positioned to solidly establish time-interleaving ADC systems in the area of high-speed, 12-/14-bit applications. Besides its matching functions, AFB also provides phase linearization and gain-flatness compensation for ADC systems. Figure 8.155 displays a basic block diagram representation of a system employing AFB.

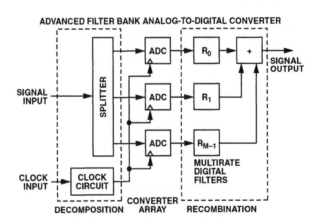

Figure 8.155: AFB Basic Block Diagram

By using a unique multirate FIR filter structure, AFB can be easily implemented into a convenient digital hardware platform, such as an FPGA or CPLD. The FIR coefficients are calculated using a patented method that involves starting with the equations seen in Figure 8.154, and then applying a variety of advanced mathematical techniques to solve for the digital calibration transfer function.

AFB enables time-interleaving ADC systems to use up to 90% of their Nyquist band, and can be configured to operate in any Nyquist zone of the converter (e.g., first, second, third, etc.) The appropriate Nyquist zone can be selected using a set of logic inputs, which control the required FIR coefficients.

AFB Design Example: The AD12400 12-Bit, 400 MSPS ADC

The AD12400 is the first member of a new family of Analog Devices products that leverage time inter-leaving and AFB. Its performance will be used to illustrate what can be achieved when state-of-the-art ADC design is combined with advanced digital post-processing technologies. Figure 8.156 illustrates the AD12400's block diagram and its key circuit functions. The AD12400 employs a unique analog front-end circuit with 400 MHz input bandwidth, two 12-bit, 200-MSPS ADC channels, and an AFB implementa-tion using an advanced field-programmable gate array (FPGA). It was designed using many of the classical matching techniques discussed above, together with a very low jitter clock distribution circuit. These key components are combined to develop a 12-bit, 400 MSPS ADC module that performs very well over 90% of the Nyquist band and over an 85°C temperature range. It has an analog input bandwidth of 400 MHz.

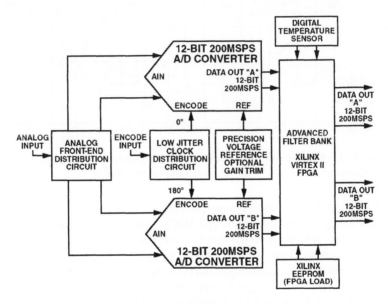

Figure 8.156: AD12400 12-Bit, 400 MSPS ADC Block Diagram

The ADCs' transfer functions are obtained using wide-bandwidth, wide-temperature range measurements during the manufacturing process. This characterization routine feeds the ADCs' measured transfer func-tions directly into the AFB coefficient calculation process. Once the ADCs have been characterized, and the required FIR coefficients have been calculated, the FPGA is programmed and the product is ready for action. Wide bandwidth matching is achieved using AFB's special FIR structure and coefficient calculation process. Wide temperature performance is achieved by selecting one of the multiple FIR coefficient sets, using an on-board digital temperature sensor.

The true impact of this technology can be seen in Figure 8.157. Figure 8.157A displays the image-spur performance across the first Nyquist zone of this system. The top curve in Figure 8.157A represents the performance of a 2-channel time-interleaved system that has been carefully designed to provide optimal matching in the layout. The behavior of the image spur in this curve makes it obvious that this system was manually trimmed at an analog input frequency of 128 MHz. A similar observation of Figure 8.157B suggests a manual trim temperature at 40°C.

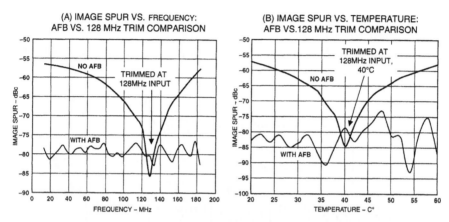

Figure 8.157: Performance of a Manually Trimmed System "Before-and-After" AFB Compensation Over Frequency and Temperature Range

Despite a careful PCB layout, tightly matched front-end circuit, tightly matched clock-distribution circuit, and common reference voltages used in the AD12400 ADC, the dynamic range degrades rapidly as the frequency and/or temperature deviates from the manual trim conditions. This rapid rate of degradation can be anticipated in any two-converter time-interleaved ADC system by analyzing some of the sensitive factors affecting this circuit. For example, the gain-temperature coefficient of a typical high performance, 12-bit ADC is 0.02%/°C. In this case, a 10°C change in temperature would cause a 0.2% change in gain, resulting in an image spur of 60 dBc (see Equation 8.23). Considering just this single ADC temperature characteristic, the predicted image spur is 3 dB worse than the 30°C performance displayed in Figure 8.157B. By contrast, the dynamic range performance shown in these figures remains solid when the AFB compensation is enabled. In fact, the dynamic range performance surpasses the 12-bit level across a bandwidth of nearly 190 MHz and a temperature range of 40°C. Another significant advantage of this approach is that the temperature range can actually be expanded from the 20°C to 60°C range shown to 0°C to 85°C by using additional FIR coefficient sets—as embodied in the AD12400.

The AD12400 achieves impressive specifications using time interleaving followed by digital post processing. The device has a full-power input bandwidth of 300 MHz. The SNR is 64 dBFS and the SFDR is 75 dBFS for a 180 MHz input signal, sampling at 400 MSPS.

Time interleaving is growing into a significant trend in performance enhancement for high speed ADC systems. Advanced digital post-processing methods, such as AFB, provide a convenient solution to the tough channel-matching requirements at resolution levels that were not previously achievable for time-interleaved systems. When combined with the best ADC architectures available, advanced DSP technologies, such as AFB, are ready to take high speed ADC systems to the next level of performance and facilitate greatly improved products and systems in demanding markets such as medical imaging, precise medicine dispensers (fluid flow measurement), synthetic aperture radar, digital beam-forming communication systems, and advanced test/measurement systems. This technology will result in many breakthroughs that will include 14-bit/400 MSPS and 12-bit/800 MSPS ADC systems in the near future.

References:
8.6 Software Radio and IF Sampling

1. Richard Groshong and Stephen Ruscak, "Undersampling Techniques Simplify Digital Radio," **Electronic Design**, May 23 1991, pp. 67–78.

2. Lackey and Upmall, "SPEAKeasy: The Military Software Radio," **IEEE Communications Magazine**, May 1995, pp. 56–61.

3. Brad Brannon, "Using Wide Dynamic Range Converters for Wideband Radios," **RF Design**, May 1995, pp. 50–65.

4. Jim Mitola, "The Software Radio Architecture," **IEEE Communications Magazine**, Vol. 33, No. 5, May 1995, pp. 26–38.

5. Jeffery Wepman, "Analog-to-Digital Converters and Their Applications in Radio Receivers," **IEEE Communications Magazine**, Vol. 33, No. 5, May 1995, pp. 39–45.

6. Brad Brannon, "Wide Dynamic Range A/D Converters Pave the way for Wideband Digital Radio Receivers," **EDN**, November 7, 1996, pp. 187–205.

7. Dave Robertson, "Selecting Mixed-Signal Components for Digital Communication Systems I: An Introduction," **Analog Dialogue**, 30-3, 1996, www.analog.com.

8. Dave Robertson, "Selecting Mixed-Signal Components for Digital Communication Systems II: Digital Modulation Schemes," **Analog Dialogue**, 30-4, 1996, www.analog.com.

9. Dave Robertson, "Selecting Mixed-Signal Components for Digital Communication Systems III: Sharing the Channel," **Analog Dialogue**, 31-1, 1997, www.analog.com.

10. Dave Robertson, "Selecting Mixed-Signal Component for Digital Communications Systems IV: Receiver Architecture Considerations," **Analog Dialogue**, 31-2, 1997, www.analog.com.

11. Dave Robertson, "Selecting Mixed-Signal Components for Digital Communications Systems V: Aliases, Images, and Spurs," **Analog Dialogue**, 31-3, 1997, www.analog.com.

12. Brad Brannon, "Digital Radio Receiver Design Requires Re-Evaluation of Parameters," **EDN**, November 6, 1998, pp. 163–170.

13. Brad Brannon, Dimitrios Efstathiou, and Tom Gratzek, "A Look at Software Radios: Are they Fact or Fiction?," **Electronic Design**, December 1, 1998.

14. Brad Brannon, "Designing a Superheterodyne Receiver Using an IF Sampling Diversity Chipset," **Application Note AN-502**, Analog Devices, 1998, www.analog.com.

15. Patrick Mannion, "Direct Conversion Prepares for Cellular Prime Time," **Electronic Design**, November 22, 1999, pp. 85–95.

16. Louis E. Frenzel, "Designers Face Tough Challenges in 3G Cellular/PCS Phone Specs," **Electronic Design**, October 2, 2000, pp. 107–122.

17. Brad Brannon and Chris Cloninger, "Redefining the Role of ADCs in Wireless," **Applied Microwave and Wireless**, March 2001, pp. 94–105.

18. Brad Brannon, "Correlating High Speed ADC Performance to Multicarrier 3G Requirements," **RF Design**, June 2003, pp. 22–28.

19. Brad Brannon, "Brad's Radio Page," www.converter-radio.com.

20. Brad Brannon, "Overcoming Converter Nonlinearities with Dither," **Application Note AN-410**, Analog Devices, 1995, www.analog.com.

21. Walt Jung, "Simple Wideband Noise Generator," Ideas for Design, **Electronic Design**, October 1, 1996.

22. Walt Jung, **Op Amp Applications**, Analog Devices, 2002, ISBN 0-916550-26-5, p. 6.165.

23. Walt Kester, "Add Noise Dither to Blow Out ADCs' Dynamic Range," **Electronic Design, Analog Applications Supplement**, November 22, 1999, pp. 20–26.

24. Doug Mercer and Joe DiPilato, "DACs are Optimized for Communication Transmit Path," **Analog Dialogue**, 30-3, 1996, www.analog.com.

25. Editor, "New TxDAC Generation," **Analog Dialogue**, Volume 33- 4, April 1999, www.analog.com.

26. W. C. Black Jr. and D. A. Hodges, "Time Interleaved Converter Arrays," **IEEE International Conference on Solid State Circuits**, February 1980, pp. 14–15.

27. W. C. Black Jr. and D. A. Hodges, "Time Interleaved Converter Arrays," **IEEE Journal of Solid State Circuits**, December 1980, Volume 15, pp. 1022–1029.

28. K. Poulton, et al., "A 20GS/s 8-b ADC with a 1MB Memory in 0.18 micron CMOS," **IEEE International Conference on Solid State Circuits**, February 2003, pp. 318–319, 496.

29. Press Release, "Agilent Technologies Introduces Industry First 6 GHz, 20 GSample/s-Per-Channel Oscilloscope and Probing Measurement System," Agilent Technologies Web Page, November 1, 2002, www.agilent.com/about/newsroom/presrel/archive.html.

30. S. Velazquez, "High-Performance Advanced Filter Bank Analog-to-Digital Converter for Universal RF Receivers," **IEEE SP International Symposium on Time-Frequency and Time-Scale Analysis**, 1998, pp. 229–232.

31. Technical Description, "Advanced Filter Bank (AFB) Analog-to-Digital Converter Technical Description," V Corp Technologies, www.v-corp.com/analogfilterbank.htm.

32. N. Kurosawa, et al., "Explicit Analysis of Channel Mismatch Effects in Time Interleaved ADC Systems," **IEEE Transactions on Circuits and Systems I – Fundamental Theory and Applications**, Volume 48, Number 3, March 2003.

33. M. Gustavsson, J. J. Wikner and N. N. Tan, **CMOS Data Converters for Communications**, Boston: Kluwer Academic Publishers, 2000, pp. 257–267.

34. Brad Brannon, "Aperture Uncertainty and ADC System Performance," **Application Note, AN-501**, Analog Devices, Inc., www.analog.com.

35. Mark Looney, "Advanced Digital Post-Processing Techniques Enhance Performance in Time-Interleaved ADC Systems," **Analog Dialogue**, 37-8, August 2003, www.analog.com.

Direct Digital Synthesis (DDS)
Walt Kester

Introduction to DDS

A frequency synthesizer generates multiple frequencies from one or more frequency references. These devices have been used for decades, especially in communications systems. Many are based upon switching and mixing frequency outputs from a bank of crystal oscillators. Others have been based upon well understood techniques utilizing phase-locked loops (PLLs). This mature technology is illustrated in Figure 8.158. A fixed-frequency reference drives one input of the phase comparator. The other phase comparator input is driven from a divide-by-N counter which is in turn driven by a voltage-controlled-oscillator (VCO). Negative feedback forces the output of the internal loop filter to a value which makes the VCO output frequency N-times the reference frequency. The time constant of the loop is controlled by the loop filter. There are many trade-offs in designing a PLL, such a phase noise, tuning speed, frequency resolution, etc., and there are many good references on the subject (References 1–5). Analog Devices has a complete selection of both integer and fractional-N PLLs as well as simulation software to aid the design process.

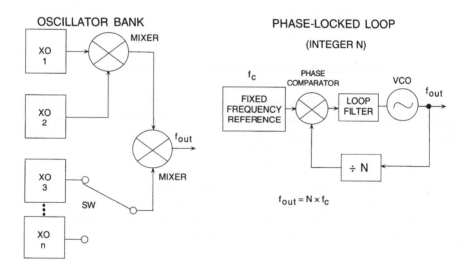

Figure 8.158: Frequency Synthesis Using Oscillators and Phase-locked Loops

With the widespread use of digital techniques in instrumentation and communications systems, a digitally-controlled method of generating multiple frequencies from a reference frequency source has evolved called Direct Digital Synthesis (DDS). The basic architecture is shown in Figure 8.159. In this simplified model, a stable clock drives a programmable-read-only-memory (PROM) which stores one or more integral number of cycles of a sinewave (or other arbitrary waveform, for that matter). As the address counter steps through each memory location, the corresponding digital amplitude of the signal at each location drives a DAC which in turn generates the analog output signal. The spectral purity of the final analog output signal is determined primarily by the DAC. The phase noise is basically that of the reference clock.

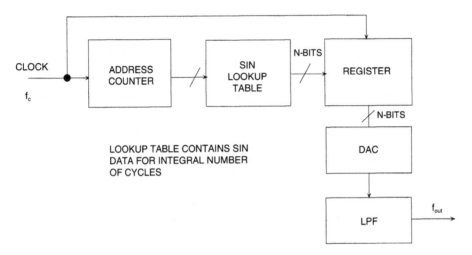

Figure 8.159: Fundamental Direct Digital Synthesis System

The DDS system differs from the PLL in several ways. Because a DDS system is a sampled data system, all the issues involved in sampling must be considered: quantization noise, aliasing, filtering, etc. For instance, the higher order harmonics of the DAC output frequencies fold back into the Nyquist bandwidth, making them unfilterable, whereas, the higher order harmonics of the output of PLL-based synthesizers can be filtered. Other considerations will be discussed shortly.

A fundamental problem with this simple DDS system is that the final output frequency can be changed only by changing the reference clock frequency or by reprogramming the PROM—making it rather inflexible. A practical DDS system implements this basic function in a much more flexible and efficient manner using digital hardware called a Numerically Controlled Oscillator (NCO). A block diagram of such a system is shown in Figure 8.160.

The heart of the system is the *phase accumulator* whose contents is updated once each clock cycle. Each time the phase accumulator is updated, the digital number, M, stored in the *delta phase register* is added to the number in the phase accumulator register. Assume that the number in the delta phase register is 00...01 and that the initial contents of the phase accumulator is 00...00. The phase accumulator is updated by 00...01 on each clock cycle. If the accumulator is 32 bits wide, 2^{32} clock cycles (over 4 billion) are required before the phase accumulator returns to 00...00, and the cycle repeats.

The truncated output of the phase accumulator serves as the address to a sine (or cosine) lookup table. Each address in the lookup table corresponds to a phase point on the sinewave from 0° to 360°. The lookup table contains the corresponding digital amplitude information for one complete cycle of a sinewave. The lookup

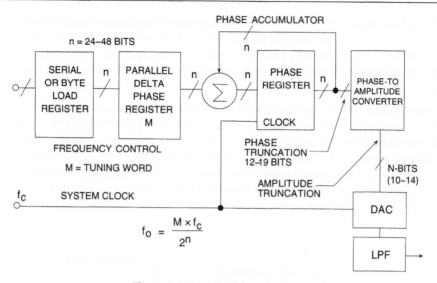

Figure 8.160: A Flexible DDS System

table therefore maps the phase information from the phase accumulator into a digital amplitude word, which in turn drives the DAC. In practice, only data for 90° is required because the quadrature data is contained in the two MSBs. In order to further reduce the size of the lookup tables, various proprietary algorithms have been developed to compute the sine values, however the fundamental concept is still the same.

Consider the case for n = 32, and M = 1. The phase accumulator steps through each of 2^{32} possible outputs before it overflows. The corresponding output sinewave frequency is equal to the clock frequency divided by 2^{32}. If M = 2, then the phase accumulator register "rolls over" twice as fast, and the output frequency is doubled. This can be generalized as follows.

For an n-bit phase accumulator (n generally ranges from 24 to 32 in most DDS systems), there are 2^n possible phase points. The digital word in the delta phase register, M, represents the amount the phase accumulator is incremented each clock cycle. If f_c is the clock frequency, then the frequency of the output sinewave is equal to:

$$f_o = \frac{M \times f_c}{2^n}$$

Eq. 8.28

This equation is known as the DDS "tuning equation." Note that the frequency resolution of the system is equal to $f_c/2^n$. For n = 32, the resolution is greater than one part in four billion. In a practical DDS system, all the bits out of the phase accumulator are not passed on to the lookup table, but are truncated, thereby reducing the size of the lookup table without affecting frequency resolution. The amount of truncation depends upon the resolution and performance of the output DAC. In general, the phase address information should have 2 to 4 bits more resolution than the DAC, but this can vary some from product to product. The objective is to use enough resolution in the lookup table address so that the overall noise and distortion of the analog output signal is limited by the DAC and not the effects of phase truncation.

The basic DDS system described above is extremely flexible and has high resolution. The frequency can be changed instantaneously with no phase discontinuity by simply changing the contents of the M-register. However, practical DDS systems first require the execution of a serial, or byte-loading sequence to get the new frequency word into an internal buffer register which precedes the parallel-output M-register. This is done to minimize package pin count. After the new word is loaded into the buffer register, the parallel-output delta phase register is clocked, thereby changing all the bits simultaneously. The number of clock cycles required to load the delta-phase buffer register determines the maximum rate at which the output frequency can be changed.

Figure 8.161 shows another way to view the operation of the phase accumulator. The sine wave oscillation is visualized as a vector rotating around a phase circle. Each designated point on the phase wheel corresponds to the equivalent point on a cycle of a sine waveform. As the vector rotates around the wheel, a corresponding output sinewave is being generated. One revolution of the vector around the phase wheel, at constant speed, results in one complete cycle of the output sinewave. The phase accumulator is utilized to provide the equivalent of the vector's linear rotation around the phase wheel. The contents of the phase accumulator corresponds to the points on the cycle of the output sinewave.

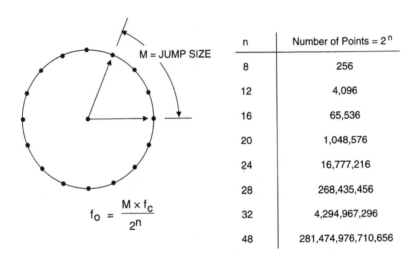

n	Number of Points = 2^n
8	256
12	4,096
16	65,536
20	1,048,576
24	16,777,216
28	268,435,456
32	4,294,967,296
48	281,474,976,710,656

$$f_o = \frac{M \times f_c}{2^n}$$

M = JUMP SIZE

Figure 8.161: Digital Phase Wheel

The number of discrete points on the phase circle is determined by the resolution of the phase accumulator. For an n-bit accumulator, there are 2^n number of points on the phase circle. The digital word in the delta phase register (M) represents the "jump size" between updates. It commands the phase accumulator to jump by M points on the phase circle each time the system is clocked.

Figure 8.162 shows the signal flow through the DDS architecture. The phase accumulator is actually a modulus M counter that increments its stored number each time it receives a clock pulse. The magnitude of the increment is determined by the binary input number or word (M) contained in the delta phase register that is summed with the overflow of the counter. The digital phase information from the phase accumulator is converted into a corresponding digital amplitude by the phase-to-amplitude converter. Finally, the DAC converts the digital amplitude into a corresponding analog signal.

When IC DDS systems became popular in the mid 1980s, the digital NCO was generally fabricated on a CMOS process, and the DAC on a bipolar process, thereby yielding a two-chip solution. Today, however, modern CMOS processes are suitable for not only the digital circuits but for the high performance DAC as well (as illustrated by the many TxDACs currently offered by Analog Devices). Modern DDS systems therefore are fully integrated and include many additional options as well.

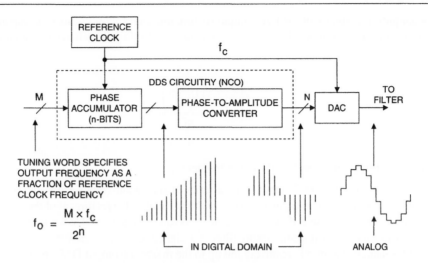

Figure 8.162: Signal Flow Through the DDS Architecture

Aliasing in DDS Systems

There is one important limitation to the range of output frequencies that can be generated from the simple DDS system. The Nyquist Criteria states that the clock frequency (sample rate) must be at least twice the output frequency. Practical limitations restrict the actual highest output frequency to about 40% of the clock frequency. Figure 8.163 shows the output of a DAC in a DDS system where the output frequency is 30 MHz and the clock frequency is 100 MSPS. An anti-imaging filter must follow the reconstruction DAC to remove the lower image frequency (100 – 30 = 70 MHz) as shown in the figure.

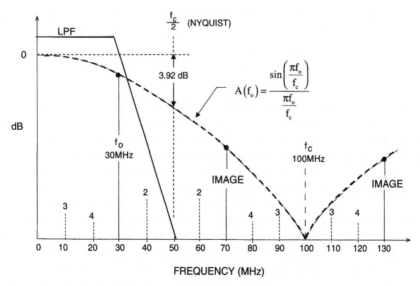

Figure 8.163: Aliasing in a DDS System Clocked at 100 MSPS with a 30 MHz Output

Note that the amplitude response of the DAC output (before filtering) follows a sin(x)/x response with zeros at the clock frequency and multiples thereof. The exact equation for the normalized output amplitude, $A(f_o)$, is given by:

$$A(f_o) = \frac{\sin\left(\frac{\pi f_o}{f_c}\right)}{\frac{\pi f_o}{f_c}}$$

Eq. 8.29

where f_o is the output frequency and f_c is the clock frequency.

This roll-off occurs because the DAC output is not a series of zero-width impulses (as in a perfect impulse resampler), but a series of rectangular pulses whose width is equal to the reciprocal of the update rate. The amplitude of the sin(x)/x response is down 3.92 dB at the Nyquist frequency (1/2 the DAC update rate). In practice, the transfer function of the antialiasing filter is designed to compensate for the sin(x)/x roll-off so that the overall frequency response is relatively flat up to the maximum output DAC frequency (generally 40% of the update rate).

Another important consideration is that unlike a PLL-based system, the higher order harmonics of the fundamental output frequency in a DDS system will fold back into the baseband because of aliasing. These harmonics cannot be removed by the antialiasing filter. For instance, if the clock frequency is 100 MSPS, and the output frequency is 30 MHz, the second harmonic of the 30 MHz output signal appears at 60 MHz (out of band), but also at 100 – 60 = 40 MHz (an inband aliased component). Similarly, the third harmonic (90 MHz) appears inband at 100 – 90 = 10 MHz, and the fourth at 120 – 100 = 20 MHz. Higher order harmonics also fall within the Nyquist bandwidth (dc to $f_c/2$). The locations of the first four harmonics are labeled in the diagram.

Frequency Planning in DDS Systems

In many DDS applications, the spectral purity of the DAC output is of primary concern. Unfortunately, the measurement, prediction, and analysis of this performance is complicated by a number of interacting factors.

It is wise to carefully choose the output frequency and the clock frequency such that the aliased harmonics discussed above do not fall close to the fundamental output frequency, and can therefore be removed with a band-pass filter.

Even an ideal N-bit DAC can produce unwanted harmonics in a DDS system. The amplitude of these harmonics is highly dependent upon the ratio of the output frequency to the clock frequency. This is because the spectral content of the DAC quantization noise varies as this ratio varies, even though its theoretical rms value remains equal to $q/\sqrt{12}$ (where q is the weight of the LSB). The assumption that the quantization noise appears as white noise and is spread uniformly over the Nyquist bandwidth is simply not true in a DDS system (it is more apt to be a true assumption in an ADC-based system, because the ADC adds a certain amount of noise to the signal which tends to "dither" or randomize the quantization error. However, a certain amount of correlation still exists). For instance, if the DAC output frequency is set to an exact sub-multiple of the clock frequency, then the quantization noise will be concentrated at multiples of the output frequency, i.e., it is highly signal dependent. If the output frequency is slightly offset, however, the quantization noise will become more random, thereby giving an improvement in the effective SFDR.

This is illustrated in Figure 8.164, where a 4096-point FFT is calculated based on digitally generated data from an ideal 12-bit DAC. In the left-hand diagram, the ratio between the clock frequency and the output

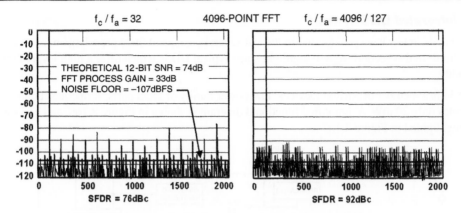

Figure 8.164: Effect of Ratio of Sampling Clock to Output Frequency on SFDR for Ideal 12-bit DAC

frequency was chosen to be exactly 32 (128 cycles of the sinewave in the FFT record length), yielding an SFDR of about 78 dBc. In the right-hand diagram, the ratio was changed to 32.25196850394 (127 cycles of the sinewave within the FFT record length), and the effective SFDR is now increased to 92 dBc. In this ideal case, we observed a change in SFDR of 16 dB just by slightly changing the frequency ratio.

Best SFDR can therefore be obtained by the careful selection of the clock and output frequencies. However, in some applications, this may not be possible. In ADC-based systems, adding a small amount of random noise to the input tends to randomize the quantization errors and reduce this effect. The same thing can be done in a DDS system as shown in Figure 8.165 (Reference 16). The pseudo-random digital noise generator output is added to the DDS sine amplitude word before being loaded into the DAC. The amplitude of the digital noise is set to about ½ LSB. This accomplishes the randomization process at the expense of a slight increase in the overall output noise floor. In most DDS applications, however, there is enough flexibility in selecting the various frequency ratios so that this type of dithering is not required.

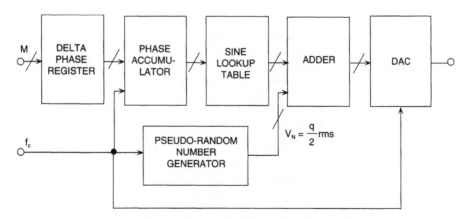

Figure 8.165: Injection of Digital Dither in a DDS System to Randomize Quantization Noise and Increase SFDR

Modern Integrated DDS Systems

DDS integrated circuits have proliferated in the last several years, and there are a large number of devices to choose from. In this section we will highlight some typical DDSs which offer a high level of integration and flexibility.

The AD9834 is a member of Analog Devices' low power family of DDS parts. It operates up to 50 MSPS and dissipates only 20 mW. A simplified functional diagram is shown in Figure 8.166, and key specifications are highlighted in Figure 8.167.

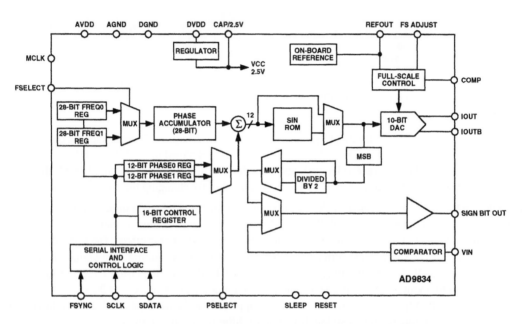

Figure 8.166: AD9834 Low Power (20 mW), 50 MSPS DDS Synthesizer

- Sinusoidal/Triangular DAC Output
- Programmable phase and frequency
- Narrow-Band SFDR >72dB
- 10-Bit DAC, 28-Bit Phase Accumulator
- 2.3V to 5.5V Operation
  - Low Power (20–35mW)
  - Power-Down Option
- Two Frequency Registers and Two Phase Registers
- Low Jitter Clock Output
- Narrow Band SFDR >72 dB
- 40Mhz SPI Serial Interface
- 25MHz (AD9833)/50 MHz (AD9834) Speed
- Serial Loading
- Extended Temperature Range: –40°C to +105°C
- 10-Lead µSOIC (AD9833) or 20-Lead TSSOP (AD9834)

Figure 8.167: AD9833/AD9834 Key Features

The 50 MSPS AD9834 contains a 10-bit TxDAC core which yields a narrowband SFDR greater than 72 dB. The sin ROM can be bypassed to produce a triangular waveform output. The phase accumulator is 28 bits wide, and the output is truncated to 12 bits at the sin ROM lookup table address input. An on-chip comparator allows a square wave output to be produced for clock generation. The AD9834 is written to via a 3-wire serial interface that can operate at clock rates up to 40 MHz and is compatible with DSP and microcontroller standards.

The AD9834 has a power-down pin that allows external control of the power-down mode. Sections of the device that are not being used can be powered down to minimize the current consumption.

Phase and frequency modulation capability is provided. The Frequency registers are 28 bits wide, and the phase registers are 12 bits wide. Because of the various output options available from the part, the AD9834 can be configured to suit a wide variety of applications. One of the areas where the AD9834 is suitable is in modulation applications. The part can be used to perform simple modulation such as FSK. More complex modulation schemes such as GMSK and QPSK can also be implemented using the AD9834. In an FSK application, the two frequency registers of the AD9834 are loaded with different values. One frequency will represent the space frequency, while the other will represent the mark frequency. The digital data stream is fed to the FSELECT pin, which will cause the AD9834 to modulate the carrier frequency between the two values. The AD9834 has two phase registers; this enables the part to perform PSK. With phase shift keying, the carrier frequency is phase shifted, the phase being altered by an amount that is related to the bit stream being input to the modulator. The AD9834 is also suitable for signal generator applications. With its low current consumption, the part is suitable for applications in which it can be used as a local oscillator.

Figure 8.168 summarizes the current low power DDS offerings from Analog Devices.

Parameter	AD9830	AD9831	AD9832	AD9833	AD9834	AD9835
Master Clock	50 MHz	25 MHz	25 MHz	25 MHz	50 MHz	50 MHz
DAC Resolution	10-bit	10-bit	10-bit	10-bit	10-bit	10-bit
Interface	Par	Par	Serial	Serial	Serial	Serial
Freq/Phase Registers	4 Phase, 2 Freq	4 Phase, 2 Freq	4 Phase, 2 Freq	2 Phase, 2 Freq	2 Phase, 2 Freq	4 Phase, 2 Freq
Supply Voltage	5V±5%	2.97V to 5.5V	2.97V to 5.5V	2.5V to 5.5V	2.5V to 5.5V	5V±5%
Power	275mW max	45mW max	45mW max	21mW	24mW	200mW max
Package	48-TQFP	48-TQFP	16-TSSOP	10-µSOIC	20-TSSOP	16-TSSOP
Comparator Output					Yes	

Figure 8.168: AD983x Low Power DDS Synthesizers

The AD9858 is a direct digital synthesizer (DDS) featuring a 10-bit DAC operating up to 1 GSPS. The AD9858 uses advanced DDS technology, coupled with an internal high speed, high performance DAC to form a digitally programmable, complete high frequency synthesizer capable of generating a frequency-agile analog output sinewave at up to 400+ MHz. The AD9858 is designed to provide fast frequency hopping and fine tuning resolution (32-bit frequency tuning word). The frequency tuning and control words are loaded into the AD9858 via parallel (8-bit) or serial loading formats. The AD9858 contains an integrated charge pump (CP) and phase frequency detector (PFD) for synthesis applications requiring the combination of a high speed DDS along with phase-locked loop (PLL) functions. An analog mixer is also provided on-chip for applications requiring the combination of a DDS, PLL, and mixer, such as frequency translation loops, tuners, and so on. The mixer can operate at frequencies up to 2 GHz.

The AD9858 also features a divide-by-two on the clock input, allowing the external clock to be as high as 2 GHz. The AD9858 is specified to operate over the extended industrial temperature range of –40°C to +85°C. A functional block diagram is shown in Figure 8.169, and key specifications are summarized in Figure 8.170.

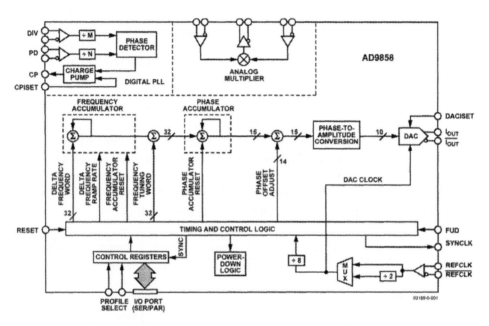

Figure 8.169: AD9858 1 GSPS DDS with Phase Detector and Analog Multiplier

- 1 GSPS Internal Clock Speed
- Integrated 10-Bit D/A Converter
- Phase Noise <130dBc/Hz @ 1 kHz Offset (DAC Output)
- 32-Bit Programmable Frequency Register
- Simplified 8-Bit Parallel and SPI Serial Control Interface
- Automatic Frequency Sweeping Capability
- 3.3 V Power Supply
- Power Dissipation < 2 Watts @ 1 GHz
- 100-Lead LQFP Surface-Mount Package
- Integrated Programable Charge Pump and Phase/Frequency Detector with Fast Lock Circuit
- Integrated 2 GHz Mixer

Figure 8.170: AD9858 1 GSPS DDS Key Specifications

Writing data to the on-chip digital registers that control all operations of the device easily configures the AD9858. The AD9858 offers a choice of both serial and parallel ports for controlling the device. Four user profiles can be selected by a pair of external pins. These profiles allow independent setting of the frequency tuning word and the phase offset adjustment word for each of four selectable configurations. The AD9858 can be programmed to operate in single-tone mode or in a frequency-sweeping mode. To save on power consumption, there is also a programmable full-sleep mode, during which most of the device is powered down to reduce current flow.

Figure 8.171 shows the AD9858 configured as an upconverter using the internal phase detector and charge pump along with external filtering and a VCO to form a high-speed PLL. The basic AD9858 DDS can generate a frequency up to 400 MHz. The PLL circuitry in conjunction with a high frequency VCO and divider is capable of multiplying the reference frequency well up into the GHz region. The reference frequency into the phase detector can be as high as 150 MHz. Further details on using DDS devices as upconverters can be found in References 10, 11, and 15.

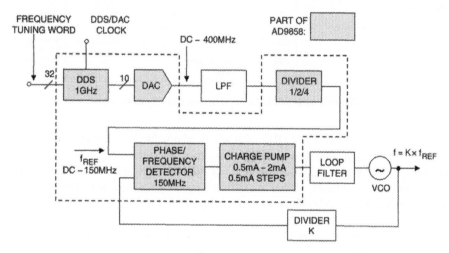

Figure 8.171: DDS Single Loop Upconversion Using the AD9858

Figure 8.172 summarizes the high-speed DDS products currently available from Analog Devices.

Parameter	AD9850	AD9851	AD9852	AD9854	AD9857	AD9858
Master Clock	125 MHz	180 MHz	300 MHz	300 MHz	200 MHz	1 GHz
DAC Resolution	10-bit	10-bit	12-bit	12-bit	14-bit	10-bit
Control Interface	Par / Serial	Par / Serial	Par / Serial	Par / Serial	Serial	Par/Serial
SFDR	>50 dBc @ 40 MHz Aout	>43 dBc @ 70 MHz Aout	80 dBc @ 100 MHz (±1 MHz) Aout	80 dBc @ 100 MHz (±1 MHz) Aout	80 dBc @ 65 MHz(±100 kHz) Aout	>50 dBc @ 360 MHz Aout
Supply Voltage	+3.3 V or +5.25 V	+2.7 V to +5.25 V	3.3 V	3.3 V	3.3 V	3.3 V
Power	155 mW @ 110MHz (+3.3V)	555 mW @ 180MHz	1.9 to 2.7 W	1.9 to 3.4 W	1 to 2 W	1.9W @ 1 GHz
Package	28-SSOP	28-SSOP	80-LQFP	80-LQFP	80-LQFP	100-EPAD TQFP
On-Chip Comparator	Yes	Yes	Yes	Yes	No	No
Notes		Clock-multiplier (6X)	Auto Freq. Sweep Clock multiplier (4-20X)	Quadrature Outputs	Modulator or Single Tone Mode	Auto Freq. Sweep, PLL, Mixer, 2 GHz input clocking

Figure 8.172: AD985x High Speed DDS Synthesizers

The AD9954 is a 400 MSPS 14-bit, 1.8 V DDS with advanced on-chip FSK modulation capability. The AD9954 is a digitally programmable, complete high frequency synthesizer capable of generating a frequency-agile analog output sinusoidal waveform at up to 200 MHz. The AD9954 is designed to provide fast frequency hopping and fine tuning resolution (32-bit frequency tuning word). The frequency tuning and control words are loaded into the AD9954 via a serial I/O port.

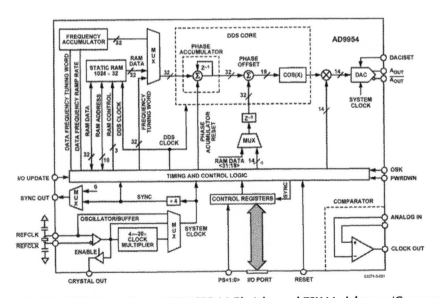

Figure 8.173: AD9954 Low Power 400 MSPS 14-Bit Advanced FSK Modulator w/Comparator

The AD9954 includes an integrated 1024 word × 32 bit static RAM to support flexible frequency sweep capability in several modes. The AD9954 also supports a user-defined linear sweep mode of operation. The device includes an on-chip high speed comparator for applications requiring a square wave output. The AD9954 is specified to operate over the extended industrial temperature range of –40° to +85°C. A simplified functional diagram of the AD9954 is shown in Figure 8.173, and key specifications are given in Figure 8.174. A summary of the 400 MSPS DDS parts from Analog Devices is given in Figure 8.175.

- 400MSPS Internal Clock Speed
- 1.8V Power Supply Operation
- Integrated 14-Bit DAC
- RAM: 1024 Word, 32-Bit
- 14-Bit Amplitude Modulation
- 32-Bit Programmable Frequency Register
- On-chip Oscillator/Buffer
- 4× – 20× Programmable Reference Clock Multiplier
- High-Speed Comparator
- SPI Serial Control Interface
- Automatic Frequency Sweeping
- Power Dissipation < 250mW @ 400MSPS
- Small 48-Lead TQFP Packaging

Figure 8.174: AD9954 Low Power 400 MSPS 14-Bit Advanced FSK Modulator Key Specifications

Parameter	AD9859	AD9951	AD9952	AD9953	AD9954
Master Clock (Max)	400 MHz	400 MHz	400 MHz	400 MHz	400 MHz
DAC Resolution	10-bit	14-bit	14-bit	14-bit	14-bit
Interface	Serial	Serial	Serial	Serial	Serial
Supply Voltage	1.8 V	1.8 V	1.8 V	1.8 V	1.8 V
Power	<250 mW	<250 mW	<250 mW	<250 mW	<250 mW
Phase, Frequency RAM				✓	✓
Automatic Frequency Sweep					✓
Comparator			✓		✓

Figure 8.175: 400 MSPS Low Power DDS Product Family

With the availability of high performance low power CMOS processes, digital quadrature modulation capability can be added to the basic DDS function. Figure 8.176 shows a simplified block diagram of the AD9857 200 MSPS 14-bit quadrature digital upconverter. The AD9857 is intended to function as a universal I/Q modulator and agile upconverter, single-tone DDS, or interpolating DAC for communications applications. The device has excellent dynamic performance with 80 dB narrowband SFDR at a 65 MHz output frequency. Figure 8.177 summarizes Analog Devices' DDS parts which are optimized as I/Q modulators.

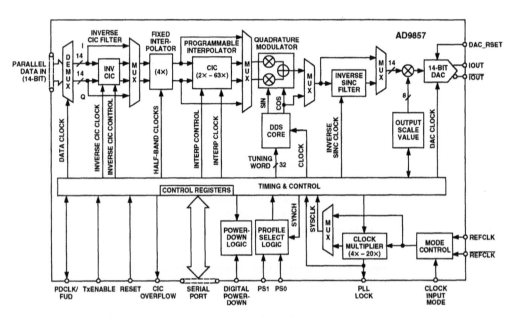

Figure 8.176: AD9857 200 MSPS, 14-Bit Quadrature Digital Upconverter

Parameter	AD9853	AD9856	AD9857
Master Clock	165 MHz	200 MHz	200 MHz
DAC Res	10-bit	12-bit	14-bit
Control Interface	Serial	Serial	Serial
SFDR	>50 dB @ 42 MHz (single-tone)	>80 dB Narrowband @ 70 MHz	80 dB @ 65 MHz (±100 kHz) Aout
Supply Voltage	3.3 V to 5 V	3 V	3.3 V
Power	750 mW @ 3.3 V	1 to 1.5 W depending on configuration	1 to 2 W depending on configuration
Package	44-MQFP	48-TQFP	80-LQFP
Notes	Complete QPSK and 16QAM Modulator	Single Tone or Modulator Mode	Single Tone or Modulator Mode

Figure 8.177: DDS I/Q Quadrature Modulators

References:
8.7 Direct Digital Synthesis

1. F. M. Gardner, **Phaselock Techniques**, Second Edition, John Wiley, 1979, ISBN 0471042943.

2. Ulrich L. Rohde, Digital PLL Frequency Synthesizers, Theory and Design, Prentice-Hall, 1983, ISBN 0-13-214239-2.

3. Mike Curtin and Paul O'Brien, "Phase-Locked Loops for High-Frequency Receivers and Transmitters, Part 1, **Analog Dialogue 33-3**, 1999, www.analog.com.

4. Mike Curtin and Paul O'Brien, "Phase-Locked Loops for High-Frequency Receivers and Transmitters, Part 2," **Analog Dialogue 33-5**, 1999, www.analog.com.

5. R. E. Best, **Phase-Locked Loops: Theory, Design and Applications**, Fourth Edition, McGraw-Hill, 1999, ISBN 0071349030.

6. Richard Cushing, "A Technical Tutorial on Digital Signal Synthesis," Analog Devices, 1999, www.analog.com/dds.

7. David Buchanan, "Choosing DACs for Direct Digital Synthesis," **Application Note AN-237**, Analog Devices, www.analog.com.

8. Richard Cushing, "Amplitude Modulation of the AD9850 Direct Digital Synthesizer," Application Note AN-423, Analog Devices, www.analog.com.

9. Colm Slattery, "Programming the AD9832/DA9835," **Application Note AN-621**, Analog Devices, www.analog.com.

10. Richard Cushing, "Single-Sideband Upconversion of Quadrature DDS Signals to the 800-to-2500-MHz Band," **Analog Dialogue**, 34-3, 2000, www.analog.com.

11. Ken Gentile, "Digital Upconverter IC Tames Complex Modulation," **Microwaves and RF**, August, 2000.

12. Jon Baird, "400 Msamples DDSs Run on Only +1.8 VDC," **Microwaves and RF**, December 2002.

13. Colm Slattery, "DDS and Converter for Signal Generator," Design Idea, **EDN**, February 20, 2003, pp. 80–82.

14. Niamh Collins, "DDS Device Produces Sawtooth Waveform," Design Idea, **EDN**, July 10, 2003, pp. 77–78.

15. David Crook, "Hybrid Synthesizer Tutorial," **Microwave Journal**, February 2003.

16. Richard J. Kerr and Lindsay A. Weaver, "Pseudorandom Dither for Frequency Synthesis Noise," **U.S. Patent 4,901,265**, filed December 14, 1987, issued February 13, 1990.

17. Henry T. Nicholas, III and Henry Samueli, "An Analysis of the Output Spectrum of Direct Digital Frequency Synthesizers in the Presence of Phase-Accumulator Truncation," **IEEE 41st Annual Frequency Control Symposium Digest of Papers**, 1987, pp. 495–502, IEEE Publication No. CH2427-3/87/0000-495.

18. Henry T. Nicholas, III and Henry Samueli, "The Optimization of Direct Digital Frequency Synthesizer Performance in the Presence of Finite Word Length Effects," **IEEE 42nd Annual Frequency Control Symposium Digest of Papers**, 1988, pp. 357–363, IEEE Publication No. CH2588-2/88/0000-357.

Precision Analog Microcontrollers
Grayson King

Introduction

Many modern sensor interfacing designs require not only precision signal conditioning and A/D conversion, but also some local embedded processing to control the ADC and perform some signal manipulation in the digital domain. Microcontrollers are ideal for this function, and the addition of nonvolatile memory allows the storage of various calibration coefficients and facilitates system reprogramming. Of course, the combination of ADC, nonvolatile memory, and microcontroller is useful in many other applications as well, including communications, medical, and handheld instrumentation to name just a few.

In addition, there has been considerable effort to define "smart sensors" which have standardized digital interfaces for connecting to various buses (References 1–4). Signal conditioning, A/D conversion, and microcontroller-based digital processing form the basis of these smart sensors.

Analog Devices has taken these three main ingredients and integrated them into a single chip called a "MicroConverter." Each product in the MicroConverter family contains high performance analog I/O, nonvolatile flash EEPROM memory, and an industry-standard microcontroller core as shown in Figure 8.178. In addition to these three basic functional blocks, many additional on-chip peripherals are included.

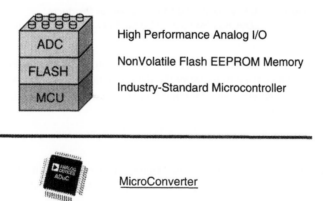

Figure 8.178: MicroConverter Definition

There are many benefits to this type of integration, including smaller overall size, reduced manufacturing cost (because of reduced parts count), reduced emissions (because data buses are kept internal to the chip), and easier software design (because interface to on-chip peripherals is already done).

Characteristics of the MicroConverter Product Family

Of the approximately 15 individual 8051-based MicroConverter devices currently available from Analog Devices, about half utilize a standard switched capacitor SAR ADC architecture with 12-bit resolution and up to 400 kSPS sampling frequency. The parts are fully specified at both 3 V and 5 V supplies. Figure 8.179 shows the ADuC842 which is representative of the SAR-based MicroConverter products. In addition to the ADC, there are two 12-bit on-chip DACs and other peripherals.

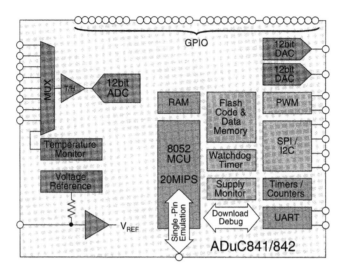

Figure 8.179: Standard SAR MicroConverter Products
ADuC812/ADuC831/ADuC832/ADuC841/ADuC842/ADuC843

All members of the SAR MicroConverter family have an 8-channel input multiplexer. In addition, the on-chip temperature monitor can be used to measure the die temperature. The flash memory is divided into code memory (8 Kbytes to 62 Kbytes) and data memory (640 bytes to 4 Kbytes). On-chip RAM is either 256 bytes or 256 bytes + 2 Kbytes.

The microcontroller core is the industry-standard 8051, and the processor speed is between 1 and 20 MIPS, depending on the particular device.

Figure 8.180 shows the lowest cost member of the MicroConverter family, the ADuC814 which, is housed in a 28-lead TSSOP package. When operating from 3 V supplies, the power dissipation for the part is below 10 mW.

The MicroConverter product family integrates precision data acquisition circuitry on the same chip as the microcontroller, without compromising performance of the analog functions. Figure 8.181 shows an FFT of the SAR ADC output with a 10 kHz input signal. Note that there is practically no degradation in ac performance whether or not the microcontroller is active. In either case, the SNR is greater than 70 dB, and the SFDR is greater than 80 dBc.

Figure 8.182 summarizes the 12-bit SAR-based MicroConverter family.

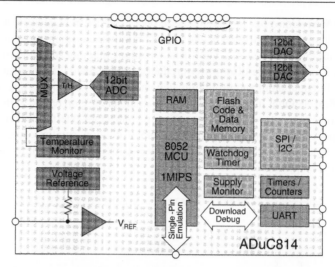

Figure 8.180: ADuC814 Reduced Pin-Count/Low Cost SAR MicroConverter

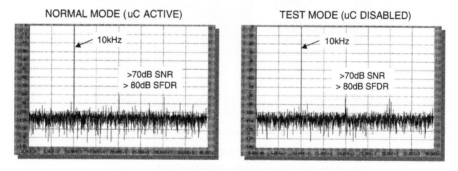

Figure 8.181: MicroConverter SAR ADC Performance

Part #	ADC	DAC	MCU	Flash/EE Code	Flash/EE Data	RAM	PKGs	Special Features
ADuC812	8-chan 12-bit	Dual 12-bit	1.3MIPS 8052	8K-byte	640-byte	256-byte	52-PQFP 56-CSP	Not Recommended for New Designs
ADuC814	6-chan 12-bit	Dual 12-bit	1.3MIPS 8052	8K-byte	640-byte	256-byte	28-TSSOP	Small, Low-Cost
ADuC831	8-chan 12-bit	Dual 12-bit +Dual PWM	1.3MIPS 8052	62K-byte	4K-byte	256-byte +2K-byte	52-PQFP 56-CSP	"Big-Memory" Upgrade to ADuC812
ADuC832	8-chan 12-bit	Dual 12-bit +Dual PWM	1.3MIPS 8052	62K-byte	4K-byte	256-byte +2K-byte	52-PQFP 56-CSP	Same As ADuC831, But With PLL Clock
ADuC841	8-chan 12-bit	Dual 12-bit +Dual PWM	20MIPS 8052	8K,32K,62 K-byte	4K-byte	256-byte +2K-byte	52-PQFP 56-CSP	"Fast-Core" Upgrade to ADuC831
ADuC842	8-chan 12-bit	Dual 12-bit +Dual PWM	16MIPS 8052	8K,32K,62 K-byte	4K-byte	256-byte +2K-byte	52-PQFP 56-CSP	"Fast-Core" Upgrade to ADuC832
ADuC843	8-chan 12-bit	Dual PWM	16MIPS 8052	8K,32K,62 K-byte	4K-byte	256-byte +2K-byte	52-PQFP 56-CSP	Stripped-Down ADuC842

Figure 8.182: Summary of 12-Bit SAR-Based MicroConverter Products

Other members of the MicroConverter family are based on a Σ-Δ ADC architecture, which is known for its low speed but very high precision. This half of the product family includes many features designed specifically for interfacing with low-level sensors. The ADuC834 block diagram shown in Figure 8.183 is well representative of these Σ-Δ based MicroConverter products. Its key analog features include a 24-bit primary ADC with buffered differential input and programmable gain, a 16-bit auxiliary ADC with unbuffered single-ended input, a flexible input multiplexing configuration, an on-chip temperature sensor accurate to about ±2°C, an on-chip voltage reference with an option to connect an external differential reference source instead, a pair of 200 μA current sources for resistive sensor excitation, smaller excitation current sources on the primary ADC's inputs that can be used to detect open-circuit conditions at the sensor, and finally a 12-bit rail-to-rail voltage output DAC.

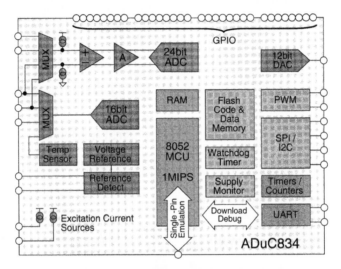

Figure 8.183: Standard Σ-Δ MicroConverter Products:
ADuC816/ADuC824/ADuC836/ADuC834

With the exception of the DAC, all these features are also found in our AD77xx stand-alone ADC products; but MicroConverter products like the ADuC834 also include an 8051 microcontroller, flash code and data memory, and a host of digital peripherals including serial communication ports, timer/counters, etc. Each MicroConverter product is supported by a suite of software development tools, enhanced by on-chip features that allow in-system programming and debugging, either through the chip's UART serial port, or using the single-pin emulation feature. This single-pin emulation is basically the same as JTAG from a functional point of view, but uses only one pin.

More recent additions to the Σ-Δ-based MicroConverter line include devices with more analog input channels and a faster 8051 core. Specifically, the ADuC845 shown in Figure 8.184 accepts up to 10 single-ended analog inputs, or five fully differential inputs, or any combination thereof. Any input can be multiplexed to the primary or the auxiliary ADC, both of which are 24-bit; but only the primary ADC includes the buffered programmable-gain differential input stage.

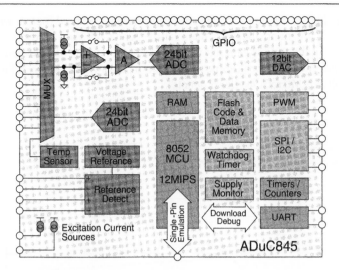

Figure 8.184: ADuC845/ADuC847/ADuC848 High Channel-Count, Fast Core Σ-Δ-Based MicroConverter

Another unique feature of these high-channel-count Σ-Δ MicroConverter products is the option to connect two completely independent differential voltage references. This is useful in multichannel applications with one or more ratiometric signal sources.

On the digital side, these high-channel-count Σ-Δ MicroConverter products also feature a faster microcontroller core. Specifically, an optimized 8052 capable of over 12 MIPS peak, compared to just over 1 MIPS peak for the ADuC834. Beyond these enhancements, the ADuC845 also includes all the features of the ADuC834. However, the ADuC834 or other MicroConverter products will often provide a more cost-effective solution if the added features of the ADuC845 are not required.

As with most of ADI's high-resolution Σ-Δ converters, the conversion speed (output word rate) of Σ-Δ MicroConverter ADCs is programmable, as is their gain. Figure 8.185 shows effective resolution as a function of output word rate at the various gain settings. Output word rate can be anywhere from around 5 Hz to about 105 Hz. The noise-free code (peak-to-peak resolution) performance shown in Figure 8.185 is characteristic of all the 24-bit Σ-Δ MicroConverter products.

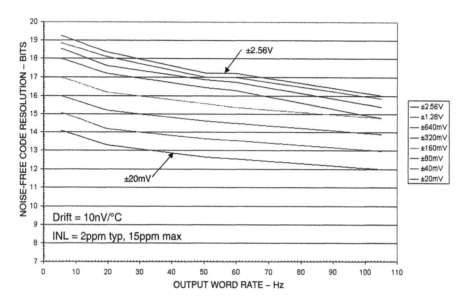

Figure 8.185: Σ-Δ ADC Performance – Normal Mode

Notice that peak-to-peak resolution is better than 19 bits at the lowest gain and slowest throughput rate. This equates to an effective rms resolution of better than 21.7 bits, and makes it pretty clear that all the digital integration in these MicroConverter products has not affected their analog precision.

Furthermore, in their normal mode of operation, these ADCs employ an ADI patented chopper stabilization technique, similar to that of chopper stabilized amplifiers, to achieve unprecedented temperature drift performance. Also, built-in self-calibration features can essentially eliminate endpoint offset and gain errors.

With the high-channel-count Σ-Δ delta MicroConverter products, the user has the option of turning the ADC's chop mode off. This allows much higher output word rates (up to 1.3 kHz), but at the cost of degraded effective resolution and significantly worse temperature drift performance as shown in Figure 8.186. This is a software selectable option, however, and in most applications it is desirable to leave chop mode enabled, resulting in the superior noise and drift performance shown previously in Figure 8.185.

The current Analog Devices' Σ-Δ MicroConverter family is summarized in Figure 8.187.

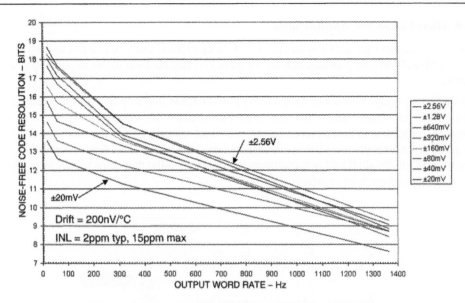

Figure 8.186: Σ-Δ ADC Performance – Chop Mode Disabled

Part #	ADC	DAC	MCU	Flash/EE Code	Flash/EE Data	RAM	PKGs	Special Features
ADuC816	Dual 16-bit	Single 12-bit	1MIPS 8052	8K-byte	640-byte	256-byte	52-PQFP 56-CSP	Lowest Cost Σ-Δ MicroConverter
ADuC824	24-bit + 16-bit	Single 12-bit	1MIPS 8052	8K-byte	640-byte	256-byte	52-PQFP 56-CSP	Pin-Compatible Upgrade to ADuC816
ADuC834	24-bit + 16-bit	Single 12-bit +Dual PWM	1MIPS 8052	62K-byte	4K-byte	256-byte +2K-byte	52-PQFP 56-CSP	"Big-Memory" Upgrade to ADuC824
ADuC836	Dual 16-bit	Single 12-bit +Dual PWM	1MIPS 8052	62K-byte	4K-byte	256-byte +2K-byte	52-PQFP 56-CSP	"Big-Memory" Upgrade to ADuC816
ADuC845	10-chan 24-bit	Single 12-bit +Dual PWM	12MIPS 8052	8K,32K,62 K-byte	4K-byte	256-byte +2K-byte	52-PQFP 56-CSP	"Fast-Core" ΣΔ with Multi-Channel Input
ADuC847	10-chan 24-bit	Dual PWM	12MIPS 8052	8K,32K,62 K-byte	4K-byte	256-byte +2K-byte	52-PQFP 56-CSP	Stripped-Down ADuC845
ADuC848	10-chan 16-bit	Dual PWM	12MIPS 8052	8K,32K,62 K-byte	4K-byte	256-byte +2K-byte	52-PQFP 56-CSP	16-bit Version of ADuC847

Figure 8.187: Summary of Σ-Δ-Based MicroConverter Products

Some Σ-Δ MicroConverter Applications

The following application examples use the ADuC834. Keep in mind that you can just as easily substitute any of the Σ-Δ MicroConverter products into any of these example configurations, depending upon system requirements.

Figure 8.188 shows a bridge transducer design using the ADuC834 that employs the ratiometric technique discussed earlier in this chapter, and therefore a voltage reference is not required. Notice that the sense lines from the bridge (connecting to the reference inputs) are wired separately from the excitation lines (going to VDD and ground). This results in a total of six wires going to the bridge. This 6-wire connection scheme is a feature of most off-the-shelf bridge transducers (such as load cells) that helps to minimize errors that would otherwise result from wire resistance.

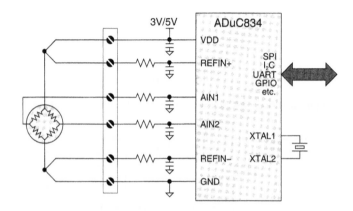

Figure 8.188: Bridge Transducer Interfacing Using a MicroConverter

This represents the complete design except for the serial connection to the rest of the system. No other support components are required, with the possible exception of overvoltage protection diodes at the terminal block inputs. This is an example of how MicroConverter products have earned the label, "system on a chip." The software design is facilitated by the excellent set of MicroConverter development tools.

Figure 8.189 shows a typical MicroConverter thermocouple application. Again, this is similar to the thermocouple interface to a standard Σ-Δ ADC as previously discussed in this chapter. The AD592 is used to sense the "cold junction" temperature. The calibration and linearization coefficients for the thermocouple are stored in the microcontroller memory. Since signals in this circuit are not ratiometric like they were for the bridge transducer circuit, a precision voltage reference is required. We could use the ADuC834's on-chip voltage reference, but to take full advantage of the high resolution ADC performance, it is beneficial to use a precision, low noise, external reference like the AD780.

Figure 8.190 shows the MicroConverter interfaced to an RTD via a 4-wire connection. Off-the-shelf RTDs are typically available in 2-wire, 3-wire, or 4-wire configurations, but the four-wire connection is the best way to minimize errors caused by lead resistance, which can otherwise be significant. Note that this is another ratiometric configuration that does not require a voltage reference. The excitation current generated in the MicroConverter (IEXC1) flows through the RTD and the R_{REF} resistor. The voltage developed across the R_{REF} resistor is used to set the reference voltage for the MicroConverter, thereby establishing ratiometric operation. The MicroConverter software is used to correct for nonlinearities in the RTD transfer function.

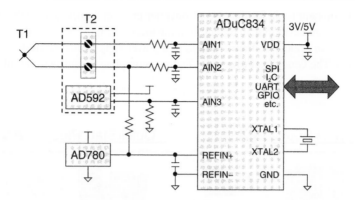

Figure 8.189: Thermocouple Interfacing Using a MicroConverter

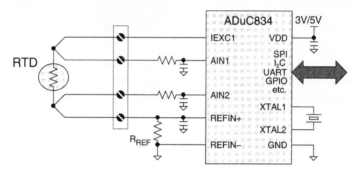

Figure 8.190: RTD Interfacing Using a MicroConverter

Finally, the table shown in Figure 8.191 summarizes current 8051-based Analog Devices MicroConverter products.

Figure 8.191: ADuC8xx Product Family Overview

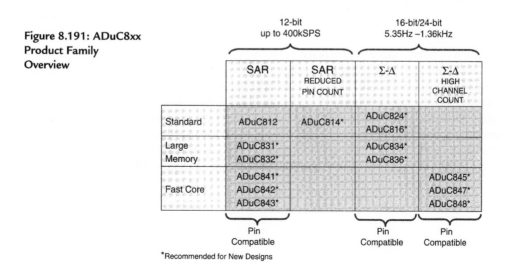

	12-bit up to 400kSPS		16-bit/24-bit 5.35Hz –1.36kHz	
	SAR	SAR REDUCED PIN COUNT	Σ-Δ	Σ-Δ HIGH CHANNEL COUNT
Standard	ADuC812	ADuC814*	ADuC824* ADuC816*	
Large Memory	ADuC831* ADuC832*		ADuC834* ADuC836*	
Fast Core	ADuC841* ADuC842* ADuC843*			ADuC845* ADuC847* ADuC848*
	Pin Compatible		Pin Compatible	Pin Compatible

*Recommended for New Designs

A complete set of development tools for the MicroConverter product family is available from Analog Devices to facilitate software development and system integration (Figure 8.192). The QuickStart™ development kit includes an evaluation board as well as a power supply, download/debug cable, and software development tools. The QuickStart Plus™ kit features complete nonintrusive emulation capability (C-Source/Assembly).

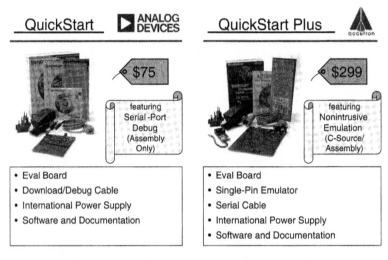

Figure 8.192: MicroConverter Development Tools

ADuC7xxx MicroConverter Products Based on the ARM7 Processor Core

Although the 8051/8052 is a popular 8-bit microcontroller core, there are applications where a more powerful core is required. Analog Devices has chosen the ARM7TDMI core for the ADuC7xxx MicroConverter product family. This popular core offers industry-standard software, and provides a 16/32-bit RISC (reduced instruction set computer) architecture. The concept for the ADuC7xxx family is similar to that of the standard ADuC8xx family, with some important added enhancements including the ARM7 core (Figure 8.193).

The first group of ADuC7xxx products is the ADuC702x. A simplified diagram of the ADuC702x series is shown in Figure 8.194. The analog section has a flexible multiplexer with up to 12 standard inputs. The inputs can be configured as single-ended, pseudo-differential, or fully differential. The SAR ADC has 12-bit resolution and a sampling rate of 1 MSPS. In addition there is an uncommitted comparator, a low drift bandgap reference, and multiple 12-bit voltage output DACs. The processor core is the ARM7TDMI, and there is 62 Kbytes of flash memory, and 8 Kbytes of SRAM. Regarding additional peripherals, there is an on-chip programmable logic array (PLA), power supply monitor (PSM), general-purpose I/O (GPIO), serial I/O, general-purpose timers, and three phase pulsewidth modulator (PWM).

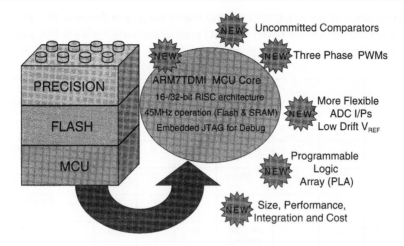

Figure 8.193: ADuC7xxx Family of ARM7-Based MicroConverter Products

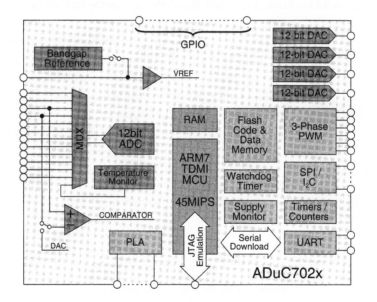

Figure 8.194: ADuC702x ARM7-Based MicroConverter Products

A summary of the members of the ADuC702x product family is given in Figure 8.195. The ADuC702x allows in-system programming via the UART or JTAG ports as shown in Figure 8.196, and Figure 9.197 summarizes the development system.

Part #	ADC Chnls	DAC Cnnls*	Flash (bytes)	SRAM (bytes)	GPIO (max)	Com-parator	PWM	Package
ADuC7020	5	4	62k	8k	14	Yes	No	40LFCSP
ADuC7021	8	2	62k	8k	13	Yes	No	40LFCSP
ADuC7022	10	0	62k	8k	13	Yes	No	40LFCSP
ADuC7024	10	2	62k	8k	30	Yes	Yes	64LFCSP
ADuC7026	12	4	62k (+Ext Memory)	8k	40	Yes	Yes	80LQFP

*Unused DAC output channels can be used as additional ADC input channels.

Figure 8.195: ADuC702x ARM7-Based MicroConverter Product Family

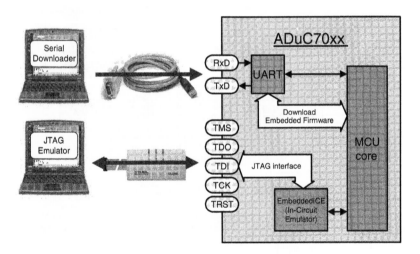

Figure 8.196: On-Chip Tools Resources

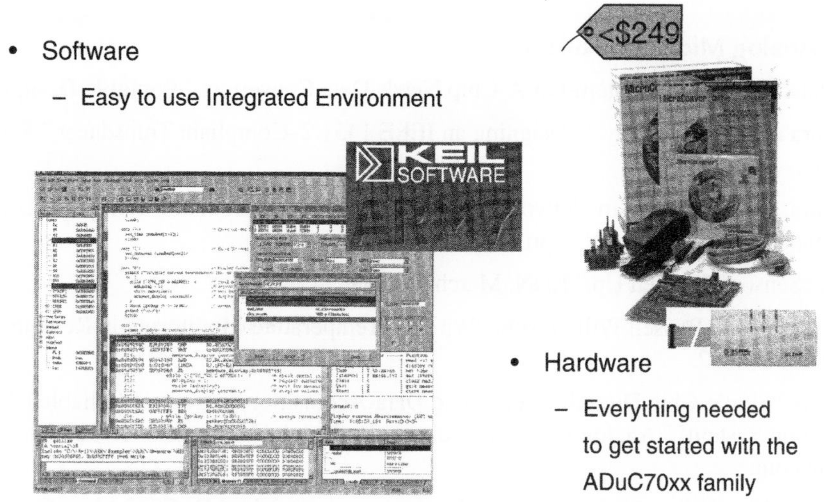
• Software
– Easy to use Integrated Environment
KEIL SOFTWARE
<$249
• Hardware
– Everything needed to get started with the ADuC70xx family

Figure 8.197: ADuC70xx Development System

Additional applications circuits for the MicroConverter product family can be found in References 5–13. General programming and architecture information for the 8051/8052 core can be found in References 14–18. The primary source for ARM7 information is the ARM website, www.arm.com (Reference 19).

The interested reader should also consult Analog Devices' MicroConverter website for additional technical notes, articles, references, sample code, etc., at www.analog.com/microconverter.

References:
8.8 Precision Analog Microcontrollers

1. Alex Mendelsohn, "A Real System-On-A-Chip Needs Data Converters," **Portable Design**, June 1998.

2. Brian O'Mara and Paul Conway, "Designing an IEEE 1451.2–Compliant Transducer," **Sensors**, August 2000.

3. Brian O'Mara, "From Sensors to Networks: Microcontrollers Are Common Components In Sensor Measurement Signal Paths," **ECN**, April 2001.

4. Bill Travis, "Sensors Smarten Up," EDN, March 4, 1999.

5. Darragh Maxwell and Russell Williamson, "Wireless Temperature Monitoring in Remote Systems," **Sensors**, October 2002.

6. Mark Malaeb, "Single-Chip Digitally Controlled Data-Acquisition as Core of Reliable DWDM Communication Systems," Analog Dialogue, Volume 36, Number 5, September-October, 2002, www.analog.com.

7. Mark Malaeb, "Closed-Loop Control Circuit Implementation of the ADuC832 MicroConverter IC and the AD8305 Logarithmic Converter in a Digital Variable Optical Attenuator," **Application Note AN-643**, Analog Devices, www.analog.com.

8. Eamon Neary, "Frequency Measurement Using Timer 2 on a MicroConverter," **Application Note AN-644**, Analog Devices, www.analog.com.

9. Eamon Neary, "Interfacing an HD44780 Character LCD to a MicroConverter," **Application Note AN-645**, Analog Devices, www.analog.com.

10. Luca Vassalli and Mark Malaeb, "Optical Module Development Platform 2.5 Gbps Transmitter with Digital Diagnostics," **Application Note AN-654**, Analog Devices, www.analog.com.

11. Nobuhiro Matsuzoe, "Tunable Laser Reference Design for Designers with the ADuC832/ADN8830/ADN2830," **Application Note AN-655**, Analog Devices, www.analog.com.

12. Brian Moss, "XY-Matrix Keypad Interface to MicroConverter," **Application Note AN-660**, Analog Devices, www.analog.com.

13. Brian Moss, "ADuC814 to ADM1032 via I²C® Interface," **Application Note AN-661**, Analog Devices, www.analog.com.

14. Thomas W. Shultz, **C and the 8051 Vol.I: Hardware, Modular Programming & Multitasking**, Second Edition, Prentice-Hall, 1998.

15. Thomas W. Shultz, **C and the 8051 Vol. II: Building Efficient Applications**, Prentice Hall, 1999.

16. I. Scott MacKenzie, **The 8051 Microcontroller**, Third Edition, Prentice-Hall, 1999.

17. James W. Stewart and Kai Miao, **The 8051 Microcontroller: Hardware, Software, and Interfacing**, Second Edition, Prentice Hall, 1999.

18. www.8052.com. (*A general reference site on the 8051, 8052 microcontrollers.*)

19. www.arm.com. (*A general reference site for the ARM-family of microcontrollers.*)

20. Sensor Interfacing Seminar, NetSeminar, www.analog.com/mcvNetSeminar.

CHAPTER 9

Hardware Design Techniques

Hardware Design Techniques

This chapter, one of the longer ones in this book, deals with topics just as important as all of those basic circuits immediately surrounding the data converter, discussed earlier. The chapter deals with various and sundry circuit/system issues which fall under the guise of system *hardware design techniques*. In this context, the design techniques may be all those support items surrounding a data converter, excluding the data converter itself. This includes issues of passive components, printed circuit design, power supply systems, protection of linear devices against overvoltage and thermal effects, EMI/RFI issues, high speed logic considerations, and finally, simulation, breadboarding and prototyping. Some of these topics aren't directly involved in the actual signal path of a design, but they are every bit as important as choosing the correct device and surrounding circuit values.

Remote sensing and signal conditioning is such a vital part of data conversion that a considerable amount of discussion is given to topics such as overvoltage protection, cable driving, shielding, and receiving—where the remote interface is often with op amps and instrumentation amplifiers. Much of this material has been extracted from a companion publication by Walter G. Jung: *Op Amp Applications Handbook*, Newnes, 2005.

Hardware Design Techniques

Passive Components
James Bryant, Walt Jung, Walt Kester

Introduction

When designing with data converters, op amps, and other precision analog devices, it is critical that users avoid the pitfall of poor passive component choice. In fact, the wrong passive component can derail even the best op amp or data converter application. This section includes discussion of some basic traps of choosing passive components for op amp and data converter applications.

So, good money has been spent for a precision op amp or data converter, only to find that, when plugged into the board, the device doesn't meet spec. Perhaps the circuit suffers from drift, poor frequency response, and oscillations—or simply doesn't achieve expected accuracy. Well, before blaming the device, closely examine your passive components—including capacitors, resistors, potentiometers, and yes, even the printed circuit boards. In these areas, subtle effects of tolerance, temperature, parasitics, aging, and user assembly procedures can unwittingly sink a circuit. All too often these effects go unspecified (or under-specified) by passive component manufacturers.

In general, if using data converters having 12 bits or more of resolution, or op amps that cost more than a few dollars, pay very close attention to passive components. Consider the case of a 12-bit DAC, where ½ LSB corresponds to 0.012% of full scale, or only 122 ppm. A host of passive component phenomena can accumulate errors far exceeding this. But, buying the most expensive passive components won't necessarily solve these problems. Often, a *correct* 25-cent capacitor yields a better-performing, more cost-effective design than a premium-grade part. With a few basics, understanding and analyzing passive components may prove rewarding, albeit not easy.

Capacitors

Most designers are generally familiar with the range of available capacitors. But the mechanisms by which both static and dynamic errors can occur in precision circuit designs using capacitors are sometimes easy to forget, because of the tremendous variety of types available. These include dielectrics of glass, aluminum foil, solid tantalum and tantalum foil, silver mica, ceramic, Teflon, and the film capacitors, including polyester, polycarbonate, polystyrene, and polypropylene types. In addition to the traditional leaded packages, many of these are now also offered in surface-mount styles.

Figure 9.1 is a workable model of a non-ideal capacitor. The nominal capacitance, C, is shunted by a resistance R_P which represents *insulation resistance* or leakage. A second resistance, R_S—*equivalent series resistance*, or ESR,—appears in series with the capacitor and represents the resistance of the capacitor leads and plates.

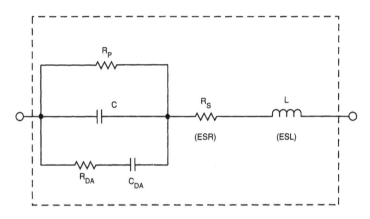

Figure 9.1: A Nonideal Capacitor Equivalent Circuit Includes Parasitic Elements

Note that capacitor phenomena aren't that easy to isolate. The matching of phenomena and models is for convenience in explanation. Inductance, L—the *equivalent series inductance*, or ESL—models the inductance of the leads and plates. Finally, resistance R_{DA} and capacitance C_{DA} together form a simplified model of a phenomenon known as *dielectric absorption,* or DA. It can ruin fast and slow circuit dynamic performance. In a real capacitor R_{DA} and C_{DA} extend to include multiple parallel sets. These parasitic RC elements can act to degrade timing circuits substantially, and the phenomenon is discussed further below.

Dielectric Absorption

Dielectric absorption, which is also known as "soakage" and sometimes as "dielectric hysteresis"—is perhaps the least understood and potentially most damaging of various capacitor parasitic effects. Upon discharge, most capacitors are reluctant to give up all of their former charge, due to this memory consequence.

Figure 9.2 illustrates this effect. On the left of the diagram, after being charged to the source potential of V volts at time t_0, the capacitor is shorted by the switch S1 at time t_1, discharging it. At time t_2, the capacitor is then open-circuited; a residual voltage slowly builds up across its terminals and reaches a nearly constant

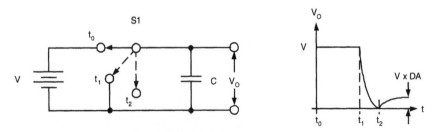

Figure 9.2: A Residual Open-Circuit Voltage After Charge/Discharge Characterizes Capacitor Dielectric Absorption

value. This error voltage is due to DA, and is shown in the right figure, a time/voltage representation of the charge/discharge/recovery sequence. Note that the recovered voltage error is proportional to both the original charging voltage V, as well as the rated DA for the capacitor in use.

Standard techniques for specifying or measuring dielectric absorption are few and far between. Measured results are usually expressed as the percentage of the original charging voltage that reappears across the capacitor. Typically, the capacitor is charged for a long period, then shorted for a shorter established time. The capacitor is then allowed to recover for a specified period, and the residual voltage is then measured (see Reference 8 for details). While this explanation describes the basic phenomenon, it is important to note that real-world capacitors vary quite widely in their susceptibility to this error, with their rated DA ranging from well below to above 1%, the exact number being a function of the dielectric material used.

In practice, DA makes itself known in a variety of ways. Perhaps an integrator refuses to reset to zero, a voltage-to-frequency converter exhibits unexpected nonlinearity, or a sample-hold amplifier (SHA) exhibits varying errors. This last manifestation can be particularly damaging in a data-acquisition system, where adjacent channels may be at voltages that differ by nearly full scale, as shown below.

Figure 9.3 illustrates the case of DA error in a simple SHA. On the left, switches S1 and S2 represent an input multiplexer and SHA switch, respectively. The multiplexer output voltage is V_X, and the sampled voltage held on C is V_y, which is buffered by the op amp for presentation to an ADC. As can be noted by the timing diagram on the right, a DA error voltage, ϵ, appears in the hold mode, when the capacitor is effectively open circuit. This voltage is proportional to the difference of voltages V1 and V2, which, if at opposite extremes of the dynamic range, exacerbates the error. As a practical matter, the best solution for good performance in terms of DA in a SHA is to use only the best capacitor.

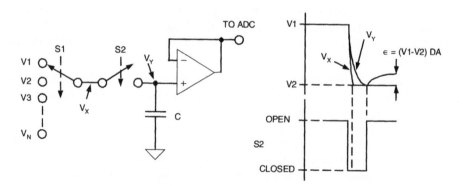

Figure 9.3: Dielectric Absorption Induces Errors in SHA Applications

The DA phenomenon is a characteristic of the dielectric material itself, although inferior manufacturing processes or electrode materials can also affect it. DA is specified as a percentage of the charging voltage. It can range from a low of 0.02% for Teflon, polystyrene, and polypropylene capacitors, up to a high of 10% or more for some electrolytics. For some time frames, the DA of polystyrene can be as low as 0.002%.

Common high-K ceramics and polycarbonate capacitor types display typical DA on the order of 0.2%, it should be noted this corresponds to ½ LSB at only 8 bits. Silver mica, glass, and tantalum capacitors typically exhibit even larger DA, ranging from 1.0% to 5.0%, with those of polyester devices falling in the vicinity of 0.5%. As a rule, if the capacitor spec sheet doesn't specifically discuss DA *within your time frame and voltage range*, exercise caution. Another type with lower *specified* DA is likely a better choice.

DA can produce long tails in the transient response of fast-settling circuits, such as those found in high-pass active filters or ac amplifiers. In some devices used for such applications, Figure 9.1's R_{DA}-C_{DA} model of DA can have a time constant of milliseconds. Much longer time constants are also quite usual. In fact, several paralleled R_{DA}-C_{DA} circuit sections with a wide range of time constants can model some devices. In fast-charge, fast-discharge applications, the behavior of the DA mechanism resembles "analog memory"; the capacitor in effect tries to remember its previous voltage.

In some designs, a user can compensate for the effects of DA if it is simple and easily characterized, and is willing to do custom tweaking. In an integrator, for instance, the output signal can be fed back through a suitable compensation network, tailored to cancel the circuit equivalent of the DA by placing a negative impedance effectively in parallel. Such compensation has been shown to improve SH circuit performance by factors of 10 or more (Reference 6).

Capacitor Parasitics and Dissipation Factor

In Figure 9.1, a capacitor's leakage resistance, R_P, the effective series resistance, R_S, and effective series inductance, L, act as parasitic elements, which can degrade an external circuit's performance. The effects of these elements are often lumped together and defined as a dissipation factor, or DF.

A capacitor's leakage is the small current that flows through the dielectric when a voltage is applied. Although modeled as a simple insulation resistance (R_P) in parallel with the capacitor, the leakage actually is nonlinear with voltage. Manufacturers often specify leakage as a megohm-microfarad product, which describes the dielectric's self-discharge time constant, in seconds. It ranges from a low of 1 second or less for high-leakage capacitors, such as electrolytic devices, to the 100s of seconds for ceramic capacitors. Glass devices exhibit self-discharge time-constants of 1,000 or more; but the best leakage performance is shown by Teflon and the film devices (polystyrene, polypropylene), with time constants exceeding 1,000,000 megohm-microfarads. For such a device, external leakage paths—created by surface contamination of the device's case or in the associated wiring or physical assembly—can overshadow the internal dielectric-related leakage.

Effective series inductance, ESL (Figure 9.1, again) arises from the inductance of the capacitor leads and plates, which, particularly at the higher frequencies, can turn a capacitor's normally capacitive reactance into an inductive reactance. Its magnitude strongly depends on construction details within the capacitor. Tubular wrapped-foil devices display significantly more lead inductance than molded radial-lead configurations. Multilayer ceramic (MLC) and film-type devices typically exhibit the lowest series inductance, while ordinary tantalum and aluminum electrolytics typically exhibit the highest. Consequently, standard electrolytic types, if used alone, usually prove insufficient for *high speed* local bypassing applications. Note however that there also are more specialized aluminum and tantalum electrolytics available, which may be suitable for higher speed uses, however, localized bypassing is still recommended. These are the types generally designed for use in switch-mode power supplies, which are covered more completely in a following section.

Manufacturers of capacitors often specify effective series impedance by means of impedance-versus-frequency plots. Not surprisingly, these curves show graphically a predominantly capacitive reactance at low frequencies, with rising impedance at higher frequencies because of the effect of series inductance.

Effective series resistance, ESR (resistor R_S of Figure 9.1), is made up of the resistance of the leads and plates. As noted, many manufacturers lump the effects of ESR, ESL, and leakage into a single parameter called *dissipation factor*, or DF. Dissipation factor measures the basic inefficiency of the capacitor. Manufacturers define it as the ratio of the energy lost to energy stored per cycle by the capacitor. The ratio of ESR to total capacitive reactance—at a specified frequency—approximates the dissipation factor, which turns out to be equivalent to the reciprocal of the figure of merit, Q. Stated as an approximation, $Q \approx 1/DF$ (with DF in numeric terms). For example, a DF of 0.1% is equivalent to a fraction of 0.001; thus the inverse in terms of Q would be 1000.

Dissipation factor often varies as a function of both temperature and frequency. Capacitors with mica and glass dielectrics generally have DF values from 0.03% to 1.0%. For ceramic devices, DF ranges from a low of 0.1 % to as high as 2.5% at room temperature. And electrolytics usually exceed even this level. The film capacitors are the best as a group, with DFs of less than 0.1 %. Stable-dielectric ceramics, notably the NP0 (also called COG) types, have DF specs comparable to films (more below).

Tolerance, Temperature, and Other Effects

In general, precision capacitors are expensive and—even then—not necessarily easy to buy. In fact, choice of capacitance is limited both by the range of available values, and also by tolerances. In terms of size, the better performing capacitors in the film families tend to be limited in practical terms to 10 µF or less (for dual reasons of size and expense). In terms of low value tolerance, ±1% is possible for NP0 ceramic and some film devices, but with possibly unacceptable delivery times. Many film capacitors can be made available with tolerances of less than ±1%, but on a special order basis only.

Most capacitors are sensitive to temperature variations. DF, DA, and capacitance value are all functions of temperature. For some capacitors, these parameters vary approximately linearly with temperature, in others they vary quite nonlinearly. Although it is usually not important for SHA applications, an excessively large *temperature coefficient* (TC, measured in ppm/°C) can prove harmful to the performance of precision integrators, voltage-to-frequency converters, and oscillators. NP0 ceramic capacitors, with TCs as low as 30 ppm/°C, are the best for stability, with polystyrene and polypropylene next best, with TCs in the 100–200 ppm/°C range. On the other hand, when capacitance stability is important, one should stay away from types with TCs of more than a few hundred ppm/°C, or in fact any TC which is nonlinear.

A capacitor's maximum working temperature should also be considered, in light of the expected environment. Polystyrene capacitors, for instance, melt near 85°C, compared to Teflon's ability to survive temperatures up to 200°C.

Sensitivity of capacitance and DA to applied voltage, expressed as *voltage coefficient*, can also hurt capacitor performance within a circuit application. Although capacitor manufacturers don't always clearly specify voltage coefficients, the user should always consider the possible effects of such factors. For instance, when maximum voltages are applied, some high-K ceramic devices can experience a decrease in capacitance of 50% or more. This is an inherent distortion producer, making such types unsuitable for signal path filtering, for example, and better suited for supply bypassing. Interestingly, NP0 ceramics, the stable dielectric subset from the wide range of available ceramics, do offer good performance with respect to voltage coefficient.

Similarly, the capacitance and dissipation factor of many types vary significantly with frequency, mainly as a result of a variation in dielectric constant. In this regard, the better dielectrics are polystyrene, polypropylene, and Teflon.

Assemble Critical Components Last

The designer's worries don't end with the design process. Some common printed circuit assembly techniques can prove ruinous to even the best designs. For instance, some commonly used cleaning solvents can infiltrate certain electrolytic capacitors—those with rubber end caps are particularly susceptible. Even worse, some of the film capacitors, polystyrene in particular, actually melt when contacted by some solvents. Rough handling of the leads can damage still other capacitors, creating random or even intermittent circuit problems. Etched-foil types are particularly delicate in this regard. To avoid these difficulties it may be advisable to mount especially critical components as the last step in the board assembly process—if possible.

Table 9.1 summarizes selection criteria for various capacitor types, arranged roughly in order of decreasing DA performance. In a selection process, the general information of this table should be supplemented by consultation of current vendor's catalog information (see References at end of section).

Designers should also consider the natural failure mechanisms of capacitors. Metallized film devices, for instance, often self-heal. They initially fail due to conductive bridges that develop through small perforations in the dielectric film. But, the resulting fault currents can generate sufficient heat to destroy the bridge, thus returning the capacitor to normal operation (at a slightly lower capacitance). Of course, applications in high-impedance circuits may not develop sufficient current to clear the bridge, so the designer must be wary here.

Tantalum capacitors also exhibit a degree of self-healing, but—unlike film capacitors—the phenomenon depends on the temperature at the fault location rising slowly. Therefore, tantalum capacitors self-heal best in high impedance circuits which limit the surge in current through the capacitor's defect. Use caution therefore, when specifying tantalums for high-current applications.

Electrolytic capacitor life often depends on the rate at which capacitor fluids seep through end caps. Epoxy end seals perform better than rubber seals, but an epoxy sealed capacitor can explode under severe reverse-voltage or overvoltage conditions. Finally, *all* polarized capacitors must be protected from exposure to voltages outside their specifications.

Table 9.1
Capacitor Comparison Chart

TYPE	TYPICAL DA	ADVANTAGES	DISADVANTAGES
Polystyrene	0.001% to 0.02%	Inexpensive Low DA Good stability (~120ppm/°C)	Damaged by temperature > +85°C Large High inductance Vendors limited
Polypropylene	0.001% to 0.02%	Inexpensive Low DA Stable (~200ppm/°C) Wide range of values	Damaged by temperature > +105°C Large High inductance
Teflon	0.003% to 0.02%	Low DA available Good stability Operational above +125 °C Wide range of values	Expensive Large High inductance
Polycarbonate	0.1%	Good stability Low cost Wide temperature range Wide range of values	Large DA limits to 8-bit applications High inductance
Polyester	0.3% to 0.5%	Moderate stability Low cost Wide temperature range Low inductance (stacked film)	Large DA limits to 8-bit applications High inductance (conventional)
NP0 Ceramic	<0.1%	Small case size Inexpensive, many vendors Good stability (30ppm/°C) 1% values available Low inductance (chip)	DA generally low (may not be specified) Low maximum values (10nF)
Monolithic Ceramic (High K)	>0.2%	Low inductance (chip) Wide range of values	Poor stability Poor DA High voltage coefficient
Mica	>0.003%	Low loss at HF Low inductance Good stability 1% values available	Quite large Low maximum values (10nF) Expensive
Aluminum Electrolytic	Very high	Large values High currents High voltages Small size	High leakage Usually polarized Poor stability, accuracy Inductive
Tantalum Electrolytic	Very high	Small size Large values Medium inductance	High leakage Usually polarized Expensive Poor stability, accuracy

Resistors and Potentiometers

Designers have a broad range of resistor technologies from which to choose, including carbon composition, carbon film, bulk metal, metal film, and both inductive and noninductive wire-wound types. As perhaps the most basic—and presumably most trouble-free—of components, resistors are often overlooked as error sources in high performance circuits.

Yet, an improperly selected resistor can subvert the accuracy of a 12-bit design by developing errors well in excess of 122 ppm (½ LSB). It is surprising what can be learned from an informed review of a resistor data sheet.

Consider the simple circuit of Figure 9.4, showing a noninverting op amp where the 100× gain is set by R1 and R2. The TCs of these two resistors are a somewhat obvious source of error. Assume the op amp gain errors to be negligible, and that the resistors are perfectly matched to a 99/1 ratio at +25°C. If, as noted, the resistor TCs differ by only 25 ppm/°C, the gain of the amplifier changes by 250 ppm for a 10°C temperature change. This is about a 1 LSB error in a 12-bit system, and a major disaster in a 16-bit system.

Temperature changes, however, can limit the accuracy of the Figure 9.4 amplifier in several ways. In this circuit (as well as many op amp circuits with component-ratio defined gains), the *absolute* TC of the resistors is less important—*as long as they track one another in ratio*. But even so, some resistor types simply aren't suitable for precise work. For example, *carbon composition* units—with TCs of approximately 1,500 ppm/°C, won't work. Even if the TCs could be matched to an unlikely 1%, the resulting 15 ppm/°C differential still proves inadequate—an 8°C shift creates a 120 ppm error.

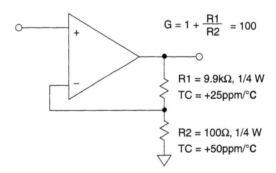

$$G = 1 + \frac{R1}{R2} = 100$$

R1 = 9.9kΩ, 1/4 W
TC = +25ppm/°C

R2 = 100Ω, 1/4 W
TC = +50ppm/°C

- Temperature change of 10°C causes gain change of 250ppm

- This is 1LSB in a 12-bit system and a disaster in a 16-bit system

**Figure 9.4: Mismatched Resistor TCs Can Induce
Temperature-Related Gain Errors**

Many manufacturers offer metal film and bulk metal resistors, with absolute TCs ranging between ±1 and ±100 ppm/°C. Beware, though; TCs can vary a great deal, particularly among discrete resistors from different batches. To avoid this problem, more expensive matched resistor pairs are offered by some manufacturers, with temperature coefficients that track one another to within 2 to 10 ppm/°C. Low-priced thin-film networks have good relative performance and are widely used.

Suppose, as shown in Figure 9.5, R1 and R2 are ¼ W resistors with identical 25 ppm/°C TCs. Even when the TCs are identical, there can still be significant errors. When the signal input is zero, the resistors dissipate no heat. But, if it is 100 mV, there is 9.9 V across R1, which then dissipates 9.9 mW. It will experience a temperature rise of 1.24°C (due to a 125°C/W, ¼ W resistor thermal resistance). This 1.24°C rise causes a resistance change of 31 ppm, and thus a corresponding gain change. But R2, with only 100 mV across it, is only heated a negligible 0.0125°C. The resulting 31 ppm net gain error represents a fullscale error of ½ LSB at 14 bits, and is a disaster for a 16-bit system.

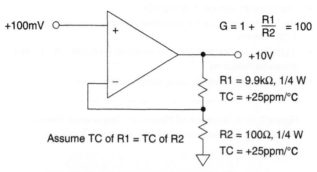

$$G = 1 + \frac{R1}{R2} = 100$$

+10V

R1 = 9.9kΩ, 1/4 W
TC = +25ppm/°C

R2 = 100Ω, 1/4 W
TC = +25ppm/°C

Assume TC of R1 = TC of R2

- R1, R2 Thermal Resistance = 125°C / W
- Temperature of R1 will rise by 1.24°C, P_D = 9.9mW
- Temperature rise of R2 is negligible, P_D = 0.1mW

- Gain is altered by 31ppm, or 1/2 LSB @ 14-bits

Figure 9.5: Uneven Power Dissipation Between Resistors With Identical TCs Can Also Introduce Temperature-Related Gain Errors

Even worse, the effects of this resistor self-heating also create easily calculable *nonlinearity errors*. In the Figure 9.5 example, with ½ the voltage input, the resulting self-heating error is only 15 ppm. In other words, the stage gain is not constant at ½ and full-scale (nor is it so at other points), as long as uneven temperature shifts exist between the gain-determining resistors. This is by no means a worst-case example; physically smaller resistors would give worse results, due to higher associated thermal resistance.

These, and similar errors, are avoided by selecting critical resistors that are accurately matched for both value and TC, are well derated for power, and have tight thermal coupling between those resistors were matching is important. This is best achieved by using a resistor network on a single substrate—such a network may either be within an IC, or it may be a separately packaged thin-film resistor network.

When the circuit resistances are very low (≤10 Ω), *interconnection stability* also becomes important. For example, while often overlooked as an error, the resistance TC of typical copper wire or printed circuit traces can add errors. The TC of copper is typically ~3,900 ppm/°C. Thus a precision 10 Ω, 10 ppm/°C wirewound resistor with 0.1 Ω of copper interconnect effectively becomes a 10.1 Ω resistor with a TC of nearly 50 ppm/°C.

One final consideration applies mainly to designs that see widely varying ambient temperatures: a phenomenon known as *temperature retrace* describes the change in resistance which occurs after a specified number of cycles of exposure to low and high ambients with constant internal dissipation. Temperature retrace can exceed 10 ppm/°C, even for some of the better thin-film components.

In summary, to design resistance-based circuits for minimum temperature-related errors, consider the points noted in Figure 9.6 (along with their cost).

- Closely match resistance TCs.
- Use resistors with low absolute TCs.
- Use resistors with low thermal resistance (higher power ratings, larger cases).
- Tightly couple matched resistors thermally (use standard common-substrate networks).
- For large ratios consider using stepped attenuators.

Figure 9.6: A Number of Points are Important Towards Minimizing Temperature-Related Errors in Resistors

Resistor Parasitics

Resistors can exhibit significant levels of parasitic inductance or capacitance, especially at high frequencies. Manufacturers often specify these parasitic effects as a reactance error, in % or ppm, based on the ratio of the difference between the impedance magnitude and the dc resistance, to the resistance, at one or more frequencies.

Wirewound resistors are especially susceptible to difficulties. Although resistor manufacturers offer wirewound components in either normal or noninductively wound form, even noninductively wound resistors create headaches for designers. These resistors still appear slightly inductive (of the order of 20 µH) for values below 10 kΩ. Above 10 kΩ the same style resistors actually exhibit 5 pF of shunt capacitance.

These parasitic effects can raise havoc in dynamic circuit applications. Of particular concern are applications using wirewound resistors with values greater than 10 kΩ. Here it isn't uncommon to see peaking, or even oscillation. These effects become more evident at low kHz frequency ranges.

Even in low frequency circuit applications, parasitic effects in wirewound resistors can create difficulties. Exponential settling to 1 ppm may take 20 time constants or more. The parasitic effects associated with wirewound resistors can significantly increase net circuit settling time to beyond the length of the basic time constants.

Unacceptable amounts of parasitic reactance are often found even in resistors that aren't wirewound. For instance, some metal-film types have significant interlead capacitance, which shows up at high frequencies. In contrast, when considering this end-to-end capacitance, carbon resistors do the best at high frequencies.

Thermoelectric Effects

Another more subtle problem with resistors is the *thermocouple effect*, also sometimes referred to as *thermal EMF*. Wherever there is a junction between two different metallic conductors, a thermoelectric voltage results. The thermocouple effect is widely used to measure temperature. However, in any low level precision op amp circuit it is also a potential source of inaccuracy, since wherever two different conductors meet, a thermocouple is formed (whether we like it or not). In fact, in many cases, it can easily produce the dominant error within an otherwise precision circuit design.

Parasitic thermocouples will cause errors when and if the various junctions forming the parasitic thermocouples are at different temperatures. With two junctions present on each side of the signal being processed within a circuit, by definition we have formed at least one thermocouple pair. If the two junctions of this thermocouple pair are at different temperatures, there will be a net temperature dependent error voltage produced. Conversely, if the two junctions of a parasitic thermocouple pair are kept at an identical temperature, then the net error produced will be zero, as the voltages of the two thermocouples effectively will be canceled.

This is a critically important point, since in practice we cannot avoid connecting dissimilar metals together to build an electronic circuit. We can, however, carefully control temperature differentials across the circuit, so that the undesired thermocouple errors cancel one another.

The effect of such parasitics is very hard to avoid. To understand this, consider a case of making connections *with copper wire only*. In this case, even a junction formed by different copper wire alloys can have a thermoelectric voltage that is a small fraction of 1 µV/°C. Taking things a step further, even such apparently benign components as resistors contain parasitic thermocouples, with potentially even stronger effects.

For example, consider the resistor model shown in Figure 9.7. The two connections between the resistor material and the leads form thermocouple junctions, T1 and T2. This thermocouple EMF can be as high as 400 µV/°C for some carbon composition resistors, and as low as 0.05 µV/°C for specially constructed resistors (see Reference 15). Ordinary metal film resistors (RN-types) are typically about 20 µV/°C.

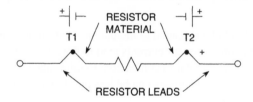

TYPICAL RESISTOR THERMOCOUPLE EMFs

* CARBON COMPOSITION ≈ 400 µV/°C
* METAL FILM ≈ 20 µV/°C
* EVENOHM OR
  MANGANIN WIREWOUND ≈ 2 µV/°C
* RCD Components HP-Series ≈ 0.05 µV/°C

**Figure 9.7: Every Resistor Contains Two Thermocouples,
Formed Between the Leads and Resistance Element**

Note that these thermocouple effects are relatively unimportant for ac signals. Even for dc-only signals, they will nicely cancel one another, if, as noted above, the entire resistor is at a uniform temperature. However, if there is significant power dissipation in a resistor, or if its orientation with respect to a heat source is nonsymmetrical, this can cause one of its ends to be warmer than the other, causing a net thermocouple error voltage. Using ordinary metal film resistors, an end-to-end temperature differential of 1°C causes a thermocouple voltage of about 20 µV. This error level is quite significant compared to the offset voltage drift of a precision op amp like the OP177, and extremely significant when compared to chopper-stabilized op amps, with their drifts of <1 µV/°C.

Figure 9.8 shows how resistor orientation can make a difference in the net thermocouple voltage. In the left diagram, standing the resistor on end in order to conserve board space will invariably cause a temperature gradient across the resistor, especially if it is dissipating any significant power. In contrast, placing the resistor flat on the PC board as shown at the right will generally eliminate the gradient. An exception might occur, if there is end-to-end resistor airflow. For such cases, orienting the resistor axis perpendicular to the airflow will minimize this source of error, since this tends to force the resistor ends to the same temperature.

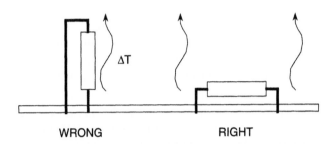

WRONG   RIGHT

**Figure 9.8: The Effects of Thermocouple EMFs Generated by Resistors
can be Minimized by Orientation that Equalizes the End Temperatures**

Note that this line of thinking should be extended to include orientation of resistors on a vertically mounted PC board. In such cases, natural convection air currents tend to flow upward across the board. Again, the resistor thermal axis should be perpendicular to convection, to minimize thermocouple effects. With tiny surface-mount resistors, the thermocouple effects can be less problematic, due to tighter thermal coupling between the resistor ends.

In general, designers should strive to avoid thermal gradients on or around critical circuit boards. Often this means thermally isolating components that dissipate significant amounts of power. Thermal turbulence created by large temperature gradients can also result in dynamic noise-like low frequency errors.

Voltage Sensitivity, Failure Mechanisms, and Aging

Resistors are also plagued by changes in value as a function of applied voltage. The deposited-oxide high-megohm type components are especially sensitive, with voltage coefficients ranging from 1 ppm/V to more than 200 ppm/V. This is another reason to exercise caution in such precision applications as high-voltage dividers.

The normal failure mechanism of a resistor can also create circuit difficulties, if not carefully considered beforehand. For example, carbon-composition resistors fail safely, by turning into open circuits. Consequently, in some applications, these components can play a useful secondary role, as a fuse. Replacing such a resistor with a carbon-film type can possibly lead to trouble, since carbon-films can fail as short circuits. (Metal-film components usually fail as open circuits.)

All resistors tend to change slightly in value with age. Manufacturers specify long-term stability in terms of change—ppm/year. Values of 50 ppm or 75 ppm/year are not uncommon among metal film resistors. For critical applications, metal-film devices should be burned in for at least one week at rated power. During burn-in, resistance values can shift by up to 100 ppm or 200 ppm. Metal film resistors may need 4–5000 operational hours for full stabilization, especially if deprived of a burn-in period.

Resistor Excess Noise

Most designers have some familiarity with thermal, or Johnson, noise which occurs in resistors. But a less widely recognized secondary noise phenomenon is associated with resistors, and it is called *excess noise*. It can prove particularly troublesome in precision op amp and converter circuits, as it is evident only when current passes through a resistor.

To review briefly, thermal noise results from thermally induced random vibration of charge resistor carriers. Although the average current from the vibrations remains zero, instantaneous charge motions result in an instantaneous voltage across the terminals.

Excess noise on the other hand, occurs primarily when dc flows in a discontinuous medium—for example the conductive particles of a carbon composition resistor. The current flows unevenly through the compressed carbon granules, creating microscopic particle-to-particle "arcing." This phenomenon gives rise to a 1/f noise-power spectrum, in addition to the thermal noise spectrum. In other words, the excess spot noise voltage increases as the inverse square-root of frequency.

Excess noise often surprises the unwary designer. Resistor thermal noise and op amp input noise set the noise floor in typical op amp circuits. Only when voltages appear across input resistors and causes current to flow does the excess noise become a significant—and often dominant—factor. In general, carbon composition resistors generate the most excess noise. As the conductive medium becomes more uniform, excess noise becomes less significant. Carbon film resistors do better, with metal film, wirewound and bulk-metal-film resistors doing better yet.

Manufacturers specify excess noise in terms of a noise index—the number of microvolts of rms noise in the resistor in each decade of frequency per volt of dc drop across the resistor. The index can rise to 10 dB (3 microvolts per dc volt per decade of bandwidth) or more. Excess noise is most significant at low frequencies, while above 100 kHz thermal noise predominates.

Potentiometers

Trimming potentiometers (trimpots) can suffer from most of the phenomena that plague fixed resistors. In addition, users must also remain vigilant against some hazards unique to these components.

For instance, many trimpots aren't sealed, and can be severely damaged by board washing solvents, and even by excessive humidity. Vibration—or simply extensive use—can damage the resistive element and wiper terminations. Contact noise, TCs, parasitic effects, and limitations on adjustable range can all hamper trimpot circuit operation. Furthermore, the limited resolution of wirewound types and the hidden limits to resolution in cermet and plastic types (hysteresis, incompatible material TCs, slack) make obtaining and maintaining precise circuit settings anything but an "infinite resolution" process. Given this background, two rules are suggested for the potential trimpot user. Rule 1: Use infinite care and infinitesimal adjustment range to avoid infinite frustration when applying manual trimpots. Rule 2: *Consider the elimination of manual trimming potentiometers altogether, if possible.* A number of digitally addressable potentiometers (RDACs or TrimDACs) are now available for direct application in similar circuit functions as classic trimpots (see Reference 17). There are also many low cost multichannel voltage output DACs expressly designed for system voltage trimming.

Table 9.2 summarizes selection criteria for various fixed resistor types, both in discrete form and as part of networks. In a selection process, the general information of this table should be supplemented by consultation of current vendor's catalog information (see References at end of section).

Table 9.2
Resistor Comparison Chart

	TYPE	ADVANTAGES	DISADVANTAGES
DISCRETE	Carbon Composition	Lowest Cost High Power/Small Case Size Wide Range of Values	Poor Tolerance (5%) Poor Temperature Coefficient (1500 ppm/°C)
	Wirewound	Excellent Tolerance (0.01%) Excellent TC (1ppm/°C) High Power	Reactance is a Problem Large Case Size Most Expensive
	Metal Film	Good Tolerance (0.1%) Good TC (<1 to 100ppm/°C) Moderate Cost Wide Range of Values Low Voltage Coefficient	Must be Stabilized with Burn-In Low Power
	Bulk Metal or Metal Foil	Excellent Tolerance (to 0.005%) Excellent TC (to <1ppm/°C) Low Reactance Low Voltage Coefficient	Low Power Very Expensive
	High Megohm	Very High Values (10^8 to $10^{14}\Omega$) Only Choice for Some Circuits	High Voltage Coefficient (200ppm/V) Fragile Glass Case (Needs Special Handling) Expensive
NETWORKS	Thick Film	Low Cost High Power Laser-Trimmable Readily Available	Fair Matching (0.1%) Poor TC (>100ppm/°C) Poor Tracking TC (10ppm/°C)
	Thin Film	Good Matching (<0.01%) Good TC (<100ppm/°C) Good Tracking TC (2ppm/°C) Moderate Cost Laser-Trimmable Low Capacitance Suitable for Hybrid IC Substrate	Often Large Geometry Limited Values and Configurations

Inductance

Stray Inductance

All conductors are inductive, and at high frequencies, the inductance of even quite short pieces of wire or printed circuit traces may be important. The inductance of a straight wire of length L mm and circular cross-section with radius R mm in free space is given by the first equation shown in Figure 9.9.

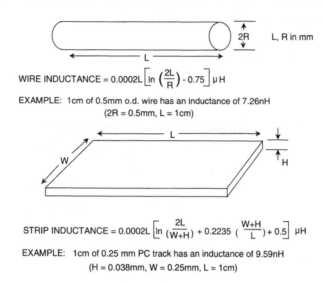

WIRE INDUCTANCE = $0.0002L \left[\ln \left(\frac{2L}{R} \right) - 0.75 \right] \mu H$

EXAMPLE: 1cm of 0.5mm o.d. wire has an inductance of 7.26nH
(2R = 0.5mm, L = 1cm)

STRIP INDUCTANCE = $0.0002L \left[\ln \left(\frac{2L}{W+H} \right) + 0.2235 \left(\frac{W+H}{L} \right) + 0.5 \right] \mu H$

EXAMPLE: 1cm of 0.25 mm PC track has an inductance of 9.59nH
(H = 0.038mm, W = 0.25mm, L = 1cm)

Figure 9.9: Wire and Strip Inductance Calculations

The inductance of a strip conductor (an approximation to a PC track) of width W mm and thickness H mm in free space is also given by the second equation in Figure 9.9.

In real systems, both these formulas turn out to be approximate, but they do give some idea of the order of magnitude of inductance involved. They tell us that 1 cm of 0.5 mm od wire has an inductance of 7.26 nH, and 1 cm of 0.25 mm PC track has an inductance of 9.59 nH—these figures are reasonably close to measured results.

At 10 MHz, an inductance of 7.26 nH has an impedance of 0.46 Ω, and so can give rise to 1% error in a 50 Ω system.

Mutual Inductance

Another consideration regarding inductance is the separation of outward and return currents. Kirchoff's Law tells us that current flows in closed paths—there is always an outward and return path. The whole path forms a single-turn inductor.

725

This principle is illustrated by the contrasting signal trace routing arrangements of Figure 9.10. If the area enclosed within the turn is relatively large, as in the upper "nonideal" picture, then the inductance (and hence the ac impedance) will also be large. On the other hand, if the outward and return paths are closer together, as in the lower "improved" picture, the inductance will be much smaller.

Note that the nonideal signal routing case of Figure 9.10 has other drawbacks—the large area enclosed within the conductors produces extensive external magnetic fields, which may interact with other circuits, causing unwanted coupling. Similarly, the large area is more vulnerable to interaction with external magnetic fields, which can induce unwanted signals in the loop.

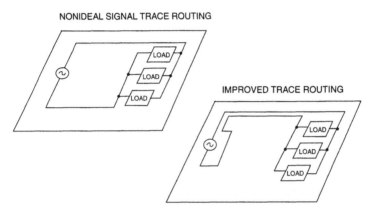

Figure 9.10: Nonideal and Improved Signal Trace Routing

The basic principle is illustrated in Figure 9.11, and is a common mechanism for the transfer of unwanted signals (noise) between two circuits.

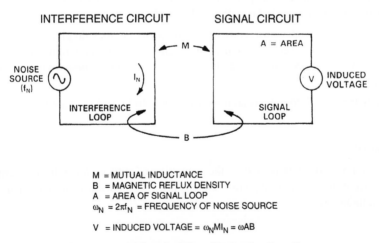

M = MUTUAL INDUCTANCE
B = MAGNETIC REFLUX DENSITY
A = AREA OF SIGNAL LOOP
$\omega_N = 2\pi f_N$ = FREQUENCY OF NOISE SOURCE

V = INDUCED VOLTAGE = $\omega_N M I_N = \omega A B$

Figure 9.11: Basic Principles of Inductive Coupling

As with most other noise sources, as soon as we define the working principle, we can see ways of reducing the effect. In this case, reducing any or all of the terms in the equations in Figure 9.11 reduces the coupling. Reducing the frequency or amplitude of the current causing the interference may be impracticable, but it is frequently possible to reduce the mutual inductance between the interfering and interfered with circuits by reducing loop areas on one or both sides and, possibly, increasing the distance between them.

A layout solution is illustrated by Figure 9.12. Here two circuits, shown as Z1 and Z2, are minimized for coupling by keeping each of the loop areas as small as is practical.

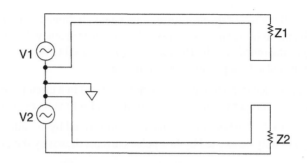

Figure 9.12: Proper Signal Routing and Layout can Reduce Inductive Coupling

As also illustrated in Figure 9.13, mutual inductance can be a problem in signals transmitted on cables. Mutual inductance is high in ribbon cables, especially when a single return is common to several signal circuits (top). Separate, dedicated signal and return lines for each signal circuit reduces the problem (middle). Using a cable with twisted pairs for each signal circuit as in the bottom picture is even better.

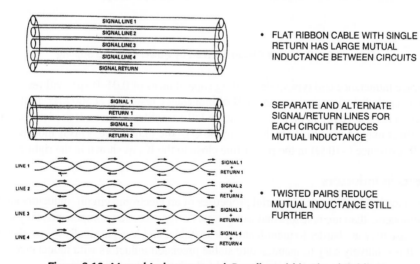

Figure 9.13: Mutual Inductance and Coupling within Signal Cabling

Shielding of magnetic fields to reduce mutual inductance is sometimes possible, but is by no means as easy as shielding an electric field with a Faraday shield (following section). HF magnetic fields are blocked by conductive material provided the skin depth in the conductor at the frequency to be screened is much less than the thickness of the conductor, and the screen has no holes. (Faraday shields can tolerate small holes, magnetic screens cannot.) LF and dc fields may be screened by a shield made of mu-metal sheet. Mu-metal is an alloy having very high permeability, but it is expensive, its magnetic properties are damaged by mechanical stress, and it will saturate if exposed to too high fields. Its use, therefore, should be avoided where possible.

Ringing

An inductor in series or parallel with a capacitor forms a resonant, or "tuned," circuit, whose key feature is that it shows marked change in impedance over a small range of frequency. Just how sharp the effect is depends on the relative Q of the tuned circuit. The effect is widely used to define the frequency response of narrow-band circuitry, but can also be a potential problem source.

If stray inductance and capacitance (which may or may not be stray) in a circuit should form a tuned circuit, then that tuned circuit may be excited by signals in the circuit, and ring at its resonant frequency.

An example is shown in Figure 9.14, where the resonant circuit formed by an inductive power line and its decoupling capacitor may possibly be excited by fast pulse currents drawn by the powered IC.

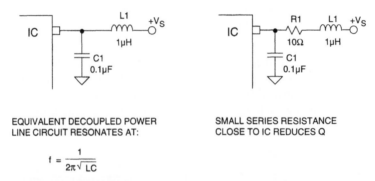

EQUIVALENT DECOUPLED POWER
LINE CIRCUIT RESONATES AT:

SMALL SERIES RESISTANCE
CLOSE TO IC REDUCES Q

$$f = \frac{1}{2\pi\sqrt{LC}}$$

Figure 9.14: Resonant Circuit Formed by Power Line Decoupling

While normal trace inductance and typical decoupling capacitances of 0.01–0.1μF will resonate well above a few MHz, an example 0.1 μF capacitor and 1 μH of inductance resonates at 500 kHz. Left unchecked, this could present a resonance problem, as shown in the left case. Should an undesired power line resonance be present, the effect may be minimized by lowering the Q of the inductance. This is most easily done by inserting a small resistance (~10 Ω) in the power line close to the IC, as shown in the right case.

Parasitic Effects in Inductors

Although inductance is one of the fundamental properties of an electronic circuit, inductors are far less common as components than are resistors and capacitors. As for precision components, they are even more rare. This is because they are harder to manufacture, less stable, and less physically robust than resistors and capacitors. It is relatively easy to manufacture stable precision inductors with inductances from nH to tens or hundreds of μH, but larger valued devices tend to be less stable, and large.

As we might expect in these circumstances, circuits are designed, where possible, to avoid the use of precision inductors. We find that stable precision inductors are rarely used in precision analog circuitry, except in tuned circuits for high frequency narrow band applications.

Of course, they are widely used in power filters, switching power supplies, and other applications where lack of precision is unimportant (more on this in a following section). The important features of inductors used in such applications are their current carrying and saturation characteristics, and their Q. If an inductor consists of a coil of wire with an air core, its inductance will be essentially unaffected by the current it is carrying. On the other hand, if it is wound on a core of a magnetic material (magnetic alloy or ferrite), its inductance will be nonlinear, since at high currents, the core will start to saturate. The effects of such saturation will reduce the efficiency of the circuitry employing the inductor and is liable to increase noise and harmonic generation.

As mentioned above, inductors and capacitors together form tuned circuits. Since all inductors will also have some stray capacity, all inductors will have a resonant frequency (which will normally be published on their data sheet), and should only be used as precision inductors at frequencies well below this.

Q or "Quality Factor"

The other characteristic of inductors is their Q (or "Quality Factor"), which is the ratio of the reactive impedance to the resistance, as indicated in Figure 9.15.

- $Q = 2\pi f\ L/R$
- The Q of an inductor or resonant circuit is a measure of the ratio of its reactance to its resistance.
- The resistance is the HF and NOT the DC value.
- The 3 dB bandwidth of a single tuned circuit is Fc/Q where Fc is the center frequency.

Figure 9.15: Inductor Q or Quality Factor

It is rarely possible to calculate the Q of an inductor from its dc resistance, since skin effect (and core losses if the inductor has a magnetic core) ensure that the Q of an inductor at high frequencies is always lower than that predicted from dc values.

Q is also a characteristic of tuned circuits (and of capacitors—but capacitors generally have such high Q values that it may be disregarded, in practice). The Q of a tuned circuit, which is generally very similar to the Q of its inductor (unless it is deliberately lowered by the use of an additional resistor), is a measure of its bandwidth around resonance. LC tuned circuits rarely have Q of much more than 100 (3 dB bandwidth of 1%), but ceramic resonators may have a Q of thousands, and quartz crystals tens of thousands.

Don't Overlook Anything

Remember, if a precision op amp or data-converter-based design does not meet specification, try not to overlook anything in your efforts to find the error sources. Analyze both active *and* passive components, trying to identify and challenge any assumptions or preconceived notions that may blind you to the facts. Take nothing for granted.

For example, when not tied down to prevent motion, cable conductors, moving within their surrounding dielectrics, can create significant static charge buildups that cause errors, especially when connected to high impedance circuits. Rigid cables, or even costly low noise Teflon-insulated cables, are expensive alternative solutions.

As more and more high precision op amps become available, and system designs call for higher speed and increased accuracy, a thorough understanding of the error sources described in this section (as well those following) becomes more important.

Some additional discussions of passive components within a succeeding power supply filtering section complements this one. In addition, the very next section on PCB design issues also complements many points within this section. Similar comments apply to the section on EMI/RFI.

References:
9.1 Passive Components

1. James E. Buchanan, "Dielectric Absorption— It Can Be a Real Problem In Timing Circuits," **EDN**, January 20, 1977, p. 83.

2. Lew Counts and Scott Wurcer, "Instrumentation Amplifier Nears Input Noise Floor," **Electronic Design,** June 10, 1982.

3. W. Doeling, W. Mark, T. Tadewald, and P. Reichenbacher, "Getting Rid of Hook: The Hidden PC-Board Capacitance," **Electronics,** October 12, 1978, p 111–117.

4. Tarlton Fleming, "Data-Acquisition System (DAS) Design Considerations," **WESCON '81 Professional Program Session Record No. 23**.

5. Walter G. Jung and Richard Marsh, "Picking Capacitors, Parts I and II," **Audio,** February and March, 1980.

6. Robert A. Pease, "Understand Capacitor Soakage to Optimize Analog Systems," **EDN**, October 13, 1982, p. 125.

7. Andy Rappaport, "Capacitors" **EDN**, October 13, 1982, p. 105.

8. Specification MIL-PRF-19978G, Capacitors, Fixed, Plastic (or Paper-Plastic) Dielectric (Hermetically Sealed in Metal, Ceramic or Glass Cases), Established and Nonestablished Reliability General Specification for, May 27, 1999.

9. Specification MIL-PRF-123B, Capacitors, Fixed, Ceramic Dielectric, (Temperature Stable and General-Purpose), High Reliability, General Specification for, August 6, 1990.

10. **Tantalum and Ceramic Surface Mount Capacitor Catalog**, Kemet Electronics Corporation, P.O. Box 5928, Greenville, SC, 29606, 864-963-6300.

11. A general capacitor information resource: www.faradnet.com

12. Southern and F-Dyne film capacitors, Southern Electronics, 215 Research Drive, Milford, CT, 06460, 203-876-7488.

13. Wesco film capacitors, Wesco Electrical Company, 201 Munson Street, Greenfield, MA, 01301, 413-774-4358.

14. Doug Grant and Scott Wurcer, "Avoiding Passive Component Pitfalls," **The Best of Analog Dialogue**, Analog Devices, 1991, p. 143–148.

15. RCD Components, Inc., 520 E. Industrial Park Drive, Manchester NH, 03109, 603-669-0054, www.rcd-comp.com.

16. Steve Sockolov and James Wong, "High Accuracy Analog Needs More Than Op Amps," **Electronic Design**, October 1, 1992, p. 53.

17. Selection guide for digital potentiometers: www.analog.com/digitalpots.

18. Precision Resistor Co., Inc., 10601 75th St. N., Largo, FL, 33777-1427, 727-541-5771, www.precisionresistor.com.

19. Ohmite Victoreen MAXI-MOX Resistors, 3601 Howard Street, Skokie, IL 60076, 847-675-2600, www.ohmite.com/victoreen/.

20. Vishay/Dale Resistors, 2300 Riverside Blvd., Norfolk, NE, 68701-2242, 402-371-0800, www.vishay.com.

21. Beyschlag Resistor Products, PO Box 1220, D-25732 Heide, Germany,www.beyschlag.com.

22. B. I. & B. Bleaney, **Electricity & Magnetism**, Oxford at the Clarendon Press, 1957, pp. 23, 24, and 52.

23. Henry W. Ott, **Noise Reduction Techniques in Electronic Systems, 2nd Edition**, John Wiley, Inc., 1988, ISBN: 0-471-85068-3.

24. G. W. A. Dummer, **Materials for Conductive and Resistive Functions**, Hayden, 1970.

Acknowledgments:

Portions of this and the following section were adapted from Doug Grant and Scott Wurcer, "Avoiding Passive Component Pitfalls," originally published in **Analog Dialogue 17-2**, 1983.

PC Board Design Issues
James Bryant, Walt Kester, Walt Jung

Printed circuit boards (PCBs) are by far the most common method of assembling modern electronic circuits. Composed of a sandwich of insulating layer (or layers) and one or more copper conductor patterns, they can introduce various forms of errors into a circuit, particularly if the circuit is operating at either high precision or high speed. PCBs then, act as "unseen" components, wherever they are used in precision circuit designs. Since designers don't always consider the PCB electrical characteristics as additional components of their circuit, overall performance can easily end up worse than predicted. This general topic, manifested in many forms, is the focus of this section.

PCB effects that are harmful to precision circuit performance include leakage resistances; spurious voltage drops in trace foils, vias, and ground planes; the influence of stray capacitance, dielectric absorption (DA), and the related "hook." In addition, the tendency of PCBs to absorb atmospheric moisture, *hygroscopicity,* means that changes in humidity often cause the contributions of some parasitic effects to vary from day to day.

In general, PCB effects can be divided into two broad categories— those that most noticeably affect the static or dc operation of the circuit, and those that most noticeably affect dynamic or ac circuit operation.

Another very broad area of PCB design is the topic of grounding. Grounding is a problem area in itself for all analog designs, and it can be said that implementing a PCB based circuit doesn't change that fact. Fortunately, certain principles of quality grounding, namely the use of ground planes, are intrinsic to the PCB environment. This factor is one of the more significant advantages to PCB based analog designs, and an appreciable amount of this section is focused on this issue.

Some other aspects of grounding that must be managed include the control of spurious ground and signal return voltages that can degrade performance. These voltages can be due to external signal coupling, common currents, or simply excessive IR drops in ground conductors. Proper conductor routing and sizing, as well as differential signal handling and ground isolation techniques enables control of such parasitic voltages.

One final area of grounding to be discussed is grounding appropriate for a mixed-signal, analog/digital environment. This topic is the subject of many application calls, and it is certainly true that interfacing with ADCs (or DACs) is a major part of the system design, and thus it shouldn't be overlooked. Indeed, the single issue of quality grounding can drive the entire layout philosophy of a high performance mixed-signal PCB design— as well it should.

Resistance of Conductors

Every engineer is familiar with resistors, although perhaps fewer are aware of their idiosyncrasies, as generally covered in Section 9.1. But too few engineers consider that all the wires and PCB traces with which their systems and circuits are assembled are also resistors. In higher precision systems, even these trace resistances and simple wire interconnections can have degrading effects. Copper is *not* a superconductor— and too many engineers appear to think it is!

Figure 9.16 illustrates a method of calculating the sheet resistance R of a copper square, given the length Z, the width X, and the thickness Y.

$$R = \frac{\rho \, Z}{XY}$$

ρ = RESISTIVITY

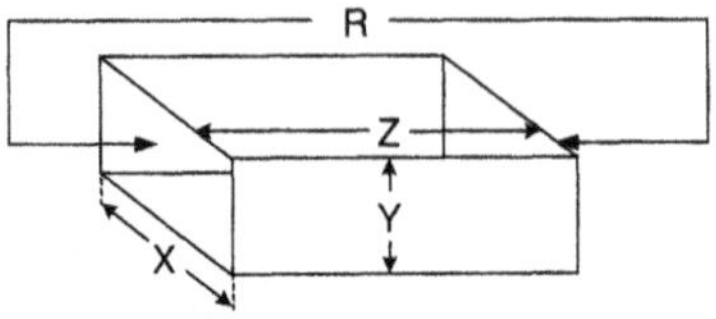

SHEET RESISTANCE CALCULATION FOR
1 OZ. COPPER CONDUCTOR:

$\rho = 1.724 \times 10^{-6}\ \Omega$ cm, Y = 0.0036cm

$R = 0.48 \, \dfrac{Z}{X} m\,\Omega$

$\dfrac{Z}{X}$ = NUMBER OF SQUARES

R = SHEET RESISTANCE OF 1 SQUARE (Z = X)
   = 0.48m Ω/SQUARE

Figure 9.16: Calculation of Sheet Resistance and Linear Resistance for Standard Copper PCB Conductors

At 25°C the resistivity of pure copper is 1.724×10^{-6} Ωcm. The thickness of standard 1 ounce PCB copper foil is 0.036 mm (0.0014"). Using the relations shown, the resistance of such a standard copper element is therefore 0.48 mΩ/square. One can readily calculate the resistance of a linear trace, by effectively "stacking" a series of such squares end-end, to make up the line's length. The line length is Z and the width is X, so the line resistance R is simply a product of Z/X and the resistance of a single square, as noted in the figure.

For a given copper weight and trace width, a resistance/length calculation can be made. For example, the 0.25 mm (10 mil) wide traces frequently used in PCB designs equates to a resistance/length of about 19 mΩ/cm (48 mΩ/inch), which is quite large. Moreover, the temperature coefficient of resistance for copper is about 0.4%/°C around room temperature. This is a factor that shouldn't be ignored, in particular within low impedance precision circuits, where the TC can shift the net impedance over temperature.

As shown in Figure 9.17, PCB trace resistance can be a serious error when conditions aren't favorable. Consider a 16-bit ADC with a 5 kΩ input resistance, driven through 5 cm of 0.25 mm wide 1 oz PCB track between it and its signal source. The track resistance of nearly 0.1 Ω forms a divider with the 5 kΩ load, creating an error. The resulting voltage drop is a gain error of 0.1/5000 (~0.0019%), well over 1 LSB (0.0015% for 16 bits).

So, when dealing with precision circuits, the point is made that even simple design items such as PCB trace resistance cannot be dealt with casually. There are various solutions to address this issue, such as wider traces (which may take up excessive space), the use of heavier copper (which may be too expensive), or simply choosing a high impedance converter. But, the most important thing is to think it all through, avoiding any tendency to overlook items appearing innocuous on the surface.

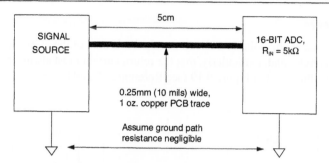

**Figure 9.17: Ohm's Law Predicts >1 LSB of Error
due to Drop In PCB Conductor**

Voltage Drop in Signal Leads—"Kelvin" Feedback

The gain error resulting from resistive voltage drop in PCB signal leads is important only with high precision and/or at high resolutions (the Figure 9.17 example), or where large signal currents flow. Where load impedance is constant and resistive, adjusting overall system gain can compensate for the error. In other circumstances, it may often be removed by the use of "Kelvin" or "voltage sensing" feedback, as shown in Figure 9.18.

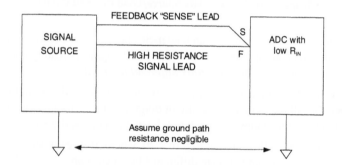

**Figure 9.18: Use of a Sense Connection
Moves Accuracy to the Load Point**

In this modification to the case of Figure 9.17, a long resistive PCB trace is still used to drive the input of a high resolution ADC, with low input impedance. In this case however, the voltage drop in the signal lead does *not* give rise to an error, as feedback is taken directly from the input pin of the ADC, and returned to the driving source. This scheme allows full accuracy to be achieved in the signal presented to the ADC, despite any voltage drop across the signal trace.

The use of separate force (F) and sense (S) connections at the load removes any errors resulting from voltage drops in the force lead, but, of course, may only be used in systems where there is negative feedback. It is also impossible to use such an arrangement to drive two or more loads with equal accuracy, since feedback may only be taken from one point. Also, in this much-simplified system, errors in the common lead source/load path are ignored, the assumption being that ground path voltages are negligible. In many systems this may not necessarily be the case, and additional steps may be needed, as noted below.

Signal Return Currents

Kirchoff's Law tells us that at any point in a circuit the algebraic sum of the currents is zero. This tells us that all currents flow in circles and, particularly, that the return current must always be considered when analyzing a circuit, as is illustrated in Figure 9.19 (see References 7 and 8).

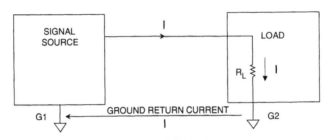

Figure 9.19: Kirchoff's Law Helps in Analyzing Voltage Drops Around a Complete Source/Load Coupled Circuit

In dealing with grounding issues, common human tendencies provide some insight into how the correct thinking about the circuit can be helpful towards analysis. Most engineers readily consider the ground return current, "I," *when they are considering a fully differential circuit.*

However, when considering the more usual circuit case, where a single-ended signal is referred to "ground," it is common to assume that all the points on the circuit diagram where ground symbols are found are at the same potential. Unfortunately, this happy circumstance "just ain't necessarily so."

This overly optimistic approach is illustrated in Figure 9.20, where, if it really should exist, "infinite ground conductivity" would lead to zero ground voltage difference between source ground G1 and load ground G2. Unfortunately this approach isn't a wise practice, and when dealing with high precision circuits, it can lead to disasters.

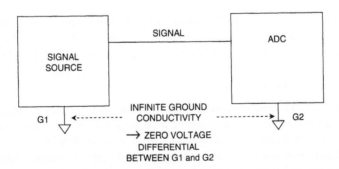

Figure 9.20: Unlike This Optimistic Diagram, it is Unrealistic to Assume Infinite Conductivity between Source/Load Grounds in a Real-World System

A more realistic approach to ground conductor integrity includes analysis of the impedance(s) involved, and careful attention to minimizing spurious noise voltages.

A more realistic model of a ground system is shown in Figure 9.21. The signal return current flows in the complex impedance existing between ground points G1 and G2 as shown, giving rise to a voltage drop ΔV in this path. But it is important to note that additional *external* currents, such as I_{EXT}, may also flow in this same path. It is critical to understand that such currents may generate uncorrelated noise voltages between G1 and G2 (dependent upon the current magnitude and relative ground impedance).

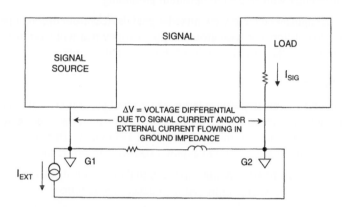

Figure 9.21: A More Realistic Source-to-Load Grounding System View Includes Consideration of the Impedance Between G1-G2, Plus the Effect of Any Nonsignal-Related Currents

Some portion of these undesired voltages may end up being seen at the signal's load end, and they can have the potential to corrupt the signal being transmitted.

Grounding in Mixed Analog/Digital Systems
Walt Kester, James Bryant, Mike Byrne

Today's signal processing systems generally require mixed-signal devices such as analog-to-digital converters (ADCs) and digital-to-analog converters (DACs) as well as fast digital signal processors (DSPs). Requirements for processing analog signals having wide dynamic ranges increases the importance of high performance ADCs and DACs. Maintaining wide dynamic range with low noise in hostile digital environments is dependent upon using good high speed circuit design techniques including proper signal routing, decoupling, and grounding.

In the past, "high precision, low speed" circuits have generally been viewed differently than so-called "high speed" circuits. With respect to ADCs and DACs, the sampling (or update) frequency has generally been used as the distinguishing speed criteria. However, the following two examples show that in practice, most of today's signal processing ICs are really "high speed," and must therefore be treated as such in order to maintain high performance. This is certainly true of DSPs, and also true of ADCs and DACs.

All sampling ADCs (ADCs with an internal sample-and-hold circuit) suitable for signal processing applications operate with relatively high speed clocks with fast rise and fall times (generally a few nanoseconds) and must be treated as high speed devices, even though throughput rates may appear low. For example, a medium-speed 12-bit successive approximation (SAR) ADC may operate on a 10 MHz internal clock, while the sampling rate is only 500 kSPS.

Sigma-delta (Σ-Δ) ADCs also require high speed clocks because of their high oversampling ratios. Even high resolution, so-called "low frequency" Σ-Δ industrial measurement ADCs (having throughputs of 10 Hz to 7.5 kHz) operate on 5 MHz or higher clocks and offer resolution to 24 bits (for example, the Analog Devices AD77xx series).

To further complicate the issue, mixed-signal ICs have both analog and digital ports, and because of this, much confusion has resulted with respect to proper grounding techniques. In addition, some mixed-signal ICs have relatively low digital currents, while others have high digital currents. In many cases, these two types must be treated differently with respect to optimum grounding.

Digital and analog design engineers tend to view mixed-signal devices from different perspectives, and the purpose of this section is to develop a general grounding philosophy that will work for most mixed signal devices, without having to know the specific details of their internal circuits.

Ground and Power Planes

The importance of maintaining a low impedance large area ground plane is critical to all analog circuits today. The ground plane not only acts as a low impedance return path for decoupling high frequency currents (caused by fast digital logic) but also minimizes EMI/RFI emissions. Because of the shielding action of the ground plane, the circuit's susceptibility to external EMI/RFI is also reduced.

Ground planes also allow the transmission of high speed digital or analog signals using transmission line techniques (microstrip or stripline) where controlled impedances are required.

The use of "buss wire" is totally unacceptable as a "ground" because of its impedance at the equivalent frequency of most logic transitions. For instance, #22 gauge wire has about 20 nH/inch inductance. A transient current having a slew rate of 10 mA/ns created by a logic signal would develop an unwanted voltage drop of 200 mV at this frequency flowing through 1 inch of this wire:

$$\Delta v = L \frac{\Delta i}{\Delta t} = 20 \text{ nH} \times \frac{10 \text{ mA}}{\text{ns}} = 200 \text{ mV}. \qquad \text{Eq. 9.1}$$

For a signal having a 2 V peak-to-peak range, this translates into an error of about 200 mV, or 10% (approximate 3.5-bit accuracy). Even in all-digital circuits, this error would result in considerable degradation of logic noise margins.

Figure 9.22 shows an illustration of a situation where the digital return current modulates the analog return current (top figure). The ground return wire inductance and resistance is shared between the analog and digital circuits, and this is what causes the interaction and resulting error. A possible solution is to make the digital return current path flow directly to the GND REF as shown in the bottom figure. This is the fundamental concept of a "star," or single-point ground system. Implementing the true single-point ground in a system which contains multiple high frequency return paths is difficult because the physical length of the individual return current wires will introduce parasitic resistance and inductance which can make obtaining a low impedance high frequency ground difficult. In practice, the current returns must consist of large area ground planes for low impedance to high frequency currents. Without a low impedance ground plane, it is therefore almost impossible to avoid these shared impedances, especially at high frequencies.

All integrated circuit ground pins should be soldered directly to the low impedance ground plane to minimize series inductance and resistance. The use of traditional IC sockets is not recommended with high-speed devices. The extra inductance and capacitance of even "low profile" sockets may corrupt the device performance by introducing unwanted shared paths. If sockets must be used with DIP packages, as in prototyping, individual "pin sockets" or "cage jacks" may be acceptable. Both capped and uncapped versions of

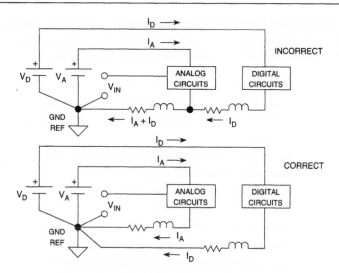

Figure 9.22: Digital Currents Flowing in Analog Return Path Create Error Voltages

these pin sockets are available (AMP part numbers 5-330808-3, and 5-330808-6). They have spring-loaded gold contacts which make good electrical and mechanical connection to the IC pins. Multiple insertions, however, may degrade their performance.

Power supply pins should be decoupled directly to the ground plane using low inductance ceramic surface-mount capacitors. If through-hole mounted ceramic capacitors must be used, their leads should be less than 1 mm. The ceramic capacitors should be located as close as possible to the IC power pins. Ferrite beads may be also required for additional decoupling.

Double-Sided versus Multilayer Printed Circuit Boards

Each PCB in the system should have at least one complete layer dedicated to the ground plane. Ideally, a double-sided board should have one side completely dedicated to ground and the other side for interconnections. In practice, this is not possible, since some of the ground plane will certainly have to be removed to allow for signal and power crossovers, vias, and through-holes. Nevertheless, as much area as possible should be preserved, and at least 75% should remain. After completing an initial layout, the ground layer should be checked carefully to make sure there are no isolated ground "islands," because IC ground pins located in a ground "island" have no current return path to the ground plane. Also, the ground plane should be checked for "skinny" connections between adjacent large areas which may significantly reduce the effectiveness of the ground plane. Needless to say, autorouting board layout techniques will generally lead to a layout disaster on a mixed-signal board, so manual intervention is highly recommended.

Systems that are densely packed with surface-mount ICs will have a large number of interconnections; therefore multilayer boards are mandatory. This allows at least one complete layer to be dedicated to ground. A simple 4-layer board would have internal ground and power plane layers with the outer two layers used for interconnections between the surface mount components. Placing the power and ground planes adjacent to each other provides additional inter-plane capacitance which helps high frequency decoupling of the power supply. In most systems, four layers are not enough, and additional layers are required for routing signals as well as power. Figure 9.23 summarizes the key issues relating to ground planes.

- Use Large Area Ground (and Power) Planes for Low Impedance Current Return Paths (Must Use at Least a Double-Sided Board)
- Double-Sided Boards:
  - Avoid High-Density Interconnection Crossovers and Vias Which Reduce Ground Plane Area
  - Keep > 75% Board Area on One Side for Ground Plane
- Multilayer Boards: Mandatory for Dense Systems
  - Dedicate at Least One Layer for the Ground Plane
  - Dedicate at Least One Layer for the Power Plane
- Use at Least 30% to 40% of PCB Connector Pins for Ground
- Continue the Ground Plane on the Backplane Motherboard to Power Supply Return

Figure 9.23: Ground Planes Are Mandatory!

Multicard Mixed-Signal Systems

The best way of minimizing ground impedance in a multicard system is to use a "motherboard" PCB as a backplane for interconnections between cards, thus providing a continuous ground plane to the backplane. The PCB connector should have at least 30%–40% of its pins devoted to ground, and these pins should be connected to the ground plane on the backplane mother card. To complete the overall system grounding scheme there are two possibilities:

1. The backplane ground plane can be connected to chassis ground at numerous points, thereby diffusing the various ground current return paths. This is commonly referred to as a "multipoint" grounding system and is shown in Figure 9.24.

2. The ground plane can be connected to a single system "star ground" point (generally at the power supply).

The first approach is most often used in all-digital systems, but can be used in mixed-signal systems provided the ground currents due to digital circuits are sufficiently low and diffused over a large area. The low ground impedance is maintained all the way through the PC boards, the backplane, and ultimately the chassis. However, it is critical that good electrical contact be made where the grounds are connected to the sheet metal chassis. This requires self-tapping sheet metal screws or "biting" washers. Special care must be taken where anodized aluminum is used for the chassis material, since its surface acts as an insulator.

The second approach ("star ground") is often used in high speed mixed-signal systems having separate analog and digital ground systems and warrants further discussion.

Separating Analog and Digital Grounds

In mixed-signal systems with large amounts of digital circuitry, it is highly desirable to *physically* separate sensitive analog components from noisy digital components. It may also be beneficial to use separate ground planes for the analog and the digital circuitry. These planes should not overlap in order to minimize

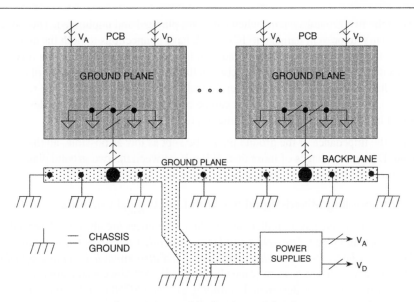

Figure 9.24: Multipoint Ground Concept

capacitive coupling between the two. The separate analog and digital ground planes are continued on the backplane using either motherboard ground planes or "ground screens" which are made up of a series of wired interconnections between the connector ground pins. The arrangement shown in Figure 9.25 illustrates that the two planes are kept separate all the way back to a common system "star" ground, generally located at the power supplies. The connections between the ground planes, the power supplies, and the "star" should be made up of multiple bus bars or wide copper braids for minimum resistance and inductance. The back-to-back Schottky diodes on each PCB are inserted to prevent accidental dc voltage from

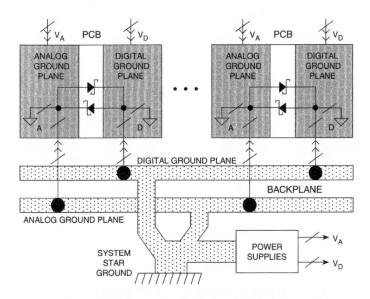

Figure 9.25: Separating Analog and Digital Ground Planes

developing between the two ground systems when cards are plugged and unplugged. This voltage should be kept less than 300 mV to prevent damage to ICs which have connections to both the analog and digital ground planes. Schottky diodes are preferable because of their low capacitance and low forward voltage drop. The low capacitance prevents ac coupling between the analog and digital ground planes. Schottky diodes begin to conduct at about 300 mV, and several parallel diodes in parallel may be required if high currents are expected. In some cases, ferrite beads can be used instead of Schottky diodes, however they introduce dc ground loops which can be troublesome in precision systems.

It is mandatory that the impedance of the ground planes be kept as low as possible, all the way back to the system star ground. DC or ac voltages of more than 300 mV between the two ground planes can not only damage ICs but cause false triggering of logic gates and possible latchup.

Grounding and Decoupling Mixed-Signal ICs with Low Digital Currents

Sensitive analog components such as amplifiers and voltage references are always referenced and de-coupled to the analog ground plane. *The ADCs and DACs (and other mixed-signal ICs) with low digital currents should generally be treated as analog components and also grounded and decoupled to the analog ground plane.* At first glance, this may seem somewhat contradictory, since a converter has an analog and digital interface and usually has pins designated as *analog ground* (AGND) and *digital ground* (DGND). The diagram shown in Figure 9.26 will help to explain this seeming dilemma.

Inside an IC that has both analog and digital circuits, such as an ADC or a DAC, the grounds are usually kept separate to avoid coupling digital signals into the analog circuits. Figure 9.26 shows a simple model of a converter. There is nothing the IC designer can do about the wirebond inductance and resistance associated with connecting the bond pads on the chip to the package pins except to realize it's there. The rapidly changing digital currents produce a voltage at point B which will inevitably couple into point A of the analog circuits through the stray capacitance, C_{STRAY}. In addition, there is approximately 0.2 pF unavoidable

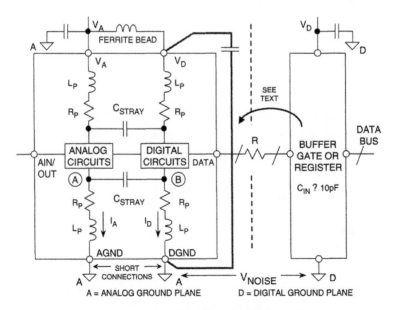

Figure 9.26: Proper Grounding of Mixed-Signal ICs with Low Internal Digital Currents

stray capacitance between every pin of the IC package. It's the IC designer's job to make the chip work in spite of this. However, in order to prevent further coupling, the AGND and DGND pins should be joined together externally to the *analog* ground plane with minimum lead lengths. Any extra impedance in the DGND connection will cause more digital noise to be developed at point B; it will, in turn, couple more digital noise into the analog circuit through the stray capacitance. *Note that connecting DGND to the digital ground plane applies V_{NOISE} across the AGND and DGND pins and invites disaster.*

The name "DGND" on an IC tells us that this pin connects to the digital ground of the IC. This does not imply that this pin must be connected to the digital ground of the system.

It is true that this arrangement may inject a small amount of digital noise onto the analog ground plane. These currents should be quite small, and can be minimized by ensuring that the converter output does not drive a large fanout (they normally can't, by design). Minimizing the fanout on the converter's digital port will also keep the converter logic transitions relatively free from ringing and minimize digital switching currents, thereby reducing any potential coupling into the analog port of the converter. The logic supply pin (V_D) can be further isolated from the analog supply by the insertion of a small lossy ferrite bead as shown in Figure 9.26. The internal transient digital currents of the converter will flow in the small loop from V_D through the decoupling capacitor and to DGND (this path is shown with a heavy line on the diagram). The transient digital currents will therefore not appear on the external analog ground plane, but are confined to the loop. The V_D pin decoupling capacitor should be mounted as close to the converter as possible to minimize parasitic inductance. These decoupling capacitors should be low inductance ceramic types, typically between 0.01 µF and 0.1 µF.

Treat the ADC Digital Outputs with Care

It is always a good idea (as shown in Figure 9.26) to place a buffer register adjacent to the converter to isolate the converter's digital lines from noise on the data bus. The register also serves to minimize loading on the digital outputs of the converter and acts as a Faraday shield between the digital outputs and the data bus. Even though many converters have three-state outputs/inputs, this isolation register still represents good design practice. In some cases it may be desirable to add an additional buffer register on the analog ground plane next to the converter output to provide greater isolation.

The series resistors (labeled "R" in Figure 9.26) between the ADC output and the buffer register input help to minimize the digital transient currents which may affect converter performance. The resistors isolate the digital output drivers from the capacitance of the buffer register inputs. In addition, the RC network formed by the series resistor and the buffer register input capacitance acts as a low-pass filter to slow down the fast edges.

A typical CMOS gate combined with PCB trace and a through-hole will create a load of approximately 10 pF. A logic output slew rate of 1 V/ns will produce 10 mA of dynamic current if there is no isolation resistor:

$$\Delta I = C \frac{\Delta v}{\Delta t} = 10 \text{ pF} \times \frac{1 \text{ V}}{\text{ns}} = 10 \text{ mA.} \qquad \text{Eq. 9.2}$$

A 500 Ω series resistors will minimize this output current and result in a rise and fall time of approximately 11ns when driving the 10 pF input capacitance of the register:

$$t_r = 2.2 \times \tau = 2.2 \times R \times C = 2.2 \times 500 \ \Omega \times 10 \text{ pF} = 11 \text{ ns.} \qquad \text{Eq. 9.3}$$

TTL registers should be avoided, since they can appreciably add to the dynamic switching currents because of their higher input capacitance.

The buffer register and other digital circuits should be grounded and decoupled to the *digital* ground plane of the PC board. Notice that any noise between the analog and digital ground plane reduces the noise margin at the converter digital interface. Since digital noise immunity is of the order of hundreds or thousands of millivolts, this is unlikely to matter. The analog ground plane will generally not be very noisy, but if the noise on the digital ground plane (relative to the analog ground plane) exceeds a few hundred millivolts, then steps should be taken to reduce the digital ground plane impedance, thereby maintaining the digital noise margins at an acceptable level. Under no circumstances should the voltage between the two ground planes exceed 300 mV, or the ICs may be damaged.

Separate power supplies for analog and digital circuits are also highly desirable, even if the voltages are the same. The analog supply should be used to power the converter. If the converter has a pin designated as a digital supply pin (V_D), it should either be powered from a separate analog supply, or filtered as shown in the diagram. All converter power pins should be decoupled to the analog ground plane, and all logic circuit power pins should be decoupled to the digital ground plane as shown in Figure 9.27.

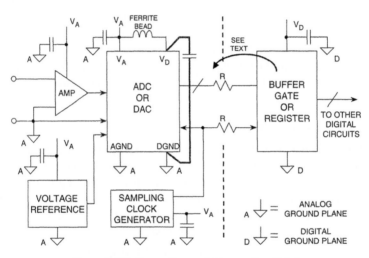

Figure 9.27: Grounding and Decoupling Points

In some cases it may not be possible to connect V_D to the analog supply. Some of the newer, high speed ICs may have their analog circuits powered by 5 V, but the digital interface powered by 3 V to interface to 3 V logic. In this case, the 3 V pin of the IC should be decoupled directly to the analog ground plane. It is also advisable to connect a ferrite bead in series with the power trace that connects the pin to the 3 V digital logic supply.

The sampling clock generation circuitry should be treated like analog circuitry and also be grounded and heavily decoupled to the analog ground plane. Phase noise on the sampling clock produces degradation in system SNR as will be discussed shortly.

Sampling Clock Considerations

In a high performance sampled data system a low phase-noise crystal oscillator should be used to generate the ADC (or DAC) sampling clock because sampling clock jitter modulates the analog input/output signal and raises the noise and distortion floor. The sampling clock generator should be isolated from noisy digital circuits and grounded and decoupled to the analog ground plane, as is true for the op amp and the ADC.

The effect of sampling clock jitter on ADC signal-to-noise ratio (SNR) is given approximately by the equation:

$$\text{SNR} = 20\log_{10}\left[\frac{1}{2\pi f\, t_j}\right]$$

Eq. 9.4

where SNR is the SNR of a perfect ADC of infinite resolution where the only source of noise is that caused by the rms sampling clock jitter, t_j. Note that f in the above equation is the analog input frequency. Just working through a simple example, if $t_j = 50$ ps rms, f = 100 kHz, then SNR = 90 dB, equivalent to about 15-bit dynamic range.

It should be noted that t_j in the above example is the root-sum-square (rss) value of the external clock jitter *and* the internal ADC clock jitter (called aperture jitter). However, in most high performance ADCs, the internal aperture jitter is negligible compared to the jitter on the sampling clock.

Since degradation in SNR is primarily due to external clock jitter, steps must be taken to ensure the sampling clock is as noise-free as possible and has the lowest possible phase jitter. This requires that a crystal oscillator be used. There are several manufacturers of small crystal oscillators with low jitter (less than 5 ps rms) CMOS compatible outputs. (For example, MF Electronics, 10 Commerce Dr., New Rochelle, NY 10801, Tel. 914-576-6570 and Wenzel Associates, Inc., 2215 Kramer Lane, Austin, Texas 78758 Tel. 512- 835-2038.)

Ideally, the sampling clock crystal oscillator should be referenced to the analog ground plane in a split-ground system. However, this is not always possible because of system constraints. In many cases, the sampling clock must be derived from a higher frequency multipurpose system clock which is generated on the digital ground plane. It must then pass from its origin on the digital ground plane to the ADC on the analog ground plane. Ground noise between the two planes adds directly to the clock signal and will produce excess jitter. The jitter can cause degradation in the signal-to-noise ratio and also produce unwanted harmonics.

This can be somewhat remedied by transmitting the sampling clock signal as a differential signal using either a small RF transformer as shown in Figure 9.28 or a high speed differential driver and receiver IC. If an active differential driver and receiver are used, they should be ECL to minimize phase jitter. In a single

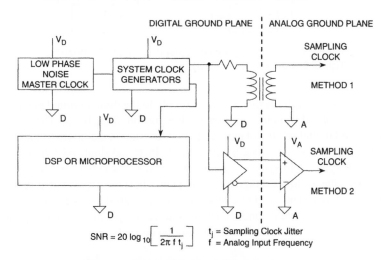

Figure 9.28: Sampling Clock Distribution From Digital to Analog Ground Planes

5 V supply system, ECL logic can be connected between ground and 5 V (PECL), and the outputs ac coupled into the ADC sampling clock input. In either case, the original master system clock must be generated from a low phase noise crystal oscillator, and not the clock output of a DSP, microprocessor, or microcontroller.

The Origins of the Confusion about Mixed-Signal Grounding: Applying Single-Card Grounding Concepts to Multicard Systems

Most ADC, DAC, and other mixed-signal device data sheets discuss grounding relative to a single PCB, usually the manufacturer's own evaluation board. This has been a source of confusion when trying to apply these principles to multicard or multi-ADC/DAC systems. The recommendation is usually to split the PCB ground plane into an analog plane and a digital plane. It is then further recommended that the AGND and DGND pins of a converter be tied together and that the analog ground plane and digital ground planes be connected at that same point as shown in Figure 9.29. This essentially creates the system "star" ground at the mixed-signal device.

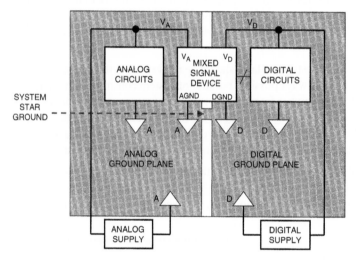

Figure 9.29: Grounding Mixed-Signal ICs : Single PC Board (Typical Evaluation/Test Board)

All noisy digital currents flow through the digital power supply to the digital ground plane and back to the digital supply; they are isolated from the sensitive analog portion of the board. The system star ground occurs where the analog and digital ground planes are joined together at the mixed-signal device. While this approach will generally work in a simple system with a single PCB and single ADC/DAC, it is not optimum for multicard mixed-signal systems. In systems having several ADCs or DACs on different PCBs (or on the same PCB, for that matter), the analog and digital ground planes become connected at several points, creating the possibility of ground loops and making a single-point "star" ground system impossible. For these reasons, this grounding approach is not recommended for multicard systems, and the approach previously discussed should be used for mixed-signal ICs with low digital currents.

Summary: Grounding Mixed-Signal Devices with Low Digital Currents in a Multicard System

Figure 9.30 summarizes the approach previously described for grounding a mixed-signal device which has low digital currents. The analog ground plane is not corrupted because the small digital transient currents flow in the small loop between V_D, the decoupling capacitor, and DGND (shown as a heavy line). The mixed-signal device is for all intents and purposes treated as an analog component. The noise V_N between the ground planes reduces the noise margin at the digital interface, but is generally not harmful if kept less than 300 mV by using a low impedance digital ground plane all the way back to the system star ground.

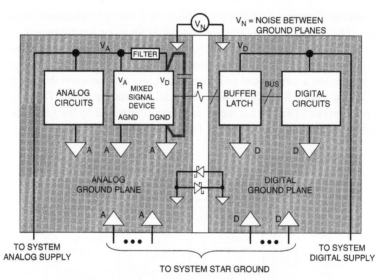

Figure 9.30: Grounding Mixed-Signal ICs with Low Internal Digital Currents: Multiple PC Boards

However, mixed-signal devices such as sigma-delta ADCs, codecs, and DSPs with on-chip analog functions are becoming more and more digitally intensive. Along with the additional digital circuitry come larger digital currents and noise. For example, a sigma-delta ADC or DAC contains a complex digital filter which adds considerably to the digital current in the device. The method previously discussed depends on the decoupling capacitor between V_D and DGND to keep the digital transient currents isolated in a small loop. However, if the digital currents are significant enough and have components at dc or low frequencies, the decoupling capacitor may have to be so large that it is impractical. Any digital current that flows outside the loop between V_D and DGND must flow through the analog ground plane. This may degrade performance, especially in high resolution systems.

It is difficult to predict what level of digital current flowing into the analog ground plane will become unacceptable in a system. All we can do at this point is to suggest an alternative grounding method which may yield better performance.

Summary: Grounding Mixed-Signal Devices with High Digital Currents in a Multicard System

An alternative grounding method for a mixed-signal device with high levels of digital currents is shown in Figure 9.31. The AGND of the mixed-signal device is connected to the analog ground plane, and the DGND of the device is connected to the digital ground plane. The digital currents are isolated from the analog ground plane, but the noise between the two ground planes is applied directly between the AGND and DGND pins of the device. For this method to be successful, the analog and digital circuits within the mixed signal device must be well isolated. The noise between AGND and DGND pins must not be large enough to reduce internal noise margins or cause corruption of the internal analog circuits.

Figure 9.31 shows optional Schottky diodes (back-to-back) or a ferrite bead connecting the analog and digital ground planes. The Schottky diodes prevent large dc voltages or low frequency voltage spikes from developing across the two planes. These voltages can potentially damage the mixed-signal IC if they exceed 300 mV because they appear directly between the AGND and DGND pins. As an alternative to the back-to-back Schottky diodes, a ferrite bead provides a dc connection between the two planes but isolates them at frequencies above a few MHz where the ferrite bead becomes resistive. This protects the IC from dc voltages between AGND and DGND, but the dc connection provided by the ferrite bead can introduce unwanted dc ground loops and may not be suitable for high resolution systems.

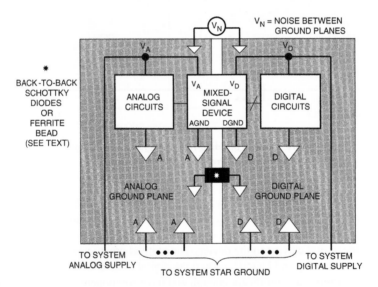

Figure 9.31: Grounding Alternative for Mixed-Signal ICs with High Digital Currents: Multiple PC Boards

Grounding DSPs with Internal Phase-Locked Loops

As if dealing with mixed-signal ICs with AGND and DGNDs wasn't enough, DSPs such as the ADSP-21160 SHARC with internal phase-locked-loops (PLLs) raise issues with respect to proper grounding. The ADSP-21160 PLL allows the internal core clock (determines the instruction cycle time) to operate at a user-selectable ratio of 2, 3, or 4 times the external clock frequency, CLKIN. The CLKIN rate is the rate at which the synchronous external ports operate. Although this allows using a lower frequency external clock, care must be taken with the power and ground connections to the internal PLL as shown in Figure 9.32.

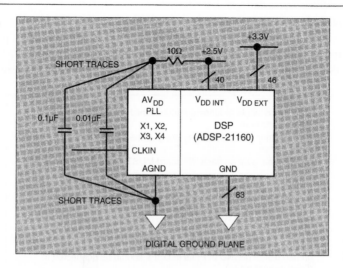

**Figure 9.32: Grounding DSPs with Internal
Phase-Locked-Loops (PLLs)**

In order to prevent internal coupling between digital currents and the PLL, the power and ground connections to the PLL are brought out separately on pins labeled AV_{DD} and AGND, respectively. The AV_{DD} 2.5 V supply should be derived from the $V_{DD\,INT}$ 2.5 V supply using the filter network as shown. This ensures a relatively noise-free supply for the internal PLL. The AGND pin of the PLL should be connected to the digital ground plane of the PC board using a short trace. The decoupling capacitors should be routed between the AV_{DD} pin and AGND pin using short traces.

Grounding Summary

No single grounding method will guarantee optimum performance 100% of the time. This section has presented a number of possible options, depending upon the characteristics of the particular mixed-signal devices in question. It is helpful, however, to provide for as many options as possible when laying out the initial PC board.

It is mandatory that at least one layer of the PC board be dedicated to ground plane. The initial board layout should provide for nonoverlapping analog and digital ground planes, but pads and vias should be provided at several locations for the installation of back-to-back Schottky diodes or ferrite beads, if required. Pads and vias should also be provided so that the analog and digital ground planes can be connected together with jumpers if required.

The AGND pins of mixed-signal devices should in general always be connected to the analog ground plane. An exception to this are DSPs such as the ADSP-21160 SHARC, which have internal phase-locked-loops (PLLs). The ground pin for the PLL is labeled AGND, but should be directly connected to the digital ground plane for the DSP. See Figure 9.33 for a general summary of grounding philosophy.

- There is no single grounding method which is guaranteed to work 100% of the time
- Different methods may or may not give the same levels of performance
- At least one layer on each PC board MUST be dedicated to ground plane
- Do initial layout with split analog and digital ground planes
- Provide pads and vias on each PC board for back-to-back Schottky diodes and optional ferrite beads to connect the two planes
- Provide "jumpers" so that DGND pins of mixed-signal devices can be connected to AGND pins (analog ground plane) or to digital ground plane. (AGND of PLLs in DSPs should be connected to digital ground plane)
- Provide pads and vias for "jumpers" so that analog and digital ground planes can be joined together at several points on each PC board
- Follow recommendations on mixed signal device data sheet

Figure 9.33: Grounding Philosophy Summary

Some General PC Board Layout Guidelines for Mixed-Signal Systems

It is evident that noise can be minimized by paying attention to the system layout and preventing different signals from interfering with each other. High level analog signals should be separated from low level analog signals, and both should be kept away from digital signals. We have seen elsewhere that in waveform sampling and reconstruction systems the sampling clock (which is a digital signal) is as vulnerable to noise as any analog signal, but is as liable to cause noise as any digital signal, and so must be kept isolated from both analog and digital systems. If clock driver packages are used in clock distribution, only one frequency clock should be passed through a single package. Sharing drivers between clocks of different frequencies in the same package will produce excess jitter and crosstalk and degrade performance.

The ground plane can act as a shield where sensitive signals cross. Figure 9.34 shows a good layout for a data acquisition board where all sensitive areas are isolated from each other and signal paths are kept as short as possible. While real life is rarely as tidy as this, the principle remains a valid one.

There are a number of important points to be considered when making signal and power connections. First of all a connector is one of the few places in the system where all signal conductors must run in parallel—it is therefore imperative to separate them with ground pins (creating a Faraday shield) to reduce coupling between them.

Multiple ground pins are important for another reason: they keep down the ground impedance at the junction between the board and the backplane. The contact resistance of a single pin of a PCB connector is quite low (of the order of 10 mΩ) when the board is new—as the board gets older the contact resistance is likely to rise, and the board's performance may be compromised. It is therefore well worthwhile to allocate extra PCB connector pins so that there are many ground connections (perhaps 30%–40% of all the pins on the PCB connector should be ground pins). For similar reasons there should be several pins for each power connection, although there is no need to have as many as there are ground pins.

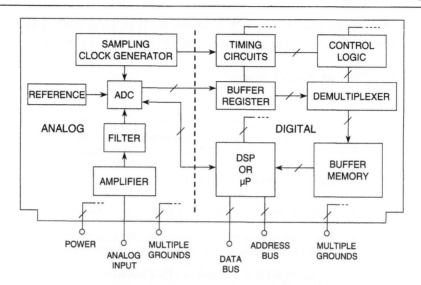

**Figure 9.34: Analog and Digital Circuits
Should be Partitioned on PCB Layout**

Analog Devices and other manufacturers of high performance mixed-signal ICs offer evaluation boards to assist customers in their initial evaluations and layout. ADC evaluation boards generally contain an on-board low jitter sampling clock oscillator, output registers, and appropriate power and signal connectors. They also may have additional support circuitry such as the ADC input buffer amplifier and external reference.

The layout of the evaluation board is optimized in terms of grounding, decoupling, and signal routing and can be used as a model when laying out the ADC PC board in the system. The actual evaluation board layout is usually available from the ADC manufacturer in the form of computer CAD files (Gerber files). In many cases, the layout of the various layers appears on the data sheet for the device.

Skin Effect

At high frequencies, also consider *skin effect*, where inductive effects cause currents to flow only in the outer surface of conductors. Note that this is in contrast to the earlier discussions of this section on dc resistance of conductors.

The skin effect has the consequence of increasing the resistance of a conductor at high frequencies. Note also that this effect is separate from the increase in impedance due to the effects of the self-inductance of conductors as frequency is increased.

Skin effect is quite a complex phenomenon, and detailed calculations are beyond the scope of this discussion. However, a good approximation for copper is that the skin depth in centimeters is $6.61/\sqrt{f}$, (f in Hz).

A summary of the skin effect within a typical PCB conductor foil is shown in Figure 9.35. Note that this copper conductor cross-sectional view assumes looking into the *side* of the conducting trace.

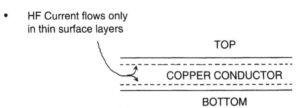

- HF Current flows only in thin surface layers

TOP

COPPER CONDUCTOR

BOTTOM

- Skin Depth: $6.61/\sqrt{f}$ cm, f in Hz

- Skin Resistance: $2.6 \times 10^{-7} \sqrt{f}$ ohms per square, f in Hz

- Since skin currents flow in both sides of a PC track, the value of skin resistance in PCBs must take account of this

Figure 9.35: Skin Depth in a PC Conductor

Assuming that skin effects become important when the skin depth is less than 50% of the thickness of the conductor, this tells us that for a typical PC foil, we must be concerned about skin effects at frequencies above approximately 12 MHz.

Where skin effect is important, the resistance for copper is $2.6 \times 10^{-7} \sqrt{f}$ ohms per square, (f in Hz). This formula is invalid if the skin thickness is greater than the conductor thickness (i.e., at dc or low frequencies).

Figure 9.36 illustrates a case of a PCB conductor with current flow, as separated from the ground plane underneath.

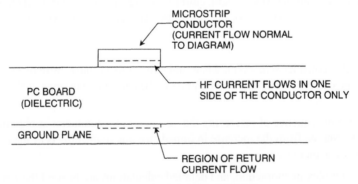

MICROSTRIP
CONDUCTOR
(CURRENT FLOW NORMAL
TO DIAGRAM)

PC BOARD
(DIELECTRIC)

HF CURRENT FLOWS IN ONE
SIDE OF THE CONDUCTOR ONLY

GROUND PLANE

REGION OF RETURN
CURRENT FLOW

**Figure 9.36: Skin Effect with PC
Conductor and Ground Plane**

In this diagram, note the (dotted) regions of high frequency current flow, as reduced by the skin effect. When calculating skin effect in PCBs, it is important to remember that current generally flows in both sides of the PC foil (this is not necessarily the case in microstrip lines, see below), so the resistance per square of PC foil may be half the above value.

Transmission Lines

We earlier considered the benefits of outward and return signal paths being close together so that inductance is minimized. As shown previously in Figure 9.36, when a high frequency signal flows in a PC track running over a ground plane, the arrangement functions as a *microstrip* transmission line, and the majority of the return current flows in the ground plane underneath the line.

Figure 9.37 shows the general parameters for a microstrip transmission line, given the conductor width, w, dielectric thickness, h, and the dielectric constant, E_r.

The characteristic impedance of such a microstrip line will depend upon the width of the track and the thickness and dielectric constant of the PCB material. Designs of microstrip lines are covered in more detail later in this chapter.

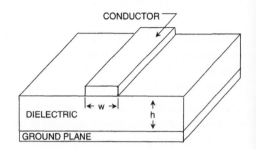

Figure 9.37: A PCB Microstrip Transmission Line is an Example of a Controlled Impedance Conductor Pair

For most dc and lower frequency applications, the characteristic impedance of PCB traces will be relatively unimportant. Even at frequencies where a track over a ground plane behaves as a transmission line, it is not necessary to worry about its characteristic impedance or proper termination if the free space wavelengths of the frequencies of interest are greater than ten times the length of the line.

However, at VHF and higher frequencies, it is possible to use PCB tracks as microstrip lines within properly terminated transmission systems. Typically the microstrip will be designed to match standard coaxial cable impedances, such as 50 Ω, 75 Ω or 100 Ω, simplifying interfacing.

Note that if losses in such systems are to be minimized, the PCB material must be chosen for low high-frequency losses. This usually means the use of Teflon or some other comparably low-loss PCB material. Often, though, the losses in short lines on cheap glass-fiber board are small enough to be quite acceptable.

Be Careful With Ground Plane Breaks

Wherever there is a break in the ground plane beneath a conductor, the ground plane return current must by necessity flow *around* the break. As a result, both the inductance and the vulnerability of the circuit to external fields are increased. This situation is diagrammed in Figure 9.38, where conductors A and B must cross one another.

Where such a break is made to allow a cross-over of two perpendicular conductors, it would be far better if the second signal were carried across both the first and the ground plane by means of

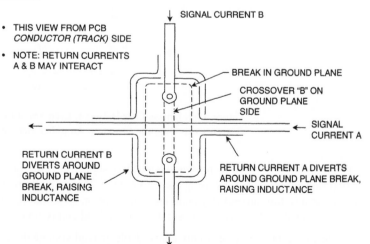

Figure 9.38: A Ground Plane Break Raises Circuit Inductance, and Increases Vulnerability to External Fields

a piece of wire or a resistor. The ground plane then acts as a shield between the two signal conductors, and the two ground return currents, flowing in opposite sides of the ground plane as a result of skin effects, do not interact.

With a multilayer board, both the crossover and the continuous ground plane can be accommodated without the need for a wire link. Multilayer PCBs are expensive and harder to trouble-shoot than more simple double-sided boards, but do offer even better shielding and signal routing. The principles involved remain unchanged but the range of layout options is increased.

The use of double-sided or multilayer PCBs with at least one continuous ground plane is undoubtedly one of the most successful design approaches for high performance mixed-signal circuitry. Often the impedance of such a ground plane is sufficiently low to permit the use of a single ground plane for both analog and digital parts of the system. However, whether or not this is possible does depend upon the resolution and bandwidth required, and the amount of digital noise present in the system.

Ground Isolation Techniques

While the use of ground planes does lower impedance and helps greatly in lowering ground noise, there may still be situations where a prohibitive level of noise exists. In such cases, the use of ground error minimization and isolation techniques can be helpful.

Another illustration of a common-ground impedance coupling problem is shown in Figure 9.39. In this circuit a precision gain-of-100 preamp amplifies a low level signal V_{IN}, using an AD8551 chopper-stabilized amplifier for best dc accuracy. At the load end, the signal V_{OUT} is measured with respect to G2, the local ground. Because of the small 700 µA I_{SUPPLY} of the AD8551 flowing between G1 and G2, there is a 7 µV ground error—about seven times the typical input offset expected from the op amp.

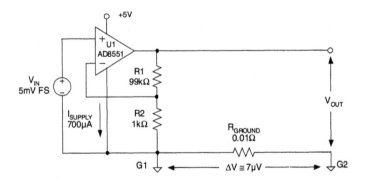

Figure 9.39: Unless Care is Taken, Even Small Common Ground Currents can Degrade Precision Amplifier Accuracy

This error can be avoided by routing the negative supply pin current of the op amp back to star ground G2 as opposed to ground G1, by using a separate trace. This step eliminates the G1-G2 path power supply current, and so minimizes the ground leg voltage error. Note that there will be little error developed in the "hot" V_{OUT} lead, so long as the current drain at the load end is small.

In some cases, there may be simply unavoidable ground voltage differences between a source signal and the load point where it is to be measured. Within the context of this "same-board" discussion, this might require rejecting ground error voltages of several tens-of-mV. Or, should the source signal originate from an "off-board" source, then the magnitude of the common-mode voltages to be rejected can easily rise into a several volt range (or even tens of volts).

Fortunately, full signal transmission accuracy can still be accomplished in the face of such high noise voltages, by employing a principle discussed earlier. This is the use of a differential-input, *ground isolation* amplifier. The ground isolation amplifier minimizes the effect of ground error voltages between stages by processing the signal in differential fashion, thereby rejecting common-mode voltages by a substantial margin (typically 60 dB or more). Note, however, that this approach is only effective for very low frequency signals.

Two ground isolation amplifier solutions are shown in Figure 9.40. This diagram can alternately employ either the AD629 to handle CM voltages up to ±270 V, or the AMP03, which is suitable for CM voltages up to ±20 V.

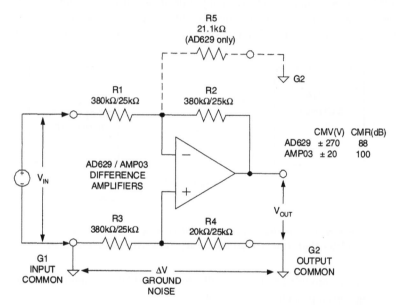

Figure 9.40: A Differential Input Ground Isolating Amplifier Allows High Transmission Accuracy by Rejecting Ground Noise Voltage Between Source (G1) and Measurement (G2) Grounds

In the circuit, input voltage V_{IN} is referred to G1, but must be measured with respect to G2. With the use of a high CMR unity-gain difference amplifier, the noise voltage ΔV existing between these two grounds is easily rejected. The AD629 offers a typical CMR of 88 dB, while the AMP03 typically achieves 100 dB. In the AD629, the high CMV rating is done by a combination of high CM attenuation, followed by differential gain, realizing a net differential gain of unity. The AD629 uses the first listed value resistors noted in the figure for R1–R5. The AMP03 operates as a precision four-resistor differential amplifier, using the 25 kΩ value R1–R4 resistors noted. Both devices are complete, one package solutions to the ground-isolation amplifier.

This scheme allows relative freedom from tightly controlling ground drop voltages, or running additional and/or larger PCB traces to minimize such error voltages. Note that it can be implemented with either the fixed gain difference amplifiers shown, or with a standard in amp IC, configured for unity gain. The AD623, for example, also allows single-supply use. In any case, signal polarity is also controllable by simple reversal of the difference amplifier inputs.

In general terms, transmitting a signal from one point on a PCB to another for measurement or further processing can be optimized by two key interrelated techniques. These are the use of high impedance, differential signal handling techniques. The high impedance loading of an in amp minimizes voltage drops, and differential sensing of the remote voltage minimizes sensitivity to ground noise.

When the further signal processing is A/D conversion, these transmission criteria can be implemented *without* adding a differential ground isolation amplifier stage. Simply select an ADC that operates differentially. The high input impedance of the ADC minimizes load sensitivity to the PCB wiring resistance. In addition, the differential input feature allows the output of the source to be sensed directly at the source output terminals (even if single-ended). The CMR of the ADC then eliminates sensitivity to noise voltages between the ADC and source grounds.

An illustration of this concept using an ADC with high impedance differential inputs is shown in Figure 9.41. Note that the general concept can be extended to virtually any signal source, driving any load. All loads, even single-ended ones, become differential-input by adding an appropriate differential input stage.

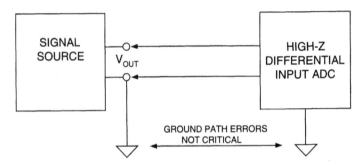

Figure 9.41: A High-Impedance Differential Input ADC Also Allows High Transmission Accuracy Between Source and Load

The differential input can be provided by either a fully developed high Z in amp or, in many cases, it can be a simple subtractor stage op amp, such as Figure 9.40.

Static PCB Effects

Leakage resistance is the dominant static circuit board effect. Contamination of the PCB surface by flux residues, deposited salts, and other debris can create leakage paths between circuit nodes. Even on well-cleaned boards, it is not unusual to find 10 nA or more of leakage to nearby nodes from 15 V supply rails. Nanoamperes of leakage current into the wrong nodes often cause volts of error at a circuit's output; for example, 10 nA into a 10 MΩ resistance causes a 0.1 V error. Unfortunately, the standard op amp pinout places the $-V_S$ supply pin next to the + input, which is often hoped to be at high impedance. To help identify nodes sensitive to the effects of leakage currents ask the simple question: If a spurious current of a few nanoamperes or more were injected into this node, would it matter?

If the circuit is already built, it is possible to localize moisture sensitivity to a suspect node with a classic test. While observing circuit operation, blow on potential trouble spots through a simple soda straw. The straw focuses the breath's moisture which, with the board's salt content in susceptible portions of the design, disrupts circuit operation upon contact. There are several means of eliminating simple surface leakage problems. Thorough washing of circuit boards to remove residues helps considerably. A simple procedure includes vigorously brushing the boards with isopropyl alcohol, followed by thorough washing with deionized water and an 85°C bakeout for a few hours. Be careful when selecting board-washing solvents, though. When cleaned with certain solvents, some water-soluble fluxes create salt deposits, exacerbating the leakage problem.

Unfortunately, if a circuit displays sensitivity to leakage, even the most rigorous cleaning can offer only a temporary solution. Problems soon return upon handling, or exposure to foul atmospheres, and high humidity. Some additional means must be sought to stabilize circuit behavior, such as conformal surface coating.

Fortunately, there is an answer to this, namely *guarding*, which offers a fairly reliable and permanent solution to the problem of surface leakage. Well-designed guards can eliminate leakage problems, even for circuits exposed to harsh industrial environments. Two schematics illustrate the basic guarding principle, as applied to typical inverting and noninverting op amp circuits.

Figure 9.42 illustrates an inverting mode guard application. In this case, the op amp reference input is grounded, so the guard is a grounded ring surrounding all leads to the inverting input, as noted by the dotted line.

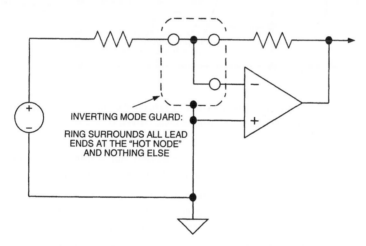

INVERTING MODE GUARD:

RING SURROUNDS ALL LEAD ENDS AT THE "HOT NODE" AND NOTHING ELSE

Figure 9.42: Inverting Mode Guard Encloses All Op Amp Inverting Input Connections Within a Grounded Guard Ring

Basic guarding principles are simple: *Completely* surround sensitive nodes with conductors that can readily sink stray currents, and maintain the guard conductors at the exact potential of the sensitive node (as otherwise the guard will serve as a leakage source rather than a leakage sink). For example, to keep leakage into a node below 1 pA (assuming 1000 MΩ leakage resistance) the guard and guarded node must be within 1 mV. Generally, the low offset of a modern op amp is sufficient to meet this criterion.

There are important caveats to be noted with implementing a true high quality guard. For traditional through-hole PCB connections, the guard pattern should appear on *both* sides of the circuit board, to be most effective. It should also be connected along its length by several vias. Finally, when either justified or required by the system design parameters, do make an effort to include guards in the PCB design process from the outset—there is little likelihood that a proper guard can be added as an afterthought.

Figure 9.43 illustrates the case for a noninverting guard. In this instance the op amp reference input is directly driven by the source, which complicates matters considerably. Again, the guard ring completely surrounds all of the input nodal connections. In this instance however, the guard is driven from the low impedance feedback divider connected to the inverting input.

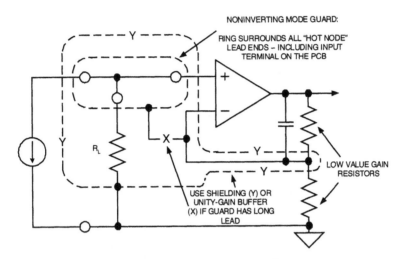

Figure 9.43: Noninverting Mode Guard Encloses All Op Amp Noninverting Input Connections within a Low Impedance, Driven Guard Ring

Usually the guard-to-divider junction will be a direct connection, but in some cases a unity gain buffer might be used at "X" to drive a cable shield, or also to maintain the lowest possible impedance at the guard ring.

In lieu of the buffer, another useful step is to use an additional, directly grounded screen ring, "Y," which surrounds the inner guard and the feedback nodes as shown. This step costs nothing except some added layout time, and will greatly help buffer leakage effects into the higher impedance inner guard ring.

Of course what hasn't been addressed to this point is just how the op amp itself is connected into these guarded islands without compromising performance. The traditional method using a TO-99 metal can package device was to employ double-sided PCB guard rings, with both op amp inputs terminated within the guarded ring.

Many high impedance sensors use the above-described method. The section immediately following illustrates how more modern IC packages can be mounted to PCB patterns, and take advantage of guarding and low leakage operation.

Sample MINIDIP and SOIC Op Amp PCB Guard Layouts

Modern assembly practices have favored smaller plastic packages such as 8-pin MINIDIP and SOIC types. Some suggested partial layouts for guard circuits using these packages is shown in the next two figures. While guard traces may also be possible with even more tiny op amp footprints, such as SOT-23, SC70, etc., the required trace separations become even more confining, challenging the layout designer as well as the manufacturing processes.

For the ADI "N" style MINIDIP package, Figure 9.44 illustrates how guarding can be accomplished for inverting (left) and noninverting (right) operating modes. This setup would also be applicable to other op amp

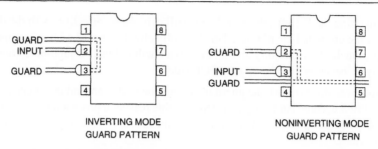

Figure 9.44: PCB Guard Patterns for Inverting and Noninverting Mode Op Amps Using 8-Pin MINIDIP (N) Package

devices where relatively high voltages occur at Pin 1 or 4. Using a standard 8-pin DIP outline for a single op amp, it can be noted that this package's 0.1" pin spacing allows a PC trace (here, the guard trace) to pass between adjacent pins. This is the key to implementing effective DIP package guarding, as it can adequately prevent a leakage path from the $-V_S$ supply at Pin 4, or from similar high potentials at Pin 1.

For the left-side inverting mode, note that the grounded guard traces connected to Pin 3 surround the op amp inverting input (Pin 2), and run parallel to the input trace. This guard would be continued out to and around the source and feedback connections of Figure 9.42 (or other similar circuit), including an input pad in the case of a cable. In the right-side noninverting mode, the guard voltage is the feedback divider voltage to Pin 2. This corresponds to the inverting input node of the amplifier, from Figure 9.43.

Note that in both of the cases of Figure 9.44, the guard physical connections shown are only partial—an actual layout would include all sensitive nodes within the circuit. In both the inverting and the noninverting modes using the MINIDIP or other through-hole style package, the PCB guard traces should be located on both sides of the board, with top and bottom traces connected with several vias.

Things become slightly more complicated when using guarding techniques with the SOIC surface mount ("R") package, as the 0.05" pin spacing doesn't easily allow routing of PCB traces between the pins. But, there is still an effective guarding answer, at least for the inverting case. Figure 9.45 shows guards for the ADI "R" style SOIC package.

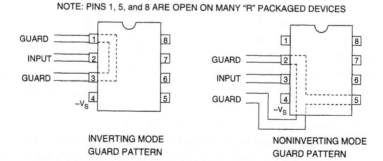

Figure 9.45: PCB Guard Patterns for Inverting and Noninverting Mode Op Amps Using 8-Pin SOIC (R) Package

Note that for many single op amp devices in this SOIC "R" package, Pins 1, 5, and 8 are "No Connect" pins. For such instances, this means that these locations can be employed in the layout to route guard traces.

In the case of the inverting mode (left), the guarding is still completely effective, with the dummy Pin 1 and Pin 3 serving as the grounded guard trace. This is a fully effective guard without compromise. Also, with SOIC op amps, much of the circuitry around the device will not use through-hole components. So, the guard ring may only be necessary on the op amp PCB side.

In the case of the follower stage (right), the guard trace must be routed around the negative supply at Pin 4, and thus Pin 4 to Pin 3 leakage isn't fully guarded. For this reason, a precision high impedance follower stage using an SOIC package op amp isn't generally recommended, as guarding isn't as effective for dual supply connected devices.

However, an exception to this caveat does apply to the use of a *single-supply* op amp as a noninverting stage. For example, if the AD8551 is used, Pin 4 becomes ground, and some degree of intrinsic guarding is then established by default.

Dynamic PCB Effects

Although static PCB effects can come and go with changes in humidity or board contamination, problems that most noticeably affect the dynamic performance of a circuit usually remain relatively constant. Short of a new design, washing or any other simple fixes can't fix them. As such, they can permanently and adversely affect a design's specifications and performance. The problems of stray capacitance, linked to lead and component placement, are reasonably well known to most circuit designers. Since lead placement can be permanently dealt with by correct layout, any remaining difficulty is solved by training assembly personnel to orient components or bend leads optimally.

Dielectric absorption (DA), on the other hand, represents a more troublesome and still poorly understood circuit-board phenomenon. Like DA in discrete capacitors, DA in a printed-circuit board can be modeled by a series resistor and capacitor connecting two closely spaced nodes. Its effect is inverse with spacing and linear with length.

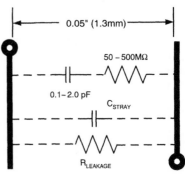

As shown in Figure 9.46, the RC model for this effective capacitance ranges from 0.1 pF to 2.0 pF, with the resistance ranging from 50 MΩ to 500 MΩ. Values of 0.5 pF and 100 MΩ are most common. Consequently, circuit-board DA interacts most strongly with high impedance circuits.

PCB DA most noticeably influences dynamic circuit response, for example, settling time. Unlike circuit leakage, the effects aren't usually linked to humidity or other environmental conditions, but rather, are a function of the board's dielectric properties. The chemistry involved in producing plated-through holes seems to exacerbate the problem. If circuits don't meet expected transient response specs, consider PCB DA as a possible cause.

Figure 9.46: DA Plagues Dynamic Response of PCB-Based Circuits

Fortunately, there are solutions. As in the case of capacitor DA, external components can be used to compensate for the effect. More importantly, surface guards that totally isolate sensitive nodes from parasitic coupling often eliminate the problem (note that these guards should be duplicated on both sides of the board, in cases of through-hole components). As noted previously, low loss PCB dielectrics are also available at higher costs.

PCB "hook," similar if not identical to DA, is characterized by variation in effective circuit-board capacitance with frequency (see Reference 1). In general, it affects high impedance circuit transient response where board capacitance is an appreciable portion of the total in the circuit. Circuits operating at frequencies below 10 kHz are the most susceptible. As in circuit board DA, the board's chemical makeup very much influences its effects.

Stray Capacitance

When two conductors aren't short-circuited together, or totally screened from each other by a conducting (Faraday) screen, there is a capacitance between them. So, on any PCB, there will be a large number of capacitors associated with any circuit (which may or may not be considered in models of the circuit). Where high frequency performance matters (and even dc and VLF circuits may use devices with high F_t and therefore be vulnerable to high frequency instability), it is very important to consider the effects of this stray capacitance.

Any basic textbook will provide formulas for the capacitance of parallel wires and other geometric configurations (see References 9 and 10). The example we need consider in this discussion is the parallel plate capacitor, often formed by conductors on opposite sides of a PCB. The basic diagram describing this capacitance is shown in Figure 9.47.

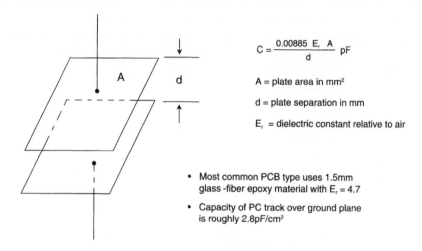

$$C = \frac{0.00885 \; E_r \; A}{d} \; pF$$

A = plate area in mm²

d = plate separation in mm

E_r = dielectric constant relative to air

- Most common PCB type uses 1.5mm glass-fiber epoxy material with $E_r = 4.7$
- Capacity of PC track over ground plane is roughly 2.8pF/cm²

Figure 9.47: Capacitance of Two Parallel Plates

Neglecting edge effects, the capacitance of two parallel plates of area A mm² and separation d mm in a medium of dielectric constant E_r relative to air is 0.00885 E_rA/d pF.

From this formula, we can calculate that for general-purpose PCB material ($E_r = 4.7$, d = 1.5 mm), the capacitance between conductors on opposite sides of the board is just under 3 pF/cm². In general, such capacitance will be parasitic, and circuits must be designed so that it does not affect their performance.

While it is possible to use PCB capacitance in place of small discrete capacitors, the dielectric properties of common PCB substrate materials cause such capacitors to behave poorly. They have a rather high temperature coefficient and poor Q at high frequencies, which makes them unsuitable for many applications. Boards made with lower loss dielectrics such as Teflon are expensive exceptions to this rule.

Capacitive Noise and Faraday Shields

There is a capacitance between any two conductors separated by a dielectric (air or vacuum are dielectrics). If there is a change of voltage on one, there will be a movement of charge on the other. A basic model for this is shown in Figure 9.48.

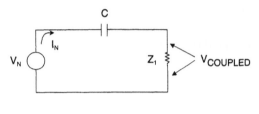

It is evident that the noise voltage, $V_{COUPLED}$ appearing across Z_1, may be reduced by several means, all of which reduce noise current in Z_1. They are reduction of the signal voltage V_N, reduction of the frequency involved, reduction of the capacitance, or reduction of Z_1 itself. Unfortunately, however, often none of these circuit parameters can be freely changed, and an alternate method is needed to minimize the interference. The best solution

Z_1 = CIRCUIT IMPEDANCE
$Z_2 = 1/j\omega C$

$$V_{COUPLED} = V_N \left(\frac{Z_1}{Z_1 + Z_2} \right)$$

Figure 9.48: Capacitive Coupling Equivalent Circuit Model

towards reducing the noise coupling effect of C is to insert a grounded conductor, also known as a *Faraday shield*, between the noise source and the affected circuit. This has the desirable effect of reducing Z_1 noise current, thus reducing $V_{COUPLED}$.

A Faraday shield model is shown by Figure 9.49. In the left picture, the function of the shield is noted by how it effectively divides the coupling capacitance, C. In the right picture the net effect on the coupled voltage across Z_1 is shown. Although the noise current I_N still flows in the shield, most of it is now diverted away from Z_1. As a result, the coupled noise voltage $V_{COUPLED}$ across Z_1 is reduced.

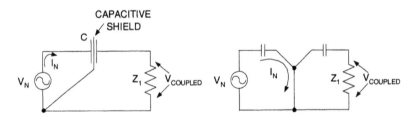

Figure 9.49: An Operational Model of a Faraday Shield

A Faraday shield is easily implemented and almost always successful. Thus capacitively coupled noise is rarely an intractable problem. However, to be fully effective, a Faraday shield must completely block the electric field between the noise source and the shielded circuit. It must also be connected so that the displacement current returns to its source, without flowing in any part of the circuit where it can introduce conducted noise.

The Floating Shield Problem

And, it is quite important to note here—*a conductor that is intended to function as a Faraday shield must never be left floating, as this almost always increases capacity and exacerbates the noise problem.*

An example of this "floating shield" problem is seen in side-brazed ceramic IC packages. These DIP packages have a small square conducting Kovar lid soldered onto a metallized rim on the ceramic package top. Package manufacturers offer only two options: the metallized rim may be connected to one of the corner pins of the package, or it may be left unconnected.

Most logic circuits have a ground pin at one of the package corners, and therefore the lid is grounded. Alas, many analog circuits don't have a ground pin at a package corner, and the lid is left floating—acting as an antenna for noise. Such circuits turn out to be far more vulnerable to electric field noise than the same chip in a plastic DIP package, where the chip is completely unshielded.

Whenever practical, it is good practice for the user to ground the lid of any side-brazed ceramic IC where the lid is not grounded by the manufacturer, thus implementing an *effective* Faraday shield. This can be done with a wire soldered to the lid (this will not damage the device, as the chip is thermally and electrically isolated from the lid). If soldering to the lid is unacceptable, a grounded phosphor-bronze clip or conductive paint from the lid to the ground pin may be used to make the ground connection,.

A safety note is appropriate at this point. Never attempt to ground such a lid without first verifying that it is unconnected. Occasionally device types are found with the lid connected to a power supply rather than to ground.

A case where a Faraday shield is impractical is between IC chip bondwires. This can have important consequences, as the stray capacitance between chip bondwires and associated leadframes is typically ≈ 0.2 pF, with observed values generally between 0.05 pF and 0.6 pF.

Buffering ADCs Against Logic Noise

If we have a high resolution data converter (ADC or DAC) connected to a high speed data bus that carries logic noise with a 2 V/ns–5 V/ns edge rate, this noise is easily connected to the converter analog port via stray capacitance across the device. Whenever the data bus is active, intolerable amounts of noise are capacitively coupled into the analog port, thus seriously degrading performance.

This particular effect is illustrated by the diagram of Figure 9.50, where multiple package capacitors couple noisy edge signals from the data bus into the analog input of an ADC.

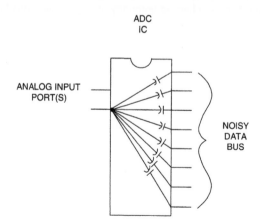

Figure 9.50: A High Speed ADC IC Sitting on a Fast Data Bus Couples Digital Noise into the Analog Port, Thus Limiting Performance

Present technology offers no cure for this problem, within the affected IC device itself. The problem also limits performance possible from other broadband monolithic mixed signal ICs with single-chip analog and digital circuits. Fortunately, this coupled noise problem can simply be avoided by *not* connecting the data bus directly to the converter.

Instead, *use a CMOS latched buffer as a converter-to-bus interface*, as shown by Figure 9.51. Now the CMOS buffer IC acts as a Faraday shield, and dramatically reduces noise coupling from the digital bus. This solution costs money, occupies board area, reduces reliability (very slightly), consumes power, and it complicates the design—but it does improve the signal-to-noise ratio of the converter. The designer must decide whether it is worthwhile for individual cases, but in general it is highly recommended.

Bus switches can also be utilized to isolate data lines from buses as described later in this chapter.

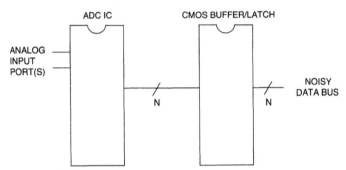

- THE OUTPUT BUFFER/LATCH ACTS AS A FARADAY SHIELD BETWEEN "N" LINES OF A FAST, NOISY DATA BUS AND A HIGH PERFORMANCE ADC

- THIS MEASURE ADDS COST, BOARD AREA, POWER CONSUMPTION, RELIABILITY REDUCTION, DESIGN COMPLEXITY AND, MOST IMPORTANTLY, IMPROVED PERFORMANCE

Figure 9.51: A High Speed ADC IC Using a CMOS Buffer/Latch at the Output Shows Enhanced Immunity of Digital Data Bus Noise

References:
9.2 PC Board Design Issues

1. W. Doeling, W. Mark, T. Tadewald, and P. Reichenbacher, "Getting Rid of Hook: The Hidden PC-Board Capacitance," **Electronics,** October 12, 1978, p. 111–117.

2. Alan Rich, "Shielding and Guarding," **Analog Dialogue,** Vol. 17, No. 1, 1983, p. 8.

3. Ralph Morrison, **Grounding and Shielding Techniques, 4th Edition**, John Wiley, Inc., 1998, ISBN: 0471245186.

4. Henry W. Ott, **Noise Reduction Techniques in Electronic Systems, 2nd Edition**, John Wiley, Inc., 1988, ISBN: 0-471-85068-3.

5. Paul Brokaw, "An IC Amplifier User's Guide to Decoupling, Grounding and Making Things Go Right for a Change," **Analog Devices AN202**.

6. Paul Brokaw, "Analog Signal-Handling for High Speed and Accuracy," **Analog Devices AN342**.

7. Paul Brokaw and Jeff Barrow, "Grounding for Low- and High-Frequency Circuits," **Analog Devices AN345.**

8. Jeff Barrow, "Avoiding Ground Problems in High Speed Circuits," **RF Design**, July 1989.

9. B. I. & B. Bleaney, **Electricity & Magnetism**, Oxford at the Clarendon Press, 1957, pp. 23, 24, and 52.

10. G. W. A. Dummer, H. Nordenberg, **Fixed and Variable Capacitors**, McGraw-Hill, 1960, pp. 11–13.

11. William C. Rempfer, *Get All the Fast ADC Bits You Pay For*, **Electronic Design, Special Analog Issue**, June 24, 1996, p. 44.

12. Mark Sauerwald, *Keeping Analog Signals Pure in a Hostile Digital World*, **Electronic Design, Special Analog Issue**, June 24, 1996, p. 57.

13. Jerald Grame and Bonnie Baker, *Design Equations Help Optimize Supply Bypassing for Op Amps*, **Electronic Design, Special Analog Issue**, June 24, 1996, p. 9.

14. Jerald Grame and Bonnie Baker, *Fast Op Amps Demand More Than a Single-Capacitor Bypass*, **Electronic Design, Special Analog Issue**, November 18, 1996, p. 9.

15. Walt Kester and James Bryant, *Grounding in High Speed Systems*, **High Speed Design Techniques**, Analog Devices, 1996, Chapter 7, p. 7–27.

16. Jeffrey S. Pattavina, *Bypassing PC Boards: Thumb Your Nose at Rules of Thumb*, **EDN**, Oct. 22, 1998, p. 149.

17. Howard W. Johnson and Martin Graham, **High-Speed Digital Design**, PTR Prentice Hall, 1993, ISBN: 0133957241.

18. Walt Kester, *A Grounding Philosophy for Mixed-Signal Systems*, **Electronic Design Analog Applications Issue**, June 23, 1997, p. 29.

19. Ralph Morrison, **Solving Interference Problems in Electronics**, John Wiley, 1995.

20. C. D. Motchenbacher and J. A. Connelly, **Low Noise Electronic System Design**, John Wiley, 1993.

21. Crystal Oscillators: MF Electronics, 10 Commerce Drive, New Rochelle, NY, 10801, 914-576-6570.

22. Crystal Oscillators: Wenzel Associates, Inc., 2215 Kramer Lane, Austin, Texas USA 78758, 512-835-2038, www.wenzel.com.

23. Mark Montrose, **EMC and the Printed Circuit Board**, IEEE Press, 1999 (IEEE Order Number PC5756).

Acknowledgments:

Portions of this section were adapted from Doug Grant and Scott Wurcer, "Avoiding Passive Component Pitfalls," originally published in **Analog Dialogue 17-2**, 1983.

Analog Power Supply Systems
Walt Jung, Walt Kester

Analog circuits have traditionally been powered from well-regulated, low noise linear power supplies. This type of power system is typically characterized by medium-to-low power conversion efficiency. Such linear regulators usually excel in terms of self-generated and radiated noise components. If the designer's life were truly simple, it might continue with such familiar designs offering good performance and minimal side effects.

But, the designer's life is hardly so simple. Modern systems may allow using linear regulators, but multiple output levels and/or polarities are often required. There may also be some additional requirements set for efficiency, which may dictate the use of dc-dc conversion techniques, and, unfortunately, their higher associated noise output.

This section addresses power supply design issues for analog systems (including op amps, analog multiplexers, ADCs, DACs, etc.), taking into account the regulator types most likely to be used. The primary dc power sources are assumed to be either rectified and smoothed ac sources (i.e., mains derived), a battery stack, or a switching regulator output. The latter example could be fed from either a battery or a mains-derived dc source.

As noted in Figure 9.52, linear mode regulation is generally recommended as an optimum starting point in all instances (first bullet). Nevertheless, in some cases, a degree of hybridization between fully linear and switching mode regulation may be required (second bullet). This could be either for efficiency or other diverse reasons.

- High performance analog power systems use linear regulators, with primary power derived from:
    - AC line power
    - Battery power systems
    - DC- DC power conversion systems
- Switching regulators should be avoided if at all possible, but if not...
    - Apply noise control techniques
    - Use quality layout and grounding
    - Be aware of EMI

Figure 9.52: Regulation Priorities for Analog Power Supply Systems

Whenever switching-type regulators are involved in powering precision analog circuits, noise control is very likely to be a design issue. Therefore some focus of this section is on minimizing noise when using switching regulators.

Linear IC Regulation

Linear IC voltage regulators have long been standard power system building blocks. After an initial introduction in 5 V logic voltage regulator form, they have since expanded into other standard voltage levels spanning from 3 V to 24 V, handling output currents from as low as 100 mA (or less) to as high as 5 A (or more). For several good reasons, linear style IC voltage regulators have been valuable system components since the early days. As mentioned above, a basic reason is the relatively low noise characteristic vis-à-vis the switching type of regulator. Others are a low parts count and overall simplicity compared to discrete solutions. But, because of their power losses, these linear regulators have also been known for being relatively inefficient. Early generation devices (of which many are still available) required 2 V or more of unregulated input above the regulated output voltage, making them lossy in power terms.

More recently however, linear IC regulators have been developed with more liberal (i.e., lower) limits on minimum input-output voltage. This voltage, known more commonly as *dropout* voltage, has led to what is termed the *Low Drop Out* regulator, or more simply, the LDO. Dropout voltage (V_{MIN}) is defined simply as that minimum input-output differential where the regulator undergoes a 2% reduction in output voltage. For example, if a nominal 5.0 V LDO output drops to 4.9 V (–2%) under conditions of an input-output differential of 0.5 V, by this definition the LDO's dropout voltage is 0.5 V.

Dropout voltage is extremely critical to a linear regulator's power efficiency. The lower the voltage allowable across a regulator while still maintaining a regulated output, the less power the regulator dissipates as a result. A low regulator dropout voltage is the key to this, as it takes a lower dropout to maintain regulation as the input voltage lowers. In performance terms, the bottom line for LDOs is simply that more useful power is delivered to the load and less heat is generated in the regulator. LDOs are key elements of power systems providing stable voltages from batteries, such as portable computers, cellular phones, etc. This is because they maintain a regulated output down to lower points on the battery's discharge curve. Or, within classic mains-powered raw dc supplies, LDOs allow lower transformer secondary voltages, reducing system shutdowns under brownout conditions, as well as allowing cooler operation.

Some Linear Voltage Regulator Basics

A brief review of three terminal linear IC regulator fundamentals is necessary before understanding the LDO variety. Most (but not all) of the general three terminal regulator types available today are *positive leg, series style* regulators. This simply means that they control the regulated voltage output by means of a pass element in series with the positive unregulated input. And, although they are fewer in number, there are also *negative leg* series style regulators, which operate in a fashion complementary to the positive units.

A basic hookup diagram of a three terminal regulator is shown in Figure 9.53. In terms of basic functionality, many standard voltage regulators operate in a series mode, three-terminal form, just as shown here. As can be noted from this figure, the three I/O terminals are V_{IN}, GND (or Common), and V_{OUT}. Note also that this regulator block, in the absence of any assigned voltage polarity, could in principle be a positive type regulator. Or, it might also be a negative style of voltage regulator—the principle is the same for both— a common terminal, as well as input and output terminals.

In operation, two power components are dissipated in the regulator, one a function of $V_{IN} - V_{OUT}$ and I_L, plus a second which is a function of V_{IN} and I_{GROUND}. The first of these is usually dominant. Analysis of the situation will reveal that as the dropout voltage V_{MIN} is reduced, the regulator is able to deliver a higher percentage of the input power to the load, and is thus more efficient, running cooler and saving power. This is the core appeal of the modern LDO type of regulator (see Chapter 7 of this book and Reference 1).

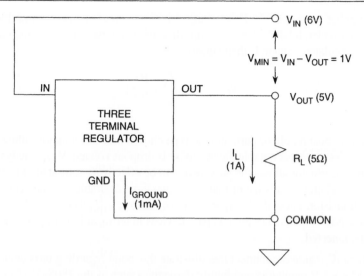

Figure 9.53: A Basic Three Terminal Regulator Hookup (Either Positive or Negative)

A more detailed look within a typical regulator block diagram reveals a variety of elements, as is shown in Figure 9.54. Note that all regulators will contain those functional components connected via solid lines. The connections shown dotted indicate options that might be available when more than three I/O pins are available.

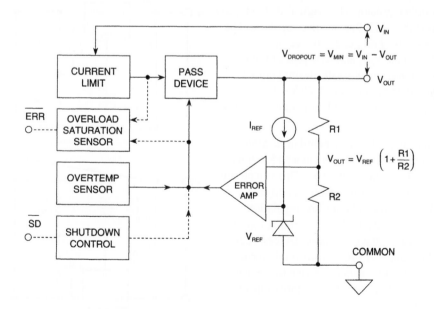

Figure 9.54: Block Diagram of a Voltage Regulator

In operation, a voltage reference block produces a stable voltage V_{REF}, which is almost always a voltage based on the bandgap voltage of silicon, typically ~1.2 V (see Reference 2). This allows output voltages of

3 V or more from supplies as low as 5 V. This voltage drives one input of an error amplifier, with the second input connected to the divider, R1-R2. The error amplifier drives the pass device, which in turn controls the output. The resulting regulated voltage is then simply:

$$V_{OUT} = V_{REF}\left(1 + \frac{R1}{R2}\right)$$

Eq. 9.5

Pass Devices

The pass device is a foremost regulator part, and the type chosen here has a major influence on almost all regulator performance issues. Most notable among these is dropout voltage, V_{MIN}. Analysis shows the use of an *inverting* mode pass transistor allows the pass device to be effectively saturated, thus minimizing the associated voltage losses. Therefore this factor makes the two most desirable pass devices for LDO use a PNP bipolar, or a PMOS transistor. These device types achieve the lowest levels of $V_{IN} - V_{OUT}$ required for LDO operation. In contrast, NPN bipolars are poor as pass devices in terms of low dropout, particularly when they are Darlington connected.

Standard fixed-voltage IC regulator architectures illustrate this point regarding pass devices. For example, the fixed-voltage LM309 5 V regulators and family derivatives such as the 7805, 7815 et. al., (and their various low and medium current alternates) are poor in terms of dropout voltage. These designs use a Darlington pass connection, not known for low dropout (~1.5 V typical), or for low quiescent current (~5 mA).

±15 V Regulator Using Adjustable Voltage ICs

Later developments in references and three-terminal regulation techniques led to the development of the *voltage-adjustable* regulator. The original IC to employ this concept was the LM317, a positive regulator. The device produces a fixed reference voltage of 1.25 V, appearing between the V_{OUT} and ADJ pins of the IC. External scaling resistors set up the desired output voltage, adjustable in the range of 1.25 V–30 V. A complementary device, the LM337, operates in similar fashion, regulating negative voltages.

An application example using standard *adjustable* three terminal regulators to implement a ±15 V linear power supply is shown in Figure 9.55. This is a circuit as might be used for powering traditional op amp supply rails. It is capable of better line regulation performance than would an otherwise similar circuit, using standard fixed-voltage regulator devices, such as for example 7815 and 7915 ICs. However, in terms of power efficiency it isn't outstanding, due to the use of the chosen ICs, which require 2 V or more of headroom for operation.

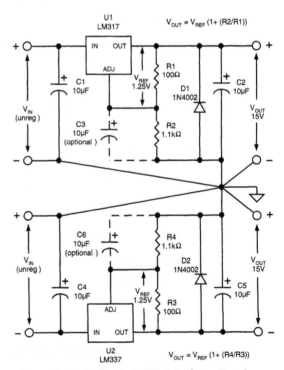

Figure 9.55: A Classic ±15V, 1 A Linear Supply Regulator Using Adjustable Voltage Regulator ICs

In the upper portion of this circuit an LM317 adjustable regulator is used, with R2 and R1 chosen to provide a 15 V output at the upper output terminal. If desired, R2 can easily be adjusted for other output levels, according to the figure's V_{OUT} equation. Resistor R1 should be left fixed, as it sets the minimum regulator drain of 10 mA or more.

In this circuit, capacitors C1 and C2 should be tantalum types, and R1-R2 metal films. C3 is optional, but is highly recommended if the lowest level of output noise is desired. The normally reverse-biased diode D1 provides a protective output clamp, for system cases where the output voltage would tend to reverse, if one supply should fail. The circuit operates from a rectified and filtered ac supply at V_{IN}, polarized as shown. The output current is determined by choosing the regulator IC for appropriate current capability.

To implement the negative supply portion, the sister device to the LM317 is used, the LM337. The bottom circuit section thus mirrors the operation of the upper, delivering a negative 15 V at the lowest output terminal. Programming of the LM337 for output voltage is similar to that of the LM317, but uses resistors R4 and R3. R4 should be used to adjust the voltage, with R3 remaining fixed. C6 is again optional, but is recommended for reasons of lowest noise.

Low Dropout Regulator Architectures

In contrast to traditional three-terminal regulators with Darlington or single-NPN pass devices, low dropout regulators employ lower voltage threshold pass devices. This basic operational difference allows them to operate effectively down to a range of 100 mV–200 mV in terms of their specified V_{MIN}. In terms of use within a system, this factor can have fairly significant operational advantages.

An effective implementation of some key LDO features is contained in the Analog Devices series of any-CAP LDO regulators. Devices of this ADP33xx series are so named for their relative insensitivity to the output capacitor, in terms of both its size and ESR. Available in power efficient packages such as the ADI Thermal Coastline (and other thermally enhanced packages), they come in both standalone LDO and LDO controller forms (used with an external PMOS FET). They also offer a wide span of fixed output voltages from 1.5 V to 5 V, with rated current outputs up to 1500 mA. User-adjustable output voltage versions are also available. A basic simplified diagram for the family is shown schematically in Figure 9.56.

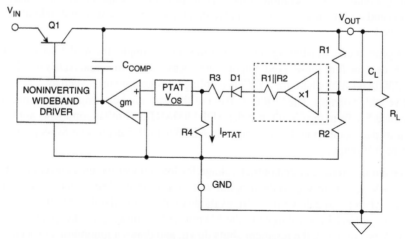

Figure 9.56: The ADP33xx anyCAP LDO Architecture Has Both DC and AC Performance Advantages

One of the key differences in the ADP33xx LDO series is the use of a high gain vertical PNP pass device, Q1, allowing typical dropout voltages for the series to be on the order of 1mV/mA for currents of 200 mA or less.

In circuit operation, V_{REF} is defined as a reference voltage existing at the output of a zero impedance divider of ratio R1/R2. In the figure, this is depicted symbolically by the (dotted) unity gain buffer amplifier fed by R1/R2, which has an output of V_{REF}. This reference voltage feeds into a series connection of (dotted) R1∥R2, then actual components D1, R3, R4, etc. The regulator output voltage is:

$$V_{OUT} = V_{REF}\left(1 + \frac{R1}{R2}\right)$$
Eq. 9.6

In the various devices of the ADP33xx series, the R1-R2 divider is adjusted to produce standard output voltages of 1.5, 1.8, 2.5, 2.7, 2.75, 2.77, 2.85, 2.9, 3.0, 3.15, 3.2, 3.3, 3.6, and 5.0 V. The regulator behaves as if the entire error amplifier has simply an offset voltage of V_{REF} volts, as seen at the output of a conventional R1-R2 divider.

While the above-described dc performance enhancements of the ADP33xx series are worthwhile, more dramatic improvements come in areas of ac-related performance. Capacitive loading and the potential instability it brings is a major deterrent to easy LDO applications. One method of providing some measure of immunity to variation in an amplifier response pole is the use of a frequency compensation technique called *pole splitting*. In the Figure 9.56 circuit, C_{COMP} functions as the pole splitting capacitor, and provides benefits of a buffered, C_L independent single-pole response. As a result, frequency response is dominated by the regulator's internal compensation, and becomes relatively immune to the value and ESR of load capacitor C_L.

This feature makes the design tolerant of virtually any output capacitor type. C_L, the load capacitor, can be as low as 0.47 µF, and it can also be a multilayer ceramic capacitor (MLCC) type, allowing a very small physical size for the entire regulation function.

Fixed-Voltage, 50/100/200/500/1000/1500 mA LDO Regulators

A basic regulator application diagram common to various fixed voltage devices of the ADP33xx device series is shown by Figure 9.57. Operation of the various pins and internal functions is discussed below. Note that all pins and all functions are not available on all parts in the series, and individual data sheets should be consulted.

This circuit is a general one, illustrating common points. For example, the ADP33xx is a 50 mA basic LDO regulator device, designed for those fixed output voltages as noted. An actual ADP3300 device ordered would be ADP3300ART-YY, where the "YY" is a voltage designator suffix such as 2.7, 3, 3.2, 3.3, or 5, for the respective table voltages. The "ART" portion of the part number designates the package (SOT-23 6-lead). To produce 5 V from the circuit, use the ADP3300ART-5. Similar comments apply to the other devices, insofar as part numbering. For example, an ADP3301AR-5 depicts an SO-8 packaged 100 mA device, producing 5 V output.

In operation, the circuit produces rated output voltage for loads under the max current limit, for input voltages above $V_{OUT} + V_{MIN}$ (where V_{MIN} is the dropout voltage for the specific device used, at rated current). The circuit is ON when the shutdown input (if available on the particular device selected) is in a HIGH state, either by a logic HIGH control input to the $\overline{SD}$ pin, or by simply tying this pin to V_{IN} (shown dotted). When $\overline{SD}$ is LOW or grounded, the regulator shuts down, and draws a minimum quiescent current.

The anyCAP regulator devices maintain regulation over a wide range of load, input voltage and temperature conditions. Most devices have a combined error band of ±1.4% (or less). When an overload condition is

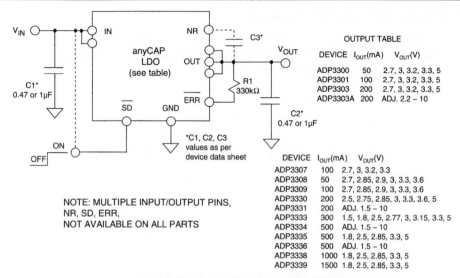

Figure 9.57: A Basic LDO Regulator Hookup Useful by Device Selection from 50 mA to 1500 mA, At Fixed or Adjustable Voltages Per Table

detected, the open collector $\overline{\text{ERR}}$ goes to a LOW state (if available on the particular device selected). R1 is a pull-up resistor for the $\overline{\text{ERR}}$ output. This resistor can be eliminated if the load provides a pull-up current.

C3, connected between the OUT and NR pins, can be used for an optional noise reduction (NR) feature (if available on the particular device selected). This is accomplished by bypassing a portion of the internal resistive divider, which reduces output noise ~10 dB. When exercised, only the recommended low leakage capacitors as specific to a particular part should be used.

The C1 input and C2 output capacitors should be selected as either 0.47 µF or 1 µF values respectively, again, as per the particular device used. For most devices of the series 0.47 µF suffices, but the ADP3335 uses the 1 µF values. Larger capacitors can also be used, and will provide better transient performance.

Heat sinking of device packages with more than five pins is enhanced, by use of multiple IN and OUT pins. All of the pins available should therefore be used in the PCB design, to minimize layout thermal resistance.

Adjustable Voltage, 200 mA LDO Regulator

In addition to the fixed output voltage LDO devices discussed above, adjustable versions are also available, to realize nonstandard voltages. The ADP3331 is one such device, and it is shown in Figure 9.58, configured as a 2.8 V output, 200 mA LDO application.

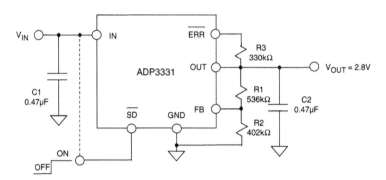

**Figure 9.58: An Adjustable 200 mA LDO
Regulator Set Up for a 2.8 V Output**

The ADP3331 is generally similar to other anyCAP LDO parts, with two notable exceptions. It has a lower quiescent current (~34 µA when lightly loaded) and most importantly, the output voltage is user-adjustable. As noted in the circuit, R1 and R2 are external precision resistors used to define the regulator operating voltage.

The output of this regulator is V_{OUT}, which is related to feedback pin FB voltage V_{FB} as:

$$V_{OUT} = V_{FB}\left(1 + \frac{R1}{R2}\right)$$

Eq. 9.7

where V_{FB} is 1.204 V. Resistors R1 and R2 program V_{OUT}, and their parallel equivalent should be kept close to 230 kΩ for best stability. To select R1 and R2, first calculate their ideal values, according to the following two expressions:

$$R1 = 230\left(\frac{V_{OUT}}{V_{FB}}\right) k\Omega$$

Eq. 9.8

$$R2 = \frac{230}{\left(1 + \dfrac{V_{FB}}{V_{OUT}}\right)} k\Omega$$

Eq. 9.9

In the example circuit, V_{OUT} is 2.8 V, which yields R1 = 534.9 kΩ, and R2 = 403.5 kΩ. As noted in the figure, closest standard 1% values are used, which provides an output of 2.8093 V (perfect resistors assumed). In practice, the resistor tolerances should be added to the ±1.4% tolerance of the ADP3331 for an estimation of overall error.

To complement the above-discussed anyCAP series of standalone LDO regulators, there is the LDO *regulator controller*. The regulator controller IC picks up where the standalone regulator stops for either load current or power dissipation, using an external PMOS FET pass device. As such, the current capability of the LDO can be extended to several amps. An LDO regulator controller application is shown later in this discussion.

The application examples above illustrate a subset of the entire anyCAP family of LDOs. Further information on this series of standalone and regulator controller LDO devices can be found in the references at the end of the section.

Charge-Pump Voltage Converters

Another method for developing supply voltage for op amp systems employs what is known as a *charge-pump* circuit (also called switched capacitor voltage conversion). Charge-pump voltage converters accomplish energy transfer and voltage conversion using charges stored on capacitors, thus the name, charge-pump.

Using switching techniques, charge pumps convert supply voltage of one polarity to a higher or lower voltage, or to an alternate polarity (at either higher or lower voltage). This is accomplished with only an array of low resistance switches, a clock for timing, and a few external storage capacitors to hold the charges being transferred in the voltage conversion process. No inductive components are used, thus EMI generation is kept to a minimum. Although relatively high currents are switched internally, the high current switching is localized, and therefore the generated noise is not as great as in inductive type switchers. With due consideration towards component selection, charge-pump converters can be implemented with reasonable noise performance.

The two common charge-pump voltage converters are the *voltage inverter* and the *voltage doubler* circuits. In a voltage inverter, a charge pump capacitor is charged to the input voltage during the first half of the switching cycle. During the second half of the switching cycle the input voltage stored on the charge pump capacitor is inverted, and is applied to an output capacitor and the load. Thus the output voltage is essentially the negative of the input voltage, and the average input current is approximately equal to the output current. The switching frequency impacts the size of the external capacitors required, and higher switching frequencies allow the use of smaller capacitors. The duty cycle—defined as the ratio of charge pump charging time to the entire switching cycle time—is usually 50%, which yields optimal transfer efficiency.

A voltage doubler works similarly to the inverter. In this case the pump capacitor accomplishes a voltage doubling function. In the first phase it is charged from the input, but in the second phase of the cycle it appears in series with the output capacitor. Over time, this has the effect of doubling the magnitude of the input voltage across the output capacitor and load. Both the inverter and voltage doubler circuits provide no voltage regulation in basic form. However, techniques exist to add regulation (discussed below).

There are advantages and disadvantages to using charge-pump techniques, compared to inductor-based switching regulators. An obvious key advantage is the elimination of the inductor and the related magnetic design issues. In addition, charge-pump converters typically have relatively low noise and minimal radiated EMI. Application circuits are simple, and usually only two or three external capacitors are required. Because there are no inductors, the final PCB height can generally be made smaller than a comparable inductance-based switching regulator. Charge-pump inverters are also low in cost, compact, and capable of efficiencies greater than 90%. Obviously, current output is limited by the capacitor size and the switch capacity. Typical IC charge-pump inverters have 150 mA maximum outputs.

On their downside, charge-pump converters don't maintain high efficiency for a wide voltage range of input to output, unlike inductive switching regulators. Nevertheless, they are still often suitable for lower current loads where any efficiency disadvantages are a small portion of a larger system power budget. A summary of general charge-pump operating characteristics is shown in Figure 9.59.

- No Inductors
- Minimal Radiated EMI
- Simple Implementation: 2 External Capacitors, 1 Diode (for Doubler), Input Capacitor
- Efficiency > 90% Achievable
- Low Cost, Compact, Low Profile (Height)
- Optimized for Inverting or Doubling Supply Voltage with Output Regulation: ADP3605, ADP3607
- Inverter/ Triplers with Three Outputs from 3V Input: ADM8830, ADM8839, ADM8840

Figure 9.59: Some General Charge-Pump Characteristics

An example of charge-pump applicability is the voltage inverter function. Inverters are often useful where a relatively low current negative voltage (i.e., –3 V) is required, in addition to a primary positive voltage (such as 5 V). This may occur in a single supply system, where only a few high performance parts require the negative voltage. Similarly, voltage doublers (and triplers) are useful in low current applications, where a voltage greater than a primary supply voltage is required. Both regulated and unregulated charge pump voltage inverters and doublers are available depending upon the particular application.

Regulated Output Charge-Pump Voltage Converters

Adding regulation to a simple charge-pump voltage converter function greatly enhances its usefulness for most applications. There are several techniques for adding regulation to a charge-pump converter. The most straightforward is to follow the charge-pump inverter/doubler with an LDO regulator. The LDO provides the regulated output, and can also reduce the charge-pump converter's ripple. This approach, however, adds complexity and reduces the available output voltage by the dropout voltage of the LDO (~200 mV). These factors may or may not be a disadvantage.

By far the simplest and most effective method for achieving regulation in a charge-pump voltage converter is to simply use a charge-pump design with an internal error amplifier, to control the on-resistance of one of the switches.

This method is used in the ADP3605 voltage inverter and the ADP3607 voltage doubler, devices offering regulated outputs for positive input voltage ranges. The output is sensed and fed back into the device via a sensing pin, V_{SENSE}. Key features of the series are good output regulation, 5% in the ADP3605, and a high switching frequency of 250 kHz, good for both high efficiency and small component size.

An simplified functional diagram of the ADP3607 voltage doubler IC from this series is shown in Figure 9.60. The application circuit shown in Figure 9.61 is a 3 V to 5 V doubler, with the output regulated ±5% for currents up to 50 mA. In normal operation, the SHUTDOWN pin is connected to ground. Alternately, a logic HIGH at this pin shuts the device down to a standby current of 150 µA.

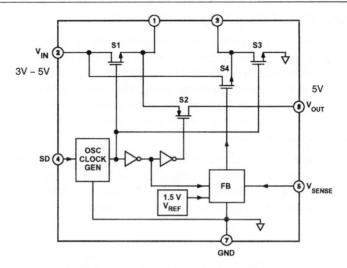

Figure 9.60: ADP3607 Regulated 5 V,
50 mA Output Charge Pump Doubler

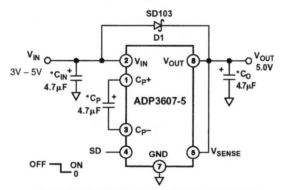

*FOR BEST PERFORMANCE, 10μF IS RECOMMENDED
C_P : SPRAGUE, 293D475X0010B2W
C_{IN}, C_O: TOKIN, 1E475ZY5UC205F

Figure 9.61: ADP3607-5 Charge
Pump Application Circuit

The capacitors for C_{IN}, C_P, and C_O should have ESRs of less than 150 mΩ and should be 10 μF. However, 4.7 μF can be used at the expense of slightly higher output ripple voltage. C_P is the most critical of the three, because of its higher current flow. The tantalum type listed is recommended for lowest output ripple.

With values as shown, typical output ripple voltage ranges up to approximately 30 mV as the output current varies over the 50 mA range.

These application examples illustrate a subset of the entire charge-pump IC family. Other charge pumps are available, including regulated charge pumps specifically for TFT displays with three output voltages (+5 V, +15 V, and –15 V) are available. For further information on these and other devices consult www.analog.com.

Linear Post Regulator for Switching Supplies

Another powerful noise reduction option which can be utilized in conjunction with a switching type supply is the option of a *linear post regulator* stage. This is at best an LDO type of regulator, chosen for the desired clean analog voltage level and current. It is preceded by a switching stage, which might be a buck or boost type inductor-based design, or it may also be a charge pump. The switching converter allows the overall design to be more power-efficient, and the linear post regulator provides clean regulation at the load, reducing the noise of the switcher. This type of regulator can also be termed *hybrid regulation*, since it combines both switching and linear regulation concepts.

An example circuit is shown in Figure 9.62, which features a 3.3 V/1 A low noise, analog-compatible regulator. It operates from a nominal 9 V supply, using a buck or step-down type of switching regulator, as the first stage at the left. The switcher output is set for a few hundred mV above the desired final voltage output, minimizing power in the LDO stage at the right. This feature may eliminate need for a heat sink on the LDO pass device.

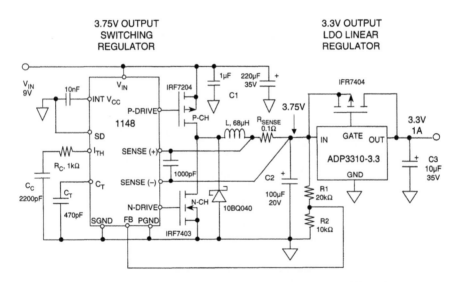

Figure 9.62: A Linear Post-Regulator Operating after a Switching/Linear Regulator is Capable of Low Noise, as Well as Good DC Efficiency

In this example the 1148 IC switcher is set up for a 3.75 V output by R1-R2, but in principle, this voltage can be anything suitable to match the headroom of the companion LDO (within specification limits, of course). In addition, the principle extends to any LDO devices and other current levels, and other switching regulators. The ADP3310-3.3 is a fixed-voltage LDO controller, driving a PMOS FET pass device, with a 3.3 V output.

The linear post regulation stage provides both noise-reduction (in this case about 14 dB), as well as good dc regulation. To realize best results, good grounding practices must be followed. In tests, noise at the 3.3 V output was about 5 mV p-p at the 150 kHz switcher frequency. Note that the LDO noise rejection for such relatively high frequencies is much less than at 100 Hz/120 Hz. Note also that C2's ESR will indirectly control the final noise output. The ripple figures given are for a general-purpose C2 part, and can be improved.

Grounding Linear and Switching Regulators

The importance of maintaining a low impedance large area ground plane is critical to practically all analog circuits today, especially high current low dropout linear regulators or switching regulators. The ground plane not only acts as a low impedance return path for high frequency switching currents but also minimizes EMI/RFI emissions. In addition, it serves to minimize unwanted voltage drops due to high load currents. Because of the shielding action of the ground plane, the circuit's susceptibility to external EMI/RFI is also reduced. When using multilayer PC boards, it is wise to add a power plane. In this way, low impedances can be maintained on both critical layers.

Figure 9.63 shows a grounding arrangement for a low dropout linear regulator such as the ADP3310. It is important to minimize the total voltage drop between the input voltage and the load, as this drop will subtract from the voltage dropped across the pass transistor and reduce its headroom. For this reason, these runs should be wide heavy traces, and are indicated by the wide interconnection lines on the diagram. The low current ground (GND) and V_{OUT} (sense) pins of the ADP3310 are connected directly to the load so that the regulator regulates the voltage at the load rather than at its own output. The IS and V_{IN} connections to the R_S current sense resistor should be made directly to the resistor terminals to minimize parasitic resistance, since the current limit resistor is typically a very low value (milliohms). In fact, for very low values it may actually consist of a PC board trace of the proper width, length, and thickness to yield the desired resistance.

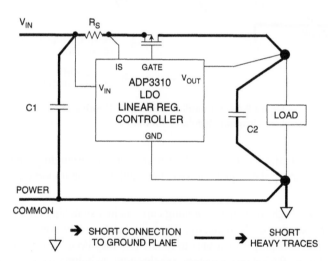

Figure 9.63: Grounding and Signal Routing Techniques for Low Dropout Regulators Method 1

The input decoupling capacitor (C1) should be connected with short leads at the regulator input in order to absorb any transients that may couple onto the input voltage line. Similarly, the load capacitor (C2) should have minimum lead length in order to absorb transients at that point and prevent them from coupling back into the regulator. The single-point connection to the low impedance ground plane is made directly at the load.

Figure 9.64 shows a grounding arrangement that is similar to that of Figure 9.63 with the exception that all ground connections are made with direct connections to the ground plane. This method works extremely well when the regulator and the load are on the same PC board, and the load is distributed around the board rather than located at one specific point. If the load is not distributed, the connection from V_{OUT} (sense) should be connected directly to the load as shown by the dotted line in the diagram. This ensures the regulator provides the proper voltage at the load regardless of the drop in the trace connecting the pass transistor output to the load.

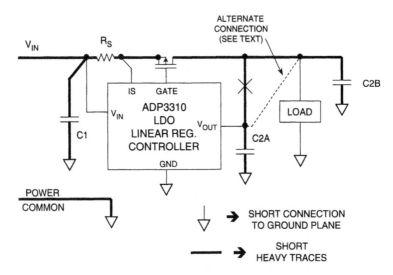

Figure 9.64: Grounding and Signal Routing for Low Dropout Regulator Method 2

Switching regulators present major challenges with respect to layout, grounding, and filtering. The discussion above on linear regulators applies equally to switchers, although the importance of dc voltage drops may not be as great.

There is no way to eliminate high frequency switching currents in a switching regulator, since they are necessary for the proper operation of the regulator. What one must do, however, is to recognize the high switching current paths and take proper measures to ensure that they do not corrupt circuits on other parts of the board or system. Figure 9.65 shows a generic synchronous switching regulator controller IC and the associated external MOSFET switching transistors. The heavy bold lines indicate the paths where there are large switching currents and/or high dc currents. Notice that all these paths are connected together at a single-point ground which in turn connects to a large area ground plane.

In order to minimize stray inductance and resistance, each of the high current paths should be as short as possible. Capacitors C1 and C2A must absorb the bulk of the input and output switching current and shunt it to the single-point ground. Any additional resistance or inductance in series with these capacitors will degrade their effectiveness. Minimizing the area of all the loops containing the switching currents prevents them from significantly affecting other parts of the circuit. In actual practice, however, the single-point concept in Figure 9.65 is difficult to implement without adding additional lead length in series with the various components. The added lead length required to implement the single-point grounding scheme tends to degrade the effects of using the single-point ground in the first place.

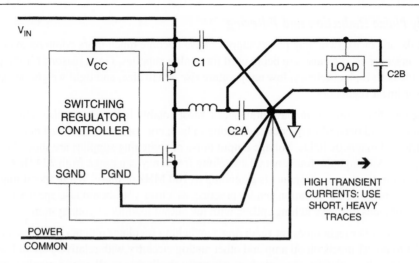

Figure 9.65: Grounding and Signal Routing Techniques for Switching Regulators Method 1

A more practical solution is to make multiple connections to the ground plane and make each of them as short as possible. This leads to the arrangement shown in Figure 9.66, where each critical ground connection is made directly to the ground plane with the shortest connection length possible. By physically locating all critical components associated with the regulator close together and making the ground connections short, stray series inductance and resistance are minimized. It is true that several small ground loops may occur using this approach, but they should not cause significant system problems because they are confined to a very small area of the overall large-area ground plane.

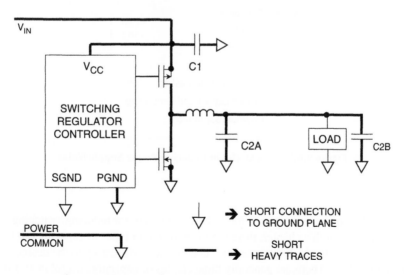

Figure 9.66: Grounding and Signal Routing Techniques for Switching Regulators Method 2

Power Supply Noise Reduction and Filtering

During the last decade or so, switching power supplies have become much more common in electronic systems. As a consequence, they also are being used for analog supplies. Good reasons for the general popularity include their high efficiency, low temperature rise, small size, and light weight, as opposed to equivalent-power linear regulators.

In spite of these benefits, switchers *do* have drawbacks, most notably high output noise. This noise generally extends over a broad band of frequencies, resulting in both conducted and radiated noise, as well as unwanted electric and magnetic fields. Voltage output noise of switching supplies are short-duration voltage transients, or spikes. Although the fundamental switching frequency can range from 20 kHz to 1 MHz, the spikes can contain frequency components extending to 100 MHz or more. While specifying switching supplies in terms of rms noise is common vendor practice, as a user you should also specify the *peak* (or peak-to-peak) amplitudes of the switching spikes, with the output loading of your system.

This section discusses filter techniques for rendering a switching regulator output *analog ready*, that is sufficiently quiet to power precision op amp and other analog circuitry with relatively small loss of dc terminal voltage. The filter solutions presented are generally applicable to all power supply types incorporating switching element(s) in their energy path. This includes charge-pump as well as other switching type converters and supplies.

This section focuses on reducing *conducted type* switching power supply noise with external post filters, as opposed to radiated type noise.

Tools useful for combating high frequency switcher noise are shown by Figure 9.67. These differ in electrical characteristics as well as practicality towards noise reduction, and are listed roughly in an order of priorities. Of these tools, L and C are the most powerful filter elements, and are the most cost-effective, as well as small in size.

> - Capacitors
> - Inductors
> - Ferrites
> - Resistors
> - Linear Post Regulation
> - Proper Layout and Grounding
> - Physical Separation

Figure 9.67: Tools Useful in Reducing Power Supply Noise

Capacitors

Capacitors are probably the single most important filter component for reducing switching-related noise. As noted in the first section of this chapter, there are many different types of capacitors. It is also quite true that understanding of their individual characteristics is absolutely mandatory to the design of effective and practical power supply filters. There are generally three classes of capacitors useful in 10 kHz to 100 MHz filters, broadly distinguished as the generic dielectric types; *electrolytic, film,* and *ceramic*. These discussions complement earlier ones, focusing on power-related concepts.

With any dielectric, a major potential filter loss element is ESR (equivalent series resistance), the net parasitic resistance of the capacitor. ESR provides an ultimate limit to filter performance, and requires more than casual consideration, because it can vary both with frequency and temperature in some types. Another capacitor loss element is ESL (equivalent series inductance). ESL determines the frequency where the net impedance characteristic switches from capacitive to inductive. This varies from as low as 10 kHz in some electrolytics to as high as 100 MHz or more in chip ceramic types. Both ESR and ESL are reduced when a leadless package is used. All capacitor types mentioned are available in surface mount packages, preferable for high speed uses.

The *electrolytic* family provides an excellent, cost-effective low frequency filter component, because of the wide range of values, a high capacitance-to-volume ratio, and a broad range of working voltages. It includes *general-purpose aluminum electrolytic* types, available in working voltages from below 10 V up to about 500 V, and in size from 1 µF to several thousand µF (with proportional case sizes). All electrolytic capacitors are polarized, and cannot withstand more than a volt or so of reverse bias without damage.

A subset of the general electrolytic family includes *tantalum* types, generally limited to voltages of 100 V or less, with capacitance of 500 µF or less (see Reference 7). In a given size, tantalums exhibit a higher capacitance-to-volume ratios than do general-purpose electrolytics, and have both a higher frequency range and lower ESR. They are generally more expensive than standard electrolytics, and must be carefully applied with respect to surge and ripple currents.

A subset of aluminum electrolytic capacitors is the *switching* type, designed for handling high pulse currents at frequencies up to several hundred kHz with low losses (see Reference 8). This capacitor type can compete with tantalums in high frequency filtering applications, with the advantage of a broader range of values.

A more specialized high performance aluminum electrolytic capacitor type uses an organic semiconductor electrolyte (see Reference 9). The *OS-CON* capacitors feature appreciably lower ESR and higher frequency range than do other electrolytic types, with an additional feature of minimal low temperature ESR degradation.

Film capacitors are available in very broad value ranges and an array of dielectrics, including polyester, polycarbonate, polypropylene, and polystyrene. Because of the low dielectric constant of these films, their volumetric efficiency is quite low, and a 10 µF/50 V polyester capacitor (for example) is actually a handful. Metalized (as opposed to foil) electrodes do help to reduce size, but even the highest dielectric constant units among film types (polyester, polycarbonate) are still larger than any electrolytic, even using the thinnest films with the lowest voltage ratings (50 V). Where film types excel is in their low dielectric losses, a factor that may not necessarily be a practical advantage for filtering switchers. For example, ESR in film capacitors can be as low as 10 mΩ or less, and the behavior of films generally is very high in terms of Q. In fact, this can cause problems of spurious resonance in filters, requiring damping components.

As typically constructed using wound layers, film capacitors can be inductive, which limits their effectiveness for high frequency filtering. Obviously, only noninductively made film caps are useful for switching regulator filters. One specific style which is noninductive is the *stacked-film* type, where the capacitor plates are cut as small overlapping linear sheet sections from a much larger wound drum of dielectric/plate material. This technique offers the low inductance attractiveness of a plate sheet style capacitor with conventional leads (see References 8 and 10). Obviously, minimal lead length should be used for best high frequency effectiveness. Very high current polycarbonate film types are also available, specifically designed for switching power supplies, with a variety of low inductance terminations to minimize ESL (see Reference 11). Dependent upon their electrical and physical size, film capacitors can be useful at frequencies to above 10 MHz. At the highest frequencies, only stacked film types should be considered. Leadless surface-mount packages are now available for film types, minimizing inductance.

Ceramic is often the capacitor material of choice above a few MHz, due to its compact size, low loss, and availability up to several μF in the high-K dielectric formulations (X7R and Z5U), at voltage ratings up to 200 V (see ceramic families of Reference 7).

Multilayer ceramic "chip caps" are very popular for bypassing and/or filtering at 10 MHz or more, simply because their very low inductance design allows near optimum RF bypassing. For smaller values, ceramic chip caps have an operating frequency range to 1 GHz. For high frequency applications, a useful selection can be ensured by selecting a value that has a self-resonant frequency *above* the highest frequency of interest.

The capacitor model and waveforms of Figure 9.68 illustrate how the various parasitic model elements become dominant, dependent upon the operating frequency. Assume an input current pulse changing from 0 A to 1 A in 100 ns, as noted in the figure, and consider what voltage will be developed across the capacitor.

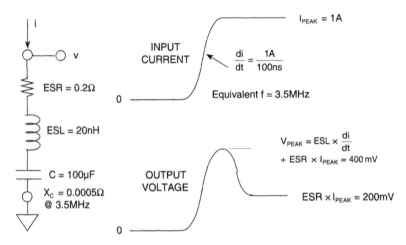

Figure 9.68: Capacitor Equivalent Circuit and Response to Input Current Pulse

The fast-rising edge of the current waveform shown results in an initial voltage peak across the capacitor, which is proportional to the ESL. After the initial transient, the voltage settles down to a longer duration level which is proportional to the ESR of the capacitor. Thus the ESL determines how effective a filter the capacitor is for the fastest components of the current signal, and the ESR is important for longer time frame components. Note that an overall time frame of a few microseconds (or even less) is relevant here. As things turn out, this means switching frequencies in the 100 kHz to 1 MHz range. Unfortunately however, this happens to be the region where most electrolytic types begin to perform poorly.

All electrolytics will display impedance curves similar in general shape to that of Figure 9.69. In a practical capacitor, at frequencies below about 10 kHz the net impedance seen at the terminals is almost purely capacitive (C region). At intermediate frequencies, the net impedance is determined by ESR, for example about 0.1 Ω to 0.5 Ω at ~125 kHz, for several types (ESR region). Above about several hundred kHz to 1 MHz these capacitor types become inductive, with net impedance rising (ESL region).

The minimum impedance within the 10 kHz to 1 MHz range will vary with the magnitude of the capacitor's ESR. This is the primary reason why ESR is the most critical item in determining a given capacitor's effectiveness as a switching supply filter element. Higher up in frequency, the inductive region will vary

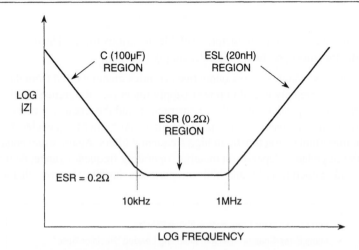

Figure 9.69: Electrolytic Capacitor Impedance versus Frequency

with ESL (which in turn is also strongly effected by package style). It should go without saying that a wide-band impedance plot for a capacitor being considered for a filter application will go a long way towards predicting its potential value, as well as for comparing one type against another.

It should be understood that all real world capacitors have some finite ESR. While it is usually desirable for filter capacitors to possess low ESR, this isn't always so. In some cases, the ESR may actually be helpful in reducing resonance peaks in filters, by supplying "free" damping. For example, in most electrolytic types, a nominally flat broad series resonance region can be noted in an impedance versus frequency plot. This occurs where |Z| falls to a minimum level, nominally equal to the capacitor's ESR at that frequency. This low Q resonance can generally be noted to cover a relatively wide frequency range of several octaves. Contrasted to the high Q sharp resonances of film and ceramic caps, electrolytic's low Q behavior can be useful in controlling resonant peaks.

Ferrites

A second important filter element is the inductor, available in various forms. The use of *ferrite* core materials is prevalent in inductors most practical for power supply filtering.

Regarding inductors, ferrites, which are nonconductive ceramics manufactured from the oxides of nickel, zinc, manganese, etc., are extremely useful in power supply filters (see Reference 12). Ferrites can act as either inductors or resistors, dependent upon their construction and the frequency range. At low frequencies (<100 kHz), inductive ferrites are useful in low-pass LC filters. At higher frequencies, ferrites become resistive, which can be an important characteristic in high frequency filters. Again, exact behavior is a function of the specifics. Ferrite impedance depends on material, operating frequency range, dc bias current, number of turns, size, shape, and temperature. Figure 9.70 summarizes a number of ferrite characteristics.

- Ferrites Good for Frequencies Above 25kHz
- Many Sizes/Shapes Available Including Leaded "Resistor Style"
- Ferrite Impedance at High Frequencies Primarily Resistive—Ideal for HF Filtering
- Low DC Loss: Resistance of Wire Passing Through Ferrite is Very Low
- High Saturation Current Versions Available
- Choice Depends Upon:
  - Source and Frequency of Interference
  - Impedance Required at Interference Frequency
  - Environmental: Temperature, AC and DC Field Strength, Size and Space Available
- Always Test the Design

Figure 9.70: A Summary of Ferrite Characteristics

Several ferrite manufacturers offer a wide selection of ferrite materials from which to choose, as well as a variety of packaging styles for the finished network (see References 13 and 14). A simple form is the *bead* of ferrite material, a cylinder of the ferrite which is simply slipped over the power supply lead to the decoupled stage. Alternately, the *leaded ferrite bead* is the same bead, premounted on a length of wire and used as a component (see Reference 14). More complex beads offer multiple holes through the cylinder for increased decoupling, plus other variations. Surface-mount beads are also available. PSpice models of Fair-Rite ferrites are available, allowing ferrite impedance estimations (see Reference 15). The models match measured rather than theoretical impedances.

A ferrite's impedance is dependent upon a number of interdependent variables, and is difficult to quantify analytically, thus selecting the proper ferrite is not straightforward. However, knowing the following system characteristics will make selection easier. First, determine the frequency range of the noise to be filtered. Second, the expected temperature range of the filter should be known, as ferrite impedance varies with temperature. Third, the dc current flowing through the ferrite must be known, to ensure that the ferrite does not saturate. Although models and other analytical tools may prove useful, the general guidelines given above, coupled with actual filter experimentation connected under system load conditions, should lead to a proper ferrite selection.

Card Entry Filter

Using proper component selection, low and high frequency band filters can be designed to smooth a noisy switching supply output so as to produce an *analog ready* supply. It is most practical to do this over two (and sometimes more) stages, each stage optimized for a range of frequencies.

A basic stage can be used to carry the entire load current, and filter noise by 60 dB or more up to a 1 MHz to 10 MHz range. Figure 9.71 illustrates this type of filter, which is used as a *card entry filter,* providing broadband filtering for all power entering a PC card.

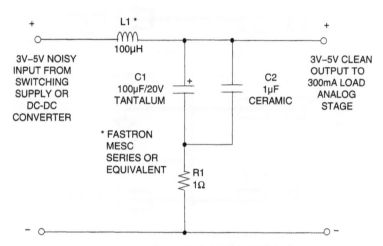

Figure 9.71: A Card-Entry Filter is Useful for Low-Medium Frequency Power Line Noise Filtering in Analog Systems

In this filter, L1 and C1 perform the primary filtering, which provides a corner frequency of about 1.6 kHz. With the corner thus placed well below typical switching frequencies, the circuit can have good attenuation up to 1 MHz, where the typical attenuation is on the order of 60 dB. At higher frequencies parasitics limit performance, and a second filter stage will be more useful.

The ultimate level of performance available from this filter will be related to the components used within it. L1 should be derated for the operating current, thus for 300 mA loads it is a 1 A type. The specified L1 choke has a typical dc resistance of 0.65 Ω, for low drop across the filter (see Reference 16). C1 can be either a tantalum or an aluminum electrolytic, with moderately low ESR. For current levels lower than 300 mA, L1 can be proportionally downsized, saving space. The resistor R1 provides damping for the LC filter, to prevent possible ringing. R1 can be reduced or even possibly eliminated, if the ESR of C1 provides a comparable impedance.

While the example shown is a single-supply configuration, obviously the same filter concepts apply for dual supplies.

Rail Bypass/Distribution Filter

A complement to the card-entry filter is the rail-bypass filter scheme of Figure 9.72. When operating from relatively clean power supplies, the heavy noise filtering of the card entry filter may not be necessary. However, some sort of low frequency bypassing with appreciable energy storage is almost always good, and this is especially true if high currents are being delivered by the stages under power.

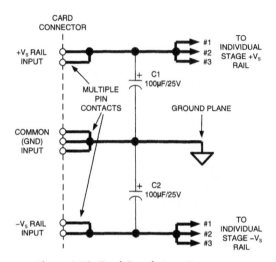

Figure 9.72: Dual-Supply Low Frequency
Rail Bypass/Distribution Filter

In such cases, some lumped low frequency bypassing is appropriate on the card. Although these energy storage filters need not be immediately adjacent to the ICs they serve, they should be within a few inches. This type of bypassing scheme should be considered a minimum for powering any analog circuit. The exact capacitor values aren't critical, and can vary appreciably. The most important thing is to avoid leaving them out.

The circuit shown uses C1 and C2 as these bypasses in a dual-rail system. Note that multiple card contacts are recommended for the I/O pins, especially ground connection. From the capacitors outward, supply rail traces are distributed to each stage as shown, in "star" distribution fashion. Note—while this is the optimum method to minimize inter-stage crosstalk, in practice some degree of "daisy chaining" is often difficult to avoid. A prudent designer should therefore carefully consider common supply currents effects in designing these PCB distribution paths.

Wider than normal traces are recommended for these supply rails, especially those carrying appreciable current. If the current levels are in the ampere region, star-type supply distribution with ultrawide traces should be considered mandatory. In extreme cases, a dedicated power plane can be used. The impedance of the ground return path is minimized by the use of a ground plane.

Local High Frequency Bypass/Decoupling

At each individual analog stage, further local, high-frequency-only filtering is used. With this technique, used in conjunction with either the card-entry filter or the low frequency bypassing network, such smaller and simpler local filter stages provide optimum high frequency decoupling. *These stages are provided directly at the power pins, of* all *individual analog stages.*

Figure 9.73 shows this technique, in both correct (left) as well as incorrect (right) example implementations. In the left example, a typical 0.1 µF chip ceramic capacitor goes directly to the opposite PCB side ground plane, by virtue of the via, and on to the IC's GND pin by a second via. In contrast, the less desirable setup at the right adds additional PCB trace inductance in the ground path of the decoupling cap, reducing effectiveness.

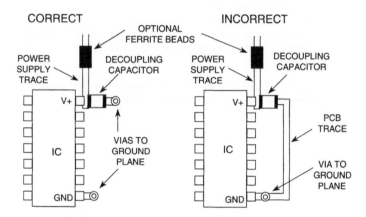

Figure 9.73: Localized High Frequency Supply Filter(s) Provides Optimum Filtering and Decoupling Via Short Low-Inductance Path (Ground Plane)

The general technique is shown here as suitable for a single-rail power supply, but the concept obviously extends to dual rail systems. Note—if the decoupled IC in question is an op amp, the GND pin shown is the $-V_S$ pin. For dual supply op amp uses, there is no op amp GND pin per se, so the dual decoupling networks should go directly to the ground plane when used, or other local ground.

All high frequency (i.e., ≥10 MHz) ICs should use a bypassing scheme similar to Figure 9.73 for best performance. Trying to operate op amps and other high performance ICs without local bypassing is almost always folly. It *may* be possible in a few circumstances, *if* the circuitry is strictly micropower in nature, and the gain-bandwidth in the kHz range. To put things into an overall perspective however, note that a pair of 0.1 µF ceramic bypass caps cost less than 25 cents. Hardly a worthy saving compared to the potential grief and lost time of troubleshooting a system without bypassing.

In contrast, the ferrite beads aren't 100% necessary, but they will add extra high frequency noise isolation and decoupling, which is often desirable. Possible caveats here would be to verify that the beads never saturate, when the op amps are handling high currents.

Note that with some ferrites, even before full saturation occurs, some beads can be nonlinear, so if a power stage is required to operate with a low distortion output, this should also be lab checked.

Figure 9.74 summarizes the previous points of this section regarding power supply conditioning techniques for high performance analog circuitry.

- Use Proper Layout and Grounding Techniques
- At HF Local Decoupling at IC Power Pins is Mandatory
- At HF Ground Planes are Mandatory
- External LC Filters Very Effective in Reducing Ripple
- Low ESR/ESL Capacitors Give Best Results
- Parallel Caps Lower ESR/ESL and Increase C
- Linear Post Regulation Effective for Noise Reduction and Best Regulation
- Completely Analytical Approach Difficult
- *Prototyping Required for Optimum Results*
- Once Design is Final, Don't Switch Vendors or Substitute Parts Without First Verifying Performance Within the Circuit

Figure 9.74: A Summary of Power Supply Conditioning Techniques for High Performance Analog Circuitry

References:
9.3 Analog Power Supply Systems

1. Walt Jung, "References and Low Dropout Linear Regulators," Section 2 within Walt Kester, Ed., **Practical Design Techniques for Power and Thermal Management**, Analog Devices, Inc., 1998, ISBN 0-916550-19-2.

2. Paul Brokaw, "A Simple Three-Terminal IC Bandgap Voltage Reference," **IEEE Journal of Solid State Circuits**, Vol. SC-9, December, 1974.

3. Frank Goodenough, "Vertical-PNP-Based Monolithic LDO Regulator Sports Advanced Features," **Electronic Design**, May 13, 1996.

4. Frank Goodenough, "Low Dropout Regulators Get Application Specific," **Electronic Design**, May 13, 1996.

5. Walt Kester, Brian Erisman, Gurgit Thandi, "Switched Capacitor Voltage Converters," Section 4 within Walt Kester, Editor, **Practical Design Techniques for Power and Thermal Management**, Analog Devices, Inc., 1998, ISBN 0-916550-19-2.

6. Walt Jung, Walt Kester, Bill Chesnut, "Power Supply Noise Reduction and Filtering," portion of Section 8 within Walt Kester, Editor, **Practical Design Techniques for Power and Thermal Management**, Analog Devices, Inc., 1998, ISBN 0-916550-19-2.

7. **Tantalum Electrolytic and Ceramic Capacitor Families**, Kemet Electronics, Box 5928, Greenville, SC, 29606, 803-963-6300.

8. Type HFQ Aluminum Electrolytic Capacitor and Type V Stacked Polyester Film Capacitor, Panasonic, 2 Panasonic Way, Secaucus, NJ, 07094, 201-348-7000.

9. **OS-CON Aluminum Electrolytic Capacitor Technical Book**, Sanyo, 3333 Sanyo Road, Forrest City, AK, 72335, 501-633-6634.

10. Ian Clelland, "Metallized Polyester Film Capacitor Fills High Frequency Switcher Needs," **PCIM**, June 1992.

11. Type 5MC Metallized Polycarbonate Capacitor, Electronic Concepts, Inc., Box 1278, Eatontown, NJ, 07724, 908-542-7880.

12. Henry W. Ott, **Noise Reduction Techniques in Electronic Systems, 2nd Edition**, John Wiley, Inc., 1988, ISBN: 0-471-85068-3.

13. **Fair-Rite Linear Ferrites Catalog**, Fair-Rite Products, Box J, Wallkill, NY, 12886, 914-895-2055.

14. Type EXCEL leaded ferrite bead EMI filter, and Type EXC L leadless ferrite bead, Panasonic, 2 Panasonic Way, Secaucus, NJ, 07094, 201-348-7000.

15. Steve Hageman, "Use Ferrite Bead Models to Analyze EMI Suppression," **The Design Center Source,** MicroSim Newsletter, January, 1995.

16. "MESC series RFI suppression chokes," FASTRON GmbH, Zum Kaiserblick 25, 83620 Feldkirchen-Westerham, Germany, www.fastron.de.

Overvoltage Protection
Walt Jung, Walt Kester, James Bryant, Joe Buxton, Wes Freeman

Data converters, op amps, and other analog ICs frequently require protection against destructive potentials at their input and output terminals. One basic reason behind this is that these ICs are by nature relatively fragile components. Although designed to be as robust as possible *for normal signals*, there are nevertheless certain application and/or handling conditions where they can see voltage transients beyond their ratings. This situation can occur for either of two instances. The first of these is *in-circuit*, that is, operating within an application circuit. The second instance is *out-of-circuit*, which might be at anytime after receipt from a supplier, but prior to final assembly and mounting of the IC. In either case, under overvoltage conditions, it is a basic fact-of-life that unless the designer limits the fault currents at the input (or possibly output) of the IC, it can be damaged or destroyed.

Obviously the designer should fully understand all of the fault mechanisms internal to those ICs that may require protection. This then allows design of networks that can protect the in-circuit IC throughout its lifetime, without undue compromise of speed, precision, etc. Or, for the out-of-circuit IC, it can help define proper protective handling procedures until it reaches its final destination. This section of the chapter examines a variety of protection schemes to ensure adequate protection for op amps and other analog ICs for in-circuit applications, as well as for out-of-circuit environments.

In-Circuit Overvoltage Protection

There are many common cases that stress op amps and other analog ICs at the input, while operating within an application, i.e., in-circuit. Since these ICs must often interface to the outside world, this may entail handling voltages exceeding their absolute maximum ratings. For example, sensors are often placed in environments where a fault condition can expose the circuit to a dangerously high voltage. With the sensor connected to a signal processing amplifier, the input then sees excessive voltages during a fault.

General Input Common Mode Limitations

Whenever the op amp or data converter input common-mode (CM) voltage goes outside its supply range, the device can be damaged, even if the supplies are turned off. Accordingly, the absolute maximum input ratings of almost all linear ICs limits the greatest applied voltage to a level equal to the positive and negative supply voltage, plus about 0.3 V beyond these voltages (i.e., $+V_S + 0.3$ V, or $-V_S - 0.3$ V). While some exceptions to this general rule might exist it is important to note this: *Most linear ICs require input protection when overvoltage of more than 0.3 V beyond the rails occurs.*

A safe operating rule is to always keep the applied CM voltage between the rail limits. Here, "safe" implies prevention of outright IC destruction. As will be seen later, there are also intermediate "danger-zone" CM conditions between the rails with certain devices, which can invoke dangerous (but not necessarily destructive) behavior.

Speaking generally, it is important to note that almost *any* op amp or data converter input will break down, given sufficient overvoltage to the positive or negative rail. Under breakdown conditions high and uncontrolled current can flow, so the danger is obvious. The exact breakdown voltage is entirely dependent on the individual op amp input stage. It may be a 0.6 V diode drop, or a process-related breakdown of 50 V or more. In many cases, overvoltage stress can result in currents over 100 mA, which destroys a part almost instantly.

Therefore, unless otherwise stated on the data sheet, input fault current should be limited to ≤5 mA to avoid damage. This is a conservative guideline, based on metal trace widths in a typical op amp or data converter input. Higher levels of current can cause *metal migration*, a cumulative effect, which, if sustained, eventually leads to an open trace. Should a migration situation be present, failure may only appear after a long time due to multiple overvoltages, a very difficult failure to identify. So, even though an amplifier may appear to withstand overvoltage currents well above 5 mA for a short time period, it is important to limit the current to 5 mA (or preferably less) for long term reliability.

Figure 9.75 illustrates an external, general-purpose op amp CM protection circuit. The basis of this scheme is the use of Schottky diodes D1 and D2, plus an external current limiting resistor, R_{LIMIT}. With appropriate selection of these parts, input protection for a great many op amps can be ensured. Note that an op amp may also have *internal* protection diodes to the supplies (as shown) which conduct at about 0.6 V forward drop above or below the respective rails. In this case however, the external Schottky diodes effectively parallel any internal diodes, so the internal units never reach their threshold. Diverting fault currents externally eliminates potential stress, protecting the op amp.

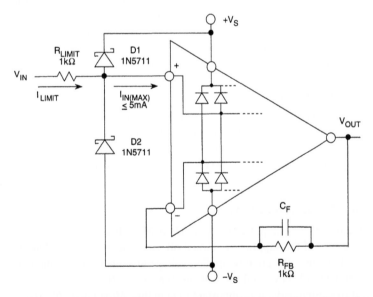

Figure 9.75: A General-Purpose Op Amp CM Overvoltage Protection Network Using Schottky Clamp Diodes with Current Limit Resistance

The external diodes also allow other degrees of freedom, some not so obvious. For example, if fault current is allowed to flow in the op amp, R_{LIMIT} must then be chosen so that the maximum current is no more than 5 mA for the worst case V_{IN}. This criterion can result in rather large R_{LIMIT} values, and the associated increase in noise and offset voltage may not be acceptable. For instance, to protect against a V_{IN} of 100 V with the

5 mA criterion, R_{LIMIT} must be ≥20 kΩ. However with external Schottky clamping diodes, this allows R_{LIMIT} to be governed by the maximum allowable D1-D2 current, which can be larger than 5 mA. However, care must be used here, for at very high currents the Schottky diode drop may exceed 0.6 V, possibly activating internal op amp diodes.

It is very useful to keep the R_{LIMIT} value as low as possible, to minimize offset and noise errors. R_{LIMIT}, in series with the op amp input, produces a bias-current-proportional voltage drop. Left uncorrected, this voltage appears as an increase in the circuit's offset voltage. Thus for op amps where the bias currents are moderate and approximately equal (most bipolar types) compensation resistor R_{FB} balances the dc effect, and minimizes this error. For low bias current op amps (I_b ≤10 nA, or FET types) it is likely R_{FB} won't be necessary. To minimize noise associated with R_{FB}, bypass it with a capacitor, C_P.

Clamping Diode Leakage

For obvious reasons, it is critical that diodes used for protective clamping at an op amp input have a leakage sufficiently low to not interfere with the bias level of the application.

Figure 9.76 illustrates how some well-known diodes differ in terms of leakage current, as a function of the reverse bias voltage, Vbias.

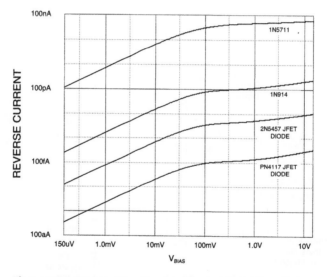

Figure 9.76: Reverse Bias Current Characteristics for Diodes Useful in Protective Clamping Networks (PSpice Simulation)

In this chart, a 25°C simulation using PSpice diode models, it is easy to see that not only is the diode type critical, so is the reverse bias. The 1N5711 Schottky type for example, has a leakage of nearly 100 nA at a reverse bias of 15 V, as it would typically be used with a ±15 V powered op amp. With this level of leakage, such diodes will only be useful with op amps with bias currents of several µA. For protection of appreciably lower bias current op amps (particularly most FET input devices) much lower leakage is necessary.

As the data of Figure 9.76 shows, not only does selecting a better diode help control leakage current, but operating it at a low bias voltage condition substantially reduces leakage. For example, while an ordinary 1N914 or 1N4148 diode may have 200 pA of leakage at 15 V, this is reduced to slightly more than 1 pA with bias controlled to 1 mV. But there is a caveat here. When used in a high impedance clamp circuit, glass diodes such as the 1N914/1N4148 families should either be shielded from incident light, or use opaque packages. This is necessary to minimize parasitic photocurrent from the surrounding light, which effectively appears as diode leakage current.

Specialty diodes with much lower leakage are also available, such as diode-connected FET devices characterized as protection diodes (see DPAD series of Reference 2). Within the data of Figure 9.76, the 2N5457 is a general purpose JFET, and the 2N4117/PN4117 family consists of parts designed for low current levels. Other low leakage and specialty diodes are described in References 3 and 4.

Finally, whenever protective diodes are used, the effects of their capacitance on circuit performance must be analyzed.

A Flexible Voltage Follower Protection Circuit

Of course, it isn't a simple matter to effectively apply protective clamping to op amp inputs, while reducing diode bias level to a sub-mV level.

The circuit of Figure 9.77 shows low leakage input clamping and other means used with a follower connected FET op amp, with protection at input and output, for both power on or off conditions.

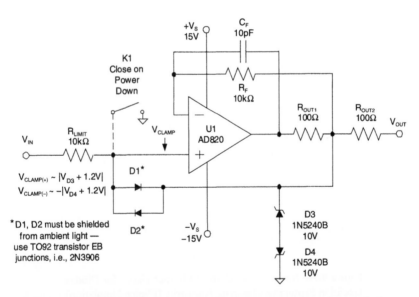

Figure 9.77: Bootstrapping the D1-D2 Protection Network Reduces Diode Leakage to Negligible Levels, and is Voltage-Programmable for Clamp Level

Momentarily disregarding the various diodes, this circuit is an output-current-limited voltage follower. With the addition of diodes D1-D2 and D3-D4, it has both a voltage-limited output, and an overvoltage protected input. Operating below the voltage threshold of output series-connected Zener diodes D3-D4, the circuit behaves as a precision voltage follower. Under normal follower operation, that is at input/output voltages $< |V_Z + 0.6|$ volts (where V_Z is the breakdown voltage of D3 or D4), diodes D1-D2 see only the combined offset and CM voltage errors of U1 as bias voltage. This reduces the D1-D2 leakage to very low levels, consistent with the pA level bias current of a FET input op amp. Note that D1-D2 *must* be prevented from photo-conduction, and one direct means of this is to use opaque package diodes, such as the 2N3906 EB junctions discussed by Pease (see References 3 and 4). If 1N914s are used they must be light shielded. In either case, bootstrapping greatly reduces the effective D1-D2 leakage.

For input/output voltage levels greater than $V_Z + 0.6$ V, Zener diodes D3-D4 breakdown. This action clamps both the V_{OUT} output node and the V_{CLAMP} node via D1-D2. The input of the op amp is clamped to either polarity of the two input levels of V_{CLAMP}, as indicated within the figure. Under clamp conditions, input voltage V_{IN} can rise to levels beyond the supply rails of U1 without harm, with excess current limited by R_{LIMIT}. If sustained high-level (~100 V) inputs will be applied, R_{LIMIT} should be rated as a 1 W–2 W (or fusible) type.

This circuit has very good dc characteristics, due to the fact that the clamping network is bootstrapped. This produces very low input/output errors below the V_{CLAMP} threshold (consistent with the op amp specifications, of course). Note that this bootstrapping has ac benefits as well, as it reduces the D1-D2 capacitance seen by the source. While the ~100 pF capacitance of D3-D4 might cause a loading problem with some op amps, this is mitigated by the isolating effect of R_{OUT1}, plus the feedback compensation of C_F. Both R_{OUT1} and R_{OUT2} protect the op amp output.

The input voltage clamping level is also programmable, and is set by the choice of Zener voltage V_Z. This voltage plus 1.2 V should be greater than the maximum input, but below the rail voltage, as summarized in the figure. The example uses 10 V ± 5% Zener diodes, so input clamping typically will occur at ±11.2 V, allowing ±10 V swings.

An important caveat to the above is that it applies for *power on* conditions. With *power off*, D1–D4 still clamp to the noted levels, but this now produces a condition whereby the U1 input and output voltage can exceed the rails.

Note that this could be dangerous, for a given U1 device. If so, an optional and simple means towards providing a lower, safe clamping level for power off conditions is to use a relay at the V_{CLAMP} node. The contacts are open with power applied, and closed with power absent. With attention paid to an overall PCB layout, this can preserve a pA level bias current of FET op amps used for U1.

Common-Mode Overvoltage Protection Using CMOS Channel Protectors

A much simpler alternative for overvoltage protection is the CMOS *channel protector*. A channel protector is a device in series with the signal path, for example preceding an op amp input. It provides overvoltage protection by dynamically altering its resistance under fault conditions. Functionally, it has the distinct advantage of affording protection for sensitive components from voltage transients, whether the power supplies are present or not. Representative devices are the ADG465/ADG466/ADG467, which are channel protectors with single, triple, and octal channel options. Because this form of protection works whether or not supplies are present, the devices are ideal for use in applications where input overvoltages are common, or where correct power sequencing can't always be guaranteed. One such example is within hot-insertion rack systems.

An application of a channel protector for overvoltage protection of a precision buffer circuit is shown in Figure 9.78. A single channel device, the ADG465 at U2, is used here at the input of the U1 precision op amp buffer, an OP777.

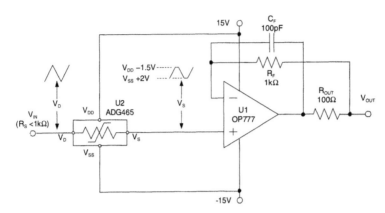

Figure 9.78: Using an ADG465 Channel Protector IC With a Precision Buffer Offers Great Simplicity of Protection and Fail-Safe Operation During Power Off

A channel protector behaves just like a series resistor of 60 Ω to 80 Ω in normal operation (i.e., nonfault conditions). Consisting of a series connection of multiple P and N MOSFETs, the protector dynamically adjusts channel resistance according to the voltage seen at the V_D terminal. Normal conduction occurs with V_D more than a threshold level above or below the rails, i.e., $(V_{SS} + 2 \text{ V}) < V_D < (V_{DD} - 1.5 \text{ V})$. For fault conditions the analog input voltage exceeds this range, causing one of the series MOSFETs to switch off, thus raising the channel resistance to a high level. This clamps the V_S output at one extreme range, either $V_{SS} + 2$ V or $V_{DD} - 1.5$ V, as shown in Figure 9.78.

A major channel protector advantage is the fact that both circuit and signal source protection are provided, in the event of overvoltage or power loss. Although shown here operating from op amp ±15 V supplies, these channel protectors can handle total supplies of up to 40 V. They also can withstand overvoltage inputs from $V_{SS} - 20$ V to $V_{DD} + 20$ V with power on (or ±35 V in the circuit shown). With power off $(V_{DD} = V_{SS} = 0 \text{ V})$, maximum input voltage is ±35 V. Maximum room temperature channel leakage is 1 nA, making them suitable for op amps and in amps with bias currents of several nA and up.

Related to the ADG46x series of channel protectors are several *fault-protected multiplexers*, for example the ADG508F/ADG509F, and the ADG438F/ADG439F families. Both the channel protectors and the fault-protected multiplexers are low power devices, and even under fault conditions, their supply current is limited to submicroampere levels. A further advantage of the fault-protected multiplexer devices is that they retain proper channel isolation, even for input conditions of one channel seeing an overvoltage, that is the remaining channels still function.

CM Overvoltage Protection Using High CM Voltage In Amp

The ultimate simplicity for analog channel overvoltage protection is achieved with resistive input attenuation ahead of a precision op amp. This combination equates to a high voltage capable in amp, such as the AD629, which is able to linearly process differential signals riding upon CM voltages of up to ±270 V. Further, and most important to overvoltage protection considerations, the on-chip resistors afford protection for either common mode or differential voltages of up to ±500 V. All of this is achieved by virtue of a precision laser-trimmed thin-film resistor array and op amp, as shown in Figure 9.79.

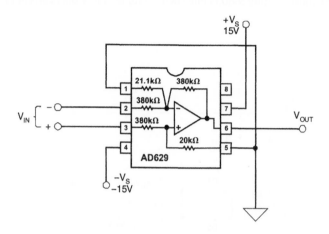

Figure 9.79: The AD629 High Voltage In Amp IC Offers ± 500 V Input Overvoltage Protection, One-Component Simplicity, and Fail-Safe Power Off Operation

Examination of this topology shows that the resistive network around the AD629's precision op amp acts to divide down the applied CM voltage at V_{IN} by a factor of 20/1. The AD629 simultaneously processes the input differential mode signal V_{IN} to a single-ended output referred to a local ground, at a gain factor of unity. Gain errors are no more than ±0.03 or 0.05%, while offset voltage is no more than 0.5 mV or 1 mV (grade-dependent). The AD629 operates over a supply range of ±2.5 V to ±18 V.

These factors combine to make the AD629 a simple, one-component choice for the protection of off-card analog inputs that can potentially see dangerous transient voltages. Due to the relatively high resistor values used, protection of the device is also inherent with no power applied, since the input resistors safely limit fault currents. In addition, it offers those operating advantages inherent to an in amp: high CMR (86 dB minimum at 500 Hz), excellent overall dc precision, and the flexibility of simple polarity changes.

On the flip side of performance issues, several factors make the AD629's output noise and drift relatively high, if compared to a lower gain in amp configuration such as the AMP03. These are the Johnson noise of the high value resistors, and the high noise gain of the topology (21×). These factors raise the op amp noise and drift along with the resistor noise by a factor higher than typical. Of course, whether or not this is an issue relevant to an individual application will require evaluation on a case-by-case basis.

Inverting Mode Op Amp Protection Schemes

There are some special cases of overvoltage protection requirements that don't fit into the more general CM protection schemes above. Figure 9.80 is one such example, a low bias current FET input op amp I/V converter.

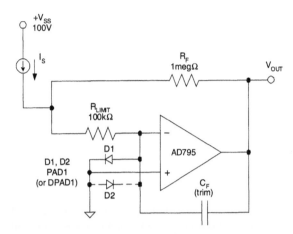

Figure 9.80: A Low Bias Current FET Input Op Amp I/V
Converter with Overvoltage Protection Network R_{LIMIT} and D1

In this circuit the AD795 1 pA bias current op amp is used as a precision inverter. Some current-source-generated signals can originate from a high voltage potential, such as the 100 V V_{SS} level shown. As such, they have the potential of developing fault voltage levels beyond the op amp rails, producing fault current into the op amp well above safe levels. To prevent this, protection resistor R_{LIMIT} is used inside the feedback loop as shown, along with voltage clamp D1 (D2).

For normal signal condition (i.e., $I_S \leq 10$ µA) the op amp's inverting node is very close to ground, with just a tiny voltage drop across R_{LIMIT}. Normal I/V conversion takes place, with gain set by R_F. For protection, D1 is a special low leakage diode, clamping any excess voltage at the (–) node to ~0.6 V, thus protecting the op amp. The value of R_{LIMIT} is chosen to allow a 1 mA max current under fault conditions. Bootstrapping the D1 (and/or D2) clamp diodes as shown minimizes the normal operating voltage across the inverting node, keeping the diode leakage low (see Figure 9.76, again). Note that for a positive source voltage as shown, only positive clamping is needed, so just one diode suffices.

Only the lowest leakage diodes ($\leq$1 pA) such as the PAD1 (or the DPAD1 dual) should be used in this circuit. As previously noted, any clamping diode used here should be shielded from light or use opaque packaging, to minimize photocurrent from ambient light. Even so, the diode(s) will increase the net input current and shunt capacitance, and feedback compensation C_F will likely be necessary to control response peaking. C_F should be a very low leakage type. Also, with the use of very low input bias current devices such as the AD795, it isn't possible to use the same level of internal protection circuitry as with other ADI op amps. This factor makes the AD795 more sensitive to handling, so ESD precautions should be taken.

Amplifier Output Voltage Phase-Reversal

As alluded to above, there are "gray-area" op amp groups that have anomalous CM voltage zones, falling between the supply rails. As such, protection for these devices cannot be guaranteed by simply ensuring that the inputs stay between the rails—they must additionally stay *entirely* within their rated CM range, for consistent behavior.

Peculiar to some op amps, this misbehavior phenomenon is called *output voltage phase-reversal*. It is seen when one or both of the op amp inputs exceeds allowable input CM voltage range. Note that the inputs may still be well within the extremes of rail voltage, but simply below one specified CM limit. Typically, this is towards the negative range. Phase-reversal is most often associated with JFET and/or BiFET amplifiers, but some bipolar single-supply amplifiers are also susceptible to it.

The Figure 9.81 waveforms illustrates this general phenomenon, with an overdriven voltage follower input on the left, and the resulting output phase-reversal at the right.

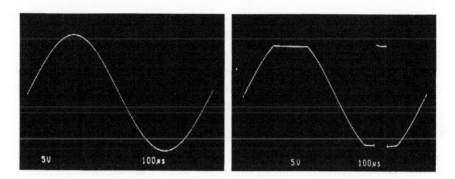

Figure 9.81: An Illustration of Input Overdriving Waveform (Left) and the Resulting Output Phase-Reversal (Right), Using a JFET Input Op Amp

While the specific details of the internal mechanism may vary with individual op amps, it suffices to say that the output phase-reversal occurs when a critical section of the amplifier front end saturates, causing the input-output sign relationship to temporarily reverse. Under this condition, when the CM range is exceeded, the negative-going input waveform in Figure 9.81 (left) does not continue going more negative in the output waveform, Figure 9.81 (right). Instead, the input-output relationship *phase-reverses*, with the output suddenly going positive, i.e., the spike. It is important to note that this is *not* a latching form of phase-reversal, as the output will once again continue to properly track the input, when the input returns to the CM range. In Figure 9.81, this can be seen in the continuance of the output sine wave, after the positive-going phase-reversal spike settles.

In most applications, this output voltage phase-reversal does no harm to the op amp, nor to the circuit where it is used. Indeed, since it is triggered when the CM limit is exceeded, noninverting stages with appreciable signal gain never see it, since their applied CM voltage is too small.

Note that with inverting applications the output phase-reversal problem is nonexistent, as the CM range isn't exercised. So, although a number of (mostly older) op amps suffer from phase-reversal, it still is rarely a serious problem in system design.

Nevertheless, when and if a phase-reversal susceptible amplifier used in a servo loop application sees excess CM voltage, the effect can be disastrous—it goes **Bang!** So, the best advice is to be forewarned.

An Output Phase-Reversal Do-it-Yourself Test

Since output phase-reversal may not always be fully described on a data sheet, it is quite useful to test for it. This is easily done in the lab, by driving a questionable op amp as a unity-gain follower, from a source impedance (R_{LIMIT}) of ~1 kΩ. It is helpful to make this a variable, 1 kΩ to 100 kΩ range resistance.

With a low resistance setting (1 kΩ), while bringing the driving signal level slowly up towards the rail limits, observe the amplifier output. If a phase-reversal mechanism is present, when the CM limit of the op amp is exceeded, the output will suddenly reverse (see Figure 9.81, again). If there is no phase-reversal present in an amplifier, the output waveform will simply clip at the limits of its swing. It may prove helpful to have a well-behaved op amp available for this test, to serve as a performance reference. One such device is the AD8610.

Note that in general, some care should be used with this test. Without a series current-limit resistor, if the generator impedance is too low (or level too high), it could possibly damage an internal junction of the op amp under test. So, obviously, caution is best for such cases.

Once a suitable R_{LIMIT} resistance value is found, well-behaved op amps will simply show a smooth, bipolar range, clipped output waveform when overdriven. This clipping will appear more like the *upper (positive swing)* portion of the waveform within Figure 9.81, right (again).

Fixes for Output Phase–Reversal

An op amp manufacturer might not always give the R_{LIMIT} resistance value appropriate to prevent output phase-reversal. But, the value can be determined empirically with the driving method mentioned above. Most often, the R_{LIMIT} resistor value providing protection against phase-reversal will also safely limit fault current through any input CM clamping diodes. If in doubt, a nominal value of 1 kΩ is a good starting point for testing.

Typically, FET input op amps will need only the current limiting series resistor for protection, but bipolar input devices are best protected with this same limiting resistor, *along with a Schottky diode* (i.e., R_{LIMIT} and D2, of Figure 9.75).

For a more detailed description of the output voltage phase-reversal effect, see References 7 and 8. Figure 9.82 summarizes a number of the key points relating to output voltage phase-reversal.

- Non-Latching Inversion of Transfer Function, Triggered by Exceeding Common Mode Limit

- Sometimes Occurs in FET and Bipolar (Single-Supply) Op Amps

- Doesn't Harm Amplifier... but Disastrous for Servo Systems

- Not Usually Specified on Data Sheet, so Amplifier Must be Checked

- Easily Prevented:

    - All op amps: Limit applied CM voltage by clamping or other means

    - BiFETs: Add series input resistance, R_{LIMIT}

    - Bipolars: R_{LIMIT} and Schottky clamp diode to rail

Figure 9.82: A Summary of Key Points Regarding Output Phase-Reversal in FET and Bipolar Input Op Amps

Alternately, any of the several previously mentioned CM clamping schemes can be used to prevent output phase reversal, by setting the clamp voltage to be less than the amplifier CM range limit where phase-reversal occurs. For example, Figure 9.77 would operate to prevent phase-reversal in FET amplifiers susceptible to it, if the negative clamp limit is set so that $V_{CLAMP(-)}$ never exceeds the typical negative CM range of -11 V on a -15 V rail.

For validation of this or any of the previous overvoltage protection schemes, the circuit should be verified on a number of op amps, over a range of conditions as suitable to the final application environment.

Input Differential Protection

The discussions thus far have been on overvoltage common-mode conditions, which is typically associated with forward biasing of PN junctions inherent in the structure of the input stage. There is another equally important aspect of protection against overvoltage, which is that due to excess *differential* voltages. Excessive differential voltage, when applied to certain op amps, can lead to degradation of their operating characteristics.

This degradation is brought about by *reverse junction breakdown,* a second case of undesirable input stage conduction, occurring under conditions of *differential* overvoltage. However, in the case of reverse breakdown of a PN junction, the problem can be more subtle in nature. It is illustrated by the partial op amp input stage in Figure 9.83

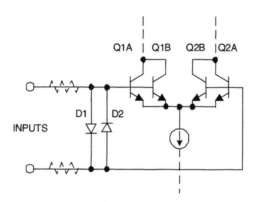

Figure 9.83: An Op Amp Input Stage With D1-D2 Input Differential Overvoltage Protection Network

This circuit, applicable to a low noise op amp such as the OP27, is also typical of many others using low noise bipolar transistors for differential pair Q1-Q2. In the absence of any protection, it can be shown that voltages above about 7 V between the two inputs will cause a reverse junction breakdown of either Q2 or Q1 (dependent upon relative polarity). Note that, in cases of emitter-base breakdown, even small reverse currents can cause degradation in both transistor gain and noise (see Reference 6). After emitter-base breakdown occurs, op amp parameters such as the bias currents and noise may well be out of specification. This is usually permanent, and it can occur gradually and quite subtly, particularly if triggered by transients. For these reasons, virtually all low noise op amps, whether NPN- or PNP-based, utilize protection diodes such as D1-D2 across the inputs. These diodes conduct for applied voltages greater than ± 0.6 V, protecting the transistors.

The dotted series resistors function as current limiters (protection for the protection diodes) but aren't used in all cases. For example, the AD797 doesn't have the resistors, simply because they would degrade the part's specified noise of $1\ \text{nV}/\sqrt{\text{Hz}}$. Note—when the resistors are absent internally, some means of external current limiting must be provided, when and if differential overvoltage conditions do occur. Obviously, this is a trade-off situation, so the confidence of full protection must be weighed against the noise degradation. Note that an application circuit itself may provide sufficient resistance in the op amp inputs, such that additional resistance isn't needed.

In applying a low noise bipolar input stage op amp, first check the chosen part's data sheet for internal protection. When necessary, protection diodes D1-D2, if not internal to the op amp, should be added to guarantee prevention of Q1-Q2 emitter-base breakdown. If differential transients of more than 5 V can be seen by the op amp in the application, the diodes are in order. Ordinary low capacitance diodes will suffice, such as the 1N4148 family. Add current limiting resistors as necessary, to limit diode current to safe levels.

Other IC device junctions, such as base-collector and JFET gate-source junctions don't exhibit the same degradation in performance upon break down, and for these the input current should be limited to 5 mA, unless the data sheet specifies a different value.

Protecting In Amps Against Overvoltage

From a protection standpoint, instrumentation amplifiers (in amps) are similar in many ways to op amps. Like op amps, their absolute maximum ratings must be observed for both common and differential mode input voltages.

A much simplified schematic of the AD620 in amp is shown in Figure 9.84, showing the input differential transistors and their associated protection parts.

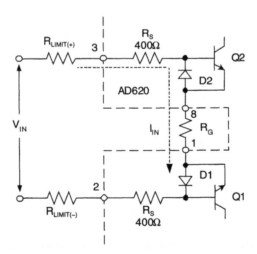

Figure 9.84: The AD620 In Amp Input Internally Uses D1-D2 and Series Resistors R$_S$ for Protection (Additional Protection Can Be Added Externally)

An important point, unique to the AD620 device, is the fact that the 400 Ω internal R_S protection resistors are *thin-film types*. Therefore these resistors don't show symptoms of diode-like conduction to the IC substrate (as would be the case were they diffused resistors). Practically, this means that the input ends of these resistors (Pins 3 and 2) can go above or below the supplies. Differential fault currents will be limited by the combination of twice the internal R_S plus the external gain resistance, R_G. Excess applied CM voltages will show current limited by R_S.

In more detail, it can be noted that input transistors Q1 and Q2 have protection diodes D1 and D2 across their base-emitter junctions, to prevent reverse breakdown. For differential voltages, analysis shows that a fault current, I_{IN}, flows through the external R_{LIMIT} resistors (if present), the internal R_S resistors, the gain-setting resistor R_G, and two diode drops (Q2, D1). For the AD620 topology, R_G varies inversely with gain, and a worst case (lowest resistance) occurs with the maximum gain of 1000, when R_G is 49.9 Ω. Therefore the lowest total internal path series resistance is about 850 Ω.

For the AD620, any combination of CM and differential input voltages should be limited to levels that limit the input fault current to 20 mA, maximum. A purely differential voltage of 17 V would result in this current level, for the lowest resistance case. For CM voltages that may go beyond either rail, an internal diode, not shown in Figure 9.84, conducts, effectively clamping the driven input to either $+V_S$ or $-V_S$ at the R_S inner end. For this overvoltage CM condition, the 400 Ω value of R_S and the excess voltage beyond the rail determines the current level. If for example V_{IN} is 23 V with $+V_S$ at 15 V, 8 V appears across R_S, and the 20 mA current rating is reached. Higher fault voltages can be dealt with by adding R_{LIMIT} resistance, to maintain fault current at 20 mA or less.

A more generalized external voltage protection circuit for an in amp like the AD620 is shown in Figure 9.85.

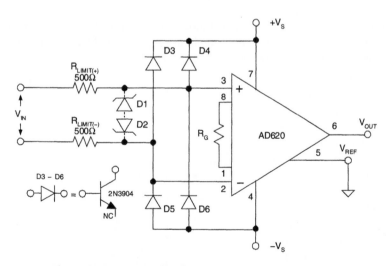

Figure 9.85: A Generalized Diode Protection Circuit for the AD620 and Other In Amps Uses D3–D6 for CM Clamping and Series Resistors R_{LIMIT} for Protection

In this circuit, low-leakage diodes D3–D6 are used as CM clamps. Since the in amp bias current may be only 1 nA or so (for the AD620), a low leakage diode type is mandatory. As can be noted from the topology, diode bootstrapping isn't possible with this configuration.

It should be noted that not only must the diodes have basically low leakage, they must also maintain low leakage at the highest expected temperature. This suggests either FET type diodes (see Figure 9.76, again), or the transistor C-B types shown. The R_{LIMIT} resistors are chosen to limit the maximum diode current under fault conditions. If additional *differential* protection is used, either back-back Zener or Transzorb clamps can be used, shown as D1-D2. If this is done, the leakage and capacitance of these diodes should be carefully considered.

The protection scheme of Figure 9.85, while effective using appropriate parts, has the downside of requiring a number of components. A much more simple in amp protection using fault protected devices is shown in Figure 9.86. Although shown with an AD620, this circuit is useful with many other dual-supply in amps with bias currents of 1 nA or more. It uses two-thirds of a triple ADG466 channel-protector for the in amp differential inputs.

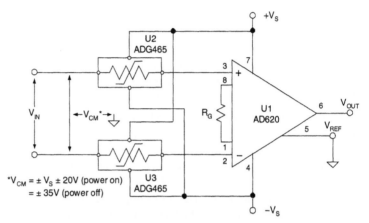

Because the nature of a channel protection device is to turn off as V_{IN} approaches either rail, the scheme of Figure 9.86 doesn't function with rail sensing single-supply in amps. If near-rail operation and protection is required in an in amp application, an alternative method is necessary. Many single-supply in amps are topologically similar to the two-amplifier in amp circuit shown within the dotted box of Figure 9.87.

Figure 9.86: A Channel Protector Device (Or Fault-Protected Multiplexer) Provides Protection for Dual-Supply In Amps With a Minimum of Extra Parts

In terms of the necessity for externally added protection components, a given in amp may or may not require them. Each case needs to be considered individually. For example, some in amps have clamp diodes as shown, but *internal to the device*. The AD623 is such a part, but it lacks the series resistors, which can be added externally when and if necessary. Note that this approach allows the R_{LIMIT} value to be optimized for protection, with negligible impact on noise for those applications not needing the protection.

Some in amp devices also have both internal protection resistors *and* clamping diodes, an example here is the AD627. In this device, the internal protection is adequate for transients up to 40 V

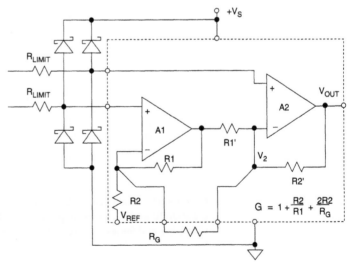

Figure 9.87: Single-Supply In Amps May or May Not Require External Protection in the Form of Resistors and Clamp Diodes—If So, They Can Be Added As Shown

beyond the supplies (a 20 mA fault current in the internal resistors). For overvoltage levels higher than this, external R_{LIMIT} resistors can be added.

The use of the Schottky diodes as shown at the two inputs is an option for in amp protection. If no clamping is specifically provided internally, then they are applicable. Their use is generally similar to the op amp protection case of Figure 9.75, with comparable caveats as far as leakage. Note that in many cases, due to internal protection networks of modern in amps, these diodes just won't be necessary. But again, there aren't hard rules on this, so always check the data sheet before finalizing an application.

ADCs whose input range falls between the supply rails can generally be protected with external Schottky diodes and a current limit resistor as shown in Figure 9.88. Even if internal ESD protection diodes are provided, the use of the external ones allows smaller values of R_{LIMIT} and lower noise and offset errors. ADCs with thin-film input attenuators, such as the AD7890-10 (see Figure 9.89), can be protected with Zener diodes or transient voltage suppressors (TVSs) with an R_{LIMIT} resistor to limit the current through them.

Figure 9.88: Input Protection for ADCs with Input Ranges Within Supply Voltages

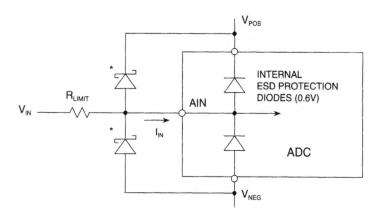

- Choose R_{LIMIT} to Limit I_{IN} Current to 5mA

- *Additional External Schottky Diodes Allow Lower Values of R_{LIMIT}

Figure 9.89: Input Protection for Single-Supply ADCs With Thin Film Resistor Input Attenuators

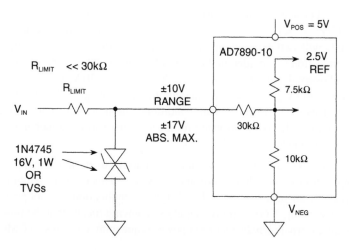

Overvoltage Protection Using CMOS Channel Protectors

The ADG465/ADG466/ADG467 are CMOS channel protectors which are placed in series with the signal path. The channel protector will protect sensitive components from voltage transients whether the power supplies are present or not. Because the channel protection works whether the supplies are present or not, the channel protectors are ideal for use in applications where correct power sequencing cannot always be guaranteed (e.g., hot-insertion rack systems) to protect analog inputs.

Each channel protector (see Figure 9.90) has an independent operation and consists of four MOS transistors—two NMOS and two PMOS. One of the PMOS devices does not lie directly in the signal path but is used to connect the source of the second PMOS device to its backgate. This has the effect of lowering the threshold voltage and so increasing the input signal range of the channel for normal operation. The source and backgate of the NMOS devices are connected for the same reason.

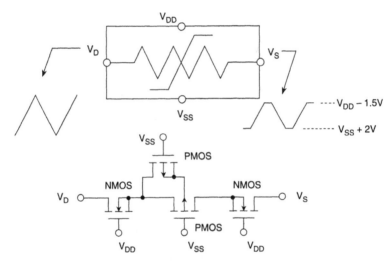

Figure 9.90: ADG465, ADG466, and ADG467
Single, Triple, and Octal Channel Protectors

The channel protector behaves just like a series resistor (60 Ω to 80 Ω) during normal operation, i.e., $(V_{SS} + 2 \text{ V}) < V_D < (V_{DD} - 1.5 \text{ V})$. When a channel's analog input voltage exceeds this range, one of the MOSFETs will switch off, clamping the output at either $V_{SS} + 2$ V or $V_{DD} - 1.5$ V. Circuitry and signal source protection is provided in the event of an overvoltage or power loss. The channel protectors can withstand overvoltage inputs from $V_{SS} - 20$ V to $V_{DD} + 20$ V with power on ($V_{DD} - V_{SS} = 44$ V maximum). With power off ($V_{DD} = V_{SS} = 0$ V), maximum input voltage is ± 35 V. The channel protectors are very low power devices, and even under fault conditions, the supply current is limited to sub microampere levels. All transistors are dielectrically isolated from each other using a trench isolation method thereby ensuring that the channel protectors cannot latch up.

Figure 9.91 shows a typical application that requires overvoltage and power supply sequencing protection. The application shows a hot-insertion rack system. This involves plugging a circuit board or module into a live rack via an edge connector. In this type of application it is not possible to guarantee correct power supply sequencing. Correct power supply sequencing means that the power supplies should be connected before any external signals. Incorrect power sequencing can cause a CMOS device to latch up. This is true of most CMOS devices, regardless of the functionality. RC networks are used on the supplies of the channel protector to ensure that the rest of the circuit is powered up before the channel protectors. In this way, the outputs of the channel protectors are clamped well below V_{DD} and V_{SS} until the capacitors are charged. The diodes ensure that the supplies on the channel protector never exceed the supply rails when it is being disconnected. Again this ensures that signals on the inputs of the CMOS devices never exceed the supplies.

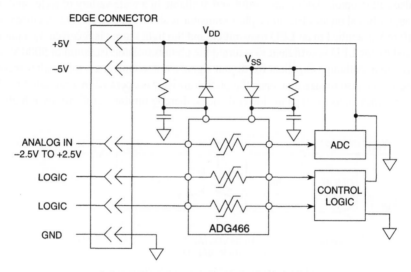

Figure 9.91: Overvoltage and Power Supply Sequencing Protection Using the ADG466

Digital Isolators

One way to break ground loops is to use digital isolation techniques. Analog isolation amplifiers find many applications where a high degree of isolation is required, such as in medical instrumentation. Digital isolation techniques offer a reliable method of transmitting digital signals over interfaces with high common-mode voltages without introducing ground noise.

Optocouplers (also called optoisolators) are useful and available in a wide variety of styles and packages. A typical optocoupler based on an LED and a phototransitor is shown in Figure 9.92. A current of approximately 10 mA is applied to an LED transmitter, and the light output is received by a phototransistor. The light produced by the LED is sufficient to saturate the phototransistor. Isolation of 5000 V rms to 7000 V rms is common. Although excellent for digital signals, optocouplers are too nonlinear for most analog applications. One should also realize that since the phototransistor is operated in a saturated mode, rise and fall-times can range from 10 µs to 20 µs in slower devices, thereby limiting applications at high speeds.

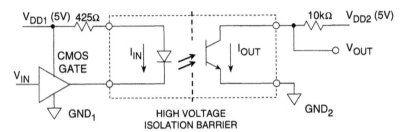

- Uses Light for Transmission Over a High Voltage Barrier
- The LED is the Transmitter, and the Phototransistor is the Receiver
- High Voltage Isolation: 5000V to 7000V RMS
- Nonlinear—Best for Digital or Frequency Information
- Rise and Fall-times can be 10µs to 20µs in Slower Devices
- Example: Siemens ILQ-1 Quad (www.siemens.com)

Figure 9.92: Digital Isolation Using LED/Phototransistor Optocouplers

A faster optocoupler architecture is shown in Figure 9.93 and is based on an LED and a photodiode. The LED is again driven with a current of approximately 10 mA. This produces a light output sufficient to generate enough current in the receiving photodiode to develop a valid high logic level at the output of the transimpedance amplifier. Speed can vary widely between optocouplers, and the fastest ones have propagation delays of 20 ns typical, and 40 ns maximum, and can handle data rates up to 25 Mbps for NRZ data. This corresponds to a maximum square wave operating frequency of 12.5 MHz, and a minimum allowable passable pulse width of 40 ns.

The ADuM1100A and ADuM1100B are digital isolators based on Analog Devices' *i*Coupler® technology. Combining high speed CMOS and monolithic air core transformer technology, these isolation components provide outstanding performance characteristics superior to the traditional optocouplers previously described.

Configured as pin-compatible replacements for existing high speed optocouplers, the ADuM1100A and ADuM1100B support data rates as high as 25 Mbps and 100 Mbps, respectively. A functional diagram of the devices is shown in Figure 9.94.

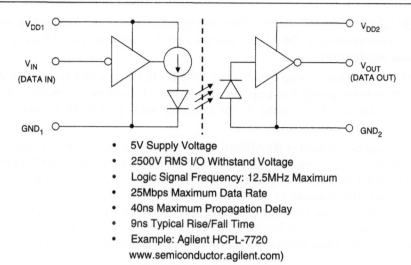

- 5V Supply Voltage
- 2500V RMS I/O Withstand Voltage
- Logic Signal Frequency: 12.5MHz Maximum
- 25Mbps Maximum Data Rate
- 40ns Maximum Propagation Delay
- 9ns Typical Rise/Fall Time
- Example: Agilent HCPL-7720
  www.semiconductor.agilent.com)

Figure 9.93: Digital Isolation Using LED/Photodiode Optocouplers

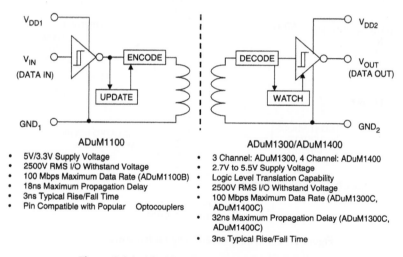

ADuM1100

- 5V/3.3V Supply Voltage
- 2500V RMS I/O Withstand Voltage
- 100 Mbps Maximum Data Rate (ADuM1100B)
- 18ns Maximum Propagation Delay
- 3ns Typical Rise/Fall Time
- Pin Compatible with Popular   Optocouplers

ADuM1300/ADuM1400

- 3 Channel: ADuM1300, 4 Channel: ADuM1400
- 2.7V to 5.5V Supply Voltage
- Logic Level Translation Capability
- 2500V RMS I/O Withstand Voltage
- 100 Mbps Maximum Data Rate (ADuM1300C, ADuM1400C)
- 32ns Maximum Propagation Delay (ADuM1300C, ADuM1400C)
- 3ns Typical Rise/Fall Time

Figure 9.94: ADuM1100A/ADuM1100B Digital Isolators

Both the ADuM1100A and ADuM1100B operate at either 3.3 V or 5 V supply voltages, have propagation delays < 18 ns, edge asymmetry of <2 ns, and rise- and fall-times <3 ns. They operate at very low power, less than 900 µA of quiescent current (sum of both sides) and a dynamic current of less than 160 µA per Mbps of data rate. Unlike common transformer implementations, the parts provide dc correctness with a patented refresh feature that continuously updates the output signal.

The ADuM1300/ADuM1301 digital isolators offer three channels of isolation, and the ADuM1400/ADuM1401 isolators offer four channels of isolation. These devices operate from 2.7 V to +5.5 V and have independent input and output power inputs allowing them to operate not only as isolators but also as logic level translators.

The AD260/AD261 family of digital isolators isolates five digital control signals to/from high speed DSPs, microcontrollers, or microprocessors. The AD260 also has a 1.5 W transformer for a 3.5 kV rms isolated external dc/dc power supply circuit.

Each line of the AD260 can handle digital signals up to 20 MHz (40 Mbps) with a propagation delay of only 14 ns which allows for extremely fast data transmission. Output waveform symmetry is maintained to within ±1 ns of the input so the AD260 can be used to accurately isolate time-based pulsewidth modulator (PWM) signals.

A simplified schematic of one channel of the AD260/AD261 is shown in Figure 9.95. The data input is passed through a schmitt trigger circuit, through a latch, and a special transmitter circuit that differentiates the edges of the digital input signal and drives the primary winding of a proprietary transformer with a "set-high/set-low" signal. The secondary of the isolation transformer drives a receiver with the same "set-hi/set-low" data that regenerates the original logic waveform. An internal circuit operates in the background which interrogates all inputs about every 5 µs and in the absence of logic transitions, sends appropriate "set-hi/set-low" data across the interface. Recovery time from a fault condition or at power-up is thus between 5 µs and 10 µs.

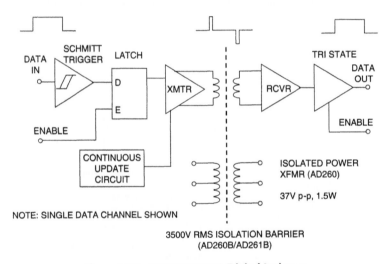

NOTE: SINGLE DATA CHANNEL SHOWN

3500V RMS ISOLATION BARRIER
(AD260B/AD261B)

Figure 9.95: AD260/AD261 Digital Isolators

The power transformer (available on the AD260) is designed to operate between 150 kHz and 250 kHz and will easily deliver more than 1 W of isolated power when driven push-pull (5 V) on the transmitter side. Different transformer taps, rectifier and regulator schemes will provide combinations of ±5 V, 15 V, 24 V, or even 30 V or higher. The output voltage when driven with a low voltage-drop drive will be 37 V p-p across the entire secondary with a 5 V push-pull drive. Key specifications for the parts are given in Figure 9.96.

To summarize, Figure 9.97 reviews the major points of the in-circuit overvoltage issues discussed in this section.

If these varied overvoltage precautions for op amps and in amps seem complex, they are. Whenever op amp (or in amp or data converter) inputs (and outputs) go outside equipment boundaries, dangerous or destructive things can happen to them. For highest reliability, these potentially hazardous situations should be anticipated.

- Isolation Test Voltage to 3500V RMS (AD260B/AD261B)
- Five Isolated Digital Lines Available in Six Input/Output Configurations
- Logic Signal Frequency: 20MHz Max.
- Data Rate: 40Mbps Max.
- Isolated Power Transformer: 37V p-p, 1.5W (AD260)
- Waveform Edge Transmission Symmetry: ±1ns
- Propagation Delay: 14ns
- Rise and Fall-Times < 5ns

Figure 9.96: AD260/AD261 Digital Isolator Key Specifications

- Input Voltages Must Not Exceed Absolute Maximum Ratings
  (Usually Specified With Respect to Supply Voltages)
- Requires $V_{IN(CM)}$ Stay Within a Range Extending to ≤0.3V Beyond Rails
  $(-V_S-0.3V \geq V_{IN} \leq + V_S + 0.3V)$
- IC Input Stage Fault Currents Must Be Limited
  (≤ 5mA Unless Otherwise Specified)
- Avoid Reverse-Bias Breakdown in Input Stage Junctions
- Differential and Common Mode Ratings Often Differ
- No Two Amplifiers are Exactly the Same
- Watch Out for Output Phase-Reversal in JFET and SS Bipolar Op Amps
- Some ICs Contain Internal Input Protection
  - Diode Voltage Clamps, Current Limiting Resistors (or both)
  - Absolute Maximum Ratings Must Still Be Observed

Figure 9.97: A Summary of In-Circuit Overvoltage Points

Fortunately, most applications are contained entirely within the equipment, and usually see inputs and outputs to/from other ICs on the same power system. Therefore clamping and protection schemes typically aren't necessary for these cases.

Out-of-Circuit Overvoltage Protection

Linear ICs such as op amps, in amps, and data converters must also be protected prior to the time that they are mounted to a printed circuit board. That is an *out-of-circuit* state. In such a condition, ICs are completely at the mercy of their environment as to what stressful voltage surges they may see. Most often the harmful voltage surges come from *electrostatic discharge,* or, as more commonly referenced, ESD. This is a single, fast, high current transfer of electrostatic charge resulting from one of two conditions. These conditions are:

1. *Direct contact transfer between two objects at different potentials (sometimes called contact discharge)*

2. *A high electrostatic field between two objects when they are in close proximity (sometimes called air discharge)*

The prime sources of static electricity are mostly insulators and are typically synthetic materials, e.g., vinyl or plastic work surfaces, insulated shoes, finished wood chairs, Scotch tape, bubble pack, soldering irons with ungrounded tips, etc. Voltage levels generated by these sources can be extremely high since their charge is not readily distributed over their surfaces or conducted to other objects. The generation of static electricity caused by rubbing two substances together is called the *triboelectric* effect. Some common examples of ordinary acts producing significant ESD voltages are shown in Figure 9.98.

- Walking Across a Carpet
  1000V – 1500V
- Walking Across a Vinyl Floor
  150V – 250V
- Handling Material Protected by Clear Plastic Covers
  400V – 600V
- Handling Polyethylene Bags
  1000V – 2000V
- Pouring Polyurethane Foam Into a Box
  1200V – 1500V
- Note: Above Assumes 60% RH. For Low RH (30%), Voltages Can Be > 10 Times

**Figure 9.98: ESD Voltages Generated
By Various Ordinary Circumstances**

ICs can be damaged by the high voltages and high peak currents generated by ESD. Precision analog circuits, often featuring very low bias currents, are more susceptible to damage than common digital circuits, because traditional input-protection structures which protect against ESD damage increase input leakage—and thus can't be used.

For the design engineer or technician, the most common manifestation of ESD damage is a catastrophic failure of the IC. However, exposure to ESD can also cause increased leakage or degrade other parameters. If a device appears not to meet a data sheet specification during evaluation, the possibility of ESD damage should be considered. Figure 9.99 outlines some relevant points on ESD-induced failures.

All ESD-sensitive devices are shipped in protective packaging. ICs are usually contained in either conductive foam or antistatic shipping tubes, and the container is then sealed in a static-dissipative plastic bag. The sealed bag is marked with a distinctive sticker, such as in Figure 9.100, which outlines the appropriate handling procedures.

- ESD Failure Mechanisms:
  – Dielectric or junction damage
  – Surface charge accumulation
  – Conductor fusing
- ESD Damage Can Cause:
  – Increased leakage
  – Degradation in performance
  – Functional failures of ICs
- ESD Damage is often Cumulative:
  – For example, each ESD "zap" may increase junction damage until, finally, the device fails.

Figure 9.99: Understanding ESD damage

All static sensitive devices are sealed in
protective packaging and marked with
special handling instructions

CAUTION

SENSITIVE ELECTRONIC DEVICES

DO NOT SHIP OR STORE NEAR STRONG
ELECTROSTATIC, ELECTROMAGNETIC,
MAGNETIC, OR RADIOACTIVE FIELDS

CAUTION

SENSITIVE ELECTRONIC DEVICES

DO NOT OPEN EXCEPT AT
APPROVED FIELD FORCE
PROTECTIVE WORK STATION

Figure 9.100: Recognizing ESD-Sensitive Devices by Package and Labeling

The presence of outside package notices such as those shown in Figure 9.100 is notice to the user that device handling procedures appropriate for ESD protection are necessary.

In addition, data sheets for ESD-sensitive ICs generally have a bold statement to that effect, as shown in Figure 9.101.

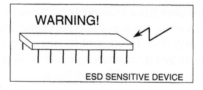

CAUTION

ESD-(Electrostatic Discharge) sensitive device. Electrostatic charges
as high as 4000 V readily accumulate on the human body and test
equipment and can discharge without detection. Although the ADxxx
features proprietary ESD protection circuitry, permanent damage may
occur on devices subjected to high energy electrostatic discharges.
Therefore,proper ESD precautions are recommended to avoid
performance degradation or loss of functionality.

Figure 9.101: ESD Data Sheet Statement for Linear ICs

Once ESD-sensitive devices are identified, protection is relatively easy. Obviously, keeping ICs in their original protective packages as long as possible is a first step. A second step is discharging potentially damaging ESD sources before IC damage occurs. Discharging such voltages can be done quickly and safely, through a high impedance.

A key component required for ESD-safe IC handling is a workbench with a static-dissipative surface, shown in the workstation of Figure 9.102. The surface is connected to ground through a 1 MΩ resistor, which dissipates any static charge, while protecting the user from electrical ground fault shock hazards. If existing bench tops are nonconductive, a static-dissipative mat should be added, along with the discharge resistor.

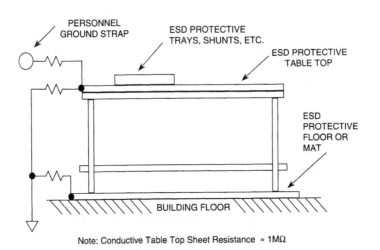

Note: Conductive Table Top Sheet Resistance » 1MΩ

Figure 9.102: A Workstation Environment Suitable for Handling ESD-Sensitive ICs

Note that the surface of the workbench has a moderately high sheet resistance. It is neither necessary nor desirable to use a low resistance surface (such as a sheet of copper-clad PC board) for the work surface. Remember, a high peak current may flow if a charged IC is discharged through a low impedance. This is precisely what happens when a charged IC contacts a grounded copper clad board. When the same charged IC is placed on the high impedance surface of Figure 9.102 however, the peak current isn't high enough to damage the device.

Several personnel handling techniques are keys to minimizing ESD-related damage. At the workstation, a conductive wrist strap is recommended while handling ESD-sensitive devices. The wrist strap ensures that normal tasks, such as peeling tape from packages, won't cause IC damage. Again, a 1 MΩ resistor, from the wrist strap to ground, is required for safety. When building prototype breadboards or assembling PC boards that contain ESD-sensitive ICs, all passive components should be inserted and soldered before the ICs. This minimizes the ESD exposure of the sensitive devices. The soldering iron must, of course, have a grounded tip.

Protecting ICs from ESD requires the participation of both the IC manufacturer and the customer. IC manufacturers have a vested interest in providing the highest possible level of ESD protection for their products. IC circuit designers, process engineers, packaging specialists and others are constantly looking for new and improved circuit designs, processes, and packaging methods to withstand or shunt ESD energy.

A complete ESD protection plan, however, requires more than building ESD protection into ICs. The users of ICs must also provide their employees with the necessary knowledge of and training in ESD handling procedures, so that protection can be built in at all key points along the way, as outlined in Figure 9.103.

ANALOG DEVICES:

- Circuit Design and Fabrication
  - Design and manufacture products with the highest level of ESD protection consistent with required analog and digital performance.

- Pack and Ship
  - Pack in static dissipative material. Mark packages with ESD warning.

CUSTOMERS:

- Incoming Inspection
  - Inspect at grounded workstation. Minimize handling.

- Inventory Control
  - Store in original ESD-safe packaging. Minimize handling.

- Manufacturing
  - Deliver to work area in original ESD-safe packaging. Open packages only at grounded workstation. Package subassemblies in static dissipative packaging.

- Pack and Ship
  - Pack in static dissipative material if required. Replacement or optional boards may require special attention.

Figure 9.103: ESD Protection Requires a Partner Relationship between ADI and the End Customer with Control at Key Points

Special care should be taken when breadboarding and evaluating ICs. The effects of ESD damage can be cumulative, so repeated mishandling of a device can eventually cause a failure. Inserting and removing ICs from a test socket, storing devices during evaluation, and adding or removing external components on the breadboard should all be done while observing proper ESD precautions. Again, if a device fails during a prototype system development, repeated ESD stress may be the cause.

The key word to remember with respect to ESD is *prevention*. There is no way to undo ESD damage, or to compensate for its effects.

ESD Models and Testing

Some applications have higher sensitivity to ESD than others. ICs located on a PC board surrounded by other circuits are generally much less susceptible to ESD damage than circuits which must interface with other PC boards or the outside world. These ICs are generally not specified or guaranteed to meet any particular ESD specification (with the exception of MIL-STD-883 Method 3015 classified devices). A good example of an ESD-sensitive interface is the RS-232 interface port ICs on a computer, which can easily be exposed to excess voltages. In order to guarantee ESD performance for such devices, the test methods and limits must be specified.

A host of test waveforms and specifications have been developed to evaluate the susceptibility of devices to ESD. The three most prominent of these waveforms currently in use for semiconductor or discrete devices are: The Human Body Model (HBM), the Machine Model (MM), and the Charged Device Model (CDM). Each of these models represents a fundamentally different ESD event, consequently, correlation between the test results for these models is minimal.

Since 1996, all electronic equipment sold to or within the European Community must meet Electromechanical Compatibility (EMC) levels as defined in specification IEC1000-4-x. Note that this does not apply to individual ICs, *but to the end equipment.* These standards are defined along with test methods in the various IEC1000 specifications, and are listed in Figure 9.104.

- IEC1000-4   Electromagnetic Compatibility EMC

- IEC1000-4-1 Overview of Immunity Tests

- IEC1000-4-2 Electrostatic Discharge Immunity (ESD)

- IEC1000-4-3 Radiated Radio-Frequency Electromagnetic Field Immunity

- IEC1000-4-4 Electrical Fast Transients (EFT)

- IEC1000-4-5 Lightening Surges

- IEC1000-4-6 Conducted Radio Frequency Disturbances above 9kHz

- Compliance Marking: $\mathsf{C}\mathsf{E}$

**Figure 9.104: A Listing of the IEC Standards Applicable
to ESD Specifications and Testing Procedures**

IEC1000-4-2 specifies compliance testing using two coupling methods, *contact discharge* and *air-gap discharge*. Contact discharge calls for a direct connection to the unit being tested. Air-gap discharge uses a higher test voltage, but does not make direct contact with the unit under test. With air discharge, the discharge gun is moved toward the unit under test, developing an arc across the air gap, hence the term air discharge. This method is influenced by humidity, temperature, barometric pressure, distance and rate of closure of the discharge gun. The contact-discharge method, while less realistic, is more repeatable and is gaining acceptance in preference to the air-gap method.

Although very little energy is contained within an ESD pulse, the extremely fast rise time coupled with high voltages can cause failures in unprotected ICs. Catastrophic destruction can occur immediately as a result of arcing or heating. Even if catastrophic failure does not occur immediately, the device may suffer from parametric degradation, which may result in degraded performance. The cumulative effects of continuous exposure can eventually lead to complete failure.

I-O lines are particularly vulnerable to ESD damage. Simply touching or plugging in an I-O cable can result in a static discharge that can damage or completely destroy the interface product connected to the I-O port (such as RS-232 line drivers and receivers).

Traditional ESD test methods such as MIL-STD-883B Method 3015.7 do not fully test a product's susceptibility to this type of discharge. This test was intended to test a product's susceptibility to ESD damage during handling. Each pin is tested with respect to all other pins. There are some important differences between the MIL-STD-883B Method 3015.7 test and the IEC test, noted as follows:

1) The IEC test is much more stringent in terms of discharge energy. The peak current injected is over four times greater.

2) The current rise time is significantly faster in the IEC test.

3) The IEC test is carried out while power is applied to the device.

It is possible that ESD discharge could induce latch-up in the device under test. This test is therefore more representative of a real-world I-O discharge where the equipment is operating normally with power applied. For maximum confidence, however, both tests should be performed on interface devices, thus ensuring maximum protection both during handling, and later, during field service.

A comparison of the test circuit values for the IEC1000-4-2 model versus the MIL-STD-883B Method 3015.7 Human Body Model is shown in Figure 9.105.

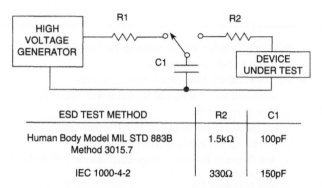

ESD TEST METHOD	R2	C1
Human Body Model MIL STD 883B Method 3015.7	1.5kΩ	100pF
IEC 1000-4-2	330Ω	150pF

NOTE: CONTACT DISCHARGE VOLTAGE SPEC FOR IEC 1000 -4-2 IS ±8kV

Figure 9.105: ESD Test Circuits and Values

The ESD waveforms for the MIL-STD-883B, METHOD 3015.7 and IEC 1000-4-2 tests are compared in Figure 9.106, left and right, respectively.

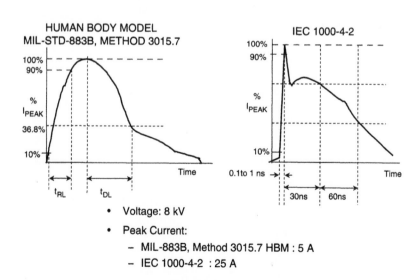

- Voltage: 8 kV
- Peak Current:
  - MIL-883B, Method 3015.7 HBM : 5 A
  - IEC 1000-4-2 : 25 A

Figure 9.106: ESD Test Waveforms

Suitable ESD-protection design measures are relatively easy to incorporate, and most of the overvoltage protection methods already discussed in this section will help. Additional protection can also be obtained. For RS-232 and RS-485 drivers and receivers, the ADMxxx-E series is supplied with guaranteed 15 kV (HBM) ESD specifications. For more general uses, the addition of TransZorbs at appropriate places in a system can provide protection against ESD (see References).

Figure 9.107 summarizes the major points about ESD prevention, from both an out-of-circuit as well as an in-circuit perspective.

```
•  Observe all Absolute Maximum Ratings on Data Sheet
•  Read ADI AN397 (See Reference 16)
•  Purchase ESD-Specified Digital Interface Devices
   –  ADMxxx-E Series of RS-232/RS-485 Drivers/Receivers
      (See Reference 18)
•  Follow General Overvoltage Protection Recommendations
   –  Add Series Resistance to Limit Currents
   –  Add Zeners or Transient Voltage Suppressors (TVS) for Extra
      Protection
      (See Reference 19)
```

Figure 9.107: A Summary of ESD Points

References:
9.4 Overvoltage Protection

1. Walt Kester, Wes Freeman, Joe Buxton, "Overvoltage Protection," portion of Section 10 within Walt Kester, Editor, **Practical Design Techniques for Sensor Signal Conditioning**, Analog Devices, Inc., 1999, ISBN 0-916550-20-6.

2. Siliconix PAD/JPAD/SSTPAD series Low leakage Pico-Amp Diodes, Vishay/Siliconix, www.vishay.com/brands/siliconix/SSFsglld.html.

3. Bob Pease, "Bounding, Clamping Techniques Improve Circuit Performance," **EDN**, November 10, 1983, p. 277.

4. Bob Pease, "Understanding Diodes and Their Problems," Chapter 6 within **Troubleshooting Analog Circuits**, Butterworth-Heinemann, 1991, ISBN 0-7506-9184-0.

5. Dov Kurz, Avner Cohen, "Bootstrapping Reduces Amplifier Input Capacitance," **EDN**, March 20, 1978.

6. C.D. Motchenbacher, J.A. Connelly, Chapter 5, within **Low Noise Electronic System Design**, John Wiley, 1993.

7. Adolfo Garcia, "Operational Amplifier Output Voltage Phase Reversal," Section 11, pp. 1–10 within Walt Kester, Editor, **1992 Amplifier Applications Guide**, Analog Devices, Inc., Norwood, MA, 1992, ISBN 0-916550-10-9.

8. Adolfo Garcia, Wes Freeman, "Overvoltage Effects On Analog Integrated Circuits," Section 7 within Walt Kester, Editor, **Practical Analog Design Techniques**, Analog Devices, Inc., Norwood, MA, 1995, ISBN 0-916550-16-8.

9. Charles Kitchin, Lew Counts, **A Designer's Guide to Instrumentation Amplifiers**, Analog Devices, Inc., 2000.

10. Walt Kester, Wes Freeman, James Bryant, "Electrostatic Discharge," portion of Section 10 within Walt Kester, Editor, **Practical Design Techniques for Sensor Signal Conditioning**, Analog Devices, Inc., 1999, ISBN 0-916550-20-6.

11. MIL-STD-883 Method 3015, "Electrostatic Discharge Sensitivity Classification." Available from Standardization Document Order Desk, 700 Robbins Ave., Building #4, Section D, Philadelphia, PA, 19111-5094.

12. EIAJ ED-4701 Test Method C-111, "Electrostatic Discharges." Available from the Japan Electronics Bureau, 250 W 34th St., New York NY 10119, Attn.: Tomoko.

13. ESD Association Standard S5.2 for "Electrostatic Discharge (ESD) Sensitivity Testing – Machine Model (MM) Component Level." Available from the ESD Association, Inc., 200 Liberty Plaza, Rome, NY 13440.

14. ESD Association Draft Standard DS5.3 for "Electrostatic Discharge (ESD) Sensitivity Testing – Charged Device Model (CDM) Component Testing." Available from the ESD Association, Inc., 200 Liberty Plaza, Rome, NY 13440.

15. **ESD Prevention Manual**, Analog Devices, Inc.

16. Niall Lyne, "Electrically Induced Damage to Standard Linear Integrated Circuits: The Most Common Causes and the Associated Fixes to Prevent Reoccurrence," **Analog Devices AN397**.

17. Mike Bryne, "How to Reliably Protect CMOS Circuits Against Power Supply Overvoltaging," **Analog Devices AN311**.

18. Data sheet for **ADM3311E RS-232 Port Transceiver**, Analog Devices, Inc., www.analog.com.

19. TransZorbs are available from General Semiconductor, Inc., 10 Melville Park Road, Melville, NY, 11747-3113, (631) 847-3000, www.gensemi.com/product/categories/tvs/tvs.htm.

Thermal Management
Walt Jung

For reliability reasons, data converter systems handling appreciable power are increasingly called upon to observe *thermal management*. All semiconductors have some specified safe upper limit for junction temperature (T_J), usually on the order of 150°C (sometimes 175°C). Like maximum power supply voltages, maximum junction temperature is a worst-case limitation which must not be exceeded. In conservative designs, it won't be approached by a less than ample safety margin. Note that this is critical, since semiconductor lifetime is inversely related to operating junction temperature. Simply put, the cooler ICs are, the more they can approach their maximum life.

This limitation of power and temperature is basic, and is illustrated by a typical data sheet statement as in Figure 9.108. In this case it is for the AD8017AR, an 8-pin SOIC device.

> The maximum power that can be safely dissipated by the AD8017 is limited by the associated rise in junction temperature. The maximum safe junction temperature for plastic encapsulated device is determined by the glass transition temperature of the plastic, approximately 150°C. Temporarily exceeding this limit may cause a shift in parametric performance due to a change in the stresses exerted on the die by the package. Exceeding a junction temperature of 175°C for an extended period can result in device failure.

Figure 9.108: Maximum Power Dissipation Data Sheet Statement for the AD8017AR, an ADI Thermally Enhanced SOIC Packaged Device

Tied to these statements are certain conditions of operation, such as the power dissipated by the device, and the package mounting specifics to the printed circuit board (PCB). In the case of the AD8017AR, the part is rated for 1.3 W of power at an ambient of 25°C. This assumes operation of the 8-lead SOIC package on a two-layer PCB with about 4 in^2 (~2500 mm^2) of 2 oz. copper for heat sinking purposes. Predicting safe operation for the device under other conditions is covered below.

Thermal Basics

The symbol θ is generally used to denote *thermal resistance*. Thermal resistance is in units of °C/watt (°C/W). Unless otherwise specified, it defines the resistance heat encounters transferring from a hot IC junction to the ambient air. It might also be expressed more specifically as θ_{JA}, for *thermal resistance, junction-to-ambient*. θ_{JC} and θ_{CA} are two additional θ forms used, and are further explained below.

In general, a device with a thermal resistance θ equal to 100°C/W will exhibit a temperature differential of 100°C for a power dissipation of 1 W, as measured between two reference points. Note that this is a linear relationship, so 1 W of dissipation in this part will produce a 100°C differential (and so on, for other powers). For the AD8017AR example, θ is about 95°C/W, so 1.3 W of dissipation produces about a 124°C junction-to-ambient temperature differential. It is, of course, this rise in temperature that is used to predict the internal temperature, in order to judge the thermal reliability of a design. With the ambient at 25°C, this allows an internal junction temperature of about 150°C. In practice most ambient temperatures are above 25°C, so less power can then be handled.

For any power dissipation P (in watts), one can calculate the effective temperature differential (ΔT) in °C as:

$$\Delta T = P \times \theta \qquad\qquad \text{Eq. 9.10}$$

where θ is the total applicable thermal resistance.

Figure 9.109 summarizes a number of basic thermal relationships.

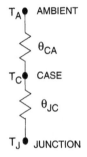

- θ = Thermal Resistance (°C/W)
- P = Total Device Power Dissipation (W)
- T = Temperature (°C)
- ΔT = Temperature Differential = P × θ
- θ_{JA} = Junction-Ambient Thermal Resistance
- θ_{JC} = Junction-Case Thermal Resistance
- θ_{CA} = Case-Ambient Thermal Resistance
- $\theta_{JA} = \theta_{JC} + \theta_{CA}$
- $T_J = T_A + (P \times \theta_{JA})$
- Note: $T_{J(Max)}$ = 150°C (Sometimes 175°C)

Figure 9.109: Basic Thermal Relationships

Note that series thermal resistances, such as the two shown at the right, model the total thermal resistance path a device may see. Therefore the total θ for calculation purposes is the sum, i.e., $\theta_{JA} = \theta_{JC}$ and θ_{CA}. Given the ambient temperature T_A, P, and θ, then T_J can be calculated. As the relationships signify, to maintain a low T_J, either θ or the power being dissipated (or both) must be kept low. A low ΔT is the key to extending semiconductor lifetimes, as it leads to lower maximum junction temperatures.

In ICs, one temperature reference point is always the device junction, taken to mean the hottest spot inside the chip operating within a given package. The other relevant reference point will be either T_C, the case of the device, or T_A, that of the surrounding air. This then leads in turn to the above mentioned individual thermal resistances, θ_{JC} and θ_{JA}.

Taking the most simple case first, θ_{JA} is the thermal resistance of a given device measured between its *junction* and the *ambient* air. This thermal resistance is most often used with small, relatively low power ICs such as op amps, which often dissipate 1 W or less. Generally, θ_{JA} figures typical of op amps and other small devices are on the order of 90°C/W–100°C/W for a plastic 8-pin DIP package, as well as the better SOIC packages.

It should be clearly understood that these thermal resistances are *highly* package-dependent, as different materials have different degrees of thermal conductivity. As a general guideline, thermal resistance of conductors is analogous to electrical resistances; that is, copper is the best, followed by aluminum, steel, and so on. Thus copper lead frame packages offer the highest performance, i.e., the lowest θ.

Heat Sinking

By definition, a *heat sink* is an added low thermal resistance device attached to an IC to aid heat removal. A heat sink has additional thermal resistance of its own, θ_{CA}, rated in °C/W. However, most current op amp packages don't easily lend themselves to heat sink attachment (exceptions are older TO-99 metal can types). Devices meant for heat sink attachment will often be noted by a θ_{JC} dramatically lower than the θ_{JA}. In this case θ will be composed of more than one component. Thermal impedances add, making a net calculation relatively simple. For example, to compute a net θ_{JA} given a relevant θ_{JC}, the thermal resistance of the heat sink, θ_{CA}, or *case* to *ambient* is added to the θ_{JC} as:

$$\theta_{JA} = \theta_{JC} + \theta_{CA} \qquad\qquad \text{Eq. 9.11}$$

and the result is the θ_{JA} for that specific circumstance.

More generally however, modern ICs *don't* use commercially available heat sinks. Instead, when significant power needs to be dissipated, such as ≥ 1 W, low thermal resistance copper PCB traces are used as the heat sink. In such cases, the most useful form of manufacturer data for this heat sinking are the boundary conditions of a sample PCB layout, and the resulting θ_{JA} for those conditions. This is, in fact, the type of specific information supplied for the AD8017AR, as mentioned earlier. Applying this approach, example data illustrating thermal relationships for such conditions is shown by Figure 9.110. These data apply for an AD8017AR mounted to a heat sink with an area of about 4 square inches on a 2-layer, 2-ounce copper PCB.

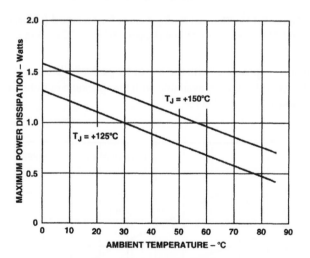

**Figure 9.110: Thermal Rating Curves
for AD8017AR Op Amp**

These curves indicate the maximum power dissipation versus temperature characteristic for the AD8017, for maximum junction temperatures of both 150°C and 125°C. Such curves are often referred to as *derating* curves, since allowable power decreases with ambient temperature.

With the AD8017AR, the proprietary ADI *Thermal Coastline* IC package is used, which allows additional power to be dissipated with no increase in the SO-8 package size. For a $T_{J(max)}$ of 150°C, the upper curve shows the allowable power in this package, which is 1.3 W at an ambient of 25°C. If a more conservative $T_{J(max)}$ of 125°C is used, the lower of the two curves applies.

A performance comparison for an 8-pin standard SOIC and the ADI Thermal Coastline version is shown in Figure 9.111. Note that the Thermal Coastline provides an allowable dissipation at 25°C of 1.3 W, whereas a standard package allows only 0.8 W. In the Thermal Coastline heat transferal is increased, accounting for the package's lower θ_{JA}.

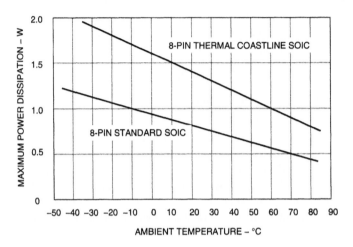

Figure 9.111: Thermal Rating Curves for Standard (Lower) and ADI Thermal Coastline (Upper) 8-Pin SOIC Packages

Even higher power dissipation is possible, with the use of IC packages better able to transfer heat from chip to PCB. An example is the AD8016 device, available with two package options rated for 5.5 W and 3.5 W at 25°C, respectively, as shown in Figure 9.112.

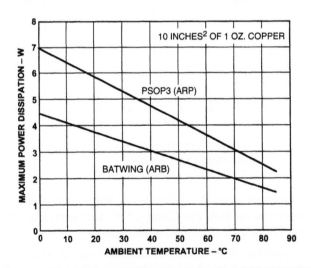

Figure 9.112: Thermal Characteristic Curves for the AD8016 BATWING (Lower) and PSOP3 (Upper) Packages, for $T_{J(Max)}$ Equal to 125°C

Taking the higher rated power option, the AD8016ARP PSOP3 package, when used with a 10 inch^2 1 oz. heat sink plane, the combination is able to handle up to 3 W of power at an ambient of 70°C, as noted by the upper curve. This corresponds to a θ_{JA} of 18°C/W, which in this case applies for a maximum junction temperature of 125°C.

The reason the PSOP3 version of the AD8016 is so better able to handle power lies with the use of a large area copper slug. Internally, the IC die rest directly on this slug, with the bottom surface exposed as shown in Figure 9.113. The intent is that this surface be soldered directly to a copper plane of the PCB, thereby extending the heat sinking.

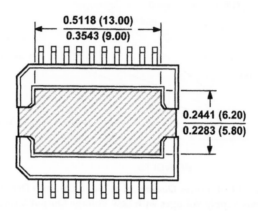

Figure 9.113: Bottom View of AD8016 20-Lead PSOP3 Package Showing Copper Slug for Aid in Heat Transfer (Central Grayed Area)

Both of the AD8016 package options are characterized for both still and moving air, but the thermal information given above applies *without* the use of directed airflow. Therefore, adding additional airflow lowers thermal resistance further (see Reference 2).

For reliable, low thermal resistance designs with op amps, several design *Do's and Don'ts* are listed below. Consider all of these points, as may be practical.

1)   *Do use as large an area of copper as possible for a PCB heat sink, up to the point of diminishing returns.*

2)   *In conjunction with 1), do use multiple (outside) PCB layers, connected together with multiple vias.*

3)   *Do use as heavy copper as is practical (2 oz. or more preferred).*

4)   *Do provide sufficient natural ventilation inlets and outlets within the system, to allow heat to freely move away from hot PCB surfaces.*

5)   *Do orient power-dissipating PCB planes vertically, for convection-aided airflow across heat sink areas.*

6)   *Do consider the use of external power buffer stages, for precision op amp applications.*

7)   *Do consider the use of forced air, for situations where several watts must be dissipated in a confined space.*

8)   *Don't use solder mask planes over heat dissipating traces.*

9)   *Don't use excessive supply voltages on ICs delivering power.*

For the most part, these points are obvious. However, one that could use some elaboration is number 9. Whenever an application requires only modest *voltage* swings (such as for example standard video, 2 V p-p) a wide supply voltage range can often be used. But, as the data of Figure 9.114 indicates, operation of an op amp driver on higher supply voltages produces a large IC dissipation, even though the load power is constant.

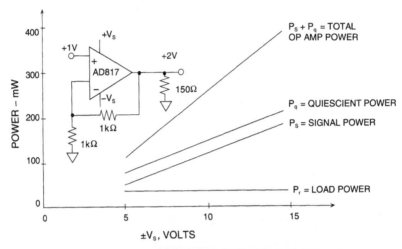

Figure 9.114: Power Dissipated in Video Op Amp Driver for Various Supply Voltages with Low Voltage Output Swing

In such cases, as long as the distortion performance of the application doesn't suffer, it can be advantageous to operate the IC on lower supplies, say ±5 V, as opposed to ±15 V. The above example data was calculated on a dc basis, which will generally tax the driver more in terms of power than a sine wave or a noise-like waveform, such as a DMT signal (see Reference 2). The general principles still hold for these ac waveforms, i.e., the op amp power dissipation is high when load current is high and the voltage low.

While there is ample opportunity for high power handling with the thermally enhanced packages described above for the AD8016 and AD8017, the increasingly popular smaller IC packages actually move in an opposite direction. Without question, it is true that today's smaller packages do noticeably sacrifice thermal performance. But, it must be understood that this is done in the interest of realizing a smaller size for the packaged op amp, and, ultimately, a much greater final PCB density for the overall system.

These points are illustrated by the thermal ratings for the AD8057 and AD8058 family of single and dual op amp devices, as is shown in Figure 9.115. The AD8057 and AD8058 op amps are available in three different packages. These are the SOT-23-5, and the 8-pin μSOIC, along with standard SOIC.

As the data shows, as the package size becomes smaller and smaller, much less power is capable of being removed. Since the lead frame is the only heatsinking possible with such tiny packages, their thermal performance is thus reduced. The θ_{JA} for the packages mentioned is 240, 200, and 160°C/W, respectively. Note this is more of a *package* than *device* limitation. Other ICs with the same packages have similar characteristics.

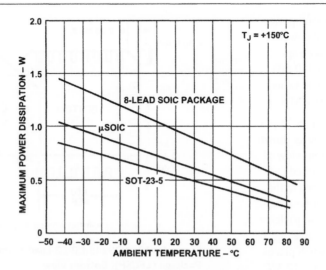

Figure 9.115: Comparative Thermal Performance
for Several AD8057/58 Op Amp Package Options

Data Converter Thermal Considerations

At first glance, one might assume that the power dissipation of an ADC or a DAC will remain constant for a given power supply voltage. However, many data converters, especially CMOS ones, have power dissipations that are highly dependent upon not only output data loading but also the sampling clock frequency. Since many of the newer high-speed converters can dissipate between 1.5 W and 2 W maximum power under the worst-case operating conditions, this point must be well understood in order to ensure that the package is mounted in such a way as to maintain the junction temperature within acceptable limits at the highest expected operating temperature.

The previous discussion in this chapter on grounding emphasized that the digital outputs of high performance ADCs, especially those with parallel outputs, should be lightly loaded (5 pF–10 pF) in order to prevent digital transient currents from corrupting the SNR and SFDR. Even under light output loading, however, most CMOS and BiCMOS ADCs have power dissipations that are a function of sampling clock frequency and in some cases, the analog input frequency and amplitude.

For example, Figure 9.116 shows the AD9245 14-bit, 80 MSPS, 3 V CMOS ADC power dissipation versus frequency for a 2.5 MHz analog input and 5 pF output loading of the data lines. The graphs show the digital and analog power supply currents

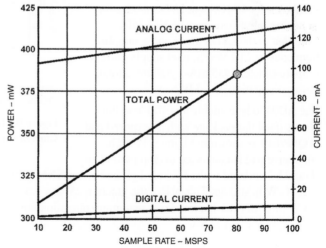

Figure 9.116: AD9245 14-Bit, 80 MSPS, 3 V CMOS ADC
Power Dissipation versus Sample Rate for 2.5 MHz Input,
5 pF Output Loads

separately as well as the total power dissipation. Note that total power dissipation can vary between approximately 310 mW and 380 mW as the sampling frequency is varied between 10 MSPS and 80 MSPS.

The AD9245 is packaged in a 32-pin leadless chip-scale package as shown in Figure 9.117. The bottom view of the package shows the exposed paddle which should be soldered to the PC board ground plane for best thermal transfer. The worst-case package junction-to-ambient resistance, θ_{JA}, is specified as 32.5°C/W, which places the junction $32.5°C \times 0.38 = 12.3°C$ above the ambient for a power dissipation of 380 mW. For a maximum operating temperature of +85°C, this places the junction at a modest $85°C + 12.3°C = 97.3°C$.

The AD9430 is a high performance 12-bit, 170/210 MSPS 3.3 V BiCMOS ADC. Two output modes are available: dual 105 MSPS demultiplexed CMOS outputs, or 210 MSPS LVDS outputs. Power dissipation as a function of sampling frequency is shown in Figure 9.118. Analog and digital supply currents are shown for CMOS and LVDS modes for an analog input frequency of 10.3 MHz. Note that in the LVDS mode and a sampling frequency of 210 MSPS, total supply current is approximately 455 mA—yielding a total power dissipation of 1.5 W.

The AD9430 is available in a 100-lead thin plastic quad flat package with an exposed pad (TQFP/EP) as shown in Figure 9.119. The conducive pad is connected to chip ground and should be soldered to the PC board ground plane. The θ_{JA} of the package when soldered to the ground plane is 25°C/W in still air. This places the junction $25°C \times 1.5 = 37.5°C$ above the ambient temperature for 1.5 W of power dissipation. For a maximum operating temperature of 85°C, this places the junction at $85°C + 37.5°C = 122.5°C$.

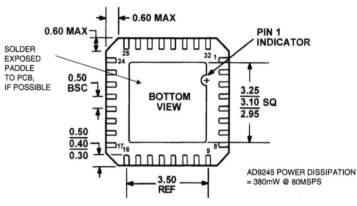

$\theta_{JA} = 32.5°C/W$, PER EIA/DESD51-1, STILL AIR

Figure 9.117: AD9245 CP-32 Lead-Frame Chip-Scale Package (LFCSP), Bottom View

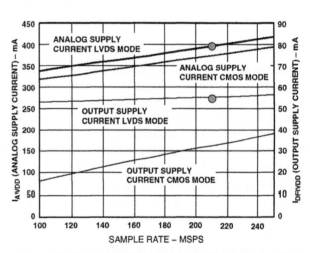

TOTAL CURRENT @ 210MSPS, LVDS MODE = 55mA + 400mA = 455mA
TOTAL POWER DISSIPATION = 3.3V × 455mA = 1.5W

Figure 9.118: AD9430 12-Bit 170/210 MSPS ADC Supply Current versus Sample Rate for 10.3 MHz Input

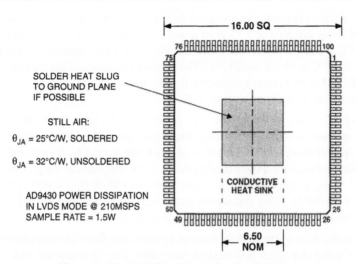

Figure 9.119: AD9430 100-Lead e-PAD TQFP

The AD6645 is a high performance 14-bit, 80/105 MSPS ADC fabricated on a high speed complementary bipolar process (XFCB), and offers the highest SFDR (89 dBc) and SNR (75 dB) currently available in 2004. Although there is little variation in power as a function of sampling frequency, the maximum power dissipation of the device is 1.75 W. The package is a thermally enhanced 52-lead PowerQuad 4® with an exposed pad as shown in Figure 9.120.

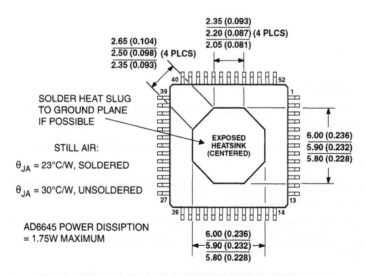

Figure 9.120: AD6645 52-Lead Power-Quad 4 (LQFP_ED) (SQ-52) Thermally Enhanced Package, Bottom View

It is recommended that the exposed center heatsink be soldered to the PC board ground plane to reduce the package θ_{JA} to 23°C/W in still air. For 1.75 W of power dissipation, this places the junction temperature 23°C × 1.75 = 40.3°C above the ambient temperature. For a maximum operating temperature of 85°C, this places the junction at 85°C + 40.3°C = 125.3°C. The thermal resistance of the package can be reduced to 17°C/W with 200 LFPM airflow, thereby reducing the junction temperature to 30°C above the ambient, or 115°C for an operating ambient temperature of 85°C.

High speed CMOS DACs (such as the TxDAC series) and DDS ICs (such as the AD985x series) also have clock-rate dependent power dissipation. For example, in the case of the AD9777 16-bit, 160 MSPS dual interpolating DAC, power dissipation is a function of clock rate, output frequency, and the enabling of the PLL and the modulation functions. Power dissipation on 3.3 V supplies can range from 380 mW (f_{DAC} = 100 MSPS, f_{OUT} = 1 MHz, no interpolation, no modulation) to 1.75 W (f_{DAC} = 400 MSPS, f_{DATA} = 50 MHz, $f_s/2$ modulation, PLL enabled). These and similar parts in the family are also offered in thermally enhanced packages with exposed pads for soldering to the PC board ground plane.

These discussions on the thermal application issues of op amps and data converters haven't dealt with the classic techniques of using clip-on (or bolt-on) type heat sinks. They also have not addressed the use of forced air cooling, generally considered only when tens of watts must be handled. These omissions are mainly because these approaches are seldom possible or practical with today's op amp and data converter packages.

The more general discussions within References 4-7 can be consulted for this and other supplementary information.

References:
9.5 Thermal Management

1. Data sheet for **AD8017 Dual High Output Current, High Speed Amplifier**, Analog Devices, Inc., www.analog.com.

2. Data sheet for **AD8016 Low Power, High Output Current, xDSL Line Driver**, Analog Devices, Inc., www.analog.com.

3. "Power Consideration Discussions," data sheet for **AD815 High Output Current Differential Driver**, Analog Devices, Inc., www.analog.com.

4. Walt Jung, Walt Kester, "Thermal Management," portion of Section 8 within Walt Kester, Editor, **Practical Design Techniques for Power and Thermal Management**, Analog Devices, Inc., 1998, ISBN 0-916550-19-2.

5. General Catalog, **AAVID Thermal Technologies, Inc.**, One Kool Path, Laconia, NH, 03246, 603-528-3400.

6. Seri Lee, "How to Select a Heat Sink," **Aavid Thermal Technologies**, www.aavid.com.

7. Seri Lee, "Optimum Design and Selection of Heat Sinks," **11th IEEE SEMI-THERM™ Symposium**, 1995, www.aavid.com.

EMI/RFI Considerations
James Bryant, Walt Jung, Walt Kester

Analog circuit performance is often adversely affected by high-frequency signals from nearby electrical activity. And, equipment containing analog circuitry may also adversely affect systems external to it. Reference 1 (page 4) describes this complementary transmission of undesirable high-frequency signals from or into local equipment as per an IEC50 definition. These corresponding aspects of the broad arena of *electromagnetic compatibility*, better known as EMC, are:

1) It describes the ability of electrical and electronic systems to operate without interfering with other systems...

2) It also describes the ability of such systems to operate as intended within a specified electromagnetic environment.

So, complete EMC assurance would indicate that the equipment under design should neither produce spurious signals, nor should it be vulnerable to out-of-band external signals (i.e., those outside its intended frequency range). It is the latter class of EMC problem to which analog equipment most often falls prey. The graceful handling of these spurious signals is emphasized within this section.

The externally produced electrical activity may generate noise, and is referred to either as electromagnetic interference (EMI), or radio frequency interference (RFI). In this section, we will refer to EMI in terms of both electromagnetic and radio frequency interference. One of the more challenging tasks of the analog designer is the control of equipment against undesired operation due to EMI. It is important to note that, in this context, *EMI and or RFI is almost always detrimental*. Once given entrance into your equipment, it can and will degrade its operation, quite often considerably.

This section is oriented heavily towards minimizing undesirable analog circuit operation due to the *receipt* of EMI/RFI. Misbehavior of this sort is also known as EMI or RFI *susceptibility*, indicating a tendency towards anomalous equipment behavior when exposed to EMI/RFI. There is, of course, a complementary EMC issue, namely with regard to spurious *emissions*. However, since analog circuits typically involve fewer pulsed, high speed, high current signal edges that give rise to such spurious signals (compared to high speed logic, for example), this aspect of EMC isn't as heavily treated here. Nevertheless, the reader should bear in mind that it can be important, particularly if the analog circuitry is part of a mixed-signal environment along with high speed logic.

Since all of these various EMC design points can be critical, *the end-of-chapter references are strongly recommended for supplementary study*. Indeed, for a thorough, fully competent design with respect to EMI, RFI and EMC, the designer will need to become intimately acquainted with one or more of these references (see References 1–6).

As for the material following, it is best viewed as an introduction to this extremely broad but increasingly important topic.

EMI/RFI Mechanisms

To understand and properly control EMI and RFI, it is helpful to first segregate it into manageable portions. Thus it is useful to remember that when EMI/RFI problems do occur, they can be fundamentally broken down into a *Source*, a *Path*, and a *Receiver*. A systems designer will have the receiver part of this landscape, and perhaps some portion of the path under direct control. But seldom will the designer have control over the actual source.

EMI Noise Sources

There are countless ways in which undesired noise can couple into an analog circuit to ruin its accuracy. Some of the many examples of these noise sources are listed in Figure 9.121.

- EMI/RFI noise sources can couple from anywhere
- Some common sources of externally generated noise:
  - Radio and TV Broadcasts
  - Mobile Radio Communications
  - Cellular Telephones
  - Vehicular Ignition
  - Lightning
  - Utility Power Lines
  - Electric Motors
  - Computers
  - Garage Door Openers
  - Telemetry Equipment

Figure 9.121: Some Common EMI Noise Sources

Since little control is possible over these sources of EMI, the next best management tool one can exercise over them is to recognize and understand the possible paths by which they couple into the equipment under design.

EMI Coupling Paths

The EMI coupling paths are actually very few in terms of basic number. Three very general paths are by:

1) *Interference due to conduction (common-impedance)*

2) *Interference due to capacitive or inductive coupling (near-field interference)*

3) *Electromagnetic radiation (far-field interference)*

Noise Coupling Mechanisms

EMI energy may enter wherever there is an impedance mismatch or discontinuity in a system. In general this occurs at the interface where cables carrying sensitive analog signals are connected to PC boards, and through power supply leads. Improperly connected cables or poor supply filtering schemes are often perfect conduits for interference.

Conducted noise may also be encountered when two or more currents share a common path (impedance). This common path is often a high impedance "ground" connection. If two circuits share this path, noise currents from one will produce noise voltages in the other. Steps may be taken to identify potential sources of this interference (see References 1 and 2, plus Section 2 of this chapter).

Figure 9.122 shows some of the general ways noise can enter a circuit from external sources.

```
•   Impedance mismatches and discontinuities
•   Common-mode impedance mismatches → Differential Signals
•   Capacitively Coupled (Electric Field Interference)
      –  dV/dt → Mutual Capacitance → Noise Current
      –  (Example: 1V/ns produces 1mA/pF)
•   Inductively Coupled (Magnetic Field)
      –  di/dt → Mutual Inductance → Noise Voltage
      –  (Example: 1mA/ns produces 1mV/nH)
```

Figure 9.122: How EMI finds Paths into Equipment

There is a capacitance between any two conductors separated by a dielectric (air and vacuum are dielectrics, as well as all solid or liquid insulators). If there is a change of voltage on one conductor there will be change of charge on the other, and a *displacement current* will flow in the dielectric. Where either the capacitance or the dV/dT is high, noise is easily coupled. For example, a 1 V/ns rate-of-change gives rise to displacement currents of 1 mA/pF.

If changing magnetic flux from current flowing in one circuit couples into another circuit, it will induce an emf in the second circuit. Such *mutual inductance* can be a troublesome source of noise coupling from circuits with high values of dI/dT. As an example, a mutual inductance of 1 nH and a changing current of 1 A/ns will induce an emf of 1 V.

Reducing Common-Impedance Noise

Steps to be taken to eliminate or reduce noise due to the conduction path sharing of impedances, or *common-impedance noise* are outlined in Figure 9.123.

```
•   Common-impedance noise
      –  Decouple op amp power leads at LF and HF
      –  Reduce common-impedance
      –  Eliminate shared paths
•   Techniques
      –  Low impedance electrolytic (LF) and local low inductance
         (HF) bypasses
      –  Use ground and power planes
      –  Optimize system design
```

Figure 9.123: Some Solutions to Common-Impedance Noise

These methods should be applied in conjunction with all of the related techniques discussed earlier within Section 2 of this chapter.

Power supply rails feeding several circuits are good common-impedance examples. Real world power sources may exhibit low output impedance, or may they not—especially over frequency. Furthermore, PCB traces used to distribute power are both inductive and resistive, and may also form a ground loop. The use of power and ground planes also reduces the power distribution impedance. These dedicated conductor layers in a PCB are continuous (ideally, that is) and as such, offer the lowest practical resistance and inductance.

In some applications where low-level signals encounter high levels of common-impedance noise it will not be possible to prevent interference and the system architecture may need to be changed. Possible changes include:

1) *Transmitting signals in differential form*

2) *Amplifying signals to higher levels for improved S/N*

3) *Converting signals into currents for transmission*

4) *Converting signals directly into digital form*

Noise Induced by Near-Field Interference

Crosstalk is the second most common form of interference. In the vicinity of the noise source, i.e., near-field, interference is not transmitted as an electromagnetic wave, and the term crosstalk may apply to either inductively or capacitively coupled signals.

Reducing Capacitance-Coupled Noise

Capacitively-coupled noise may be reduced by reducing the coupling capacity (by increasing conductor separation), but is most easily cured by shielding. A conductive and grounded shield (known as a *Faraday shield*) between the signal source and the affected node will eliminate this noise, by routing the displacement current directly to ground.

With the use of such shields, it is important to note that it is always *essential* that a Faraday shield be grounded. A floating or open-circuit shield almost invariably increases capacitively-coupled noise. For a brief review of this shielding, consult Section 2 of this chapter again, and see References 2 and 3 at the end of this section.

Methods to eliminate capacitance-coupled interference are summarized in Figure 9.124.

- Reduce Level of High dV/dtNoise Sources
- Use Proper Grounding Schemes for Cable Shields
- Reduce Stray Capacitance
  - Equalize Input Lead Lengths
  - Keep Traces Short
  - Use Signal-Ground Signal-Routing Schemes
- Use Grounded Conductive Faraday Shields to Protect
- Against Electric Fields

Figure 9.124: Methods to Reduce Capacitance-Coupled Noise

Reducing Magnetically-Coupled Noise

Methods to eliminate interference caused by magnetic fields are summarized in Figure 9.125.

- Careful Routing of Wiring
- Use Conductive Screens for HF Magnetic Shields
- Use High Permeability Shields for LF Magnetic Fields
- (mu-Metal)
- Reduce Loop Area of Receiver
  - Twisted Pair Wiring
  - Physical Wire Placement
  - Orientation of Circuit to Interference
- Reduce Noise Sources
  - Twisted Pair Wiring
  - Driven Shields

Figure 9.125: Methods to Reduce Magnetically-Coupled Noise

To illustrate the effect of magnetically-coupled noise, consider a circuit with a closed-loop area of A cm² operating in a magnetic field with an rms flux density value of B gauss. The noise voltage V_n induced in this circuit can be expressed by the following equation:

$$V_n = 2 \pi f B A \cos\theta \times 10^{-8} V$$

Eq. 9.12

In this equation, f represents the frequency of the magnetic field, and θ represents the angle of the magnetic field B to the circuit with loop area A. Magnetic field coupling can be reduced by reducing the circuit loop area, the magnetic field intensity, or the angle of incidence. Reducing circuit loop area requires arranging the circuit conductors closer together. Twisting the conductors together reduces the loop net area. This has the effect of canceling magnetic field pickup, because the sum of positive and negative incremental loop areas is ideally equal to zero. Reducing the magnetic field directly may be difficult. However, since magnetic field intensity is inversely proportional to the cube of the distance from the source, physically moving the affected circuit away from the magnetic field has a very great effect in reducing the induced noise voltage. Finally, if the circuit is placed perpendicular to the magnetic field, pickup is minimized. If the circuit's conductors are in parallel to the magnetic field the induced noise is maximized because the angle of incidence is zero.

There are also techniques that can be used to reduce the amount of magnetic-field interference, *at its source*. In the previous paragraph, the conductors of the receiver circuit were twisted together, to cancel the induced magnetic field along the wires. The same principle can be used on the source wiring. If the source of the magnetic field is large currents flowing through nearby conductors, these wires can be twisted together to reduce the net magnetic field.

Shields and cans are not nearly as effective against magnetic fields as against electric fields, but can be useful on occasion. At low frequencies magnetic shields using high-permeability material such as Mu-metal can provide modest attenuation of magnetic fields. At high frequencies simple conductive shields are quite effective provided that the thickness of the shield is greater than the skin depth of the conductor used (at the frequency involved). Note—copper skin depth is $6.6/\sqrt{f}$, with f in Hz.

Passive Components: Your Arsenal Against EMI

Passive components, such as resistors, capacitors, and inductors, are powerful tools for reducing externally induced interference when used properly.

Simple RC networks make efficient and inexpensive one-pole, low-pass filters. Incoming noise is converted to heat and dissipated in the resistor. But note that a fixed resistor does produce thermal noise of its own. Also, when used in the input circuit of an op amp or in amp, such resistor(s) can generate input-bias-current induced offset voltage. While matching the two resistors will minimize the dc offset, the noise will remain. Figure 9.126 summarizes some popular low-pass filters for minimizing EMI.

LP Filter Type	ADVANTAGE	DISADVANTAGE
RC Section	Simple Inexpensive	Resistor Thermal Noise $I_B \times R$ Drop $\rightarrow$ Offset Single-Pole Cutoff
LC Section (Bifilar)	Very Low Noise at LF Very Low IR Drop Inexpensive Two-Pole Cutoff	Medium Complexity Nonlinear Core Effects Possible
π Section (C-L-C)	Very Low Noise at LF Very Low IR Drop Pre-packaged Filters Multiple-Pole Cutoff	Most Complex Nonlinear Core Effects Possible Expensive

**Figure 9.126: Using Passive Components
Within Filters to Combat EMI**

In applications where signal and return conductors aren't well-coupled magnetically, a common-mode (CM) choke can be used to increase their mutual inductance. Note that these comments apply mostly to in amps, which naturally receive a balanced input signal (whereas op amps are inherently unbalanced inputs—unless one constructs an in amp with them). A CM choke can be simply constructed by winding several turns of the differential signal conductors together through a high-permeability (> 2000) ferrite bead. The magnetic properties of the ferrite allow differential-mode currents to pass unimpeded while suppressing CM currents.

Capacitors can also be used before and after the choke, to provide additional CM and differential-mode filtering, respectively. Such a CM choke is cheap and produces very low thermal noise and bias current-induced offsets, due to the wire's low dc resistance. However, there is a field around the core. A metallic shield surrounding the core may be necessary to prevent coupling with other circuits. Also, note that high current levels should be avoided in the core as they may saturate the ferrite.

The third method for passive filtering takes the form of packaged π-networks (C-L-C). These packaged filters are completely self-contained and include feedthrough capacitors at the input and the output as well as a shield to prevent the inductor's magnetic field from radiating noise. These more expensive networks offer high levels of attenuation and wide operating frequency ranges, but the filters must be selected so that for the operating current levels involved the ferrite doesn't saturate.

Reducing System Susceptibility to EMI

The general examples discussed above and the techniques illustrated earlier in this section outline the procedures that can be used to reduce or eliminate EMI/RFI. Considered on a *system* basis, a summary of possible measures is given in Figure 9.127.

> - Always Assume That Interference Exists
> - Use Conducting Enclosures Against Electric and HF Magnetic Fields
> - Use mu-Metal Enclosures Against LF Magnetic Fields
> - Implement Cable Shields Effectively
> - Use Feedthrough Capacitors and Packaged PI Filters

Figure 9.127: Reducing System EMI/RFI Susceptibility

Other examples of filtering techniques useful against EMI are illustrated later in this section, under "Reducing RFI rectification within op amp and in amp circuits."

The section immediately below further details shielding principles.

A Review of Shielding Concepts

The concepts of shielding effectiveness presented next are background material. Interested readers should consult References 4–9 cited at the end of the section for more detailed information.

Applying the concepts of shielding effectively requires an understanding of the source of the interference, the environment surrounding the source, and the distance between the source and point of observation (the receiver). If the circuit is operating close to the source (in the *near*, or induction-field), the field characteristics are determined by the source. If the circuit is remotely located (in the *far*, or radiation-field), the field characteristics are determined by the transmission medium.

A circuit operates in a near-field if its distance from the source of the interference is less than the wavelength (λ) of the interference divided by 2π, or $\lambda/2\pi$. If the distance between the circuit and the source of the interference is larger than this quantity, the circuit operates in the far field. For instance, the interference caused by a 1 ns pulse edge has an upper bandwidth of approximately 350 MHz. The wavelength of a 350 MHz signal is approximately 32 inches (the speed of light is approximately 12"/ns). Dividing the wavelength by 2π yields a distance of approximately 5 inches, the boundary between near- and far-field. If a circuit is within 5 inches of a 350 MHz interference source, then the circuit operates in the near-field of the interference. If the distance is greater than 5 inches, the circuit operates in the far-field of the interference.

Regardless of the type of interference, there is a characteristic impedance associated with it. The characteristic, or wave impedance of a field is determined by the ratio of its electric (or E-) field to its magnetic (or H-) field. In the far field, the ratio of the electric field to the magnetic field is the characteristic (wave impedance) of free space, given by $Z_o = 377 \ \Omega$. In the near field, the wave-impedance is determined by the nature of the interference and its distance from the source. If the interference source is high-current and low voltage (for example, a loop antenna or a power line transformer), the field is predominately magnetic and exhibits a wave impedance which is less than 377 Ω. If the source is low current and high voltage (for example, a rod antenna or a high speed digital switching circuit), then the field is predominately electric and exhibits a wave impedance which is greater than 377 Ω.

Conductive enclosures can be used to shield sensitive circuits from the effects of these external fields. These materials present an *impedance mismatch* to the incident interference, because the impedance of the shield is lower than the wave impedance of the incident field. The effectiveness of the conductive shield depends on two things: First is the loss due to the *reflection* of the incident wave off the shielding material. Second is the loss due to the *absorption* of the transmitted wave *within* the shielding material. The amount of reflection loss depends upon the type of interference and its wave impedance. The amount of absorption loss, however, is independent of the type of interference. It is the same for near- and far-field radiation, as well as for electric or magnetic fields.

Reflection loss at the interface between two media depends on the difference in the characteristic impedances of the two media. For electric fields, reflection loss depends on the frequency of the interference and the shielding material. This loss can be expressed in dB, and is given by:

$$R_e(dB) = 322 + 10\log_{10}\left[\frac{\sigma_r}{\mu_r f^3 r^2}\right]$$ Eq. 9.13

where σ_r = relative conductivity of the shielding material, in Siemens per meter;
μ_r = relative permeability of the shielding material, in Henries per meter;
f = frequency of the interference, and
r = distance from source of the interference, in meters

For magnetic fields, the loss depends also on the shielding material and the frequency of the interference. Reflection loss for magnetic fields is given by:

$$R_m(dB) = 14.6 + 10\log_{10}\left[\frac{fr^2\sigma_r}{\mu_r}\right]$$ Eq. 9.14

and, for plane waves ($r > \lambda/2\pi$), the reflection loss is given by:

$$R_{pw}(dB) = 168 + 10\log_{10}\left[\frac{\sigma_r}{\mu_r f}\right]$$ Eq. 9.15

Absorption is the second loss mechanism in shielding materials. Wave attenuation due to absorption is given by:

$$A(dB) = 3.34t\sqrt{\sigma_r \mu_r f}$$ Eq. 9.16

where *t* = thickness of the shield material, in inches. This expression is valid for plane waves, electric and magnetic fields. Since the intensity of a transmitted field decreases exponentially relative to the thickness of the shielding material, the absorption loss in a shield one skin-depth (δ) thick is 9 dB. Since absorption loss is proportional to thickness and inversely proportional to skin depth, increasing the thickness of the shielding material improves shielding effectiveness at high frequencies.

Reflection loss for plane waves in the far field decreases with increasing frequency because the shield impedance, Z_s, increases with frequency. Absorption loss, on the other hand, increases with frequency because skin depth decreases. For electric fields and plane waves, the primary shielding mechanism is reflection loss, and at high frequencies, the mechanism is absorption loss.

Thus for high-frequency interference signals, lightweight, easily worked high conductivity materials such as copper or aluminum can provide adequate shielding. At low frequencies however, both reflection and absorption loss to magnetic fields is low. It is thus very difficult to shield circuits from low frequency magnetic fields. In these applications, high permeability materials that exhibit low reluctance provide the best protection. These low reluctance materials provide a magnetic shunt path that diverts the magnetic field away from the protected circuit.

To summarize the characteristics of metallic materials commonly used for shielded purposes: Use high conductivity metals for HF interference, and high permeability metals for LF interference.

A properly shielded enclosure is very effective at preventing external interference from disrupting its contents as well as confining any internally-generated interference. However, in the real world, openings in the shield are often required to accommodate adjustment knobs, switches, connectors, or to provide ventilation. Unfortunately, these openings may compromise shielding effectiveness by providing paths for high frequency interference to enter the instrument.

The longest dimension (not the total area) of an opening is used to evaluate the ability of external fields to enter the enclosure, because the openings behave as slot antennas. Equation Eq. 9.17 can be used to calculate the shielding effectiveness, or the susceptibility to EMI leakage or penetration, of an opening in an enclosure:

$$\text{Shielding Effectiveness (dB)} = 20 \log_{10} \left(\frac{\lambda}{2 \times L} \right) \qquad \text{Eq. 9.17}$$

where λ = wavelength of the interference and
$\quad$ L = maximum dimension of the opening

Maximum radiation of EMI through an opening occurs when the longest dimension of the opening is equal to one-half-wavelength of the interference frequency (0 dB shielding effectiveness). A rule-of-thumb is to keep the longest dimension less than 1/20 wavelength of the interference signal, as this provides 20 dB shielding effectiveness. Furthermore, a few small openings on each side of an enclosure is preferred over many openings on one side. This is because the openings on different sides radiate energy in different directions, and as a result, shielding effectiveness is not compromised. If openings and seams cannot be avoided, then conductive gaskets, screens, and paints alone or in combination should be used judiciously to limit the longest dimension of any opening to less than 1/20 wavelength. Any cables, wires, connectors, indicators, or control shafts penetrating the enclosure should have circumferential metallic shields physically bonded to the enclosure at the point of entry. In those applications where unshielded cables/wires are used, then filters are recommended at the shield entry point.

General Points on Cables and Shields

Although covered in detail elsewhere, it is worth noting that the improper use of cables and their shields can be a significant contributor to both radiated and conducted interference. Rather than developing an entire treatise on these issues, the interested reader should consult References 2, 3, 5, and 6 for background.

As shown in Figure 9.128, proper cable/enclosure shielding confines sensitive circuitry and signals *entirely within the shield*, with no compromise to shielding effectiveness.

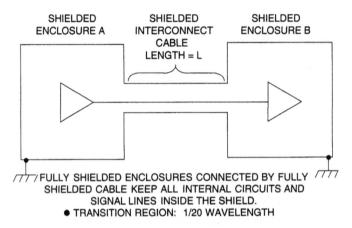

Figure 9.128: Shielded Interconnect Cables Are Either Electrically Long or Short, Depending Upon the Operating Frequency

As can be noted by this diagram, the enclosures and the shield must be properly grounded, otherwise they can act as an antenna, thereby making the radiated and conducted interference problem worse (rather than better).

Depending on the type of interference (pickup/radiated, low/high frequency), proper cable shielding is implemented differently and is very dependent on the length of the cable. The first step is to determine whether the length of the cable is *electrically short* or *electrically long* at the frequency of concern. A cable is considered electrically short if the length of the cable is less than 1/20 wavelength of the highest frequency of the interference. Otherwise it is considered to be electrically long.

For example, at 50 Hz/60 Hz, an electrically short cable is any cable length less than 150 miles, where the primary coupling mechanism for these low frequency electric fields is capacitive. As such, for any cable length less than 150 miles, the amplitude of the interference will be the same over the entire length of the cable.

In applications where the length of the cable is electrically long, or protection against high frequency interference is required, then the preferred method is to connect the cable shield to low impedance points, *at both ends*. As will be seen shortly, this can be a direct connection at the driving end, and a capacitive connection at the receiver. If left ungrounded, unterminated transmission line effects can cause reflections and standing waves along the cable. At frequencies of 10 MHz and above, circumferential (360°) shield bonds and metal connectors are required to main low impedance connections to ground.

In summary, for protection against low frequency (<1 MHz), electric-field interference, grounding the shield at one end is acceptable. For high frequency interference (>1 MHz), the preferred method is grounding the shield at both ends, using 360° circumferential bonds between the shield and the connector, and maintaining metal-to-metal continuity between the connectors and the enclosure.

However, in practice, there is a caveat involved with directly grounding the shield at both ends. When this is done, it creates a low frequency ground loop, shown in Figure 9.129.

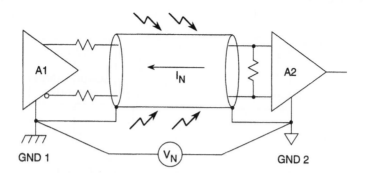

- V_N Causes Current in Shield (Usually 50/60Hz)

- Differential Error Voltage is Produced at Input of A2 Unless:
    - A1 Output is Perfectly Balanced and
    - A2 Input is Perfectly Balanced and
    - Cable is Perfectly Balanced

Figure 9.129: Ground Loops in Shielded Twisted Pair Cable Can Cause Errors

Whenever two systems A1 and A2 are remote from each other, there is usually a difference in the ground potentials at each system, i.e., V_N. The frequency of this potential difference is generally the line frequency (50 Hz or 60 Hz) and multiples thereof. But, if the shield is directly grounded at both ends as shown, noise current I_N flows in the shield. In a perfectly balanced system, the common-mode rejection of the system is infinite, and this current flow produces no differential error at the receiver A2. However, perfect balance is never achieved in the driver, its impedance, the cable, or the receiver, so a certain portion of the shield current will appear as a differential noise signal, at the input of A2. The following illustrates correct shield grounding for various examples.

As noted above, cable shields are subject to both low and high frequency interference. Good design practice requires that the shield be grounded at both ends if the cable is electrically long to the interference frequency, as is usually the case with RF interference.

Figure 9.130 shows a remote passive RTD sensor connected to a bridge and conditioning circuit by a shielded cable. The proper grounding method is shown in the upper part of the figure, where the shield is grounded at the receiving end.

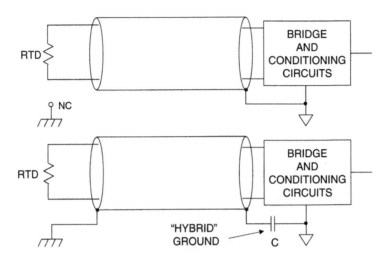

Figure 9.130: Hybrid Grounding of Shielded Cable With Passive Sensor

However, safety considerations may require that the remote end of the shield also be grounded. If this is the case, the receiving end can be grounded with a low inductance ceramic capacitor (0.01 µF to 0.1 µF), still providing high frequency grounding. The capacitor acts as a ground to RF signals on the shield but blocks low frequency line current to flow in the shield. This technique is often referred to as a *hybrid ground*.

A case of an active remote sensor and/or other electronics is shown Figure 9.131. In both of the two situations, a hybrid ground is also appropriate, either for the balanced (upper) or the single-ended (lower) driver case. In both instances the capacitor "C" breaks the low frequency ground loop, providing effective RF grounding of the shielded cable at the A2 receiving end at the right side of the diagram.

There are also some more subtle points that should be made with regard to the source termination resistances used, R_S. In both the balanced as well as the single-ended drive cases, the driving signal seen on the balanced line originates from a net impedance of R_S, which is split between the two twisted pair legs as twice $R_S/2$. In the upper case of a fully differential drive, this is straightforward, with an $R_S/2$ valued resistor connected in series with the complementary outputs from A1.

In the bottom case of the single-ended driver, note that there are still two $R_S/2$ resistors used, one in series with both legs. Here the grounded dummy return leg resistor provides an impedance-balanced ground connection drive to the differential line, aiding in overall system noise immunity. Note that this implementation is only useful for those applications with a balanced receiver at A2, as shown.

Coaxial cables are different from shielded twisted pair cables in that the signal return current path is through the shield. For this reason, the ideal situation is to ground the shield at the driving end and allow the shield to float at the differential receiver (A2) as shown in the upper portion of Figure 9.132. For this technique to work, however, the receiver must be a differential type with good high frequency CM rejection.

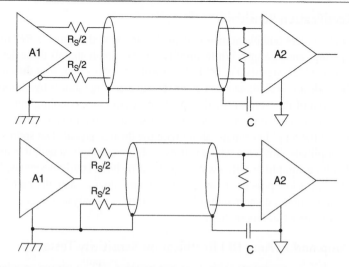

**Figure 9.131: Impedance-Balanced Drive of Balanced Shielded Cable Aids
Noise-Immunity with Either Balanced or Single-Ended Source Signals**

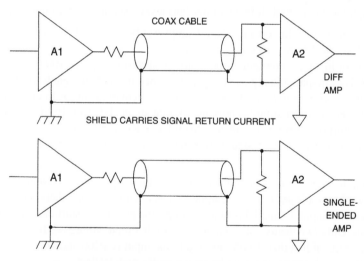

**Figure 9.132: Coaxial Cables Can Use Either Balanced
or Single-Ended Receivers**

However, the receiver may be a single-ended type, such as typical of a standard single op amp type circuit. This is true for the bottom example of Figure 9.132, so there is no choice but to ground the coaxial cable shield at both ends for this case.

Input-Stage RFI Rectification Sensitivity

A well-known but poorly understood phenomenon in analog integrated circuits is *RFI rectification*, specifically as it occurs in op amps and in amps. While amplifying very small signals these devices can rectify large-amplitude, out-of-band HF signals, i.e., RFI. As a result, dc errors appear at the output in addition to the desired signal. The undesired HF signals can enter sensitive analog circuits by various means. Conductors leading into and out of the circuit provide a path for interference coupling into a circuit. These conductors pick up noise through capacitive, inductive, or radiation coupling, as discussed earlier. The spurious signals appear at the amplifier inputs, along with the desired signal. The spurious signals can be several tens of mV in amplitude, however, which causes problems. Simply stated, it cannot be assumed that a sensitive, low bandwidth dc amplifier will always reject out-of-band spurious signals. While this would be the case for a simple linear low pass filter, op amp and in amp devices actually rectify high-level HF signals, leading to nonlinearities and anomalous offsets. Methods of analysis for as well as the prevention of RFI rectification are discussed in this section.

Background: Op Amp and In Amp RFI Rectification Sensitivity Tests

Just about all in amp and op amp input stages use emitter-coupled BJT or source-coupled FET differential pairs of some type. Depending on the device operating current, the interfering frequency and its relative amplitude, these differential pairs can behave as high-frequency detectors. As will be shown, the detection process produces spectral components at the harmonics of the interference, as well at dc. It is the detected dc component of the interference that shifts amplifier bias levels, leading to inaccuracies.

The effect of RFI rectification within op amps and in amps can be evaluated with relatively simple test circuits, as described for the *RFI Rectification Test Configuration* (see pages 1–38 of Reference 10). In these tests, an op amp or in amp is configured for a gain of –100 (op amp), or 100 (in amp), with dc output measured after a 100 Hz low-pass filter, preventing interference from other signals. A 100 MHz, 20 mVp-p signal is the test stimulus, chosen to be well above test device frequency limits. In operation, the test evaluates dc output shift observed under stimulus presence. While an ideal dc shift for this measurement would be zero, the actual dc shift of a given part indicates the relative RFI rectification sensitivity. Devices using both BJT and FET technologies can be tested by this method, as can devices operating at either low or high supply current levels.

In the original op amp test device set of Reference 10, some FET-input devices (OP80, OP42, OP249 and AD845) exhibited no observable shift in their output voltages, while several others showed shifts of less than 10 µV referred to the input. Of the BJT-input op amps, the amount of shift decreased with increasing device supply current. Only two devices showed no observable output voltage shift (AD797 and AD827), while others showed shifts of less than 10 µV referred to the input (OP200 and OP297). For other op amps, it is to be expected that similar patterns would be shown under such testing.

From these tests, some generalizations on RFI rectification can be made. First, device susceptibility appears to be inversely proportional to supply current; that is, devices biased at low quiescent supply currents exhibit greatest output voltage shift. Second, ICs with FET-input stages appeared to be less susceptible to rectification than those with BJTs. Note that these points are independent of whether the device is an op amp or an in amp. In practice this means that the lower power op amps *or* in amps will tend to be more susceptible to RFI rectification effects. And, FET-input op amps (or in amps) will tend to be *less* susceptible to RFI, especially those operating at higher currents.

Based on these data and from the fundamental differences between BJTs and FETs, we can summarize what we know. Bipolar transistor action is controlled by a forward-biased p-n junction (the base-emitter junction) whose I-V characteristic is exponential and quite nonlinear. FET behavior, on the other hand, is controlled by voltages applied to a reverse-biased p-n junction diode (the gate-source junction). The I-V characteristic of FETs is a square-law, and thus it is inherently more linear than that of BJTs.

For the case of the lower supply current devices, transistors in the circuit are biased well below their peak f_T collector currents. Although the ICs may be constructed on processes whose device f_ts can reach hundreds of MHz, charge transit times increase, when transistors are operated at low current levels. The impedance levels used also make RFI rectification in these devices worse. In low-power op amps, impedances are on the order of hundreds to thousands of kΩs, whereas in moderate supply-current designs impedances might be no more than just a few kΩ. Combined, these factors tend to degrade a low power device's RFI rectification sensitivity.

Figure 9.133 summarizes these general observations on RFI rectification sensitivity, and is applicable to both op amps and in amps.

> - BJT input devices rectify readily
>   - Forward-biased B-E junction
>   - Exponential I-V Transfer Characteristic
> - FET input devices less sensitive to rectifying
>   - Reversed-biased p-n junction
>   - Square-law I-V Transfer Characteristic
> - Low I_{supply} devices versus High I_{supply} devices
>   - Low I_{supply} ⇒ Higher rectification sensitivity
>   - High I_{supply} ⇒ Lower rectification sensitivity

Figure 9.133: Some General Observations on Op Amp and In Amp Input Stage RFI Rectification Sensitivity

An Analytical Approach: BJT RFI Rectification

While lab experiments can demonstrate that BJT-input devices exhibit greater RFI rectification sensitivity than comparable devices with FET inputs, a more analytical approach can also be taken to explain this phenomenon.

RF circuit designers have long known that p-n junction diodes are efficient rectifiers because of their nonlinear I-V characteristics. A spectral analysis of a BJT transistor current output for a HF sinewave input reveals that, as the device is biased closer to its "knee," nonlinearity increases. This, in turn, makes its use as a detector more efficient. This is especially true in low power op amps, where input transistors are biased at very low collector currents.

A rectification analysis for the collector current of a BJT has been presented in Reference 10, and will not be repeated here except for the important conclusions. These results reveal that the original quadratic second-order term can be simplified into a frequency-dependent term, Δi_c (AC), at twice the input frequency and a dc term, Δi_c(DC). The latter component can be expressed as noted in Eq. 9.18, the final form for the rectified dc term:

$$\Delta i_c\left(DC\right)=\left(\frac{V_X}{V_T}\right)^2 \times \frac{I_C}{4} \qquad\qquad \text{Eq. 9.18}$$

This expression shows that the dc component of the second-order term is directly proportional to the *square* of the HF noise amplitude V_X, and, also, to I_C, the quiescent collector current of the transistor. To illustrate this point on rectification, note that the change in dc collector current of a bipolar transistor operating at an I_C of 1 mA with a spurious 10 mV$_{peak}$ high frequency signal impinging upon it will be about 38 µA.

Reducing the amount of rectified collector current is a matter of reducing the quiescent current, or the magnitude of the interference. Since the op amp and in amp input stages seldom provide adjustable quiescent collector currents, reducing the level of interfering noise V_X is by far the best (and almost always the only) solution. For example, reducing the amplitude of the interference by a factor of 2, down to 5 mV$_{peak}$ produces a net 4-to-1 reduction in the rectified collector current. Obviously, this illustrates the importance of keeping spurious HF signals away from RFI sensitive amplifier inputs.

An Analytical Approach: FET RFI Rectification

A rectification analysis for the drain current of a JFET has also been presented in Reference 10, and isn't repeated here. A similar approach was used for the rectification analysis of a FET's drain current as a function of a small voltage V_X, applied to its gate. The results of evaluating the second-order rectified term for the FET's drain current are summarized in Eq. 9.19. Like the BJT, an FET's second-order term has an ac and a dc component. The simplified expression for the dc term of the rectified drain current is given here, where the rectified dc drain current is directly proportional to the square of the amplitude of V_X, the spurious signal. However, Eq. 9.19 also reveals a very important difference between the *degree* of the rectification produced by FETs relative to BJTs.

$$\Delta i_D(DC) = \left(\frac{V_X}{V_P}\right)^2 \times \frac{I_{DSS}}{2}$$

Eq. 9.19

Whereas in a BJT the change in collector current has a direct relationship to its quiescent collector current level, the change in a JFET's drain current is proportional to its drain current at zero gate-source voltage, I_{DSS}, and inversely proportional to the square of its channel pinch-off voltage, V_P—parameters that are geometry and process dependent. Typically, JFETs used in the input stages of in amps and op amps are biased with their quiescent current of ~0.5 × I_{DSS}. Therefore, the change in a JFET's drain current is independent of its quiescent drain current; hence, independent of the operating point.

A quantitative comparison of second-order rectified dc terms between BJTs and FETs is illustrated in Figure 9.134. In this example, a bipolar transistor with a unit emitter area of 576 µm² is compared to a unit-area JFET designed for an I_{DSS} of 20 µA and a pinch-off voltage of 2 V. Each device is biased at 10 µA and operated at $T_A = 25°C$.

The important result is that, under identical quiescent current levels, the change in collector current in bipolar transistors is about 1500 times greater than the change in a JFET's drain current. This explains why FET-input amplifiers behave with less sensitivity to large amplitude HF stimulus. As a result, they offer more RFI rectification immunity.

- BJT:

  Emitter area = 576µm²

  I_C = 10µA

  V_T = 25.68mV @ 25°C

  $$\Delta i_C = \left(\frac{V_X}{V_T}\right)^2 \times \frac{I_C}{4}$$

  $$= \frac{V_X^2}{264}$$

- JFET:

  I_{DSS} = 20µA (Z/L=1)

  V_P = 2V

  I_D = 10µA

  $$\Delta i_D = \left(\frac{V_X}{V_T}\right)^2 \times \frac{I_{DSS}}{2}$$

  $$= \frac{V_X^2}{400 \times 10^3}$$

- Conclusion: BJTs ~1500 more sensitive than JFETs

Figure 9.134: Relative Sensitivity Comparison – BJT versus JFET

What all this boils down to is this: Since a user has virtually no access to the amplifier's internal circuitry, the prevention of IC circuit performance degradation due to RFI is essentially left to those means which are external to the ICs.

As the analysis above shows, regardless of the amplifier type, *RFI rectification is directly proportional to the square of the interfering signal's amplitude*. Therefore, to minimize RFI rectification in precision amplifiers, the level of interference must be reduced or eliminated, *prior to the stage*. The most direct way to reduce or eliminate the unwanted noise is by proper filtering.

This topic is covered in the section immediately following.

Reducing RFI Rectification Within Op amp and In Amp Circuits

EMI and RFI can seriously affect the dc performance of high accuracy analog circuits. Because of their relatively low bandwidth, precision op amps and in amps simply won't accurately amplify RF signals in the MHz range. However, if these out-of-band signals are allowed to couple into a precision amplifier through either its input, output, or power supply pins, they can be internally rectified by various amplifier junctions, ultimately causing an undesirable dc offset at the output. The previous theoretical discussion of this phenomenon has shown its basic mechanisms. The logical next step is to show how proper filtering can minimize or eliminate these errors.

Elsewhere in this chapter we have discussed how proper supply decoupling minimizes RFI on IC power pins. Further discussion is required with respect to the amplifier inputs and outputs, *at the device level*. It is assumed at this point that system level EMI/RFI approaches have already been implemented, such as an RFI-tight enclosure, properly grounded shields, power rail filtering, etc. The steps following can be considered as circuit-level EMI/RFI prevention.

Op Amp Inputs

The best way to prevent input stage rectification is to use a low-pass filter located close to the op amp input as shown in Figure 9.135. In the case of the inverting op amp at the left, filter capacitor C is placed between equal-value resistors R1-R2. This results in a simple corner frequency expression, as shown in the figure. At very low frequencies or dc, the closed loop gain of the circuit is –R3/(R1+R2). Note that C cannot be connected directly to the inverting input of the op amp, since that would cause instability. The filter bandwidth can be chosen at least 100 times the signal bandwidth to minimize signal loss.

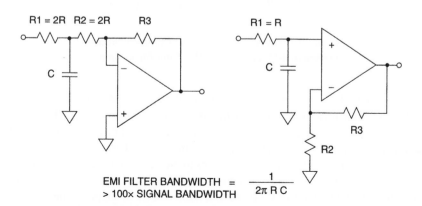

$$\text{EMI FILTER BANDWIDTH} = \frac{1}{2\pi R C} > 100 \times \text{SIGNAL BANDWIDTH}$$

Figure 9.135: Simple EMI/RFI Noise Filters for Op Amp Circuits

For the noninverting case on the right, capacitor C can be connected directly to the op amp input as shown, and an input resistor with a value "R" yields the same corner frequency as the inverting case. In both cases low inductance chip-style capacitors such as NP0 ceramics should be used. The capacitor should in any case be free of losses or voltage coefficient problems, which limits it to either the NP0 mentioned, or a film type.

It should be noted that a ferrite bead can be used instead of R1; however, ferrite bead impedance is not well controlled and is generally no greater than 100 Ω at 10 MHz to 100 MHz. This requires a large value capacitor to attenuate lower frequencies.

In Amp Inputs

Precision in amps are particularly sensitive to dc offset errors due to the presence of CM EMI/RFI. This is very much like the problem in op amps. And, as is true with op amps, the sensitivity to EMI/RFI is more acute with the lower power in amp devices.

A general-purpose approach to proper filtering for device level application of in amps is shown in Figure 9.136. In this circuit the in amp could in practice be any one of a number of devices. The relatively complex balanced RC filter preceding the in amp performs all of the high frequency filtering. The in amp would be programmed for the gain required in the application, via its gain-set resistance (not shown).

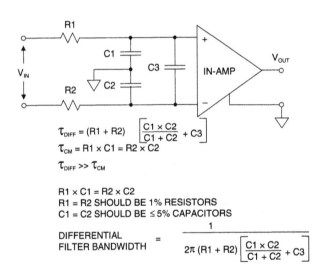

$$\tau_{DIFF} = (R1 + R2) \left[\frac{C1 \times C2}{C1 + C2} + C3 \right]$$

$$\tau_{CM} = R1 \times C1 = R2 \times C2$$

$$\tau_{DIFF} \gg \tau_{CM}$$

R1 $\times$ C1 = R2 $\times$ C2
R1 = R2 SHOULD BE 1% RESISTORS
C1 = C2 SHOULD BE $\leq$ 5% CAPACITORS

$$\text{DIFFERENTIAL FILTER BANDWIDTH} = \frac{1}{2\pi (R1 + R2) \left[\frac{C1 \times C2}{C1 + C2} + C3 \right]}$$

Figure 9.136: A General-Purpose Common-Mode/
Differential-Mode RC EMI/RFI Filter for In Amps

Within the filter, note that fully balanced filtering is provided for both CM (R1-C1 and R2-C2) as well as differential mode (DM) signals (R1+R2, and C3 || the series connection of C1-C2). If R1-R2 and C1-C2 aren't well matched, some of the input common-mode signal at V_{IN} will be converted to a differential mode signal at the in amp inputs. For this reason, C1 and C2 should be matched to within at least 5% of each other. Also, R1 and R2 should be 1% metal film resistors to aid this matching. It is assumed that the source resistances seen at the V_{IN} terminals are low with respect to R1-R2, and matched. In this type of filter, C3 should be chosen much larger than C1 or C2 (C3 $\geq$ C1, C2), in order to suppress spurious differential signals due to CM$\Rightarrow$DM conversion resulting from mismatch of the R1-C1 and R2-C2 time constants.

The overall filter bandwidth should be at least 100 times the input signal bandwidth. Physically, the filter components should be symmetrically mounted on a PC board with a large area ground plane and placed close to the in amp inputs for optimum performance.

Figure 9.137 shows a family of these filters, as suited to a range of different in amps. The RC components should be tailored to the different in amp devices, as per the table. These filter components are selected for a reasonable balance of low EMI/RFI sensitivity and a low increase in noise (vis-à-vis that of the related in amp, without the filter).

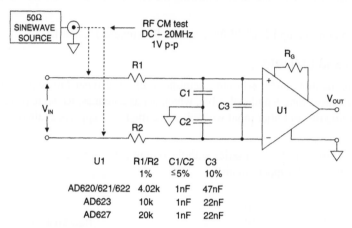

U1	R1/R2	C1/C2	C3
	1%	≤5%	10%
AD620/621/622	4.02k	1nF	47nF
AD623	10k	1nF	22nF
AD627	20k	1nF	22nF

Figure 9.137: Flexible Common-Mode and Differential-Mode RC EMI/RFI Filters Are Useful with the AD620 Series, the AD623, AD627, and Other In Amps

To test the EMI/RFI sensitivity of the configuration, a 1 Vp-p CM signal can be applied to the input resistors, as noted. With a typically used in amp such as the AD620 working at a gain of 1000, the maximum RTI input offset voltage shift observed was 1.5 µV over the 20 MHz range. In the AD620 filter example, the differential bandwidth is about 400 Hz.

Common-mode chokes offer a simple, one-component EMI/RFI protection alternative to the passive RC filters, as shown in Figure 9.138.

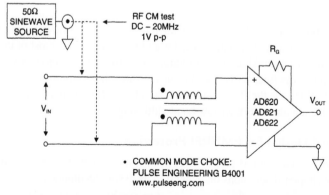

Figure 9.138: For Simplicity as Well as Lowest Noise EMI/RFI Filter Operation, a Common-Mode Choke is Useful with the AD620 Series In Amp Devices

In addition to being a low component count approach, choke-based filters offer low noise, by dispensing with the resistances. Selecting the proper common-mode choke is critical, however. The choke used in the circuit of Figure 9.138 is a Pulse Engineering B4001. The maximum RTI offset shift measured from dc to 20 MHz at G = 1000 was 4.5 µV. Either an off-the-shelf choke such as the B4001 can be used for this filter, or, alternately one can be constructed. Since balance of the windings is important, bifilar wire is suggested. The core material must of course operate over the expected frequency band. Note that, unlike the Figure 9.137 family of RC filters, a choke-only filter offers no differential filtration. Differential mode filtering can be optionally added, with a second stage following the choke, by adding the R1-C3-R2 connections of Figure 9.137.

For further information on in amp EMI/RFI filtering, see References 10, and 12–15.

Amplifier Outputs and EMI/RFI

In addition to filtering the input and power pins, amplifier *outputs* also need to be protected from EMI/RFI, especially if they must drive long lengths of cable, which act as antennas. RF signals received on an output line can couple back into the amplifier input where it is rectified, and appears again on the output as an offset shift.

A resistor and/or ferrite bead, or both, in series with the output is the simplest and least expensive output filter, as shown in Figure 9.139 (upper circuit).

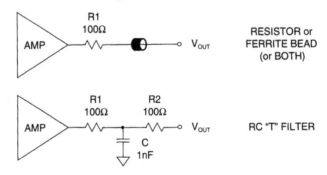

Figure 9.139: Op Amp and In Amp Outputs Should be Protected Against EMI/RFI, Particularly if They Drive Long Cables

Adding a resistor-capacitor-resistor "T" circuit as shown in Figure 9.139 (lower circuit) improves this filter with just slightly more complexity. The output resistor and capacitor divert most of the high frequency energy away from the amplifier, making this configuration useful even with low power active devices. Of course, the time constant of the filter parts must be chosen carefully, to minimize any degradation of the desired output signal. In this case the RC components are chosen for an approximate 3 MHz signal bandwidth, suitable for instrumentation or other low bandwidth stages.

Printed Circuit Board Design for EMI/RFI Protection

This section summarizes general points on EMI/RFI with respect to the printed circuit board (PCB) layout. It complements earlier chapter discussions on general PCB design techniques.

When a PCB design has not been optimized in terms of EMI/RFI, system performance can be compromised. This is true not only for signal-path performance, but also for the system's susceptibility to EMI, plus the degree of EMI radiated by the system. Failure to implement sound PCB layout techniques will very likely lead to system/instrument EMC failures.

To summarize earlier points of this section, a real-world PCB layout may allow multiple paths through which high-frequency noise can couple/radiate into and/or out of the circuit. This is especially true for digital circuitry, operating at high *edge rates*. It is the rapid changes of logic state ($1 \Rightarrow 0$ or $0 \Rightarrow 1$), i.e., the edge rate which contains the HF energy which can easily radiate as EMI. While similar points are applicable to precision high speed analog or mixed analog/digital circuits, logic devices are by far the worst potential EMI offenders. Identifying critical circuits and paths helps in designing the PCB for both low emissions and susceptibility to radiated and conducted external and internal noise sources.

Choose Logic Devices Carefully

Logic family speaking, a key point in minimizing system noise problems is to *choose devices no faster than actually required by the application*. Many designers assume that faster is always better—fast logic is better than slow, high bandwidth amplifiers better than low bandwidth ones, and fast DACs and ADCs are better, even if the speed isn't required by the system. Unfortunately, faster is *not* better, and actually may be worse for EMI concerns.

Many fast DACs and ADCs have digital inputs and outputs with edge rates in the 1 V/ns region. Because of this wide bandwidth, the sampling clock and the digital inputs can respond to any form of high frequency noise, even glitches as narrow as 1 ns to 3 ns. These high speed data converters and amplifiers are thus easy prey for the high frequency noise of microprocessors, digital signal processors, motors, switching regulators, hand-held radios, electric jackhammers, etc. With some of these high speed devices, a small amount of input/output filtering may be required to desensitize the circuit from its EMI/RFI environment. A ferrite bead just before the local decoupling capacitor is very effective in filtering high frequency noise on supply lines. Of course, with circuits requiring bipolar supplies, this technique should be applied to both positive and negative supply lines.

To help reduce emissions generated by extremely fast moving digital signals at DAC inputs or ADC outputs, a small resistor or ferrite bead may be required at each digital input/output.

Design PCBs Thoughtfully

Once the system's critical paths and circuits have been identified, the next step in implementing sound PCB layout is to partition the printed circuit board according to circuit function. This involves the appropriate use of power, ground, and signal planes. Good PCB layouts also isolate critical analog paths from sources of high interference (I/O lines and connectors, for example). High frequency circuits (analog and digital) should be separated from low frequency ones. Furthermore, automatic signal routing CAD layout software should be used with extreme caution. Critical signal paths should be routed by hand, to avoid undesired coupling and/or emissions.

Properly designed multilayer PCBs can reduce EMI emissions and increase immunity to RF fields, by a factor of 10 or more, compared to double-sided boards. A multilayer board allows a complete layer to be used for the ground plane, whereas the ground plane side of a double-sided board is often disrupted with signal crossovers, etc. If the system has separate analog and digital ground and power planes, the analog ground plane should be underneath the analog power plane, and similarly, the digital ground plane should be underneath the digital power plane. There should be no overlap between analog and digital ground planes, nor analog and digital power planes.

Designing Controlled Impedances Traces on PCBs

A variety of trace geometries are possible with controlled impedance designs, and they may be either integral to or allied to the PCB pattern. In the discussions below, the basic patterns follow those of the IPC, as described in standard 2141 (see Reference 16).

Note that the figures below use the term "ground plane." It should be understood that this plane is in fact a large area, low impedance *reference* plane. In practice it may actually be either a ground plane or a power plane, both of which are assumed to be at zero ac potential.

The first of these is the simple wire-over-a-plane form of transmission line, also called a *wire microstrip*. A cross-sectional view is shown in Figure 9.140. This type of transmission line might be a signal wire used within a breadboard, for example. It is composed simply of a discrete insulated wire spaced a fixed distance over a ground plane. The dielectric would be either the insulation wall of the wire, or a combination of this insulation and air.

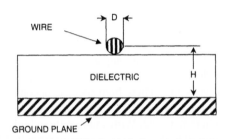

Figure 9.140: A Wire Microstrip Transmission Line with Defined Impedance is Formed by an Insulated Wire Spaced From a Ground Plane

The impedance of this line in ohms can be estimated with Eq. 9.20. Here D is the conductor diameter, H the wire spacing above the plane, and ε_r the dielectric constant.

$$Z_0\left(\Omega\right)= \frac{60}{\sqrt{\varepsilon_r}} \ln\left[\frac{4H}{D}\right]$$

Eq. 9.20

For patterns integral to the PCB, there are a variety of geometric models from which to choose, single-ended and differential. These are covered in some detail within IPC standard 2141 (see Reference 16), but information on two popular examples is shown here.

Before beginning any PCB-based transmission line design, it should be understood that there are abundant equations, all claiming to cover such designs. In this context, "Which of these are accurate?" is an extremely pertinent question. The unfortunate answer is, *"None are perfectly so."* All of the existing equations are approximations, and thus accurate to varying degrees, depending upon specifics. The best known and most widely quoted equations are those of Reference 16, but even these come with application caveats.

Reference 17 has evaluated the Reference 16 equations for various geometric patterns against test PCB samples, finding that predicted accuracy varies according to target impedance. Reference 18 also evaluates the Reference 16 equations, offering an alternative and even more complex set (see Reference 19). The equations quoted below are from Reference 16, and are offered here as a starting point for a design, subject to further analysis, testing and design verification. The bottom line is, study carefully, and take PCB trace impedance equations with a proper dose of salt.

Microstrip PCB Transmission Lines

For a simple two-sided PCB design where one side is a ground plane, a signal trace on the other side can be designed for controlled impedance. This geometry is known as a *surface microstrip,* or more simply, *microstrip.*

A cross-sectional view of a two-layer PCB illustrates this microstrip geometry as shown in Figure 9.141.

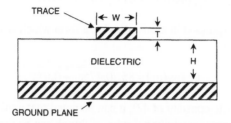

Figure 9.141: A Microstrip Transmission Line with Defined Impedance is Formed by a PCB Trace of Appropriate Geometry, Spaced From a Ground Plane

For a given PCB laminate and copper weight, note that all parameters will be predetermined except for W, the width of the signal trace. Eq. 9.21 can then be used to design a PCB trace to match the impedance required by the circuit. For the signal trace of width W and thickness T, separated by distance H from a ground (or power) plane by a PCB dielectric with dielectric constant ε_r, the characteristic impedance is:

$$Z_O\left(\Omega\right) = \frac{87}{\sqrt{\varepsilon_r + 1.41}} \ln\left[\frac{5.98H}{\left(0.8W + T\right)}\right]$$

Eq. 9.21

Note that in these expressions, measurements are in common dimensions (mils).

These transmission lines will have not only a characteristic impedance, but also capacitance. This can be calculated in terms of pF/in as shown in Eq. 9.22.

$$C_O\left(pF/in\right) = \frac{0.67\left(\varepsilon_r + 1.41\right)}{\ln\left[5.98H/\left(0.8W + T\right)\right]}$$

Eq. 9.22

As an example including these calculations, a 2-layer board might use 20 mil wide (W), 1 ounce (T=1.4) copper traces separated by 10 mil (H) FR-4 (ε_r = 4.0) dielectric material. The resulting impedance for this microstrip would be about 50 Ω. For other standard impedances, for example the 75 Ω video standard, adjust "W" to about 8.3 mils.

Some Microstrip Guidelines

This example touches an interesting and quite handy point. Reference 17 discusses a useful guideline pertaining to microstrip PCB impedance. For a case of dielectric constant of 4.0 (FR-4), it turns out that when W/H is 2/1, the resulting impedance will be close to 50 Ω (as in the first example, with W = 20 mils).

Careful readers will note that Eq. 9.21 predicts Z_O to be about 46 Ω, generally consistent with accuracy quoted in Reference 17 (>5%). The IPC microstrip equation is most accurate between 50 Ω and 100 Ω, but is substantially less so for lower (or higher) impedances. Reference 20 gives tabular results of various PCB industry impedance calculator tools.

The propagation delay of the microstrip line can also be calculated, as per Eq. 9.23. This is the one-way transit time for a microstrip signal trace. Interestingly, for a given geometry model, *the delay constant in ns/ft is a function only of the dielectric constant, and not the trace dimensions* (see Reference 21). Note that this is quite a convenient situation. It means that with a given PCB laminate (and given ε_r), the propagation delay constant is fixed for various impedance lines.

$$t_{pd}\,(ns/ft) = 1.017\sqrt{0.475\varepsilon_r + 0.67} \qquad\qquad \text{Eq. 9.23}$$

This delay constant can also be expressed in terms of ps/in, a form which will be more practical for smaller PCBs. This is:

$$t_{pd}\,(ps/in) = 85\sqrt{0.475\varepsilon_r + 0.67} \qquad\qquad \text{Eq. 9.24}$$

Thus for an example PCB dielectric constant of 4.0, it can be noted that a microstrip's delay constant is about 1.63 ns/ft, or 136 ps/in. These two additional rules of thumb can be useful in designing the timing of signals across PCB trace runs.

Symmetric Stripline PCB Transmission Lines

A method of PCB design preferred from many viewpoints is a multilayer PCB. This arrangement *embeds* the signal trace between a power and a ground plane, as shown in the cross-sectional view of Figure 9.142. The low impedance ac ground planes and the embedded signal trace form a *symmetric stripline* transmission line.

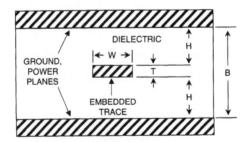

Figure 9.142: A Symmetric Stripline Transmission Line with Defined Impedance is Formed by a PCB Trace of Appropriate Geometry Embedded between Equally Spaced Ground and/or Power Planes

As can be noted from the figure, the return current path for a high frequency signal trace is located directly above and below the signal trace on the ground/power planes. The high frequency signal is thus contained entirely inside the PCB, minimizing emissions, and providing natural shielding against incoming spurious signals.

The characteristic impedance of this arrangement is again dependent upon geometry and the ε_r of the PCB dielectric. An expression for Z_0 of the stripline transmission line is:

$$Z_0\,(\Omega) = \frac{60}{\sqrt{\varepsilon_r}}\ln\left[\frac{1.9(B)}{(0.8W + T)}\right] \qquad\qquad \text{Eq. 9.25}$$

Here, all dimensions are again in mils, and B is the spacing between the two planes. In this symmetric geometry, note that B is also equal to 2H + T. Reference 17 indicates that the accuracy of this Reference 16 equation is typically on the order of 6%.

Another handy guideline for the symmetric stripline in an $\varepsilon_r = 4.0$ case is to make B a multiple of W, in the range of 2 to 2.2. This will result in an stripline impedance of about 50Ω. Of course this rule is based on a further approximation, by neglecting T. Nevertheless, it is still useful for ballpark estimates.

The symmetric stripline also has a characteristic capacitance, which can be calculated in terms of pF/in as shown in Eq. 9.26.

$$C_O\left(pF/in\right) = \frac{1.41\left(\varepsilon_r\right)}{1n\left[3.81H/\left(0.8W + T\right)\right]} \qquad \text{Eq. 9.26}$$

The propagation delay of the symmetric stripline is shown in Eq. 9.27.

$$t_{pd}\left(ns/ft\right) = 1.017\sqrt{\varepsilon_r} \qquad \text{Eq. 9.27}$$

or, in terms of ps:

$$t_{pd}\left(ps/in\right) = 85\sqrt{\varepsilon_r} \qquad \text{Eq. 9.28}$$

For a PCB dielectric constant of 4.0, it can be noted that the symmetric stripline's delay constant is almost exactly 2 ns/ft, or 170 ps/in.

Some Pros and Cons of Embedding Traces

The above discussions allow the design of PCB traces of defined impedance, either on a surface layer or embedded between layers. There of course are many other considerations beyond these impedance issues.

Embedded signals do have one major and obvious disadvantage— the debugging of the hidden circuit traces is difficult to impossible. Some of the pros and cons of embedded signal traces are summarized in Figure 9.143.

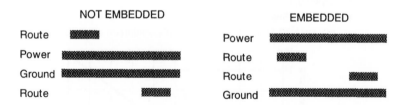

- Advantages
  - Signal traces shielded and protected
  - Lower impedance, thus lower emissions and crosstalk
  - Significant improvement > 50MHz
- Disadvantages
  - Difficult prototyping and troubleshooting
  - Decoupling may be more difficult
  - Impedance may be too low for easy matching

Figure 9.143: The Pros and Cons of Not Embedding versus the Embedding of Signal Traces in Multilayer PCB Designs

Multilayer PCBs can be designed *without* the use of embedded traces, as is shown in the left-most cross-sectional example. This embedded case could be considered as a doubled two-layer PCB design (i.e., four copper layers overall). The routed traces at the top form a microstrip with the power plane, while the traces at the bottom form a microstrip with the ground plane. In this example, the signal traces of both outer layers are readily accessible for measurement and troubleshooting purposes. But, the arrangement does nothing to take advantage of the shielding properties of the planes.

This nonembedded arrangement will have greater emissions and susceptibility to external signals, vis-a-vis the embedded case at the right, which uses the embedding, and does take full advantage of the planes. As in many other engineering efforts, the decision of embedded versus not-embedded for the PCB design becomes a trade-off, in this case one of reduced emissions versus ease of testing.

Dealing with High-Speed Logic

Much has been written about terminating PCB traces in their characteristic impedance, to avoid signal reflections. A good guideline to determine when this is necessary is as follows: *Terminate the transmission line in its characteristic impedance when the one-way propagation delay of the PCB track is equal to or greater than one-half the applied signal rise/fall time (whichever edge is faster).* For example, a 2 inch microstrip line over an $E_r = 4.0$ dielectric would have a delay of ~270 ps. Using the above rule strictly, termination would be appropriate whenever the signal rise time is < ~500 ps. A more conservative rule is to use a 2 inch (PCB track length)/nanosecond (rise/fall time) rule. If the signal trace exceeds this trace-length/speed criterion, then termination should be used.

For example, PCB tracks for high-speed logic with rise/fall time of 5 ns should be terminated in their characteristic impedance if the track length is equal to or greater than 10 inches (where measured length *includes* meanders).

As an example of what can be expected today in modern systems, Figure 9.144 shows typical rise/fall times for several logic families including the SHARC DSPs operating on 3.3V supplies. As would be expected, the rise/fall times are a function of load capacitance.

- GaAs: 0.1ns
- ECL: 0.75ns
- ADI SHARC DSPs: 0.5 ns to 1 ns (Operating on +3.3V Supply)

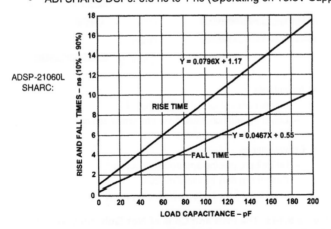

Figure 9.144: Typical DSP Output Rise Times and Fall Times

In the analog domain, it is important to note that this same 2-inch/nanosecond guideline should also be used with op amps and other circuits, to determine the need for transmission line techniques. For instance, if an amplifier must output a maximum frequency of f_{max}, then the equivalent rise time t_r is related to this f_{max}. This limiting rise time, t_r, can be calculated as:

$$t_r = 0.35/f_{max}$$

Eq. 9.29

The maximum PCB track length is then calculated by multiplying t_r by 2-inch/nanosecond. For example, a maximum frequency of 100 MHz corresponds to a rise time of 3.5 ns, so a 7-inch or more track carrying this signal should be treated as a transmission line.

The best ways to keep sensitive analog circuits from being affected by fast logic are to physically separate the two by the PCB layout, and to use no faster logic family than is dictated by system requirements. In some cases, this may require the use of several logic families in a system. An alternative is to use series resistance or ferrite beads to slow down the logic transitions where highest speed isn't required. Figure 9.145 shows two methods.

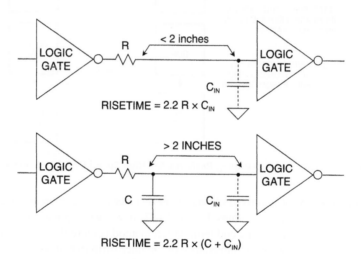

**Figure 9.145: Damping Resistors Slow Down
Fast Logic Edges to Minimize EMI/RFI Problems**

In the first, the series resistance and the input capacitance of the gate form a low-pass filter. Typical CMOS input capacitance is 5 pF to 10 pF. Locate the series resistor close to the driving gate. The resistor minimizes transient currents and may eliminate the necessity of using transmission line techniques. The value of the resistor should be chosen such that the rise and fall times at the receiving gate are fast enough to meet system requirement, but no faster. Also, make sure that the resistor is not so large that the logic levels at the receiver are out of specification because of the voltage drop caused by the source and sink currents that flow through the resistor. The second method is suitable for longer distances (>2 inches), where additional capacitance is added to slow down the edge speed. Notice that either one of these techniques increases delay and increases the rise/fall time of the original signal. This must be considered with respect to the overall timing budget, and the additional delay may not be acceptable.

Figure 9.146 shows a situation where several DSPs must connect to a single point, as would be the case when using read or write strobes bidirectionally connected from several DSPs. Small damping resistors shown in Figure 9.146A can minimize ringing provided the length of separation is less than about 2 inches. This method will also increase rise/fall times and propagation delay. If two groups of processors must be connected, a single resistor between the pairs of processors as shown in Figure 9.146B can serve to damp out ringing.

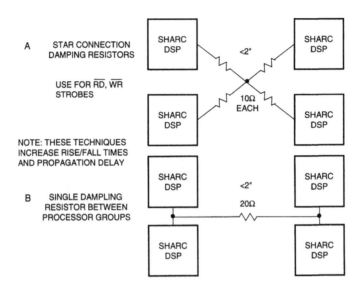

Figure 9.146: Series Damping Resistors for High Speed DSP Interconnections

The only way to preserve 1 ns or less rise/fall times over distances greater than about 2 inches without ringing is to use transmission line techniques. Figure 9.147 shows two popular methods of termination: end termination, and source termination. The end termination method (Figure 9.147A) terminates the cable at its terminating point in the characteristic impedance of the microstrip transmission line. Although higher impedances can be used, 50 Ω is popular because it minimizes the effects of the termination impedance mismatch due to the input capacitance of the terminating gate (usually 5 pF to 10 pF).

In Figure 9.147A, the cable is terminated in a Thevenin impedance of 50 Ω terminated to 1.4 V (the midpoint of the input logic threshold of 0.8 V and 2.0 V). This requires two resistors (91 Ω and 120 Ω), which add about 50 mW to the total quiescent power dissipation to the circuit. Figure 9.147A also shows the resistor values for terminating with a 5V supply (68 Ω and 180 Ω). Note that 3.3 V logic is much more desirable in line driver applications because of its symmetrical voltage swing, faster speed, and lower power. Drivers are available with less than 0.5 ns time skew, source and sink current capability greater than 25 mA, and rise/fall times of about 1 ns. Switching noise generated by 3.3 V logic is generally less than 5 V logic because of the reduced signal swings and lower transient currents.

The source termination method, shown in Figure 9.147B, absorbs the reflected waveform with an impedance equal to that of the transmission line. This requires about 39 Ω in series with the internal output impedance of the driver, which is generally about 10 Ω. This technique requires that the end of the transmission line be terminated in an open circuit, therefore no additional fanout is allowed. The source termination method adds no additional quiescent power dissipation to the circuit.

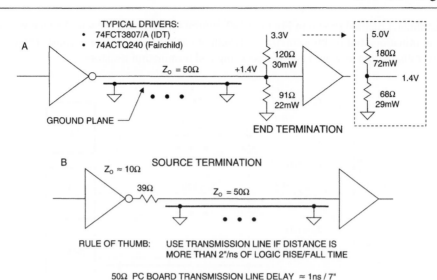

Figure 9.147: Termination Techniques for Controlled Impedance Microstrip Transmission Lines

Figure 9.148 shows a method for distributing a high speed clock to several devices. The problem with this approach is that there is a small amount of time skew between the clocks because of the propagation delay of the microstrip line (approximately 1 ns /7"). This time skew may be critical in some applications. It is important to keep the stub length to each device less than 0.5" in order to prevent mismatches along the transmission line.

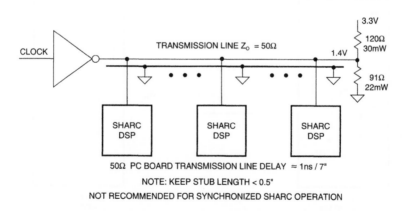

Figure 9.148: Clock Distribution Using End-of-Line Termination

The clock distribution method shown in Figure 9.149 minimizes the clock skew to the receiving devices by using source terminations and making certain the length of each microstrip line is equal. There is no extra quiescent power dissipation as would be the case using end termination resistors.

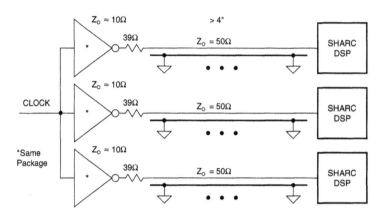

Figure 9.149: Preferred Method of Clock Distribution Using Source Terminated Transmission Lines

Figure 9.150 shows how source terminations can be used in bidirectional link port transmissions between SHARC DSPs. The output impedance of the SHARC driver is approximately 17 Ω, and therefore a 33 Ω series resistor is required on each end of the transmission line for proper source termination.

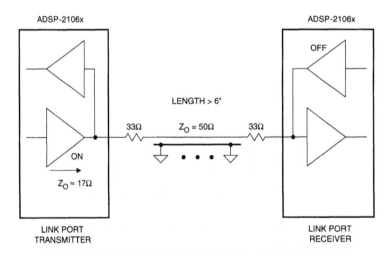

Figure 9.150: Source Termination for Bidirectional Transmission Between SHARC DSPs

The method shown in Figure 9.151 can be used for bidirectional transmission of signals from several sources over a relatively long transmission line. In this case, the line is terminated at both ends, resulting in a dc load impedance of 25 Ω. SHARC drivers are capable of driving this load to valid logic levels.

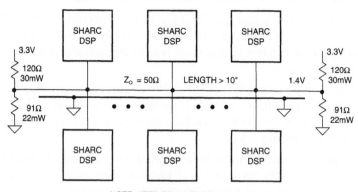

NOTE: KEEP STUB LENGTH < 0.5"

NOT RECOMMENDED FOR CLOCKS IN SYNCHRONIZED SHARC OPERATION!

Figure 9.151: Single Transmission Line Terminated at Both Ends

Emitter-coupled-logic (ECL) has long been known for low noise and its ability to drive terminated transmission lines with rise/fall times less than 2 ns. The family presents a constant load to the power supply, and the low level differential outputs provide a high degree of common-mode rejection. However, ECL dissipates lots of power.

Recently, low-voltage-differential-signaling (LVDS) logic has attained widespread popularity because of similar characteristics, but with lower amplitudes and lower power dissipation than ECL. The defining LVDS specification can be found in Reference 23, and References 24 and 25 should also prove useful. The LVDS logic swing is typically 350 mV peak-to-peak centered about a common-mode voltage of 1.2 V. A typical driver and receiver configuration is shown in Figure 9.152. The driver consists of a nominal 3.5 mA current source with polarity switching provided by PMOS and NMOS transistors as in the case of the AD9430 12-bit, 170/210 MSPS ADC. The output voltage of the driver is nominally 350 mV peak-to-peak at each output, and can vary between 247 mV and 454 mV. The output current can vary between 2.47 mA and

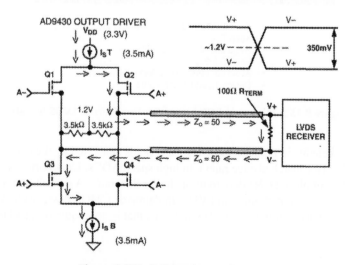

Figure 9.152: LVDS Driver and Receiver

4.54 mA. The LVDS receiver is terminated in a 100 Ω line-to-line. According to the LVDS specification, the receiver must respond to signals as small as 100 mV, over a common-mode voltage range of 50 mV to 2.35 V. The wide common-mode receiver voltage range is to accommodate ground voltage differences up to ±1 V between the driver and receiver.

The LVDS edge speed is defined as the 20% to 90% rise/fall time (as opposed to 10% to 90% for CMOS logic) and specified to be less than < 0.3 t_{ui}, where t_{ui} is the inverse of the data signaling rate. For a 210 MSPS sampling rate, t_{ui} = 4.76 ns, and the 20% to 80% rise/fall time must be less than 0.3 × 4.76 = 1.43 ns. For the AD9430, the rise/fall time is nominally 0.5 ns.

LVDS outputs for high performance ADCs should be treated differently than standard LVDS outputs used in digital logic. While standard LVDS can drive 1 to 10 meters in high speed digital applications (dependent on data rate), it is not recommended to let a high performance ADC drive that distance. It is recommended to keep the output trace lengths short (< 2 in.), minimizing the opportunity for any noise coupling onto the outputs from the adjacent circuitry, which may get back to the analog inputs. The differential output traces should be routed close together, maximizing common-mode rejection, with the 100 Ω termination resistor close to the receiver. Users should pay attention to PCB trace lengths to minimize any delay skew. A typical differential microstrip PCB trace cross section is shown in Figure 9.153 along with some recommended layout guidelines.

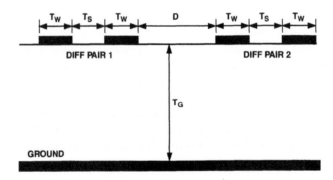

- Keep T_W, T_S, and D constant over the trace length
- Keep T_S ~ < $2T_W$
- Avoid use of vias if possible
- Keep D > $2T_S$
- Avoid 90° bends if possible
- Design T_W and T_G for ~ 50Ω

Figure 9.153: Microstrip PCB Layout for Two Pairs of LVDS Signals

LVDS also offers some benefits in reduced EMI. The EMI fields generated by the opposing LVDS currents tend to cancel each other (for matched edge rates). In high speed ADCs, LVDS offers simpler timing constraints compared to demultiplexed CMOS outputs at similar data rates. A demultiplexed data bus requires a synchronization signal that is not required in LVDS. In demuxed CMOS buses, a clock equal to one-half the ADC sample rate is needed, adding cost and complexity, that is not required in LVDS.

References:
9.6 EMI/RFI Considerations

1. Tim Williams, **EMC for Product Designers, 2nd Ed.**, Newnes, Oxford, 1996, ISBN: 0-7506-2466-3.

2. Henry Ott, **Noise Reduction Techniques In Electronic Systems, 2nd Ed.**, John Wiley & Sons, New York, 1988, ISBN 0-471-85068-3.

3. Mark Montrose, **EMC and the Printed Circuit Board**, IEEE Press, 1999, ISBN 0-7803-4703-X.

4. Ralph Morrison, **Grounding And Shielding Techniques in Instrumentation, 3rd Ed.**, John Wiley & Sons, New York, 1986, ISBN 0-471-83805-5.

5. Daryl Gerke and William Kimmel, "Designer's Guide to Electromagnetic Compatibility," **EDN**, January 20, 1994.

6. **Designing for EMC** *(Workshop Notes),* Kimmel Gerke Associates, Ltd., 1994.

7. Daryl Gerke and William Kimmel, "EMI and Circuit Components," **EDN**, September 1, 2000.

8. Alan Rich, "Understanding Interference-Type Noise," **Analog Dialogue**, Vol. 16, No. 3, 1982, pp. 16–19. *(Also available as application note AN346.)*

9. Alan Rich, "Shielding and Guarding," **Analog Dialogue**, Vol. 17, No. 1, 1983, pp. 8–13. *(Also available as application note AN347.)*

10. James Wong, Joe Buxton, Adolfo Garcia, James Bryant, "Filtering and Protection Against EMI/RFI" and "Input Stage RFI Rectification Sensitivity," Chapter 1, pp. 21–55 of **Systems Application Guide**, 1993, Analog Devices, Inc., Norwood, MA, ISBN 0-916550-13-3.

11. Adolfo Garcia, "EMI/RFI Considerations," Chapter 7, pp. 69–88 of **High Speed Design Techniques**, 1996, Analog Devices, Inc., Norwood, MA, 1993, ISBN 0-916550-17-6.

12. Walt Kester, Walt Jung, Chuck Kitchin, "Preventing RFI Rectification," Chapter 10, pp. 10.39–10.43 of **Practical Design Techniques for Sensor Signal Conditioning**, Analog Devices, Inc., Norwood, MA, 1999, ISBN 0-916550-20-6.

13. Charles Kitchin, Lew Counts, **A Designer's Guide to Instrumentation Amplifiers**, Analog Devices, Inc., 2000.

14. **B4001 and B4003 Common-Mode chokes**, Pulse Engineering, Inc., 12220 World Trade Drive, San Diego, CA, 92128, 619-674-8100, www.pulseeng.com.

15. **Understanding Common-Mode Noise**, Pulse Engineering, Inc., 12220 World Trade Drive, San Diego, CA, 92128, 619-674-8100, www.pulseeng.com.

16. Standard IPC-2141, "Controlled Impedance Circuit Boards and High Speed Logic Design," 1996, Institute for Interconnection and Packaging Electronic Circuits, 2215 Sanders Road, Northbrook, IL, 60062-6135, 847-509-9700, www.ipc.org.

17. Eric Bogatin, "Verifying the Accuracy of 2D Field Solvers for Characteristic Impedance Calculation," Ansoft Seminar, October 11, 2000, www.bogatinenterprises.com.

18. Andrew Burkhardt, Christopher Gregg, Alan Staniforth, "Calculation of PCB Track Impedance," Technical Paper S-19-5, presented at the IPC Printed Circuits Expo '99 Conference, March 14–18, 1999.

19. Brian C. Wadell, **Transmission Line Design Handbook,** Artech House, Norwood, MA, 1991, ISBN: 0-89006-436-9.

20. Eric Bogatin, "No Myths Allowed, "Impedance Calculations," a Chip Center column, November 1, 1999, www.chipcenter.com/signalintegrity.

21. William R. Blood, Jr., **MECL System Design Handbook** (HB205/D, Rev. 1A May 1988), ON Semiconductor, August, 2000.

22. Paul Brokaw, "An IC Amplifier User Guide To Decoupling, Grounding, and Making Things Go Right For A Change," **Analog Devices AN202**.

23. TIA/EIA-644-A Standard, **Electrical Characteristics of Low Voltage Differential Signaling (LVDS) Interface Circuits**, January 30, 2001.

24. IEEE Std. 1596.3-1996, **IEEE Standard for Low Voltage Differential Signals (LVDS) for Scalable Coherent Interface**, IEEE, 1996.

25. Cindy Bloomingdale and Gary Hendrickson, "LVDS Data Outputs for High Speed Analog-to-Digital Converter," **Application Note AN-586**, Analog Devices, 2002.

Some useful EMC and signal integrity related URLs:

Eric Bogatin website, www.bogatinenterprises.com.

Chip Center's "Signal Integrity" page, www.chipcenter.com/signalintegrity.

Kimmel Gerke Associates website, www.emiguru.com.

Henry Ott website, www.hottconsultants.com.

IEEE EMC website, www.ewh.ieee.org/soc/emcs.

Mark Montrose website, www.montrosecompliance.com/index.html.

Tim Williams website, www.elmac.co.uk.

Acknowledgments:

Eric Bogatin made helpful comments on this section, which were very much appreciated.

Low Voltage Logic Interfacing

Walt Kester, Ethan Bordeax, Johannes Horvath, Catherine Redmond, Eva Murphy

For nearly 20 years, the standard V_{DD} for digital circuits was 5 V. This voltage level was used because bipolar transistor technology required 5 V to allow headroom for proper operation. However, in the late 1980s, Complementary Metal Oxide Semiconductor (CMOS) became the standard for digital IC design. This process did not necessarily require the same voltage levels as TTL circuits, but the industry adopted the 5 V TTL standard logic threshold levels to maintain backward compatibility with older systems (Reference 1).

The current revolution in supply voltage reduction has been driven by demand for faster and smaller products at lower costs. This push has caused silicon geometries to drop from 2 µm in the early 1980s to 0.18 µm used in today's latest microprocessor and IC designs. As feature sizes have become increasingly smaller, the voltage for optimum device performance has also dropped below the 5 V level. This is illustrated in the current microprocessors for PCs, where the optimum core operating voltage is programmed externally using voltage identification (VID) pins, and can be as low as 1.3 V.

The strong interest in lower voltage DSPs is clearly visible in the shifting sales percentages for 5 V and 3.3 V parts. Sales growth for 3.3 V DSPs has increased at more than twice the rate of the rest of the DSP market (30% for all DSPs versus more than 70% for 3.3 V devices). This trend will continue as the high volume/high growth portable markets demand signal processors that contain all of the traits of the lower voltage DSPs.

On the one hand, the lower voltage ICs operate at lower power, allow smaller chip areas, and higher speeds. On the other hand, the lower voltage ICs must often interface to other ICs which operate at larger V_{DD} supply voltages thereby causing interface compatibility problems. Although lower operating voltages mean smaller signal swings, and hence less switching noise, noise margins are lower for low supply voltage ICs. A summary of key points relating to low voltage logic is summarized in Figure 9.154.

- Lower Power for Portable Applications
- 2.5V ICs Can Operate on Two "AA" Alkaline Cells
- Faster CMOS Processes, Smaller Geometries, Lower Breakdown Voltages
- Multiple Voltages in System: 5V, 3.3V, 2.5V, 1.8V DSP Core Voltage (VID), Analog Supply Voltage
- Interfaces Required Between Multiple Logic Types
- Lower Voltage Swings Produce Less Switching Noise
- Lower Noise Margins
- Less Headroom in Analog Circuits Decreases Signal Swings and Increases Sensitivity to Noise

---

Figure 9.154: Low Voltage Mixed-Signal ICs

The popularity of 2.5 V devices can be partially explained by their ability to operate from two AA alkaline cells. Figure 9.155 shows the typical discharge characteristics for an AA cell under various load conditions (Reference 2). Note that at a load current of 15 mA, the voltage remains above 1.25 V (2.5 V for two cells in series) for nearly 100 hours. Therefore, an IC that can operate effectively at low currents with a supply voltage of 2.5 V ±10% (2.25 V – 2.75 V) is very useful in portable designs.

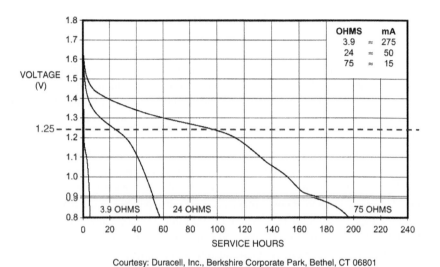

Courtesy: Duracell, Inc., Berkshire Corporate Park, Bethel, CT 06801
www.duracell.com

Figure 9.155: Duracell MN1500 "AA"
Alkaline Battery Discharge Characteristics

In order to understand the compatibility issues relating to interfacing ICs operated at different V_{DD} supplies, it is useful to first look at the structure of a typical CMOS logic stage as shown in Figure 9.156.

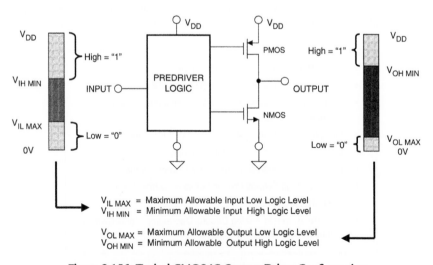

$V_{IL\ MAX}$ = Maximum Allowable Input Low Logic Level
$V_{IH\ MIN}$ = Minimum Allowable Input High Logic Level

$V_{OL\ MAX}$ = Maximum Allowable Output Low Logic Level
$V_{OH\ MIN}$ = Minimum Allowable Output High Logic Level

Figure 9.156: Typical CMOS IC Output Driver Configuration

Note that the output driver stage consists of a PMOS and an NMOS transistor. When the output is high, the PMOS transistor connects the output to the $+V_{DD}$ supply through its low on-resistance (R_{ON}), and the NMOS transistor is off. When the output is low, the NMOS transistor connects the output to ground through its on-resistance, and the PMOS transistor is off. The R_{ON} of a CMOS output stage can vary between 5 Ω and 50 Ω depending on the size of the transistors, which in turn, determines the output current drive capability.

A typical logic IC has its power supplies and grounds separated between the output drivers and the rest of the circuitry (including the predriver). This is done to maintain a clean power supply, which reduces the effect of noise and ground bounce on the I/O levels. This is increasingly important, since added tolerance and compliance are critical in I/O driver specifications, especially at low voltages.

Figure 9.156 also shows "bars" that define the minimum and maximum required input and output voltages to produce a valid high or low logic level. Note that for CMOS logic, the actual output logic levels are determined by the drive current and the R_{ON} of the transistors. For light loads, the output logic levels are very close to 0 V and $+V_{DD}$. The input logic thresholds, on the other hand, are determined by the input circuit of the IC.

There are three sections in the "input" bar. The bottom section shows the input range that is interpreted as a logic low. In the case of 5 V TTL, this range would be between 0 V and 0.8 V. The middle section shows the input voltage range where it is interpreted as neither a logic low nor a logic high. The upper section shows where an input is interpreted as a logic high. In the case of 5 V TTL, this would be between 2 V and 5 V.

Similarly, there are three sections in the "output" bar. The bottom range shows the allowable voltage for a logic low output. In the case of 5 V TTL, the IC must output a voltage between 0 V and 0.4 V. The middle section shows the voltage range that is not a valid high or low—the device should never transmit a voltage level in this region except when transitioning from one level to the other. The upper section shows the allowable voltage range for a logic high output signal. For 5 V TTL, this voltage is between 2.4 V and 5 V. The chart does not reflect a 10% overshoot/undershoot also allowed on the inputs of the logic standard.

A summary of the existing logic standards using these definitions is shown in Figure 9.157. Note that the input thresholds of classic CMOS logic (series 4000, for example) are defined as 0.3 V_{DD} and 0.7 V_{DD}.

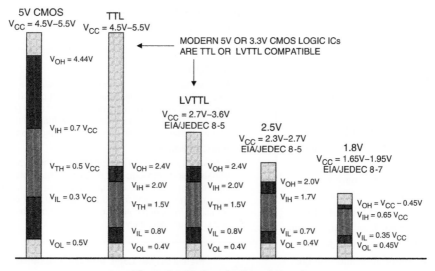

Figure 9.157: Standard Logic Levels

However, most CMOS logic circuits in use today are compatible with TTL and LVTTL levels which are the dominant 5 V and 3.3 V operating standards for DSPs. Note that 5 V TTL and 3.3 V LVTTL input and output threshold voltages are identical. The difference is the upper range for the allowable high levels.

The international standards bureau JEDEC (Joint Electron Device Engineering Council) has created a 2.5 V standard (JEDEC standard 8-5) and a 1.8 V standard (Reference 3). There are also a wide range of other low voltage standards, such as GTL (Gunning Transceiver Logic), BTL (Backplane Transceiver Logic), ECL (Emitter-Coupled-Logic) PECL (Positive ECL Logic), and LVDS. However, most of these standards are aimed at application specific markets and not for general purpose semiconductor systems.

From this chart (Figure 9.157), it is possible to visualize some of the possible problems in connecting together two ICs operating on different standards. One example would be connecting a 5 V TTL device to a 3.3 V LVTTL IC. The 5 V TTL high level is too high for the LVTTL to handle (> 3.3 V). This could cause permanent damage to the LVTTL chip. Another possible problem would be a system with a 2.5 V IC driving a 5 V CMOS device. The logic high level from the 2.5 V device is not high enough for it to register as a logic high on the 5 V CMOS input ($V_{IH\ MIN}$ = 3.5 V). These examples illustrate two possible types of logic level incompatibilities—either a device being driven with too high a voltage or a device not driving a voltage high enough for it to register a valid high logic level with the receiving IC. These interfacing problems introduce two important concepts: *voltage tolerance* and *voltage compliance*.

Voltage Tolerance and Voltage Compliance

A device that is *voltage tolerant* can withstand a voltage greater than its V_{DD} on its I/O pins. For example, if a device has a V_{DD} of 2.5 V and can accept inputs equal to 3.3 V and can withstand 3.3 V on its outputs, the 2.5 V device is called 3.3 V tolerant. The meaning of *input* voltage tolerance is fairly obvious, but the meaning of *output* voltage tolerance requires some explanation. The output of a 2.5 V CMOS driver in the high state appears like a small resistor (R_{ON} of the PMOS FET) connected to 2.5 V. Obviously, connecting its output directly to 3.3 V is likely to destroy the device due to excessive current. However, if the 2.5 V device has a three-state output which is connected to a bus which is also driven by a 3.3 V IC, then the meaning becomes clearer. Even though the 2.5 V IC is in the off (third-state) condition, the 3.3 V IC can drive the bus voltage higher than 2.5 V, potentially causing damage to the 2.5 V IC output.

A device that is *voltage compliant* can receive signals from and transmit signals to a device which is operated at a voltage greater than its own V_{DD}. For example, if a device has a 2.5 V V_{DD} and can transmit and receive signals to and from a 3.3 V device, the 2.5 V device is said to be 3.3 V compliant.

The interface between the 5 V CMOS and 3.3 V LVTTL parts illustrates a lack of voltage tolerance; the LVTTL IC input is overdriven by the 5 V CMOS device output. The interface between the 2.5 V JEDEC and the 5 V CMOS part demonstrates a lack of voltage compliance; the output high level of the JEDEC IC does not comply to the input level requirement of a the 5 V CMOS device. The definitions of voltage compliance and voltage tolerance are repeated in Figure 9.158.

- Voltage Tolerance:
  - A device that is *Voltage Tolerant* can withstand a voltage greater than its V_{DD} on its input and output pins. If a device has a V_{DD} of 2.5V and can accept inputs of 3.3V (±10%), the 2.5V device is 3.3V tolerant on its input. Input and output tolerance should be examined and specified separately.
- Voltage Compliance:
  - A device that is *Voltage Compliant* can transmit and receive signals to and from logic that is operated at a voltage greater than its own V_{DD}. If a device has a 2.5V V_{DD} and can properly transmit signals to and from 3.3V logic, the 2.5V device is 3.3V compliant. Input and output compliance should be examined and specified separately.

Figure 9.158: Logic Voltage Tolerance and Compatibility Definitions

Interfacing 5 V Systems to 3.3 V Systems using NMOS FET "Bus Switches"

When combining ICs that operate on different voltage standards, one is often forced to add additional discrete elements to ensure voltage tolerance and compliance. In order to achieve voltage tolerance between 5 V and 3.3 V logic; for instance, a bus switch voltage translator such as the ADG3257 can be used (also see References 4, 5). The bus switch limits the voltage applied to an IC. This is done to avoid applying a larger input high voltage than the receiving device can tolerate.

As an example, it is possible to place a bus switch between a 5 V CMOS and 3.3 V LVTTL IC, and the two devices can then transmit data properly as shown in Figure 9.159. The bus switch is basically an NMOS FET. If 4.3 V is placed on the gate of the FET, the maximum passable signal is 3.3 V (approximately 1 V less than the gate voltage). If both input and output are below 3.3 V, the NMOS FET acts as a low resistance ($R_{ON} \approx 2\ \Omega$). As the input approaches 3.3 V, the FET on-resistance increases, thereby limiting the signal output. The ADG3257 is a quad 2:1 Mux/Demux bus switch with a gate drive enable as shown in the lower half of Figure 9.159. The V_{CC} of the ADG3257 sets the high level for the gate drive.

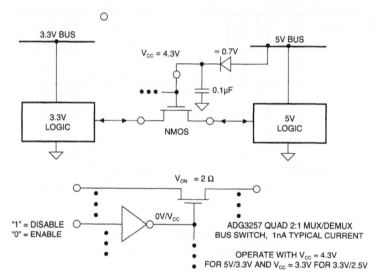

Figure 9.159: 5 V/3.3 V Bidirectional Interface Using NMOS FET Achieves Voltage Tolerance

One way of creating a 4.3 V supply on a 5 V/3.3 V system board is to simply place a silicon diode between the 5 V supply and V_{CC} on the bus switch as shown in Figure 9.159. For 3.3 V/2.5 V applications, the V_{CC} pin can be connected directly to the 3.3 V supply. Some bus switches are designed to operate on either 3.3 V or 5 V directly and generate the internal gate bias level internally.

A bus switch removes voltage tolerance concerns in this mixed logic design. One convenient feature of bus switches is that they are bidirectional; this allows the designer to place a bus translator between two ICs and not have to create additional routing logic for input and output signals.

A bus switch increases the total power dissipation along with the total area required to lay out a system. Since voltage bus switches are typically CMOS circuits, they have very low power dissipation ratings. An average value for added continuous power dissipation is 5 mW per package (10 switches), and this is independent of the frequency of signals that pass through the circuit. Bus switches typically have 8–20 I/O pins per package and take up approximately 25 to 50 mm² of board space.

One concern when adding interface logic into a circuit is a possible increase in propagation delay. Added propagation delay can create many timing problems in a design. Bus switches have very low propagation delay values.

The bus switch contributes practically no propagation delay (0.1 ns typical for the ADG3257) other than the RC delay of the typical R_{ON} of the switch and the load capacitance when driven by an ideal voltage source. Since the time constant is typically much smaller than the rise/fall times of typical driving signals, bus switches add very little propagation delay to the system. Low R_{ON} is therefore critical for bus switches, since the switch on-resistance in conjunction with the bus capacitance creates a single-pole filter which can add delay and reduce the maximum data rate. The typical on-capacitance of the ADG3257 is 10 pF, and this capacitance in conjunction with an R_{ON} of 4 Ω yields a rise/fall time of approximately 90 ps. Figure 9.160 shows the ADG3257 on-resistance as a function of input voltage for 5.5 V, 5 V, 4.5 V, 3.3 V, 3.0 V, and 2.7 V supplies. Maximum pass voltage as a function of input voltage is shown in Figure 9.161.

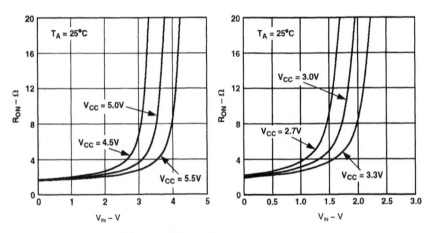

**Figure 9.160: ON Resistance versus
Input Voltage for ADG3257 Bus Switch**

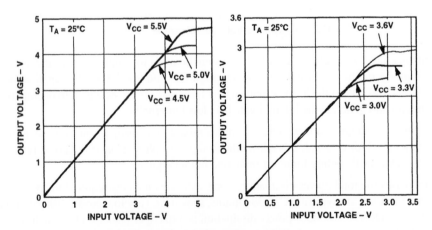

**Figure 9.161: Maximum Pass Voltage versus
Input Voltage for ADG3257 Bus Switch**

Eye diagrams for the ADG3257 operating at 622 Mbps and 933 Mbps are shown in Figure 9.162.

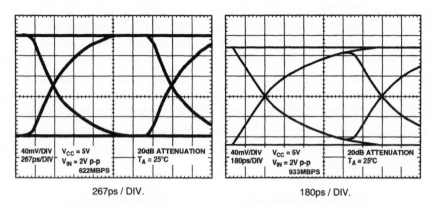

40mV/DIV	$V_{CC} = 5V$	20dB ATTENUATION	40mV/DIV	$V_{CC} = 5V$	20dB ATTENUATION
267ps/DIV	$V_{IN} = 2V$ p-p	$T_A = 25°C$	180ps/DIV	$V_{IN} = 2V$ p-p	$T_A = 25°C$
	622MBPS			933MBPS	

267ps / DIV.   180ps / DIV.

Figure 9.162: Eye Diagrams for 622 Mbps and 933 Mbps Data Rates

3.3 V/2.5 V Interfaces

Figure 9.163 shows two possibilities for a 3.3 V to 2.5 V logic interface. The top diagram (A) shows a direct connection. This will work provided the 2.5 V IC is 3.3 V tolerant on its input. If the 2.5 V IC is not 3.3 V tolerant, a low voltage bus switch such as the ADG3231 can be used. In most cases, the connection between 3.3 V and 2.5 V systems can be bidirectional, even though the V_{OH} of 2.5 V logic is specified as 2.0 V which is the same as the V_{IH} specification of 3.3 V logic (refer back to Figure 9.157). This point deserves further discussion.

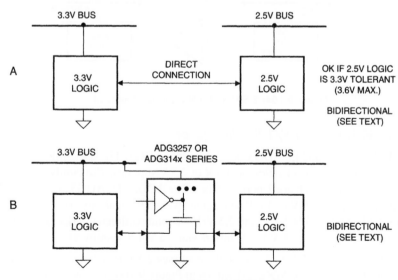

Figure 9.163: 3.3 V to 2.5 V Interface

Figure 9.164A shows a direct connection between 2.5 V and 3.3 V logic. In order for this to work, the 2.5 V output must be at least 2 V minimum per the JEDEC specifications. With no loading on the 2.5 V output, the 3.3 V IC input is connected directly to 2.5 V through the on-resistance of the PMOS transistor driver. This provides 0.5 V noise margin for the nominal supply voltage of 2.5 V. However, the tolerance on the 2.5 V bus allows it to drop to a minimum of 2.3 V, and the noise margin is reduced to 0.3 V. This may still work in a relatively quiet environment, but could be marginal if there is noise on the supply voltages.

Adding a 1.6 kΩ pull-up resistor as shown in Figure 9.164B ensures the 2.5 V output will not drop below 2.5 V due to the input current of the 3.3 V device, but the degraded noise margin still exists for a 2.3 V supply. With a 50% duty cycle, the resistor adds about 3.4 mW power dissipation per output.

A more reliable interface between 2.5 V and 3.3 V logic is shown in Figure 9.164C, where a logic translator such as the ADG3231 is used. This solves all noise margin problems associated with (A) and (B) and requires about 2 µA maximum per output.

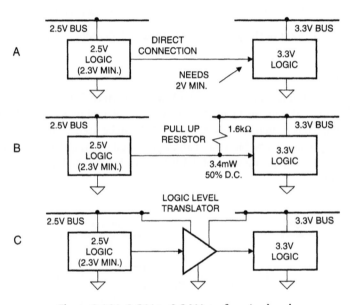

Figure 9.164: 2.5 V to 3.3 V Interface Analyzed

3.3 V/2.5 V, 3.3 V/1.8 V, 2.5 V/1.8 V Interfaces

The ADG3241, ADG3242, ADG3243, ADG3245, ADG3246, ADG3247, ADG3248, and ADG3249 are low voltage bus switches optimized for operation on 3.3 V or 2.5 V supplies. The family includes 1-, 2-, 8-, 10- and dual 8-bit switches, all of which are 2-port switches. The ADG3241, ADG3242, ADG3245, ADG3246, ADG3247, and ADG3249 have 2.5 V or 1.8 V selectable level shift capability. The family offers a fast, low power solution for 3.3 V/2.5 V, 3.3 V/1.8 V, and 2.5 V/1.8 V unidirectional interfaces. Figure 9.165 shows the ADG32xx-family used as 3.3 V/1.8 V level shifters and 2.5 V/1.8 V shifters.

Translating from 1.8 V to 2.5 V, 1.8 V to 3.3 V, (and sometimes 2.5 V to 3.3 V as previously discussed) requires a logic translator such as the ADG3231 shown in Figure 9.166. The two voltage buses can be any value between 1.65 V and 3.6 V. The ADG3231 is a single-channel translator in a SOT-23 package, and the ADG3232 is a 2:1 multiplexer/level translator also in a SOT-23 package.

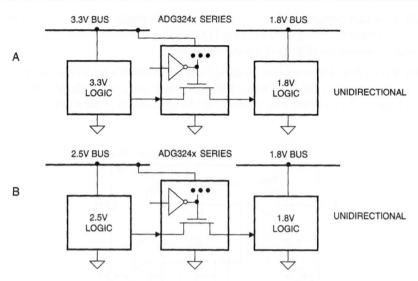

Figure 9.165: 3.3 V to 1.8 V, 2.5 V to 1.8 V Unidirectional Interfaces

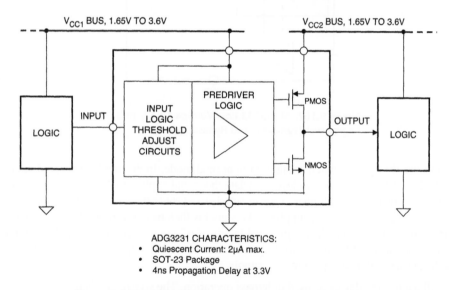

ADG3231 CHARACTERISTICS:
- Quiescent Current: 2µA max.
- SOT-23 Package
- 4ns Propagation Delay at 3.3V

Figure 9.166: ADG3231 Low Voltage Logic Level Translator

The ADG3233 is a bypass switch designed on a submicron process that operates from supplies as low as 1.65 V. The device is guaranteed for operation over the supply range 1.65 V to 3.6 V. It operates from two supply voltages, allowing bidirectional level translation, i.e., it translates low voltages to higher voltages and vice versa. The signal path is unidirectional, meaning data may only flow from A to Y. This type of device may be used in applications that require a bypassing function. It is ideally suited to bypassing devices in a JTAG chain or in a daisy-chain loop. One switch could be used for each device or a number of devices, thus allowing easy bypassing of one or more devices in a chain. This may be particularly useful in reducing the time overhead in testing devices in the JTAG chain or in daisy-chain applications where the user does not wish to change the settings of a particular device.

The bypass switch is packaged in two of the smallest footprints available for its required pin count. The 8-lead SOT-23 package requires only 8.26 mm × 8.26 mm board space, while the MSOP package occupies approximately 15 mm × 15 mm board area. A functional block diagram of the ADG3233 is shown in Figure 9.167.

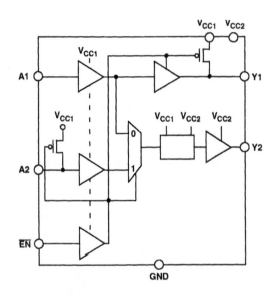

**Figure 9.167: ADG3233 Low Voltage 1.65 V to 3.6 V,
Logic Level Translator and Bypass Switch**

Figure 9.168 shows the bypass switch being used in normal mode. In this mode, the signal paths are from A1 to Y1 and A2 to Y2. The device will level translate the signal applied to A1 to a V_{CC1} logic level (this level translation can be either to a higher or lower supply) and route the signal to the Y1 output, which will have standard V_{OL}/V_{OH} levels for V_{CC1} supplies. The signal is then passed through Device 1 and back to the A2 input pin of the bypass switch. The logic level inputs of A2 are with respect to the V_{CC1} supply. The signal will be level translated from V_{CC1} to V_{CC2} and routed to the Y2 output pin of the bypass switch. Y2 output logic levels are with respect to the V_{CC2} supply.

Figure 9.169 illustrates the device as used in bypass operation. The signal path is now from A1 directly to Y2, thus bypassing Device 1 completely. The signal will be level translated to a V_{CC2} logic level and available on Y2, where it may be applied directly to the input of Device 2. In bypass mode, Y1 is pulled up to V_{CC1}. The three supplies in Figures 9.168 and 9.169 may be any combination of supplies, i.e., V_{CC0}, V_{CC1}, and V_{CC2} may be any combination of supplies, for example, 1.8 V, 2.5 V, and 3.3 V.

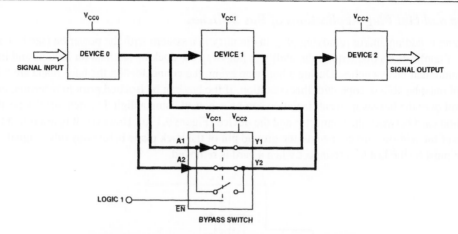

Figure 9.168: ADG3233 Bypass Switch in Normal Mode

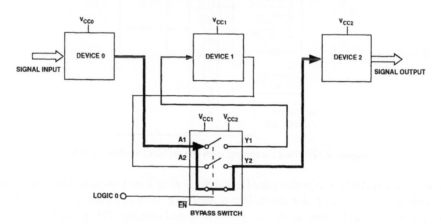

Figure 9.169: ADG3233 Bypass Switch in Bypass Mode

Hot Swap and Hot Plug Applications of Bus Switches

Hot swapping is adding and/or removing plug-in circuitry in a system with the power on (see Reference 8). Examples of applications that require the ability to hot swap are docking stations for laptops and line cards for telecommunications switches. During a hot swap event, the connectors on the back plane are "live"; the add-on card must be able to cope with this condition. If the bus can be isolated prior to insertion, one has more control over the hot-swap event. Isolation can be achieved using a digital switch, ideally positioned on the add-on card between the connector and the device (Figure 9.170). However, it is important that the ground pin of the add-on card connect to the ground pin of the back plane before any other signal or power pins, and it must be the last to disconnect when a card is removed.

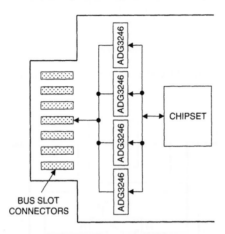

Figure 9.170: Hot Swapping with the ADG3246 Bus Switch

Critical systems, such as ADSL (Asynchronous Digital Subscriber Line), manufacturing controls, servers, and airline reservations must not be shut down. If new hardware, such as a plug-in modem, needs to be added to the system, it has to be done while the system is up and running. This process of adding hardware during mandatory continuous operation is known as hot plug (see Reference 9). To ensure smooth execution of the process, a digital switch can be wired between the connector and the internal bus (Figure 9.171). During the hot-plug event, the switch is turned off to provide isolation of the specific circuit location.

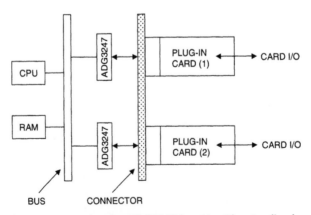

Figure 9.171: Using the ADG3247 in a Hot-Plug Application

Internally Created Voltage Tolerance / Compliance

Modern high performance CMOS DSPs and microprocessors typically operate on core voltages between 1 V and 2 V. These low voltages yield optimum speed-power performance. However, the logic levels in the core are not compatible with standard 2.5 V or 3.3 V I/O interfaces. This problem is typically solved as shown in Figure 9.172, where the logic core operates at a reduced voltage, but the output drivers operate at a standard supply voltage level of 2.5 V or 3.3 V.

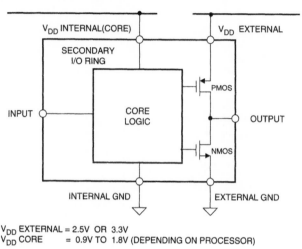

V$_{DD}$ EXTERNAL = 2.5V OR 3.3V
V$_{DD}$ CORE = 0.9V TO 1.8V (DEPENDING ON PROCESSOR)
INPUT 2.5V OR 3.3V TOLERANT

Figure 9.172: Internal Compliance and Tolerance in a CMOS IC with Secondary I/O Ring

The technique followed by many IC manufacturers is to provide a secondary I/O ring, i.e., the I/O drivers are driven by the 2.5 V or 3.3 V power supply, hence the device is compatible with 2.5 V or 3.3 V logic levels. Note that the inputs must be compliant and tolerant to the I/O supply voltage. There are several issues to consider in a dual-supply logic IC design of this type:

- *Power-Up Sequencing*: If two power supplies are required to give an IC additional tolerance/compliance, what is the power-up sequence? Is it a requirement that the power supplies are switched on simultaneously or can the device only have a voltage supplied on the core or only on the I/O ring? This problem can be easily solved if the core voltage is generated from the I/O supply voltage using a low dropout linear regulator.

- *Process Support and Electro-Static Discharge (ESD) Protection*: The transistors created in the IC's fabrication process must be able to both withstand and drive high voltages. The high voltage transistors create additional fabrication costs since they require more processing steps to build in high voltage tolerance. Designs with standard transistors require additional circuitry. The I/O drivers must also provide ESD protection for the device. Most current designs limit the overvoltage to below one diode drop (0.7 V) above the power supply. Protection for larger overvoltage requires more diodes in series.

- *Internal High Voltage Generation*: The PMOS transistors need to be placed in a substrate well that is tied to the highest on-chip voltage to prevent lateral diodes from turning on and drawing excessive current. This high voltage can either be generated on-chip using charge pumps, or from an external supply. This requirement can make the design complex, since one cannot efficiently use charge pumps to generate higher voltages and also achieve low standby current. In most cases, the voltage is supplied externally.

- *Chip Area*: Die size is a primary factor in reducing costs and increasing yields. Tolerance and compliance circuitry may require either more or larger I/O devices to achieve the desired performance levels.

- *Testing*: Since the core and the I/O can be at different voltages, testing the device for all possible combinations of voltages can be complicated, adding to the total cost of the IC.

References:
9.7 Low Voltage Logic Interfacing

1. P. Alfke, *Low-Voltage FPGAs Allow 3.3V/5V System Design*, **Electronic Design**, pp. 70–76, August 18, 1997.

2. AA Alkaline Battery Discharge Characteristics, Duracell Inc., Berkshire Corporate Park, Bethel, CT 06801, www.duracell.com.

3. Joint Electron Device Engineering Council (JEDEC), www.jedec.org, Standard JESD8-5, October 1995, and Standard JESD8-7, February, 1997.

4. QS3384 Data Sheet, Integrated Device Technology (IDT), Inc., 2975 Stender Way, Santa Clara, CA 95054, www.idt.com.

5. Pericom Semiconductor Corporation, 2380 Bering Drive, San Jose, CA 95131, www.pericom.com.

6. H. Johnson, M. Graham, **High Speed Digital Design**, Prentice Hall, 1993.

7. Eva Murphy and Catherine Redmond, "Bus Switches for Speed, Safety, and Efficiency: What They Are and What You Should Know about Them," **Analog Dialogue** 36-06, Analog Devices, Inc., 2002.

8. Compact PCI Hot Swap Specification R1.0.

9. PCI Hot-Plug Specification R1.0, October 6, 1997.

Breadboarding and Prototyping
Walt Kester, James Bryant, Walt Jung

A basic principle of a breadboard or a prototype structure is that it is a *temporary* one, designed to test the performance of an electronic circuit or system. By definition it must therefore be easy to modify, particularly so for a breadboard.

There are many commercial prototyping systems, but unfortunately for the analog designer, almost all of them are designed for prototyping *digital* systems. In such environments, noise immunities are hundreds of millivolts or more. Prototyping methods commonly used include noncopper-clad Matrix board, Vectorboard, wire-wrap, and plug-in breadboard systems. Quite simply, these are all unsuitable for high performance or high frequency analog prototyping, because of their excessively high parasitic resistance, inductance, and capacitance levels. Even the use of standard IC sockets is inadvisable in many prototyping applications (more on this, below).

Figure 9.173 summarizes a number of key points on selecting a useful analog breadboard and/or prototyping system, which are further discussed below.

- Always Use a Large Area Ground Plane for Precision or High Frequency Circuits
- Minimize Parasitic Resistance, Capacitance, and Inductance
- If Sockets Are Required, Use "Pin Sockets" ("Cage Jacks")
- Pay Equal Attention to Signal Routing, Component Placement, Grounding, and Decoupling in Both the Prototype and the Final Design
- Popular Prototyping Techniques:
  - Freehand "Deadbug" Using Point-to-point Wiring
  - "Solder-mount"
  - Milled PC Board From CAD Layout
  - Multilayer Boards: Double-sided With Additional Point-to-point Wiring
- Modern Surface-Mount ICs in Small Packages Require Special Techniques—Usually a Preliminary Multilayer PC Board Layout

---

Figure 9.173: A Summary of Analog Prototyping System Key Points

One of the more important considerations in selecting a prototyping method is the requirement for a large-area ground plane. This is required for high frequency circuits as well as low speed precision circuits, especially when prototyping circuits involving ADCs or DACs. The differentiation between *high speed* and *high precision* mixed-signal circuits is difficult to make. For example, 16+ bit ADCs (and DACs) may operate on high speed clocks (>10 MHz) with rise and fall times of less than a few nanoseconds, while the effective throughput rate of the converters may be less than 100 kSPS. Successful prototyping of these circuits requires that equal (and thorough) attention be given to good high speed and high precision circuit techniques.

Several years ago, many ICs were offered in both DIP and surface-mount packages, so breadboarding and prototyping could be done using the user-friendly DIP package. Today, however, most high performance data converters are not available in DIP packages and, if they were, the added package parasitics would limit performance in many cases.

Breadboarding and prototyping in today's environment is especially difficult, because modern surface-mounted ICs in small packages can be extremely difficult to solder into any type of PC board using manual techniques. Ball grid array (BGA) packages are nearly impossible to solder manually. Sockets—very expensive if available—are generally out of the question because of added parasitics so, in many cases, an actual multilayer PC board must be designed and fabricated. This trend has placed an even greater responsibility on the IC manufacturer to supply a variety of high quality well documented evaluation boards to assist in the initial design phases of a project.

"Deadbug" Prototyping

A simple technique for analog prototyping where DIP ICs are available uses a solid copper-clad board as a ground plane (see References 1 and 2). In this method, the ground pins of the ICs are soldered directly to the plane, and the other components are wired together above it. This allows HF decoupling paths to be very short indeed. All lead lengths should be as short as possible, and signal routing should separate high level and low level signals. Connection wires should be located close to the surface of the board to minimize the possibility of stray inductive coupling. In most cases, 18-gauge or larger insulated wire should be used. Parallel runs should not be "bundled" because of possible coupling. Ideally the layout (at least the relative placement of the components on the board) should be similar to the layout to be used on the final PCB. This approach is often referred to as *deadbug* prototyping, because the ICs are often mounted upside down with their leads up in the air (with the exception of the ground pins, which are bent over and soldered directly to the ground plane). The upside-down ICs look like deceased insects, hence the name.

Figure 9.174 shows a hand-wired "deadbug" analog breadboard. This circuit uses two high speed op amps, and in fact gives excellent performance in spite of its lack of esthetic appeal. The IC op amps are mounted upside down on the copper board with the leads bent over. The signals are connected with short point-to-point wiring. The characteristic impedance of a wire over a ground plane is about 120 Ω, although this may

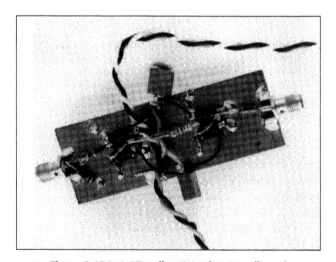

Figure 9.174: A "Deadbug" Analog Breadboard

vary as much as ±40% depending on the distance from the plane. The decoupling capacitors are connected directly from the op amp power pins to the copper-clad ground plane. When working at frequencies of several hundred MHz, it is a good idea to use only one side of the board for ground. Many people drill holes in the board and connect the sides together by soldering short pieces of wire. If care isn't taken, this may result in unexpected ground loops between the two sides of the board, especially at RF frequencies.

Pieces of copper-clad board may be soldered at right angles to the main ground plane to provide screening, or circuitry may be constructed on both sides of the board (with through-hole connections) with the board itself providing screening. For this, the board will need corner standoffs to protect underside components from being crushed.

When the components of a breadboard of this type are wired point-to-point in the air (a type of construction strongly advocated by Bob Pease (see Reference 2) and sometimes known as "bird's nest" construction) there is always the risk of the circuitry being crushed and resulting short circuits. Also, if the circuitry rises high above the ground plane, the screening effect of the ground plane is diminished, and interaction between different parts of the circuit is more likely. Nevertheless, the technique is very practical and widely used because the circuit may easily be modified (this of course assumes the person doing the modifications is adept with soldering techniques).

Another prototype breadboard variation is shown in Figure 9.175. Here the single-sided copper-clad board has predrilled holes on 0.1" centers (see Reference 3). Power buses are used at the top and bottom of the board. The decoupling capacitors are used on the power pins of each IC. Because of the loss of copper area due to the predrilled holes, this technique does not provide as low a ground impedance as a completely covered copper-clad board of Figure 9.174, so be forewarned.

In a variation of this technique, the ICs and other components are mounted on the noncopper-clad side of the board. The holes are used as vias, and the point-to-point wiring is done on the copper-clad side of the board. Note that the copper surrounding each hole used for a via must be drilled out, to prevent shorting. This approach requires that all IC pins be on 0.1" centers. For low frequency circuits, low profile sockets can be used, and the socket pins then will allow easy point-to-point wiring.

Figure 9.175: A "Deadbug" Prototype Using 0.1" Predrilled Single-Sided, Copper-Clad Printed Board Material

Solder-Mount Prototyping

There is a commercial breadboarding system with most of the advantages of the above techniques (robust ground, screening, ease of circuit alteration, low capacitance and low inductance) and several additional advantages: it is rigid, components are close to the ground plane and, where necessary, node capacitances and line impedances can be calculated easily. This system is made by Wainwright Instruments and is available in Europe as "Mini-Mount" and in the USA (where the trademark "Mini-Mount" is the property of another company) as "Solder-Mount" (see References 4 and 5).

Solder-Mount consists of small pieces of PCB with etched patterns on one side and contact adhesive on the other. These pieces are stuck to the ground plane, and components are soldered to them. They are available in a wide variety of patterns, including ready-made pads for IC packages of all sizes from 8-pin SOICs to 64-pin DILs, strips with solder pads at intervals (which intervals range from 0.040" to 0.25", the range includes strips with 0.1" pad spacing which may be used to mount DIL devices), strips with conductors of the correct width to form microstrip transmission lines (50 Ω, 60 Ω, 75 Ω or 100 Ω) when mounted on the ground plane, and a variety of pads for mounting various other components. Self-adhesive tinned copper strips and rectangles (LO-PADS) are also available as tie-points for connections. They have a relatively high capacitance to ground and therefore serve as low inductance decoupling capacitors. They come in sheet form and may be cut with a knife or scissors.

The main advantage of Solder-Mount construction over "bird's nest" or "deadbug" is that the resulting circuit is far more rigid, and, if desired, may be made far smaller (the latest Solder-Mounts are for surface-mount devices and allow the construction of breadboards scarcely larger than the final PCB, although it is generally more convenient if the prototype is somewhat larger). Solder-Mount is sufficiently durable that it may be used for small quantity production as well as prototyping.

Figure 9.176 shows an example of a 2.5-GHz phase-locked-loop prototype, built with Solder-Mount techniques. While this is a high speed circuit, the method is equally suitable for the construction of high resolution low frequency analog circuitry.

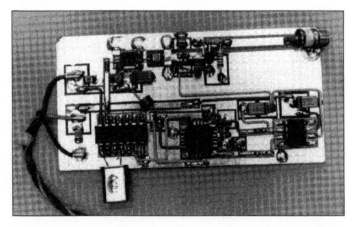

Figure 9.176: A "Solder-Mount" Constructed Prototype Board

A particularly convenient feature of Solder-Mount at VHF is the relative ease with which transmission lines can be formed. As noted earlier, if a conductor runs over a ground plane, it forms a microstrip transmission line. The Solder-Mount components include strips which form microstrip lines when mounted on a ground plane (they are available with impedances of 50 Ω, 60 Ω, 75 Ω, and 100 Ω). These strips may be used as transmission lines for impedance matching, or alternately, more simply as power buses. Note that glass fiber/epoxy PCB is somewhat lossy at VHF/ UHF, but losses will probably be tolerable if microstrip runs are short.

Milled PCB Prototyping

Both "deadbug" and "Solder-Mount" prototypes become tedious for complex analog circuits, and larger circuits are better prototyped using more formal layout techniques.

There is a prototyping approach that is but one step removed from conventional PCB construction, described as follows. This is to actually lay out a double-sided board, using conventional CAD techniques. PC-based software layout packages offer ease of layout as well as schematic capture to verify connections (see References 6 and 7). Although most layout software has some degree of autorouting capability, this feature is best left to digital designs. The analog traces and component placements should be done by hand, following the rules discussed elsewhere in this chapter. After the board layout is complete, the software verifies the connections per the schematic diagram net list.

Many designers find that they can make use of CAD techniques to lay out simple boards. The result is a pattern-generation tape (or Gerber file) which would normally be sent to a PCB manufacturing facility where the final board is made.

Rather than use a PCB manufacturer, however, automatic drilling and milling machines are available which accept the PG tape directly (see References 8 and 9). An example of such a prototype circuit board is shown in Figure 9.177 (top view).

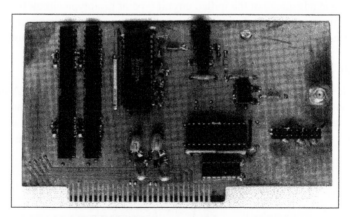

**Figure 9.177: A Milled Circuit Construction
Prototype Board (Top View)**

These systems produce either single or double-sided circuit boards directly, by drilling all holes and using a milling technique to remove conductive copper, thus creating the required insulation paths, and finally, the finished prototype circuit board. The result can be a board functionally quite similar to a final manufactured double-sided PCB.

However, it should be noted that a chief caveat of this method is that there is no "plated-through" hole capability. Because of this, any conductive "vias" required between the two layers of the board must be manually wired and soldered on both sides.

Minimum trace widths of 25 mils (1 mil = 0.001") and 12 mil spacing between traces are standard, although smaller trace widths can be achieved with care. The minimum spacing between lines is dictated by the size of the milling bit used, typically 10 to 12 mils.

A bottom-side view of this same milled prototype circuit board is shown in Figure 9.178. The accessible nature of the copper pattern allows access to the traces for modifications.

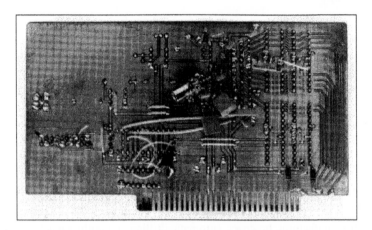

Figure 9.178: A Milled Circuit Construction Prototype Board (Bottom View)

Perhaps the greatest single advantage of the milled circuit type of prototype circuit board is that it approaches the format of the final PCB design most closely. By its very nature however, it is basically limited to only single or double-sided boards—rendering it virtually useless for surface-mount designs.

Beware of Sockets

IC sockets can degrade the performance of high speed or high precision analog ICs. Although they make prototyping easier, even *low profile* sockets often introduce enough parasitic capacitance and inductance to degrade the performance of a high speed circuit. If sockets must be used, a socket made of individual *pin sockets* (sometimes called *cage jacks*) mounted in the ground plane board may be acceptable, as in Figure 9.179.

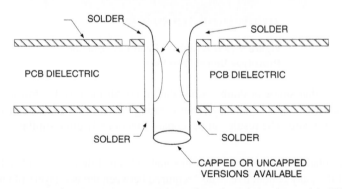

Figure 9.179: When Necessary, Use Pin Sockets for Minimal Parasitic Effects

To use this technique, clear the copper (on both sides of the board) for about 0.5 mm around each ungrounded pin socket, Then solder the grounded socket pins to ground, on both sides of the board.

Both capped and uncapped versions of these pin sockets are available (AMP part numbers 5-330808-3, and 5-330808-6, respectively). The pin sockets protrude through the board far enough to allow point-to-point wiring interconnections.

Because of the spring-loaded gold-plated contacts within the pin socket, there is good electrical and mechanical connection to the IC pins. Multiple insertions, however, may degrade the performance of the pin socket, so this factor should be kept in mind.

Note also that the uncapped versions allow the IC pins to extend out the bottom of the socket. This feature leads to an additional useful function. Once a prototype using the pin sockets is working and no further changes are to be made the IC pins can be soldered directly to the bottom of the socket. This establishes a rugged, permanent connection.

Some Additional Prototyping Points

The prototyping techniques discussed so far have been limited to single- or double-sided PCBs. Multilayer PCBs do not easily lend themselves to standard prototyping techniques. If multilayer board prototyping is required, one side of a double-sided board can be used for ground and the other side for power and signals. Point-to-point wiring can be used for additional runs which would normally be placed on the additional layers provided by a multilayer board. However, it is difficult to control the impedance of the point-to-point wiring runs, and the high frequency performance of a circuit prototyped in this manner may differ significantly from the final multilayer board.

Other difficulties in prototyping may occur with op amps or other linear devices having bandwidths greater than a few hundred megahertz. Small variations in parasitic capacitance (< 1 pF) between the prototype and the final board can cause subtle differences in bandwidth and settling time.

Sometimes, prototyping is done with DIP packages (if available), when the final production package is an SOIC. *This is not recommended.* At high frequencies, small package-related parasitic differences can account for different performance, between prototype and final PCB. To minimize this effect, always prototype with the final packages.

Evaluation Boards

Most manufacturers of analog ICs provide *evaluation boards* usually at a nominal cost. These boards allow customers to evaluate ICs without constructing their own prototypes. Regardless of the product, the manufacturer has taken proper precautions regarding grounding, layout, and decoupling to ensure optimum device performance. Where applicable, the evaluation PCB artwork is usually made available free of charge, should a customer wish to copy the layout directly or make modifications to suit an application.

General-Purpose Op Amp Evaluation Board from the Mid-1990s

Evaluation boards can either be dedicated to a particular IC, or can be general-purpose. With op amps the most universal linear IC, it is logical that evaluation boards be developed for them, to aid easy applications. However, it is also important that a good quality evaluation board avoid the parasitic effects discussed above. An example is the general purpose dual amplifier evaluation board of in Figure 9.180 (see Reference 10).

**Figure 9.180: A Mid-1990s General-Purpose Op Amp Evaluation Board
Allowed Fast, Easy Configuration of Low Frequency Op Amps in DIL Packages**

This board uses pin sockets for any standard dual op amp pinout device, and a flexible set of component jumper locations allows it to be setup for inverting or noninverting amplifiers. Various gains can be configured by choice of the component values, in either ac- or dc-coupled configurations.

The card design provides signal coupling via BNC connectors at input and output. It also uses external lab power supplies, which are wired to the lug terminals at the top. The card does however contain local supply voltage decoupling and bypassing components.

These general-purpose boards are intended for medium to high precision uses at frequencies below 10 MHz, with moderate op amp input currents. For higher operating speeds, a dedicated, device-specific evaluation board is a better choice.

Dedicated Op Amp Evaluation Boards

In high speed/high precision ICs, special attention must be given to power supply decoupling. For example, fast slewing signals into relatively low impedance loads produce high speed transient currents at the power supply pins of an op amp. The transient currents produce corresponding voltages across any parasitic impedance that may exist in the power supply traces. These voltages, in turn, may couple to the amplifier output, because of the op amp's finite power supply rejection at high frequencies.

The AD8001 high speed current-feedback amplifier is a case in point, and a dedicated evaluation board is available for it. A bottom side view of this SOIC board is shown in Figure 9.181. A triple decoupling scheme was chosen, to ensure a low impedance ground path at all transient frequencies. Highest frequency transients are shunted to ground by dual 1000 pF/0.01 μF ceramic chip capacitors, located as close to the power supply pins as possible to minimize series inductance and resistance. With these surface-mount components, there is minimum stray inductance and resistance in the ground plane path. Lower frequency transient currents are shunted by the larger 10 μF tantalum capacitors.

Figure 9.181: A High Speed Op Amp Such as the AD8001 Requires a Dedicated
Evaluation Board With Suitable Ground Planes and Decoupling (Bottom View)

The input and output signal traces of this board are 50 Ω microstrip transmission lines, as can be noted towards the right and left. Gain-set resistors are chip-style film resistors, which have low parasitic inductance. These can be seen in the center of the photo, mounted at a slight diagonal.

Note also that there is considerable continuous ground plane area on both sides of the PCB. Plated-through holes connect the top and bottom side ground planes at several points, in order to maintain lowest possible impedance and best high frequency ground continuity.

Input and output connections to the card are provided via the SMA connectors as shown, which terminate the input/output signal transmission lines. The board's power connection from external lab supplies is made via solder terminals, which are seen at the ends of the broad supply line traces.

Some of these points are more easily seen in a topside view of the same card, which is shown in Figure 9.182. This AD8001 evaluation board is a noninverting signal gain stage, optimized for lowest parasitic capacitance. The cutaway area around the SOIC outline of the AD8001 provides lowest stray capacitance, as can be noted in this view.

In this view is also seen the virtually continuous ground plane and the multiple vias, connecting the top/bottom planes.

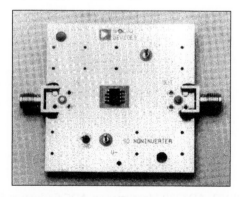

Figure 9.182: The AD8001 Evaluation Board Uses a Large Area Ground
Plane as well as Minimal Parasitic Capacitance (Top View)

Data Converter Evaluation Boards

A well designed manufacturers' evaluation board is a powerful tool that can greatly simplify the integration of an ADC or DAC into a system. Probably the best feature of an evaluation board is that its layout is designed to optimize the performance of the data converter. Analog Devices provides a complete electrical schematic and parts list as well as a PC board layout of its evaluation boards on the data sheet for most ADCs and DACs. Each layer of the multilayer board is also shown, and Analog Devices will supply the CAD layout files (Gerber format) for the board if needed. Many system level problems related to layout can be avoided simply by studying the evaluation board layout and using it as a guide in the system board layout—perhaps even copying critical parts of the layout directly if needed.

Evaluation boards typically have input/output connectors for the analog, digital, and power interfaces to facilitate interfacing with external test equipment. Any required support circuitry such as voltage references, crystal oscillators for clock generation, etc., are generally included as part of the board.

Many modern data converters have a considerable amount of on-chip digital logic for controlling various modes of operation, including gain, offset, calibration, data transfer, etc. These options are set by loading the appropriate words into internal control registers, usually via a serial port. In some converters, especially sigma-delta ADCs, just setting the basic options requires considerable knowledge of the internal control registers and the interface. For this reason, most ADC/DAC evaluation boards have interfaces (either parallel, serial, or USB) and software to allow easy menu-driven control of the various internal options from an external PC. In many cases, configuration files created in the evaluation software can be downloaded into the final system design.

Figure 9.183 shows the evaluation board for the AD7730 24-bit bridge transducer sigma-delta ADC. This ADC has an on-chip PGA and is designed to interface directly to a variety of bridge transducers. A load cell with a 10-mV full-scale output can be connected directly to the ADC input, and the output is read by the PC

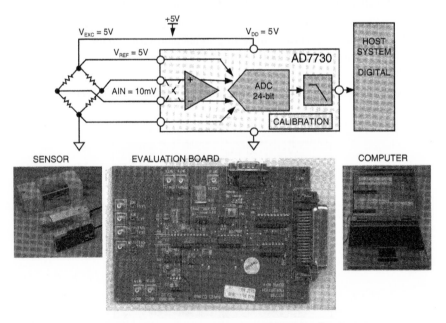

Figure 9.183: AD7730 Measurement ADC Evaluation System

via the parallel port interface. The evaluation board software allows the system designer to see the effects of sample rate, gain, filter bandwidth, and output data averaging on the overall effective resolution. The software also provides histogram displays for direct evaluation of system noise.

Figure 9.184 shows the evaluation board for the AD5535 32-channel 14-bit high voltage DAC. The evaluation board interfaces with an external PC via a parallel port connector. The software provided with the board allows data to be easily loaded into the individual DAC registers via the 3-wire serial interface.

Figure 9.184: AD5535 32-Channel, 14-Bit, 200 V Output DAC Evaluation Board

Figure 9.185 shows an ADC evaluation board for the AD7450 12-bit 1 MSPS ADC connected to an Evaluation/Control board. The ADC evaluation board (right side of diagram) is product-specific, but the evaluation/control board interfaces to a variety of ADC evaluation boards and has an on-board 16-bit buffer memory and control logic which interfaces to a PC via the parallel port. The software provided includes an

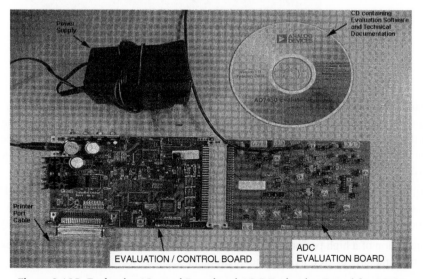

Figure 9.185: Evaluation/Control Board and ADC Evaluation Board for ADCs

FFT routine which allows evaluating the ADC under dynamic conditions. The evaluation/control board can handle ADCs with sampling rates up to several MHz.

Figure 9.186 shows Analog Devices' high-speed ADC FIFO evaluation kit which interfaces to a variety of high-speed ADC evaluation boards, such as the AD9430 12-bit, 210 MSPS ADC. The FIFO evaluation kit includes a memory board to capture blocks of data from the ADC as well as ADC Analyzer software. The FIFO board can be connected to the parallel port of a PC through a standard printer cable and used with the ADC Analyzer software to quickly evaluate the performance of the high speed ADC. The FIFO board contains two 32 K, 16-bit wide FIFOs, and data can be captured at clock rates up to 133 MSPS on each channel. Memory upgrades are available to increase the size of the FIFO to 64 K, 132 K, or 256 K. Two versions of the FIFO are available—one version is used with dual ADCs or ADCs with demultiplexed digital outputs, and the other version is used with single-channel ADCs. Users can view the FFT output and analyze SNR, SINAD, SFDR, THD, and harmonic distortion information.

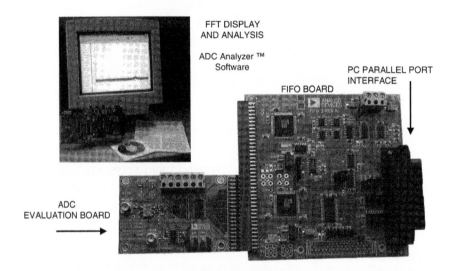

FFT DISPLAY
AND ANALYSIS

ADC Analyzer ™
Software

PC PARALLEL PORT
INTERFACE

FIFO BOARD

ADC
EVALUATION BOARD

Figure 9.186: Analog Devices' High Speed ADC FIFO Evaluation Kit

Summary

The prototyping techniques described earlier in this section are quite useful for ICs that are in DIP packages, and it is well worth the effort to prototype at least some of the critical analog circuitry before going to a final board layout. However, modern high performance ADCs and DACs are often provided in small surface-mount packages that do not lend themselves to simple prototyping techniques. In the system, multi-layer PC boards are required which further complicates the prototyping process.

In many cases, the only effective prototype for high performance analog systems is an actual PC board layout, especially if a multilayer is required in the final design. Evaluation boards are not only useful in the initial evaluation phases, but their layouts can be used as guides for the actual system board layout.

Successful integration of a high performance data converter into a system therefore requires excellent support from the manufacturer as well as care and attention to detail by the user.

References:
9.8 Breadboarding and Prototyping

1.   Jim Williams, "High Speed Amplifier Techniques," Linear Technology AN-47, August, 1991.

2.   Robert A. Pease, **Troubleshooting Analog Circuits**, Butterworth-Heinemann, 1991, ISBN 0-7506-9184-0.

3.   Vector Electronic Company, 12460 Gladstone Ave., Sylmar, CA 91342, Tel. 818-365-9661.

4.   Wainwright Instruments Inc., 69 Madison Ave., Telford, PA, 18969-1829, 215-723-4333, www.rdi-wainwright.com.

5.   Wainwright Instruments GmbH, Widdersberger Strasse 14, DW-8138 Andechs-Frieding, Germany. +49-8152-3162.

6.   PADS Software, Advanced CAM Technologies, Inc., 16450 Los Gatos Blvd., Suite 110, Los Gatos, CA 95032, www.ecam.com.

7.   ACCEL Technologies, Inc., 17140 Bernardo Center Drive, Suite 100, San Diego, CA 92128, www.acceltech.com.

8.   LPKF Laser & Electronics, 28220 SW Boberg Rd., Wilsonville, OR 97020, 800-345-LPKF or 503-454-4200, www.lpkfcadcam.com.

9.   T-Tech, Inc., 5591-B New Peachtree Road, Atlanta, GA, 34341, 800-370-1530 or 770-455-0676, www.T-Tech.com.

10. Adolfo Garcia, "Evaluation Boards for Single, Dual and Quad Operational Amplifiers," **Analog Devices AN398**, January 1996.

INDEX

- Subject Index
- Analog Devices' Parts Index

Printed and bound by CPI Group (UK) Ltd, Croydon, CR0 4YY
03/10/2024
01040335-0011